Library of
Davidson College

INDUSTRIAL ELECTROCHEMICAL PROCESSES

INDUSTRIAL ELECTROCHEMICAL PROCESSES

EDITED BY

A. T. KUHN

Lecturer in Electrochemistry
University of Salford, England

ELSEVIER PUBLISHING COMPANY
AMSTERDAM – LONDON – NEW YORK
1971

ELSEVIER PUBLISHING COMPANY
335 JAN VAN GALENSTRAAT
P.O. BOX 211, AMSTERDAM, THE NETHERLANDS

ELSEVIER PUBLISHING CO. LTD.
BARKING, ESSEX, ENGLAND

AMERICAN ELSEVIER PUBLISHING COMPANY, INC.
52 VANDERBILT AVENUE
NEW YORK, NEW YORK 10017

LIBRARY OF CONGRESS CARD NUMBER: 70-118254

ISBN: 0-444-40885-1

WITH 144 ILLUSTRATIONS AND 56 TABLES

COPYRIGHT © 1971 BY ELSEVIER PUBLISHING COMPANY, AMSTERDAM
ALL RIGHTS RESERVED. NO PART OF THIS PUBLICATION MAY BE REPRODUCED, STORED IN A RETRIEVAL SYSTEM, OR TRANSMITTED IN ANY FORM OR BY ANY MEANS, ELECTRONIC, MECHANICAL, PHOTOCOPYING, RECORDING, OR OTHERWISE, WITHOUT THE PRIOR WRITTEN PERMISSION OF THE PUBLISHER,
ELSEVIER PUBLISHING COMPANY, JAN VAN GALENSTRAAT 335, AMSTERDAM

PRINTED IN THE NETHERLANDS

DEDICATION

To the many who have inspired and guided me; most of all my father, and my many teachers, Gerd Sommerhoff, Jim Gauntlett, Freddy Yorke and Peter Church, Ronnie Bell and Alec Waters. Also to Professor Orville-Thomas, who has created at the University an environment so conducive to creativity. Lastly, to my wife, Sylvia.

List of Contributors

B. A. COOKE, Imperial Chemical Industries Ltd., Paints Division, Slough, Bucks.
D. GILROY, Electricity Council Research Centre, Capenhurst, Chester.
P. F. HART, G. E. C., Hirst Research Centre, Wembley, Middlesex.
A. W. D. HILLS, Metallurgy Department, Sheffield Polytechnic, Sheffield, Yorkshire.
G. ISSERLIS, Department of Metal Science, Polytechnic of the South Bank, London.
C. JACKSON, Imperial Chemical Industries Ltd., Mond Division, Runcorn, Cheshire.
R. J. KENDRICK, Electroformers Ltd., Staplehurst, Kent.
A. T. KUHN, Department of Chemistry and Applied Chemistry, University of Salford, Lancashire.
D. LAWSON, European Chemical News, London.
N. M. NESS, Imperial Chemical Industries Ltd., Paints Division, Slough, Bucks.
A. L. L. PALLUEL, Imperial Chemical Industries Ltd., Paints Division, Slough, Bucks.
A. J. RUDGE, Research and Development Department, Imperial Chemical Industries Ltd., Mond Division, Runcorn, Cheshire.
D. H. SMITH, CJB (Projects) Limited, Portsmouth, Hants.
G. S. SOLT, Wm. Boby Ltd., Rickmansworth, Herts.
J. M. STEELE, Capper Pass and Son Ltd., Melton Works, N. Ferriby, Yorkshire.
E. V. TUCK, Machining Development, Manufacturing Development Department, Rolls-Royce Ltd., Bristol Engine Division, Filton, Bristol.
B. J. WOODHALL, Imperial Chemical Industries Ltd., Petrochemical and Polymer Laboratory, Runcorn, Cheshire.
P. M. WRIGHT, Department of Chemistry and Applied Chemistry, University of Salford, Lancashire.

Foreword

Over ten years have passed since the last work on the subject of Industrial Electrochemical Processes was written. In that interval, we have seen nuclear energy grow from an exciting experimental stage to the situation where it now produces a significant fraction of the world's electrical energy. This fraction will unquestionably grow to the point where fossil fuels assume only minor importance in power generation, and we are promised that in this coming situation, electricity will be substantially cheaper than it is today, to the point where countries endowed with hydro-electric power schemes will lose most of the advantages they currently possess. This same decade has been equally interesting in that it has witnessed a veritable explosion of research on the fuel cell, together with the engineering of such systems to the point where they formed a crucial part of the machinery in which man first landed on the moon. This achieved, it now seems most probable that fuel cells will recede again in importance, until they remain more than a mere scientific curiosity, but of value only in very specialised situations. Yet this decade of research has resulted in the training of many electrochemists and electrochemical engineers, as well as the creation of a rich store of ideas, materials, and hardware, much of which is available as "spin-off" for the Electrochemical Industry. The manner in which this "spin-off" may be used is partly seen in this book, though the potential has hardly been tapped, as yet.

In this work, the authors have endeavoured to present a picture of electrochemical processes which have been in the past, are now, or may one day become an important part of our industrial technology. They have attempted to examine the scientific work which underlies these processes, and to assess the economic importance of each process, on a nation-by-nation basis, as well as discussing the manner in which processes compete with other, non-electrochemical, rivals. It might be suggested that the earlier, now discarded, processes deserve a specially close study, for in the light of changing costs, new ideas or materials, who is to say whether they might not now be attractive once more. And yet, this is a study requiring special courage, for not only is there much in-built prejudice against early failures, but also one knows that there is no chance of patent protection, except as regards detailed aspects or features of cell design. Right from the start, it was decided to omit any fundamental treatment of electrochemistry, though most of the thirty or more books dealing with batteries and fuel cells at present in print include such a treatment. Some of our readers may be electrochemists, while others may be technologists or even economists. We are satisfied that they will already possess as much electrochemical knowledge as they require.

Who is best qualified to write a work such as this? Unquestionably, it is the man in industry, the worker in the front line, as it were. However, such experts are not always free to undertake this type of venture or to speak uninhibitedly. Then too, there is a tendency in some fields for these men to be specialists in a very limited area of their technology. The authors who agreed to contribute to this volume are, it is hoped, the best of those who could or would be considered for such a task. They have been assisted in their work by literally hundreds of private communications, both written and verbal, from their colleagues in the field, and these have been acknowledged wherever possible. In such a way, we have sought to overcome presentation of a given technology from the viewpoint of one firm only. At the same time, the fruits of this exercise in seeking opinions and advice from industry have been thought-provoking. Thus, all letters written to German firms (in that tongue) were promptly acknowledged, often with much helpful information. One only wishes the same could be said of more countries. Indeed, the German- and Russian-speaking countries are noteworthy for the freedom with which their technology is described in the literature. One wonders how much longer a situation can continue in which, at international meetings, technologists from one nation present detailed papers, while their counterparts from countries with equally well-developed industries sit mutely, with orders to remain thus, or at best, present "review" papers, containing no new information; as far as electrochemical processes are concerned, the British have no cause for self-congratulation.

Our technology is an upwards-growing helix, whose growth is best achieved by spiralling through the industry itself, the neighbouring technologies, the seats of learning and so back to the industry concerned once more. In an atmosphere where discussion is inhibited, all parties lose, for the research specialist outside industry is deprived of the stimulus which results from knowing his work is important, while he forgoes any opportunity of seeing his ideas come to fruition on a large scale. Nor can he learn in what areas most profitably to apply his thoughts. His counterpart in industry loses equally, and forgoes the satisfaction of being known and accepted as one of the international fraternity of science, and the fact that he is the equal of his academic counterpart, both in his grasp of the facts, and in his achievement potential, can pass unrecognised. Such thoughts are not unrelated to a problem whose existence is slowly being acknowledged in Britain and the U.S.A. – and in all probability elsewhere. This concerns the young and more intelligent members of society. How these should be trained or educated, how many, and why, are questions which have yet to be fully faced. To do this, and then to find, for these people, intellectually satisfying tasks for most of their working life span is a matter which has not been solved. Indeed, at this time, and speaking of the U.K. and the U.S.A., the prospects appear if anything worse than they were. It would be pretentious to suggest a simple solution for this, but the writer is convinced that, whatever solution is found, it will be preceded by a "breaking-down" of the barriers between the three worlds of industry, learning and the research institute. And this can only be done by facilitating the movement of scientists on a temporary visiting, or permanent, basis between them. U.S. industry has been much praised for its use of junior consultants, where U.K. industry tends in so many cases to favour a limited number of very eminent "prestige consultants". The British Scientific Civil Service has a commendable scheme for "vacation consultancies" in its own establishments, and some universities appoint visiting lecturers or professors. But there is still far too much lip-service and too little real action in this area.

How can one view the electrochemical industry? Because the electrochemistry is merely a means to an end, and not the only means, in most cases, we can consider the situation as an interface at which electrochemical and non-electrochemical processes are competitive. An example of this might well be found in the technology of electrochemical machining. The interface is an ill-defined zone, and frequently there will be advances in technology – such as the recently announced titanium-skinned tools for machining, which push back the interfacial zone one way or another – in this case by permitting much higher throughputs using conventional machine tools, while leaving unchallenged the technology as a whole. The recent advances in the technique of lost-wax moulding, itself as old as Christianity, provide another example of competition not just for electrochemical machining, but for many forms of mechanical shaping too. The interface of competitiveness can sometimes move in such a way that it utterly swamps a completely established area of technology, and the rapid disappearance of heavy water electrolysis might be cited in this context.

Electricity is a reactant, and in particular, a uniquely clean and versatile oxidising or reducing agent. However it is made (ignoring the once proud boast of the fuel cell world), it is subject to the inefficiencies of the Carnot cycle, which in current practice means that no more than 40% of the thermal energy can be converted. A tragedy of our society is that, in so many cases, the residual 60% of the energy goes to waste. Only the resourceful few, large manufacturers of chemicals have been able to utilise this otherwise wasted energy for process steam or similar purposes, while District Heating schemes, in the U.K. at least, can be numbered on the fingers of one hand. Here lies the first challenge to the successful future use of electricity. If we consider the processes occurring in an electrochemical cell, and the voltage associated with them, as the author of Chapter 16 on Cell Design has done, we see that catalysis and cell design are two approaches which may reduce a cell voltage to a term in which only the thermodynamically reversible voltage term is left unconsidered. And herein lies the second challenge. This reversible voltage can be reduced by means of air-depolarised electrodes or analogous fuel cell type technology. But even more desirable would be the development of processes in which both anode and cathode products were useful. There are, of course, examples of such processes. In the Downs process, both the sodium metal and the chlorine are desirable products, and the same is true in the electrolytic production of magnesium. The electrolysis of aqueous brines to yield three products – caustic soda, chlorine and hydrogen – all of which are desirable, though in varying degrees, provides a last example. At the same time, it may serve to illustrate one problem inevitably associated with the concept, namely the maintenance of a balance between the variable demands for the different products. Chlorine is the most sought after, while hydrogen commands a high premium so long as it can be sold into the glass industry or the fat industry, or similar users. But thereafter, it is relegated to serve as a boiler fuel, a poor return on the energy expended in making it. The caustic soda has, up to the present time, been saleable easily enough. But it is understood that Solvay, in Belgium, now carbonate this caustic, and this, too, is tantamount to downgrading it energetically. It may or may not be coincidental that the processes mentioned here are among the most secure and unchallenged in the electrochemical hierarchy. (The fact might also stem from the absence of any serious non-electrochemical process for effecting these chemical reactions.) The idea of conducting two chemically unrelated processes in a single cell is not novel, and Soviet workers in particular have devoted much thought to the idea. Such a cell, if successful, would result in halving, not only the operating, but also the capital costs associated with each process. One can enumerate dozens of ideas which might be transformed from failure to success by such an approach. And yet, the concept has not been followed up. Clearly, there would be problems. The demand for the product from one half of the cell would rarely be equal to that from the other half. Then, too, the optimal reactant concentrations and temperatures for one process might not be those for the other, and a compromise would have to be called for. There are answers to all these drawbacks, and the others which might be raised – but surely the reward is worth devoting further study to this concept.

The third challenge, like the preceding one, awaits the chemical engineer. The few highly successful electrochemical processes such as chlorine electrolysis or aluminium smelting, and also the highly developed, though now less important, electrolysis of water share the common feature that the reacting species, whatever it may be, is present at high concentrations. Thus the electrochemical engineer in such situations has not been overly concerned with problems of mass transport of the reactant. But for every single one of these processes, a dozen other processes, patents or ideas exist where the reactant is not present in high concentrations. The first hundred years of electrochemical technology were marked by the situation where the electrochemist was content to accept this fact, and to operate his plant at low rates to accommodate problems of mass transport in a cell whose geometry conceded nothing to the situation. It is slowly being appreciated that the investment of a little extra capital, or design sophistication, or energy expended to promote mass transport, can pay enormous dividends in the shape of overall performance. In this context the psychological importance of the much-discussed fluidised bed electrode should not be overlooked. Its validity as a patent has been doubted, its merits too strongly praised and, indeed, disputed. But it has initiated a renaissance of thought, and cells with rotating electrodes or "windscreen wipers" are now considered much more seriously than hitherto.

Apart from the outward form and concept of the cell itself, the materials available for its construction have become far more numerous in the last decade. The advances in both polymer technology and metallurgy have presented the cell designer with a choice that can lead to more efficient and less costly cells. In the selection of the electrode materials themselves in many cases a wider choice obtains, and the penultimate chapters in the book are conceived in the spirit of a "shopping guide" for the basic components of an industrial cell. But looking only slightly further ahead, one sees a period in which many new electrocatalysts could appear, and research on electrode materials such as the tungsten bronzes, slightly platinum doped, or the carbides, borides and nitrides of the refractory metals, or even magnetite, doped with precious metal oxide promoter could culminate in a product in which precious metal capital cost or operating loss ceased to be important. Here then, are many grounds for optimism and for thinking that there may be a continuous growth in the importance of the industry.

Technologists everywhere are bowing before the forces which seek to prevent further pollution of our world before it is too late. Industrial electrochemical processes must play their part in this movement, and at least one large mercury cell chlorine plant in the U.S.A. has stood idle for many months, accused of polluting inland waters with mercury. But by and large, it is true to say that electrochemical processes, especially those in aqueous solutions, are less guilty of such pollution than their purely thermal counterparts and this is true both for oxidations and reductions. At a price, there is no doubt that we shall be able to solve the problems of pollution, and in this changed economic setting the merits of established processes may suffer a complete transformation. But looking beyond even this, the prophets of doom have foreseen a graver disaster. Just as few people formerly realised the gravity of the pollution problem, today just as few are aware of the rate at which we are consuming our natural resources. By the end of the present century – and that is, after all, only three books away at the rate of one per decade – there are dangers that many of our basic resources will be in gravely short supply. In practice, of course, this will mean that less and less attractive sources of the desired commodity will be tapped, both as primary sources and reclaimed materials. This, too, will transform the entire industrial structure as we know it today. Three years ago, the Chairman of Britain's largest Chemical company was urging that natural gas should be treated as a chemical feedstock, not merely as a fuel. While coal and oil resources are both considered to have a lifetime limited to the measurable future, there is one energy resource, at least, for which no gloomy predictions have been made – electricity. True, the world's uranium stocks are believed to be comparable, in longevity, to those of the hydrocarbon fuels. But beyond these, there lies the breeder reactor, already a reality, and beyond that still, the ultimate goal of fusion energy, truly the limitless source of power. Should this state of affairs be reached, then all reductions might be electrochemical, either directly or indirectly, all oxidations anodic, and the fuels which we so cavalierly burn or crack, will become infinitely precious primary building blocks. It matters not, while this book is in press, that an oil bonanza is discovered under the Yorkshire moors or in Northern Europe. Such a discovery will only delay for a decade or two the inevitable reckoning. Changed conditions such as these would transform economics as we know them today, and at this time there will be a reappraisal of older processes and ideas with a view to their possible revival.

Thus it is, whether amid the optimistic predictions of cheaper electrical power, or of a threatened world, that the electrochemists who have written this book, from industry, the research institutes or the world of learning share a common feeling that their discipline will became more, not less, important with the passage of time.

Who can predict the course that chemistry and its technology will take in the future? Restricting oneself to a single guess, it would be that the next generation of chemical products, as well as the technology required for their manufacture, will be more sophisticated. This is a challenge which electrochemistry is well fitted to meet, and the authors, in their various fields look forward to this challenge. Nor are they ashamed to look back, and they would all of them wish here to pay a tribute to some of the great electrochemists who have built the industry as we know it today. Let us not forget Faraday, Jacobi, Billiter, Hamilton Young Castner, Gibbs, Downs, Moissan, Hall and Heroualt, and the many others who have left their mark in the field, while the writings of Engelhardt, Eger and Mantell are quoted often enough for our high esteem of them to be apparent. To all of these, and to the oldest and most eminent of electrochemists, A. N. Frumkin, the contributors to this book record their deepest respects.

A. T. KUHN

Contents

List of Contributors vii

Foreword ix

Chapter 1. PRODUCTION OF ELEMENTAL FLUORINE BY ELECTROLYSIS
BY A. J. RUDGE 1
 1. Historical 1
 2. The electrolyte 7
  2.1 Melting point and hydrogen fluoride partial pressures of solutions of potassium fluoride in hydrogen fluoride, 7; 2.2. Electrical conductivity, 10; 2.3. Specific gravity, 10; 2.4. Surface tension of KF/HF mixtures, 10; 2.5. Electrolyte manufacture, 13; 2.6. Electrolyte specification, 13.
 3. Anode phenomena 14
  3.1 Polarisation, 14; 3.2. The anode/electrolyte contact angle, 15; 3.3. Mechanism of fluorine evolution, 16; 3.4. Mechanism of polarisation, 18; 3.5. Anode effect, 19.
 4. Methods of avoiding or overcoming polarisation 21
  4.1. Influence of structure of anode material on polarisation, 21; 4.2. Influence of electrolyte purity and additives on polarisation, 24; 4.3. The use of low permeability carbon anodes, 25.
 5. The reversible voltage 26
 6. Fluorine electrode kinetics 27
 7. The working voltage 29
 8. Heat produced in the cell and its removal 34
 9. Mechanism of fluorine and hydrogen evolution in relation to cell design. 35
10. Influence of the mechanism of fluorine evolution on the hydrogen fluoride content of the fluorine 38
11. Removal of hydrogen fluoride from fluorine 39
12. The anode connection 44
13. Energy efficiency and energy usage 45
14. Fluorine cell design and operating characteristics 45
  14.1 I.C.I., 50; 14.2. U.K.A.E.A., 52; 14.3. Imperial Smelting Corporation, 53; 14.4. Union Carbide, 53; 14.5. Allied Chemical Corporation, 55; 14.6. Pennsalt Chemicals Corporation, 56; 14.7. Kali-Chemie, 56; 14.8. Montecatini-Edison, 57; 14.9. La Société des Usines Chimiques de Pierrelatte, 57.
15. Fluorine compression 57
16. Liquid fluorine 57
17. Disposal of fluorine and hydrogen fluorine-containing hydrogen . . 58
18. Analysis of elemental fluorine 59
19. Fluorine in the laboratory 60

20. Materials of construction for use with elemental fluorine 60
21. The cost of elemental fluorine 63
22. Uses of fluorine 64
    Acknowledgements 65
    References 65

Chapter 2. ELECTROCHEMICAL FLUORINATION
BY A. J. RUDGE 71
1. Introduction 71
2. Historical 71
3. Scope of the process 72
4. Mechanism of electrolytic fluorination 74
5. Perfluorocompounds of industrial importance produced by electrochemical fluorination 76
    5.1. Perfluorocarboxylic acids and by-products, 76; 5.2. Perfluorosulphonic acids, 80.
6. Other routes to perfluoroacids 81
7. Other compounds made by electrochemical fluorination 81
8. Commercial availability and cost of products made by electrochemical fluorination 82
9. Uses of perfluorocompounds produced by electrochemical fluorination 84
    9.1. Surfactants, 84; 9.2. Surface treatments, 85.
    Acknowledgements 87
    References 87

Chapter 3. THE CHLOR-ALKALI INDUSTRY
BY A. T. KUHN 89
1. Introduction 89
2. Hydrochloric acid electrolysis 89
    2.1. General, 89; 2.2. Basic electrochemistry of the HCl system, 90; 2.3. Industrial cell design, 90; 2.4. Cell operation, 91; 2.5. Tolerance to organics, 91; 2.6. Process developments, 91.
3. Chlorates 92
    3.1. Introduction, 92; 3.2. Cell design, 93; 3.3. Other references, 96; 3.4. Chlorites, 96.
4. Perchlorates 96
    4.1. Introduction, 96; 4.2. General industrial, 97; 4.3. Mechanism of perchlorate formation, 97; 4.4. Cell design and operation, 97; 4.5. Other perchlorate processes, 98.
5. Hypochlorites 99
6. Sodium 99
    6.1. Introduction, 99; 6.2. The Downs process, 100; 6.3. Cell operating data, 102; 6.4. Constitution of the melt, 102; 6.5. Other processes, 102; 6.6. Further reading – sodium manufacture, 104.
7. Chlorine from brine 106
    7.1. Introduction, 106; 7.2. Comparison of diaphragm and mercury cells, 108; 7.3. Energetics of mercury and diaphragm cells, 109; 7.4. Industrial cells, 109; 7.5. Future developments, 117; 7.6. Further reading – chlorine electrolysis, 118.

8. Iodine electrochemistry 120
9. Bromates 121
    9.1. Introduction, 121; 9.2. Basic reaction, 121; 9.3. Commercial cells, 121; 9.4. Other processes, 122.
10. Bromine 122
    References 123

Chapter 4. INDUSTRIAL WATER ELECTROLYSIS
BY D. H. SMITH 127
1. Introduction 127
2. The electrolytic decomposition of water 128
    2.1. Cell voltage, 128; 2.2. Gas evolution and gas purity, 130; 2.3. Energy consumption, 130.
3. General features of industrial water electrolysis plant 131
    3.1. Feed water preparation, 132; 3.2. Electrolyte preparation and circulation, 132; 3.3. Cell construction, 133; 3.4. Tank electrolysers, 133; 3.5. Filter-press electrolysers, 134; 3.6. Pressure electrolysers, 135.
4. Gas separation, purification and drying 135
5. Safety and control 135
6. Commercial water electrolysis plant 136
    6.1. Tank electrolysers, 136; 6.2. Filter-press electrolysers, 138; 6.3. Pressure electrolysers, 147; 6.4. Miscellaneous and future developments, 153.
7. Uses of electrolytic hydrogen and oxygen 154
    7.1. Economics of hydrogen manufacture, 155.
8. Literature of the water electrolysis process 155
    Acknowledgement 156
    References 156

Chapter 5. ELECTROLYTIC HEAVY WATER MANUFACTURE
BY C. JACKSON 159
1. Introduction 159
2. Manufacture by electrolysis 160
    2.1. Norsk Hydro plant, 160; 2.2. U.S. Atomic Energy Commission plant at Trail, B.C., 161.
3. Pre-enrichment by electrolysis 162
    3.1. Norsk Hydro plant, Norway, 162; 3.2. Fertilizer Corporation of India plant at Nangal, 163; 3.3. Emser Werke plant, Domat Ems, Switzerland, 163; 3.4. Junta Energía Nuclear plant, Madrid, 164.
4. Final enrichment by electrolysis 164
    4.1. Manhattan District Project, Morgantown, W.Va., 164; 4.2. Savannah River plant, 165.
5. Electrolytic upgrading of diluted heavy water 166
    5.1. Atomic Energy of Canada Ltd. unit, Chalk River, Ontario, 166; 5.2. United Kingdom Atomic Energy Authority unit at A. E. A. Winfrith, 167; 5.3. Showa Denko K.K. unit, Japan, 167; 5.4. Atomic Energy Establishment, India, 167.
6. Other electrolysis methods 167
    6.1. Reversible electrolysis, 167; 6.2. Direct electrolysis for deuterium, 168.

xviii   CONTENTS

  7. Trends 168
  8. Economic assessment 169
  9. Conclusion 170
10. The literature of electrolytic heavy water production 170
     Acknowledgement 171
     References 171

Chapter 6. THE ELECTROWINNING OF METALS
BY D. GILROY 175
  1. Introduction 175
     1.1. Electrowinning in aqueous media, 176; 1.2. Electrowinning in molten salts, 177.
  2. Aqueous processes 178
     2.1. Copper, 178; 2.2. Cobalt, 181; 2.3. Nickel, 182; 2.4. Zinc, 183; 2.5. Cadmium, 186; 2.6. Chromium, 188; 2.7. Manganese, 189; 2.8. Gallium, 190; 2.9. Thallium and indium, 191.
  3. Molten salt processes 192
     3.1. Aluminium, 192; 3.2. Magnesium, 199; 3.3. Titanium, 201; 3.4. Niobium and tantalum, 203; 3.5. Rare earth metals, 205; 3.6. Lithium, 207; 3.7. Beryllium, 208; 3.8. Boron, 209.
  4. Other metals 209
  5. Summary and conclusions 211
     Acknowledgements 212
     References 213

Chapter 7 (Part 1). ELECTROREFINING IN AQUEOUS ELECTROLYTES
BY J. M. STEELE 219
  1. Introduction 219
     1.1. History of electrorefining, 219; 1.2. Basic principles, 220.
  2. Process summary 222
  3. Electrorefining compared with other methods of refining 223
     3.1. Copper, 224; 3.2. Nickel, 224; 3.3. Cobalt, 224; 3.4. Lead, 225; 3.5. Silver and gold, 225.
  4. Electrorefining circuits 225
     4.1. Simple layout, 225; 4.2. Electrolytic cells, 226; 4.3. Electrolyte circulation, 227; 4.4. Electrodes, 228; 4.5. Electrical connection, 229; 4.5. Slimes recovery, 230.
  5. Anodes 231
     5.1. Production of anodes, 231; 5.2. Classification of anodes, 231; 5.3. Slime characteristics, 232; 5.4. Anode life, 233.
  6. Electrolyte 233
     6.1. Function, 233; 6.2. Purification of electrolyte, 234; 6.3. Addition agents, 235; 6.4. Temperature, 236; 6.5. Metal ion balance, 237.
  7. Cathode 237
     7.1. Positioning of electrodes in cells, 237; 7.2. Flow of electrolyte through the cell, 238; 7.3. Maintenance of an adequate supply of addition agent, 238; 7.4. Absence of solids in the electrolyte, 238; 7.5. Elimination of short circuits between anode and cathode, 238.

8. Refinery costs 239
   8.1. Power costs, 239; 8.2. Labour costs, 240; 8.3. Fuel and reagent costs, 240; 8.4. Maintenance, 240.
9. Future developments 241
10. Electrolytic production of metal powders 242
    References. 243

Chapter 7 (Part 2). ELECTROREFINING IN MOLTEN SALTS
BY P. F. HART AND A. W. D. HILLS 245
1. Introduction 245
2. Aluminium 246
3. Lead 248
4. Plutonium 250
5. Beryllium 251
6. Niobium 252
7. Titanium 253
8. Vanadium 253
9. Zirconium, tungsten and molybdenum 254
10. Uranium 254
11. Tin and antimony 254
12. Other metals 255
13. Future possibilities 256
    References. 256
    Bibliography. 260

Chapter 8. ELECTROCHEMICAL MACHINING
BY E. V. TUCK 263
1. Introduction 263
2. Plant 264
   2.1. The machine tool, 265; 2.2. Electrolyte supply system, 266; 2.3. Clarification of electrolyte, 267; 2.4. Hydroxide sludge disposal, 268; 2.5. Hydrogen extraction, 268; 2.6. Pumps, 268.
3. Power supplies 269
4. Dimensional accuracy and surface finish 273
   4.1. Dimensional accuracy, 273; 4.2. Surface finish, 274.
5. Operating conditions 274
   5.1. Factors affecting the efficiency, 278.
6. Design of tools 280
   6.1. Materials for tools and fixtures, 280; 6.2. Power connections, 281; 6.3. Tool insulation, 282; 6.4. Tool design, 283; 6.5. Starting on surfaces which do not conform to the tool shape, 289; 6.6. Tool correction, 291.
7. Electrolytes 295
   7.1. Post-ECM treatment, 298.
8. Electrochemical grinding 298
9. Economics 301

10. Applications 304
   10.1. Electrochemical forming, 305; 10.2. Electrochemical turning, 306; 10.3. Electrochemical drilling, 306; 10.4. Trepanning, 308.
11. Future developments 309
   Acknowledgements 311
12. Key patents (UK numbers) 311
13. Bibliography 323
   References 326

Chapter 9. THE ELECTROLYTIC FINISHING OF METALS
BY G. ISSERLIS 327
1. Introduction 327
2. Electrodeposition of metals 327
   2.1. Introduction, 327; 2.2. Mechanism and kinetics of electrodeposition, 328; 2.3. Mode of growth of electrodeposits, 331; 2.4. Throwing power of an electrolyte, 331; 2.5. Metal distribution formulae, 332; 2.6. Micro throwing power and levelling, 333; 2.7. Codeposition of foreign substances, 334; 2.8. Deposition of bright metal coatings, 335; 2.9. Effect of dissolved hydrogen, 337; 2.10. Internal stresses in electrodeposits, 338; 2.11. Measurement of internal stress, 338; 2.12. Hardness and wear resistance, 339; 2.13. Porosity, 339; 2.14. Metal surface preparation for electroplating, 339; 2.15. Electrodeposition of copper, 344; 2.16. Electrodeposition of nickel, 348; 2.17. Electrodeposition of iron and cobalt, 352; 2.18. Electrodeposition of chromium, 354; 2.19. Protection offered by decorative nickel–chromium coatings against corrosion of the basis metal, 358; 2.20. Electrodeposition of zinc, 360; 2.21. Electrodeposition of cadmium, 364; 2.22. Electrodeposition of tin, 365; 2.23. Electrodeposition of noble metals, 369; 2.24. Electroplating on aluminium, 371; 2.25. Plating of zinc-base alloy diecastings, 372; 2.26. Electrodeposition of alloys, 372; 2.27. Electrodeposition of metals from non-aqueous electrolytes, 376; 2.28. Electrodeposition of refractory metals, 377; 2.29. Electrodeposition of metals on plastics, 378.
3. Anodic oxidation 378
   3.1. Aluminium anodising, 379; 3.2. Copper anodising, 379; 3.3. Theoretical background of anodic film formation, 380.
4. Electrocolouring 380
5. Plant for electroplating 381
   Bibliography and references 382

Chapter 10. ELECTROFORMING
BY R. J. KENDRICK 385
1. Introduction and historical 385
2. Advantages of electroforming as a production process 387
3. Problems encountered during electroforming 387
   3.1. Internal stress, 387; 3.2. Non-uniform metal distribution, 389; 3.3. Corner weakness defects, 390; 3.4. Low deposition speeds, 391.
4. Electroforming solutions and deposit properties 392
   4.1. Copper, 392; 4.2. Nickel, 393; 4.3. Precious metals, 395.
5. Anode materials 396
6. Equipment for electroforming 397

7. Measuring properties of electroforms 398
   7.1. Hardness and tensile strength, 398; 7.2. Internal stress, 398; 7.3. Ductility, 399.
8. Preparation of mandrels 399
   8.1. Mandrel types for plain metal products, 399; 8.2. Mandrels for perforated metal products, 402.
9. Applications of electroforming 403
   9.1. Record stampers, 403; 9.2. Seamless perforated tubes, 404; 9.3. Wave guides, 405; 9.4. Plain and perforated metal foils, 405; 9.5. Endless plain or perforated nickel bands, 406; 9.6. Copper spark machining electrodes, 407; 9.7. Razor foils, 407; 9.8. Moulds and dies, 407; 9.9. Metals bellows, 408; 9.10. Diamond cutting tools, 408; 9.11. Aerospace components, 408; 9.12. Consumer items, 408; 9.13. Apertured components, 409; 9.14. Metal fibres, 409; 9.15. Silverware, 410; 9.16. Gold, 411; 9.17. Electrotyping, 411.
10. The future of electroforming 411
    10.1. Metal and alloys, 411; 10.2. Refractory metals, 415; 10.3. Engineering development, 415.
    References 416

Chapter 11. THE ELECTRODEPOSITION OF PAINT
BY B. A. COOKE, N. M. NESS AND A. L. L. PALLUEL 417

1. Introduction 417
2. General nature of the process 417
   2.1. Advantages and limitations of paint electrodeposition, 419.
3. The composition of electrocoat paints 423
   3.1. Vehicles for electrodeposition, 425; 3.2. Other features of electrocoat paint composition, 430.
4. The physical chemistry of paint electrodeposition 434
   4.1. Electrolytic properties of the aqueous dispersion, 434; 4.2. The anodic process, 435; 4.3. The cathodic process, 437; 4.4. The stoichiometry of deposition, 437; 4.5. Initiation of film growth, 438; 4.6. Film conduction, 440; 4.7. Throwing power, 446.
5. The technology of paint electrodeposition 449
   5.1. Approaches to the technology, 449; 5.2. Ancillary operations, 457; 5.3. Electrocoating as part of a complete metal finishing system, 459.
   Addendum 462
   Acknowledgements 463
   References 463
   Bibliography 465

Chapter 12. ELECTRODIALYSIS
BY G. S. SOLT 467

1. Introduction 467
2. General details of construction 469
3. Polarisation 471
   3.1. Introduction, 471; 3.2. Effect of temperature change, 474; 3.3. Anomalous polarisation, 475; 3.4. Polarity reversal, 476.
4. Construction and design 477
   4.1. Introduction, 477; 4.2. Electrodes, 479; 4.3. The flow sheet, 480.

5. Economics and plant design 483
   5.1. Introduction, 483; 5.2. Optimisation calculations, 484; 5.3. Capital costs, 485; 5.4. Other costs, 487; 5.5. Fields of economic application, 488.
6. Variations on the process 490
   6.1. The filled cell, 490; 6.2. Transport depletion, 490.
7. Examples in the field 491
   7.1. Desalination for potable water, 491; 7.2. Salt production from sea water, 494; 7.3. Desalination of cheese whey, 495.
   References 495

Chapter 13. MISCELLANEOUS INDUSTRIAL PROCESSES
BY C. JACKSON AND A. T. KUHN 497
1. Inorganic oxidation and reduction 497
   1.1. Hydrogen peroxide and persalts, 497; 1.2. Manganese dioxide and permanganate, 503; 1.3. Metal oxides, 505; 1.4. Miscellaneous processes, 506.
2. Organic oxidation and reduction processes 507
   2.1. Electrohydrodimerisation of acrylonitrile to adiponitrile, 507; 2.2. Electrolytic lead alkyls, 508; 2.3. Methyl ethyl ketone, 509; 2.4. Propylene oxide, 509; 2.5. Sorbitol and mannitol, 510; 2.6. Reduction of nitrobenzene, 510; 2.7. Reduction of salicylic acid, 512; 2.8. Piperidine, 512; 2.9. Iodoform and chloroform, 513; 210. Dimethyl sulphoxide, 513.
3. Processes involving electrolytic regeneration 513
   3.1. Dialdehyde starch, 513; 3.2. Chromic acid regeneration, 514; 3.3. Bromide regeneration, 516; 5.4. Caustic soda regeneration, 516.
4. Use of electrolytic amalgam as reducing agent 518
   4.1. Benzidine and analogues, 518; 4.2. Sodium sulphide, 518; 4.3. Sodium hydrosulphite, 519; 4.4. Sodium alcoholate, 519; 4.5. Hydrodimerisation of acrylonitrile to adiponitrile, 519.
5. Electrochemistry and glass manufacture 520
   Acknowledgements 520
   References 521

Chapter 14. ELECTRODES FOR INDUSTRIAL PROCESSES
BY A. T. KUHN AND P. M. WRIGHT 525
1. Introduction 525
2. The lead dioxide anode 525
3. The magnetite anode 533
4. Lead and lead alloy anodes and the "Chilex" anode 535
5. Carbon and graphite anodes 538
6. Noble metal coated anodes 545
   6.1. Introduction, 545; 6.2. Properties of noble metal coated titanium anodes, 548; 6.3. Anode design, 551; 6.4. Application of noble metal coated titanium anodes in particular industrial electrochemical processes, 556; 6.5. Conclusions, 561.
7. Cathode materials 561
   References 566

Chapter 15. DIAPHRAGMS AND ELECTROLYTES
BY C. JACKSON, B. A. COOKE AND B. J. WOODHALL 575

1. Diaphragms 575
   1.1. Introduction, 575; 1.2. Mechanical diaphragms, 575; 1.3. Examples of commercially available electrolytic diaphragms, 579.
2. Ion-exchange membranes 584
3. Electrolytes for electrochemical processes 593
   3.1. The criteria for a suitable electrolyte, 593; 3.2. Aqueous electrolytes, 594; 3.3. Molten salts, 595; 3.4. Non-aqueous media, 595.
   Acknowledgement 595
   References 596

Chapter 16. CELL DESIGN
BY C. JACKSON 599
1. General discussion of cell design 599
   1.1. Current efficiency, 600; 1.2. Cell voltage, 601.
2. More detailed consideration of cell design 601
   2.1. Electrode design, 601; 2.2. Materials of construction, 603; 2.3. Cell heating and cooling, 604.
   References 604

Chapter 17. AN INTERNATIONAL SURVEY OF INDUSTRIAL ELECTROCHEMICAL PROCESSES
BY A. T. KUHN AND D. LAWSON 607
1. Economic information on a product group 607
   1.1. Chlorine, 607; 1.2. Aluminium, tin, copper, zinc, magnesium, 608; 1.3. Hydrogen peroxide, 608; 1.4. Perchlorates, 608; 1.5. Manganese dioxide, 608; 1.6. Metals (Misc.), 608.
2. Economic information on a national basis 608
   2.1. The Soviet Union and Comecon group countries, 609; 2.2. United States, 613; 2.3. Canada, 614; 2.4. Australia, 614; 2.5. India, 614; 2.6. Israel, 615; 2.7. Japan, 616; 2.8. Sweden, 617; 2.9. Norway, 618; 2.10. Finland, 618; 2.11. The Benelux countries, 619; 2.12. France, 619.
3. Bibliography (U.S.S.R.) 620
   References 620

Index 623

Chapter 1

Production of Elemental Fluorine by Electrolysis

A. J. RUDGE

Research and Development Department, Imperial Chemical Industries Ltd., Mond Division, Runcorn, Cheshire

Fluorine has been manufactured on an industrial scale for less than 30 years. Commensurate with this late development of fluorine technology, published information is relatively sparse and even the fundamentals of the subject have received very limited attention.

Because fluorine is possibly much less familiar than other electrochemical products and processes discussed in this book, brief mention is made of certain aspects which would not normally be referred to in a text book of electrochemistry, such as fluorine compression, analysis of fluorine and the like, in order to provide references for those who require more detailed information.

Where reference is made to British or other patents, it should be borne in mind that equivalent patents in countries other than those quoted exist in many cases.

All patent dates are dates of publication.

1. Historical

There are several reviews which provide accounts of the production of elemental fluorine by the electrolysis of solutions of potassium fluoride in hydrogen fluoride and which cover the period from the classical work of Moissan* in 1886 to the mid 1950's[1-5]. The last two references include tables summarising the characteristics of the more important cells. Two short accounts deal particularly with more recent historical developments in this country[6,7]. The most recent review, though brief, appears to be that of Kwasnik[8].

It seems desirable to give here a condensed version of the early history of elemental fluorine in order that the reasons for the current universal acceptance of the medium temperature carbon anode cell may be appreciated. Some historical information, not hitherto published, has been included.

All work involving the production of elemental fluorine carried out between 1886 and 1919, including that of the pioneer fluorine chemist Otto Ruff, made use

---

* Whereas Moissan was certainly the first to develop a method capable of preparing fluorine in appreciable quantity, Domange has pointed out that Fremy was successful in producing a small quantity of fluorine some 30 years before Moissan[9].

of low temperature cells, with an electrolyte containing 60% or more of hydrogen fluoride and platinum or platinum alloy anodes, which were essentially copies of or variants on Moissan's apparatus. The anhydrous hydrogen fluoride required had to be made by the tedious thermal decomposition of potassium bifluoride, KF·HF, and it was necessary to cool the cell to below room temperature to reduce the partial pressure of the hydrogen fluoride (b.p. 19.5°) over the electrolyte*. The anode was heavily corroded with a resultant low current efficiency. Particularly during this early period, the extremely hazardous nature of anhydrous hydrogen fluoride was not recognised so that, as Ruff pointed out in the introduction to his book *Die Chemie des Fluors*, "Scarcely anybody has obtained results in this field without also acquiring honourable scars. The battle cost Louyet his life, while others suffered severe injury to their health"[10].

Although it is somewhat surprising that such an arrangement had not been tried before, it was not until 1919 that Argo and co-workers** described a high temperature cell using molten, anhydrous potassium bifluoride (m.p. 239°) as the electrolyte with a graphite anode[11]. The cathodic container was of copper. Such a cell represented a notable simplification, insofar as the need to prepare anhydrous hydrogen fluoride and to use platinum apparatus was avoided, but there remained two attendant difficulties. The melting point of the electrolyte rises rapidly as the hydrogen fluoride is consumed by electrolysis and anodic polarisation† is a recurrent problem. Cady makes the very pertinent comment "The numerous investigators who described cells of the high temperature type before 1942 used graphite anodes and sometimes wrote enthusiastically about successful operation of their cells. In general, they did not say much about the many hours spent trying to make their cells work properly"[3]. As has been stated elsewhere, up to the 1940's "more time was taken up in getting the cell to work than in carrying out research using the fluorine produced"[7].

Leech has pointed out that between 1919 and 1944 forty out of the fifty or so cells described in over fifty papers and patents were of the high temperature type[1].

The cell used by Lebeau and Damiens in 1925 may be considered to be the precursor of the modern medium temperature type[13]. With an electrolyte of composition KF·3HF (51% HF), having a convenient melting point†† of 65.8°, it was found

---

* Anhydrous hydrogen fluoride was not available commercially before 1931 in America, when the output was some 500 tons per year, and about 1942 in Britain. The need for chlorofluoromethanes for use as low toxicity, non-inflammable refrigerants was the original justification for producing the anhydrous product in America: in Britain it was the production of fluorine and uranium tetrafluoride for the "Tube Alloy Project" (see p. 3).

** Argo and his associates were interested in the potentialities of fluorine as a war gas and are on record as having made the following incautious statement "The free fluorine is not poisonous. With no problems whatever we have worked in a noticeable concentration for several weeks without ill effects."

† The term "polarisation", as applied to fluorine cells, is defined and discussed on p. 14 *et seq*.

†† Lebeau and Damiens give the melting point of KF·3HF as 56°. The correct value is 65.8° as given by Leech from the data of Cady[14].

that a nickel anode could be used and was free from polarisation troubles. Little attention was paid to this work until 1942, by which time Cady had investigated the potassium fluoride/hydrogen fluoride system in detail and recognised the outstanding merits of an electrolyte approximating in composition to KF·2HF (41 % HF) with a melting point of 71.7°[14, 15].

Cady's cell made use of either a nickel or non-graphitic carbon anode, but the former was preferred because of the absence of polarisation and other difficulties[15]. Graphitic carbon swells up and disintegrates in electrolyte of composition KF·2HF[16]. This work, and Cady's phase studies of the potassium fluoride/hydrogen fluoride system, together with his measurements of the hydrogen fluoride partial pressures for the same system, provide a complete rationale for the adoption of the medium temperature cell and all other types are now obsolete.

Until 1940* all the work carried out with elemental fluorine was essentially of an academic character and no commercial applications for the element (or for compounds which could only be derived therefrom) either existed or were foreseen. The discovery of nuclear fission with the concomitant possibilities of nuclear energy and the "atom bomb" altered this position dramatically and resulted in a requirement for fluorine on tonnage scale for the preparation of uranium hexafluoride, used in the diffusional separation of ^{235}U and ^{238}U.

On May 5th, 1940 the writer was asked to prepare a small quantity of uranium hexafluoride for Professor J. Chadwick (later Sir James Chadwick) at Liverpool University. Approximately 1 g was handed over on July 10th, 1940, while about a month later 7 g was passed to Dr. O. R. Frisch (now Jacksonian Professor of Natural Philosophy, Cambridge) who was then at Liverpool. Subsequently, uranium hexafluoride was supplied to Professor W. N. Haworth (later Sir Norman Haworth) and his school at Birmingham University and to Professor F. Simon (later Sir Francis Simon) and his team at the New Clarendon Laboratory, Oxford.

By the beginning of 1941, uranium hexafluoride was being prepared at the rate of some 300 to 400 g per month, but only with the expenditure of much effort using temperamental nickel anode, low temperature cells (Fig. 1). During 1943, 150 kg of uranium hexafluoride was produced using nickel anode medium temperature cells, while in 1944 the quantity made was 200 kg.

The above work was carried out in the Research Department of the former General Chemicals Division (now incorporated in Mond Division) of I.C.I. at Castner-Kellner Works, Weston Point, Runcorn, Cheshire. Initially it was sponsored by the Ministry of Aircraft Production, but it later became part of the so-called "Tube Alloy" project. This aspect of the General Chemicals Division's work for the "Tube Alloy" project ceased in December, 1945.

According to Neumark and Siegmund[17] the investigation of commercial

---

* No precision is claimed for this date, since details of American activities at about this time are lacking.

Fig. 1.
Early low temperature, nickel anode fluorine cells used by the General Chemicals Division of I.C.I. at Weston Point, Runcorn, Cheshire for the production of uranium hexafluoride during the early 1940's.

fluorine cells in America was started by the Harshaw Chemical Co., Du Pont and the Hooker Chemical Corporation in 1942 at the request of the U.S. Government. Pennsalt Chemical Corporation and Allied Chemical Corporation independently started fluorine production in 1946.

Both for the "Manhattan Project" in America and the "Tube Alloy" project in this country early supplies of fluorine on a laboratory or semi-technical scale were produced using nickel anode cells. This avoided the considerable difficulties involved in the use of carbon anodes, but such cells had no commercial future because heavy corrosion of the anodes limited the current efficiency to 60 to 70%, and accumulation of nickel fluoride in the electrolyte ultimately necessitated their being shut down for removal of the sludge[6].

Despite the initial impetus given to fluorine cell development by the "Tube Alloy" project, production of the element did not reach an appreciable scale in Britain during the war years because of the decision, following a visit of Sir John Anderson to America in 1943, to concentrate the major military atomic energy effort in that country. A short account of the build up and organisation of the British effort in the sphere is recorded in a Stationery Office publication[18].

Research on fluorine in the General Chemicals Division of I.C.I. continued after the war and ultimately resulted in the development of reliable carbon anode medium temperature laboratory cells having special features and on which the design of works scale cells was ultimately based.

American war-time activities relating to the design and operation of fluorine cells were first disclosed in 1947 in the March number of Industrial and Engineering Chemistry[19], which was devoted entirely to fluorine chemistry and, together with other information in this field, was ultimately published as part of the massive National Nuclear Energy Series[20]. The Massachusetts Institute of Technology[21], The Hooker Electrochemical Co.[22], The Harshaw Chemical Co.[23], Du Pont[24], The Pennsylvania Salt Manufacturing Co.[25], and Johns Hopkins University[26] all designed and operated fluorine cells. These ranged in capacity from 50 A (M.I.T.) to 2 kA (Hooker) and were all medium temperature carbon anode cells with the exception of those at Johns Hopkins University which were of the high temperature type. The most successful cell appears to have been the 2 kA unit developed by Hooker which formed the basis of design of the fluorine cells subsequently operated by Union Carbide[27]. In all cases difficulty was experienced in making a reliable electrical connection to the carbon anode (see p. 44). Polarisation was also a source of trouble which was overcome by a variety of empirical procedures (see p. 14).

Considerable developments in fluorine technology also took place in Germany during the war years[16]. According to Neumark*, who investigated the German war-time fluorine industry under the auspices of the U.S. Chemical Warfare Service in 1945, the impetus for these developments derived both from atomic energy investigations and the decision to produce the recently discovered chlorine trifluoride as an incendiary agent. Although this work received the highest priority it is now evident that it was chlorine trifluoride rather than the atomic energy aspect which provided the principal *raison d'être*.

Supporting this contention is the fact that the only sizable plant in Germany was one with a capacity of some 600 to 700 tons per year, which was erected at Falkenhagen near Berlin, and which produced fluorine for conversion to chlorine trifluoride. Falkenhagen is now in East Germany and much of the plant from this location was removed to Stuln in Lower Bavaria to prevent its falling into the hands of the advancing Russians.

In addition to the above, Kwasnik designed and operated, from 1942 onwards, a single 2 kA medium temperature carbon anode cell which was located at Leverkusen[16]. A noteworthy feature of this cell, the body of which was made of magnesium alloy, was the absence of a diaphragm and the fact that with an electrode separation of approximately 2.5 in (6.6 cm) the partition separating the fluorine and hydrogen compartments dipped little more than an inch into the

---

* Leech[1] gives references to reports of British teams which studied the German fluorine industry following the military collapse of Germany in 1945.

electrolyte. Since the current efficiency was only 75 to 80% it is evident that appreciable recombination of fluorine and hydrogen occurred.

In view of the successful operation of a medium temperature cell, its simplicity and its undemanding requirements as to materials of construction, it is surprising that the Falkenhagen plant consisted of 60 high temperature cells which had silver cathodes. The explanation probably lies in the importance the Germans attached to the project with the consequent necessity of freezing designs at an early stage.

Since 1950 only seven accounts of significance giving new information relating to the technology of fluorine production have appeared in the literature and two of these, published in 1958, merely bring up-to-date earlier accounts by the same producer (Union Carbide, formerly Carbide and Carbon Chemical Co.) which were published in 1955[7, 17, 27-31].

Following the 1943 decision to transfer the greater part of the "Tube Alloy" effort to the United States, fluorine production in Britain did not pass beyond the semi-technical scale and it was not until it was decided to build a British diffusional separation plant that a demand arose for fluorine production on a larger scale. The first of six medium temperature carbon anode fluorine cells of nominal 1kA capacity was started up at the Rocksavage (Runcorn, Cheshire) Works of I.C.I. in July, 1948. The fluorine from this plant was used to prepare a variety of products required for the initiation of the British Atomic Energy programme* including uranium hexafluoride, various fully fluorinated organic compounds (model compounds, lubricating and sealing liquids) sulphur hexafluoride and halogen fluorides, particularly chlorine trifluoride, over the period 1948–51. In 1952 a redesigned fluorine plant with twelve cells of nominal 1.4 kA capacity and a chlorine trifluoride plant of 60 tons per year capacity was commissioned. Subsequent increases in capacity were as follows: In 1955 there were 16 cells of 1.4 kA nominal capacity. In 1957 the 16 1.4 kA cells were up-rated to 2.2 kA. In 1959, 14 cells of nominal 5.0 kA capacity were installed. All the fluorine was converted to chlorine trifluoride.

Practically all the chlorine trifluoride produced at Rocksavage Works between 1952 and 1963 was consumed by the U.K.A.E.A. for the production of uranium hexafluoride at their Springfield Works, though a little was used in the manufacture of the I.C.I. chlorofluorocarbon ("Florube") lubricating oils and greases, now discontinued. In 1963 the demand for chlorine trifluoride by the U.K.A.E.A. ceased and this organisation now manufactures its own fluorine for uranium hexafluoride production using cells of basic I.C.I. design[32]. I.C.I. at present manufacture fluorine solely for conversion to sulphur hexafluoride.

---

* The United Kingdom Atomic Energy Authority was formed in 1954 and the first annual report published by H.M.S.O. covers the period 1954–55.

2. The electrolyte

Modern fluorine cells use an electrolyte consisting of a molten mixture of potassium fluoride and hydrogen fluoride approximating in composition to KF·2HF (40.8% HF), the usual working limits corresponding to a hydrogen fluoride content of 38 to 42% and temperatures ranging from 80° to 110°C.

2.1 MELTING POINT AND HYDROGEN FLUORIDE PARTIAL PRESSURES OF SOLUTIONS OF POTASSIUM FLUORIDE IN HYDROGEN FLUORIDE

A phase study of the system potassium fluoride/hydrogen fluoride was carried out by Cady[14] in 1934. The melting point/composition diagram is shown in Fig. 2 while the melting points of compounds and eutectics are given in Table 1. Cady also measured the partial pressure of hydrogen fluoride over the potassium fluoride/hydrogen fluoride system and its variation with temperature[14]. This information is of great importance in connection with the rational design and operation of fluorine cells.

In Fig. 3 hydrogen fluoride partial pressures at temperatures near to the melting points are plotted against composition, from which it is clear that an electrolyte of approximate composition KF·2HF has the lowest possible hydrogen fluoride

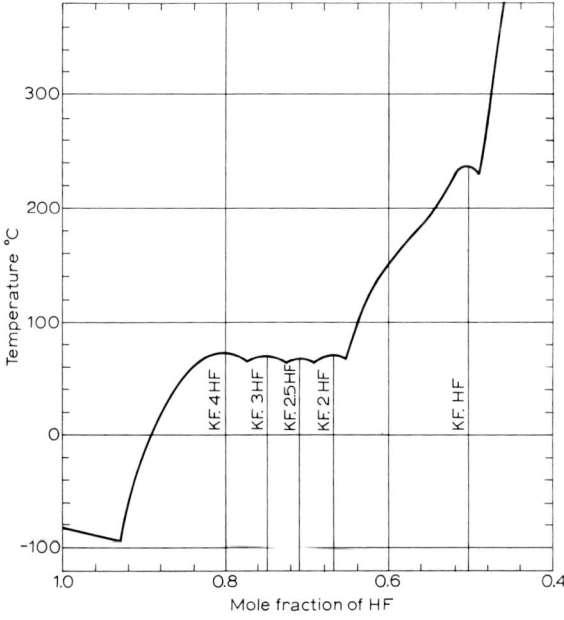

Fig. 2.
Melting point/composition diagram for the potassium fluoride/hydrogen fluoride system.

TABLE 1

MELTING POINTS OF COMPOUNDS AND EUTECTICS IN THE KF/HF SYSTEM[1, 14]

Solid phase	M.p. (°C)	% HF	Solid phase	M.p. (°C)	% HF
HF	−83.7	100	KF·2.5 HF	64.3	46.2
Eutectic	−97	88.3	Eutectic	61.8	44.2
KF·4HF	72.0	57.9	KF·2HF	71.7	40.8
Eutectic	63.6	53.7	Eutectic	68.3	38.5
KF·3HF	65.8	50.8	KF·HF	239.0	25.6
Eutectic	62.4	47.9	Eutectic	229.5	24.6

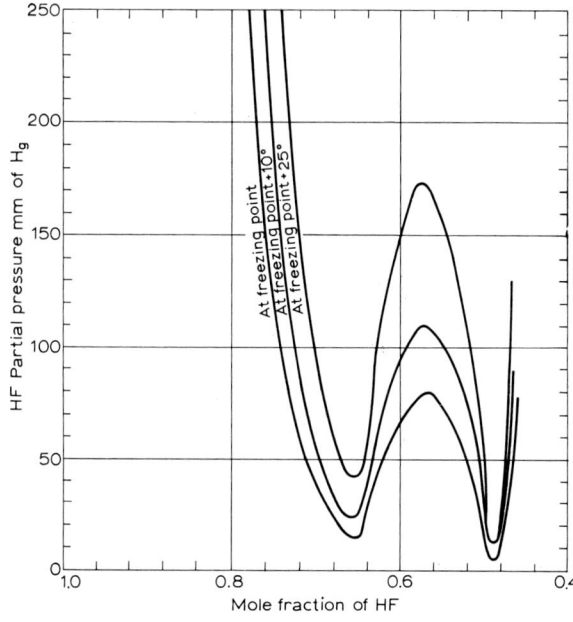

Fig. 3.
Partial pressures of hydrogen fluoride over the potassium fluoride/hydrogen fluoride system at temperatures near to the melting point.

partial pressure associated with a convenient operating temperature[2].

A statistical analysis of Cady's vapour pressure figures, involving multiple regression analysis, has been carried out by Baines and Davies, from which an equation relating partial pressure, temperature and composition was derived which is considered to give results which are accurate to ± 5% over the region of importance in fluorine cell operation[33].
The equation is:

$$\log p = 2.0733 - 4244\,(1/T) + 0.2975\,C + 47.94\,(C/T) - 0.003785\,C^2$$

where p is the partial pressure (mm Hg), T is the absolute temperature and C is the concentration of hydrogen fluoride expressed as weight per cent. Partial pres-

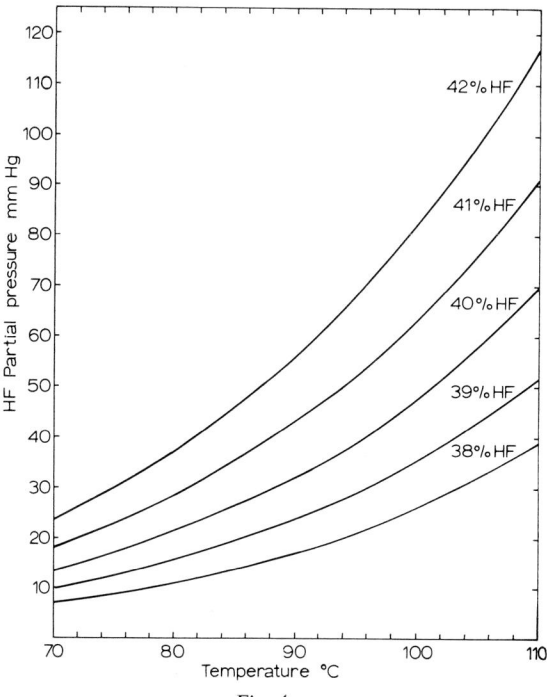

Fig. 4.
Partial pressures of hydrogen fluoride over the potassium fluoride/hydrogen fluoride system over the range 38 to 42% hydrogen fluoride and 70° to 110°.

sure figures derived from the above equation are plotted in Fig. 4 for the range 38 to 42% hydrogen fluoride and 70° to 110°.

Table 2 shows the variation of the partial pressure of hydrogen fluoride with temperature over the same concentration range together with the corresponding *equilibrium* concentrations to be expected in the hydrogen and fluorine. The hydrogen fluoride concentration in the hydrogen usually corresponds to the equilibrium partial pressure over the electrolyte at the temperature and composition existing in the main bulk of the electrolyte, but this may not apply in the case of the fluorine. As is not surprising in view of the high anodic overvoltage, the anode surface may be appreciably hotter than the bulk of the electrolyte, particularly where most of the fluorine is not discharged in free bubbles (see pp. 16, 38).

From the figures in Table 2 the advantage of operating cells at low hydrogen fluoride concentrations and low temperatures in minimising hydrogen fluoride losses in the product gases is apparent. A change from 38% hydrogen fluoride at 80° to 42% at 110° represents a ten-fold increase in the loss. Low temperatures and hydrogen fluoride concentrations are, of course, associated with a decrease in electrolyte conductivity and hence an increase in the working voltage of the cells (p. 29) so that it is necessary to strike an economic balance.

TABLE 2

EQUILIBRIUM CONCENTRATIONS OF HYDROGEN FLUORIDE OVER FLUORINE CELL ELECTROLYTE[14,33]

Temp. (°C)	HF concentration in electrolyte (wt. %)									
	38		39		40		41		42	
	P (mm)	Vol. (%)	P (mm)	Vol. (%)	P (mm)	Vol. (%)	P (mm)	Vol. (%)	P (mm)	Vol. (%)
80	11.2	1.5	15.6	2.1	21	2.8	28	3.8	37	4.9
85	14.0	1.8	19.4	2.5	26	3.4	35	4.6	46	6.1
90	17.4	2.3	24	3.1	32	4.2	43	5.7	56	7.4
95	21	2.8	29	3.8	39	5.1	52	6.8	68	9.1
100	26	3.4	36	4.7	48	6.3	63	8.3	82	10.8
105	32	4.2	43	5.7	70	9.2	76	10.0	98	12.9
110	39	5.1	52	6.9	83	10.9	91	12.0	117	15.4

2.2 ELECTRICAL CONDUCTIVITY

The only recorded measurements of the specific conductivity of potassium fluoride/hydrogen fluoride mixtures are those of Schumb, et al.[21], which cover the range 37.4 to 44.7% hydrogen fluoride at temperatures of 90°, 95° and 100°. Temperature control was accurate to ± 0.05°. The accuracy of hydrogen fluoride concentration is not stated but is given to the nearest 0.1%.

Since British fluorine cells commonly operate with the bulk of the electrolyte at temperatures of 80 to 90°, Fig. 5 gives extrapolated curves for 80° and 85° in addition to those covered by Schumb et al. Values of the specific conductivity at 80° and 85° read from these curves should be considered to be approximations.

2.3 SPECIFIC GRAVITY

The variation of the specific gravity of electrolyte with composition and temperature was also measured by Schumb et al.[21] and their results are shown graphically in Fig. 6. These investigators used the density of electrolyte at a specific temperature as a method of checking electrolyte composition.

2.4 SURFACE TENSION OF KF/HF MIXTURES

The surface tension of fluorine cell electrolyte is of importance in connection with the phenomenon which occurs at the surface of carbon anodes (see p. 14). An

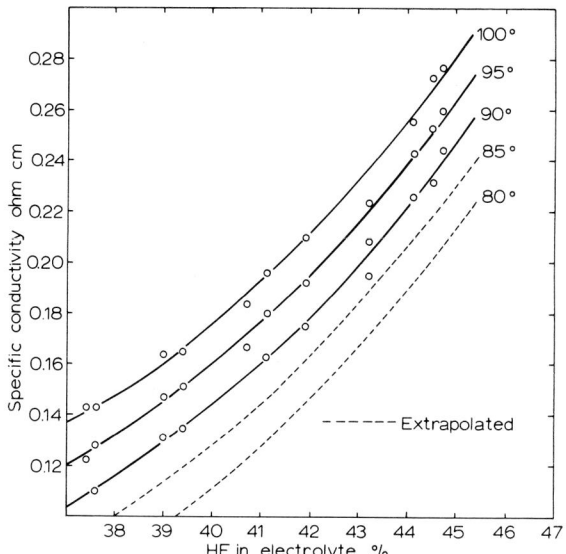

Fig. 5.
Specific conductivity of the potassium fluoride/hydrogen fluoride system over the range 37 to 45 % hydrogen fluoride at 80° to 100°.

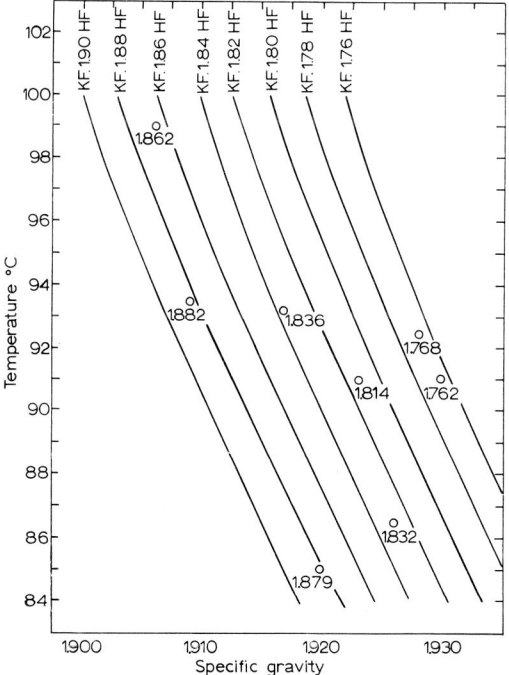

Fig. 6.
Specific gravity of the potassium fluoride/hydrogen fluoride system over the range 38 to 39.6 % hydrogen fluoride at 85° to 100°. o = Experimental molar ratio KF : HF.

investigation of the relationship between surface tension and electrolyte composition and temperature was carried out by Crowle in 1947–48[34] using the maximum bubble pressure method with two platinum capillary tubes of different diameter.

Calibration of the apparatus with benzene between 20° and 50° and with water between 17° and 90° gave results which agreed with published values to within ± 0.8%. The average accuracy of the measurement with KF/HF mixtures was estimated to be ± 0.5 dyne/cm. The results of measurements with KF/HF mixtures containing 0.2% of water are shown in Fig. 7.

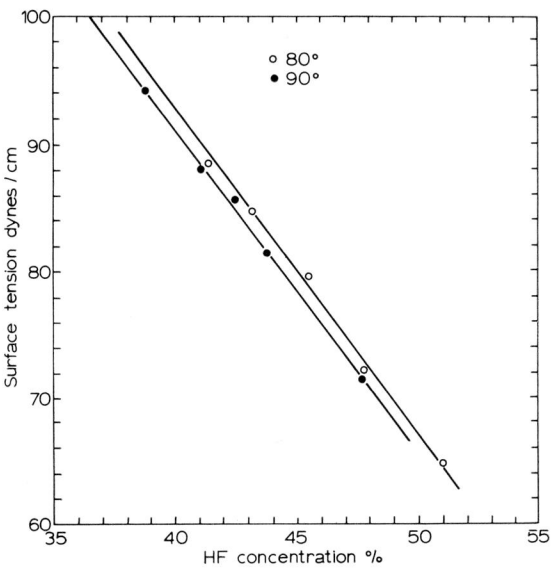

Fig. 7.
Surface tension of the potassium fluoride/hydrogen fluoride system over the range 39 to 51 % hydrogen fluoride at 80° and 90°.

There is a linear relationship between surface tension and composition corresponding to a decrease of about 2.5 dyne/cm for each 1% increase in HF concentration. Though it is impossible to give an accurate figure, the temperature coefficient appears to be about 0.1 dyne/°C. An isolated measurement of the surface tension of KF·HF at 245° gave a value of 118 dyne/cm.

Water in concentrations up to about 3% had only a small effect on the surface tension of mixtures containing 38.5 to 52.1% of HF at 80° and measured changes are probably not significant. The addition of 2% of lithium fluoride to solutions containing 39 to 51% of HF and about 1% of water at 80° produced small and erratic changes (−2.5 to +7.0 dyne/cm).

2.5 Electrolyte manufacture

Electrolyte is manufactured by adding potassium bifluoride to a quantity of the pre-formed product contained in a steam-jacketed steel vessel, fitted with an agitator and heated bottom run-off, and then passing liquid hydrogen fluoride into the molten charge *via* a dip-pipe until the required composition is reached. Provision is made for the removal of hydrogen fluoride vapour from the reactor *via* a duct to an absorption system. The process is repeated until the vessel is full, after which the product is run off directly into any cells which require to be filled.

If the preparation of electrolyte has to start from scratch, then liquid hydrogen fluoride is added cautiously to solid potassium bifluoride while cooling water is passing through the reactor jacket, since the reaction is exothermic. Having prepared sufficient product to cover the hydrogen fluoride dip-pipe the procedure follows that described above.

Since electrolysis removes only hydrogen fluoride from the electrolyte, which is replaced by appropriate additions during the operation of the cells, only sufficient electrolyte is required to fill the cells initially and to replace "drag out" losses, or losses resulting from inadvertent flooding of a cell by water should a leak develop in the cooling system when the wet electrolyte must be disposed of.

2.6 Electrolyte specification

An exacting specification is not required for British fluorine cells, except that a reasonably low water content is desirable and the sulphate content must not be too high, otherwise attack or disintegration of the anodes will occur. An acceptable specification for the potassium bifluoride starting material would be K_2SiF_6 2.5% max., K_2SO_4 0.5% max., H_2O 0.3% max. American fluorine cells are said to require a very low water content in the hydrogen fluorine feed so that the water content of the electrolyte should also be as low as possible. Allied Chemical Corporation's specification for potassium bifluoride used for electrolyte manufacture is KF·HF 99.0% min., KCl 0.02% max., K_2SiF_6 0.5% max., SO_4 0.01% max., Fe 0.02% max., heavy metals as Pb 0.005% max., H_2O 0.1% max.[17].

Union Carbide Corporation use potassium bifluoride containing less than 0.01% of sulphate while the electrolyte is conditioned before use by bubbling fluoride through it until the water content is between 0.001 and 0.003%[27].

While the above statements apply to electrolyte before it is used in the cells, it has now been shown that the behaviour of cells is strongly influenced by the presence of nickel in the electrolyte resulting from corrosion of nickel or "Monel" skirts or diaphragms. The effect of this nickel is dependent on the water concentration of the electrolyte (see p. 24). Other deliberate additions may be made such as lithium or sodium fluorides (see p. 24).

It is necessary for the sulphate content of the electrolyte to be reasonably low; 2% caused some attack of the anode while 5% resulted in complete disintegration[31].

Allied Chemical Corporation specification for hydrogen fluoride used for electrolyte manufacture and cell feed calls for HF 99.5% min., SO_2 0.005% max., H_2SiF_6 0.001% max., H_2O 0.02% max., H_2SO_4 (including HSO_3F) 0.005% max. and nonvolatile acids (as H_2SO_4) 0.01% max.[17]. Since Union Carbide Corporation[28] specifies 99.5% pure hydrogen fluoride it would appear that they obtain this from Allied Chemical Corporation.

The specification for hydrogen fluoride used in I.C.I. cells is: HF 99.6% min., SO_2 0.03% max., H_2O 0.2% max., H_2SO_4 0.005% max., H_2SiF_6 0.14% max.

3. Anode phenomena

3.1 POLARISATION

From the earliest years the recurrent incidence of "polarisation" was a serious difficulty in the development of both medium and high temperature, carbon anode fluorine cells.

Though the term "polarisation" is used in a number of different senses in electrochemistry, in the present connection it is restricted to the condition under which, *at a fixed voltage*, a sudden or gradual decrease occurs in the current flowing through the cell to a value which is a small fraction of that passing when the cell is operating normally*. It is essentially an anodic phenomenon. Though it has been stated that cathodic polarisation can occur in fluorine cells the cause of a high resistance developing at this electrode has no relation to the type of process which occurs at the anode[21]. A high resistance at the cathode is most likely to result from such a mundane cause as a low hydrogen fluoride content in the electrolyte with resultant deposition of the solid KF·2HF (m.p. 71.7°) or the eutectics 2KF·3HF (m.p. 68.3°) or 2KF·4.5HF (m.p. 61.8°) depending on electrolyte composition, on the electrode (see p. 8).

A detailed historical account of the difficulties experienced by early investigators would do little to illuminate the subject now that the basic reason for polarisation is understood, but some idea of the confusion which existed as late as 1949 is provided by Leech[1].

---

* In this section and elsewhere the word "polarisation" as used in connection with fluorine cells has the above defined meaning. It does not refer to overvoltages as usually defined which result from phenomena common to all electrolytic processes, such as the series of steps, ions → molecules → gas bubbles. The ohmic resistance and resulting potential gradient due to these causes is a small fraction of that caused by "polarisation" in this special sense. Use of the term "polarisation" in this way is retained because it has been so employed since the earliest days of carbon anode fluorine cells.

3.2 THE ANODE/ELECTROLYTE CONTACT ANGLE

Polarisation was first reported by Argo and co-workers in 1919 during the operation of a high temperature cell with a graphite anode[11]. In the discussion which followed their papers, Mathers stated that the graphite anode was not "wetted"* by the electrolyte, and this observation was confirmed by many subsequent users of the high temperature cell[12]. This non-wetting effect was re-encountered when the medium temperature cell became the preferred choice and Cady pointed out that anodes removed from the cell and washed with water retained this characteristic[15]. Water adhered to the whole cathode surface but only to a few spots on the anode.

If a piece of ungraphitised carbon is inserted into molten $KF \cdot 2HF$ (or graphite into molten $KF \cdot HF$) it is "wetted" by the melt and the contact angle is substantially zero. If, now, the carbon is made anodic with respect to a second electrode the contact angle increases within seconds to a high value which approximates to 150° and (as ordinarily understood) the carbon is no longer "wetted" by the melt (Fig. 8).

There was much speculation as to the reason for this behaviour until it was suggested by Ruff that it might be due to the presence of a film of the intercalation compound carbon monofluoride, $(CF)_n$[40,41]. The correctness of this hypothesis was demonstrated by Rüdorff and co-workers using X-ray analysis[42]. It has also been

---

* It should be noted that the relationship between the specific free surface energy of a solid surface, the surface tension of a liquid and the contact angle is not the simple, uncontroversial matter which those unfamiliar with the subject might suppose[35]. In particular, considerable confusion still exists, on a semantic plane, as what is meant by "wetting", and misunderstanding will result in the absence of a clear definition[36].

The term "wetted" is commonly used in a restricted sense with reference to the spreading of a liquid as a continuous film on a solid. Under such conditions the contact angle is zero.

The work of adhesion, W_{SL}, between a solid and a liquid is related to the surface tension of the liquid in air or any other gas, γ_{LA}, and the contact angle θ by the expression:

$$W_{SL} = \gamma_{LA}(1 + \cos\theta)$$

when $\theta = 0$, $\cos\theta = 1$ and $W_{SL} = 2\gamma_{LA}$

Thus the contact angle becomes zero when W_{SL} attains a value which is equal to twice the surface tension of the liquid.

It follows that, if the work of adhesion between the solid and the liquid is less than twice the surface tension, the liquid will make a finite contact angle with the solid. When, however, $\theta = 180°$, $\cos\theta = 1$ and $W_{SL} = 0$. This represents a limiting case where there is complete absence of interaction between the molecules of the liquid and of the solid. It is highly improbable that such a state is realisable and no examples of *true* contact angles of 180° are known[36,37]. ("Whether, in the case of a liquid displacing its vapour from a solid surface, but at a finite rate, the dynamic advancing contact angle could ever reach 180° is uncertain, at the moment, but improbable.") Because roughness of the surface of a solid increases the *apparent* value when the true value exceeds 90°, examples are known where the contact angle *appears* to be 180°[37].

The above considerations make it possible to define "wetting" more accurately as follows: *"Complete wetting" of a solid by a liquid is represented by a contact angle of zero, and "complete non-wetting" by a contact angle of 180°. All intermediate values of the contact angle represent varying degrees of wetting or "partial" wetting*[39].

Fig. 8.
Illustration shows large contact angle *(reversed meniscus)* at the surface of a carbon anode in medium temperature cell electrolyte (approximately KF·2HF). Note "wetted" gauze cathode and evolution of fluorine at air/electrolyte/carbon interface as shown by paper soaked in potassium iodide solution.

shown that carbon monofluoride in bulk is wetted neither by water nor by molten KF·2HF, *i.e.* these liquids show a large contact angle when in contact with it[43].

3.3 Mechanism of fluorine evolution

In spite of the early recognition of the existence of a large contact angle at the electrode/electrolyte/gas interface of a working carbon or graphite anode, the effect of this on the mechanism of gas evolution at the surface of the carbon anodes of fluorine cells does not seem to have been fully appreciated until 1947[44]. In 1933 Kabanov and Frumkin[45] showed that the shape and size of gas bubbles produced electrolytically at a mercury surface depends on the electrode/electrolyte/gas contact angle and that where the contact angle is equal to or greater than 90° the shape of the bubbles must, of necessity, be lenticular and they will adhere to the electrode surface.

Where lenticular bubbles are produced on a vertical low permeability electrode bubble breakaway cannot readily occur in the usual way, and transfer of gas to the electrolyte surface more often takes place by slipping of the bubbles up the electrode under the influence of buoyancy forces, or by a process of coalescence and passing on of gas from bubble to bubble—as with rain drops running down a not-too-clean window pane (Fig. 9).

Rüdorff et al.[42] observed the same phenomena under the conditions of "anode effect" (see p. 19) in a high temperature cell. "At the moment of the onset of the voltage increase there occur at the electrode individual points of light. As the voltage increases they increase rapidly in number and size and finally rise *perpendicularly* from the anode in large, white, shining gas bubbles. If the anode is inclined in the melt *the luminous gas bubbles wander from the lower side of the anode to the upper and creep preferentially along this until the surface of the melt is reached*". (A.J.R.'s italics).

Much later, in 1961, the above phenomena were described independently by Watanabe et al. who appeared to be unaware of the prior work described above[46].

Fig. 9.
Lenticular fluorine bubbles on surface of low permeability carbon anode. Note electrolyte meniscus near top of photograph.

At the (steel) cathode of a fluorine cell the contact angle is, or approximates to, zero and the hydrogen bubbles are spherical and relatively small and readily break away from the electrode surface as in normal aqueous electrolysis. Hydrogen bubbles produced at the start of electrolysis are frequently much smaller than those evolved at a later stage, which may indicate a small progressive increase in contact angle.

3.4 Mechanism of polarisation

The occurrence of polarisation in carbon anode fluorine cells is a direct consequence of a large anode/electrolyte/gas contact angle and, in the sense in which the term is used here, is unknown in aqueous electrolysis. At a constant value of the contact angle, growth of bubbles on an impervious or low permeability anode can occur only by increase in the length of the bubble perimeter with a resultant reduction in the area of the anode in contact with the electrolyte (Fig. 9). However, this cannot be the situation when the anode is in the polarised state since the presence of lenticular bubbles represents a transient, unstable condition, which can usually only be maintained by careful control of current density, and normally the current falls spontaneously and rapidly to a very low value and the bubbles shrink and disappear. Under these circumstances no bubbles are visible even when the surface of the anode is examined with the aid of a microscope.

While it cannot be considered to be proven, the following explanation of polarisation appears to be in accordance with the known facts. Figure 9 shows that an appreciable area of the electrode is apparently in contact with the electrolyte and it would appear that bubble coverage has not reached its ultimate extent over the whole of the electrode surface. It seems probable that when this does occur the effective anode area is so reduced that the rate of fluorine production drops sharply and the gas contained in the large bubbles is discharged into the interior of the carbon, most samples of which show at least some permeability (see Table 3). As the bubbles collapse the electrolyte approaches the electrode surface until only surface micro-asperities are in physical contact with the electrolyte, and the small residual current which continues to flow, even in the polarised state, corresponds to that small part of the surface which remains in contact with the electrolyte. Any increase in current beyond that corresponding to the rate at which gas can escape through the anode (or *via* interconnected channels on the anode face) results simply in a further reduction in the effective surface and an equilibrium state is established[31]. The shining, light-reflecting surface of the right-hand receding face of the anode in Fig. 9 is consistent with the above-described role of surface asperities and would explain the apparent absence of bubbles when the surface is viewed in a direction normal to the anode surface.

It is particularly to be noted that the existence of such a light-reflecting gas film neither implies nor requires the invocation of a contact angle of 180° and the existence of a complete "gas sheath" round the anode, as has been claimed by a number of investigators, including Kroll[47].

3.5 ANODE EFFECT

The so-called "anode effect" is readily produced by increasing the voltage across a fluorine cell having an anode in the polarised state, so that the two phenomena are obviously related. Judged from accounts in text books of industrial electrochemistry, misunderstanding as to the cause of anode effect appears to be widespread.*

Arndt and Probst[50] and Taylor[51] have provided general descriptions and discussed the conditions which give rise to anode effect in fused salt electrolysis, other than in the production of fluorine. This is discussed in particular detail by Rüdorff et al.[42] while Watanabe et al.[52] have studied the critical current density at which anode effect occurs in fluorine cells under a variety of conditions.

Arndt and Probst carried out most of their work in fused chloride melts and showed clearly that the critical current density at which anode effect commenced is strongly dependent on the oxygen content of the melt; the lower the oxygen (oxide) content, the lower the current density at which anode effect occurred. It thus appears that in the electrolysis of oxygen-free fused salts the work of adhesion between carbon or graphite anodes and the electrolyte is relatively low, with an associated large contact angle. The reason for the low work of adhesion is known in the case of fused fluoride electrolytes (see p. 15) but this is not so in the case of fused chlorides.**

Anode effect may be explained as follows. On raising the voltage across a cell which is in the polarised state, a point is reached at which the current increases in spite of the high electrical resistance due to the reduced effective area of the anode, *i.e.* that part of the anode which is in actual physical contact with the electrolyte at surface asperities. The current is now restricted to those parts of the anode which are not masked by gas. These electrical "bridges" become overheated as a result of the very high local current density and break down, possibly by vaporisation. In the act of breaking the contact, minute arcs or sparks are produced. This gives rise to an unstable condition and temporary equilibrium is re-established by the electrolyte coming into contact with the anode at other spots where the

---

* Mantell[48] has this to say by way of explanation of anode effect: "Assume first that the anode effect is the formation of a gas film surrounding substantial portions of the anode. *The film prevents the electrolyte from wetting the anode*: the result is observed in the formation of myriads of tiny arcs, *since current must pass from anode to electrolyte through the gas film*. (A.J.R.'s italics). Hampel[49] does not mention anode effect or polarisation as used in the present connection.
** Recent work in connection with fuel cells has shown that molten lithium chloride does not "wet" carbon or graphite but no explanation for the phenomenon appears to be forthcoming[53].

temperature is lower. These events are repeated with high frequency over the anode surface and explain the shifting arcs and sparks which are a feature of the phenomenon, while local liberation of energy may be so great that the gas becomes incandescent, or, more probably in the case of fluorine cells, the incandescence results from the heat of reaction of the anode gas with the anode.

If the gas evolved is capable of reacting readily with carbon or graphite at the temperature at the anode surface, a "clean", easily-wetted surface is produced and electrolysis resumes its normal course, until the carbon surface is re-contaminated by a lyophobic film.*

It has been erroneously assumed that a rise in cell voltage is a *result* rather than a cause of anode effect. Whether or not the voltage across a cell rises spontaneously, or remains sensibly constant in the event of an increase in internal resistance, is obviously decided solely by the electrical characteristics of the circuit. If these are such that the voltage cannot rise appreciably above the normal operating voltage, polarisation will occur: if a sufficient increase in voltage takes place, then polarisation will pass over into anode effect.

Medium temperature fluorine cells are unusual in industrial electrochemistry in that the anode wear rates are very low except in the presence of appreciable quantities of water and anodes appear to retain their original dimensions indefinitely, provided that burning in fluorine does not occur as a result of breakage or the development of hot spots at defective electrical connections (see p. 44). In the electrolytic production of aluminium and magnesium, in normal operation the anodes burn to carbon oxides because of attack by oxygen liberated as a result of the electrolysis of dissolved alumina and magnesia respectively.

Polarisation and anode effect are particularly well known in aluminium production[55] if there is a deficiency of alumina in the molten cryolite electrolyte,** and have been reported to occur, though infrequently, in Downs cell operation. That polarisation has been a particularly troublesome feature with fluorine cells is a consequence of the substantial absence of oxygen in the system with resulting absence of progressive attack of the anode.

---

* Rüdorff et al.[42] carried out their investigations using a high temperature cell with an electrolyte of molten KF·HF but most of their observations are relevant to medium temperature cells. It was noted that a new graphite anode became covered with a sooty layer after a short period of operation in KF·HF which, they said, resulted from the conversion of graphite asperities to carbon monofluoride. After operating at an increased voltage, such that anode effect occurred, this powdery material was burnt off with the formation of carbon tetrafluoride and higher fluorocarbons and the graphite had a highly polished appearance, previously and erroneously attributed to a glazed coating of impurities (Dennis et al.[54] went so far as to claim that it was a siliceous coating). This polished surface carried an invisible coating of carbon monofluoride.

Because polarisation could be temporarily cured by burning off the powdery surface coating by inducing anode effect, Rüdorff et al. concluded that anode effect was promoted by rough and decreased by smooth surfaces, but this is an over-simplification.

** As Mantell[48] points out, anode effect occurs most readily with molten fluorides and least readily with iodides.

4. Methods of avoiding or overcoming polarisation

The foregoing description of the mechanism of polarisation leads directly to two possible ways of overcoming the difficulty: (a), where a large anode/electrolyte contact angle exists, means must be provided whereby accumulation of gas at the anode surface is prevented or (b) additions to the electrolyte must be made, or other procedures adopted, which will have the effect of reducing the magnitude of the anode/electrolyte contact angle to a value which is considerably less than 90°.

4.1 INFLUENCE OF STRUCTURE OF ANODE MATERIAL ON POLARISATION

When the anode/electrolyte contact angle is greater than 90° escape of gas from the electrode surface is facilitated by increasing the permeability of the material of the electrode, provided that the interconnecting pores are not too large in relation to the surface tension of the electrolyte, the contact angle and the hydrostatic head of electrolyte[31,39].

The pressure P with which a liquid of surface tension γ is driven into a capillary of radius r by surface tension forces is given by the expression: $P = 2\gamma\cos\theta/r$, where θ is the contact angle. For all values of θ less than 90° P is positive and the liquid will spontaneously flood the capillary. For values of θ greater than 90° P is negative and there is a resistance to flooding. The condition which defines the depth of immersion at which flooding tendency is just balanced by hydrostatic forces, where θ is greater than 90°, is, of course, $h\varrho g = 2\cos\theta/r$, where h, ϱ and g have the usual significance.

It has been shown that there is a relationship between the permeability of various samples of carbon and their resistance to polarisation when used in a single large tank of molten electrolyte so that the conditions of test were the same for all samples (Table 3)[31,56].

In considering the results in Table 3, it should be borne in mind that a strictly quantitative relationship between permeability and resistance to polarisation cannot be expected because the permeability measured was the *total* permeability resulting from pores of all sizes, whereas the *effective* permeability relates only to those pores which are not so large as to be flooded with electrolyte under the conditions of operation. An additional factor to be taken into account in applying such results to the design of fluorine cells is that the surface-to-volume ratio is obviously of importance and carbons with permeabilities at the lower end of the scale may give results on a large scale which are superior to those obtained with small sized test anodes.

From the equation relating critical capillary size to hydrostatic head, it can be shown that the maximum pore diameter for a 1 cm depth of electrolyte is 846 μ and for a depth of 25.4 cm (10 in.) is 33μ, assuming a contact angle of 150°, a

TABLE 3

RELATIONSHIP BETWEEN THE AIR PERMEABILITY OF CARBONS AND THEIR TENDENCY TO POLARISE WHEN USED AS FLUORINE CELL ANODES[31]

Grade of carbon	Mean pore size (μ)	Permeability*	Period of operation before polarisation (hr) ø		
			Current density		
			A/in.² 0.5	1.0	2.0
			A/cm² 0.074	0.148	0.296
"Carbocell" 20+	140	18.0	⎫	⎫	24
"Carbocell" 30+	99	16.0	⎪ No polarisatn.	⎪ No polarisatn.	24
"Carbocell" 40+	69	19.0	⎬ in 242 h	⎬ in 99 h	
"Carbocell" 50+	48	3.2	⎭	⎭	
"Carbocell" 60+	33	2.3	181.0		
"Carbocell" C+	5	0.08	3.4		
National Carbon Co. CA.1260		0.06	1.2		
British Acheson		0.05	1.3		

Test conditions. Anode size 0.75 in. diam., 1.9. in long; electrolyte composition, KF 60%, HF 40% ($\pm$ 0.3%); initial water content of electrolyte 0.24% ($\pm$ 0.05%); electrolyte temp. 85.5°C ($\pm$ 2.5°).

* Permeability expressed as number of ft³ of air/min passing through 1 ft² of carbon 1 in. thick under a pressure equivalent to 2 in. of water. Measurements made on specimens 1 in. diameter and 1 in. long. *These units are those used in the manufacturer's specifications for porous carbon and graphite.*

+ Product of the National Carbon Co. of America.

ø In practice it is not always easy to measure precisely the critical current density at which polarisation (or anode effect) occurs since the results are seldom reproducible even with the same anode under the same conditions. It has been found that anodes in molten KF·2HF show a "memory effect" in that the critical current density for anode effect depends on the time which has elapsed since the phenomenon was previously induced[59].

density of 1.905 g/cc and an electrolyte surface tension of 90.6 dyne/cm. It is evident, therefore, that the pore size chosen must necessarily be a compromise so far as an acutal anode is concerned. It would appear that, ideally, anode material should have the highest possible permeability derived from pores not larger than that corresponding to the critical size for the maximum depth of immersion of the anode.

Means are available for measuring the pore size distribution of porous solids[57] and it is of interest to note that the distribution has been measured for "Graphicell" 50 and 60, which have the same mean pore size as the corresponding grades of "Carbocell"[58]. The pore size distribution is shown in Fig. 10 from which it may be seen that the mean pore size agrees closely with the manufacturer's figures (Table 3), the distribution being roughly symmetrical.

Advances have recently been made in the production of carbon and graphite having controlled pore size for use in fuel cells and some of this work may be

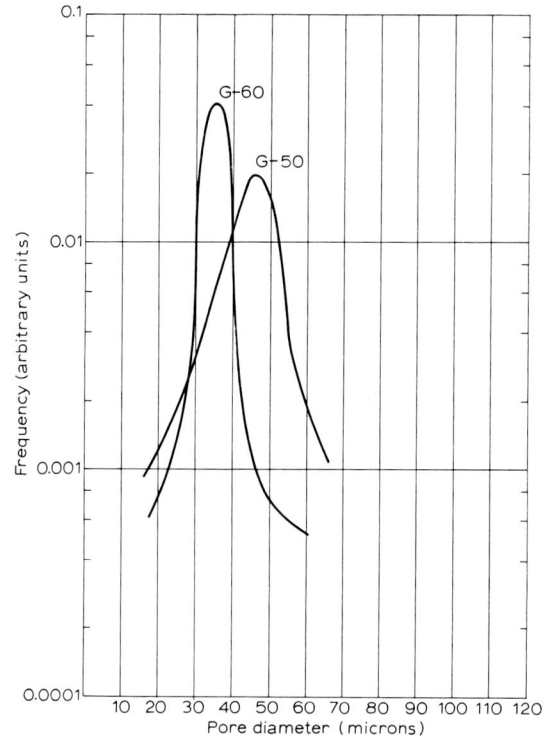

Fig. 10.
Pore size distribution for "Graphicell" carbon, grades 50 and 60.

relevant to the production of high permeability anodes for fluorine cells[60, 61]. Similarly, many of the theoretical considerations which apply to non-wetted fuel cell electrodes, including the study of mass transport through porous media, will also be applicable to the study of the behaviour of high permeability fluorine cell anodes[62].

The beneficial effect of using permeable anodes in reducing the critical current density at which anode effect occurs in fluoride-containing melts has been confirmed by Watanabe and co-workers[63] while a Japanese patent relates to the production of special types of carbon suitable for fluorine cell anodes[64].

McLaren of Oak Ridge National Laboratory has reviewed previous work relating to polarisation in fluorine cell anodes and described experimental investigations designed to overcome it, but without providing any explanation of the basic cause[65]. This investigator also refers to porous carbon anodes being free from polarisation at current densities up to 7 A/in.^{2} (1.04 A/cm^2) but no practical use appears to have been made of this information.

American investigators[66] have also used porous anodes in the high temperature electrolysis of molten fluorides to the same end.

4.2 INFLUENCE OF ELECTROLYTE PURITY AND ADDITIVES ON POLARISATION

Whereas cells which use high permeability carbon anodes are insensitive to electrolyte composition and the presence of minor impurities so far as polarisation is concerned, the reason why it is also possible to use dense, very low permeability anodes remains to be explained. The explanation almost certainly lies in the presence of deliberately-added, or adventitious, impurities in the electrolyte which have the effect of bringing about a reduction in the anode/electrolyte contact angle[31].

No certain method of preventing polarisation was known up to the time when intensive development of the fluorine cell began in the early years of World War II. Though early post-war accounts of American fluorine development make it clear that polarisation was no longer such a problem as to make reliable operation practically impossible, the subject remained confused and some of the alleged requirements for avoiding the trouble are conflicting[31]. Polarisation was still a feature of cells using low permeability anodes as late as 1955, and presumably still is[17, 28].

The water content of the electrolyte was long-suspected as being important in relation to polarisation[11, 15, 22, 24, 67-71]. It was also claimed that the addition of lithium fluoride, which is only sparingly soluble in the electrolyte, would cure polarisation though the effect was transient[23]. Sodium and aluminium fluorides were stated to have the same effect [23]. Watanabe *et al.*[72] found that the impregnation of carbon anodes with molten lithium fluoride was more effective than simple additions to the electrolyte, which were only operative when the lithium fluoride was present in colloidal suspension. It was also found that the incidence of polarisation was reduced when the electrolyte was pre-electrolysed using a nickel anode, an effect which was attributed to a reduction of the water content of the electrolyte[23, 25]. This conclusion as it stands is, however, an over-simplification.

It has recently been reported that nickel salts in the electrolyte are very effective in reducing the anode/electrolyte contact angle *provided that the nickel is in suspension and in a higher valency state*[*31]. If the concentration of water in the electrolyte is above about 0.1 % it is not possible to oxidise the nickel to the higher valency state. *The effects of water and nickel are thus interdependent* and it is necessary to

---

* Klemm and Huss[73] state that the compound K_2NiF_6 results on fluorinating a mixture of KF and $NiCl_2$ in two to one molar ratio. This would mean that Ni^{IV} was present. This compound is described as red and is decomposed by water with the evolution of a gas, presumed to be oxygen difluoride. Electrolyte containing high concentrations of nickel adhering to the surface of anodes removed from cells is initially pink in colour, but this fades on exposure to atmospheric moisture. Elsewhere reference is made to the compound K_3NiF_6[74] and it is not clear from the literature whether nickel can exist in both tri- and tetravalent states in fluorine cell electrolyte[75]. Neumark[17] has stated that NiF_3 is produced on electrolysing $KF \cdot 2HF$ with a nickel anode, but the existence of this compound has not been confirmed.

reduce the water content to a low level before the nickel is effective in causing "wetting" of the anode. It has also been shown that nickel present in a high oxidation state in the electrolyte migrates to the anode surface.

Thus an anode operating in an electrolyte containing only 7 p.p.m. of nickel in suspension on eventual withdrawal had acquired a coating of electrolyte containing 105 p.p.m. of nickel. In another case the concentration of nickel in electrolyte adhering to the anode was as high as 1,500 p.p.m. Under such conditions the anode is extensively "wetted" by the electrolyte (*i.e.* the contact angle is reduced to well below 90°) and break-away of fluorine bubbles occurs as in normal electrolysis. Low concentrations of water are effective in reversing this effect. These observations make it clear why preliminary electrolysis with a nickel anode is effective in preventing polarisation with carbon anodes of low permeability. The process not only introduces nickel into the electrolyte but also sufficiently reduces the water content to allow the nickel to be effective in promoting "wetting" of anodes. The influence of nickel, added *in this way*, in promoting "wetting" of anodes is, however, usually transient[31].

Even with low concentrations of water, copper fluoride was ineffective in causing wetting of the anode. Sodium, lead and calcium fluoride had a weak effect, which was trivial as compared with that of nickel fluoride[31]. Lithium fluoride probably behaves in a similar manner, while the same may be true of aluminium and magnesium fluorides.

4.3 THE USE OF LOW PERMEABILITY CARBON ANODES

Since, in the absence of electrolyte additives, the use of anodes made of low permeability carbon makes the incidence of polarisation inevitable, except at low current densities, it remains to be explained how this material is used in practice. Union Carbide cells[27] are fitted with Monel* metal skirts and screens and these suffer slow bipolar corrosion so that low concentrations of nickel will be introduced *continuously* into the electrolyte and it is to be expected that this will result in a reduction of the anode/electrolyte contact angle. The very low water content of the electrolyte (0.001 to 0.003%) will ensure that suspended nickel salts will be converted to, and remain in, a high valency state once formed. Polarisation which is commonly experienced when starting up new cells is usually cured by operation for 2 to 10 min at 30 to 40 V. This will have the effect of inducing anode effect, with consequent superficial combustion of the anode surface and removal of the carbon monofluoride coating; it will also accelerate corrosion of the Monel metal skirts and diaphragms to provide the nickel required to promote "wetting"

---

* "Monel" is a registered trade mark.

of the anode. When Monel metal is made anodic in KF·2HF the current/voltage curve shows a very large increase in the current flowing with increase in voltage above about 2V[29]. From 2 to 5 V the current increases by a factor of 10^4.

An alternative start-up procedure is to operate the cell at a very low current density 0.15 to 0.17 A/in.^{2} (0.022 to 0.025 A/cm^2) for as long as 48 to 72 h before increasing the current to the normal value. By operating below the critical polarisation current density in this way it is presumed that water is removed and that sufficient nickel is transferred from the skirt and diaphragm to the anode to promote wetting by the electrolyte so that operation at higher current densities then becomes possible.

Allied Chemical Corporation's cells have a machined magnesium alloy top with integral gas separation skirts. Polarisation is cured by applying a high voltage for 3 to 5 min as described above[17]. Though it is not specifically stated, it is presumed that some bipolar corrosion of the magnesium alloy skirts occurs. In the absence of further information it would appear probable that the continued introduction of electrolytically produced magnesium fluoride has the same effect in promoting "wetting" of the anodes as does the addition of lithium and other fluorides, which have a low solubility in the electrolyte, when these are present in a sufficiently fine state of subdivision.

While it would be desirable to have further information on the above subject, it is of interest to note that, at least in the case of the Union Carbide Corporation's cells, a rational explanation can be provided as to why a very low water content in the electrolyte is considered necessary.

5. The reversible voltage

The reversible voltage of the standard fluorine electrode was first calculated by Latimer[76] who derived a value of 2.85 ± 0.04 V from the heat of formation of HF in solution and the free energy of the resulting ions in hypothetical $1M$ dilution, giving a free energy of formation of HF of 65.6 kcal. The above value of the reversible voltage, which has often been quoted, refers to aqueous solutions and has sometimes been erroneously related to water-free HF solutions[22,39]. Fredenhagen and Kreft[77] reported an experimentally derived value of 2.678 V* in anhydrous HF containing KF at 0° while v. Wartenberg[77] stated that he had been unable to obtain an e.m.f. with fluorine and hydrogen electrodes in a melt of composition KF·1–3 HF. Garner and Yost[78] similarly failed to obtain a figure for the reversible potential of the fluorine electrode.

From measurements on eight different molten mixtures of chlorides and fluorides

---

* Such a value suggests that the electrolyte used was far from anhydrous.

Neumann and Richter[79] obtained a mean figure of 1.923 V, as compared with 1.929 V by direct measurements in molten KF·HF using a fluorine/graphite anode. The methods used by Neumann and Richter have, with justice, been criticised by Ruff and Busch[80]. This work must now be considered of historical interest only.

Accepting Latimer's value of 2.85 V for the standard potential of the fluorine electrode (in aqueous solution), Watanabe *et al.* calculated the value of the reversible potential referred to anhydrous hydrogen fluoride by making allowances for the different solvation energies of the ions under the two conditions[81].

$H^+ + H_2O \rightarrow H_3O^+$ = 255 kcal $H^+ + HF \rightarrow H_2F^+$ = 283 kcal
$F^- + H_2O \rightarrow F^-(H_2O)$ = 123 kcal $F^- + HF \rightarrow HF^-_2$ = 70 kcal

378 kcal   353 kcal

Difference = 25 kcal = 1.08 V.

Since the total solvation energy in hydrogen fluoride is *less* than that in water, the reversible potential of a fluorine cell under standard conditions becomes 2.85 − 1.08 = 1.77 V at 298° K.

Recent direct measurements of the reversible potential of the fluorine electrode have been carried out by Arvia and de Cusminsky[82] and by Watanabe and co-workers[81]. Using an electrolyte of compcsition KF·2HF, Watanabe and co-workers determined the reversible anode potential as 1.90 V at an unspecified temperature (probably in the region 80°–100°). Using molten KF·HF at 250°–280°, Arvia and de Cusminsky obtained 1.75 ± 0.025 V as the reversible potential which is in excellent, but possibly fortuitous, agreement with the calculated value of 1.77 V at 298°K.

In view of the unusual conditions existing at the surface of the carbon anodes of working fluorine cells (discussed under "Anode phenomena", p. 14) it would appear that further measurements of the reversible potential are required, particularly with respect to anodes of porous carbon at which no free fluorine bubbles are released. Since all modern fluorine cells use an electrolyte approximating in composition to KF·2HF, Watanabe's experimentally determined value of 1.90 V for this electrolyte is used in subsequent calculations (see p. 34).

6. Fluorine electrode kinetics

There appear to have been only two attempts to elucidate the kinetics of the evolution of fluorine at the surface of anodes of various materials[81–83]. Of these one deals with an electrolyte of molten KF·HF at temperatures in the range 251° to 256°, while the second (Watanabe *et al.*[81]) relates to KF·2HF at temperatures which are unspecified but pobably in the range 80° to 100°. Both make use of open circuit decay techniques with both carbon and graphite anodes. Watanabe *et al.* use steady state data in addition. Watanabe *et al.* studied platinum and Arvia and de Cusminsky[82] report very briefly on nickel electrodes. Values of i_o, a, the constant b in Tafel's equation and the double layer capacity are summarised in Table 4.

TABLE 4

KINETIC CONSTANTS FOR THE FLUORINE ELECTRODE[81, 82]

Constant	Watanabe et al.	Arvia et al.
i_o (graphite) (A/cm^2)	3×10^{-6}	10×10^{-3}
(carbon) (A/cm^2)	9×10^{-5}	10×10^{-6}
(platinum) (A/cm^2)	2×10^{-5}	—
a	0.8	0.82
b	2.303 RT/aF	2.303 RT/aF
Double layer capacity, C (μF/cm^2)	≈ 30*	50

* The above figure applies to carbon, graphite and platinum in KF·2HF. In the presence of 1% of water, carbon has a value of ≈ 65 μF/cm^2.

Based on the reported results, there is no change in the mechanism of fluorine evolution over the range 80° to 260° with electrolytes ranging from KF·2HF to KF·HF in composition and there is agreement that the transfer coefficient, a, has a value of 0.8. The value of b is consistent with a slope of 2.303 RT/F, assuming the above values of a. Using the hydrogen evolution reaction as an analogy, the two sets of authors propose a similar general mechanism but differ in the step which they consider to be rate-determining.

$$HF_2^- + C \rightarrow HF + C \ldots F + e \qquad (i)$$
$$C \ldots F + C \ldots F \rightarrow C + F_2 \qquad (ii)$$
$$C \ldots F + F^- \rightarrow C + F_2 + e \qquad (iii)$$

Arvia and de Cusminsky favour (i) as the rate-determining step, which is analogous to the Volmer reaction in hydrogen evolution, while Watanabe *et al.* favour (iii), which is analogous to the "ion-atom" hydrogen evolution (Heyrovsky) mechanism. However, neither set of authors has sufficient evidence to be categorical and this admission is implicit in their papers.

Work in this field has uncovered some inexplicable anomalies in the ohmic polarisation observed. Watanabe's results show deviations from the linear Tafel plot which indicate the ohmic resistance of the electrolyte to be 5.1 Ω, whereas direct measurement with a Wheatstone bridge gave a value of 1.4 Ω. The current density at which this result was obtained is not stated. Where the mechanism of fluorine evolution in these anhydrous melts is completely analogous to that encountered in aqueous electrolysis one would assume that the ohmic resistance due to gas bubbles would be current-dependent as the size of the bubbles changes. The mechanism of fluorine evolution is, however, anomalous and highly dependent on the nature of the anode material, and the presence of trace impurities in the electrolyte (see p. 16 *et seq.*).

The ohmic overpotential measurements of Arvia and de Cusminsky are more sophisticated and reveal that there are, in fact, two regions of ohmic overpotential. At low current densities the ohmic overpotential indicates a resistive component of 15 Ω. This component appears to cease increasing at a current density of approximately 75 mA/cm^2 and to remain constant at this value. At this point a different (far smaller) resistance of approximately 0.8 Ω becomes evident. The authors provide a satisfactory explanation for this second, smaller ohmic overpotential in terms of solution resistance between anode and reference electrode. The very closeness of their prediction for this leaves open the question of "bubble overvoltage" which would be expected to be prominent at these higher current densities. Their explanation for the first period of ohmic overpotential is less than satisfactory, and they state only that the value of this is dependent on the shape and size of the electrodes, as well as the presence of foreign ions in the melt. They claim that it is independent of electrode material. This latter fact appears to preclude surface phenomena from being a cause. Another possible explanation is that this first period is the manifestation of bubble overvoltage, and that this reaches a limiting value for some reason.

The only additional data come from de Cusminsky[83] in which (his Fig. 3) the extent of the first resistance is extended in the melt KF·HF (0.25% LiF) compared with that in a melt with 0.50% LiF.

It is evident that some aspects of the electrode kinetics are more complex than is the case in aqueous electrolysis and this may well be related to the large contact angle which exists at the surface of carbon anodes in both KF·HF and KF·2HF electrolytes in the absence of contact angle-reducing additives.

7. The working voltage

As in any other electrolytic process the greater part of the working voltage of the fluorine cell is made up of the sum of the reversible voltage, the voltage drop due to the ohmic resistance of the electrolyte and the overvoltages at the anode and cathode respectively. In addition, there is a small contribution from the ohmic resistance of the electrodes and the leads thereto and from the presence of gas bubbles in the electrolyte. Neglecting the effect of gas bubbles, the voltage drop due to the ohmic resistance of the electrolyte is obviously a linear function of the current flowing through the cell, while the electrode overvoltages depend on the respective current densities.

While it is, in theory, possible to calculate the voltage drop through the electrolyte from the current load for a particular electrolyte composition and temperature (Fig. 5), this is frequently difficult or impossible in practice because of complications due to cell geometry. No measurements of electrode overvoltages and their variation with current density have been made which are universally applicable to working fluorine cells, and it appears doubtful whether meaningful figures will ever be available that take into account variation of anode material, and the subtleties and variability in the mechanism of fluorine discharge which are known to occur at the anode (see p. 16). An additional difficulty arises from the fact that the anode surface temperature frequently does not correspond to that in the bulk of the electrolyte and is dependent, not only on the mechanism of fluorine discharge at the anode, but also on the amount of agitation brought about by the discharge of hydrogen bubbles at the cathode. This, in turn, depends on the electrode separation (see p. 35).

For these reasons it is impossible to predict with any precision the working voltage of a projected design of cell.*

Experiments involving two laboratory scale cells of nominal 60 A capacity give some idea of how the working voltage was made up *under the conditions which applied during these experiments*[84].

Both cells had 3 in. (7.62 cm) diameter cylindrical anodes of "Carbocell" Grade 20 porous carbon and were virtually identical except that in one the anode/cathode separation was 0.6 in. (1.52 cm) and in the other 3.0 in. (7.62 cm) .Variations of voltage with anode current density were measured for each cell corresponding to an electrolyte containing 42% of hydrogen fluoride at a temperature of 80° (measured in the cathode compartment of the cell). The geometry of the cells was such that it was possible to calculate with a reasonable degree of accuracy the ohmic resistance of the electrolyte between the electrodes.** For the above electrolyte composition and temperature the specific conductivity is 0.145 mhos/cm corresponding to a resistivity of 6.9Ω/cm (see Fig. 5).

The results of the measurements are given in Table 5, in which the combined overvoltage at anode and cathode, V_{oac}, was calculated from the expression $V_{oac} = V_w - (1.90 + V_e)$, where V_w is the working voltage and V_e is the voltage drop due to the ohmic resistance of the electrolyte. The current density at the anode was the same for each cell and, though the cathodic current densities differed by a factor of 2 : 1 it is known that the anodic overvoltage is much greater than that at the cathode.

The last two columns in Table 5 show the difference between the working voltages due to the calculated ohmic resistance of the electrolyte compared with

---

* This, of course, is not peculiar to fluorine cells. Cell development involves a continuing *experimental* effort to minimise energy usage by reducing the working voltage while maintaining, or increasing, current efficiency.

** The resistance of the annulus between the electrodes per unit length is equal to:

$$\frac{\omega}{2\pi} \int_{r_1}^{r_2} \frac{l}{r} \cdot dr \; \Omega$$

so that the total resistance for an annular length of l is $\omega/2\pi l \cdot \log r_2/r_1 \, \Omega$. For the narrow gap cell l was 9.875 in. (25.1 cm) and for the wide gap cell 10.875 in. (27.6 cm). This results in voltage drops due to the electrolyte of 0.0148 C and 0.0437 C respectively where C is the current in amperes.

TABLE 5
MAKE-UP OF WORKING VOLTAGE FOR TWO EXPERIMENTAL LABORATORY SCALE CELLS[84]

Current (A)	Anodic c.d.		(a) Electrode separation 0.6 in. (1.525 cm)			(b) Electrode separation 3.0 in. (7.619 cm)			Voltage Diff. $V_w(b) - V_w(a)$	
	(A/in^2)	(A/cm^2)	V_w	V_e	V_{oac}	V_w	V_e	V_{oac}	Calc.	Obs.
9	0.1	0.015	5.30	0.133	3.17	5.95	0.394	3.66	0.261	0.65
18	0.2	0.031	5.75	0.267	3.58	6.70	0.787	4.01	0.520	0.95
27	0.3	0.047	6.15	0.400	3.85	7.40	1.18	4.32	0.78	1.25
36	0.4	0.062	6.55	0.533	4.12	8.00	1.57	4.53	1.04	1.45
45	0.5	0.078	6.90	0.666	4.34	8.55	1.97	4.68	1.31	1.65
54	0.6	0.093	7.25	0.800	4.55	9.10	2.36	4.84	1.56	1.85
60*	0.66	0.102	7.50	0.888	4.71	9.40	2.62	4.88	1.73	1.90
63	0.7	0.108	7.60	0.935	4.76	9.55	2.76	4.89	1.82	1.95
72	0.8	0.124	7.90	1.07	4.93	10.00	3.15	4.95	2.08	2.10

* Normal maximum load. Reversible voltage taken as 1.90.

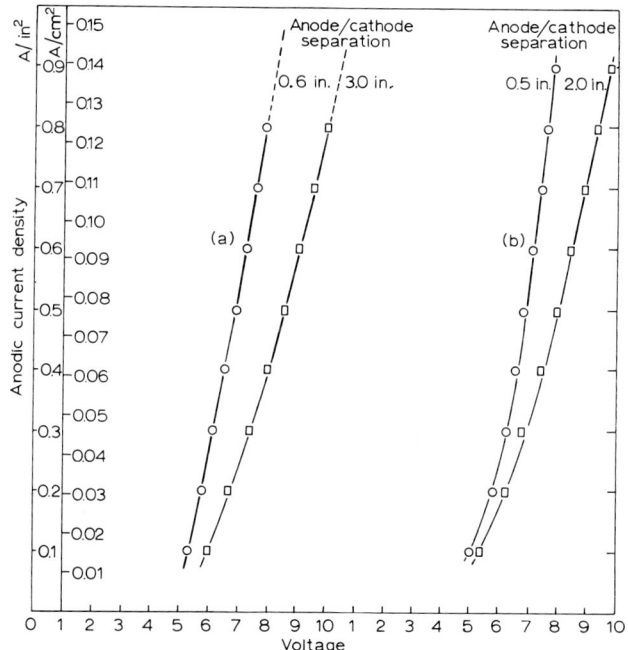

Fig. 11.
Current/voltage relationship for four experimental fluorine cells. (a) Cells with 3 in. diameter cylindrical anodes having electrode separations of 3.0 in. and 0.6 in., respectively; (b) cells with rectangular anodes 13.5 in. by 11 in. by 2¾ in., having electrode separations of 2.0 in. and 0.5 in., respectively. In all cases the electrolyte contained 40% hydrogen fluoride at a temperature of 80°.

the observed differences. The variation of V_w with current density is plotted in Fig. 11, which shows also the current/voltage relationship for two experimental cells both fitted with single works-scale anode blocks ("Carbocell", Grade 20, 13.5 in. by 11 in. by 2.75 in. or 34.2 cm by 27.9 cm by 7.0 cm) which differed chiefly in that one had an electrode separation of 2 in. (5.08 cm) and the other 0.5 in. (1.27 cm). The anode/skirt gaps were 0.5 in. (1.27 cm) and 0.0625 in. (0.159 cm) respectively.

It is seen that the agreement in the last two columns of Table 5 is poor at low current densities but improves with increasing current density. The same is true for the values of V_{oac} for the two different electrode separations, while a plot shows that the two run roughly parallel up to a current density of about 0.5 A/in.² (0.074 A/cm²) and thereafter tend to converge, as is indeed evident from the Table. The significance of this is by no means clear. Table 5 shows, however, that, particularly for small electrode separations, overvoltages at the electrodes represent the biggest single contribution to the working voltage and that the combined figure is very close to 5 V at an anodic current density of 0.8 A/in.² (0.119 A/cm²). It is necessary, however, to emphasise once again that this figure is probably correct only for the experimental conditions defined and that constancy of the individual electrode overvoltages at higher current densities cannot be assumed.

There is only one report in the published literature of an attempt being made to measure directly the anodic and cathodic overvoltages in a working fluorine cell using probes[21]. In this case the anode was of dense, ungraphitised carbon and the electrolyte contained 38.3% of hydrogen fluoride at a temperature of 110°, corresponding to a specific conductivity of about 0.180Ω/cm as compared with 0.145 Ω/cm in the experiments just described. At a current density of 0.8 A/in.² (0.119 A/cm²) the voltage drop between a probe inserted near the anode was about 5.0 V while the corresponding figure at the cathode was about 1.5 V, making V_{oac} 6.5 V. This is about 1.5 V higher than V_{oac} for the two cells using cylindrical porous "Carbocell" anodes at the same current density (Table 5); the difference could be due largely to the different types of anode material used in the two cases, or it could indicate incipient anodic polarisation in the cell used by Schumb *et al.* The overall working voltage of the American cell under these conditions was about 9 V, the electrode separation being 6 in. (15.24 cm) to facilitate the use of probes.*

Table 6 shows current/voltage relationships for two individual twelve anode (six anode assembly) experimental work scale cells, which differed only in that one had an anode/cathode separation of $1\frac{3}{8}$ in. (3.492 cm) while in the other the separation was $2\frac{11}{32}$ in. (5.952 cm) with varying electrolytic composition and temperature[85].

It is seen that the voltage saving for a decrease in electrode separation of

---

* The reported voltage drop due to the electrolyte at 0.8 A/in², (0.119 A/cm²) is much lower (≈ 2 V) than that calculated from the specific conductivity of the electrolyte.

TABLE 6

RELATIONSHIP BETWEEN CURRENT AND VOLTAGE FOR TWO EXPERIMENTAL
WORKS SCALE CELLS WITH DIFFERENT ELECTRODE SEPARATIONS[85]

	ø Cell A. electrode Separation 1⅜ in. (3.492 cm)				ø Cell B. electrode Separation 2¹¹/₃₂ in. (5.592 cm)			
Electrolyte temp. (°C)	85	83	80	80	84	83	82	83
Electrolyte comp. (% HF)	40.8	41.4	42.4	42.7	40.0	40.4	40.6	40.7
Specific conductivity (Ω/cm)	0.140	0.153	0.156	0.160	0.125	0.128	0.129	0.131
Cell load A	Cell voltage				Cell voltage			
200	5.35	5.6	5.3	5.3	5.45	5.5	5.5	5.6
400	6.25	6.4	6.0	5.9	6.4	6.4	6.35	6.6
600	6.85	6.95	6.6	6.55	7.15	7.25	7.2	7.4
800	7.3	7.45	7.05	7.0	7.85	7.95	7.9	8.0
1000	7.8	7.9	7.5	7.45	8.5	8.65	8.65	8.6
1200	8.3	8.45	7.95	7.9	9.2	9.35	9.25	9.2
1400	8.8	8.9	8.4	8.3	9.85	10.0	9.95	9.8
1600	8.35	9.4	8.9	8.75	10.55	10.7	10.65	10.4
1800	9.75	9.8	8.35	9.15	11.15	11.35	11.3	10.9
2000*	10.3	10.1	9.8	9.65	12.1	12.0	11.95	11.5

* Anode current density 0.9 A/in.² (0.134 A/cm²)
ø Anode material "Carbocell" Grade 20

approximately 1 in. (2.54 cm) varied from 1.2 to 2.3 V (10.5 to 19%), depending on electrolyte composition and temperature, at a load of 2 kA. Where the conditions of electrolyte composition and temperature were approximately the same (40 to 40.8% HF and 84 to 85°) the voltage saving was 1.8 V or 14.9%. It may also be noted that, for a given electrode separation, the working voltage is not always strictly proportioned to the electrolyte conductivity. The reasons for this are not understood but may be due to variations in electrolyte circulation, temperature inhomogeneities or variations in overvoltages.

Murray et al.[22] have published current/voltage figures for a 2 kA cell using dense, ungraphitised (low permeability) carbon anodes when operating with an electrolyte containing 39.5% of hydrogen fluoride at a temperature of 105° (specific conductivity 0.185 Ω/cm). At a current density of 0.5 A/in.² (0.078 A/cm²) the working voltage was 8.7 V. The electrode separation is not specified but appears to be 1.5 in. (3.81 cm), which is close to that of Cell A in Table 6. At the same current density of 0.5 A/in.² (0.075 A/cm²), Cell A working with an electrolyte having a conductivity about 14% lower than that of the American cell, had a working voltage of approximately 7.7 V. The difference of 1V may be due to differences in cell geometry, to differences in overvoltages and other factors already discussed.

As pointed out elsewhere (see p. 20) the anodes of fluorine cells do not normally suffer wear so that there is no progressive increase in voltage with age resulting from an increase in the electrode separation. For this reason progressive increase in cell voltage with time is most commonly the result of deterioration of the electrical connections to the anodes. Partial or complete fracture of one or more anodes can, of course, lead to the same result.

8. Heat produced in the cell and its removal

For a fluorine cell working at 100% current efficiency the heat produced, H, is equal to $H = 0.86\ A\ (V_w - V_r)$ tonne cal/h where A is the current in kA, V_w is the working voltage and V_r is the reversible voltage. The reversible voltage (minimum potential or rest potential) for a fluorine cell using molten $KF \cdot 2HF$ as electrolyte as measured by Watanabe et al.[81] was 1.90 V. Accepting this value, the heat produced by a cell working at 6 V, 1 kA and 100% current efficiency is 6.966 tonne cal/h or 27,650 B.T.U./h. If the load is 5 kA at the same voltage the corresponding figures are 34.830 tonne cal/h or 138,250 B.T.U./h*.

Since, however, cells more often than not work at lower current efficiencies, and because current efficiency is lost solely by recombination of fluorine and hydrogen, due allowance must be made for this.** Heat produced in this way is not related to the reversible voltage but to the heat of formation of hydrogen fluoride which is $-64,827$ cal at $100°$. When allowance is made for the current efficiency loss the total heat produced becomes:

$H = 0.86\ A\ (V - 1.90) + 0.86\ A\ [2.81\ (100 - E)/100]$ tonne cal/h
$\ \ \ = 0.86\ A\ \{(V - 1.90) + [2.81\ (100 - E)/100]\}$ tonne cal/h

where E is the current efficiency in percent. Thus each 10% decrease in current efficiency corresponds to extra heat production at the rate of 0.242 tonne cal/h at 1 kA and *pro rata* for higher currents.

Table 7 shows the heat balance for an I.C.I. experimental cell working at loads up to 5.1 kA in which the cathodes were cooled with water having an inlet temperature of 67 to 69.5°.

The Union Carbide cell[27] was fitted with an external water jacket and three internal cooling tubes† having a total cooling area of 38 ft² (approximately 3.5 m²). The cooling water rate is given as 40 to 80 lb/min (1.2 to 2.2 m³/h) with inlet temperature ranging from 32° to 49° and outlet temperatures 46° to 63°. For a cell working at 4 kA at 9.5 V and a water inlet temperature of 120°F (49°C) the

---

* Dykstra et al.[27] state that a cell operating at 4kA and 9.5 V has a heat output of 90,800 B.T.U./h (28.90 tonne cal/h). This figure is obtained taking V_r as 2.85 and assuming 100% current efficiency and is some 3.4 tonne cal, or about 13% *too low*.
** Assuming no abnormal conditions exist, such as combustion of anodes in fluorine.
† Later increased to twelve in a larger cell body[29].

TABLE 7

HEAT REMOVAL IN A WORKS SCALE FLUORINE CELL[85]

Cell conditions and heat liberated					Heat removed at cathodes						
					Cooling water				Transfer coeff.		
V	kA	C.E. (%)	Cell temp. (°C)	Heat produced (tonne cal/h)	In (°C)	Out (°C)	Rate (m³/h)	Heat removed (tonne cal/h)	a	b	Proportion removed by cathodes (%)
8.8	2.58	82	82.0	16.6	67.2	71.0	2	7.6	5.6	60.4	45.8
9.3	3.50	89	81.5	23.2	67.7	69.9	6	13.2	9.8	106	56.8
9.6	4.01	74	82.1	29.2	68.1	72.5	4	17.6	14.1	152	60.3
9.1	4.01	74	81.6	27.3	68.4	72.3	4.5	17.6	14.9	161	64.5
9.9	4.56	82	82.2	33.4	67.2	72.0	4	19.2	14.4	155	57.5
9.8	4.56	75	84.1	33.8	69.5	73.6	5	20.5	15.4	166	59.8
9.9	4.56	82	83.4	33.7	69.3	72.5	6	19.2	14.5	156	57.0
10.3	4.56	85	83.1	34.7	69.0	71.9	7	20.3	15.2	164	58.5
10.0	5.10	85	84.9	37.5	68.8	72.3	7	24.5	16.2	175	60.0

* Area of cathode cooling coils 105.5 ft² (9.8 m²)
(a) Kcal/ft²/degree (c)/h
(b) Kcal/m²/degree (c)/h

overall heat transfer coefficient is given as 28.5 B.T.U./ft²/degree (F)/h or 12.9 kcal/ft²/degree (C)/h (approx. 139 kcal/m²/degree (C)/h), which is of the same order as that for the I.C.I. cell in which a higher proportion of the total heat produced is dissipated by natural convection from the cell sides and top.

9. Mechanism of fluorine and hydrogen evolution in relation to cell design

Where there exists a large anode/electrolyte contact angle at the surface of *high permeability* carbon anodes operating at an appropriate current density, no fluorine is evolved from the anode as free bubbles; all the fluorine leaves the anode face *via* interconnected pores. Under these conditions it is possible to devise a very simple design of cell (Fig. 12)[31, 86]. Under appropriate conditions a similar arrangement may be used with anodes of relatively low permeability[87].

Such an arrangement is not, however, suitable for producing fluorine on a commercial scale but is useful for laboratory preparation and for lecture demonstration purposes. Because of its innate simplicity and because it is possible to use an ordinary tin as the electrolyte container and cathode such a device has been called a "tin-can fluorine cell"[31].

Other designs of cell have been evolved which take advantage of the absence of fluorine bubble evolution from anodes of permeable carbon[88].

Fig. 12.
Components of "tin can" fluorine cell. The anode consists of a block of high permeability carbon with an undersized hole into which is forced a copper tube which acts as the anode connection. The tin can is the cathode. On lowering the anode into the molten electrolyte with a potential difference of about 10 V between anode and cathode all the fluorine issues from the open end of the copper tube which becomes anodically passive when it is submerged in the electrolyte.

With *low permeability* anodes, in the absence of additives which promote "wetting" and in the absence of polarisation, lenticular bubbles do not leave the anode face but reach the electrolyte surface as has already been described (p. 16). In the case of *low permeability* anodes in the presence of depolarising agents which promote reduction of the contact angle the incidence of bubble break-away is greatly increased and it is probable that almost all the fluorine is discharged as free bubbles.

It has been shown that, after a short period of operation during which very small bubbles are produced, the hydrogen bubbles which are freely evolved from a steel cathode do not spread far through the electrolyte in a horizontal direction.

By measuring the amount of hydrogen evolved from a 12 in. (30.48 cm) deep cathode at various distances from the cathode in a horizontal direction it was found that 95% of the bubbles did not spread more than ¾ in. (1.91 cm) from the cathode face[89].

This convenient behaviour of the hydrogen bubbles and of the fluorine in the case of *high permeability* carbon anodes under the conditions described above, makes it possible to dispense with a conventional diaphragm to keep the two gases apart and all that is required is a "skirt" dipping a short distance into the electrolyte in order to divide the cell into fluorine and hydrogen compartments. On the basis of these observations and supporting experimental work it was possible to define minimum anode/cathode separations in relation to anode/skirt gap and effective electrode length[90].

That it is possible to dispense with a diaphragm with cells having *low permeability* anodes, where bubble break-away must occur if contact angle-reducing additives are present (*e.g.* as in the Allied Chemical Corporation cell)[17] is evidence that fluorine bubbles spread only a short distance from the anode face under these conditions.* Union Carbide cells have 6-mesh Monel diaphragms. "The screens are placed between the anode and cathode assemblies to prevent broken carbon anodes from causing a short circuit between the anode and cathode and to prevent hydrogen and fluorine from mixing in the cell"[29, 30].

From the above considerations, it is evident that anode/cathode separations and the location of the skirt with respect to these is dependent on the precise way in which fluorine is evolved at the anode. Thus, with *high permeability* anodes operating with nickel present in the electrolyte in the substantial absence of water so that some bubble break-away occurs, the skirt must be located at an adequate distance from the anode if there is not to be excessive loss of current efficiency due to fluorine entering the hydrogen compartment of the cell**[91].

If, on the other hand, nickel is either absent, or is present in association with a sufficient quantity of water such that the effect of the nickel in promoting "wetting" of the anode is neutralised, the skirt may be placed nearer to the anode without loss of current efficiency due to the above-mentioned cause[92]. In all cases it is, of course, necessary to adjust the anode/cathode position in relation to the skirt position and the anode/skirt gap to prevent hydrogen entering the fluorine compartment, particularly where the anode/skirt gap has to be appreciable to take account of the effect of nickel or other contact angle-reducing additives[31].

---

* This evidence is circumstantial. The small spread could indicate that the bubbles were relatively large (large buoyancy effect) and it is possible that the contact angle, though much smaller than 90°, is not zero under such conditions.
** It has been found that fluorine bubbles, which enter the hydrogen side of the cell react immediately and quietly and burn without causing explosions, provided that there is no air present in the hydrogen, or that it is present below the explosive limit. Explosions in the cathode compartment are prevented by purging the hydrogen side of the cell with nitrogen before start up[31].

10. Influence of the mechanism of fluorine evolution on the hydrogen fluoride content of the fluorine

Because of the large anodic overvoltage, the temperature of the surface of a working carbon anode may under certain circumstances be substantially higher than that of the bulk of the electrolyte in the cell. With *high permeability* anodes under conditions such that *all* the fluorine is discharged through the interconnected pores the hydrogen fluoride content of the fluorine may be abnormally high, corresponding closely to the partial pressure of hydrogen fluoride over the electrolyte at the temperature of the anode face, and may reach 20 to 30%*. This high temperature is attributed to lack of local turbulence in the electrolyte with poor heat transfer from the face of the anode. The temperature of the anode face and the concentration of hydrogen fluoride may be reduced by using water-cooled cathodes with a reduced interelectrode separation[93].

However, with nickel present in the electrolyte in the substantial absence of water a proportion of the fluorine is evolved as free bubbles even with *high permeability* anodes (pp. 21-25) and this provides sufficient turbulence to reduce substantially the anode face temperature and hence the concentration of hydrogen fluoride in the fluorine. Under such conditions the hydrogen fluoride in the fluorine is normally in the range 5 to 7% by volume when using an electrolyte containing 38 to 40% HF at 80°.**

With no nickel in the electrolyte, which can be achieved by using steel skirts, the hydrogen fluoride content of the fluorine may rise to 25% or more. Since no fluorine is evolved as free bubbles at the anode under these conditions a current efficiency of 100% can be achieved even with a relatively small anode-skirt gap, provided that the cathode is not too close to the skirt[31].

With *low permeability* carbon anodes operating under conditions such that substantially all the fluorine leaves the anodes as free bubbles, it is to be expected that the hydrogen fluoride content of the fluorine will be that corresponding to the hydrogen fluoride partial pressure of the main bulk of the electrolyte. For Union Carbide cells with an electrolyte containing 41 to 42% HF at 88° to 104° the concentration of hydrogen fluoride in the fluorine leaving the cells would be expected to be 5 to 13% but no actual figures have been reported[29]. The reported figure for Allied Chemical Corporation's cells[17], which probably operate under similar conditions, is 5 to 11% of hydrogen fluoride in the fluorine, though no specific information is given as to electrolyte composition and temperature.

Special arrangements have been proposed to overcome the difficulty resulting

---

* The temperature of the anode surface has been measured by means of thermocouples passing through the bulk of the anode with the junction located at the anode face.
** The mean hydrogen fluoride content of the hydrogen leaving I.C.I. cells is 4% which corresponds closely to the equilibrium concentration of hydrogen fluoride in the cathode compartment.

from the high hydrogen fluoride content of the fluorine from cells using *highly permeable* anodes in the absence of contact angle-reducing additives. Either the fluorine from the several anode compartments is bubbled through electrolyte in a separate compartment in the cell where the hydrogen fluoride partial pressure is lower[94] or a composite anode can be used in which the lower portion has a high permeability and the upper a low permeability, so causing the fluorine to bubble through the upper inch or so of electrolyte[95].

11. Removal of hydrogen fluoride from fluorine

The major impurity in fluorine as obtained direct from the cells is hydrogen fluoride. In addition, oxygen, nitrogen and carbon tetrafluoride may be present. Oxygen can result either from the presence of water in the electrolyte (hence the need for a low water content in the hydrogen fluoride feed) or, in association with nitrogen, from the ingress of air into the system. In the ordinary way, fluorine from the cells should be free from carbon tetrafluoride, but it will be present should burning of one or more anodes result because of anode breakage, or the development of hot spots which may occur as a consequence of defective anode connections (see p. 44). While the presence of hydrogen fluoride in fluorine from the cells is inevitable the presence of the other impurities can, in principle, be prevented.

Depending on the electrolyte composition and temperature, and the design and operating conditions of the cells, the fluorine will contain from 5 to 20% or more of hydrogen fluoride (see p. 38). The most obvious way of reducing the hydrogen fluoride content and facilitating its recovery is by refrigeration, but this is an inefficient and costly method because of the high vapour pressure of hydrogen fluoride, even at low temperatures, and the fact that it is associated under some conditions.

Figure 13 shows the variation of the vapour pressure of hydrogen fluoride with temperatures based on the equation, $\log p = 6.37 - 1315/T$[96].

Figures 14 and 15 show the variation in the degree of association of hydrogen fluoride vapour with pressure at temperatures above the boiling point[97] and the variation of the degree of association of the saturated vapour at temperatures from $0°$ to $-90°$ respectively[98].

If p is the saturated vapour pressure of hydrogen fluoride at temperature t, then the percentage by volume of hydrogen fluoride at equilibrium in a carrier gas at this temperature is $100p/760$. However, even at low partial pressures, hydrogen fluoride vapour is associated at low temperatures (Fig. 15) so that when the refrigerated gas warms up to room temperature, under which conditions low partial pressures of hydrogen fluoride are unassociated, the final concentration becomes $100p\alpha/760\%$ by volume, where α is the degree of association at the refrigeration temperature.

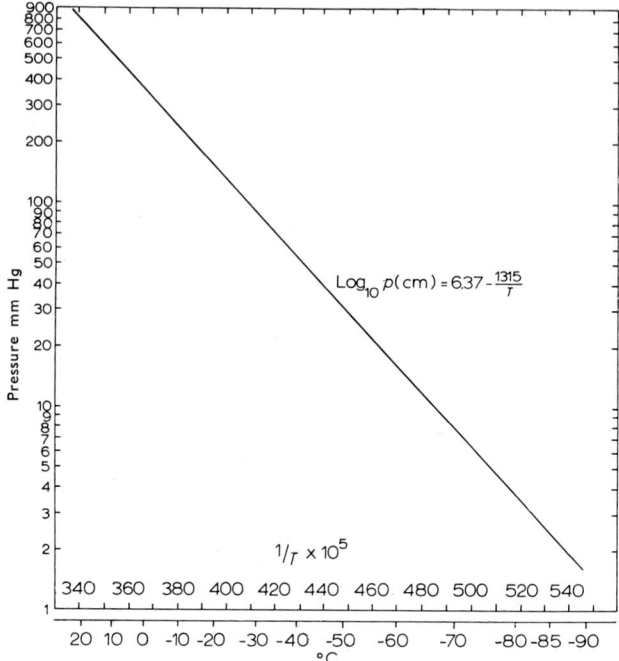

Fig. 13.
Variation of vapour pressure of hydrogen fluoride with temperature.

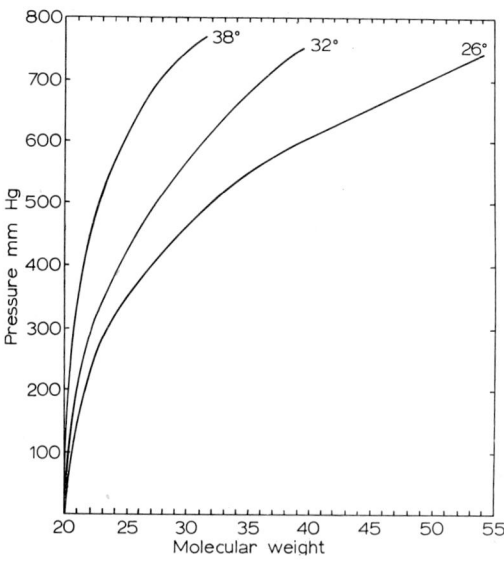

Fig. 14.
Degree of association of hydrogen fluoride vapour and its variation with pressure at temperatures above the normal boiling point.

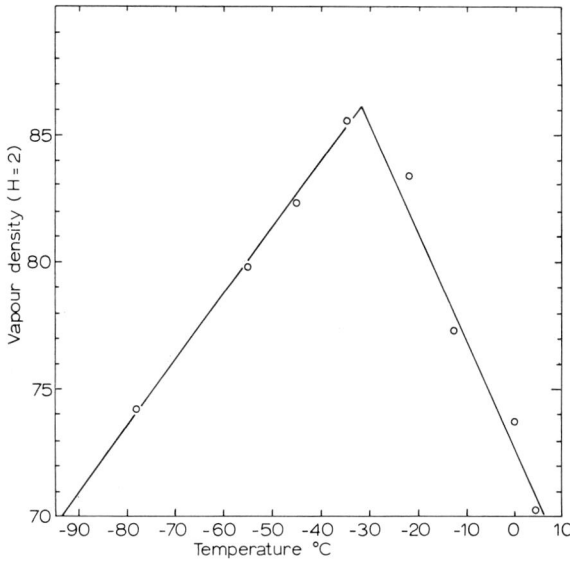

Fig. 15.
Variation of degree of association of saturated hydrogen fluoride vapour at normal pressure and temperatures between 0° and –90°.

Figure 16 shows the equilibrium concentration of hydrogen fluoride in a carrier gas after refrigeration to varying temperatures between –50° and –90° and subsequent warming to room temperature, together with the percentage recovery from inlet gases containing from 5 to 20% of hydrogen fluoride as calculated from the formula:

$$\text{Recovery \%} = 100 \left[\frac{1 - (100 - V)\, p a}{V(760 - p)} \right]$$

where V is the concentration by volume of hydrogen fluoride in the inlet gas, p is the saturated vapour pressure of hydrogen fluoride at the refrigeration temperature and a is the degree of association of the vapour at this temperature.

It is evident that, even if the effective refrigeration temperature is –80°, about 2% of hydrogen fluoride will still remain and that modest levels of refrigeration will be very inefficient even with quite high concentrations of hydrogen fluoride in the inlet gas. Froning and co-workers[99] have shown that, quite apart from the practical difficulties, combined compression and refrigeration is likewise an inefficient method for the removal of hydrogen fluoride from fluorine.

The Union Carbide Nuclear Co. at Oak Ridge, Tennessee[30] have used refrigeration to reduce the hydrogen fluoride concentration of fluorine to 4–5%. It is stated that the refrigeration temperature was –70°F (–57°C) but this temperature would correspond to an equilibrium concentration of hydrogen fluoride of 9% by

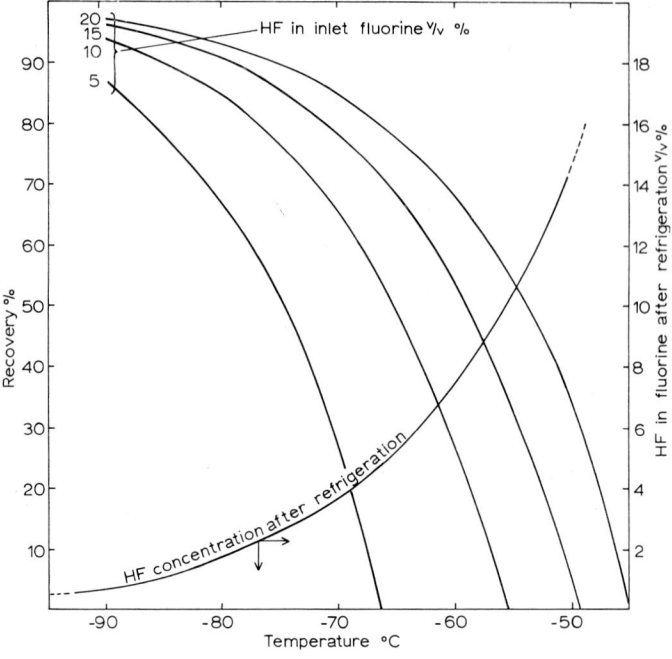

Fig. 16.

Effect of refrigeration on the concentration of hydrogen fluoride in a carrier gas previously saturated with it and the variation of recovery efficiency with temperature for various inlet concentrations.

volume. A refrigeration temperature of $-70°$ *centigrade* would correspond to a final hydrogen fluoride concentration of 4%.

In order to reduce the concentration of hydrogen fluoride to a low level, an efficient absorbent is required. Potassium fluoride is an obvious possibility but it is not used because of the tendency to produce liquid products and sodium fluoride is preferred since it shows this property to a reduced degree. Froning et al.[99] have measured the partial pressure of hydrogen fluoride over NaF·HF at various temperatures with the results shown in Table 8.

Provided that efficient contact can be ensured, sodium fluoride at room temperature should be a highly efficient reagent for the present purpose.

Froning et al. used sodium fluoride pellets in a tower which was heated to 100° to avoid the tendency to produce liquid or semi-liquid products with consequent risk of blockage. Such an arrangement was capable of reducing the hydrogen fluoride content of fluorine to about 0.02% from an inlet concentration of 4% and regeneration of the spent absorbent was possible by heating to about 300°. It was further shown that very little decrease in the efficiency of absorption occurred until about 80% of the sodium fluoride was converted to bifluoride. The preparation of porous pellets of sodium fluoride by heating sodium bifluoride and their

TABLE 8

PARTIAL PRESSURE OF HF OVER NaF·HF[99]

Temp. (°C)	Partial pressure of HF over NaF·HF mm Hg	Equilibrium HF concn. (vol. %)
25	0.01*	Negligible
100	1.4*	0.18
200	87	11.5
250	422	55.6
275	706	93.0
278	760*	100

* Extrapolated values

use at 100° to absorb hydrogen fluoride from a gas stream is claimed in a patent assigned to the Harshaw Chem. Co.[100].

Union Carbide Nuclear Co. employed a 2 in. (5.08 cm) depth of sodium fluoride pellets in trays to reduce the hydrogen fluoride content of fluorine from 4-5% to approximately 2%. I.C.I. practice involves the use of sodium fluoride powder in trays in a multi-pass arrangement with sheathed electrical heating elements beneath the trays to permit regeneration. In spite of caking the sodium fluoride maintains its absorptive capacity after repeated regeneration. A regenerative scrubber containing 450 lb (205 kg) of sodium fluoride has a power loading of 120 kW, and requires 25 kWh for regeneration, and is capable of treating 4,000 lb (1,820 kg) of fluorine before regeneration is required[101]. Such an arrangement may be backed by a tower containing pellets to reduce further the hydrogen fluoride content. Because of the low hydrogen fluoride inlet concentration, long-term operation without blockages is possible[7].

Only two other methods have been proposed for removing or reducing the hydrogen fluoride content of fluorine. In cases where conditions in the cells are such that the hydrogen fluoride content of the fluorine leaving the anode compartment(s) of the cells is well above that corresponding to the partial pressure of hydrogen fluoride in the main part of the cell (say 10-20%) it has been proposed that the gas should be bubbled through the electrolyte in a special compartment at the end of each cell so as to equilibrate with it. Alternatively, the fluorine is treated in an absorption tower with a molten mixture of potassium fluoride and hydrogen fluoride of appropriate composition with subsequent stripping of the hydrogen fluoride in a second tower. Using as an absorbent a composition containing 38% of hydrogen fluoride at 80° the hydrogen fluoride content of the fluorine would be reduced to a concentration approaching 2%[94] (Table 2).

It has been shown that fluorosulphonic acid is unaffected by concentrated fluorine and is effective in removing hydrogen fluoride. The hydrogen fluoride may either be stripped out of the fluorosulphonic acid by heating or could be converted to fluorosulphonic acid by addition of sulphur trioxide. In laboratory experiments it was shown that fluorine containing 15% of hydrogen fluoride contained 0.3% when treated with fluorosulphonic acid at 20° and greater efficiency should be possible with adequate design[102].

Though these last two methods of reducing the hydrogen fluoride content are practical propositions, economic evaluation is required to demonstrate that they would be worthwhile and would probably only be so in a large fluorine installation.

All the above hydrogen fluoride removal processes would, of course, be applicable to treating hydrogen from the cells, but, since the concentration is usually much lower than in the fluorine, it may be considered preferable to scrub with a 10–15% solution of caustic potash as is the practice in some installations. Caustic soda is inconvenient because of the low solubility of sodium fluoride.

12. The anode connection

In 1947, Murray, Osborne and Kircher of the Harshaw Chemical Co.[22] wrote "The crux of the whole problem in the carbon anode cell is the electrical contact between the carbon anode and its support." This remains a vital feature of cell design and is the factor which almost invariably determines the life of the cell.

The most usual way of making the connection is to clamp the anode between a pressure plate and its support. Using this method with the chrome-molybdenum steel pressure plates and anode support bars half submerged in electrolyte,* Union Carbide Nuclear Co.[27] obtained initially an average cell life of 1,250 h or approximately 7.5 weeks.** By 1958 Union Carbide had developed a larger cell in which the effective anode area was increased from 32 to 42 ft² (2.96 to 3.9 m²) while the pressure plate was changed from steel to copper which resulted in an increase in cell life to 24 weeks at 4 kA or about 16 weeks at 6 kA[29]. The primary cause of cell failure was said to be, not anode breakage, but increased contact resistance at the anode joint with resultant over-heating at this point. "After the metal parts have failed because of corrosion, the loss in contact caused the carbon anodes to break which resulted in carbon over-heating and burning."

Five years later it was claimed that cell life had been further increased to approximately 48 weeks (at 4 kA) by recessing the heads of the clamping bolts and covering with carbon plugs "but because other cell components deteriorated occasional desludging was needed here also"[103, 104]. Union Carbide also tried out carbon, as opposed to metal, support bars but these were inferior to the arrangement described above[105].

---

* It is not clear why Union Carbide use a submerged anode connection since, in the writer's view, this would be expected to make the anode connection problem more difficult.
** Union Carbide (and Allied Chemical Corporation) quote anode life in amp hours, which is not very informative. Anode lives in weeks have been calculated assuming a cell load of 4kA except where stated otherwise.

Siegmund and Neumark of Allied Chemical Corporation[17, 106], state that the anode connection is still the most serious design problem. Details of the Allied design of anode connection are lacking but the principle is evidently basically similar to that used by Union Carbide with the exception that the connection itself is well clear of the electrolyte surface. Anode life is stated to be "in excess of 20×10^6 A h" which at 6 kA is equivalent to about 20 weeks. The capacity of the Allied cell is not stated unequivocally but reference is made to the heat output at 4 kA. If this is taken as the cell capacity then the cell life is approximately 30 weeks.

I.C.I. cells have an average life of about a year, while in special cases individual cells have operated for as long as four years with only one individual anode replacement.

Methods other than those described above have been claimed in the patent literature[107, 108]. A recent Russian patent refers to a method whereby copper rods are inserted into the anode which is then heated to 90–100° and immersed in an impregnating composition preheated to 90–100° and comprising "the usual amounts of epoxy-resin, dibutylphthalate and maleic anhydride".[109]

13. Energy efficiency and energy usage

Since the energy efficiency of any electrolytic cell is equal to $V_r \times E/V_w$ where V_r is the reversible voltage (1.90), E is the current efficiency expressed as a decimal and V_w is the working voltage, the energy efficiency of a fluorine cell operating at 10 V and 95% current efficiency is $1.90 \times 0.95/10 = 18\%$. This compares with 60 to 70% for a brine cell producing chlorine and caustic soda liquor and about 40% for a Down's cell producing chlorine and metallic sodium.

The energy usage for a fluorine cell is 0.640 V_w/E kWh/lb (1.41 V_w/E kWh/kg) or 1434 V_w/E kWh/ton, where V_w and E have the same significance as above. Thus, a cell working at 10 V with a current efficiency of 95% will have an energy consumption of 6.74 kWh/lb (14.83 kWh/kg) or 15,100 kWh/ton. All commercial fluorine cells will have energy usages of about this figure, which is over five times that required for the production of chlorine in a diaphragm cell. This high energy usage, together with the relatively high cost of anhydrous hydrofluoric acid, makes it clear why elemental fluorine must remain an expensive product as compared with chlorine.

14. Fluorine cell design and operating characteristics

The only fluorine plants for which reasonably comprehensive details of design and operating conditions are available are those of I.C.I. and Union Carbide and these are summarised in Tables 9 and 10.

TABLE 9

SUMMARISED COMPARISON OF MAIN CONSTRUCTIONAL FEATURES AND
OPERATING CHARACTERISTICS OF BRITISH (I.C.I.) AND
AMERICAN (UNION CARBIDE) FLUORINE CELLS

Information on I.C.I. Cells From Rocksavage Data (see refs. 27–30 inclusive for Union Carbide Cells)

Constructional features or operating characteristic	Units	British cell unit	American cell unit
A. Constructional details			
1. Material of construction of (a) cell body		Mild Steel	Monel liner, steel jacket
(b) skirt		Monel	Monel
(c) diaphragm		None required	6 Mesh Monel gauze
(d) cathode		Mild steel	Mild steel insulated from body
(e) anode		Ungraphitised carbon	Ungraphitised carbon
2. Dimensions of cell body (a) length, (b) width, (c) height	in.	(Internal) (a) 119, (b) 28, (c) 22	(External) (a) 89, (b) 32, (c) 39 (1958)
	cm	(a) 302, (b) 71, (c) 56	(a) 226, (b) 96.5, (c) 104 (1958)
3. Electrolyte capacity of cell	lb	3,800	3,400 (?) (1958)
	kg	1,730	1,550 (?)
4. Number of anode blocks per cell		24	32 (1958)
5. Number of anode blocks per anode assembly		2	16 (1958)
6. Method of anode attachment to support			Clamped between contact plates of AISI-4140 steel bars and copper pressure plate with steel cap screws[a]
7. Size of anode blocks (a) length, (b) width, (c) thickness	in.	(a) 13.5, (b) 11, (c) 2.75	(a) 20.625, (b) 8, (c) 2 (1958)
	cm	(a) 34.3, (b) 27.9, (c) 6.99	(a) 52.5, (b) 20.4, (c) 5.1
8. Specification of anode carbon (a) Supplier			National Carbon Co., U.S.A.
(b) Trade name or grade			(a) GAA, (b) YAA or (c) YBD[b]
(c) Apparent density	g/cm^3		(a) 1.55, (b) 1.70, (c) ?
(d) True density	g/cm^3		Not stated
(e) Specific resistance	Ω cm		(a) 0.0007, (b) 0.0005, (c) ?
	lb/in.^{2}		(a) 3000, (b) 3100, (c) ?
(f) Flexural strength	kg/cm^2		(a) 210, (b) 218, (c) ?
9. Anode/cathode separation	in.		1.5 (1955)
10. Anode/skirt gap	in.		Not stated, but probably about 0.625
11. Cathode/skirt gap	cm		Not stated, but probably about 1.59
	in.		Not stated, but probably about 0.625
	cm		Not stated, but probably about 1.59
12. Cooling arrangements		Water cooled coils, natural convection from cell walls	Water jacket on cell walls and central cooling tubes
13. Method of HF addition		Intermittent manual addition of liquid every 12 h	Continuous addition of vapour automatic with control?

1. Normal current capacity	kA	5.0	4.0–6.0 (1958)
2. Maximum current capacity	kA	6.5	6.0? (1958)
3. Fluorine output at normal load	lb/h	7.25	At 4 kA 5.6, at 6 kA 8.82 (1958)
	kg/h	3.3	At 4 kA 2.54, at 6 kA 3.47 (1958)
4. Fluorine current efficiency (measured average)	%	93	90–95 (1955. No figures for later cells)
5. Voltage per cell at full normal load	V	10 at 5 kA	8 at 4kA, 12 at 6 kA (1958)
6. Anode current density at normal load	$A/in.^2$	1.15	0.65 at 4 kA, 0.97 at 6 kA (1958)
	A/cm^2	0.18	0.10 at 4 kA, 0.15 at 6kA (1958)
7. Cathode current density at normal load (approx.)	$A/in.^2$	0.23	Not stated, but probably about 0.3–0.5
	A/cm^2	0.036	Not stated, but probably about 0.046–0.077
8. Electrolyte temp.	°C	80–85	88–104 at 4 kA and 6 kA (1958)
9. HF content of electrolyte	%	40–41.5	40–42
10 Special requirements as regards electrolyte purity etc.		Purity of electrolyte not critical Water content of HF not critical	HF impurities <0.1%. SO_4 in KF·HF <0.01%. All electrolyte pretreated with fluorine to give 0.001 to 0.003% H_2O (1955)
11. Life of cell bodies (a) average	Years	} No failures in 11 years	} Greater than 1.2 years (1955)
(b) maximum	Years		
12. Life of diaphragm (a) average	Years	4^c	} About 1.2 (1955)
(b) maximum	Years	7^c	
13. Life of anodes (a) average	Years	1.5	0.2 (1955) 0.9? (1960)
(b) maximum	Years	2^d	0.6 (1955)
14. Life of cell unit (a) average	Years	1	} 0.1 (1955) 0.92 at 4 kA (1960)
(b) maximum	Years	2.0^d	
15. Frequency of occurrence and method of overcoming "polarisation"		No "polarisation" has ever occurred	Slow start-up or high voltage conditioning required. Occasional polarisation. When due to water, LiF added and operated at 500 A for 1 h.
16. HF in fluorine leaving cell at full load	% by vol.	Approx. 6	Not stated, probably 4–13
17. HF in hydrogen leaving cell at full load	% by vol.	4.0	4–13 (calcd. from temp. and composition of electrolyte)
18. Cooling water temp. at full load (a) inlet	°C	Approx. 65	32–49 (1955)
(b) outlet	°C	Approx. 67	46–63 (1955)

[a] Improved connection reported in 1960 (see refs. 103, 104)
[b] Later cells have YBD carbon anodes
[c] Skirt life. No diaphragm is used in I.C.I. cells.
[d] One I.C.I. cell with anodes of special carbon had a life of 4 years with loss of only one anode.

TABLE 10

SUMMARISED COMPARISON OF MAIN FEATURES AND OPERATING CHARACTERISTICS
OF BRITISH (I.C.I.) AND
AMERICAN (UNION CARBIDE) FLUORINE PLANTS AND AUXILIARY EQUIPMENT

Information on I.C.I. Plant from Rocksavage Data (see refs. 27–30 inclusive for Union Carbide cells)

Constructional feature, method of operation or operating characteristic	Units	British cell room	American cell room
Nominal maximum fluorine output (a) per year	ton	330	650[a]
(b) per hour	lb	94.25	167[a]
	kg	42.9	76
Total number of cells installed		14	36[a]
Maximum number of cells normally working at full load		12	18[a]
Electrical supply		2 air-cooled germanium rectifiers each operated at 2.5 kA; maximum voltage of 120	Two 2 kA, 400 V ignition tube rectifiers connected in parallel. Voltage variation 60–400 V. Two 2 kA 30 V selenium oxide rectifiers. Voltage variation 0–60 V (1955 plant)
Arrangement of electrical connections		14 cells in series	18 cells in series
Method of removing HF from fluorine		Contacting with anhydrous sodium fluoride (p. 43)	In 1955 plant refrigeration to –117°F In 1958 plant to –70°F followed by passage over NaF pellets in trays (regenerative)
Method of removing HF from hydrogen		Scrubbing with 10% caustic potash liquor	Scrubbed with water and neutralised with lime
HF in fluorine entering removal system	% by vol.	6	Not stated, but probably 4–13 from electrolyte composition and temp.
HF in fluorine leaving removal system	% by vol	0.15 to 0.20[b]	4–5 after refrigeration; 2% after NaF absorber
HF capacity of HF removal system	lb	Primary 200	
	kg	Primary 91	
Frequency of recharging HF removal system		Primary regenerative. Secondary non-regenerative and recharged every 10 days	

PRODUCTION OF ELEMENTAL FLUORINE

HF in hydrogen entering scrubbing system or vented to waste direct	% by vol.	4
HF in hydrogen leaving scrubbing system		Not stated, but will be 4–13 from electrolyte composition and temp.
Method of pressure control on fluorine side of cells		Negligible Automatic pressure control linked to by-pass on booster fan. Pressure on cells controlled to 0.25 in. (0.63 cm) water Automatic control *via* booster fan
Method of pressure control on hydrogen side of cells		Simple lute into 10% caustic potash liquor Automatic control *via* booster fan
Method of disposal of waste fluorine, etc.		Absorption in tower circulating 10% caustic potash Vented to 80 ft (24.4 m) stack with no scrubbing (1955 plant)
HF supply to cells		Ex drums as liquid direct to cells Ex tank wagons as liquid to storage vessels and thence as vapour to cells
Provisions made for removal of electrolyte spray in fluorine		Small catch pot and cyclone to protect automatic control valve. Product collected is essentially iron fluoride and quantity is small Entrainment separators over each cell and "catch cells" in headers to mains. Results only partially successful (1955 plant).
Fluorine surge tank capacity	ft³ m³	None required 500 (1955 plant) 14
Fluorine booster		Two conventional centrifugal fans. Deep conventional gland packed with chlorofluorocarbon oil-impregnated asbestos, with air purge. Two positive displacement lobe type blowers. Gland has four shaft seals buffered with nitrogen. Copper-impregnated "Teflon" packing (1955 plant)
Pressure of boosted fluorine	in. of water cm of water	24 61 55 (1955 plant) 140

[a] The designed output of American plant in 1955 is given as 4,000 lb/day, but this would require most of the cells to be operating simultaneously at full load. It is, however, stated that one cell series only is working while the other is under repair. It is possible that the actual normal output of the American plant is no more than half that given above. In a paper published in 1958 a plant with a capacity of 7,600 lb/day is described but it is stated that "This facility does not represent any of the actual plants in operation today".[115]

[b] The primary regenerative scrubber uses NaF powder in trays. Regeneration by electric heating. Secondary scrubber contains NaF pellets in a tower (p. 43).

Union Carbide operate their plants under contract to the U.S. Atomic Energy Commission using the name Union Carbide Nuclear Co. They are located at Oak Ridge, Tennessee and Paducah, Kentucky. It would appear that Goodyear also operate a fluorine plant under the designation Goodyear Atomic Corporation at Portsmouth, Ohio but no details are available[110].

14.1 I.C.I.

The cells are of nominal 5kA capacity (7.25 lb or 3.29 kg of fluorine/h at 93% current efficiency) though individual cells have been operated for long periods at loads up to 7.5 kA (10.5 lb or 4.8 kg of fluorine/h).* (Fig. 17). With the exception of the "skirts" the material of construction of the cells is mild steel throughout. The cell body is jacketed on the sides and separately on the bottom, which is water-cooled. Heating, either water at 80° or steam, is applied to the side jackets only when the cells are shut down. Twenty-four pancake coils connected to inlet and exit headers divide the cell transversely and function as water-cooled cathodes[93]. The cell cover has twelve rectangular openings into which anode assemblies fit so that the anodes are interposed between pairs of cathode coils. Each anode assembly consists of a flat plate of mild steel to the underside of which is attached the rectangular Monel metal skirt and inside which are located a pair of anode blocks. The anodes are insulated from the skirt assembly and from the cell top by means of neoprene gaskets. The skirt dimensions are such that they dip about 2 in. (5.08 cm) into the electrolyte and so divide the cell into twelve fluorine compartments and a single hydrogen compartment. The total effective anode surface is 30 ft^2 (2.787 m^2) and the cathode surface 105.5 ft^2 (9.8 m^2). Fluorine from the twelve anode assemblies is collected in a common header while the hydrogen leaves at an offtake located at one end of the cell.

A liquid hydrogen fluoride feed pipe passes through the cover half-way along the cell in the hydrogen compartment and dips 6 in. (15.24 cm) into the electrolyte and permits liquid hydrogen fluoride to be added at the rate of 1 lb (0.455 kg)/min. A sampling tube 1 in. (2.54 cm) in diameter, which can be closed at the upper end and also dips into the electrolyte for several inches in the hydrogen side and allows samples of electrolyte to be taken while the cell is in operation. Provision is made to purge both the hydrogen side of the cell and the individual anode compartments with dry nitrogen. Electrical connections from the several pairs of anodes are connected to the positive busbar running the length of the cell. The negative busbar is connected to the cell body which is thus at the same potential as the cathodes. All gas exits and service connections are provided with insula-

---

* Because of the basic design no difficulty is envisaged in considerably increasing cell capacity by increasing the size of the cell body and the number of cathode coils and anode assemblies.

Fig. 17
I.C.I. fluorine cell of nominal 5.0 kA capacity.

tion breaks so that the cells can be connected in electrical series.

Because of the type of carbon used for the anodes there are no polarisation problems and no conditioning is required except in so far as cells fitted with fresh electrolyte, which may contain up to 0.5% of water, require to be started up at a

Fig. 18.
General view of part of cell room at Rocksavage works of Mond Division, I.C.I. at Weston Point, Runcorn, Cheshire.

low load because the resistance is abnormally low until the water is removed by electrolysis.

Little attention is required in routine operation except to check the load, gas pressures, electrolyte temperature and level and to add hydrogen fluoride as required to maintain the latter. Routine measurements are made of the current to each anode using a portable "clip on" type ammeter ("Tong tester") in order to detect actual or incipient anode failure. An advantage of having the anodes arranged in pairs in individual assemblies is that breakage of one or more anodes in an assembly can be dealt with without disturbing the remaining anodes. With suitable provision for ventilation, the defective anode assembly is replaced with the cell *in situ* once it has been isolated from the rest of the series.

14.2 U.K.A.E.A.

Originally, the U.K.A.E.A., which was formed in 1954, was supplied with chlorine trifluoride by I.C.I. for conversion of uranium tetrafluoride to uranium hexafluoride which was fed to the diffusional separation plant at Capenhurst, Cheshire[111].

By 1962 the military requirement for enriched uranium, for which the Capenhurst plant was originally built, ceased and part of the factory was closed in 1963, the remainder operating for civil purposes[112]. As a result of this I.C.I. ceased to manufacture chlorine trifluoride.

In 1967, as a consequence of a Government decision to expand the civil atomic energy programme, that part of the Capenhurst plant which had been shut down since 1963 was recommissioned. To supply the uranium hexafluoride for this a prototype plant was built and operated at Springfield Works of the U.K.A.E.A. in Lancashire and a larger plant was designed and completed in February, 1968. This plant converts uranium tetrafluoride to the hexafluoride using elemental fluorine produced in fluorine cells of basic I.C.I. design[113].

It has been estimated that by 1975 the European need for enriched feed will amount to 5,000 tons of "separative work"[114]. This would require 8,800 to 10,300 tons of uranium, depending on the degree of enrichment of ^{235}U. The *theoretical* requirement of elemental fluorine for this is roughly 960 to 1,100 tons. Current (1969) U.K.A.E.A. fluorine capacity is about 500 tons per annum.

14.3 IMPERIAL SMELTING CORPORATION

This Company operated fluorine cells on a relatively small scale in the post-war years but no specific information is available as to design and operating characteristics. It is believed that these may have been based on Pennsalt experience in the field[25] (see p. 56).

14.4 UNION CARBIDE

The Union Carbide cell is a development of that operated by Hooker Electrochemical Co. at Niagara Falls during the war years[22, 27]. The latest design employs a Monel body with steel jacket on the sides[29, 30]. Twelve cooling tubes run longitudinally through the middle of the cell. The two anode assemblies which are fitted to the cell top each consist of eight pairs of anodes (total 32 anodes) each anode being $20\frac{5}{8}$ in. by 8 in. by 2 in. (52.4 cm by 20.3 cm by 5.08 cm). The two cathode assemblies each consist of three $\frac{1}{4}$ in. (0.635 cm) parallel plates joined at the ends to form two open-ended "boxes" which surround the anode assemblies and are supported by four steel posts which also serve as cathode connections. Monel gas separation skirts surround each pair of anode assemblies and 6 mesh Monel gauze diaphragm screens are interposed between the anodes and cathodes. The method of attaching anodes to their supports has already been described (see p. 44). An "exploded" drawing of the anode/diaphragm/cathode assembly is shown in ref. 30. A feature of the design is that the whole of the cell top must be removed to gain access to any anode assembly. The cell head is electrically insulated from the anode and cathode so that the cell body is "floating".

All cells are conditioned to prevent polarisation before they are installed in the series, each half of the cell being treated separately. This treatment consists of operation at a load of 0.5 kA which is then raised to 3.5 kA when, after 15 to 20 min., polarisation occurs. After working at high voltage for a few minutes polarisation ceases and operation at normal voltage (8 to 9 V) is possible and may be continued at 4 or 6 kA until the voltage increases to 12 V when the cell is considered to have failed. Polarisation during actual operation is said to be "infrequent".

Details of plant layout and method of operation are given by Jacobson *et al.*[28].

The actual existing capacity of the fluorine plants at Oak Ridge and Paducah is not absolutely clear. In 1955, Union Carbide, under contract to the U.S. Atomic Energy Commission, was operating a 36 cell, 4,000 lb (approx. 1820 kg)/day (1,500 tons/year approx.) plant though it is not stated whether this was at Oak Ridge or Paducah[28]. In 1958 a 25 lb (approx. 11.4 kg)/h (95 tons/year) plant at Oak Ridge was used to produce fluorine at 25 to 75 p.s.i.g.[30]. A plant with an output of approximately 3,000 tons/year described by Oak Ridge personnel in 1958 was said "not to represent any of the actual plants in operation today"[115].

Though they should be considered as speculative the following conclusions appear to fit the known facts. It would seem that the large fluorine plant described by Jacobson *et al.*[28] was operated by the U.S.A.E.C. in the years before 1959 (when Allied commenced operations (see p. 56) principally for the production of uranium hexafluoride and that thereafter most, if not all, uranium hexafluoride was produced by Allied. This is supported by the statement that Allied's process for uranium hexafluoride production, and possibly some fluorine production "know-how", is based on U.S.A.E.C. studies[116]. The 25 lb (11.4 kg)/h plant at Oak Ridge[30], together with some fluorine capacity at Paducah, is probably used for "pickling" the various units of the diffusion plants at both locations to remove absorbed moisture and other impurities before admission of the hexafluoride. For this reason there is probably some fluorine capacity at Portsmouth, Ohio, where there is a diffusion plant operated by the Goodyear Atomic Corporation, though no details are available*[110].

It has recently been reported that more separating capacity than is available from the three existing American diffusion plants referred to will be required before 1980 and that a proposal for private ownership of the new plant(s) is under

---

* It should be noted, however, that Dykstra *et al.*[30] state that "the Oak Ridge gaseous diffusion plant produces and uses large quantities of fluorine. For certain auxiliary uses a small proportion of the fluorine is compressed to 25 to 75 lb/in.^{2} (1.77 to 5.34 kg/cm^2). The fluorine is produced *in a separate facility* which has a capacity of 25 lb/h." A possible inference is that there is, or was, a large fluorine plant at Oak Ridge, which may be that described by Jacobson *et al.*[28].

consideration. The cost of the original plants has been given[110] as $ 2.3 × 10^9$.

The gross power loading of the three American diffusion plants is reported to be 6,000 MW, that at Capenhurst, England, 300 MW, while the French plant at Pierrelatte is also about 300 MW[117, 118]. It seems reasonable to suppose, therefore, that the fluorine capacities for uranium hexafluoride production is in roughly the same ratio, though British fluorine capacity for this purpose is likely to be increased (see p. 53).

14.5 ALLIED CHEMICAL CORPORATION

Unfortunately, relatively little information is available as to the details of design and operating characteristics of Allied Chemical's fluorine cells[17]. The body of the cell is of mild steel, with an external water jacket and is divided into two parts by a longitudinal partition. The cell cover is a machined magnesium alloy casting on the underside of which are two gas separation skirts arranged longitudinally and located appropriately with respect to the sides of the body and the longitudinal partition. Two rows of carbon anodes are suspended centrally within the skirts, the anodes being held between copper clamps and back-up plates by means of bolts. These connections are above the surface of the electrolyte[119].

No figure is given for cell capacity but it would appear that this is not likely to be less than 4 kA or more than 6 kA. Cell life has been reported as "greater than 20×10^6 A h". If this is taken as the actual figure it corresponds to a life of 20 or 30 weeks, depending on the assumed cell capacity[106]. Reference 17 contains a photograph of the Allied cell room.

The cells are mounted on wheels and it is assumed that they are removed for servicing and are "conditioned" (see p. 25) at a location remote from their normal operating position.

An Allied data sheet on fluorine[120] states that "General Chemical (Division of Allied Chemical Corporation) is the nation's primary supplier of fluorine, in both gaseous and liquid forms for commercial and research applications", and this Company was the first to develop the bulk shipment of liquid fluorine on a regular schedule to locations as far apart as New York and California. Allied also supply most, if not all, the fluorine which is used for research purposes in America as compressed gas in cylinders (see p. 60).

In America, "private industry entered the field of large scale fluorine production when Allied contracted to produce 5,000 tons/year of uranium hexafluoride requiring about 1,000 tons/year of fluorine"* and over the last twenty years more

---

* The theoretical *total* fluorine requirement for 5,000 tons of uranium hexafluoride is about 1,620 tons, of which roughly 540 tons would be *elemental* fluorine.

than 100 × 10⁶ pounds of fluorine have been handled"[106]. This corresponds to an average rate of approximately 2,500 tons/year. Elsewhere [116]it is reported that Allied owns the only American commercial plant for the production of uranium hexafluoride. This, located at Metropolis, Illinois, started up in 1959 and produced more than 32,000 tons of uranium hexafluoride over a five year period ending in 1964, when it was shut down. The current commercial nuclear power situation required the plant to be restarted in 1968 and existing capacity is expected to meet the requirements for several years. However, three producers of uranium concentrates, Union Carbide, Kerr-McGee Corporation and United Nuclear Co. are reported to be interested in uranium hexafluoride production but without revealing any specific plans.

14.6 PENNSALT CHEMICALS CORPORATION

According to Neumark and Siegmund[17] this Company commenced commercial fluorine production in 1946, simultaneously with Allied Chemical Corporation. Little further information is available except for a report (1951) of the operation of two 2 kA medium temperature cells[121]. The anodes consisted of eight pairs of 24 in. by 5 in. by 2 in. (61 cm by 12.6 cm by 5.1 cm) or four pairs of 24 in. by 10 in. by 2 in. (61 cm by 25.2 cm by 5.1 cm) ungraphitised electrode carbon blocks. The anode connection consisted of "shoes" shrunk onto the carbon. Cathodes were of louvred mild steel while a steel skirt dipped about 3 in. (7.6 cm) into the electrolyte. The operating temperature was 100° to 105° (nickel thermocouple wells were used) and the anode current density was 0.485 A/in.^{2} (0.075 A/cm^2) at 10 to 12 V. The current efficiency was 95% and fluorine leaving the cells contained 4 to 5% of hydrogen fluoride. It is stated that continuous operation was achieved corresponding to 4,700 kAh "except for infrequent shut downs for repair, for anode replacement or because of polarisation". If both cells were running simultaneously the above amount of electricity corresponds to about seven weeks operation.

14.7 KALI-CHEMIE

Kali-Chemie has a fluorine installation with a capacity of 100 tons/year made up of three Allied Chemicals-type cells of approximately 5 kA capacity which was started up in January, 1966. Most of the output is used to manufacture sulphur hexafluoride but some fluorine is sold as compressed gas in cylinders containing a nominal 1 and 2.5 kg. No details of cell performance and general operating characteristics are available.

14.8 Montecatini-Edison

Montecatini started up a plant for the production of sulphur hexafluoride at Linate, Milan in 1963/64. No details of the fluorine plant are available.

14.9 La Société des Usines Chimiques de Pierrelatte

It has recently been disclosed that the above organisation, which operates France's enriched uranium facilities, will head uranium hexafluoride production but no information is available as to the design of cells which are to be used[122, 123].

A one page "hand out" available at a recent exhibition discloses that the above organisation was founded in 1961, Ugine Kuhlmann being the principal shareholder. In addition, it is stated that the organisation has facilities capable of converting 5,000 to 6,000 tons/annum of natural uranium into "a product in conformity with the United States A.E.C.'s specifications, whatever the purity of the initial products to be fluorinated". U.C.P. is also a producer of sulphur hexafluoride "and can, on request, fluorinate any product".

15. Fluorine compression

In the I.C.I. fluorine plant fluorine is not stored under pressure and is simply boosted to a few inches of water pressure, using a conventional centrifugal fan, to feed the plant(s) consuming it. However, Union Carbide compress at least some of the fluorine produced and store it at 25 to 75 lb/in.^{2} (1.77 to 5.34 kg/cm^2). Details of the equipment used for this purpose may be found in refs. (28) and (30).

To obtain fluorine at higher pressures, the gas is first liquefied and then allowed to vapourize into suitably designed vessels.

16. Liquid fluorine

A discussion of the production of liquid fluorine is beyond the scope of this account but basic information and relevant references are available elsewhere[17]. It is of interest to note that nitrogen (b.p. $-196°$) cannot be used for purging and "padding" purposes since it is soluble in liquid fluorine (b.p. $-188°$) and helium must be employed.

Large quantities of liquid fluorine are shipped by road by the Allied Chemical Corporation of America while limited quantities were at one time transported by I.C.I. in this way. Allied's specification, for liquid fluorine on a helium-free basis, is as follows:[17]

Fluorine (wt. % min.)	99.0
HF + CF_4 (wt. % max.)	0.3
O_2 + N_2 (wt. % max.)	1.0
HF + CF_4 + O_2 + N_2 (wt. % max.)	1.0

The I.C.I. specification for the gas obtained from containers of liquid fluorine was as follows:

Fluorine (wt. % min.)	99.0
HF (wt. % less than)	0.1
CF_4 (wt. % less than)	0.01

Warning. If fluorine direct from the cells, after removal of hydrogen fluoride, is liquefied without pretreatment violent explosions can occur as the last of the liquid evaporates. The reason for this is not known but it may result from the spontaneous violent decomposition of traces of liquid ozone or higher oxides of fluorine such as F_2O_2, F_2O_3 and F_2O_4. The difficulty may be avoided by heating the fluorine to 300° to 400° before liquefaction.

17. Disposal of fluorine and hydrogen fluorine-containing hydrogen

In the early days of the production of fluorine on a large scale considerable difficulty was experienced in devising a suitable method for its disposal and a variety of methods were investigated. The reaction of fluorine with both hydrogen and with water is inhibited at times and can result in explosions, though the addition of alcohol to the latter results in a smooth reaction. The recommended method resulting from early studies involved injection of the fluorine into the flame of a natural gas burner. The resultant mixture of carbon tetrafluoride and hydrogen fluoride was then treated in an aqueous caustic scrubber[124].

It has been shown, however, that fluorine can be effectively absorbed in a solution of caustic potash of about 10% concentration.* This method has been used by I.C.I. for many years to dispose of both concentrated fluorine and fluorine-containing tail gas, and exits containing less than 0.05 mg of fluorine/m³ have been achieved. Hydrogen containing approximately 4% by volume of hydrogen fluoride is treated in a similar system with a similar level of hydrogen fluoride in the scrubbed gas.

At the Oak Ridge, Tennessee and Paducah, Kentucky plants of Union Carbide it would appear that emergency venting of fluorine from tanks holding fluorine under pressure, or purging of cells, took place, without absorption, *via* an 80 ft disposal stack into which air was fed by two 50,000 ft³/min blowers[28]. It seems probable that current American concern with environmental pollution would

---

* The tendency to form oxygen difluoride when aqueous alkalies react with elemental fluorine depends critically on the conditions, particularly contact time and alkali concentration[125].

not permit this procedure to continue and the above may not represent present practice. At both these installations hydrogen fluoride was absorbed in water and the dilute acid disposed of in a lime neutralisation system[28].

The disposal of high strengths of fluorine in an emergency by reaction with charcoal has been studied both in America and this country. The method is satisfactory except that an appreciable concentration of hydrogen fluoride (20 to 30%) may be present in the products of reaction even after vacuum drying of the charcoal at 100°[126, 127].

18. Analysis of elemental fluorine

Early methods for the analysis of fluorine were cumbersome and probably not very accurate[124, 128]. The obvious method of reacting with aqueous potassium iodide and titrating the liberated iodine is unsatisfactory because concentrated fluorine further oxidises the iodine, though the procedure works reasonably well with dilute fluorine ($< 10\%$). Hydrogen fluoride may be estimated by passing a known volume of the gas through weighed nickel absorption tubes containing sodium fluoride[124] or catherometrically provided that gases other than hydrogen fluoride and fluorine are substantially absent[129].

It has been found the fluorine concentration of gas from the cells, after removal of hydrogen fluoride, can be measured volumetrically with a reasonable degree of accuracy by absorption in 10% aqueous sodium sulphite solution, stabilised with a small quantity of *p*-amino phenol; or in a solution containing 14 to 15 g of potassium chloride and 5 g of potassium hydroxide per 100 g of water provided that the correct technique is used. This involves *rapidly* flushing the solution into the top of a gas burette containing the sample[130, 131].

Allied Chemical Corporation recommend as a method of analysis passing the sample through a bed of sodium fluoride to retain hydrogen fluoride and then over granular sodium chloride. The chlorine produced is reduced to chloride which is determined by the Volhard method. Residual gases, oxygen, nitrogen, helium are analysed by gas/liquid chromatography. Hydrogen fluoride and carbon tetrafluoride are measured on a separate sample using IR spectroscopy[17].

Whereas the gas burette methods using aqueous solutions described above are rapid and useful for plant control purposes they are probably not very accurate and the percentage of "inerts" is likely to be somewhat high because of the liberation of a small amount of oxygen. For the most accurate results, gas/liquid chromatography is to be preferred though the methods which have been developed are to some extent more cumbersome than is usual in the analysis of less reactive gases[132-134].

Rochefort[133] reacted the fluorine with dry sodium chloride and analysed the chlorine produced using "Kel-F" oil supported on solid "Kel-F" polymer in a nickel tube. More recently it has been shown that improved results, particularly in the measurement of hydrogen fluoride content, are obtained using a perfluorocarbon oil (approximately $C_{21}F_{44}$) on polytetrafluoroethylene as the stationary phase in polytetrafluoroethylene tubes. It is claimed that less chemisorption occurs under these conditions[135].

Typical analyses of gaseous fluorine from the cells, after hydrogen fluoride removal, are as follows:[30]

	I.C.I.	Union Carbide
Fluorine %	> 99	97.0
Hydrogen fluoride %	0.2	2
Oxygen %		0.5
Nitrogen %		0.5

For the specification of fluorine from liquid fluorine tanks see p. 58.

19. Fluorine in the laboratory

Because of its special characteristics it is not out of place in an account dealing with the industrial production of fluorine to make brief mention of the availability of fluorine in the laboratory.

In America virtually all the fluorine used in research is supplied by Allied Chemicals in cylinders at a pressure of 400 p.s.i.g. Two sizes are available holding 4.5 and 6 lb (2.04 and 2.72 kg) of fluorine respectively[17]. Kali-Chemie[137] also supply gaseous fluorine in cylinders holding 1.0 and 2.5 kg at a pressure of 400 p.s.i.g. The specification is:

Fluorine	98–99%
Hydrogen fluoride (Max.)	0.5%
Balance (chiefly oxygen and nitrogen) about	1%

In Great Britain the use of compressed fluorine for laboratory use is not favoured, though at one time supplies were available from Imperial Smelting Corporation.

For many years I.C.I. supplied very reliable, laboratory scale fluorine cells to universities and to industry[7, 39]. These were of two sizes, 10 A and 60 A, and produced approximately 7 and 40 g of fluorine/h respectively (Figs. 19 a and b and 20 a and b). This business has now been taken over by Sherman Chemicals Ltd. of Tottenham, who offer a full range of spares and servicing[136] (see also p. 35).

20. Materials of construction for use with elemental fluorine

Details of materials of construction suitable for the handling of fluorine are to be found in a number of references[2, 4, 120, 137]. In brief, almost any of the common constructional metals (other than lead and tin and their alloys) are satisfactory at

Fig. 19.
(a) and (b) I.C.I. 10A laboratory fluorine cell producing approx. 7 g of fluorine/h.

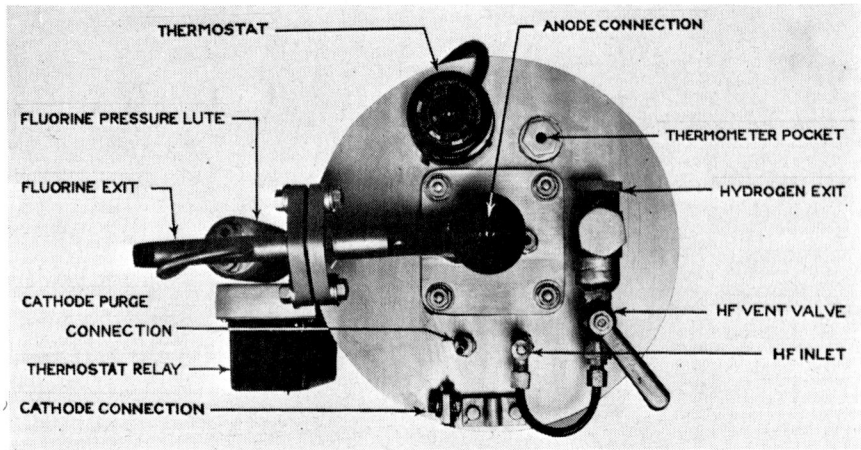

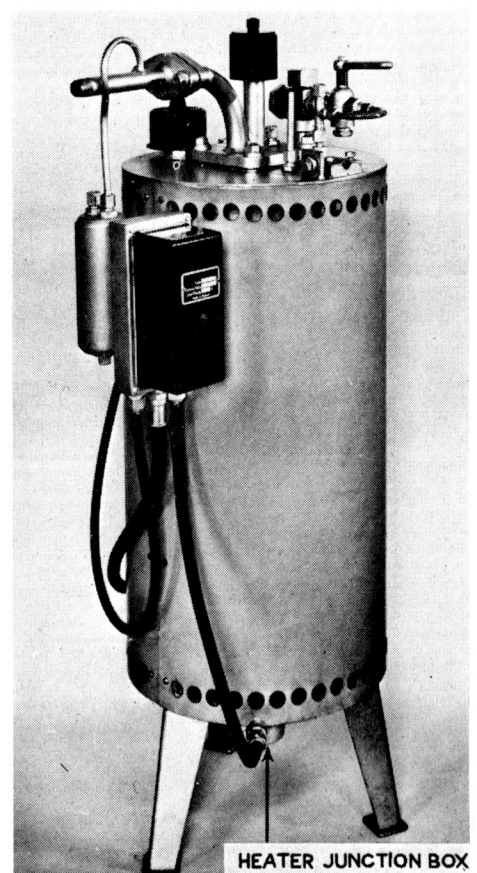

Fig. 20.
(a) and (b) I.C.I. 60A laboratory fluorine cell producing approx. 40 g of fluorine/h.

temperature and pressures not greatly above ambient. For higher temperatures, up to at least 400°, nickel and Monel metal are greatly superior to all others. As jointing materials, appropriate grades of Neoprene and rubber are satisfactory provided that they are used in thin sheets so as to minimise the surface exposed to the gas. The life of such joints is, however, finite. More resistant is polytetrafluoroethylene, though some degradation may occur in long-term use.

In all installations handling fluorine and reactive fluorides it is essential that all items are thoroughly degreased before putting into operation.

Subject to the above reservations, fluorine at close to atmospheric pressure is not difficult to handle. In 1947 Cady expressed the opinion that the hazards of working with fluorine and its compounds* had previously been greatly over-rated and that it is "just another substance"[138]. Handling fluorine at high pressure is a very different matter.

21. The cost of elemental fluorine

Leech quotes some figures for the make-up of the cost of fluorine which has been given in an American paper[111, 139] and "which might be achieved on a 1,200 ton/year scale where the capital cost is estimated at £1,400,000".

	Cost/ unit of fluorine		
	d/lb	£/lb	£/kg
Hydrogen fluoride (17d/lb)	20.5	0.0855	0.188
Potassium bifluoride (22d/lb)	1.3	0.0054	0.0119
Electrical energy (0.326 d/kWh)	3.4	0.0142	0.032
Steam, air, water, nitrogen etc.	1.7	0.0071	0.0156
Labour (includes 118% overheads)			
operation	2.1	0.0088	0.0194
maintenance	3.4	0.0142	0.032
laboratory and supervision	3.4	0.0142	0.032
Maintenance materials	3.4	0.0142	0.032
Operating cost	39.2	0.1632	0.360
Amortisation (over 20 years)	6.2	0.0259	0.056
Total cost	45.4	0.1891	0.416

Several comments may be made about the above cost. The energy consumption corresponds to some 10 kWh/lb and so must include electrical energy for all

---

* But see p. 58.

purposes (see p. 45). However, most of this will be DC energy required for electrolysis and in many circumstances would cost at least three times the figure quoted above. In addition, the capital cost will have increased appreciably since 1958 and, at the present time, few chemical plants would be amortised over as long a period as twenty years. If it is assumed that the cost of energy is 1d/kWh, that the capital cost is now £2,000,000, and that this is amortised over, say, seven years, the total cost on the above basis would be 7s/lb (£0.35/lb or £0.77/kg).

It may be noted, that on the above large scale, energy cost, hydrogen fluoride and capital are large items in the total. As with all plants, labour costs and capital costs are sensitive to scale and would increase considerably for smaller outputs. Additionally, it may be mentioned that the above cost does not include any "return on capital". The cost-conscious reader may care to revise the above figure in the light of such considerations as are discussed by Davies and McCarthy[140].

22. Uses of fluorine

To the writer, elemental fluorine has been a disappointment to the extent that very few commercial outlets of importance have developed. At the time of writing they number three only and of these two could, perhaps, be described as somewhat esoteric. These are the use of fluorine in atomic energy, both military and civil, and in rocketry. The third, and purely civil use, is in the production of sulphur hexafluoride.

Fluorine is essential for the production of uranium hexafluoride by the reaction $UF_4 + F_2 \rightarrow UF_6$, the hexafluoride being fed to the diffusional separation plants for the separation or enrichment of ^{235}U contained in natural uranium[111]. With current plans for the expansion of the civil atomic energy programme throughout the world the requirement for elemental fluorine may be expected to increase (see p. 53), though about the year 2,000 the demand may decrease because of the accumulation of increasing stocks of plutonium from fast reactors[141]. Smaller quantities of fluorine are used to "pickle", i.e. pre-treat, components of diffusional separation plants. Some methods of processing spent fuel elements from nuclear reactors involve the use of halogen fluorides, such as bromine trifluoride, for the manufacture of which elemental fluorine is required[111].

The efficacy of a rocket propellant is measured in terms of the "specific impulse" which is proportioned to the heat developed per unit weight of the reactants and inversely proportional to the molecular weight of the combustion products. The large heat of reaction of fluorine with many substances and its relatively low atomic weight makes it a potential oxidiser of high efficiency and a considerable amount of work has been carried out in this field[142-146]. Most, if not all, this work has been sponsored by the National Aeronautical and Space Administration of the

U.S. The chemical literature over the last ten years or so also shows that N.A.S.A. has sponsored a large amount of fundamental work relating to the development of fluorine-rich oxidisers other than elemental fluorine itself. If this has not led to developments of practical importance it has at least resulted in this disclosure of important new areas of chemistry, for example that of fluorine and nitrogen[147-149].

North American Rockwell's Rocketdyne Division is reported to be building a small plant near Los Angeles to produce the recently discovered chlorine pentafluoride (Trade name "Fluoridyne") for use as a high energy oxidiser[150, 151].

Sulphur hexafluoride is made by the combustion of sulphur in fluorine and finds application as a gaseous dielectric in electrical equipment, particularly switchgear, because of its high dielectric strength (about two and a half times that of air), its non-inflammability and complete absence of toxicity. There is every reason to suppose that this is an expanding market.

Though it is possible to fluorinate linear and cyclic hydrocarbons using elemental fluorine without too much breakdown, compounds containing functional groups lose these groups on fluorination and other methods have to be used to prepare such compounds as perfluoroacid fluorides (see p. 76). Perfluoroparaffins are expensive to manufacture and they have as yet no practical applications of importance

"Napalm" has for obvious reasons superseded chlorine trifluoride as an incendiary agent in warfare for which purpose it was prepared in Germany during World War II[16].

23. Acknowledgements

The author wishes to thank Mr. G. Wallace, Mr. A. Davies and Dr. J. R. Case of Mond Division, I.C.I. who were kind enough to read the original manuscript and who have made helpful comments. Dr. A. T. Kuhn, the editor of this volume, provided invaluable assistance with the sections dealing with electrode kinetics and the reversible potential of the fluorine cell.

The author is indebted to Dr. R. B. Kehoe of the U.K.A.E.A. for information relating to the significance of "separative work" and for details of La Société des Usines Chimiques de Pierrelatte".

24. References

1 H. R. LEECH, *Quart. Rev. (London)*, 3 (1949) 22.
2 H. R. LEECH in J. W. Mellor (Ed.), *Comprehensive Treatise on Inorganic and Theoretical Chemistry*, Supplement II, Part I, Longmans Green and Co., London, 1956, p. 1 *et seq.*
3 G. H. CADY in J. H. SIMONS (Ed.), *Fluorine Chemistry*, Vol. I, Academic Press Inc., New York, 1950, p. 293 *et seq.*
4 H. C. MILLER and F. D. LOOMIS in R. E. KIRK and D. F. OTHMER (Eds.), *Encyclopedia of Chemical Technology*, Vol. VI, The Interscience Encyclopedia, New York, 1951, p. 656 *et seq.*

5 *Gmelins Handbuch der Anorganischen Chemie*, System-Nummer 5, "Fluor", Verlag Chemie, G.m.b.H., Weinheim/Bergstrasse, 1959, p. 66 *et seq.*
6 A. J. RUDGE, *Chem. Ind. (London)*, (1949) 427.
7 A. J. RUDGE, *Chem. Ind. (London)*, (1956) 504.
8 W. KWASNIK, *Fortschr. Chem. Forsch.*, 8 (1967) 309.
9 L. DOMANGE, *Proc. Chem. Soc.*, June/July, (1959) 172.
10 O. RUFF, *Die Chemie des Fluors*, Julius Springer, Berlin, 1920.
11 W. L. ARGO, F. C. MATHERS, B. HUMISTON and C. O. ANDERSON, *Trans. Electrochem. Soc.*, 35 (1919) 335.
12 F. C. MATHERS, *Trans. Electrochem. Soc.*, 36 (1919) 207.
13 P. LEBEAU and A. DAMIENS, *Compt. Rend.*, 181 (1925) 917.
14 G. H. CADY, *J. Am. Chem. Soc.*, 56 (1934) 1431.
15 G. H. CADY, D. A. RODGERS and C. A. CARLSON, *Ind. Eng. Chem.*, 34 (1942) 443.
16 H. R. NEUMARK, *Trans. Electrochem. Soc.*, 91 (1947) 367.
17 H. R. NEUMARK and J. M. SIEGMUND in R. E. KIRK and D. F. OTHMER (Eds.), *Encyclopedia of Chemical Technology*, Vol. IX, Interscience Encyclopedia Inc., New York, 2nd ed., 1966, p. 506 *et seq.*
18 *Statements Relating to the Atom Bomb*, H.M.S.O., 1945.
19 Various authors, *Ind. Eng. Chem.*, 39 (1947) 235 *et seq.*
20 *The Preparation, Properties and Technology of Fluorine and Organic Fluorine Compounds*, C. SLESSER and S. R. SCHRAMM (Eds.), *Nat. Nucl. Energy Ser., Div. VII*, Vol. I, McGraw-Hill Book Co., New York, Toronto, London, 1951.
21 W. C. SCHUMB, R. C. YOUNG and K. J. RADIMIR, *Ind. Eng. Chem.*, 39 (1947) 244.
22 R. L. MURRAY, S. G. OSBORNE and M. S. KIRCHER, *Ind. Eng. Chem.*, 39 (1947) 249.
23 J. T. PINKSTON, *Ind. Eng. Chem.*, 39 (1947) 255.
24 R. C. DOWNING, A. F. BENNING, F. B. DOWNING, R. C. MCHARNESS, M. K. RICHARDS and T. W. TOMKOWIT, *Ind. Eng. Chem.*, 39 (1947) 259.
25 F. J. GALL and H. C. MILLER, *Ind. Eng. Chem.*, 39 (1947) 262.
26 R. D. FOWLER, W. B. BURFORD, H. C. ANDERSON, J. M. HAMILTON and C. E. WEBER, *Ind. Eng. Chem.*, 39 (1947) 266.
27 J. DYKSTRA, S. KATZ, C. B. CLIFFORD, E. W. POWELL and G. H. MONTILLON, *Ind. Eng. Chem.*, 47 (1955) 883.
28 J. JACOBSON, W. K. HENDERSON, T. P. FLEMING, R. W. LEVIN and J. A. MARSHALL, *Ind. Eng. Chem.*, 47 (1955) 878.
29 S. P. VALVALIDES, R. E. CABLE, W. K. HENDERSON and C. A. POWELL, *Ind. Eng. Chem.*, 50 (1958) 178.
30 J. DYKSTRA, B. H. THOMPSON and W. C. PARIS, *Ind. Eng. Chem.*, 50 (1958) 181.
31 A. J. RUDGE, *Chem. Ind. (London)*, (1966) 482.
32 *13th Report of the U.K.A.E.A.*, 1966–67, H.M.S.O.
33 A. BAINES and A. DAVIES, *J. Appl. Chem. (London)*, 5 (1955) 242.
34 D. P. CROWLE, (I.C.I. Ltd.), unpublished work.
35 *"Wetting"*, *Soc. Chem. Ind. (London), Monograph*, 25 (1967).
36 *Ibid.*, p. 52.
37 *Ibid.*, p. 433.
38 *Ibid.*, p. 31.
39 A. J. RUDGE, *The Manufacture and Use of Fluorine and its Compounds*, Oxford University Press, London, 1962.
40 O. RUFF, *Ber.*, (1936) A 185.
41 D. E. PALIN and K. D. WADSWORTH, *Nature*, 162 (1948) 925.
42 W. RÜDORFF, U. HOFMANN, G. RÜDORFF, J. ENDELL and G. RUESS, *Z. Anorg. Allgem. Chem.*, 256 (1948) 125.
43 A. J. RUDGE and A. DAVIES, (I.C.I. Ltd.), unpublished work.
44 A. J. RUDGE, W. N. HOWELL and H. HILL, *Nature*, 160 (1947) 504.
45 B. KABANOV and A. FRUMKIN, *Z. Physik. Chem. (Frankfurt)*, 165 (1933) 433.
46 N. WATANABE, M. ISHII and S. YOSHIZAWA, *J. Electrochem. Soc. Japan*, 28 (1961) E 180.
47 W. J. KROLL, *J. Electrochem. Soc.*, 106 (1959) 8C.

48 C. L. Mantell, *Electrochemical Engineering*, McGraw-Hill Book Co. Inc., New York, Toronto, London, 4th ed., 1960, p. 359.
49 C. A. Hampel, *Encyclopedia of Electrochemistry*, Reinhold Publishing Corp., New York; Chapman and Hall, London, 1964.
50 K. Arndt and H. Probst, *Z. Elektrochem.*, 29 (1923) 323.
51 C. S. Taylor, *Trans. Electrochem. Soc.*, 47 (1925) 301.
52 N. Watanabe, M. Ishii and S. Yoshizawa, *J. Electrochem. Soc. Japan*, 29 (1961) E 177.
53 D. A. J. Swinkels and R. N. Seafurth, *J. Electrochem. Soc.*, 115 (1968) 994.
54 L. M. Dennis, J. K. Veeder and E. G. Rochow, *J. Am. Chem. Soc.*, 53 (1931) 3263.
55 T. G. Pearson, *The Chemical Background of the Aluminium Industry*, Roy. Inst. Chem., (London), Lectures, Monographs Rept., 3 (1955).
56 W. N. Howell and H. Hill, (I.C.I. Ltd.), Brit. pat., 655,098 of 11.7.51.
57 P. K. C. Wiggs, *Industrial Carbon and Graphite*, Soc. Chem. Ind., London, 1958, p. 252.
58 C. Sheer, J. A. Cooney, D. L. Rothacker and F. R. Sileo, *Energy Distribution in the High Intensity Arc Plasma*, Technical Note No. 1, Vitro Laboratories, Division of Vitro Corporation of America, West Orange, New Jersey, AFOSR-860, June 16, 1961.
59 A. Davies, H. Hill and A. J. Rudge, (I.C.I. Ltd.), unpublished work.
60 R. R. Paxton, J. F. Demendi, G. J. Young and R. B. Rozelle, *J. Electrochem. Soc.*, 110 (1963) 933.
61 M. W. Reed and W. C. Schwerner, *J. Electrochem. Soc.*, 114 (1967) 852.
62 C. Berger, *Handbook of Fuel Cell Technology*, Prentice-Hall Inc., New Jersey, 1968.
63 N. Watanabe, Y. Fujii and S. Yoshizawa, *J. Electrochem. Soc. Japan*, 31 (1963) 131.
64 Daikin Inds. Co. Ltd., Japan pat., 11457 of 23.6.64.
65 J. McLaren, *U.S.A.E.C. Report* CF-51-6-157, Dec. 7, 1956.
66 F. Olstowski and J. J. Newport, (Dow Chemical Co.), Brit. pat., 890,605 of 7.3.62.
67 J. Simons, *J. Am. Chem. Soc.*, 46 (1924) 2175.
68 W. D. Bancroft and N. C. Jones, *Trans. Electrochem. Soc.*, 55 (1929) 183.
69 L. M. Dennis, J. M. Veeder and E. G. Rochow, *J. Am. Chem. Soc.*, 53 (1931) 3263.
70 W. C. Schumb and E. L. Gamble, *J. Am. Chem. Soc.*, 52 (1930) 4302.
71 A. L. Henne, *J. Am. Chem. Soc.*, 60 (1939) 96.
72 N. Watanabe, M. Inoue and S. Yoshizawa, *J. Electrochem. Soc. Japan*, 31 (1963) 113.
73 K. Klemm and E. Huss, *Z. Anorg. Allgem. Chem.*, 258 (1949) 221.
74 R. D. Peacock and D. W. A. Sharpe, *J. Chem. Soc.*, (1959) 2762.
75 A. G. Sharpe in M. Stacey, J. C. Tatlow and A. G. Sharpe, (Eds.), *Advances in Fluorine Chemistry*, Vol. I, Butterworths Scientific Publications, London, 1960, p. 53.
76 W. M. Latimer, *J. Am. Chem. Soc.*, 48 (1926) 2868.
77 H. K. Fredenhagen and O. T. Kreft, *Z. Elektrochem.*, 35 (1929) 670.
78 C. S. Garner and D. M. Yost, *J. Am. Chem. Soc.*, 59 (1937) 2738.
79 H. B. Neumann and H. Richter, *Z. Elektrochem.*, 31 (1925) 481.
80 O. Ruff and W. Busch, *Z. Elektrochem.*, 31 (1925) 614.
81 N. Watanabe, M. Inoue and S. Yoshizawa, *J. Electrochem. Soc. Japan*, 31 (1963) 168.
82 A. J. Arvia and J. B. de Cusminsky, *Trans. Faraday Soc.*, 58 (1962) 1019.
83 J. B. de Cusminsky, *Rev. Fac. Cienc. Quim. Univ. Nac. La Plata*, 34 (1961) 183.
84 A. J. Rudge and H. Hill, (I.C.I. Ltd.), unpublished work.
85 A. Davies, (I.C.I. Ltd.), unpublished work.
86 W. N. Howell and H. Hill, (I.C.I. Ltd.), Brit. Pat., 642,812 of 13.9.50.
87 W. H. Wilson and H. Hill, (I.C.I. Ltd.) Brit. Pat., 731,066 of 1.6.55.
88 A. J. Rudge, H. Hill and W. N. Howell, (I.C.I. Ltd.) Brit. Pat., 675,209 of 9.7.52.
89 A. J. Rudge and H. Hill, (I.C.I. Ltd.), unpublished work.
90 A. J. Rudge and A. Davies, (I.C.I. Ltd.), Brit. Pat., 852, 369 of 26.10.60.
91 A. J. Rudge and A. Davies, (I.C.I. Ltd.), Brit. Pat., 957,168 of 6.5.64.
92 A. J. Rudge and A. Davies, (I.C.I. Ltd.), Brit. Pat., 957,603 of 6.5.64.
93 A. J. Rudge and A. Davies, (I.C.I. Ltd.), Brit. Pat., 861,978 of 1.3.61.
94 A. J. Rudge and H. Hill, (I.C.I. Ltd.), Brit. Pat., 778,248 of 3.7.57.
95 A. Davies, (I.C.I. Ltd.), Brit. Pat., 925,870 of 8.5.63.
96 J. Simons, *J. Am. Chem. Soc.*, 46 (1924) 2179.

97 K. Fredenhagen, W. Klatt and H. Kunz, *Z. Anorg. Allgem. Chem.*, 218 (1934) 161.
98 K. Fredenhagen, *Z. Anorg. Allgem. Chem.*, 210 (1933) 24.
99 J. F. Froning, M. K. Richards, T. W. Stricklin and S. G. Turnbull, *Ind. Eng. Chem.*, 39 (1947) 275.
100 K. E. Long and H. W. Cromer, (Harshaw Chemical Co.), U.S. Pat., 2,426,558 of 26.8.47.
101 A. W. Ravenscroft and A. J. Rudge, (I.C.I. Ltd.), Brit. Pat., 825,185 of 21.5.58.
102 H. R. Leech and W. H. Wilson, (I.C.I. Ltd.), Brit. Pat., 824,427 of 2.12.59.
103 C. A. Powell, R. E. Cable and W. K. Henderson, *Ind. Eng. Chem. (Intern. Edition)*, 52 (1960) 46A.
104 R. E. Cable, W. B. Goode and W. K. Henderson, (U.S.A.E.C.), U.S. Pat., 3,041,266 of 26.6.62.
105 W. B. Goode, C. R. King, W. K. Henderson and S. Bernstein, U.S.A.E.C., KY-452. *C. A.*, 61 (1964) 3927.
106 J. M. Siegmund, *Chem. Eng .Prog.*, 63 (1967) 88.
107 A. J. Rudge and W. N. Howell, (I.C.I. Ltd.), Brit. Pat., 651,107 of 14.3.51.
108 A. J. Rudge and W. N. Howell, (I.C.I. Ltd.), Brit. Pat., 685,461 of 7.1.53.
109 A. V. Uvarov and A. V. Bogolypov, Russian Pat., 202,885 of 28.9.67.
110 Anon., *New Scientist*, 39 (1968) 380.
111 H. R. Leech, *Chem. Ind. (London)*, (1960) 242.
112 U.K.A.E.A., 8th and 9th Annual Reports and Accounts, H.M.S.O.
113 U.K.A.E.A., 14th Annual Report and Accounts, H.M.S.O.
114 B. Silcock, *Sunday Times (Business News Section)*, January 19th, 1969, p. 25.
115 A. P. Huber, J. Dykstra and B. H. Thompson, *Peaceful Uses of Atomic Energy*, 2nd U.N. Intern. Conf. Geneva, 1958, 15/P/524.
116 N. P. Chopey, *Chem. Eng.*, 73 (1966) 52.
117 G. Leach and A. Wilson, *The Observer*, February 9th, 1969. p. 6.
118 Anon., *Chem. Eng. News*, 46 (Sept. 2nd) (1968) 11.
119 Ref. 17, p. 513.
120 Allied Chemical Corporation, General Chemical Division, *Product Data Sheet*, PD-TA-85413, 15.1.58.
121 F. R. Lowdermilk, R. G. Danehower and H. C. Miller, *J. Chem. Educ.*, 28 (1951) 246.
122 Anon., *Chem. Eng. News*, 47 (June 30th) (1969) 17.
123 Anon., *New Scientist*, 43 (1969) 20.
124 S. G. Turnbull, A. F. Benning, G. W. Feldmann, A. L. Linch, R. C. McHarness and M. K. Richards, *Ind. Eng. Chem.*, 39 (1947) 286.
125 H. R. Leech in ref. 2, pp. 186–195.
126 W. H. Schmidt, *NASA Memo.*, 1-27-59E.
127 A. Davies and G. Wallace, (I.C.I. Ltd.), unpublished information.
128 W. T. Miller and L. A. Bidgelow, *J. Am. Chem. Soc.*, 58 (1936) 1585.
129 I.C.I. Ltd., unpublished information.
130 A. F. Williams, U.K.A.E. BDDA. 107, 1947 (quoted in ref. 133).
131 E. Norcott, (I.C.I. Ltd.), unpublished information.
132 J. J. Mayo, W. R. Rossmassler and E. L. Williamson, A.E.C. Accession No. 43369, No. CONF-721-1; *C. A.*, 62 (1965) 11122.
133 O. Rochefort, *Anal. Chim. Acta*, 29 (1963) 350 and references therein.
134 H. V. Dayan and B. C. Neale, *Advan. Chem. Ser.*, 54 (1965) 223; *C. A.*, 65 (1966) 4664.
135 W. S. Pappas and J. G. Million, *Anal. Chem.*, 40 (1968) 2176.
136 Anon., *Chem. Age*, 98 (May 4th) (1968) 27.
137 Kali-Chemie Brochure, "KC-Fluor F_2", undated but received 13.8.68.
138 G. H. Cady, *Ind. Eng. Chem.*, 39 (1947) 10A.
139 J. S. Nairn et al., *Peaceful Uses of Atomic Energy*, 2nd U.N. Intern. Conf. Geneva, 1958, 15/P/524.
140 D. S. Davies and C. McCarthy, *Introduction to Technological Economics*, John Wiley and Sons, London, New York and Sydney, 1967.
141 T. Johnson, *Sunday Times (Business News Section)*, February 23rd, 1969, p. 37.
142 J. Silverman and W. T. Webor, *Jet Propulsion*, 26 (1956) 874; *C. A.*, 51 (1957) 1580.

143 J. F. GALL, *Ind. Eng. Chem.*, 49 (1957) 1331; *C. A.*, 51 (1957) 18611.
144 S. GORDON, *Natl. Advisory Comm. Aeron. Res. Memo.*, No. E57/122, 1958; *C. A.*, 54 (1960) 18961.
145 S. GORDON and V. N. HUFF, *Natl. Advisory Comm. Aeron. Res. Memo.*, No. E56/A13a, 1956; *C. A.*, 54 (1960) 18961.
146 H. R. NEUMARK and F. L. HOLLOWAY, *Missiles and Rockets*, 2 (1957) 97; *C. A.*, 52 (1958) 9534.
147 C. B. COLBURN, *J. Chem. Educ.*, 38 (1961) 180.
148 C. J. HOFFMAN and R. G. NEVILLE, *Chem. Rev.*, 62 (1962) 1.
149 E. W. LAWLESS and I. C. SMITH, *Inorganic High-Energy Oxidisers*, Edward Arnold (Publishers) Ltd., London; Marcel Dekker Inc., New York, 1968.
150 Anon., *Chem. Eng.*, 76 (1969) 73.
151 Anon., *Chem. Eng. News*, 47 (Jan. 20th) (1969) 37.

Chapter 2

Electrochemical Fluorination

A. J. RUDGE

Research and Development Department, Imperial Chemical Industries Ltd., Mond Division, Runcorn, Cheshire

1. Introduction

Many organic compounds have an appreciable solubility in anhydrous hydrogen fluoride and form conducting solutions[1,2] and such solutions can be electrolysed in a diaphragmless cell having nickel anodes interleaved with steel cathodes. Hydrogen is evolved at the cathode while highly fluorinated products are formed at the anode, it being a characteristic of the process that the voltage is controlled to such a value that no free fluorine is liberated. Provided a substance such as sodium or potassium fluoride is added to the hydrogen fluoride to provide conductivity, the process may be extended to organic compounds of low solubility, to those which do not form conducting solutions in hydrogen fluoride, and to certain gaseous inorganic compounds.

Such a process is known as "electrochemical fluorination" and the apparatus in which it is carried out as a "Simons Cell", after the inventor J. H. Simons. Further practical details are given by Simons (ref. 2, p. 414) and by Burdon and Tatlow[3] but much of the information is of limited industrial significance.

Two excellent reviews of the process are available, the first written by Burdon and Tatlow of Birmingham University in 1960[3] and a much more recent one by Nagase[4]. The writer has been heavily dependent on these for basic sources of information and references. Watanabe has also reviewed the field but this account is of limited use to occidental readers since it is written in Japanese[5].

Because of the existence of the above reviews and the scope of this book the present account deals only briefly with the process in general and the emphasis is on the production of highly fluorinated products which are of established or marginal industrial importance. For the same reason the number of references herein has been limited to those of special significance in the present connection and the reviews cited should be consulted for a comprehensive list. Patent dates given are publication dates.

2. Historical

The first published account of the electrochemical fluorination process is to be found in a series of papers by Simons and co-workers[6-11]. The second paper of the above series was submitted to the Journal of the American Chemical Society in 1941 but was withdrawn "for security reasons". In addition to the above papers a general account of the apparatus and method of operation is given in the first volume of Simons *Fluorine Chemistry*[12].

The first cells used were little more than "test-tube size" but a 2 kA cell was in operation in 1949 and possibly earlier. A photograph of this appears in ref. 12. It is of interest to note that, in this early work, the fluorination of acetic acid or acetyl chloride is reported as giving a miscellaneous range of products including partially and fully fluorinated methane but no trifluoroacetyl fluoride was identified, nor was pentafluoropropionyl fluoride obtained from propionic acid. The fluorinated products were recovered both by condensation from the exit gases and by evaporation of the hydrogen fluoride contained in the cell. The first report of the preparation of perfluorocarboxylic acids by electrochemical fluorination is that of Kauck and Desslin[13,14] who obtained members of the homologous series containing up to thirteen carbon atoms by starting with the hydrocarbon acid anhydrides (Table 1).

TABLE 1

PERFLUOROCARBOXYLIC ACIDS, $CF_3(CF_2)_n \cdot COOH$, PREPARED BY KAUCK AND DESSLIN

n	0	1	2	3	4	5	6	8	12
B.p. of acid (°C)	73	96	120	139	157	175	189	218	270

Note: Boiling points are approximate since, in some cases, the pressure differed somewhat from 760 mm.

That there was little understanding of the nature of the process is suggested by the fact that they give the following somewhat improbable equation for the formation of trifluoroacetyl fluoride:

$$(CH_3CO)_2O + 10HF \rightarrow 2CF_3 \cdot COF + F_2O + 8H_2$$

It is noteworthy that no yields or current efficiencies are given, with the obvious inference. Guenthner[15] was the first to report the production of a dibasic perfluoroacid fluoride, $(CF_2)_8 \cdot (COF)_2$, together with $CF_3(CF_2)_7COF$ by the electrolysis of a 4% solution of sebacic acid in hydrogen fluoride. Again, no yields are given.

3. Scope of the process

From an academic point of view, electrochemical fluorination is of very wide application. Not only liquids may be fluorinated but also solids such as sebacic acid provided that they are soluble in anhydrous hydrogen fluoride[15]. Gases such as methane[16], ethylene, ethane and propane[17-19], chloromethanes and chlorofluoromethanes[20] may be similarly fluorinated if they are bubbled into the cell below the electrodes.

It must be emphasised, however, that except in special cases single products are rarely obtained and yields and current efficiencies are frequently very low. In the case of gases, typical reported current efficiencies and yields have not exceeded 50% for the sum of all products, while the maximum concentration of specific compounds in the fluorinated product ranged from 8 to 57%[16]. Hydrogen sulphide

TABLE 2

TYPICAL PRODUCTS OBTAINED BY ELECTROCHEMICAL FLUORINATION

Starting material	Principal products	Remarks
$CH_3(CH_2)_nCH_3$	$CF_3(CF_2)_nCF_3$	Poor yields
$CF_3(CH_2)_nCH_3$ or $CH_3(CF_2)(CH_2)_{n-1}CH_3$	$CF_3(CF_2)_nCF_3$	Moderate to good yields
Cyclohexane, benzene, naphthalene	—	No fluorination
$CCl_2 : CCl_2$	$CFCl_2 . CFCl_2$	Yield and C.E. high
$CH_3(CH_2)_n OH$	Perfluorocarbons, perfluoro- acid fluorides, perfluoroethers	Poor yields
n- and cyclo ethers, polyethers	Perfluoro-analogues	Moderate to poor yields
n-Carboxyclic acids and acid anhydrides	Perfluoroacid fluorides, per- fluorocarbons, perfluoro- ethers*	Poor yields
n-Acid chlorides & fluorides	Perfluorocarbons, perfluoro- acid fluorides, perfluoro- ethers*	Yields moderate to very good
n-Diacid fluorides	Perfluorodiacid fluorides, per- fluoroacid fluorides, per- fluorocarbons	Poor yields
Aldehydes and ketones	Perfluoroacid fluorides, per- fluorocarbons	No perfluoroaldehydes or ketones
Esters	Perfluoroacid fluorides, etc.	No perfluoroesters
Primary and secondary amines	N-fluorinated perfluoroamines and NF_3	Moderate yields
Tertiary amines	Perfluorotertiary amines and NF_3	Moderate yields
n-Alkyl sulphides	$(R_f)_2SF_4$, R_fSF_5	
$CH_3(CH_2)_nSO_3H$	Perfluorocarbons, $CF_3(CF_2)_{n-1}CF_3$	Yield of perfluorosulphonyl fluorides poor
$CH_3(CH_2)_nSO_2Cl(F)$	$CF_3(CF_2)_nSO_2F$	Yields better than perfluoro- carboxylic acid fluorides
NH_3**	NF_3	
H_2O	F_2O	
CS_2	CF_3SF_5, SF_6, CF_4, etc.	
K_2SO_4	SO_2F_2, SOF_2, F_2O	

* n-Carboxylic acids containing six or more carbon atoms give perfluoro*cyclic* ethers, frequently as major products (see pp. 78-79).
** *Caution:* NF_2H and NFH_2 may be among the products. These are very unstable and NF_2H at least may detonate spontaneously when in the liquid state.

has been fluorinated to sulphur hexafluoride with a current efficiency of 40 to 42%[18, 21, 22].

The general scope of the process may, perhaps, best be illustrated by listing the types of product obtained with various starting materials (Table 2). It should be noted, however, that Table 2 is far from comprehensive and reference to "good" and "moderate" yields are purely relative and in many conventional processes would be considered to be "poor".

4. Mechanism of electrolytic fluorination

Possible mechanisms are discussed at some length in Burdon and Tatlow's interesting review of 1960[3]. Seven years later Nagase[4] says simply that the mechanism is "not completely understood" but "it is based on an anodic reaction, very certainly free radical in type".

In considering possible mechanisms it is useful to list the principal characteristics of the process.

1. The reaction takes place at low temperatures, say $-10°$ to $+25°$, and is characterised by the ability to retain functional groups such as .COF and .SO$_2$F, in contrast to direct fluorinations at higher temperatures.

2. There are no indications that any anode material is as good as nickel.

3. No free fluorine is produced although the preferred working voltage, 5 to 6 V, is higher than that corresponding to the reversible voltage of a fluorine cell (see p. 26).

4. The process is slow and the maximum usable current density is very low as compared with most electrolytic processes.

5. There is an induction period when new anodes are used and it has been stated that free fluorine is evolved, even at low voltages, during this period.

It is difficult to avoid the conclusion that nickel possesses properties which are key features of the process. Burdon and Tatlow point out the significant fact that both Ni^{III} and Ni^{IV} are known in compounds of the type K_3NiF_6 and K_2NiF_6. The latter compound is red in colour and is prepared by treating a two to one molar mixture of potassium fluoride and nickel difluoride with fluorine at elevated temperature. Nothing is known about its chemical properties. It is to be noted, however, that during the operation of nickel anode fluorine cells the anode suffers severe corrosion and the electrolyte becomes progressively contaminated with a red sludge (see p. 4). Though the composition of the red component of the sludge has not been elucidated it is known to react rapidly with water with discharge of the red colour and formation of a gas with an ozone-like odour (oxygen difluoride?). There is, therefore, strong circumstantial evidence that a higher valency nickel fluoride (probably NiF_4^{2-}) participates in the electrochemical fluorination process, particularly since such species are, as would be expected, highly labile.

Neither NiF_3 nor NiF_4 are known except as complexes with alkali metal fluorides so that the composition of active higher valency nickel species is not obvious in cases where alkali metal fluoride have not been added to the hydrogen fluoride to promote conductivity.

A serious dilemma arises when one considers the precise species which is fluorinated at the anode. It is usually considered that substances such as hydrocarbonacid fluorides dissolve in hydrogen fluoride to give protonated species and fluoride ion. Such a positively charged moiety is unlikely to be fluorinated at the anode. It is much more likely to occur with an anion or a neutral molecule. It is suggested that the presence of a low concentration of an anion is not an impossibility since it is known that alkali metal fluorides such as potassium or caesium fluoride react with perfluoroacid fluorides to give such a species[23].

$$R_f.COF + MF \rightarrow R_f.CF_2.O^-M^+$$

There is thus the possibility of an analogous equilibrium existing between a perfluoroacid fluoride and hydrogen fluoride, and the situation when a hydrocarbonacid fluoride is dissolved in hydrogen fluoride may possibly be represented as follows:

However, the fact that compounds incapable of forming ions may be fluorinated in a Simons cell suggests that an ionic mechanism is unlikely in such cases.

Two recent papers by Donohue et al.[24, 25] describe the pulsed electrolysis of wet hydrogen fluoride by which means the current efficiency on fluorine monoxide (oxygen difluoride) production was raised from 2 to 5% to a little over 40%. The presence of potassium fluoride in the electrolyte in concentrations greater than 1 mole% decreased the yield. The mechanism of the process is also discussed.

Based on measurements of the decay of the anode voltage on open circuit Donohue et al.[25] concluded "that the anode is in an oxidised state as an electrode in a battery. Further, by analogy with the nickel oxide battery electrode, we assume that the accumulation of higher valent nickel (Ni^{3+} or Ni^{4+}) species is the cause."

A surprising feature of electrochemical fluorination, so far as long-chain compounds are concerned, is the almost total absence of partially fluorinated material. This suggests some sort of "zipper" mechanism which is almost certainly free-radical in character. Further work in this field is obviously desirable.

5. Perfluorocompounds of industrial importance produced by electrochemical fluorination

Because there have been no disclosures of the details of the large scale production of perfluoro compounds of established commercial importance it is inevitable that the information contained in this section will have serious limitations. The best that can be attempted is to present such information as is available in the published literature together with the results of laboratory-scale work with which the writer is personally familiar.

5.1 Perfluorocarboxylic acids and by-products

A real advance in the production of perfluorocarboxylic acids by electrolytic fluorination occurred with the finding that carboxylic acid fluorides are much better starting materials than carboxylic acids themselves or their anhydrides. Alternatively, carboxylic acid chlorides may be used since anhydrous hydrogen fluoride converts these to the acid fluorides with evolution of hydrogen chloride[26].

The reason for this is to be found in the behaviour of carboxylic acids or their anhydrides in solution in hydrogen fluoride.

$$(R.CO)_2O + HF \rightleftharpoons R.COF + R.COOH$$
$$R.COOH + HF \rightleftharpoons R.COF + H_2O*$$

Only the acid fluoride will give perfluoroacid fluoride on fluorination since the carboxylic acid itself would be expected to yield the very unstable fluoroxyacid $R_f.COOF$, resulting in decomposition to perfluorocarbon and carbon dioxide. Additionally, and assuming the same amount of reaction in each case, acid anhydrides would be expected to yield more acid fluoride than the carboxylic acids themselves.

According to Scholberg and Brice[26] "the yields of trifluoroacetyl fluoride can be substantially doubled and the yields of higher compounds can be improved in an even greater ratio, as compared with the yields obtained when using the anhydrides of hydrocarbon acids as starting compounds. Furthermore, the acid fluoride product yield per unit of electrical energy (electrical efficiency) is more than doubled." The results obtained by these investigators are given in Table 3.

In 1956, Muray[27] prepared a range of perfluoroacid fluorides starting with the

---

* *Caution:* Any water in the system will be converted, at least in part, to oxygen difluoride on electrolysis. A number of violent detonations have occurred as a result of condensing liquid fluorine monoxide together with products with which it forms a thermodynamically unstable system. Almost any carbon/fluorine compound other than carbon tetrafluoride would give a potentially dangerous system with oxygen difluoride.

TABLE 3

PRODUCTION OF PERFLUOROCARBOXYLIC ACIDS BY SCHOLBERG AND BRICE[26]

Starting material	Weight		Conc.	Time	Total products		Desired product		
	(lb)	(kg)	(%)	(h)	(lb)	(kg)	(lb)	(kg)	Yield (%)
$CH_3.COF$	1 750.5	796	4.5	1 145	—	—	2 277	1 040	71
$CH_3(CH_2)_2.COF$	444	202	5.0	711	416	189	387	176	35.8
$CH_3(CH_2)_6.COCl$	477	216	9.0	—	745	338	134	61	11

Notes: 1. Electrolysis was carried out in a cell with 110 ft² (10.2 m²) of anode surface, the current ranging from 1650 to 1955 A at 5.45 to 5.9 V giving a current density of about 15 to 18 A/ft² (0.016 to 0.019 A/cm²).
2. Products were collected by condensation from the exit hydrogen and/or were run off from the bottom of the cell, depending on boiling points.
3. Sodium fluoride was added to the HF except in the case of octanoyl chloride, which gave a solution of adequate conductivity.

acid chlorides using a Simons cell with an anode area of 314 in.² (2024 cm²) at currents ranging from 3 to 25 A and at a voltage which was usually about 5.5. The results are given in Table 4, from which it may be seen that the yield of perfluoroheptanoic acid at 10% was almost identical with that obtained by Scholberg and Brice[26], but the yields of perfluoroacids of shorter chain length were a little better.

In 1959 Barr[28] investigated the fluorination of hexanoyl chloride in a Simons cell with and without stirring and at temperatures ranging from −5° to +20°. Low temperatures gave poor results. With stirring the yields of perfluoroacid fluoride were 9 and 18% and without stirring 17, 25 and 22%. A mass balance on the products accounted for 87.2% of the input carbon in the best case the distribution being: perfluoroacid fluoride 21.6%, caustic-insoluble liquid products 53.8%, gaseous products (mainly carbon dioxide and carbon tetrafluoride with a little

TABLE 4

PRODUCTION OF PERFLUOROCARBOXYLIC ACIDS BY MURAY[27]

Starting material	Wt. (g)	Total products (g)	Perfluoroacid (g)	Yield of acid (%)
$CH_3(CH_2)_3 . COCl$	380	53	18	2*
$CH_3(CH_2)_4 . COCl$	1 221	1 362	420	15
$CH_3(CH_2)_5 . COCl$	507	835	175	16
$CH_3(CH_2)_6 . COCl$	406	500	85	10
$CH_3(CH_2)_{10} . COCl$	440	≈ 270	2	0.5
$CH_3(CH_2)_{14} . COCl$	138	Nil	Nil	Nil**
$(CH_2)_4 . (COCl)_2$	275	Nil	Nil	Nil*
$(CH_2)_8 . (COCl)_2$	835	657	234	11

* Products probably lost because of volatility.
** Increase in cell resistance due to solid deposits on anode prevented continued electrolysis.

carbonyl fluoride) 9.5%, unidentified gaseous product 2.3%, unaccounted for 12.8%.

In all fluorinations of hydrocarbon acid fluorides, particularly in the case of the longer chain compounds, an appreciable amount of caustic insoluble material is formed indicating loss of the acid fluoride functional group. It was originally assumed that this material consisted of linear perfluorocarbons and/or fluorohydrocarbons. However, Kauck and Simons[29] showed that the liquid, caustic-insoluble products obtained from the fluorination of acid fluorides containing six or more carbon atoms was largely a mixture of a perfluorocarbon and isomeric perfluorocylic ethers.

In 1957 Muray[27] who, at the time, was unaware of the findings of Kauck and Simons, independently investigated the caustic-insoluble liquid products obtained in the electrochemical fluorination of hexanoyl chloride, $C_5H_{11}.COCl$ and heptanoyl chloride, $C_6H_{13}.COCl$, and found that these materials consisted largely of perfluorocyclic ethers, having either five or six atoms in the ring, together with a minor proportion of linear perfluorocarbon having one carbon less than the starting material. He was, however, unable to elucidate unequivocally the structure of the perfluorocyclic ethers and to decide whether the oxygen atom was in the ring or the side chain. Muray's figures for the physical properties of $C_6F_{12}O$ and $C_7F_{14}O$ are in excellent agreement with those of Kauck and Simons[29]. Mass balances based on one mole of starting material are shown in Table 5.

The figures given in Table 5 show that the weight of perfluoroether and perfluorocarbon was roughly twice the weight of perfluoroacid fluoride obtained, which was no more than 11 mole% in the case of heptanoyl chloride.

Kauck and Simons[29] showed that the cyclic ethers prepared as above have the oxygen atom as part of either a five- or six-membered ring and that the product from the electrochemical fluorination of a particular hydrocarbonacid fluoride containing six or more carbon atoms consists of a virtually inseparable mixture of

TABLE 5

MATERIAL BALANCE IN THE ELECTROCHEMICAL FLUORINATION OF HYDROCARBONACID CHLORIDES[27]

	Hexanoyl chloride			Heptanoyl chloride		
	Formula	Weight (g)	Moles	Formula	Weight (g)	Moles
Starting material	$C_5H_{11}.COCl$	134.5	1.00	$C_6H_{13}.COCl$	148.5	1.0
Total liquid products	Mixture	188.0	—	Mixture	285.0	—
Loss on KOH treatment*	—	36.0	—	—	109.0	—
Perfluoroacid fluoride	$C_5F_{11}.COF$	47.0	0.15	$C_6F_{13}.COF$	41.0	0.11
Perfluorocarbon	C_5F_{12}	21.0	0.07	C_6F_{14}	27.0	0.08
Perfluoroether	$C_6F_{12}O$	74.0	0.24	$C_6F_{14}O$	95.0	0.26
Other compounds	Mostly $C_6F_{12}O$	10.0	~ 0.03	Mostly $C_7F_{14}O$	13.0	~ 0.04

* Material other than perfluoroacid fluoride.

the two isomers. Thus, octanoyl fluoride yields a mixture of the following ethers:

$$\begin{array}{c} CF_2\!-\!CF_2 \\ | \quad\quad | \\ CF_2 \quad CF\cdot CF_2\cdot CF_2\cdot CF_2\cdot CF_3 \\ \diagdown O \diagup \end{array} \quad \text{and} \quad \begin{array}{c} CF_2 \\ CF_2 \diagup \diagdown CF_2 \\ | \quad\quad | \\ CF_2 \quad CF\cdot CF_2\cdot CF_2\cdot CF_3 \\ \diagdown O \diagup \end{array}$$

Such a mixture is commercially available and is sold by the Minnesota Mining and Manufacturing Co. (3M's) under the description "FC-75". It is thus, presumably, the major product obtained in the production of perfluorooctanoic acid which is also produced commercially by 3M's. Table 6 gives the physical properties of FC-75, together with those of the inert liquid FC-43 which is perfluorotributylamine.

TABLE 6

SOME PHYSICAL PROPERTIES OF 3M'S FC-43 AND FC-75

	FC-43	FC-75
Molecular weight	671	≈ 338
Boiling range (°C)	170-180	99–107 (95%)
Pour point (°C)	−50	−62 max.
Density (g/cc at 25°C)	1.872	1.77
Viscosity (cstokes at 25°C)	2.74	0.65 min.
Refractive index (25°C)	1.2910	1.277
Surface tension (dyne/cm at 25°C)	16.1	13.5–18.0
Specific heat (cal/g 25–40°C)	0.27	0.248
Heat of vaporisation (cal/mole at b.p.)	11,100	8,700
Vapour pressure at (mm at 25°C)	0.3	—
Trouton constant	24.6	23.3
Dielectric strength (ASTM–D877) (kV)	40	35 min.
Dielectric constant (100 Hz at 25°C)	1.86	1.86
Power factor (100 Hz at 25°C)	< 0.0005	0.0005
Volume resistivity (Ω cm at 25°C)	10^{14}–10^{16}	6×10^{14}

Notes: 1. FC-43 is heptacosafluorotributylamine or perfluorotributylamine, $(C_4F_9)_3N$. FC-75 consists largely of isomers of $C_8F_{16}O$ and is a by-product obtained in the electrochemical fluorination of octanoyl chloride[29].
2. Physical properties are taken from 3M's Technical Information Brochure dated September, 1960.

A 3M's patent relates to the production of perfluoro(alkoxypropionic) acids of general formula $F(CF_2)_n O.CF_2.CF_2.COOH$ from the analogous hydrocarbon acid chlorides or fluorides[30] but no information is given as to the yields obtained. A recent patent (March, 1969) assigned to 3M's relates to the preparation of perfluorosulphonic acid fluorides of the type:

$$\begin{array}{c} Q \\ \diagdown \\ \quad\quad N(CF_2)_n.SO_2F \\ \diagup \\ Q_1 \end{array}$$

where Q and Q_1 may be linear perfluoroalkyl groups or form part of a cyclic structure such as a morpholine ring[31]. All the examples given involve the electrochemical fluorination of materials in which the nitrogen atom forms part of a six-membered ring system and it is claimed that the products are useful as additives in chromium plating baths of the self-regulating type.

5.2 Perfluorosulphonic acids

It would appear that Gramstad and Haszeldine[32] in England and Brice and Trott[33] of 3M's in America found independently that the electrolytic fluorination of alkane sulphonyl chlorides or fluorides gave good to moderate yields of the corresponding perfluorosulphonyl fluorides. The results obtained by Brice and Trott are shown in Table 7.

TABLE 7

PRODUCTION OF PERFLUOROSULPHONYL FLUORIDES BY BRICE AND TROTT[33]

Starting material	Weight (g)	Conc. (%)	Time (h)	Temp. (°C)	Total products (g)	Desired product		
						Wt. (g)	Yield (%)	C. Effic. (%)
$CH_3.SO_2Cl$	740	4	46	15–17	1 018	634	55	≈ 35
$CH_3.CH_2.SO_2Cl$	735	4	41	15–17				
$(CH_3)_2.CH(CH_2)_2SO_2Cl$	?	6	106	18	695	380*		
$CH_3(CH_2)_5.SO_2Cl$	?	10	50	20–24	522	329		
$CH_3.(CH_2)_7.SO_2Cl$	?	10–20	80	17–19	498			
$CH_3.(CH_2)_7.SO_2Cl$	470	10	69	18–20	509	298	≈ 25	≈ 20

Notes: 1. The cell current was approx. 40–50 A and the voltage 5–6 V in all cases.
2. Products were collected by condensation from the exit hydrogen and/or were run off from the bottom of the cell depending on the boiling points.
3. The current density, where stated, was 20 A/ft² (0.018 A/cm²).
* Product stated to be perfluoro-n-pentane-sulphonyl fluoride.

Gramstad and Haszeldine used a current density of 0.0014 A/cm² and obtained a 96% yield of trifluoromethylsulphonyl fluoride from $CH_3.SO_2F$ and an 87% yield from $CH_3.SO_2Cl$. In a later paper[34] they describe the preparation of perfluorosulphonyl fluorides containing from two to eight carbon atoms. As with the perfluoroacid fluorides, the yield decreased with increase in the length of the carbon chain as shown in Table 8, but in general, higher yields were obtained.

Illustrative of the variable results obtained by different operators in this field it may be noted that Burdon et al.[35] were able to obtain $CF_3.SO_2F$ in only 50 to 60% yields while the corresponding yield of C_8 and C_{10} acids was 10 to 15%.

TABLE 8

YIELDS OF PERFLUOROSULPHONYL FLUORIDES, $CF_3(CF_2)_n \cdot SO_2F$, OBTAINED BY GRAMSTAD AND HASZELDINE[34]

n	0	1	2	3	4	5	6	7
Yield (%)	87–96	79	68	58	45	36	31	25

By analogy with the perfluoroethers obtained as major products in the fluorination of fatty acid chlorides and fluorides it might be expected that by-products in the fluorination of hydrocarbon sulphonyl chlorides and fluorides would include five- and six-membered ring sulphones, but there is no indication in the literature that these have been identified. In fact, no information appears to be available as to the nature of the by-products but doubtless these include linear perfluorocarbons.

6. Other routes to perfluoroacids

It is perhaps, advisable to point out that perfluorocarboxylic acids and related compounds can be prepared by processes other than electrochemical fluorination. Most of these involve, as the initial step, telomerisation between tetrafluoroethylene and a suitable telogen. Thus, tetrafluoroethylene and trifluoromethyl or pentafluoroethyl iodide give a mixture of linear perfluoroakyl iodides of varying chain length[36, 37] and these may be converted to perfluorocarboxylic acids by a variety of methods. When methanol is used as the telogen with tetrafluoroethylene, alcohols of the general formula $H(C_2F_4)_n \cdot CH_2 \cdot OH$ result and these may be oxidised to give *omega*-hydroperfluoroacids which are suitable for some (but not all) applications of perfluoroacids[38, 39]. The disadvantage of processes based on telomerisation is that the telomers have a wide range of molecular weight. However, Du Pont, Pennsalt and others are known to be active in this field as illustrated by the publication of a number of recent patents.

More recently, highly branched oligomers of tetrafluoroethylene prepared by fluoride ion catalysis[40] have been converted to highly fluorinated branched chain sulphonic acids and their salts, which show excellent surfactant properties and are potentially of relatively low cost[41].

7. Other compounds made by electrochemical fluorination

So far as is known, no products of established industrial importance, other than those already described, are currently produced by electrochemical fluorination. It may be noted, however, that electrolysis of solutions of ammonium fluoride in hydrogen fluoride provides what appears to be the preferred route to nitrogen trifluoride[42], appreciable quantities of which have been made and which is available commercially.* Air Products supply nitrogen trifluoride in experimental quantities as do British Drug Houses, but at a price in excess of £90/100 g.

It is probable that the electrolysis of a dilute solution of water in hydrogen fluoride would be the preferred method for the production of oxygen difluoride should this material ever be required on an industrial scale. Pulsed electrolysis greatly improves the yield (see p. 75) and Kuhn who has studied the economics of this process concludes that it could be commercially attractive in the case of products in the higher cost range[43].

Sulphur hexafluoride has been made by electrolytic fluorination of hydrogen sulphide but the low current density used (0.023 A/cm^2) makes it very unlikely that the process as it stands would be commercially attractive[22].

3M's have marketed perfluorotributylamine under the designation FC-43 but there is no indication that this material is of commercial significance (see Table 6).

8. Commercial availability and cost of products made by electrochemical fluorination

Without question, perfluorooctanoic acid, C_7F_{15}.COOH, and perfluorooctylsulphonic acid, C_8F_{17}.SO$_3$H, and their derivatives are, from a commercial point of view, the most important materials made by electrochemical fluorination. Until recently, these were produced exclusively by the Minnesota Mining and Manufacturing Co. of St. Paul, Minnesota**. Unfortunately, there is very little information relating to the details of the process as operated on a commercial scale. The only information available is that already given.

Theoretical raw material and energy usages for the production of perfluorooctanoyl fluoride and perfluorooctyl sulphonyl fluoride are given at the top of Table 9, while a cost estimate for the manufacture of the latter compound on a 100 tons/year scale is given in the lower part of the Table.

---

* Schmeisser and Hubler have recently claimed that the electrochemical fluorination of urea provides a better route to nitrogen trifluoride[42a].
** Perfluorooctanoic acid has been manufactured for not less than three years by the Italian firm of Rimar (Italian Research Marzotto), owned by Count Marzotto who is active in the wool and textile business. The acid is used captively for textile treatment (see p. 85) but is also being marketed as such. The process used is electrochemical fluorination and it is understood that five times as much perfluoroether as perfluoroacid is obtained and that the perfluoroether is being sold to the computer industry. Kingsley and Keith are world agents for Rimar.

ELECTROCHEMICAL FLUORINATION 83

TABLE 9

THEORETICAL RAW MATERIAL AND ENERGY REQUIREMENTS FOR
PERFLUOROOCTANOYL FLUORIDE AND PERFLUOROOCTYL SULPHONYL FLUORIDE

Starting material	Units/ unit product (wt.)	Energy requirements *	
		(kWh/lb)	(kWh/kg)
$C_7H_{15}.COCl$	0.414	5.070	11.16
HF	0.740		
$C_8H_{17}.SO_2Cl$	0.434	5.532	12.18
HF	0.718		

* Voltage assumed = 6 V

COST ESTIMATE FOR THE PRODUCTION OF PERFLUOROOCTYL SULPHONYL FLUORIDE
AT THE RATE OF 100 TONS PER YEAR (101.6 TONNES/YEAR.)

The estimate is based on the conversion of $C_8H_{17}Br$ to $C_8H_{17}SO_3Na$ followed by conversion of the sodium salt to $C_8H_{17}SO_2Cl$ and electrochemical fluorination of this.

Basic data:				
	Plant capital	£350,000		
	Working capital	£100,000		
	Total capital	£450,000		
	n-Octyl bromide	36d/lb	£0.33/kg	
	Sodium sulphite	5d/lb	£0.046/kg	
	Anhydrous HF	17d/lb	£0.156/kg	
	Electrical energy	1d/kWh	£0.0042/kWh	

			Cost	
Raw materials		Pence/lb	£/lb	£/kg
n-Octyl bromide		126.0	0.525	1.155
Sodium sulphite		35.0	0.146	0.321
Anhydrous HF		68.0	0.284	0.624
Others		50.0	0.208	0.458
		279.0	1.163	2.558
Services				
Electrical energy		50.0	0.208	0.459
Others		5.0	0.021	0.046
Labour (includes 118% overheads)				
Operation		66.0	0.276	0.608
Maintenance		41.0	0.171	0.376
Laboratory and supervision		22.0	0.092	0.202
Plant cost		463.0	1.931	4.249
Amortisation over 7 years		53.5	0.224	0.494
Works cost		516.5	2.155	4.743

Note: The above cost estimate has the same general basis as that on p. 63 (q.v.) and contains no "return on capital".

Table 10 gives prices for some 3M's products.

TABLE 10

PRICES OF SOME 3M'S' PRODUCTS PRODUCED BY ELECTROCHEMICAL FLUORINATION

Code No.	Composition or probable composition	Date	Cost (£) Lots of 1 lb (0.455 kg)		500 lb or more (227 kg)	
			Per lb	Per kg	Per lb	Per kg
FC–21	$CF_3.COOH$	4.3.63	5.4	11.9	—	—
FC–22	$CF_3.CF_2.COOH$	4.3.63	17.4	38.2	—	—
FC–23	$CF_3.CF_2.CF_2.COOH$	4.3.63	8.9	19.6	—	—
FC–26	$CF_3(CF_2)_6.COOH$	4.3.63	41.6	91.6	—	—
FC–75	$C_8F_{16}O$ isomeric ethers	4.3.63	10.5	23.1	—	—
FC–95	$C_7F_{15}.SO_3K$ + shorter chains	15.9.67	12.9	28.4	10.7	23.6
FC–98	$C_6F_{13}.SO_3K$ (+ branched chain?)	15.9.67	7.2	15.9	5.55	12.2
FC–126	$C_7F_{15}.COONH_4$	15.9.67	19.7	43.4	17.8	39.2
FC–128	$C_8F_{17}SO_2N(R)CH_2.COOK$	15.9.67	13.6	29.9	11.8	26.0
FC–134	$(C_8F_{17}.SO_2.NH.CH_2.CH_2.CH_2N(CH_3)_4)^+I^-$	15.9.67	17.3	38.1	15.5	34.2
FC–170	$C_8F_{15}.SO_2N(C_2H_5)(CH_2.CH_2.O)_{14}H$	15.9.67	7.15	15.7	5.35	11.8
FC–172	$C_7F_{15}CO.NH.(CH_2)_3N^+(CH_2)_3.CH_2.CH_2COO^-$	15.9.67	12.5	27.5	10.7	23.6
FC–176	Not known	15.9.67	7.15	15.8	5.45	12.0

Notes: 1. Prices have been taken from 3M's' sales literature of the date indicated and may, therefore, not be current.
2. 4.3.63 and 15.9.67 prices have been converted from dollars at the rate of £1 = $2.80.
3. The composition of items from FC–95 onwards have been deduced from chemical analysis and by comparing the surface tension/concentration curves for aqueous solutions with those published[44]. The accuracy of the composition is not, therefore, guaranteed.

9. Uses of perfluorocompounds produced by electrochemical fluorination

A useful survey of the properties and actual and potential applications of perfluoro compounds produced by electrochemical fluorination is to be found in Kirk–Othmer[44]. In particular, this discusses the characteristics of fluorosurfactants and the formation of low free surface energy surfaces. Fluorochemical properties and applications have also been discussed by Guenthner and Vieter[45], who give reaction schemes showing how various types of surfactants are derived from $C_7F_{15}.COF$ and $C_8F_{17}.SO_2F$, and by Klaus[46].

9.1 SURFACTANTS

Fluorocarbon-based surfactants are generally more effective than hydrocarbon surfactants in aqueous solution in that certain of the former can lower the surface tension of water from 72 dyne/cm to 15 to 20 dyne/cm at only 0.01 % concentration. With hydrocarbon surfactants ten to one hundred times the concentration is required to lower the surface tension of water to 30 to 35 dyne/cm, the minimum attainable with these materials according to Klaus.*

There is little doubt that currently the principal outlet for a perfluorosurfactant is in the emulsion polymerisation of tetrafluoroethylene. In the preparation of such emulsions an emulsifier, typically ammonium perfluorooctanoate, is used at a concentration of about 0.5% with respect to the polytetrafluoroethylene emulsion polymer[48]. At a rough estimate the world production of polytetrafluoroethylene in 1968 was 9,000 tons/year, of which about 4,000 tons was emulsion polymer. Thus, the world usage of perfluorosurfactant for this purpose would be about 45,000 lb, equivalent to about £900,000/year with surfactant at £20/lb.

Pennsalt (Pennsylvania Salt Manufacturing Co.) are also offering a perfluorosurfactant which is claimed to be suitable for use in the emulsion polymerisation of tetrafluoroethylene. This is the ammonium salt of a perfluoroacid of the general structure $(CF_3)_2.CF.(C_2F_4)_n.CF_2.COOH$ produced from tetrafluoroethylene with $(CF_3)_2CFI$ as a telogen[49] (see also footnote p. 82).

Another important application of surfactants derived from electrochemically produced perfluorocompounds is as a constituent of "Light Water". This is a fire-fighting foam developed by the U.S. Naval Research Laboratory for dealing with petroleum fires, such as may occur in oil refineries and in aircraft crashes. The foam spreads rapidly over the surface of the burning fuel smothering the flames and, as the foam collapses, the water spreads as a thin film on the fuel surface forming a stable vapour seal, thus preventing re-ignition[50-53]. "Light Water" is compatible with "Purple K" (potassium bicarbonate) which is used as firefighting powder under similar circumstances.

Less important uses of fluorosurfactants are as corrosion inhibitors, as additives to increase reaction rates in immiscible systems (fluorosurfactants show surface activity in organic solvents) as spray inhibitors in chromium plating baths and as levelling agents in polishes (Guenthner and Vieter[45]).

9.2 Surface treatments

A most important application of perfluorocarboxylic and sulphonic acids is as starting materials for the production of textile treating agents which confer water and oil repellent (dirt repellent) properties on the product.

The fundamental reason for the efficiency of such textile finishes is the finding of Zisman and co-workers that close-packed CF_3 groups exhibit very low free

---

* This statement appears to be incorrect since surface tensions in water as low as 26.3 dyne/cm, and possibly lower, have been obtained with conventional surfactants[17].

surface energies. The free surface energy of a surface is conveniently defined in terms of the "critical surface tension", γ_c, which corresponds to the surface tension of a liquid which will just spread on a substrate as a continuous film[44].

The lower the value of γ_c, the lower the free surface energy of the surface and the less readily it is "wetted" (see p. 15). The lowest known value of γ_c, 6 dyne/cm, corresponds to close-packed CF_3 groups, the value for $.CF_2H$ groups is 15 dyne/cm, for $.CF_2.$ groups (P.T.F.E.) 18 dyne/cm and for $.CH_2.$ groups (polyethylene) 31 dyne/cm. However, the minimum value for γ_c is obtained only when the length of the perfluorocarbon chain exceeds a certain value. This is shown in Table 11, which gives the variation in oil repellency (in arbitrary units) with increasing perfluorocarbon chain length. The same Table also shows the catastrophic effect of a single hydrogen on the *omega* carbon atom and the uselessness of a chlorofluorocarbon chain[46].

TABLE 11

EFFECT OF CHAIN LENGTH AND STRUCTURE ON OIL REPELLENCY

Group	$.CF_3$	$.C_2F_5$	$.C_3F_7$	$.C_5F_{11}$	$.C_7F_{15}$	$.C_9F_{19}$	$.(CF_2)_7.CF_2H$	$.CF_2(CF_2.CFCl)_3.Cl$
Oil repellency	0	60	90	100	120	130	50	0

The actual textile finishes consist basically of a perfluorocarbon moiety attached *via* a wide variety of organic groupings to a polymerisable group, such as acrylate or methacrylate, which is then polymerised and made up as an aqueous emulsion or as a solution in a solvent such as $CF_2Cl.CFCl_2$ or $CCl_3.CH_3$. Invariably, other additives and modifiers are present and the formulation of a successful product involves a considerable amount of technological "know-how".

The principal producers of fluorocarbon textile finishes are, at present, 3M's, whose product is known as "Scotchgard"[54] and Du Pont who market "Zepel". The Italian firm of Rimar (see p. 82) is also active in this field but the scale of operations is believed to be limited and, so far as is known, the usage is captive. Most, if not all, 3M's textile treatments appear to be based on the C_8 perfluorosulphonic acid (or sulphonyl fluoride) while in the case of Du Pont's "Zepel" the basic raw material is a straight chain perfluoroalkyl iodide produced by a telomerisation process with tetrafluoroethylene as the principal raw material. Perfluorocompounds having geminal trifluoromethyl groups, as produced by Pennsalt (see p. 85) are inferior to straight-chain compounds in the formulation of textile finishes.

The production of fluorochemical textile finishes is of considerable importance, the actual and projected total sales in millions of dollars being as shown below[55]:

1959	1965	1970
3.0	10.0	15.0

10. Acknowledgements

The writer wishes to thank W. R. Deem, H. C. Fielding, J. Hutchinson and S. J. Webster of I.C.I., Mond Division for supplying information and checking the original manuscript; S. Sheratt, I.C.I., Plastics Division for supplying information; P. J. Craven, I.C.I., Mond Division for the cost estimate and J. R. Case, I.C.I., Mond Division for helpful discussions and advice.

11. References

1 H. R. LEECH in J. W. MELLOR (Ed.), *Comprehensive Treatise on Inorganic and Theoretical Chemistry*, Supplement II, Part I, Longmans, Green and Co., London, 1956, pp. 120–129.
2 J. H. SIMONS in J. H. SIMONS (Ed.), *Fluorine Chemistry*, Vol. I, Academic Press Inc., New York, 1950, p. 225 *et seq.*
3 J. BURDON and J. C. TATLOW in M. STACEY, J. C. TATLOW and A. G. SHARPE (Eds.), *Advances in Fluorine Chemistry*, Vol. I, Butterworth's Scientific Publications, London, 1960, p. 129.
4 S. NAGASE in P. TARRANT (Ed.), *Fluorine Chemistry Reviews*, Vol. I, Edward Arnold Ltd., London; Marcel Dekker Inc., New York, 1967, p. 77.
5 I. WATANABE, *Denki Kagaku*, 36 (1968) 172.
6 J. H. SIMONS, *J. Electrochem. Soc.*, 95 (1949) 47.
7 J. H. SIMONS, H. T. FRANCIS and J. A. HOGG, *J. Electrochem. Soc.*, 95 (1949) 53.
8 J. H. SIMONS and W. J. HARLAND, *J. Electrochem. Soc.*, 95 (1949) 55.
9 J. H. SIMONS, J. H. PEARLSON, W. H. BRICE, W. A. WATSON and R. D. DRESDNER, *J. Electrochem. Soc.*, 95 (1949) 59.
11 J. H. SIMONS and R. D. DRESDNER, *J. Electrochem. Soc.*, 95 (1949) 64.
12 Ref. 2, pp. 414–420.
13 E. A. KAUCK and A. R. DESSLIN, *Ind. Eng. Chem.*, 43 (1951) 2332.
14 E. A. KAUCK and A. R. DESSLIN, (3M's), U.S. Pat., 2,567,011 of 4.9.51.
15 R. A. GEUNTHNER, (3M's) U.S. Pat., 2,606,206 of 5.8.52.
16 S. NAGASE, U. TANAKA and H. BABA, *Bull. Chem. Soc. Japan*, 38 (1965) 834.
17 S. NAGASE, K. TANAKA and T. ABE, *Bull. Chem. Soc. Japan*, 39 (1966), 219.
18 M. SCHMEISSER and P. SARTORI, *Chem. Ing. Tech.*, 36 (1964) 9.
19 P. SARTORI, *Angew. Chem.*, 75 (1963) 417.
20 S. NAGASE, H. BABA and T. ABE, *Bull. Chem. Soc. Japan*, 39 (1966) 2304.
21 L. HEINRICH, *Z. Anorg. Allgem. Chem.*, 346 (1966) 44.
22 Inst. Für Silicon und Fluorkarbon Chemie, Ger. Pat., 1,279,002 of 3.10.68, Brit. Pat. 1,075,525 of 12.7.67.
23 M. E. REDWOOD and C. J. WILLIS, *Can. J. Chem.*, 43 (1965) 1893.
24 J. A. DONOHUE and A. ZLETZ, *J. Electrochem. Soc.*, 115 (1968) 1039.
25 J. A. DONOHUE, A. ZLETZ and R. J. FLANNERY, *J. Electrochem. Soc.*, 115 (1968) 1042.
26 H. M. SCHOLBERG and H. G. BRICE, (3M's) U.S. Pat., 2,717,871 of 13.9.55.
27 J. MURAY, (I.C.I.), unpublished information.
28 J. BARR, (I.C.I.), unpublished information.
29 E. A. KAUCK and J. H. SIMONS, (3M's), U.S. Pat., 2,644,823 of 7.7.53, Brit. Pat., 718,318 of 10.11.54.
30 T. J. BRICE, W. H. PEARLSON and H. M. SCHOLBERG, (3M's) U.S. Pat., 2,713,593 of 19.7.55.
31 R. L. HOUSEN, (3M's) Brit. Pat., 1,146,312 of 26.3.69.
32 T. GRAMSTAD and R. N. HASZELDINE, *J. Chem. Soc.*, (1956) 173.
33 T. H. BRICE and P. W. TROTT, (3M's), U.S. Pat., 2,732,398 of 24.1.56.
34 T. GRAMSTAD and R. N. HASZELDINE, *J. Chem. Soc.*, (1957) 2460.
35 J. BURDON, I. FAMAYAND, M. STACEY and J. C. TATLOW, *J. Chem. Soc.*, (1957) 2574.

36 R. N. HASZELDINE, *J. Chem. Soc.*, (1949) 2856.
37 R. N. HASZELDINE, *J. Chem. Soc.*, (1953) 3761.
38 R. M. JOYCE, (Du Pont), U.S. Pat., 2,559, 628 of 10.7.51. *C. A.*, 46 (1952) 3063.
39 K. L. BERRY, (Du Pont), U.S. Pat., 2,559,629 of 10.7.51. *C. A.*, 46 (1952) 3063.
40 H. C. FIELDING and A. J. RUDGE, (I.C.I.), Brit. Pat., 1,082,127 of 6.9.67.
41 H. C. FIELDING, (I.C.I.), Brit. Pat., 1,155,607 of 18.6.69.
42 E. W. LAWLESS and I. C. SMITH, *Inorganic High-Energy Oxidisers*, Edward Arnold (Publishers) Ltd., London; Marcel Dekker Inc., New York, 1968. 42a M. SCHMEISSER and F. HUBLER, *Z. Anorg. Allgem. Chem.*, 367 (1969) 62.
43 A. T. KUHN, *Chem. Process Eng.*, 49 (1968) 56.
44 N. L. JARVIS and W. A. ZISMAN in R. E. KIRK and D. F. OTHMER (Eds.), *Encyclopedia of Chemical Technology*, Vol. 9, Interscience Publishers Inc., New York, London, Sydney, 2nd ed., 1966, pp. 707–737.
45 R. A. GUENTHNER and M. L. VIETER, *Ind. Eng. Chem. Prod. Res. Develop.*, 1 (1962) 165.
46 C. G. KLAUS, *Chem. Eng. Progr.*, 62 (1966) 98.
47 M. K. BERNETT and W. A. ZISMAN, *J. Phys. Chem.*, 65 (1961) 448.
48 D. L. SCHINDLER, (Du Pont), Brit. Pat., 1,071,992 of 14.6.67.
49 M. HAUPTSCHEIN and M. BRAID, (Pennsalt), Brit. Pat., 1,002,324 of 28.8.65.
50 R. L. TUVE, H. B. PETERSON, E. J. JABLONSKI and R. R. NEIL, U.S.N.R.L., Report N.R.L. 6057, 13.3.64.
51 R. L. TUVE, H. B. PETERSON, E. J. JABLONSKI, R. N. NEIL and R. L. GIPE, U.S.N.R.L., Report N.R.L. 6573, 15.3.67.
52 R. L. TUVE and E. J. JABLONSKI, (Secretary of U.S. Navy), U.S. Pat., 3,258,423 of 28.6.66.
53 D. W. FITTES and P. NASH, *Fire*, 61 (Sept.) (1968) 170.
54 E. J. GRAJECK and W. H. PETERSON, *Textile Res. J.*, 32 (1962) 320.
55 Anon., *Chem. Eng. News*, 43 (Oct. 4th) (1965) 29.

Chapter 3

The Chlor–Alkali Industry

A. T. KUHN

Department of Chemistry and Applied Chemistry, University of Salford, Lancashire

1. Introduction

As we interpret the meaning of the term here, the title of this chapter embraces virtually every industrial electrochemical process involving the reaction of halide ions or anions containing the halogens (fluorine is treated in a separate chapter). The bromate sector is small in its scale of operation, and it will thus be obvious that to all intents and purposes we are concerned with the reactions of sodium chloride. Chlorine, chlorate and perchlorate plants are often found together on the same site. Since the volume of potash salts treated is small in comparison with that of sodium chlorides, the former are usually produced "in a corner" of a sodium chloride consuming plant where they may be advantageously supplied with the required services. The various areas of the industry are now considered individually, but Chapter 14 on electrodes should be consulted for certain details.

2. Hydrochloric acid electrolysis

2.1 General

The electrolysis of HCl to give gaseous hydrogen and chlorine is essentially a by-product recovery reaction. The hydrochloric acid feed is mainly an end-product of organic chlorination reactions. The growing tendency to use oxychlorinations in place of chlorinations on the industrial scale, implies that recovery of waste HCl will become less significant than it is at present (in the United Kingdom the process is not carried on at all). Although HCl production is not expected to grow rapidly, if indeed at all, of all the electrochemical chlor-alkali processes, hydrochloric acid electrolysis is most strongly in competition with non-electrochemical processes such as those developed by Kellogg or Shell, all of which are derivations of the former Deacon Process for decomposition of HCl. The relative merits of the electrochemical and non-electrochemical processes depend on individual circumstances, and are reviewed by Janson[1]. In favour of the electrochemical process are its efficient use of floor space and the high purity of both H_2 and Cl_2 evolved, which enable the former to be used for catalytic hydrogenation and the latter to be fed straight into existing chlorine mains. Lastly, in contrast to other electrolytic chlor-alkali processes, HCl electrolysis is highly flexible and can be run from 20% to 100% load without significant loss of efficiency. In this way, it can serve as a "buffer" in a chlorine manufacturing and consuming complex.

2.2 Basic electrochemistry of the HCl system

The reversible voltage of the system is quoted by Hölemann[2] as 1.36 V. Nichols[3] gives 1.02 V, while Gallone and Messner[4] give what are probably the best values, ranging from 1.0 V (25% HCl, 100°C) to 1.16 V (15% HCl, 70°C). Both Hölemann and Gallone and Messner give values for anodic and cathodic overvoltages, as well as ohmic losses due to the electrolyte and also gas bubble resistance (Gallone and Messner). An overall energy balance is given by Hölemann, who also cites the data of Haber and Grinberg[5] for current efficiency as a function of HCl concentration.

2.3 Industrial cell design

Cell designs are discussed in several works, though these are largely out of date and should be consulted for mainly historical reasons. They include Mellor[6], Mantell[7] and Sconce[8] and the *Encyclopaedia of Electrochemistry*[3]. In Sconce the earliest cell designs including the I.G. Farben design and the former de Nora design are among those discussed. The latter had an interesting anode design in which the chlorine discharge took place on lumps of graphite in contact with a ribbed graphite back-plate. These lumps could be replaced as they were consumed without dismantling the cell. The convenience of this design was, however, outweighed by the increase in cell voltage, and de Nora have discontinued this design. The two best known modern cell designs are those of Hoechst and de Nora. Both have vertical graphite anodes and cathodes, both are of filter-press construction. Both use PVC diaphragms whose main purpose is to minimise gas bubble leakage from anode to cathode and *vice-versa* which has a deleterious effect on current efficiency. Unlike the so-called "diaphragm cell" in brine electrolysis, there is no significant pH drop across this diaphragm.

The Hoechst cell is described in a number of papers which duplicate one another to a certain extent. The present Hoechst cell capacity is at least 120 tonnes/day, and has an electrode area of 2.5 m^2, operating at 0.4–4.0 kA/m^2 current density. These details can be found in Hölemann[2], Dönges and Janson[9-11] and Grosselfinger[12]. The de Nora Cell is described by Gallone and Messner[4]. Further details of cell design may be found in: Fr. pat., 1,438,213; U.S. pat. 3,312,614; Belg. pat.

640,396 (Solvay); Neth. Appl. 6,400,085 (Hooker); Neth. pat. 286,119 (Solvay); U.S. pat. 3,242,065; U.S. pat. 3,236,760 (de Nora); Neth. Appl. 6,510,493 (Hooker); Brit. pat. 1,004,207 (Asahi). Eastern Bloc practice is described in ref. 54.

2.4 Cell operation

Cells operate in the range 70°–90°C on acids of 30–35% strength. Below this concentration, the increased side reaction of oxygen evolution results not only in loss of current efficiency but also in increased rate of graphite wear. These two factors are in fact related (see Chapter 14, section 5). The spent acid is passed out from the cell at about 15–20%. The general operation costs are detailed in Gallone and Messner, and Grosselfinger[12] but the main costs are electrical energy—approx. 1650 kWh/tonne (Gallone gives the equation, E (AC) = $0.38(i{-}1000) + 1520$ kWh/tonne chlorine, for 80°C) while graphite wear is less than 100 g/tonne chlorine (Gallone and Messner) though higher values—1 kg/tonne Cl_2— are found in Sconce. The PVC diaphragms reinforced with fluorocarbon polymers or similar type materials have a life of well over two years[4]. The same authors reach a manufacturing cost of 23 dollars/tonne Cl_2, based on free acid supply and no value attributed to the hydrogen formed.

2.5 Tolerance to organics

It will be seen that the HCl used for this process can be expected to contain large quantities of dissolved organics whose removal would be expensive. Sconce takes a fairly optimistic view of the tolerance of the process to such species, but Janson[1] discusses this more fully and illustrates with microphotographs the effect of organics on anode wear, and the circumstances when this may be expected to become severe.

2.6 Process developments

In the light of the fairly static production situation, no manufacturer is definitely known to be devoting much effort to further development of this process. The following areas have been discussed or brought to pilot plant stage.

1. *Anode improvements.* The use of DSA (dimensionally stable anodes) with a resultant lowering of overvoltage, both initially and increasingly as graphite wear increases the ohmic resistance. Activation of anodes with precious metal electrocatalysts is discussed (and dismissed) by Gallone and Messner, who show results

for a Pt–Ir alloy improver. Hass[13] quotes a private communication from de Nora in which the lowering of energy requirements to 1500 kWh/tonne Cl_2 is envisaged. This could reflect either of the preceding two developments or the following one.

2. *Cathode overvoltage*. Here again, the use of electrocatalysts deposited on the graphite could be expected to lead to a reduction in the energy used.

3. *Use of fuel cell cathodes to reduce oxygen*. This is discussed elsewhere in the chapter and also by Bianchi and Traini[14] in reference to this process. However the loss of hydrogen, which would not thus be formed, might be unacceptable in many plants. This system is also studied in refs. 15 and 16.

In addition to these basic modifications of the simple electrolysis, two other completely different electrolytic processes have been suggested. Both involve recycling of a "redox" catalyst.

1. *The Westvaco System*. In this process, the reactions

$$2\ CuCl_2 \xrightarrow{\text{electrol.}} 2\ CuCl + Cl_2$$

$$2CuCl + 2HCl + \tfrac{1}{2}O_2 \longrightarrow 2CuCl_2 + H_2O$$

occur. The process was scaled up to a large pilot plant scale, and the fullest operational and cost data are available (Sconce[17, 18]). Authorities all speak well of the process which, while it does not reduce capital costs of a plant, lowers power consumption by some 20%.

2. *The Schroeder Process* is based on the redox cycle:

$$NiCl_2 \xrightarrow{\text{electrol.}} Ni + Cl_2$$

$$Ni + 2HCl \longrightarrow NiCl_2 + H_2$$

This process was never scaled up, but Sconce again quotes similar advantages for it.

3. Chlorates

3.1 Introduction

In contrast to the manufacture of perchlorates (q.v.) which is not a rapidly expanding market, chlorates are produced in rapidly growing tonnages, a fact largely connected with their use in the pulp and paper industry. The technology of their manufacture does not differ much from that of perchlorate manufacture.

Mechanism of oxidation. The anode reaction, in which chlorate is formed from chloride ions, has been extensively investigated. The overall reaction in the cell is:

$$NaCl + 3H_2O \rightarrow NaClO_3 + 3H_2$$

hydrogen is liberated at the cathode, while chlorine is formed at the anode. The chlorine reacts with water to form hypochlorous acid and hypochlorite anions:

$$Cl_2 + H_2O \rightarrow HClO + H^+ + Cl^-$$
$$Cl_2 + 2OH^- \rightarrow ClO^- + H_2O + Cl^-$$

with the succeeding slow step:

$$2HClO + ClO^- \rightarrow ClO_3^- + 2H^+ + 2Cl^-$$

which is favoured by low temperatures and somewhat acid solutions. The anode reaction:

$$6\,ClO^- + 3H_2O - 6e \rightarrow 2ClO_3^- + 4Cl^- + 6H^+ + \tfrac{3}{2}O_2$$

also occurs. Detailed studies have been made by Ibl and Landolt[19] and de Valera[20], though in each case the authors draw on several of their earlier papers in the field. The effect of Cl^- on conversion of available chlorine was reported by Jaksic et al.[21].

Reference to earlier papers can be found in the work of Ibl and Landolt and de Valera and also in the review paper by Schumacher[22].

3.2 Cell design

Cell designs have remained unchanged for many years, and details are given in the usual reference books: Mantell[7], Kirk–Othmer[18] and Hampel[3]. These cells are similar to perchlorate cells, they have no diaphragm and are often fitted with cooling coils. There is the same choice of anode material: graphite, magnetite, lead dioxide, platinum and now, platinised titanium or related titanium base anodes[23]. Cathodes are of mild steel though graphite and stainless steel have also been reported. Cells are made of steel, lined with ceramic or similar materials.

In the past few years new cell designs have been disclosed; some are monopolar, others are bipolar and these have graphite electrodes. Of these the Krebs NC 12 and the Olin Mathieson designs might be singled out. The former is fitted with Pt–Ir titanium anodes and electrolyte circulation is achieved by natural convection. Extensive use of PVC is made in the cell body. The process is shown in Fig. 1. There are three main parts, consisting of the cell itself, the pipe connecting the cell to the gas separator, and the gas separator. The makers claim that this design allows a very compact cell. They outline the design philosophy of their cell as follows:

(a) Highest possible conversion of ClO^- ions to ClO_3^- ions, outside the cell.

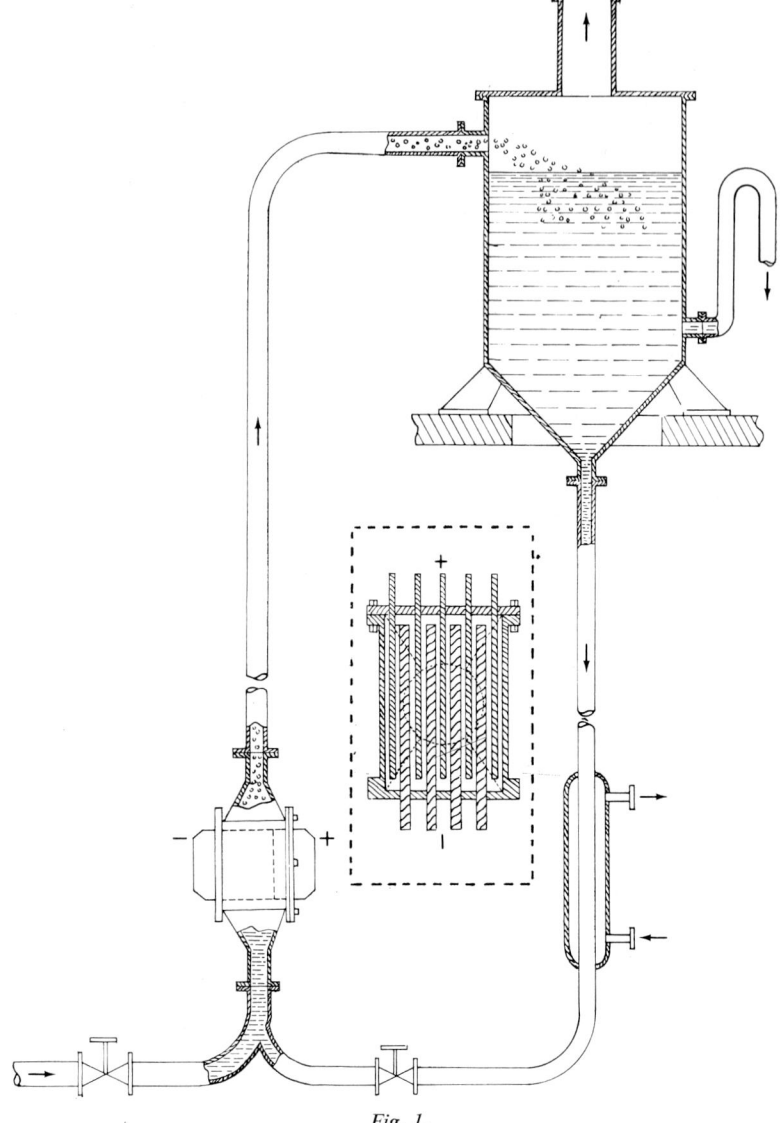

Fig. 1.
The Krebs NC 12 chlorate cell, showing electrochemical cell, reactor and heat exchanger. Inset—plan view of electrochemical cell (courtesy Messrs. Krebs et Cie).

(b) Recycling of the Cl^- ions formed in the above reaction.

(c) Cell operation at sufficiently low $[ClO^-]$ to prevent their discharge.

It is in the pipe leading from the cell that the chlorate is largely formed. This cyclic operation permits ClO_3^- : Cl^- ratios of 2.5 in contrast to the figure of 0.2 normally reached without cycling. The cell and equipment as shown are operated in trains, each stage passing on a fraction of the electrolyte to the succeeding one. Krebs quote[24] the following data.

Average cell volts	3.4
Current efficiency	94%
O_2 in H_2	< 3%
Energy	5625 kWh/tonne chlorate
Electrode surface	Pt 70%–Ir 30%
HCl (100%)	14 kg/tonne
Sodium bichromate	1 kg/tonne
Barium sulphate	2 kg/tonne

In addition, the following data is quoted:

Cell voltage $V = 2.58 + (0.311 - 0.0061(t-60)) \times I$

where I is in kA/m². Densities up to 75 A/dm² have been used, though the table above relates to 25 A/dm² at 60°C with a final concentration of 575 g/l chlorate and 125 g/l brine. Precious metal consumption was 500 mg/tonne chlorate though this is believed to be conservative, and longer term measurements indicate a value closer to 200 mg/tonne. Needless to say, other anodes are actively under consideration. Krebs have installed this plant in Finland at the Oulu Osakeythio paper mills and its ultimate size will be 10,000 tonnes p.a.

The Olin Mathieson cell with steel cathodes and platinised titanium anodes has the following characteristics, at 110°C and 160 mA/cm²:

Energy required	7000 kWh/tonne
Cell feed	190 g NaCl/l
	330 g $NaClO_3$/l
Cell effluent	110 g/l NaCl
	580 g/l $NaClO_3$
pH	6.7

Cell voltages with various inter-electrode spacings and at various temperatures are quoted. The process does not appear to have run as long on test as the Krebs cell, and details are found in refs. 25 and 26. From Eastern Europe, details of another chlorate cell design are given by Jaksic et al.[27]. Their criteria are similar to those of the Krebs engineers and their results doubtless similar. They give details of anode fastening and cell designs.

The Chemech cell is described in ref. 28. It is a bipolar, graphite electrode cell and its operation is described in the above reference and in ref. 29 with details of operating costs. The plant at Bellingham, Washington, U.S.A., has five cells which produce over 4000 tonnes p.a.

The first two cells described use noble metal anodes, while the Yugoslav design and Chemech cells have graphite anodes. Both magnetite and lead dioxide have also been used as anode materials, the former is described in refs. 31 and 32.

3.3 OTHER REFERENCES

Patents: Can. 791,699, 741,312; Brit. 1,056,889, 1,104,078; Neth. Appl. 6,509,556 (PbO_2); Fr. 1,444,557, 1,402,590, refer to cell designs. Other mechanistic studies both on laboratory and pilot scale cells) include ref. 33 (all mainly laboratory work), with ref. 34 on the use of PbO_2. On the pilot scale, Yokoyama[35] discusses the effect of HCl concentration and Rousar[36] a mathematical model of the cell. The Electrochemical Society devotes sections of its meetings periodically to chlorates, etc. From the most recent symposium (Boston, May, 1968), of which not all papers have appeared in print. Beck and Brännland[37] discussed cells with a separate reactor (*e.g.* Krebs) from a mathematical point of view; Rousar[38] presented a paper on calculated current density distributions in bipolar electrolysers; Claus[39] discussed computerised predictions of chlorate cell reactions, with optima; two speakers[40, 41] discussed cathode mechanisms and the effect of additives such as bichromate; several papers were given on the behaviour of graphite[42, 43] while Crawford *et al.*[44] discussed conductivities of the high temperature $NaCl$/$NaClO_3$ system.

A broad view of industrial practice is given in ref. 29.

3.4 CHLORITES

The production of chlorites is not truly an electrochemical process. In the Kesting process, chlorates are reacted with hydrochloric acid to give ClO_2 and chlorine. The former reacts with NaOH to form a solution of sodium chlorite. Details can be found in refs. 45 and 46.

4. Perchlorates

4.1 INTRODUCTION

This term is generally understood to mean the anodic oxidation of aqueous solutions of sodium chlorate to sodium perchlorate. Where other cations are called for, they are usually made by double decomposition reactions with the sodium perchlorate. This applies particularly to ammonium and potassium salts. The significance of this route to the perchlorates is that plants are rarely found far from chlorate cells. According to Schumacher[47] the use of perchlorates has reached a peak and may be declining. Schumacher's monograph[48] is the best source of all information relating to the chemistry, preparation, manufacture and use-patterns of perchlorates. In this section, the main points will be brought out and later developments and work will be described.

4.2 General industrial

Among plants where perchlorates are made, can be listed:
American Potash and Chem. Corp., Henderson, Nevada
Hooker Chem. Corp., Niagara Falls, N.Y.
Pacific Engineering, Henderson, Nevada
H.E.F. Inc. (U.S.)
Pennwalt Corp. Tacoma, Washington
G. F. Smith Chemical Co., Columbus, Ohio
Japan Carlite Co., Sanwa Chemical Co., Tokyo
Pechiney St Gobain, Chedde, H.Sav., France
with other plants in Scandinavia, Spain and Germany.
Two general reviews give a historical picture[49, 22].

4.3 Mechanism of perchlorate formation

This has been studied over several years, and the following studies should be consulted: Oechsli[50], Bennett and Mack[51], Knibbs and Palfreeman[52], Grotheer and Cook[53], Il'in and Skripchenko[33], Ramachandran et al.[55], Nagalingam et al.[56] Narasinham[57], de Nora, et al.[58].

4.4 Cell design and operation

Schumacher has a very complete bibliography but appears to omit one very valuable paper by Legendre[59], which presents a "European" point of view. The Japanese Sanwa cell is stated to be almost identical to that in which bromates are made[60] referring to ref. 61; patents giving cell designs include ref. 62 while recent Indian work is described in ref. 63.

The design of perchlorate cells starts with the anodes. Smooth platinum has been most widely used (Schumacher[48] states that platinised platinum leads to excess oxygen formation) though the present trend[47] is to electrodes of the platinised titanium type. A separate school of thought favours PbO_2 anodes. In addition, there has been much work on composite anodes, in which platinum is coated on a less precious metal. For details of this work refer to Chapter 14, *Electrodes for Industrial Processes*. The choice between PbO_2 and platinum anodes is largely dictated by economic factors, for PbO_2 anodes run at rather higher overvoltages.

Legendre and de Nora, among others, give a good idea of the quantitative aspects. Choice of anode dictates choice of cathode for though mild steel is the commonest solution, it is given extra protection by the addition of sodium chromate solutions to the working electrolyte. However, where PbO_2 anodes are chosen, these are incompatible with such chromate solutions[64] and in these cases nickel or its alloys are selected for cathodes. This is discussed in several other papers all referred to in Schumacher's monograph.

The cell incorporating the chosen electrodes is usually fitted with cooling coils or cooling boxes. In the latter case, these are of stainless steel and serve as additional cathodes (Osuga, Fujii et al.). Pechiney use cathodes of bronze[59].

The electrolyte used varies considerably, depending in part on whether batch runs or continuous runs are operated, and also on the premium placed on purity of product or current efficiency. Concentrations of sodium chlorate from 300–600 g/l are used. A high level of perchlorate is also maintained in most cases, again around 400 g/l though Legendre refers to working without perchlorate in the cell feed. Sodium chloride and sodium chromate or dichromates are also often added, the latter from 0.5–5.0 g/l. The purpose of the chromates is to protect the cathode, and this is stated to be achieved by formation of a semi-permeable membrane at the cathode. The temperature at which the cell is run is the compromised optimum of cell voltage (favouring high temperature) current efficiency (favoured by low temperatures and low current densities) precious metal loss (increasing with temperature) and cooling costs. Obviously, the optima will vary widely. De Nora and Legendre should again be consulted. Current efficiencies range from 95% down, and at low chlorate concentrations ozone is evolved. Schumacher gives details of platinum loss figures, though the use of newer anodes may change this situation. Cell voltages could range from 5.0–7.0 V depending on temperatures and electrode gaps, and overall energies required appear to be about 4400 kWh/tonne (all plant services included) though again such figures depend on the cell operating conditions, and figures as low as 3000 kWh have been quoted (cell consumption only) according to Schumacher's monograph. Bauer[65] quotes 2800 kWh/tonne and 92% C.E. at 65°C and 450 mA/cm².

4.5 OTHER PERCHLORATE PROCESSES

The lower solubility of potassium perchlorate provides good reasons for not operating electrolysis cells for its manufacture, for it crusts out at the electrodes. At the same time, it renders the double decomposition reaction a favourable one. A recent paper on ammonium perchlorate manufacture is found in ref. 65.

Legendre describes a process for manufacture of perchloric acid on plant scale, with silver cathodes, platinum anodes running at 4 V and 3500 A. He cites Ger. pat. 1,031,288 (Merck). Schumacher discusses the manufacture of perchlorate starting from sodium chloride and also from hypochlorites. However, high energies are required for these processes, which are not pursued on a large scale.

5. Hypochlorites

As far as can be ascertained, the electrolytic manufacture of sodium hypochlorite is moribund, at least in the sense of an industrial process. In former times it was prepared by electrolysis of brine solutions in a cell where the anode and cathode products were allowed to mix freely. The cells thus resembled the chlorate cells. A description of such cells is found in Regner[66] and Brit. pat. 1,104,747. With the widespread growth of the chlorine industry, it was found more economic to make hypochlorite by the mixing of chlorine and caustic soda in separate plants. The chlorine used is the so-called "tail gas" which emerges from the chlorine liquefaction plants. It is passed upwards through a bed of graphite or similar material packed in a tower, through which caustic flows counter to it. Apart from occasional problems resulting from "black" hypo in which graphite particles are suspended, the process is straightforward. Sodium hypochlorites are electrolytically prepared (often by sea water electrolysis) for *in-situ* generation of a sterilant. Such plants have been used for sewage sterilisation (a process in which Constructors John Brown have been successful) and on board ships (where Patterson Candy have installed many plants) as well as the protection of power station cooling water from mussel fouling, etc. In all the above devices, electrodes of platinised titanium have been used with reasonable success except in those cases where the water temperature fell below 5°C or so, when extensive platinum loss took place. This is possibly due to higher overvoltages resulting from constant current devices[67].

Pacific Engineering have investigated the use of PbO_2 electrodes for this application (see Chapter 14 on electrodes) and apart from the above uses, all of which relate to maritime usage, it is possible that the process is used for similar purposes in certain S. American mines.

6. Sodium

6.1 INTRODUCTION

All the sodium manufactured in the world today comes from the electrolysis of sodium chloride melts in a cell based on a design of Downs, and named after him. Important though this area of industrial processing is, the dearth of information is unbelievable. Apart from the patent literature, which can serve only as an overall guide to current thinking, there is the excellent monograph by Sittig[68] and the very recent account by Lemke[69]. Of these, the first is comprehensive though now outdated in certain ways, the second account is rather brief. For scientific data relating to the process, the reader must turn mainly to the Russian literature, of which some is in translation but only the minority. Data on production is also hard to come by, though the U.S. figure of approx. 150,000 t yearly is often quoted (see Chapter 17).

6.2 THE DOWNS PROCESS

This process is described in Sittig[68] who draws largely on older published work, and by Lemke[69]. The cell shown by Lemke is a good representation of modern industrial practice. It shows (Figs. 2, 3, 4) a cylindrical iron cell lined with ceramic materials, with four vertical graphite anodes inserted from below. Nothing can be stated categorically as to the shape of these anodes, though the patent literature indicates that the anodes have "wells" drilled in their centres, these wells communicating with the bulk electrolyte through vertical slots. According to the most recent patent[107] the well should be in the shape of an inverted cone to reduce ohmic losses in the graphite. These patents claim an energy consumption of 2320 kWh/tonne of sodium, which may or may not be a true reflection of modern operating efficiency. The exterior of the cell is thermally insulated, while

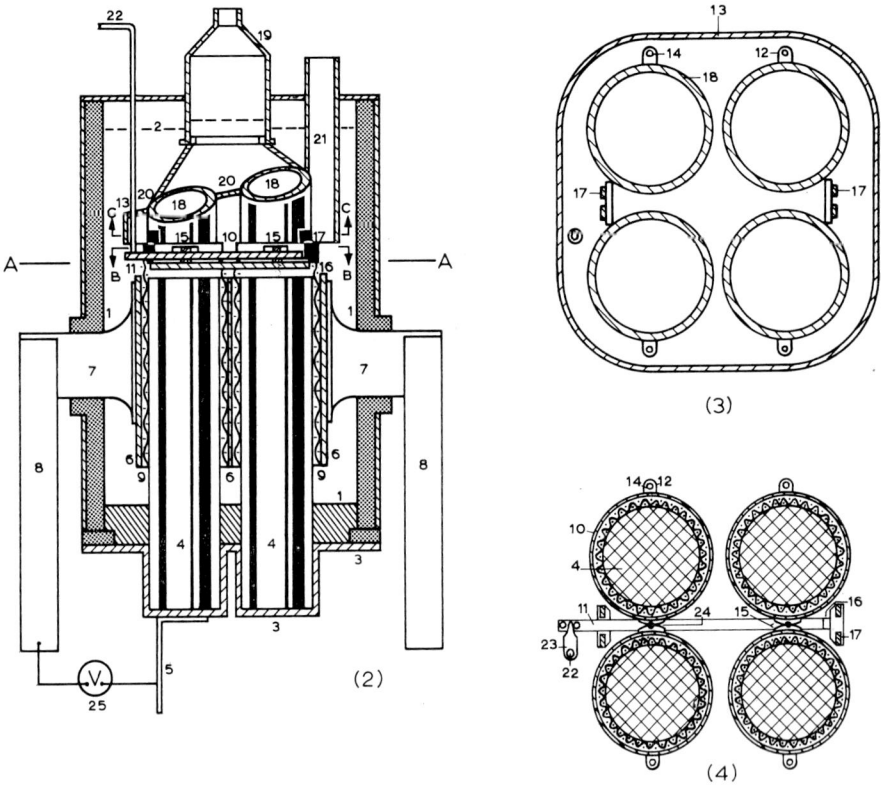

Figs. 2, 3, 4.
Plan and section view of Downs cell (from U.S. pat. 2,924,558). (1) Insulation lining; (2) electrolyte level; (3) housing for anodes; (4) anodes; (5) anode busbar; (6) cathode; (7) cathode lead-in; (8) cathode busbar; (9) diaphragms; (18) chlorine collector (hood); (19) chlorine collector dome; (20) sodium collector; (21) riser pipe (Na). The other numbers describe a mechanical diaphragm agitator which is a special feature in this patent.

iron cathodes enter the cell horizontally from the sides. An iron mesh diaphragm is used, while the chlorine formed at the anodes is collected under a series of hoods. The whole cell is accessible at two levels with a floor level just below the top of the cell itself. The sodium formed at the cathodes rises up through the melt and is collected in an annular inverted trough, which has a riser pipe through which the metal can float upwards. At the top of this, is a "two-hour" receiver, whose purpose is obvious, and "twenty-four" hour receivers are periodically wheeled in to empty the smaller receivers.

6.2.1 Operation of Downs cell

The Downs cell is notoriously difficult and dangerous to operate. The graphite anodes wear away, as might be expected. They also suffer more drastic damage on occasions. The presence of iron, in particular, is known to be highly deleterious, probably forming ferric chloride inside the graphite. Oxygen-containing species are likewise harmful to anode life. The second main problem area in the Downs cell is the diaphragm, which serves to reduce the back-reaction of metallic sodium with chlorine. The iron mesh used here suffers from both corrosion and clogging with metals and salts. The former problem has been tackled by using ceramic clad diaphragms, of which details are found under *Further Reading*, though most operators are not yet using these clad diaphragms, as far as is known. The problem of freeing diaphragms from clogging is described in U.S. pat. 2,913,381 which makes use of a mechanical shaker. As the cell grows older, after each overhaul, it becomes less and less efficient. Thus a "new" cell is heavily insulated on the outside walls and as it ages, and becomes less efficient, this insulation is removed as the internal inefficiencies provide the heat to maintain the cell at operating temperature. When a cell is too inefficient, it is shorted out with water-cooled shorting bars. Though the cell itself is very little different from plant to plant, the manner of product collection forms the subject of many patents and probably, of some actual modifications. As the molten metal rises (with its density much lower than that of the melt) calcium metal crystallises out, tending to block the riser pipe. This is prevented by "tickling" the pipe, which frees the blockage and also restores the composition of the melt to something approaching its correct value. In this way, only salt need be added to replace losses, and melt analyses can be made less often so that the Ca or Ba levels remain satisfactory. The "tickling" operation lends itself to automation, and again, while patents exist, an informed guess is that very few if any plants (in the U.S.A.) have such automatic ticklers, while in European countries one would guess that a few pilot ticklers have been installed for evaluation. The mode of sodium transfer and handling is similarly undisclosed. Again, an informed guess would predict elimination of the smaller "two-hour" vessel, with possible introduction of gas blanketing (not widely used at present).

The technology of nuclear reactors (in particular those based on liquid sodium coolant) places all manner of pumps and vessels at the disposal of the chemical industry which will doubtless adopt some of these in due course.

6.3 Cell operating data

As Lemke states, no data on this are available. Current efficiencies, judged from the Soviet literature, would be around 90–95% for large well-run cells.

6.4 Constitution of the melt

Among Downs cell operators, this is perhaps the most sensitive question there is, since from it stem all questions relating to product purity, cell temperature, current efficiency and ease of operation. Readers of Lemke's account, up to date though it is, would be forgiven for thinking that a Na–Ca melt is used, though in fact this is rarely if ever the case. Patents exist for ternary melts of Na–Ca–Ba and even quaternary melts and some of these are cited at the end of this section. The worth of these patents is very mixed. Some are clearly attempts to evade previous filings and all the patents should be judged in the context of the published, but little-known, work of Alabyshev and co-workers in the U.S.S.R., which pre-dates them by some considerable time. Several patents disclose melts in which sodium fluoride is added (approx. 1%) to increase current efficiency. Whether this is used in fact, is not known. It can be stated with some confidence that most plants use the Na–Ca–Ba melt first described by Alabyshev.

6.5 Other processes

Sittig[68] is the best guide to the numerous other processes (both electrolytic and otherwise) which have been proposed and operated at various times. One of the more obvious methods is the electrolysis of a caustic soda melt, and this was indeed the main method in earlier days. Under conditions where demand for caustic soda and chlorine (the twin products of the chlor-alkali industry) is equal, there is no special merit in sodium manufacture based on caustic soda. However, under conditions of real caustic surplus, there is a thermodynamic advantage in making the metal from the hydroxide, and the author considers that this route remains a real possibility. This is a view not shared by the author in ref. 74, who produces cell data to support his argument. However it appears that the comparison is not a meaningful one, and it is possible that the performance of modern, efficient Downs cells is being compared with that of older, less efficient (or overrun)

Castner caustic cells. The other processes mentioned in Sittig are too numerous to mention completely. However two strands of thought run through the years. The first considers the possibility of recovering metallic sodium (as distinct from its hydroxide) from the amalgam it forms in mercury cells for brine electrolysis.

The second line of thought, based on the fact that most sodium made is used in the manufacture of petrol anti-knocks, seeks to form electrolytically not sodium, but the sodium–lead alloy which can be directly used in anti-knock manufacture, though sodium itself (or potassium) can also be recovered from these lead alloys.

6.5.1 Sodium from amalgam processes

These are described in Sittig. At least one which got as far as pilot plant stage has added interest in that Dr (later Lord) Fleck was associated with it at the Runcorn works of the Castner Kellner Alkali Company. The electrolyte used in this process was liquid ammonia at approx. 5 atm. with sodium cyanide added as supporting electrolyte. The metal appears to have dissolved in the ammonia and was recovered in a second stage of the process. The process ran for a year or so but was closed down about 1923; it never progressed beyond pilot plant stage. During the years 1940–54, the Germans operated a process for recovery of sodium from mercury cell amalgam, and a 1000 A plant operating at the Gersthofen works of I.G. Farben is described in ref. 70 and later sources. The process was essentially an electrolysis of the amalgam (itself as anode) in a ternary molten salt bath of sodium hydroxide, iodide and bromide, operating at 250°C. The cathode was a slowly rotating nickel disc. The problem encountered was that of mercury solubility in the melt, both in the loss of this material and seen as a contaminant of the sodium. While the process itself was viable, the cost of purification of sodium from mercury was high (but not inordinately so, according to ref. 70).

In recent years, further thought has been devoted to this approach with Japanese workers studying melt systems similar to those used by the Germans at Gersthofen. Details of this work are listed under *Further Reading* but it is not believed to have led to any process worth scaling up, or indeed very different from its German predecessors.

A different approach, made possible by the remarkable properties of β alumina, involved a concentration cell using this material as diaphragm. Details of this are given in refs. 71–73 and the method is currently being studied on a laboratory scale in several organisations. The success of the idea will depend largely on the properties and price of the ceramic material, which leave much to be desired on both counts at this time.

The electrolysis of sodium chloride melts using lead cathodes has been operated since before 1900. The so-called "Acker" process operated at the turn of the century in the U.S.A. and was then licensed to I.G. Farben in Germany, where an annual capacity of 5000 tonnes was still operating at the end of 1945. These plants were then dismantled[75]. Prior to this, in 1938, a reappraisal of this process led to a modification of the cell described in ref. 76. Descriptions of the plant as well as full operating costs are given in ref. 77. It is clear that the process was not an easy one to operate but according to the data given (and the simple fact that the plants were operated for so many years) it is surprising that the process was abandoned. After this time, interest in the process revived in the U.S.A., and the "Szechtmann" cell[78] was offered under licence by the Philblack Corp. However nothing seems to have come of the scheme for reasons unknown, though it has been suggested to the author that the process was unable to offer the very precisely controlled lead–sodium alloy required for anti-knock manufacture. This again seems to be something which modern analytical and control techniques could overcome.

Sittig[68] describes and lists numerous other ideas for the manufacture of sodium, mainly by electrolysis in a variety of electrolytes. However since none of these, apart from the ones described above, is known to have advanced beyond the laboratory stage, and since none is presently active, the reader should refer to this monograph.

6.6 Further reading – sodium manufacture

Molten electrolytes for Downs cells
U.S. pat. 2,850,442; 3,020,221 (to DuPont).
U.S. pat. 3,119,756 (to Ethyl Corp.).
Ger. pat. 1,162,575 (to Ethyl Corp.).
Jap. pat. 10,004/63 (describes $NaCl/ZnCl_2$ system).
Brit. pat. 918,809 (quaternary melt, to I.C.I.).
U.S. pat. 3,072,544; 3,051,635.
M. F. LANTRATOV, *Zh. Prikl. Khim.*, 34 (1961) 1249.
S. V. VASILIEV and A. D. CHEKUSHKIN, *Khim. i Shkole*, (6) (1962) 65.
K. YA. GRACHEV and V. I. EVGLEVSKAYA, *Zh. Prikl. Khim.*, 35 (1962) 1141; 37 (1964) 2061.
A. F. ALABYSHEV and M. F. LANTRATOV, *Tr. Leningr. Tekhnol. Inst. im. Leningrad Soveta*, 12 (1946) 141.
A. F. ALABYSHEV and N. I. KULAKOVSKAYA, *ibid.*, p. 152.
M. F. LANTRATOV and A. F. ALABYSHEV, *8th Mendeleev Congr. Gen. Appl. Chem.*, 13 (1958) 118.
V. ARAMUTHAN *et al.*, *J. Proc. Inst. Chemists India*, 39 (1967) 27.
S. A. ZARETSKII, L. S. YURKOVA and V. B. BUSSE-MACHUKAS, *Zh. Prikl. Khim.*, 36 (1963) 506.
S. A. ZARETSKII, V. B. BUSSE-MACHUKAS and A. A. KHARAKHANOV, *Zh. Prikl. Khim.*, 34 (1961) 2478.
K. YA. GRACHEV and V. YA. ZHURYUTINA, *Zh. Prikl. Khim.*, 32 (1959) 214.

Current efficiency studies
K. YA. GRACHEV and V. YA. DENISCHENKO, *Zh. Prikl. Khim.*, 40 (1967) 668.
V. Z. GREBENIK and K. YA. GRACHEV, *ibid.*, 41 (1968) 203.
V. Z. GREBENIK and V. B. BUSSE-MACHUKAS, *ibid.*, 40 (1967) 1970.
K. YA. GRACHEV and B. V. KARTOLOV, *ibid.*, 33 (1960) 1834.

Description of Downs cell
U.S. pat. 2,924,558; 2,919,238 (to Du Pont).
Belg. pat. 578,671; 579,776 (rotating cathode, to Solvay).
U.S.S.R. pat. 185,083; Fr. pat. 1,563,518.
Ger. pat. 1,121,340 (rectangular cell, Dow).
Ger. pat. 1,128,673 (to CIBA).

Operation of Downs cells
U.S. pat. 2,913,381 (start-up technique).
U.S.S.R. pat. 186,710 (start-up technique).
Ger. pat. 1,162,574 (assembly of Downs Cell, to Ethyl Corp.).
I. E. VENERAKI *et al.*, *Vestn. Kiev Politech. Inst. Ser. Teploenerg*, (2) (1965) 154 (Heat balance on 5000 A cell).
U.S. pat. 3,245,899 (salt feed to Downs cell, to Ethyl Corp.).

Automated tickler
U.S. pat. 2,861,938 (to I.C.I.); 2,944,955 (to Ethyl Corp.).
U.S. pat. 3,037,927.

Use of ceramic materials in Downs cells
V. L. BALKEVICH and R. K. KORDONSKAYA, *Tr. Vses. Nauchn.-Issled. Inst. Stroit. Keram.*, (10) (1955) 170.
U.S. pat. 3,248,311 (coated diaphragms).

Anode study
L. M. MONASTYRSKII and L. G. GALKIN, *Tr. po Khim. i Khim. Tekhnol.*, 3 (1960) 147.
K. YA. GRACHEV and E. I. ADAEV, *Zh. Prikl. Khim.*, 33 (1960) 2368 (with diaphragm wear study included).

Sodium manufacture by other routes
Fr. pat. 1,457,562 (recovery from amalgam, liq. NH_3, to Showa Denko).
Belg. pat. 615,693 (liq. Pb cathode, Montecatini).
M. F. LANTRATOV, *Zh. Neorgan. Khim.*, 4 (1959) 2043 (thermodynamics of Na/Pb).
Brit. pat. 1,013,004; 1,007,778 (to Philblack for Na or K manufacture using Pb cathode).

Ger. pat. 1,214,007 (to Chlormetals, as above).
U.S. pat. 3,265,490 (Japanese work on recovery from amalgam).
U.S.S.R. pat. 186,138 (liquid Pb cathode).
U.S.S.R. pat. 196,731 (liquid Pb cathode).
U.S. pat. 3,006,824 (3 layer cell).
U.S. pat. 3,234,115; Ger. pat. 1,114,330; 1,114,490 (to K. Ziegler, organo sodium compds.).
S. OKADA, S. YOSHIZAWA, N. WATANABE and S. TOKUDA, *Kogyo Kagaku Zasshi*, 60 (1957) 666 (detailed aspects of recovery from Na/Hg with ternary melt).
Fr. pat. 1,390,872 (to Chlormetals).
S. A. ZARETSKII *et al.*, *Tr. Vses. Soveshch po Fiz. Khim. Rasplavlen Solei 2nd Kiev*, (1963) 338 (Na/Pb cathode process).

General reading
M. SITTIG, *Chem. Eng. Progr. Symp. Ser.*, 53 (1957) 35 (Review); A. V. ARAMUTHAN, *Trans. Soc. Adv. Electrochem. Sci. Technol. (India)*, 2 (1967) 9.
A. V. ARAMUTHAN, *Chem. Age India*, 15 (1964) 510.
K. YA. GRACHEV and V. A. NOVOSELOV, *Khim. Prom.*, 1 (1965) 57.
Sodium Metal – Process Developed at C.E.R.I., C.E.R.I. Karaikudi, India.

7. Chlorine from brine

7.1 INTRODUCTION

Though chlorine is also formed in the electrolytic manufacture of sodium and magnesium (*q.v.*) in these cases it is in the nature of a by-product. The manufacture of chlorine by electrolysis of HCl has already been considered in this chapter.

The electrolysis of brine, an aqueous solution of sodium or potassium chloride, is one of the oldest and certainly the most important and widespread of the indutrial electrochemical processes. The overall reaction is:

$$2NaCl + 2H_2O = 2NaOH + Cl_2 + H_2 \tag{1}$$

and reversible voltage for this is given[8] as 2.15 V. With the exception of reaction (2) using air-depolarised cathodes

$$2NaCl + \tfrac{1}{2}O_2 + H_2O = 2NaOH + Cl_2 \tag{2}$$

all electrolytic processes must use the reactants shown, and there is no way of avoiding caustic soda production while making chlorine. The reversible voltage for (2) has been quoted[89] as 0.95 V. In practice reversible voltages such as the ones quoted above bear little relation to what is actually observed, both because conditions are non-standard and also because of the interfering effect of side reactions. The difference between the two values is, however, a valid indication of the savings achievable with air cathodes. The difference between electrolyses of potassium and sodium brines is minimal and mainly a question of different optima being reached in the operating conditions as a result of different solubilities and conductivities.

Aqueous brines are electrolysed in two basic cell types. The diaphragm cell has graphite or noble metal oxide anodes and mild steel cathodes (usually mesh-formed) with anodes and cathodes separated by an asbestos membrane. In this type of cell the reaction is basically (1) though chlorates (and small amounts of hypochlorite) are found as a result of the reaction sequence:

$$Cl_2 + OH^- = Cl^- + HOCl$$
$$HOCl = H^+ + OCl^-$$
$$2HOCl + OCl^- = ClO_3^- + 2Cl^- + 2H^+$$

which occurs mainly in the anode compartment. The products are swept (with the general liquid flow through the cell) into the cathode compartment and out of the cell. Further details are found in ref. 80.

In the mercury cell (see Fig. 5) there are two separate compartments. In the brine cell, the reaction:

$$2NaCl + Hg = Cl_2 + 2Na/Hg \qquad (3)$$

occurs (E_{rev} = 3.05V). The amalgam so formed ranges from 0.25% Na through 0.3% (a common value) to the higher figures 0.5–0.6% described in patents assigned to Solvay. At concentrations above 0.7% a solid phase forms, which would prevent operation of the process. This amalgam is passed into a "decomposer" or "denuder" where it is treated with water:

$$2Na/Hg + 2H_2O = 2NaOH + H_2 + Hg \qquad (4)$$

a reaction which, though it is not apparent, is a pair of simultaneous anodic and cathodic reactions which do not occur at any finite rate except on the surface of an electrocatalyst such as the graphite blocks or balls used in actual denuders.

In both mercury and diaphragm cells, the brine is fed into the cells as hot as possible. In both cases, high temperatures carry certain penalties such as increased breakdown of rubber seals and linings used in construction of the cells. Mercury cells are often operated at a brine inlet temperature of approx. 80°C (higher inlet temperatures lead to increased wear of cell components). As the brine flows down the cell it is depleted of Cl^- ions by some 12%, which could lead to adverse effects on the electrode kinetics. However, the ohmic losses in the cell cause the brine to heat up (some 10°C) as it flows down the cell, and these two effects compensate one another with the result that little change is observed between cell voltage at the top (beginning) and bottom (end) of the cell. Brines are always fed in close to saturation and the emergent depleted brine undergoes a vacuum dechlorination. After this it is recirculated to a resaturator and back to the cell. In a few very rare cases, the brine is sent to waste after vacuum dechlorination, but this is only done in areas where brine is very inexpensive. It is worth noting here that sea water is insufficiently strong[108] to be of use in the cells described here and countries such as Japan lacking indigenous brine supplies use sea brines after concentration.

7.2 COMPARISON OF DIAPHRAGM AND MERCURY CELLS

The two processes are simple to compare. Mercury cells are costly to install (not least because of the cost of the mercury) and deliver a purer caustic soda. Their energy requirements are higher than those of a diaphragm cell, which is also cheap to install but produces a caustic with up to 50% of NaCl.

A further perennial problem in mercury cell operation stems from mercury loss. All operators have run trials with radioactive mercury, which has pinpointed certain sources of loss. Whether a recirculatory brine system, as described above, reduces mercury loss or facilitates its recovery is open to speculation. Few plants are without legends such as that of the man whose bicycle frame was filled with mercury as it was wheeled out of the gates. The "hold-up" or "inventory" of mercury required to operate a cell is an item has been successfully reduced over the years. Details of this are found in *Further Reading*. Just which cells are installed depends very largely on the "caustic–chlorine demand" balance which appears to be slowly swinging towards surplus of caustic soda. Manufacturers have always fought to equalise the demand/use for the two main chemical products which have completely different use-patterns and demand cycles. As the chlorine demand escalates, a much discussed solution is to carbonate the NaOH and thus dispense with the older Solvay ammonia–soda process. Though cost of electricity, steam, money, and local demand patterns are never the same in any two places, an interesting comparison of the two types of process can be found in ref. 81.

Another factor affecting the comparative economics is the availability of process steam. If power is generated on-site, L.P. steam may be used for the evaporation of cell liquor from diaphragm cells. Where such steam is not available, the energy required to raise it must be offset against the other energy savings achieved with diaphragm cells.

It was true until recent years that the use of diaphragm or mercury cells, in a particular plant or country, reflected historical facts rather than deliberate choice. Recently, mercury cells have gained in favour over diaphragm cells, but in the long run the lower capital costs, the lessening of demand for pure caustic, and the technological advances mentioned in this section would all seem to favour the diaphragm cell.

7.3 Energetics of mercury and diaphragm cells

The reason for the greater consumption of energy in the mercury cell is that the formation of amalgam requires more energy than that of caustic soda and hydrogen. Though both cells produce the same products from the same reactants, the extra energy that has been "invested" in the formation of the amalgam is dissipated as heat in the (short-circuited) simultaneous reactions which take place in the denuder. The formation of this amalgam, and its subsequent destruction uses the mercury as a "moving-burden" vehicle, which is selective in that it allows no chlorine species into the denuder. This approach has led to many experiments in which ion-exchange membranes are used in a diaphragm cell. These lead to a slight increase of energy requirements, which would be acceptable. But the main drawbacks have been the inability of the membranes to withstand the fierce conditions (pH 14 on one face, pH 2, 1 atm. Cl_2 on the other face) coupled with the reduction in permissible current densities. While there is no sign of any breakthrough in this area, it would certainly provide the last word in electrolytic manufacture of chlorine. One suggestion for such a diaphragm is interesting[30] because it illustrates the relationship between diaphragm and mercury cells. This is to operate a diaphragm cell with a diaphragm of mercury. The diaphragm thus formed would not be an ion-exchange one, nor would it be a mechanical diaphragm (see Chapter 15 on diaphragms) but such an idea is not known to have been scaled-up and there are obviously many difficulties. British Patents 1,067,447 and 1,076,783 illustrate another application of this idea.

7.4 Industrial cells

Although brine electrolysis cells fall into the two types described, there are hundreds if not thousands of patents and variations on them. Some designs of cell originate from hardware manufacturers such as Krebs, De Nora or Uhde, while certain other large chemical manufacturers such as Olin Mathieson and Hooker

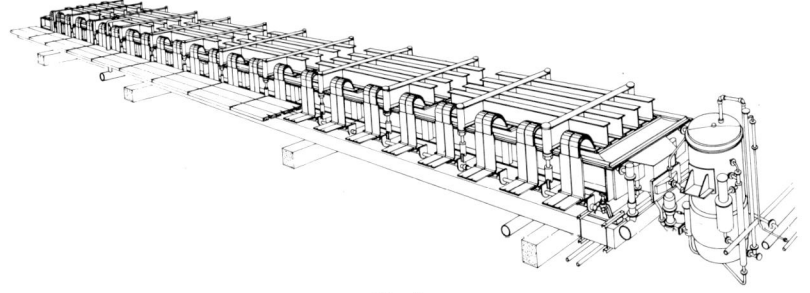

Fig. 5.
De Nora mercury cell, Type 24H5, rated capacity 300 kA (courtesy Oronzio De Nora).

sell their own cell designs to others. Details of these cells are known and compared in several papers. Yet other users of cells (such as I.C.I.) are believed to use their own cell designs and release no details of these. A study of the patent literature is rewarding to a very limited extent[82], since it provides little clue as to equipment actually in use by the holder of the patent.

7.4.1 Mercury cells

These consist of a long trough of mild steel. Sometimes it is made in two separate sections at midway, sometimes the walls are integral with the baseplate, sometimes bolted on. Inside the "trough" so formed, the walls are sometimes rubber-lined while the baseplate, on which the mercury runs, is sometimes polished, sometimes machined or both. The whole assembly is of course inclined, to enable the mercury to flow down under gravity. A lid is fitted over this trough, and the anodes come down through the centre of the lid. Lids can be made of a variety of materials, including rubber-lined mild steel or laminates with chlorine resistant plastics to face the inside.

(a) Brine feed

The brine is pretreated to remove certain heavy metal anions which would promote hydrogen evolution at the mercury cathode. This would lead to impure chlorine, current inefficiencies and at worst, explosions. Vanadium is especially deleterious in this context. Sulphate anions are also removed as they lead to oxygen evolution at the anode, with excessive anode wear (on graphite anodes only).

(b) Electrodes

Present usage is for graphite anodes. These are often ribbed (with ribs downwards and usually parallel to the brine flow) with holes drilled vertically between the ribs for exhaust of the chlorine. Illustrations of these can be found in ref. 83.

These graphite anodes wear away (see Chapter 14) as a result of anodic oxidation to CO_2 and the mechanical erosion associated with this process. The result is an increase in the inter-electrode gap with fall-off in cell performance. Alternatively, an electrode adjustment mechanism can be incorporated in the cell design, but the cost of providing and operating this has been one of the major incentives in the search for a "permanent" anode such as the DSA. The design of precious metal anodes might well be illustrated by the drawing in ref. 84, though the design based on expanded mesh titanium is possibly the most advantageous.

An important advantage claimed for the use of DSA is the reduction in the amount of brine purification required, in particular as regards removal of anions such as SO_4^{2-}. The DSA's have such a low overvoltage for chlorine, and such a high one for oxygen, that there is no concern that the latter gas is formed, even when considerable amounts of SO_4^{2-} are present.

Apart from these two anodes, only magnetite is usable, and this has been abandoned for many years now (see Chapter 14). The anodes come down through the lid (up to five abreast on the largest cells) each one on its own stem. Each stem (hence each anode) is capable of individual adjustment up and down in such a way that the inter-electrode gap is optimised. In certain cell designs such adjustment is coupled with a system whereby an entire rack of twelve or more anodes can be raised or lowered. At the end of the cell the chlorine gas is exhausted under slight vacuum. If the newer metallic anodes (see Chapter 14) are successful, the design of this anode adjustment gear will have to be reconsidered, for routine periodic adjustment will no longer be required and there may be scope for saving here on plant costs. On the other hand it is still uncertain how resistant the anodes are to massive short-circuit (*i.e.* when immersed in the mercury). Some form of "quick-release" gear is fitted to De Nora Cells. At least one cell room exists at Dormagen (W. Germany) with computer-controlled, fully automatic anode adjustment (Fr. Pat. 1,563,322 (Bayer)).

(c) Denuder design

The treatment of the amalgam with water is accomplished either in a tower denuder—which is simply a mild steel drum filled with pellets of graphite over which the amalgam trickles together with water—or in a trough denuder, which is a mild steel trough of the same length as the main cell and lying next to the cell or slung beneath it. The trough denuder is very much smaller in its cross-section than the main cell. It is filled with slabs of graphite. The relative merits of these two designs are well treated in ref. 79. Denuders too have been of vertical design,

TABLE 1

CHARACTERISTICS OF DE NORA AMALGAM CELLS WITH GRAPHITE ANODES

By courtesy of Oronzio de Nora

Characteristics	Basis 90 tonne/day chlorine capacity		Basis 270 tonne/day chlorine capacity
	14 H2	12 H6	20 H6
Number of cells installed	50	16	30
Number of cells in operation	50	16	30
Rated (A)	60,000	180,000	300,000
Maximum (A)	72,000	200,000	330,000
Anodic area (m^2)	7.5	19.4	32.3
Anodic current density at rating (A/m^2)	8,000	9,300	9,300
Cathodic current density at rating (A/m^2)	7,600	8,900	8,800
Avg. overall cell voltage at rating	4.1	4.3	4.3
Avg. overall voltage (A max.)	4.30	4.45	4.45
Current efficiency (NaOH) (%)	95–97	95–97	95–97

(continued on p. 112)

TABLE 1 *(continued)*

Characteristics	Basis 90 tonne/day chlorine capacity		Basis 270 tonne/day chlorine capacity
	14 H2	12 H6	20 H6
Current efficiency (KOH) %	94–96	94–96	94–96
Short-circuiting switches	14 unit	12 unit	20 unit
Mercury pump power (kW)	1.2	3	4.4
Power DC, at rating (kWh/tonne Cl_2)	3,225	3,375	3,375
A max. (kWh/tonne Cl_2)	3,375	3,495	3,495
Cathode width (m)	0.82	2.10	2.10
Cathode length (m)	9.65	9.60	16.20
Cathodic area (m²)	7.9	20.2	34
Cell dimensions, overall without aisles			
width (m)	1.10	2.25	2.30
length (m)	10	10.8	17
Floor space/cell, incl. aisles (m²)	19	43.5	61
Floor space (m²/tonne Cl_2)	10.9	8.0	6.7
Headroom required for piping etc., under cells (approx. m)	2.70	2.70	2.70
Height above 2nd floor level (m)	0.30	0.50	0.50
Automatic anode adjustment	Yes	Yes	Yes
Anode stem diaphragm material	Rubber—entire cell top	Rubber—entire cell top	Rubber—entire cell top
Cell cleaned under load	No	No	No
Slope of cell bottom (mm/m)	10	10	10
Decomposer, type	Vertical	Vertical	Vertical
location	at end of cell	at end of cell	at end of cell
lining	None	None	None
Sodium content entering decomposer (%)	0.2–0.3	0.2–0.3	0.2–0.3
Cell construction, bottom	steel	steel	steel
sides	rubber-lined steel	rubber-lined steel	rubber-lined steel
top	(3.2 mm) flexible rubber	(5.3 mm) flexible rubber	(5.3 mm) flexible rubber
Method of securing top	clamps	clamps	clamps
Anodes across cell	3	6	6
Anodes/cell	28	72	120
Stems/anode	2	2	2
Anode (kg/cell) (incl. stems)	1,270	3,360	5,800
Decomposer graphite (kg/cell)	150	450	745
Mercury (kg/cell)	990	2,210	3,280
Copper wt of cell incl. busbars (kg)	1,180	4,100	6,800
Total wt of cell in operation (kg)	18,000	38,500	59,000
Anode life (KOH) (months)	9	8	8
Anode life (NaOH) (months)	12	10	10
Graphite consumption			
(KOH) (kg/tonne)	2.5	2.5	2.5
(NaOH) (kg/tonne)	2.0	2.0	2.0
Mercury consumption (kg/tonne Cl_2)	0.1–0.15	0.1–0.15	0.1–0.15
Cl_2(KOH), dry basis (%)	99	99	99
H_2 in Cl_2(KOH), dry basis (%)	0.2–0.5	0.2–0.5	0.2–0.5
Cl_2 (NaOH), dry basis (%)	99	99	99
H_2 in Cl_2 (NaOH), dry basis (%)	0.1–0.4	0.1–0.4	0.1–0.4
H_2 (NaOH), dry basis (%)	99.9	99.9	99.9

and one in which the mercury amalgam runs down an expanded metal mesh is noteworthy because it would lend itself well to a fuel cell configuration[105].

Considerable amounts of heat are evolved in the reaction, and tower denuders are usually fitted with cooling equipment. Figure 5 shows the design of a De Nora mercury cell and among the features clearly visible are the girder racks on top of the cell, with cam-operated adjustment for racking up or down groups of anodes. The anode current feeds are seen looping over at the side of the cell. They connect the anodes to the cell baseplate on the adjacent cell. The drum-like structure is the denuder in which the amalgam is treated with water to form caustic and hydrogen. The performance of the de Nora cell is shown in Tables 1 (graphite anodes) and 2 (DS anodes).

(d) Other mercury cell designs

Two other cell designs are notable since neither conform with the foregoing general principles. Both are Japanese. The Kureha H.D. cell is the more conventional of the two and resembles the conventional mercury cell bent round on itself, hairpin fashion. It also incorporates a ferris wheel mercury pump for lifting the mercury from the bottom of the tower denuder back to the top of the cell.

At the time of its introduction, the Kureha H.D. cell was a major advance in cell design and performance. Its "hairpin" configuration reduced the mercury inventory (by elimination of the long mercury return pipe) and resistive losses in

TABLE 2

CHARACTERISTICS OF DE NORA AMALGAM CELLS WITH DIMENSIONALLY STABLE ELECTRODES (DSE)

Characteristics	Basis 90 tonne/day chlorine capacity		Basis 270 tonne/day chlorine capacity	Basis 900 tonne/day chlorine capacity
	10 M2	14 M2	21 M2	33 M2
Number of cells installed	20	15	30	60
Number of cells in operation	20	15	30	60
Rated (A)	150,000	200,000	300,000	500,000
Maximum (A)	180,000	240,000	360,000	600,000
Anodic area (m^2)	11	15.4	23	36.1
Anodic current density at rating (A/m^2)	13,600	13,000	13,000	13,850
Cathodic current density at rating (A/m^2)	13,000	12,500	12,500	13,300
Avg. overall cell voltage at rating	4	3.95	3.95	4
Avg. overall cell voltage (A max.)	4.15	4.10	4.12	4.18
Current efficiency (NaOH) (%)	97–98	97–98	97–98	97–98
Current efficiency (KOH) (%)	96–97	96–97	96–97	96–97

(continued on p. 114)

TABLE 2 (continued)

Characteristics	Basis 90 tonne/day chlorine capacity		Basis 270 tonne/day chlorine capacity	Basis 900 tonne/day chlorine capacity
	10 M2	14 M2	21 M2	33 M2
Short-circuit switches	10 units	14 units	14 units	22 units
Mercury pump power (kW)	3	3	4.4	4.4
Power DC, at rating (kWh/tonne Cl_2)	3,100	3,050	3,050	3,100
Power DC, A max. (kWh/tonne Cl_2)	3,210	3,180	3,190	3,230
Cathode width (m)	1.42	1.61	2.1	2.1
Cathode length (m)	8.1	9.9	11.4	17.7
Cathodic area, (m²)	11.5	16	24	37.5
Cell dimensions, overall without aisles				
width (m)	1.6	1.7	2.3	2.3
length (m)	9.1	11.1	12.5	19.2
Floor space/cell, incl. aisles (m²)	27.5	33	45.5	62.3
Floor space (m²/tonne Cl_2)	6.1	5.3	4.9	4.0
Headroom required for piping etc., under cells (approx. m)	2.7	2.7	2.7	2.7
Automatic anode protection device	Yes	Yes	Yes	Yes
Anode system diaphragm material	rubber entire cell top	rubber entire cell top	rubber entire cell top	rubber entire cell top
Cell cleaned under load	No	No	No	No
Slope of cell bottom (mm/m)	15	15	15	15
Decomposer, type	Vertical	Vertical	Vertical	Vertical
location	at end of cell	at end of cell	at end of cell	at end of cell
lining	None	None	None	None
Sodium content entering decomposer, %	0.2–0.3	0.2–0.3	0.2–0.3	0.2–0.3
Cell construction, bottom	steel	steel	steel	steel
sides	rubber lined steel	rubber lined steel	rubber lined steel	rubber lined steel
top	3.2 mm flexible rubber	5.3 mm flexible rubber	5.3 mm flexible rubber	5.3 mm flexible rubber
Method of securing top	clamps	clamps	clamps	clamps
Anodes across cell	2	2	3	3
Anodes/cell	20	28	42	66
Stems/anode	4	4	4	4
Decomposer graphite (kg/cell)	360	500	750	1,250
Mercury (kg/cell)	1,550	1,930	2,620	4,000
Copper wt. of cell incl. busbars (kg)	3,400	4,300	6,300	10,400
Total wt. of cell in operation (kg)	29,400	38,500	54,300	76,000
Anode coating life (KOH) (months)	18–24	18–24	18–24	18–24
(NaOH) (months)	18–24	18–24	18–24	18–24
Graphite consumption	None	None	None	None
Mercury consumption (kg/tonne Cl_2)	0.05–0.075	0.05–0.075	0.05–0.075	0.05–0.075
Cl_2(KOH), dry basis (%)	99.5	99.5	99.5	99.5
H_2 in Cl_2(KOH), dry basis (%)	0.2–0.3	0.2–0.3	0.2–0.3	0.2–0.3
Cl_2(NaOH), dry basis (%)	99.6	99.6	99.6	99.6
H_2 in Cl_2(NaOH), dry basis (%)	0.1–0.2	0.1–0.2	0.1–0.2	0.1–0.2
H_2(NaOH), dry basis (%)	99.9	99.9	99.9	99.9

the anode stems and lead-ins were reduced. The cell served to spur other cell designers to improve their performance.

An entirely revolutionary design was the Asahi mercury cell. A circular mild steel cathode revolved slowly, and the whole cell was drum-like in shape. The Asahi cell performed well but is not in use outside its native country.

Still other types of mercury cell are described in Sconce and these include cells with vertical mercury cathodes and cells in which the brine is forced through at high rates. The latter concept seeks to eliminate or reduce bubble overvoltage, and extensive engineering trials were made both in the U.S.A. and the U.K. However the power required to pump the brine, as well as sealing problems led to the abandonment of the idea.

A recent Japanese patent[85] covers a bipolar mercury cell, using explosion bonded mild steel–titanium laminates. Several layers of mercury cells thus lie one on top of the other. Such techniques of marrying titanium to mild steel must find application sooner or later in any bipolar cell design using precious metal anodes, whether a chlorine cell or not.

7.4.2 Diaphragm cells

The mercury cells whose performance is described in the previous section come as close to perfection as technologists and users can conceive. With diaphragm cells the case is a very different one, and the cells hitherto available have always left obvious room for improvements. The successful adoption of DSA's will, if anything, make a more profound change on diaphragm cells than on mercury cells. This can be seen by looking at present day diaphragm cell design[8]. All designs except for Dow are monopolar, with "finger" anodes interleaving with cathodes of similar shape. The Dow design is bipolar, but such designs with graphite electrodes have never been wholly successful. None of the present diaphragm cells permits any adjustment for anode wear. All use a diaphragm of (white, chrysotile) asbestos, which is formed either by laying on sheets of this material, or by vacuum deposition of an aqueous slurry of asbestos fibres. This is sucked down onto the cathode, which is often a mild steel mesh. Concrete is the commonest cell construction material for diaphragm cells. Figures 6 and 7 show the Hooker S4 cell, the latest of a long established line of similar designs. Columbia Southern and Diamond Shamrock cells are similar, and these makers have recently offered a design incorporating DSA's. In all cells, chlorine is removed from the anode compartment, while hydrogen at very low overpressures comes off from the cathode and is removed from inside the steel mesh "fingers". Sconce gives full descriptions of these and other cells, with performance figures. If the patent literature is any guide, there is much interest currently in bipolar diaphragm cell designs with mild-steel DSA composite electrodes. These might be formed by explosive bonding[86] or other techniques[109]. Such a cell would have vastly superior

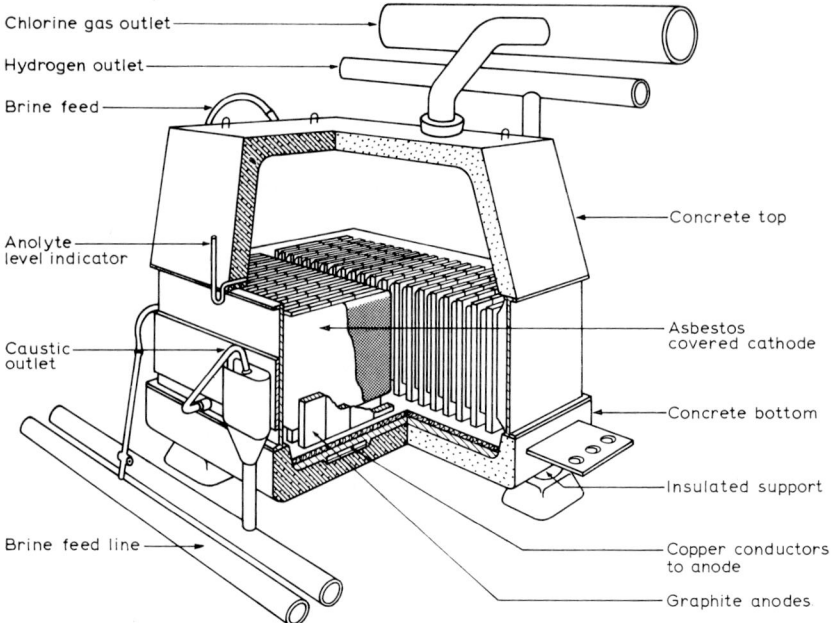

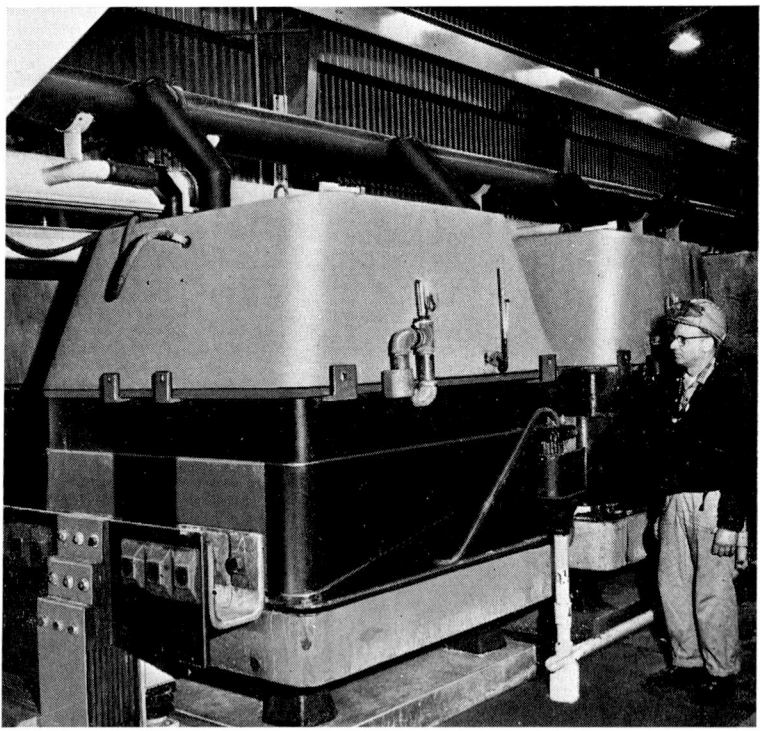

Figs. 6, 7.
Hooker S4 diaphragm cell (60,000 A), the largest commercially available model (courtesy Hooker Chem. Corp.).

performance to those presently in use. Its economic use would still be restricted to some extent by the weaknesses of asbestos as a diaphragm material. Normal lifetime for these diaphragms is little over 90 days, though the figure depends on brine composition, operating temperature, current density and also on the amount of load-shedding (if any) which is often called for in the case of cells supplied from a public authority supply. Considerable "know-how" resides in operating diaphragm cells, and diaphragm failure (flow rate too high or low) can be postponed to some extent by addition of HCl to the brine feed (to dissolve precipitates in the pores)[87] or alkali earth metal hydroxides which have the opposite effect. Diaphragm cells fitted with DSA's may have a bonus here, since less debris will be formed from the anodes and a cleaner diaphragm should result. Hooker have recently disclosed a cell in which extra circulation of anolyte liquor takes place. This is said to extend diaphragm life (Neth. pat. 6,802,179).

A continuing search for better diaphragm materials is taking place. Numerous patents, mainly Japanese, cover the use of ion exchange membranes, but no actual use of these is known. Of the newer plastics, PTFE is the only one not attacked under diaphragm cell conditions. Materials such as polypropylene are not apparently attacked in their bulk form, but as soon as exposed to the environment of a chlorine cell as high area porous materials, sufficient chlorination occurs to cause blockage. Whether work has taken place to chlorinate previously oversize pored diaphragms in a pretreatment plant, is not known. PTFE suffers the drawback that it is not wettable, and to permit sufficient flow through and avoid formation of minute gas bubbles (Jamin effect) in the pores which block them, wetting agents are required [88]. An alternative approach would be to modify the PTFE by addition of a second component which favoured wetting. Of the non-plastic materials, the rapidly growing technology of carbon fibres and cloths offers the brightest hopes for longer lasting diaphragms (see, for example, Fr. pat. 1,406,529).

7.5 Future developments

Leaving aside the question of precious metal anodes and any other anode material which might emerge, as well as novel diaphragms whether of the ion-exchange or conventional variety, the main developments will undoubtedly relate to energy savings in the chlorine industry. Two different principles are applied to the diaphragm and mercury cell respectively. In the former, reaction (1) is replaced by (2) with the appropriate voltage savings and the concept is illustrated in Fig. 1 ref. 89 where either air or oxygen would be used at the cathode (in the case of air, the nitrogen has to be vented out). Chlorine manufacturers and fuel cell manufacturers are combining to solve this problem, and test rigs on the 1 m² scale are known to exist. The problems (apart from the sheer economics of the extra fuel cell electrode cost) are largely due to the presence of NaCl crystallising out and clogging the electrode pores. Nevertheless, limited success has already been attained with such systems. In the case of the mercury cell, the principle is not one of saving energy but of recovering and re-using it. Thus a fuel cell is built which consists of Na/Hg–O_2, with the amalgam coming from the cell and returning to it. Much data exist on this type of fuel cell, which was studied by Yeager for the U.S. Navy, and the problems can be seen from this work. They include problems of how to feed the recovered energy back into the system as a whole, problems of preventing short-circuits flowing through the mercury, which requires mechanical circuit breakers of some type, and the problems of optimisation. Clearly, as the amalgam becomes denuded of its sodium it becomes less and less rewarding to recover the energy from it. Many believe that after a given fraction of the sodium has been removed in a fuel cell, it would be most economic to remove the rest in a conventional denuder system. A general picture of these problems is found in ref. 89 while a detailed design of a mercury fuel cell denuder is given in ref. 90.

7.6 FURTHER READING – CHLORINE ELECTROLYSIS

The A.C.S. Monograph *Chlorine* edited by J. S. Sconce, Reinhold, New York, 1962, provides easily the fullest and most comprehensive account available. Other descriptions are found in the *Encyclopaedia of Electrochemistry*[3] and *Caustic Soda Production Techniques*, 1962, Noyes Development Corp., which describes largely the use, under license, of equipment in India. A culling from recent patents is found in ref. 82. Other specialised references are listed below:

Brine pretreatment
U.S. pat. 3,312,609 (Hooker); 2,954,333 (Columbia Southern) see also ref. 82.
K. HARADA and K. YAMORI, *Electrochem. Technol.*, 5 (1967) 137.
M. YAMAHA, H. SUZUKI and K. TSUKAMOTO, *Electrochem. Technol.*, 5 (1967) 257.
P. H. RALSTON and R. R. MITCHELL, *Electrochem. Technol.*, 5 (1967) 262.
I. YAGI, T. MATSUNO and M. YAMAHA, *Electrochem. Technol.*, 5 (1967) 415.
G. HAUCK and W. DÜRR, *Chem. Ing.–Tech.*, 39 (1967) 720.

Process economics and design comparisons
A. SCHMIDT, *Chem. Ing.–Tech.*, 39 (1967) 692.
W. WATSON-SMITH, *Chem. Engr.*, Mar. (1968) CE54.
F. HINE, *Electrochem. Technol.*, 6 (1968) 69.
H. SOMMERS, *Electrochem. Technol.*, 5 (1967) 108.
and earlier surveys by Sommers referred to by him.

Design of mercury cell anodes and stems
H. A. SOMMERS, *Electrochem. Technol.*, 6 (1968) 124.
Fr. pat. 1,563,222 (Bayer).

Denuder designs and processes
M. M. JAKSIC and I. M. CSONKA, *Electrochem. Technol.*, 4 (1966) 49.

Busbar design
W. D. WHINEY and C. T. ABBOTT, *Electrochem. Technol.*, 5 (1967) 130.
L. T. GUESSE and G. K. CROWN, *Electrochem. Technol.*, 5 (1967) 133.

Chlorine drying and liquefaction and handling
L. COJOCARU and K. TIMM, *Chem. Ing.-Tech.*, 39 (1967) 85.
J. A. A. KETELAAR, *Electrochem. Technol.*, 5 (1967) 143.
H. NOTTEBOHM, *Chem. Ing.-Techn.*, 37 (1965) 581.
H. HAGEMANN, *Chem. Ing.-Tech.*, 39 (1967) 744.
J. MISCHKE and S. PAYER, *ibid.*, 39 (1967) 734.
K. D. MICHELS and R. HASE, *Chem. Tech. (D.D.R.)*, 10 (1969) 614, see also ref. 82.

Details of Kureha cell
H. SHIBATA and Y. YAMAZAKI, *Electrochem. Technol.*, 5 (1967) 239.
H. A. SOMMERS, *Chem. Eng. Progr.*, 61 (1965) 94.

Details of Asahai cell
M. MUROZUMI, *Electrochem. Technol.*, 5 (1967) 236.

Use of plastics in chlor-alkali industry
H. SCHIFFERDECKER, *Chem. Ing.-Tech.*, 39 (1967) 737.

Hydrogen evolution in mercury cells
J. BALEJ and I. PASEKA, *Chem. Ing.-Tech.*, 39 (1967) 725.

General surveys (economic/technological)
K. HASS, *Electrochem. Technol.*, 5 (1967) 246.
K. HASS, *Chem. Ing.-Tech.*, 39 (1967) 689.
K. HASS, *ibid.*, 40 (1968) 557.

Anodes (graphite)
T. C. JEFFERY, *Electrochem. Technol.*, 5 (1967) 124 (in diaphragm cells).
R. PROFT and S. RICHTER, *Chem. Techn. (D.D.R.)*, 10 (1969) 611 (porosity and c. eff.).

Diaphragm cell design
G. GRÜNZIG, W. SCHNAUBELT and M. HORX, *Chem. Tech., (D.D.R.)*, 10 (1969) 604 (Eastern Bloc designs); see also ref. 82.
Neth. pat. 6,802,179 (forced circulation in diaphragm cells).

V. V. STENDER, *Zh. Prikl. Khim.*, 40 1293. pilot studies.
O. S. KSENZHEK, *Elektrokhimya*, 4 (1968) 1439, pilot studies.

Diaphragm cells with ion exchangers
E. Germ. pat. 59,777 (20/1/68); 62,040 (8/6/68).
U.S. pat. 3,438,879.
Jap. pat. 17563/67.
M. MINAGAWA and T. KEDA, *Denki Kagaku*, 34 (1966) 495.
Fr. pat. 1,510,265; F. WOLF and H. WYSNOMIOSKI, *Werkstoff Korrosion*, 18 (1967) 898.

Literature of the chlor -alkali industry
Approximately every two years, the Gesellschaft Deutscher Chemiker (Fachgruppe Angewandte Chemie) hold symposia on this subject, and the material given is published shortly afterwards in *Chemie Ingenieur-Technik*. The Electrochemical Society of New York have similar symposia, published in their own journals. In the volume *Literature of Chemical Technology (Advan. Chem. Ser. of A.C.S.* 78 (1968)) Chapter 1 is devoted to the literature of the chlor-alkali industry. The Chlorine Institute of New York is a non-profit making organisation publishing a most valuable series of pamphlets including one with details of commercially available cells.

8. Iodine electrochemistry

This cannot be said to be an important field—no large tonnages are involved. Two main reactions will be discussed, namely the manufacture of iodates from iodine, and the oxidation of iodates to periodates.

Manufacture of iodates is well treated in a paper by Udupa et al.[91]. These authors use a lead dioxide anode for the work. The reaction proceeds well, but it would seem from this work that crystallisation of the iodate formed would make the process a difficult one to operate other than in batchwise fashion. The work has a good bibliography, and although this does not form the basis of any industrial process at the present time, its future is largely connected with the fate of the periodate process (see below).

The oxidation of iodates to periodates has received much attention, in particular from Mantell and co-workers. The aim of their work was to use the periodic acid as an oxidising agent for certain organic reactions, and to regenerate the iodic acid so formed. Details of this work will be found in Chapter 13, *Miscellaneous Electrochemical Processes*, and an entire chapter is devoted to the work in ref. 92 while another account is found in ref. 3. Mantell[93] described cell design for iodates and periodates. His estimate for their joint production totals (in the U.S.A.) was 25 tonnes p.a.

9. Bromates

9.1 Introduction

The electrolytic manufacture of bromates of the alkali metals is a steadily growing process. Although the process has much in common with the manufacture of chlorates (q.v.), there are differences, including more severely corrosive conditions in the cell. Mantell[7] refers to the use of graphite electrodes at 300–800 A/m² and cell voltages of 3.6 V with cell operating temperatures below 40°C. Dichromate is added to the electrolyte. In addition to the electrolysis of bromide solutions, Mantell refers to the electrolysis of solutions of bromine and caustic soda. This process is not thought to be of any commercial value today. Sugino[3] describes the process at greater length than Mantell, placing considerable emphasis on his own work using PbO_2 anodes. He also discusses the use of graphite and platinised titanium anodes (and thus presumably others of the noble-metal oxide/titanium type). The use of graphite is discouraged on account of a yellow "mud" which forms at the anode and discolours the product[94].

9.2 Basic reaction

Early work by Kretzschmar[95] and by Bray[96] enables us to postulate the following alternative reactions:

$$6\ Br^- \rightarrow 3Br_2 + 6e \qquad (1)$$

$$2Br_2 + 2OH^- \rightarrow 2HBrO + 2Br^- \qquad (2)$$

$$Br_2 + 2OH^- \rightarrow BrO^- + H_2O + Br^- \qquad (3)$$

$$2HBrO + BrO^- \rightarrow BrO_3^- + 2HBr \qquad (4)$$

which is a sequence similar to that postulated for chlorate formation (q.v.) although reaction (4) has a rate constant some two orders of magnitude greater than that of the analogous reaction with chlorine-containing species, and reactions (3) and (2) do not proceed as rapidly. In fact liquid bromine can be observed in the cell under certain conditions of poor circulation and somewhat acidic conditions. A good study of the experimental variables is that of Udupa et al.[97]. Larger scale work, especially with lead dioxide electrodes, is described by Sugino[98, 99].

9.3 Commercial cells

One cell design is illustrated in Sugino[98], another in U.S. pat. 2,191,574; Jap. pat. 9864 (1957) also describes PbO_2 electrodes in bromate cells. Current densities in the Japanese work are 20 A/dm² (referred to anode) with a cell voltage around 4 V, operating around 70°C at pH 9.5–10. With current efficiencies of 90%, the energy consumption is 4000 kWh/tonne bromate of the $KBrO_3$. For the sodium salt Sugino[98] quotes similar conditions but with a current efficiency of only 70% resulting in a consumption of 6000 kWh/tonne of the sodium bromate. Mantell[93] discussed bromate cell design in 1968 and quoted annual production (U.S.A.) at 300 tonnes. An interesting account of bromate preparation is also that by Radford[100] who quotes the very early work of Muller[101]. Radford stresses the role of dichromate as an additive in the process and its effect on raising the current efficiency from 75–95% which he attributes to the formation of a layer on the cathode which hinders reduction of the intermediate hypobromite. As to cell design, he quotes U.S. pat. 1,919,721, and also refers to the use of perforated cathodes (of copper or stainless steel). The hydrogen gas liberated should be used to the fullest extent for stirring of the electrolyte in the cell. Too small an overvoltage at the cathode results in corrosion (especially with copper) to form cupric oxides. Radford quotes the use of graphite, platinum-sheathed copper, lead dioxide and platinised titanium anodes. On graphite anodes, he refers to techniques of impregnation. For details of all these, the reader should refer to Chapter 14 on electrodes. In an interesting reference to his own unpublished work, Radford notes the dangers of corrosion of platinum anodes when the bromide feed approaches exhaustion. This, coupled with his observation of ozone at the gas exits, would indicate very high overpotentials, and he also mentions sludges of platinum oxides. A similar effect results from the presence of silicates which, Radford states, probably form a semi-permeable membrane on the anode, permitting concentration polarisation to set in.

9.4 OTHER PROCESSES

The purely chemical process reacts potassium hydroxide with bromine and Radford quotes details. As he points out, the formation of KBr as a by-product would seem to limit this product to the level of demand of that salt. It would not appear to be industrially important. Work by Sarghel[102] on electrolysis of calcium salts would not seem to be significant, leading as it does to the formation of insoluble salts at the cathode. Details of current efficiencies and cell voltages at a number of C.D.'s on smooth and platinised platinum are found in Engelhardt[106].

10. Bromine

Bromine is no longer prepared by direct electrolysis though the process has been used in the past. For plants where high bromide concentrations are not available, the electrolytic process was never attractive, and there are only a few sites in the world where bromide brines are available. The last pilot plant to close down was located in Israel, and its operation is described in ref. 103.

Although direct electrolysis always held out the promise of lower energy consumption than the chlorine replacement processes which superseded it, unforeseeable difficulties outweighed this fact, and the 0.37 V difference in the reversible potentials of chlorine and bromine is, in fact sacrificed. An excellent historical summary is found in ref. 104. where the Wünsche process (D.R. pat. 140,274) and the Kossuth Process (D.R. pat. 103,644) are described. Both processes were operated in Germany before the war. In the Wünsche process, carbon anodes and cathodes were separated by a cylindrical ceramic diaphragm and the cell operated at 3.4 V and 115 A/m^2. Separate anode and cathode streams were used, the former going to a stripping tower to be debrominated and returning then to the cathode compartment where magnesium hydroxide was deposited from it. Yaron states that 50 cells, 80–90% used at the Westeregeln Works, produced 300 kg of pure bromine daily, at 70% current efficiency. Clogging of the diaphragm by the hydroxides was a major operating problem.

In the Kossuth process, a bipolar cell design with graphite electrodes and no diaphragms was used. Periodic current reversal minimised hydroxide scale formation, but only at the expense of current efficiency which ran at around 50%. The energy required was 3000 kWh/tonne of bromine. The American chemical industry suffered from not having a concentrated bromide source and thus experienced all the above mentioned problems as well as others stemming from the dilution of their brines. The electrolytic cell of Dow in 1889 was similar to the Kossuth cell in its conception[106].

In conclusion, it is doubtful—unless there is a sudden upsurge in bromine consumption—whether there will be a return to the electrolytic processes. The Israeli work[104] (and also in Isr. pat. 10,390 and U.S. pat. 2,825,685) will have to be superseded by novel processes or materials if further work is to be anticipated.

Many of the foregoing details are much amplified in Engelhardt[106] where current–voltage curves for various concentrations of bromides are found as well as very much fuller descriptions of the processes referred to above, and others. A good picture of production in Germany and the U.S.A. and various other commercial factors is also presented.

11. References

1 H. G. JANSON, *Chem. Ing.-Tech.*, 39 (1967) 729.
2 H. HÖLEMANN, *Chem. Ing.-Tech.*, 34 (1962) 371.
3 J. H. NICHOLS in C. HAMPEL (Ed.), *Encyclopaedia of Electrochemistry*, Reinhold, New York, 1964.
4 P. GALLONE and O. MESSNER, *Electrochem. Technol.*, 3 (1965) 321.

5 F. HABER and S. GRINBERG, *Z. Anorg. Allgem. Chem.*, 16 (1897/8) 198, 329.
6 F. MELLOR, *Comprehensive Treatise on Inorganic Chemistry*, Vol. II, Suppl. I, Longmans Green, London, 1956, p. 314.
7 C. F. MANTELL, *Electrochemical Engineering*, McGraw-Hill, New York, 4th ed., 1960.
8 J. S. SCONCE, *Chlorine A.C.S. Monograph*, Reinhold, New York, 1962.
9 E. DÖNGES and H. G. JANSON, *Chem. Ing.-Tech.*, 38 (1966) 443.
10 E. DÖNGES and J. SCHÜCKER, *Chem. Ing.-Tech.*, 37 (1965) 498.
11 H. G. JANSON, *Chem. Ing.-Tech.*, 40 (1968) 562.
12 F. B. GROSSELFINGER, *Chem. Eng.*, 71 (1964) 172.
13 K. HASS, *Chem. Ing.-Tech.*, 40 (1968) 557.
14 G. BIANCHI and C. TRAINI, *Chim. Ind. Milan*, 46 (1964) 363.
15 F. HINE, S. YOSHIZAWA and Y. NAKANE, *Electrochem. Technol.*, 4 (1966) 555.
16 Fr. pat., 1,438,213.
17 C. P. ROBERTS, *Chem. Eng. Progr.*, Sept. (1960) 456.
18 KIRK–OTHMER (Ed.), *Encyclopaedia of Chemical Technology*, Vol. 5, 2nd ed., Wiley, New York, 1963.
19 N. IBL and D. LANDHOLT, *Electrochim. Acta*, 15 (1970) 1165.
20 V. DE VALERA, *Ext. Abstr. of Boston Mtg. of. Electrochem. Soc.*, May, 1968.
21 M. M. JAKSIC et al., *J. Electrochem. Soc.*, 116 (1969) 684, 1316.
22 J. C. SCHUMACHER, *J. Electrochem. Soc.*, 116 (1969) 68C.
23 Anon., *Platinum Metals Rev.*, 13 (1969) 103.
24 *Informations Chimie*, No. 66, Dec. 1968.
25 U.S. pat., 3,043,757; 3,055,821.
26 J. R. NEWBERRY and W. C. GARDINER, *J. Electrochem. Soc.*, 116 (1969) 114.
27 M. M. JAKSIC et al., *Electrochem. Technol.*, 6 (1968) 397.
28 D. G. ELLIOT, *Tappi*, 51 (1968) 88A.
29 R REMIREZ, *Chem. Eng. (N.Y.)*, 74 (1967) 136.
30 V. P. GLADYSHEV, L. M. RUBAN et al., (see *Chem. Abstr.*, 67 (1967) 39547s).
31 T. MATSUMURA et al., *J. Electrochem. Soc.*, 115 (1968) 402.
32 T. MATSUMURA, *Electrochem. Technol.*, 4 (1966) 402.
33 E. A. VAINRIB, E. A. BARANOV and G. L. MEDRISH, *Elektrokhimya*, 3 (1967) 97; K. G. IL'IN and V. I. SKRIPCHENKO, *Izv. Vysshykh Uchebn. Zavedenii Khim. i Khim. Tekhnol.*, 7 (1964) 572; V. A. SHLYAPMIKOV and T. S. FILLIPOV, *Elektrokhymya*, 4 (1968) 20.
34 H. V. UDUPA, *Indian J. Technol.*, 4 (1966) 305.
35 T. YOKOYAMA, *Denki Kagaku*, 29 (1961) 697.
36 I. ROUSAR, *Chem. Premysl*, 17 (1967) 468.
37 T. R. BECK and R. BRÄNNLAND, *J. Electrochem. Soc.*, 115 (1968) 886.
38 I. ROUSAR, *J. Electrochem. Soc.*, 116 (1969) 676, 683.
39 J. CLAUS, *Ext. Abstr. Electrochem. Soc. Mtg. May, 1968*.
40 J. A. MCLAREN, *Ext. Abstr. Electrochem. Soc. Mtg. May, 1968*.
41 M. M. JAKSIC et al., *J. Electrochem. Soc.*, 116 (1969) 394.
42 M. JANES, *Ext. Abstr. Electrochem. Soc. Mtg. May, 1968*.
43 W. A. NYSTROM, *Ext. Abstr. Electrochem. Soc. Mtg. May, 1968*.
44 R. A. CRAWFORD, W. B. DARLINGTON and L. B. KLIEVER, *Ext. Abstr. Electrochem. Soc. Mtg. May, 1968*.
45 Anon., *Ind. Chemist*, 37 (1961) 21.
46 B. POPYANKOV and N. KOLAROV, *Tekstilna Prom. (Sofia)*, 14 (1965) 26.
47 J. C. SCHUMACHER, private communication, 9.7.69.
48 J. C. SCHUMACHER, *Perchlorates, A.C.S. Monograph 146*, Reinhold, New York, 1960.
49 W. WALLACE, *J. Electrochem. Soc.*, 109 (1962) 309C.
50 W. OECHSLI, *Z. Elektrochem.*, 9 (1903) 807.
51 C. W. BENNETT and E. L. MACK, *Trans. Electrochem. Soc.*, 29 (1916) 323.
52 N. V. S. KNIBBS and H. PALFREEMAN, *Trans. Faraday Soc.*, 16 (1920) 402.
53 M. P. GROTHEER and E. H. COOK, *Electrochem. Technol.*, (1968) 221.
54 K. BERNDT, V. DÖLLE and G. KREUZBERGER, *Chem. Tech. (D.D.R.)*, 10 (1969) 607.
55 N. RAMACHANDRAN et al., *Ind. Chem. Engr.*, 8 (1966) 6.

56 N. NAGALINGAM et al., Chem. Age India, 16 (1965) 491.
57 K. G. NARASINHAM, Bull. Nat. Inst. Sci. India, 29 (1965) 279.
58 O. DE NORA, P. GALLONE, C. TRAINI and G. MENEGHINI, J. Electrochem. Soc., 116 (1969) 146.
59 A. LEGENDRE, Chem. Ing.-Tech., 34 (1962) 379.
60 T. OSUGA, S. FUJII, K. SUGINO and T. SEKINE, J. Electrochem. Soc., 116 (1969) 203.
61 T. OSUGA and K. SUGINO, J. Electrochem. Soc., 104 (1957) 448.
62 U.S. pat. 2,868,711; Brit. pat. 1,104,078; Fr. pat. 1,444,557; Can. pat. 741,312; Fr. pat. 1,402,590.
63 Anon., Ind. Chem. Engr., 9 (1967) 28.
64 G. ANGEL and H. MELLQUIST, Z. Elektrochem., 40 (1934) 702.
65 R. BAUER, Ext. Abstr. Electrochem. Soc. Mtg. May, 1968.
66 A. REGNER, Industrial Electrochemistry, Artia Press, Constable, 1957/8.
67 C. MARSHALL and J. P. MILLINGTON, J. Appl. Chem., 19 (1969) 298.
68 M. SITTIG, Sodium, A.C.S. Monograph, 133, Reinhold, New York, 1956.
69 KIRK–OTHMER (Ed.), Sodium in Encyclopaedia of Chemical Technology, 2nd ed., Wiley, New York, 1963.
70 FIAT, Final Report 819, 1946.
71 Brit. pat. 1,155,927.
72 Brit. pat. 1,200,103.
73 Belg. pat. 686,359.
74 C. HAMPEL (Ed.), Sodium in Encyclopaedia of Electrochemistry, Reinhold, New York, 1964.
75 R. MORGENTHALER, Erzmetall, 7 (1954) 358.
76 FIAT Final Report 830, 1946.
77 G. EGER, Handbuch der Technischen Elektrochemie, Band III, Akad Verlag, Leipzig, 2nd ed., 1955.
78 F. R. THOMPSON (Ed.), Chem. Eng., 62 (1955) 101.
79 H. HUND, Chem. Ing.-Tech., 39 (1967) 702.
80 T. R. BECK, J. Electrochem. Soc., 116 (1969) 1038.
81 T. A. LIEDERBACH, Ext. Abstr. Electrochem. Soc. Mtg. May, 1966; R. D. BURT and W. W. LAWRENCE, ibid.
82 R. POWELL, Chlorine and Caustic Soda Manufacture ,Recent Developments, Noyes Development Co., 1969. This book is essentially a patent survey reflecting ideas as much as actual practice.
83 H. SOMMERS, Electrochem. Technol., 6 (1968) 124.
84 O. DE NORA, Chem. Ing.-Techn., 42 (1970) 222.
85 Brit. pat. 1,121,676 (to Asahi).
86 Brit. pat. 1,099,434 (to Asahi).
87 Brit. pat. 1,092,167; Dutch pat. 66. 04328.
88 L. M. MONASTYRSKII, Khim. Prom., 42 (1966) 788; U.S.S.R. pat. 172,726.
89 A. T. KUHN, New Scientist, 36 (Oct. 5th) (1967) 21.
90 J. A. LEDUC, J. G. KOURILO and C. LURIE, J. Electrochem. Soc. 116 (1969) 546.
91 M. S. VENKATACHALAPATHY, S. KRISHNAN, M. RAMACHANDRAN and H. V. K. UDUPA, Electrochem. Technol., 5 (1967) 399, 404.
92 C. L. MANTELL, Electro-organic Chemical Processing, Noyes Development Corp., 1968.
93 C. L. MANTELL, Ext. Abstr. 276 Electrochem. Soc. Mtg. May, 1968.
94 D. T. EWING and H. W. SCHMIDT, Trans. Am. Electrochem. Soc., 47 (1925) 117.
95 H. KRETZSCHMAR, Z. Elektrochem., 10 (1904) 789.
96 W. C. BRAY, J. Am. Chem. Soc., 32 (1910) 932; 33 (1911) 1485.
97 S. SUNDARAJAN, K. C. NARASINHAM and H. V. K. UDUPA, Chem. Process Eng., Sept. (1962) 438, 447.
98 T. OSUGA and K. SUGINO, J. Electrochem. Soc., 104 (1957) 448.
99 K. SUGINO, Bull. Chem. Soc. Japan, 23 (1950) 115.
100 P. J. RADFORD in Z. E. JOLLES (Ed.), Bromine and its Compounds, Benn, London, 1966.
101 E. MÜLLER, Z. Elektrochem., 5 (1899) 469.
102 J. SARGHEL, Z. Elektrochem., 6 (1899) 149, 163.
103 O. SCHACHTER, Bull. Res. Council Israel, 7A (1958) 204.

104 Z. E. JOLLES (Ed.), *Bromine and its Compounds*, Benn, London, 1966, Chapt. I.
105 Patent of Addition to Brit. pat. 1,165,244.
106 V. ENGELHARDT, *Handbuch der Technischen Elektrochemie*, Akad Verlag, Leipzig, 1934 (and see ref. 77).
107 Fr. pat. 1,563,518 (DuPont).
108 Ger. Offen. 1,934,082 contains data on efficiency of electrolysis of dil. brines with graphite and D.S. anodes.
109 *Brit. Chem. Eng.*, 16 (1971) 153.

Chapter 4

Industrial Water Electrolysis

D. H. SMITH

CJB (Projects) Limited, Portsmouth, Hants.

1. Introduction

The production of hydrogen and oxygen by the electrolysis of water has been practised on the industrial scale since the beginning of this century. Some dates of importance in the development of the technology are shown in Table 1. Whilst most installations are comparatively small in size, there are notable exceptions. At the Aswan High Dam site, for instance, 40,000 m^3 h^{-1} of hydrogen is produced by electrolysis, using Demag electrolysers.

TABLE 1

IMPORTANT DATES IN THE DEVELOPMENT OF WATER ELECTROLYSIS

1800—Nicholson and Carlisle—Electrolytic Decomposition of Water into Hydrogen and Oxygen

1888—Latschinoff—Investigation of Water Electrolysis in Pressure Cells

1902—Oerlikon—Manufacture of a small Filter-Press Electrolyser

1910—International Electrolytic—Design of Knowles Cell.

1912—International Electrolytic—First Knowles Cell installed

1925—Noegerath and Lavaczek—Design of Pressure Cell

1926—Zdansky—Design of Filter Pressure Electrolyser

1931—Newitt and Sen—Design of Pressure Cell

1935—Diez—Design of Pressure Cell

1937—Oerlikon—Manufacture of large Filter-Press Electrolyser

1948—Zdansky-Lonza—Design of Filter-press Pressure Electrolyser

1951—Lonza A.G.—Installation of First Filter-press Pressure Electrolyser

1955—Treadwell—Pressure Electrolyser for Submarines

1962—Lurgi—4,700 Nm^3 h^{-1} Hydrogen Pressure Electrolyser installed

1963—CJB—Pressure Electrolyser for Submarines

1967—CJB—Small Fully Automatic Pressure Electrolyser

The electrolytic method has certain intrinsic advantages over others. High purity gases are readily produced, the equipment can be made fully automatic and considerable flexibility of output can be obtained.

The economics of hydrogen and oxygen production are dictated by the costs of electric power and these vary widely from area to area. At the present time, for general large scale production, the electrolytic method is only economic in regions where cheap power is available. Special situations exist however, in which water electrolysis is preferred.

This Chapter describes the electrolytic decomposition of water and the basic features of Industrial Water Electrolysers. Brief details of some commercially available units are then given, together with the uses made of such plant.

2. The electrolytic decomposition of water

Water is decomposed into its elements, when a direct current is passed between a pair of electrodes immersed in a suitable aqueous electrolyte. In order to obtain the gases evolved in a sensibly pure, and hence safe, condition, a semi-permeable membrane or diaphragm is placed between the electrodes to prevent the gases mixing. The components of the cell are thus two electrodes, a diaphragm and a suitable electrolyte.

2.1 Cell voltage

The voltage, V, applied across the electrodes can be divided into three components. These are the decomposition voltage, E_D, the overvoltage at the electrodes, E_O and the ohmic loss in the interelectrode gap which is the product of the cell current, I, and the electrical resistance of this gap, R. Thus,

$$V = E_D + E_O + IR$$

At 25°C and a pressure of one atmosphere, the reversible decomposition voltage for water is 1.23 V. For isothermal operation, the energy required results in an equivalent cell voltage of 1.48 V and further energy is absorbed in producing the products in gaseous form. No gas evolution is observed in practice until voltages of 1.65–1.7 V are applied. Practical cells operate at voltages of 1.8–2.6 V, as a result of overvoltage and ohmic losses.

Overvoltage is of two kinds. Activation overvoltage results from the slowness of the electrode reactions. It varies with the metal used as the electrode and its surface condition. It is reduced by operating at elevated temperatures and, to a lesser extent, at elevated pressures[1,2], and increases with the rate or current density of the electrode reaction. Figure 1 shows the relationships between overvoltage and current density for a number of metals.

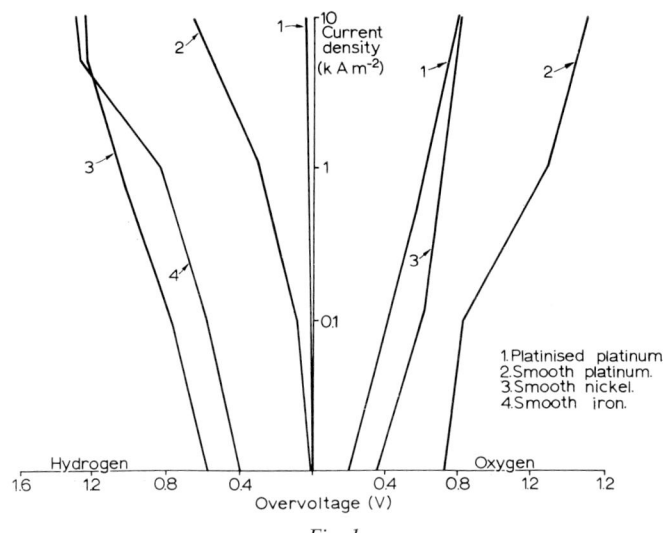

Fig. 1.
Overvoltage/current density for various metals.

Concentration overvoltage arises as a result of changes in the composition of layers of electrolyte close to the electrodes. Ions are discharged and hydrogen and oxygen are formed, the replenishment of the former and the removal of the latter being diffusion-controlled. The energy barriers which result increase the cell voltage. These effects are minimised by using elevated temperatures and by agitation of the electrolyte, either by forced circulation or by the use of favourable electrode geometry.

Ohmic losses occur both in electrolyte and diaphragm. An electrolyte is chosen which combines a maximum electrical conductivity with a minimum rate of attack upon materials of construction. For water electrolysis, solutions of sodium or potassium hydroxide are chosen, the variation of conductivity of these with concentration and temperature[3] being shown in Fig. 2.

During electrolysis the solution close to the cathode becomes more concentrated, and that close to the anode less concentrated, because of the different mobilities of the ions present and this effect is also reduced by agitation of the electrolyte. Operating at high temperature results in increased electrical conductivity.

The interelectrode gap is partially filled with gases during electrolysis, which increase its electrical resistance. Electrolyte agitation, favourable electrode geometry and operation at elevated pressure, can be employed to minimise this effect.

The diaphragm material used should have low ohmic resistance in the electrolyte and should be an effective barrier to the transfer of gas bubbles and electrolyte carrying entrained or dissolved gas. It should be resistant to chemical attack by the electrolyte, have good mechanical strength and its cost should be acceptable.

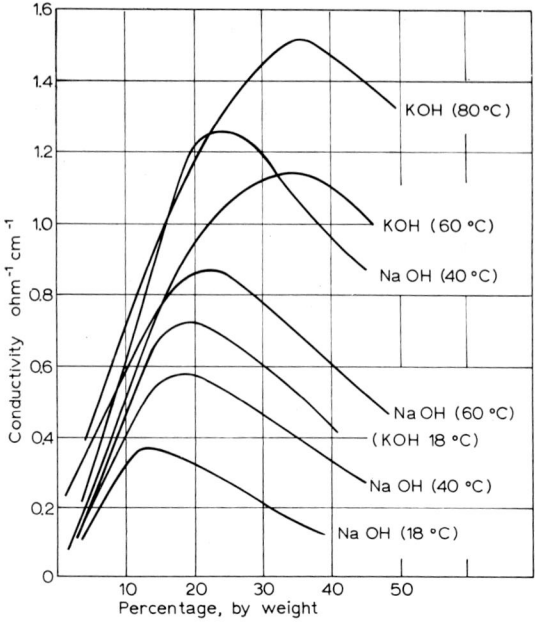

Fig. 2.
Variation of conductivity of electrolyte with concn. and temp.

Of the materials available for this service, woven asbestos-fibre cloth is almost invariably chosen, although fine metal-mesh has also been used. For diaphragms of large area, asbestos cloth is interwoven with fine nickel wire to give increased strength.

2.2 Gas evolution and gas purity

The electrolyte passing between the electrodes of a water electrolyser is rapidly expanded by the gas evolved and leaves the interelectrode gap as a froth. Vertical electrodes allow improved circulation because of gas lift. Corrugated, expanded and perforated electrodes are used to encourage gas release. Gas purities in well-designed cells are always better than 99%, the hydrogen being purer than the oxygen.

2.3 Energy consumption

Current efficiencies in commercial water electrolysers are close to 100% since no side reactions occur. The theoretical energy consumption is the product of the reversible cell voltage and the electrochemical equivalent for the desired gas (1,000 Ah per 0.418 Nm³ of hydrogen and 0.209 Nm³ of oxygen). As has been stated previously, power is used in producing both the products in a gaseous form, some water vapour and, because of ohmic losses, waste heat. Based upon the theoretical cell voltage of 1.23 V, energy efficiencies of 45–65% are observed in practice so that hydrogen is produced at 4–6 kWh Nm⁻³.

3. General features of industrial water electrolysis plant

Figure 3 shows the possible components of an industrial plant for the production of hydrogen and oxygen by the electrolysis of water.

The electrolytic cells, 1, are supplied with DC power from the rectifier, 8, and generate the gases which leave the cell together with entrained or circulating electrolyte. In the gas separators, 2, gross separation of gas and liquid takes place. The electrolyte is returned *via* a filter, 14, and pump, 15, to the cells.

The gases pass through coolers, 3, and purifiers, 4, and hence to low pressure gas holders, 5. If the gas is required at high pressure, compressors, 6, and high pressure storage, 7, are provided. Electrolyte is prepared in tank 12 which also

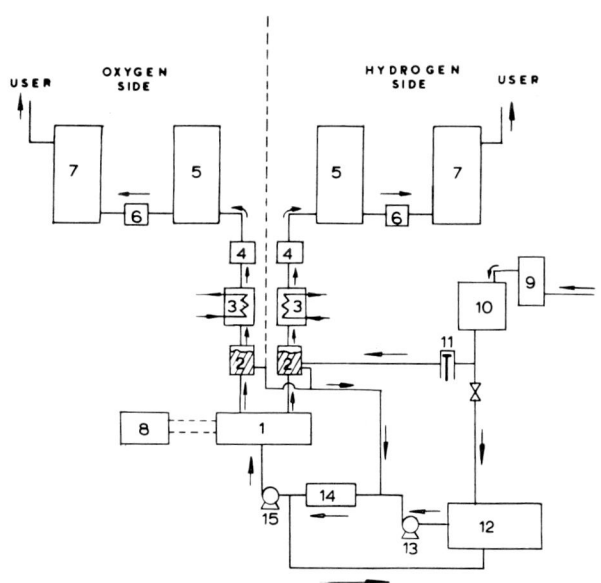

Fig. 3.
Flowsheet of industrial water electrolysis plant. (1) cell pack; (2) gas/electrolyte separators; (3) gas coolers; (4) gas purifiers; (5) low pressure gasholders; (6) compressors; (7) high pressure gas holders; (8) rectifier; (9) water purifier; (10) feed water tank; (11) feed water pump; (12) electrolyte store/preparation tank; (13) electrolyte pump; (14) filter; (15) circulating pump.

serves as a dump tank. Feed water is purified in 10 and pumped, 11, at a controlled rate, to the hydrogen gas drum, 2.

3.1 Feed water preparation

Good quality feed water is used at the rate of 0.8–1 1 Nm^{-3} of hydrogen produced. Any impurities present in this water will accumulate in the electrolyte. In particular, chloride and sulphate ions must be absent since these species promote corrosion in the cell. Depending upon the quality of the raw water and the size of unit, either distilled or deionised water is provided. The rate at which pure water is added to the electrolyte may be controlled in a number of ways depending upon the degree of sophistication desired. Manual pumping to maintain the electrolyte level in a calibrated sight-glass may be used, whilst for fully automatic operation, a piston-type metering pump, the length of whose stroke is continuously adjusted against a level-sensing device in one of the gas drums, for example, may be employed.

All parts contacting the feed water are fabricated from corrosion-resistant steel or inert non-metallic material. The quality of the feed water may be monitored by measurement of its electrical conductivity.

3.2 Electrolyte preparation and circulation

Sodium or potassium hydroxide of good quality is used. For water electrolysis at elevated pressure a higher quality of electrolyte is required. The absence of metal ions, which might be cathodically deposited, and traces of organic matter, which might poison the electrodes, are of particular importance. Although aqueous potassium hydroxide has greater conductivity than aqueous sodium hydroxide of a similar strength, it is more expensive and shows a more rapid rate of attack upon materials of construction.

Water and hydroxide are agitated in the mixing tank and circulated through the cells and filter until a clean and homogenous solution has been obtained. The electrolyte circulation system is fabricated in mild steel or cast iron for units operating at atmospheric pressure and in stainless steels for elevated pressure operation. Seals between component parts are fabricated of bitumen, asbestos-fibre sheet or organic polymers. Flexible metal seals have been used. In pressure electrolysers, halogenated hydrocarbon polymers are employed.

The electrolyte must be protected from prolonged contact with air since it absorbs carbon dioxide. Carbonate thus formed increases electrolyte resistance and eventually forms a solid carbonate phase. In some cells, the electrolyte surface is blanketed with nitrogen gas. Analysis of the electrolyte is simple and is part of the regular maintenance procedure.

3.3 Cell construction

In all cells, the electrodes are placed as close to one another and to the diaphragm as is consistent with the free flow of electrolyte and release of evolved gas. By using perforated, expanded or mesh metal, the diaphragm may be allowed to contact the electrodes, gas escaping through the holes into the electrolyte behind them. Electrodes are of sufficient thickness to retain rigidity and parallelism and should be manufactured from sound, pure metal, since inclusions and impurities may result in local corrosive attack.

For operation at atmospheric pressure, mild steel cathodes are chosen together with anodes of nickel-plated mild steel. A variety of surface treatments is used, ranging from simple mechanical roughening to the deposition of precious metals. These treatments reduce the overvoltage but increase the cost of the plant. In electrolysers operating at elevated pressures, nickel is used for both electrodes.

Diaphragms of asbestos cloth operate satisfactorily for periods in excess of ten years. They should be free to move to a limited extent, since the stresses arising from the movement of electrolyte and gas can result in rupture if they are too taut. Damage may result if they are allowed to dry out, this resulting from the disruptive effect of hydroxide crystallising with the fibre structure.

Metallic materials, other than the electrodes themselves, which are exposed to the electrolyte, are coated with insulating material.

Two basic designs of electrolyser are available. The first is commonly called the Tank Electrolyser and consists of a series of parallel, monopolar electrodes and diaphragms hung in a tank containing the electrolyte. The second is the Filter-press Electrolyser (so called because its design is similar to that of a plate and frame filter-press), in which the electrodes are bipolar and electrolyte is circulated between them.

3.4 Tank electrolysers

A series of electrodes, anode and cathode alternately, is suspended, vertical and parallel to one another, in a tank (Fig. 4). Alternate electrodes are surrounded by a diaphragm and the whole assembly is hung from a series of bells which serve to collect the evolved gases.

All the anodes and all the cathodes in one tank are connected to a pair of common conductors so that one tank operates across a 2–2.5 V DC supply. Costs of conductors rise as the current load rises whilst the cost of rectification equipment of a given output falls as the voltage of the output rises. Voltage must be supplied at a safe level. A series of tanks is therefore connected electrically so as to give the optimum current/voltage relationship consistent with the above considerations.

The tank electrolyser is rugged, simple to assemble and maintain and the component parts are fabricated cheaply. It tends, however, to occupy a large floor area.

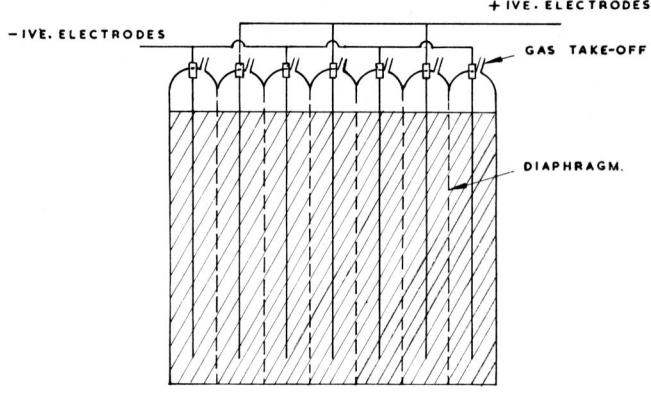

Fig. 4.
General arrangement of tank-electrolyser cell.

3.5 Filter-press electrolysers

A series of electrodes is arranged, vertical and parallel to one another, diaphragms being placed between alternate electrodes. The whole assembly is then clamped up, seals being provided between each electrode and diaphragm. Inlets and outlets are provided for electrolyte to enter and leave each electrode–diaphragm space (Fig. 5). These seals must not only prevent leakage of electrolyte, but also provide electrical insulation. The voltage is applied across the two end electrodes only, being distributed evenly down the cell pack. Each electrode is bipolar, that is, operates as an anode on one side and a cathode on the other.

Filter-press electrolysers are more compact than tank electrolysers of the same capacity. However, the necessity for multiple seals between components demands

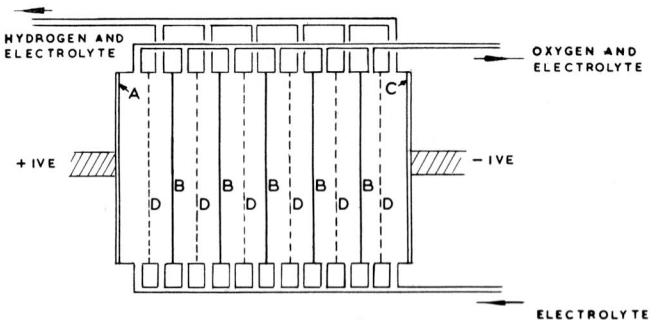

Fig. 5.
General arrangement of filter-press electrolyser cell. (A) anode; (B) bi-polar electrode; (C) cathode; (D) diaphragm.

a higher quality in fabrication and assembly. The provision and hydraulic balance of multiple electrolyte inlets and outlets presents design problems and replacement of faulty parts can also be more difficult. The whole unit therefore tends to be more expensive than a similar tank electrolyser. Units often operate across quite high voltages, this being divided equally about a central point in the cell pack which is at earth potential.

3.6 Pressure electrolysers

Some of the special features of the electrolysis of water at elevated pressure have already been described. Pressure water electrolysers have, to date, been used only to a limited extent and a number of different approaches to their design has been made. These will be described in detail later (Section 6.3).

4. Gas separation, purification and drying

In tank electrolysers the problem of separation of gas from electrolyte is dealt with within the cell unit, the gas rising from the electrolyte surface into the gas bells. In filter-press electrolysers, gas lift or electrolyte circulation carries a gas–liquid mixture out of the cells. Changes in direction and velocity are brought about in the gas separators and are sufficient to give gross removal of the electrolyte which is returned to the cell free of all but dissolved gas. Since current leakage is possible between adjacent electrolyte paths, the gas–electrolyte inlets and outlets have to be of reasonable length and in some electrolysers are fabricated of insulating material such as glass or plastics.

Removal of traces of electrolyte from the gases is achieved by passage through the pure water fed to the cells, which also serves to cool the gases. Demister filters, containing zeolite or asbestos-fibre, may be used. It is often desirable to dry the gas before storage by cooling, passage through sulphuric acid and, finally, through vessels packed with silica gel or alumina. Since the gases are pure, in most cases no further treatment is demanded. Catalytic purifiers may be fitted to remove the last traces of oxygen in hydrogen (or *vice versa*) if so desired. Alternatively, hydrogen may be purified using palladium diffusion membrane devices.

5. Safety and control

The principle safety problem is the prevention of the formation of explosive mixtures of hydrogen with oxygen or air. The avoidance of hot spots or electric sparks where hydrogen gas is present, the avoidance of hydrogen leakage and the problems of handling high voltages and large currents must also be catered for.

Within the cell itself, mixing of gas across the diaphragm is prevented by ensuring that the pressures on each side are equal. This may be achieved by allowing only a certain pressure of gas to be reached, any excess being released through a liquid seal or pressure relief valve. A more sophisticated system, used in pressure electrolysers, is a differential pressure transmitter, linked to pressure release valves in one or both of the gas vessels. Pressure build-up can occur due to blockage of electrolyte and gas lines in filter-press and pressure electrolysers. The provision and maintenance of electrolyte filters is thus essential. In pressure electrolysers, continuous automatic analysis of gas purity is often demanded.

Hot spots and electrical sparks may occur if there is electrolyte leakage and crystalline deposits of hydroxide form. Sound seals are therefore an essential safety feature. Electrical equipment should be isolated from the cells themselves, rectifiers and control gear often being placed in a separate room. Motors, driving pumps, etc., should be of flame-proof design, and the DC supply should be provided with devices to prevent overload and reversal of polarity. Electrolytic cells are mounted on insulated supports.

Electrolyte level may be controlled in the cell or the gas separators by float valves or level sensors which control the rate of supply of feed water. Electrolyte temperature is to some extent self controlling, heat losses occurring to the surroundings and in the production of gases with a variable content of water vapour. Cooling coils are fitted in the tank electrolyser and in the gas separator drums of the filter-press units.

The output of the electrolyser may be controlled by stepwise or continuous variation of current, whilst output gas pressure may be controlled in the gas holder.

6. Commercial water electrolysis plant

A number of commercial units is described in this section.

6.1 Tank electrolysers

6.1.1 The Knowles cell (The International Electrolytic Plant Co. Ltd.)

A partial section of this cell is shown in Fig. 6 and a photograph of an actual installation in Fig. 7. The block of gas collecting bells, DD, is supported on the edge of the tank, A, containing the electrolyte. The internal parts of the cell may be removed in one piece after disconnecting the gas off-take pipes, MM, and the current conductors, K. The electrodes, BB, alternately positive and negative, are

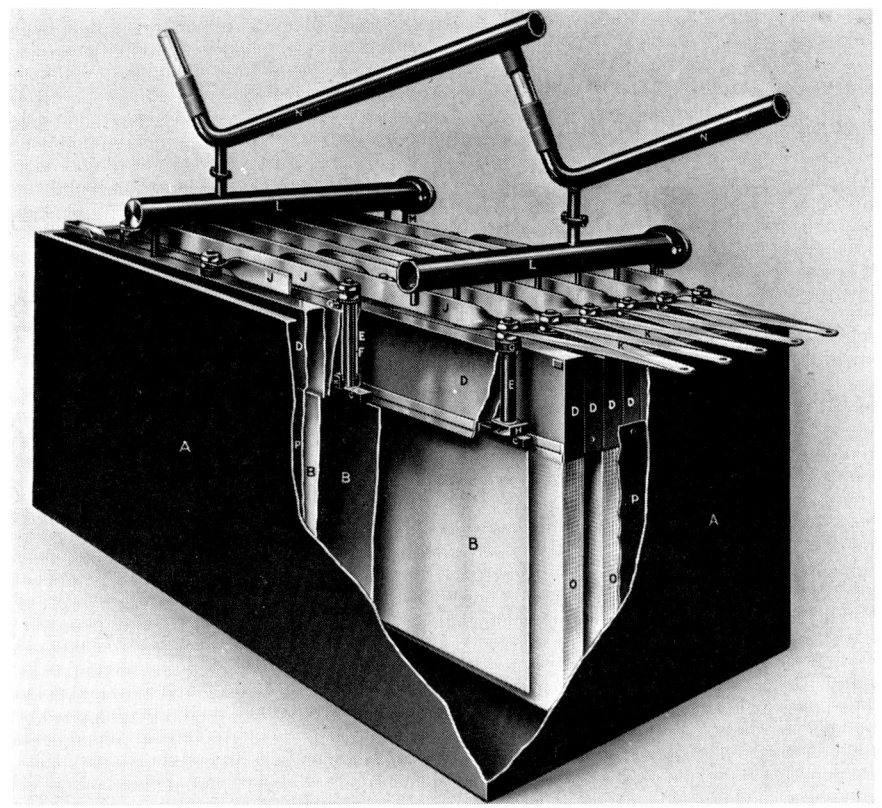

Fig. 6.
Partial section of Knowles cell.

Fig. 7.
Knowles cell installation.

suspended from the bells by electrode leads which pass through steel tubes, EE, insulated by sleeves, FF. The gases are collected through pipes, LL, one for each gas, and these are connected to S-shaped tubes, NN, which run along a row of tanks. Finely divided electrolyte spray is trapped in NN and returned to the cell *via* pipes MM. Electrodes are separated by diaphragms, OO, which are open at their lower edges. Tubes, inside the tank, carry cooling or heating water to maintain the electrolyte at constant temperature.

Special electrodes, consisting of series of parallel plates, can be used to reduce interelectrode spacing and hence power consumption. Operating conditions for a unit to produce 2.06 Nm^3 hydrogen h^{-1} are as follows.

Current	4,500 A
Voltage	1.9 V
Electrolyte	28% w/v aqueous KOH
Temperature	80°C
Gas purities	Hydrogen 99.75% v/v
	Oxygen 99.5 % v/v
Power consumption	4.15 kWh Nm^{-3} hydrogen

6.1.2 The Stuart cell (Efco Royce Furnaces Ltd.)

The Stuart cell is a compact form of tank electrolyser which makes use of electrodes comprising assemblies of vertical metal strips, closely spaced. In this way, an available electrode area, 30 times that of the superficial area is produced and although low current densities are used, high currents can be carried by each electrode. The electrodes are placed close against the diaphragm, supporting it and thereby increasing its life.

Operating conditions for a Stuart cell producing 2.4 Nm^3 hydrogen h^{-1} are as follows.

Current	5,250 A
Voltage	2.04 V
Electrolyte	28% w/v aqueous KOH
Temperature	85°C
Gas purities	Hydrogen 99.9% v/v
	Oxygen 99.7 % v/v
Power consumption	4.9 kWh Nm^{-3} hydrogen

6.2 Filter-press electrolysers

Most of these electrolysers are similar in concept, minor variations in design having been made over the years.

6.2.1 The CJB electrolyser (Constructors John Brown Ltd.)

In the CJB electrolyser, a number of cells is compressed together to form a battery, the voltage being applied across the end electrodes. Each cell of this battery (Fig. 8) consists of a frame to which is attached an asbestos cloth diaphragm. On each side of the diaphragm is placed an electrode, with perforated pre-electrodes attached, one to each side. Electrodes and frames are sealed to, and insulated from, one another by asbestos fibre gaskets. The electrodes on the hydrogen side are of mild steel whilst those on the oxygen side are heavily nickel-plated. Cell frames are of steel in the larger units, the parts exposed to the electrolyte being coated with a special cement.

The gases evolved pass through the perforations in the pre-electrodes and are carried away in the circulating electrolyte, through insulated gas off-take pipes, into channels connecting all the hydrogen and all the oxygen streams. Gas domes, one for each gas, are placed above the battery and these, in turn, are connected to gas separation and cooling drums. The electrolyte passes through a filter and back to the battery. The evolved gases are bubbled through cooled distilled-water puri-

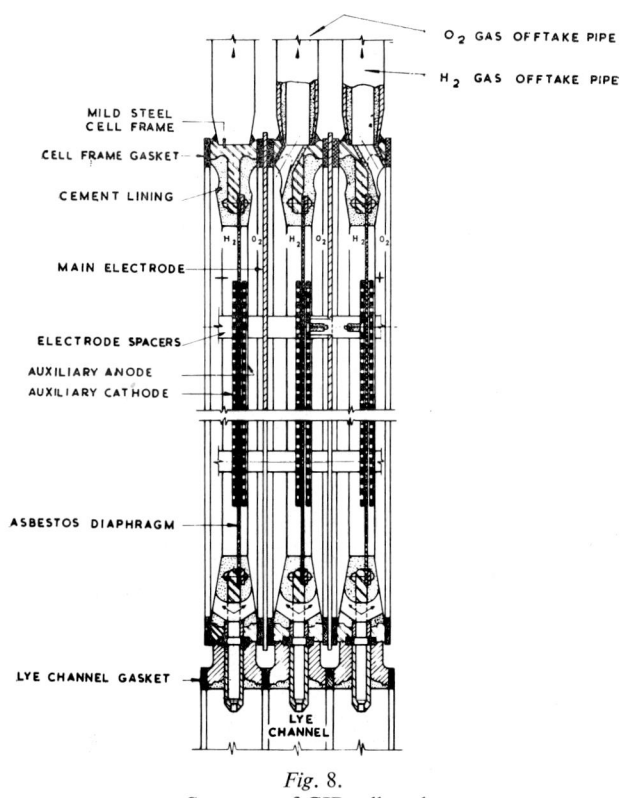

Fig. 8.
Structure of CJB cell pack.

fiers. The distilled-water feed to the battery is maintained at a constant level in these purifiers which also contain a device for equalisation of pressures. The battery of cells is clamped together with four tie-bars across heavy-section end plates and the whole battery is mounted on insulators (Fig. 9).

Operating conditions for a unit to produce 240 Nm³ hydrogen h^{-1} are as follows.

Current	6,700 A
Voltage	171 V
Electrolyte	28% w/v aqueous KOH
Temperature	80°C
Gas purities	Hydrogen 99.9% v/v
	Oxygen 99.7% v/v
Power consumption	4.75 kWh Nm^{-3} hydrogen

Fig. 9.
CJB electrolyser.

6.2.2 The Demag electrolyser (Demag Elektrometallurgie GMBH)

The design of this unit is similar to that of other filter-press electrolysers, but care has been taken to ensure uniform electrolyte circulation within the cells (Fig. 10).

Fig. 10.
Demag electrolyser.

Batteries are supplied as "blocks" of cells, a number of which may be assembled to give a range of outputs. Operating conditions for a unit to produce 150 Nm^2 hydrogen h^{-1} are as follows.

Current	7,600 A
Voltage	87–90 V
Electrolyte	28–30% w/v aqueous KOH
Temperature	80°C
Gas purities	Hydrogen 99.9% v/v
	Oxygen 99.7% v/v
Power consumption	4.3–4.5 kWh Nm^{-3} hydrogen

6.2.3 The Oerlikon electrolyser (Maschinenfabrik Oerlikon)

A somewhat different design is used in this unit (Fig. 11). Electrolyte is pumped continuously through the cells and filter and then overflows, with entrained gas, through pipes with transparent sections into separator vessels placed one on each

Fig. 11.
Oerlikon electrolyser.

side of the battery top. The gases pass to two scrubber-cooler units, in which they are washed with feed water which is kept cool in a separate heat exchanger.

Activated cathodes are available which give lower cell voltages. Operating conditions for a unit producing 210 Nm³ hydrogen h⁻¹ are as follows.

Current	6,600 A
Voltage	126 V
Electrolyte	25% w/v aqueous KOH
Temperature	75°C
Gas purities	Hydrogen 99.8% v/v
	Oxygen 99.6% v/v
Power consumption	4.3–4.4 kWh Nm⁻³ hydrogen

6.2.4 The Pintsch Bamag electrolyser (Pintsch Bamag A.G.)

This unit (Fig. 12) relies upon gas-lift for electrolyte circulation, methods similar to those used in other filter-press electrolysers being employed for gas separation, purification and cooling. New forms and activation of electrode surface reduce power consumption and operating conditions for a unit to produce 100 Nm³ hydrogen h^{-1} are as follows.

Fig. 12.
Pintsch Bamag electrolyser.

Current	6,600 A
Voltage	70 V
Electrolyte	25% w/v aqueous KOH
Temperature	78–80°C
Gas purities	Hydrogen 99.9% v/v
	Oxygen 99.8% v/v
Power consumption	4.5 kWh Nm^{-3} hydrogen

6.2.5 The Moritz oxyhydrolyser (Moritz Chemical Engineering Co. Ltd.)

In this unit the evolved gases are cooled and purified in two vertical columns at the end of the battery (Fig. 13). The base of each column contains a cooling system, whilst feed water in the top section of each column scrubs the gases free of entrained electrolyte. Level is maintained constant by a float-valve in the oxygen column.

The cell frames are electrically connected to the electrodes, the seals being formed by the diaphragm, the edges of which are impregnated with an elastomer. Very close electrode spacing is preserved and although a minor penalty is thereby paid in terms of gas purity, a low cell voltage results. Typical operating conditions for a unit producing 40 Nm^3 hydrogen h^{-1} are as follows.

Current	800 A
Voltage	220 V
Electrolyte	25% w/v aqueous KOH
Temperature	60°C
Gas purities	Hydrogen 99.2–99.5% v/v
	Oxygen 98.0–98.5% v/v
Catalytic purifiers are available which can increase these purities to 99.95%	
Power consumption	4.4 kWh Nm^{-3} hydrogen

6.2.6 The De Nora electrolyser (Oronzio De Nora)

This filter-press unit, similar in most respects to other units, uses double diaphragms so as to ensure high gas purities, the space between being filled with electrolyte and vented (Fig. 14). De Nora electrolyser systems have been especially designed for the associated production of deuterium. Operating conditions for a unit producing 18 Nm^3 hydrogen h^{-1} are as follows.

Current	2,500 A
Voltage	34 V
Electrolyte	29% w/v aqueous KOH

INDUSTRIAL WATER ELECTROLYSIS

Fig. 13.
Moritz oxyhydrolyser.

Temperature   75°C
Gas purities   Hydrogen 99.8% v/v
                   Oxygen   99.5% v/v
Power consumption 4.6 kWh Nm^{-3} hydrogen

Fig. 14.
De Nora electrolyser.

Fig. 15.
Zdansky–Lonza electrolyser.

6.3 Pressure electrolysers

6.3.1 The Zdansky–Lonza electrolyser (Lurgi GMBH)

This design of industrial electrolyser (Fig. 15), the only large pressure water electrolyser at present available, was developed as the result of a prolonged study of ways of improving the economics of water electrolysis at atmospheric pressure, and was designed by Mr. Zdansky who had previously conceived the Bamag filter-press electrolyser[4,5].

It was decided that the optimum operating pressure was 30 atm. and that the filter-press design could be modified to give closer electrode spacing which, together with the reduction of gas volume in the cell at elevated pressure, would result in appreciable power savings. Gaskets between adjacent cells were designed so as both to seal and insulate with a large pressure difference across them.

Figure 16 shows the construction of individual cells and their assembly into a cell pack. Each cell consists of two embossed, nickel-plated steel plates, 1, which act as the cell walls and carry nickel-plated and activated steel wire-mesh gauze electrodes, 3. The steel plates are sealed into ring-shaped frames, 2. Anode and

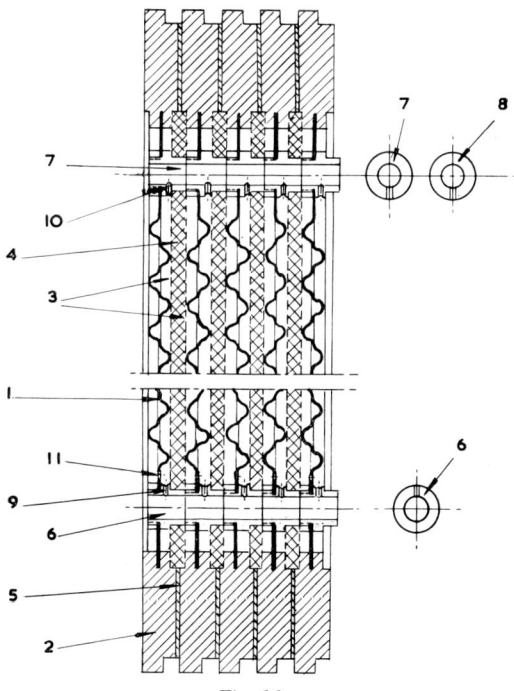

Fig. 16.
Section through Zdansky-Lonza cell.

cathode compartments are separated by a laminated plate of pure asbestos, 4, and individual cell frames are isolated, one from another, by sealing rings covered with PTFE, 5.

Electrolyte enters the cell pack through channel 6, which is composed of individual units each the width of one cell, bolted to the partition walls, 1. Transverse bore holes, 9, allow the electrolyte to enter the individual cathode compartments and thence, through holes, 11, into the anode compartments. Electrolyte–gas mixtures leave the cells through bore-holes, 10, into two collecting channels, 7 and 8, one for hydrogen and the other for oxygen.

Individual cells are very thin so that a large number may be assembled in a small volume. The general arrangement of the electrolyser is shown in the flowsheet in Fig. 17. The cell pack is clamped up between heavy end plates, 12, and produces gas/electrolyte mixtures which pass, 13, into the separators, 14, where the gas rises away from the electrolyte. Both separators are water-cooled, 15, and the gas leaves them through demisters, 16, and a control valve, 17. A further indirect cooler, 18, is used for each gas together with a water trap, 19.

Feed water, 22, of demineralised quality, enters the oxygen separator. The electrolyte streams from the separators are recycled through pump 20 and filter 21, to the cell pack. Operating data for a unit to produce 145 Nm^3 h^{-1} of hydrogen are as follows.

Current	3,000 A
Voltage	217 V
Electrolyte	25% w/v aqueous KOH
Temperature	90°C
Gas purities	Hydrogen 99.8–99.9% v/v
	Oxygen 99.3–99.5% v/v
Power consumption	4.5 kWh Nm^{-3} hydrogen

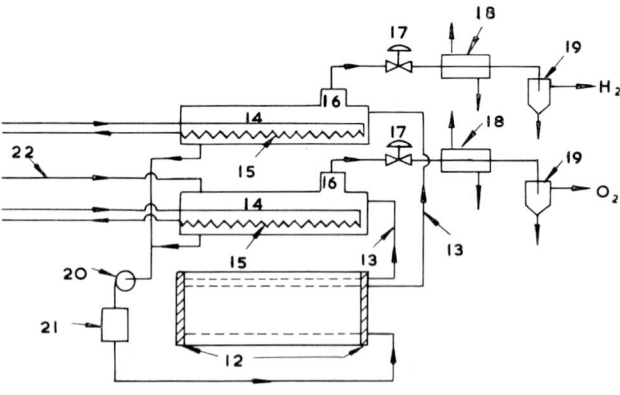

Fig. 17.
Simplified flowsheet of Zdansky–Lonza electrolyser.

Fig. 18.
CJB pressure electrolyser.

6.3.2 The CJB electrolyser (Constructors John Brown Ltd.)

This pressure electrolyser, originally developed for use in submarines but now modified for other special applications, employs a different method for the containment of pressure. The unit consists of a filter-press type of cell pack, contained in a steel pressure vessel, the space between pressure vessel and cell pack being filled with demineralised feed water which acts as an additional insulator.

Figure 18 shows one such unit which operates at pressures of up to 30 atm. Similar units have been designed to operate at pressures of up to 200 atm.

In Fig. 19 are shown the components of the cell pack. Cell frames consist of

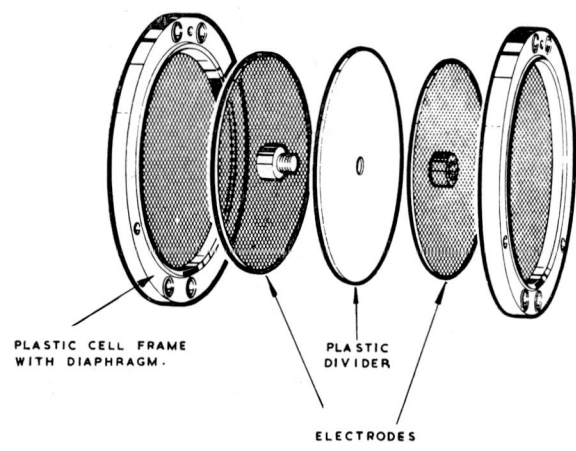

PLASTIC CELL FRAME
WITH DIAPHRAGM.

PLASTIC
DIVIDER

ELECTRODES

Fig. 19.
Cell assembly detail.

plastic mouldings with asbestos cloth diaphragm inserts, alternating with sheet plastic dividers. Electrodes of perforated nickel sheet are assembled, one on each side of the divider. The plastic cell frames have pairs of holes at top and bottom which when the cell pack is assembled form the electrolyte entry and exit channels, transverse holes being drilled from them into anode and cathode compartments.

The cell pack is assembled and clamped down on the end plate of the earthed pressure vessel, an insulated conductor carrying current to the other end of the cell pack and passing out of the pressure vessel through an insulating seal in the end plate.

Figure 20, shows the general arrangement and the basis of the control system. Electrolyte is pumped into the base of the cell pack where it divides and passes to anode and cathode compartments. The electrolyte–gas mixtures leaving the cell

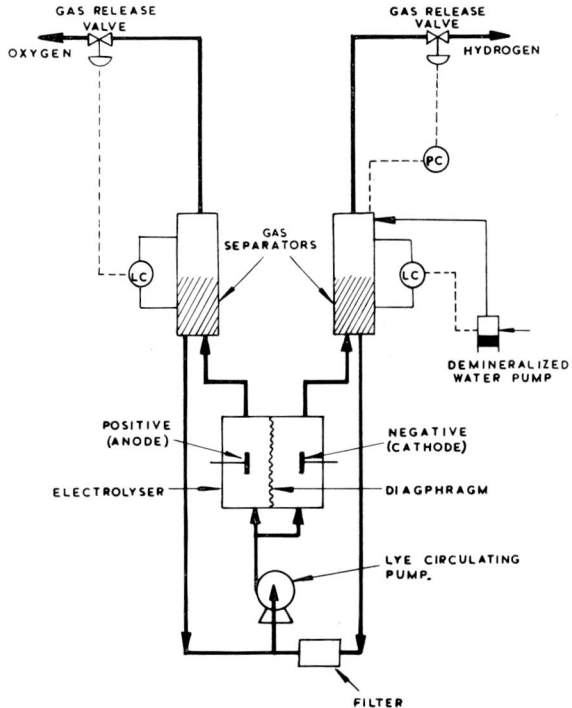

Fig. 20.
Simplified flowsheet of CJB pressure electrolyser.

pack pass to two separator vessels, the electrolyte returning to the cell pack through a pump and filter. The overall pressure in the system is controlled by a pressure regulating valve on the hydrogen outlet, whilst oxygen release is controlled by a level-sensing device in the oxygen separator. Feed water is supplied at a rate controlled by a level sensor in the hydrogen separator. Additional features include fully automatic stop and start, safety and fail safe devices.

This unit has interest as a means of effecting a variety of electrosyntheses at elevated pressure[6]. Operating conditions for a unit producing 2 Nm^3 hydrogen h^{-1} are as follows.

Current	125 A
Voltage	70 V
Electrolyte	30% w/v aqueous KOH
Temperature	65°C
Gas purities	Hydrogen 99.9% v/v
	Oxygen 99.5% v/v
Power consumption	4.4 kWh Nm^{-3} hydrogen

6.3.3 The Treadwell oxygen generator (Treadwell Corporation)

This unit was developed to produce oxygen for life-support on submarines[5]. Such service demands operation under conditions of shock and eccentric motion and the unit is contained in a severely restricted space. The minimum quantity of hydrogen must be present at any one moment and this gas must be discharged overboard whilst the boat is submerged at depth. Pressures of up to 200 atm. are achieved and the unit is fully automatic and safe.

The Treadwell design is based upon a number of individual monopolar cells, each in its own pressure vessel, coupled up in electrical series. Figure 21 shows the

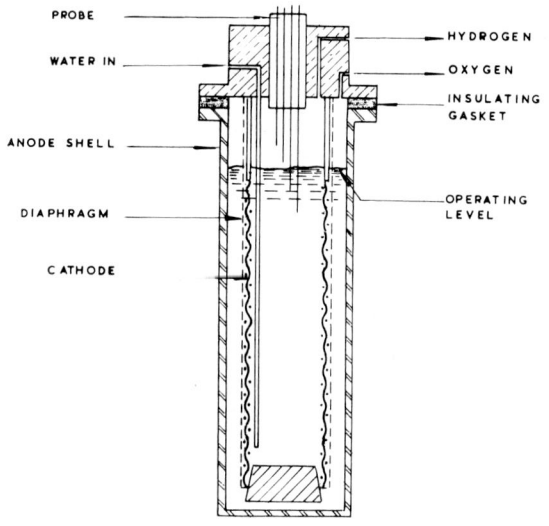

Fig. 21.
High pressure hydrogen/oxygen cell.

design of one such cell. The cylindrical nickel pressure vessel is the anode of the cell, the cathode consisting of a nickel-plated copper wire mesh cylinder, wrapped externally with asbestos cloth and placed centrally in the pressure vessel. Waste heat is removed by blowing cold air over the surface of the pressure vessels, or by placing water jackets around the cell.

The flow sheet (Fig. 22) shows the oxygen leaving the cells through a condensate trap and control valve which is actuated from a pressure transmitter; the hydrogen flowing from the cells leaves through a similar system. The difference in pressure between these gases is controlled so as to prevent mixing. Water is fed to the cells at a rate controlled by a level-sensing system in each cell.

The safety system necessary involves the superposition of three tiers of controls operating from level, pressure, temperature and purity sensors, together with fail

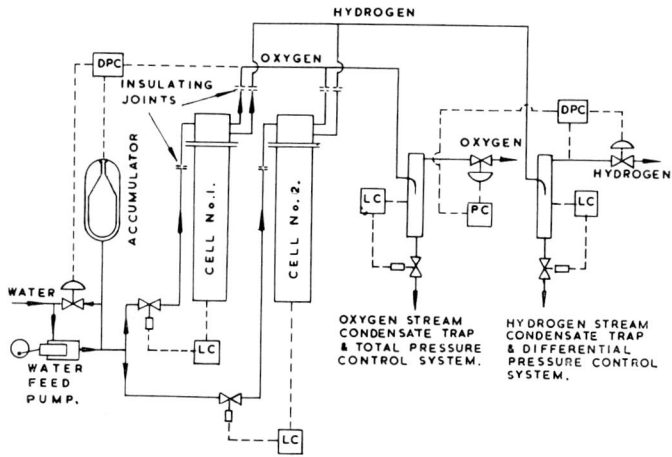

Fig. 22.
Flowsheet. High pressure electrolytic oxygen generator.

safe devices. The actuation of any safety device results in automatic alarm, shutdown and nitrogen blanketing.

Operating conditions for a cell producing 0.31 Nm³ hydrogen h⁻¹ are as follows.

Current	750 A
Voltage	3.0 V
Electrolyte	30% w/v aqueous KOH
Temperature	90°C
Gas purities	Hydrogen 99.4–99.8% v/v
	Oxygen 98.8–99.2% v/v
Power consumption	7.2 kWh Nm⁻³ hydrogen

6.4 MISCELLANEOUS AND FUTURE DEVELOPMENTS

6.4.1 *The Gaspack H. hydrogen generator (Metals Research Ltd.)*

Small quantities of pure hydrogen gas (up to 9 l/h⁻¹) for laboratories and on-line process control gas chromatographs are produced by a small unit now available. The hydrogen which is purified subsequent to electrolysis by passage through the Pd alloy cathode of the cell, has an impurity level of 1 part in 10⁹.

6.4.2 *Future developments*

The expense of further development of large-scale units may not be justified in the light of the present economics of tonnage hydrogen and oxygen production.

Concentration upon special applications in the aero-space, submarine and meteorological fields has led to the development of improved cell materials and control technology, which await large-scale exploitation.

As an example, one cell is based upon porous anodes and cathodes clamped up against the faces of a porous matrix through which the electrolyte permeates. The evolved gases emerge at the backs of the electrodes[8,9]. Modifications of such a cell may result in units producing only one gas or producing the gases at different pressures and with reduced power consumptions.

Other devices[10,11] have been described which combine catalytic and electrolytic components and which potentially solve in one step the problems of atmospheric pollution and control in closed environments. Advances of this type have resulted from fuel cell developments.

7. Uses of electrolytic hydrogen and oxygen

Unless cheap electric power is available, large scale production of hydrogen by the reforming of natural or refinery gases and of oxygen by air liquefaction, provide cheaper products. Water electrolysis has certain advantages, however, which often make it preferable on the medium and small scale. These include the intrinsic purity of the gases produced especially the absence of sulphur and organics, flexibility of output, the suitability of the electrolytic method for fully automatic operation, rapid start-up and shut-down, reliability and small labour requirement. Water electrolysers can be operated either as a steady electrical base load or intermittently, using cheap off-peak power.

It is almost always the hydrogen which is required and the main uses are as follows.
   (a) Synthesis gas for the production of ammonia, hydrogen chloride, sorbitol, etc.
   (b) Protective or reducing atmosphere in the non-ferrous metal industry, the semi-conductor industry and in the manufacture of electric lamps.
   (c) Catalytic hydrogenation in edible oils production.
   (d) Manufacture of hard fats and fatty acids.
   (e) Cooling of turbo-alternators in power stations.
   (f) Filling meteorological balloons.
   (g) Metal welding and cutting.
   (h) Synthesis of alcohol, acetic acid, etc.
   (i) Combustion gas in high temperature processes, *e.g.* manufacture of synthetic gem stones.
   (j) Heat treatment of stainless steel components to give ultra clean surfaces.

The oxygen is usually bottled and sold as a by-product.

7.1 Economics of hydrogen manufacture

Electrolysis as a source of hydrogen is a route which rapidly loses its economic advantage as the size of plant required increases. Any such statement depends on the cost of electrical energy. Nonetheless, even in Norway where power costs of 3 mills ($ 0,003)/kWh are often quoted, there is a planned switch away from electrolytic hydrogen. There are several papers in the literature on this aspect, and although somewhat out of date, insertion of up-to-date costs can give an accurate picture. Von Pichler[12] or Karwat[13] and Twist and Sagar[14] are all relevant on general costs of hydrogen manufacture. The latter reference contains an appendix with details of costs of Zdansky Lonza plants. In addition, Johnson et al.[15], Charlesworth and Schmidt[16], Chemical Week[17], Truman Lee[18] and Bowen[19, 20] give a broadly based view on costs.

Among the various types of electrolysers themselves, the main economic differences lie between the pressure electrolysers and the tank electrolyser plants. The latter tend to be less costly although in situations where the gas is required at an appreciable pressure, one can offset the cost of compressors, which are not usually necessary with pressure electrolysers. These too, are often more efficient in their use of floor space, although the Knowles "Column Cell" marketed by International Electrolytic Plant Co., succeeds in reducing this disadvantage. Other factors which act in favour of the tank electrolyser are simplicity of operation, ease of maintenance and lower down-time factors, since maintenance can be achieved with only a small proportion of the plant capacity out of action. The capital cost of a tank electrolyser, such as a Knowles cell, varies from £500/m³/h hydrogen for smaller plants (5 m³/h), down to £200/m³/h for a plant producing 200 m³/h[21]. The above costs exclude electrical equipment, gas storage and compression plant.

8. Literature of the water electrolysis process

For a historical survey, the reader is referred to Engelhardt[22] or Billiter[23]. The changes which have occurred in the intervening years, are not large. The following references are classified according to their interest.

1. *Cells designed to operate in zero-gravity conditions*, mostly with ion-exchange membranes or other devices such as porous electrodes or Pd–Ag electrodes. See refs. 24, 25, 26, 27.

2. *Pressure electrolysers:* Scharlan[29] (application to the fat industry), Lapin[30] (a review article), Zdansky[31], Zdansky–Lonza[28, 32, 33], Ibarz and Diez (a laboratory study of a pressure cell)[2] and ref. 34.

3. *Construction of electrodes:* Friese et al.[35] (porous electrodes for pressure elec-

trolysers), Pintsch Bamag[36] (activation of electrodes) and gauze electrodes[37], manufacture of Raney type electrodes[38], Houwalt and others on carbon powder electrodes[39], electroformed electrodes[40], use of precious metals to increase activity[41], electrodes with holes[42], bipolar electrodes of sintered porous nickel[43].

4. *Cell construction:* Shirikin[44] (design of a Russian cell), Asahi[45], Zdansky-Lonza[46] (battery of cells), Pintsch Bamag[47] (bipolar cell design), Demag[48] (use of rubber diaphragms), a bipolar filter-press electrolyser from the U.S.S.R.[49], a cooling unit[50] from Demag.

5. *Varia:* Mrochek and Kibbey[51] (electrolysis of water at 200°C), Koshechkin and Ponomarev[52] (thermodynamic and experimental data on pressure electrolysis), White, Schuldiner *et al.*[53] (a new electrolytic route to hydrogen and oxygen involving the cycle: $Cd(OH)_2 \rightarrow Cd + H_2O + \frac{1}{2}O_2$ with $Cd + 2H_2O \rightarrow Cd(OH)_2 + H_2$), De Nora[54] (the use of higher alkali concentrations), use of unsmooth DC[55], prevention of foaming[56] and fog removal[57], use of wetting agents[58], a study of the effect of bubbles on cell operation[59-61] and a review of water electrolysis cells[62].

6. Most cell manufacturers publish at least some literature on the construction and applications of their products. Among these, few are more detailed than those of Demag[63]. Among their "Information" series (published in several languages), DE 54 describes locations and users of plants sold by them, and is kept up to date; DE 55 describes applications in the oil and fat industry; DE 57 covers use as a protective gas in the metal and electrical industries, while other, unnumbered leaflets cover similar topics.

Acknowledgement

The author wishes to thank the Companies supplying the commercial units described in this Chapter, who kindly provided design and operating data, photographs and diagrams, and Constructors John Brown Ltd., for permission to prepare and publish this chapter.

9. References

1 D. M. NEWITT and H. K. SEN, *Trans. Inst. Chem. Engrs. London*, 10 (1932) 22.
2 J. IBARZ and A. DIEZ, C.I.T.C.E. 8th Meeting, Madrid, 1956, pp.145–157.
3 C. E. BOWEN, *J. Inst. Elec. Engrs. London*, 90 (1943) 474.
4 Brit. Pat., 679,334 (30.11.1950).
5 Brit. pat., 681,695 (30.7.1951).
6 J. T. ANDERSON, *Chemical Processing*, 12 (1966) S13.
7 R. SPITZER, Electrochem. Soc. Philadelphia Meeting, May, 1959.
8 Brit. pat., 1,139,614 (15.2.1961).
9 Brit. pat., 1,139,615 (17.2.67).
10 U.S.Govt. Rept., AD 648.505; AD 664.361; N68 34.044. 33–39.

11 D. C. POPMA and V. G. COLLINS, *Chem. Eng. Prog. Symp. Ser.*, 62 (1966) 24.
12 A. VON PICHLER, *Chem. Ing.-Tech.*, 33 (1961) 95.
13 E. KARWAT, *The Chem. Engineer*, 193 (1965) 294.
14 D. R. TWIST and K. J. SAGAR, *The Chem. Engineer*, 193 (1965) 252.
15 E. S. JOHANSON, M. MACKLES and S. C. SCHUMAN, *Hydrocarbon Process Petrol. Refiner*, 43 (1964) 119.
16 P. L. CHARLESWORTH and G. SCHMIDT, *The Chem. Engineer*, 193 (1965) 259.
17 Anon., *Chemical Week*, 97 (26) (1965) 18.
18 G. TRUMAN LEE et al., *Petrol. Refiner*, 42 (1963) 125.
19 C. E. BOWEN, *J. Inst. Elec. Engrs. London*, 90 (1943) 35.
20 C. E. BOWEN, *Chemical Processing* Feb. 1966 (suppt. *New Chemicals*).
21 F. J. KNOWLES, Private communication.
22 V. ENGELHARDT, *Handbuch der Technischen Elektrochemie*, Akad. Verlagsgesellschaft, Leipzig, 1933, Vol. II, Pt. I, pp 1–196.
23 J. BILLITER, *Die Technische Elektrolyse der Nichtmetalle*, Springer Verlag, Berlin, 1954.
24 J. E. CLIFFORD, E. S. KOLIC and C. L. FAUST, *Chem. Eng. Progr. Symp. Ser.*, 62 (1966) 43.
25 Anon., *Chem. Eng. Progr. Symp. Ser.*, 62 (1966) 24.
26 J. E. CLIFFORD, E. S. KOLIC and C. L. FAUST, *Sci. Tech. Aerospace Rept.*, 2 (1964) 2616.
27 J. E. CLIFFORD, E. S. KOLIC and C. L. FAUST, *ibid*, 1 (1963) 42.
28 U.S. pat., 2,871,179; Swiss pat., 334,616 (to Zdansky Lonza); 334,614; 330,814.
29 A. SCHARLAN, *Fette, Seifen Anstrichsmittel*, 62 (1960) 185.
30 YA. S. LAPIN, *Khim. Prom.*, (8) (1963) 600.
31 A. E. ZDANSKY, *Dechema Monograph.*, 33 (1959) 92.
32 Ger. pat., 1,007,746 (to Zdansky Lonza).
33 Ger. pat., 935,727 (to Zdansky Lonza).
34 Ger. pat., 877,750.
35 K. H. FRIESE, E. JUSTI and H. WINSEL, Ger. pat., 1,200,903; 1,167,406.
36 Brit. pat., 989,003 (to Pintsch Bamag).
37 Ger. pat., 1,198,332 (to Pintsch Bamag).
38 Ger. pat., 1,183,892; 1,065,821. Jap. pat., 6611 (1956).
39 W. TOMASSI, *Przemysl Chem.*, 39 (1960) 160; W. TOMASSI and W. ZIELENIEWSKI, *Przemysl Chem.*, 39 (1960) 421; A. HOUWALT, *Przemysl Chem.*, 41 (1962) 367, 545; 42 (1963) 285.
40 Ger. pat., 941,843.
41 Swiss pat., 331,841 (to Zdansky Lonza).
42 Ger. pat., 1,014,529 (to Demag).
43 Ger. pat., 1,094,247.
44 N. M. SHIRIKIN, *Energetik*, 6 (1958) 18.
45 Jap. pat., 309,310 (1959) (to Asahi).
46 U.S. pat., 2,881,123 (to Zdansky Lonza).
47 U.S. pat., 2,862,864 (to Pintsch Bamag).
48 Ger. pat., 1,100,604 (to Demag).
49 U.S.S.R. pat., 137,896 (1960).
50 Ger. pat., 1,112,050 (to Demag).
51 J. E. MROCHEK and A. H. KIBBEY, *Nucl. Sci. Abstr.*, 21 (1967) 5845.
52 V. V. KOSCHECHKIN and V. D. PONOMAREV, *Ref. Zh. Kim.*, 1966 (8) Pt II Abstr 8L 246.
53 U.S. pat., 3,216,919; 2,951,802; 2,984,607.
54 Fr. pat., 1,379,177 (to Oronzio de Nora).
55 S. TAJIMA and N. BABA, *Denki Kagaku*, 28 (1960) 110.
56 Jap. pat., 15,682 (1960).
57 Jap. pat., 8927 (1960).
58 Ger. pat., 1,022,194 (to Demag).
59 T. MATSUNO, *Bull. Fac. Eng. Yokohama Natl. Univ.*, 8 (1959) 155.
60 A. L. ROTINYAN and V. M. ALOITS, *Zh. Prikl. Khim.*, 30 (1957) 1781.
61 T. MURAKAWA, *Denki Kagaku*, 25 (1957) 280.
62 K. KANZAKI, *Noguchi Kenkyusho Jiho*, 4 (1955) 29.
63 Available from: Demag Elektrometallurgie, Duisburg.

Chapter 5

Electrolytic Heavy Water Manufacture

C. JACKSON

Imperial Chemical Industries Ltd., Mond Division, The Heath, Runcorn, Cheshire

1. Introduction

When water is electrolysed, deuterium separates more slowly at the cathode than does the lighter hydrogen and consequently the electrolyte becomes enriched in deuterium. The separation of the isotopes in the electrolytic cell can be defined by the equation:

$$a = \frac{(HDO)}{(H_2O)} \frac{(H_2)}{(HD)}$$

where $\frac{(HDO)}{(H_2O)}$ is the ratio of the number of moles in the electrolyte, $\frac{(H_2)}{(HD)}$ is the ratio of the number of moles in the gas and a is known as the Separation Factor.

Since the equilibrium constant for the reaction:

$$H_2O(l) + HD(g) \rightleftharpoons HDO(l) + H_2(g)$$

is 3.88 at 25°C and 2.73 at 100°C[1] it is apparent that the separation factors achieved in practice (from three to eight) must be due to some mechanism other than establishment of equilibrium in this reaction at the cathode surface. One explanation is that the hydrogen ion is discharged more readily at the cathode than the deuterium ion, thus accounting for the tendency of hydrogen to concentrate in the gas. This explanation is given support by laboratory findings[2] that separation factors up to 7.5 may be obtained under conditions of irreversible electrolysis with a high current density or with electrode materials that do not catalyse the reaction. At low current densities or with electrodes which catalyse the reaction, separation factors in electrolysis equal to the equilibrium constant are observed.

The separation factor is influenced by several factors including temperature, current density, electrode material, cleanliness of electrode surface, and vaporisation, entrainment or foaming of the electrolyte[3,4]. Reduction of the electrolyte temperature increases the separation factor as does an increase in the current density. The latter however, is usually avoided because of the increased power consumption in the cell due to the increased cell voltage.

The closer a process approaches reversibility, the more likely it is to be commercially successful since this implies high thermodynamic efficiency[5]. The electrolytic separation of heavy water is highly irreversible in spite of the high separation factor, and requires about 57 kcal/mole. The reversibility can be increased by making the process a multistage cascade process with the size of each stage approximately inversely proportional to the concentration of the isotope. However, to produce tonnage quantities of heavy water from natural water, a complex and therefore expensive plant is required to handle the quantities of raw material needed, since natural water contains only about 150 ppm heavy water. Many types of cascade system have been dealt with theoretically and presented in refs. 6–11.

Where cheap electrical power is available the irreversibility is not a constraint and electrolytic plants have been built and operated. In electrolysis more than 90% of the energy expended occurs in reaching a concentration of about 1%. Hence, since the power required is much reduced after this concentration is reached, the concentration of water pre-enriched by another process can economically be increased by electrolysis. That is to say electrolytic concentration is feasible when another process absorbs the cost, the only limitation being that the heavy water capacity may be governed by the capacity of the other process.

Since 1943 electrolysis has been employed solely as a finishing stage, or under conditions of cheap power, as a pre-enrichment stage, the bulk of the concentration being carried out by much cheaper processes. Comprehensive reviews of the methods of heavy water manufacture are given by Benedict[1] and Becker[18].

2. Manufacture by electrolysis

Electrolysis was used to produce heavy water by utilisation of existing facilities for electrolytic hydrogen by both Norsk Hydro in Norway and the U.S.A.E.C. at Trail B.C.

2.1 Norsk hydro plant

Electrolytic manufacture of heavy water by Norsk Hydro at Rjukan, Norway commenced in 1934 utilising the company's plant for the electrolytic production of hydrogen for ammonia synthesis[12–14].

Separation by electrolysis was carried out by using a cascade series of modified Pechkrantz electrolysers fitted with diaphragms to prevent mixing of hydrogen and oxygen. Nine stages of parallel-connected cells were used with the number of cells per stage decreasing as the deuterium content increased. The stages were connected in a series cascade which was operated in steady flow. The electrolyte was 25% KOH solution.

Natural water containing 0.0135 mole% D_2O was fed to the first stage of the cascade where the heavy water concentration of the electrolyte increased to about

0.03%. About 73% of the water fed to each stage was electrolysed, the remaining 27% being carried from the stage as water vapour (containing the same amount of heavy water as the electrolyte) in the hydrogen and oxygen produced. After being condensed from the gas streams, the condensate entered the next stage of the cascade as electrolyte where the concentration of heavy water was increased to about 0.06%. This continued until the concentration of heavy water was increased to about 2%. The separation factor obtained in the cells was four at a temperature of 80°C.

The further concentration was continued in electrolytic cells arranged into a small nine stage unit where a separation factor between ten and twelve was obtained by electrolysing at 5–10°C. At this stage of the process, the hydrogen was also burned with the oxygen from the cell, the resultant water being condensed and returned to the cascade at a stage operating at the same concentration. This was because the hydrogen produced contained so much deuterium that it would have been uneconomic to let it pass into the ammonia plant.

When the concentration had reached about 10% the liquid was fed to the electrolytic cells of a high concentration plant and the heavy water concentrated to 99.8% purity.

Norsk Hydro started its heavy water production by this method and by 1942 was the world's largest producer, the two hydrogen plants at Vemork and Saheim at Rjukan producing 1.5 tonne/year as a by-product of the production of 18,000 m^3 of electrolytic hydrogen per hour. The average power consumption of this plant was 91,000 kW or 54 kWh/mole of hydrogen. This low yield, equivalent to only 5–10% recovery of the heavy water passing with the feed water to the electrolytic cells, was due to incomplete electrolytic separation resulting in most of the deuterium being lost with the hydrogen gas. It was impractical to burn all the hydrogen produced since this would have meant loss of hydrogen for the ammonia plant and therefore higher heavy water operating costs.

2.2 U.S. ATOMIC ENERGY COMMISSION PLANT AT TRAIL, B.C.[1,3,12]

The Trail Plant was built by the A.E.C. under the Manhattan Project (see later) and utilised the existing electrolytic hydrogen plant for ammonia synthesis operated by the Consolidated Mining and Smelting Company. The C.M. and S. plant was suitably modified by the addition of H_2/H_2O exchange towers to avoid the need to burn hydrogen. Approximately 1000 kg/day of 2.3% heavy water were produced by this primary plant and a secondary plant was also built to further concentrate by electrolysis this heavy water to 99.7%.

The primary plant consisted of four stages of electrolytic diaphragm cells, 2687 in the first stage, 378 in the second, 94 in the third and 30 in the fourth, producing a total of 14,500 m^3/h of hydrogen. Steel cathodes were used and the cell temperature maintained at 60–70°C. Since the principal method for enrichment in the primary plant was the exchange reaction, no attempts were made to increase the separation factors of the electrolysers.

The hydrogen produced passed through the catalytic exchange towers to meet a countercurrent flow of water, the resultant enriched water being fed as electrolyte to a suitable stage in the cascade.

The secondary plant consisted of three electrolytic stages containing 126, 20 and 4 cells respectively and produced 500–550 kg/month of 99.7% heavy water. Operation was batchwise with a cell temperature of 23°C and both cell design and operation were identical to the Morgantown Plant to be described later.

Construction of the plant commenced in September 1942 and was completed by June 1943. The cost of construction was approximately U.S.$2,600,000 and operating costs were U.S.$132/kg in 1953, including overhead and profit to C.M. and S.

The plant at Trail has now been closed down because of excessive operating costs compared to other processes (notably H_2S/H_2O exchange) but Benedict[1] has suggested ways in which the process might have been improved including discharging the hydrogen from the first stage directly to the ammonia plant, rearranging the exchange columns and electrolytic cells and redesigning the cells to have a higher separation factor.

3. Pre-enrichment by electrolysis

The heavy water plants of the Fertilizer Corporation of India at Nangal, Emser Werke A.G., Switzerland, the Junta Energia Nuclear Madrid and the modern plant of Norsk Hydro at Rjukan all use electrolysis as a pre-enrichment technique because of the availability of cheap power, final concentration being achieved by cheaper processes.

3.1 NORSK HYDRO PLANT, NORWAY[13, 14]

As described above, the burning of the hydrogen produced in the Secondary plant at Rjukan in order to recover the deuterium content meant a loss of hydrogen for the ammonia plant. From 1945 the burners were gradually replaced by catalytic exchange towers where deuterium is exchanged between the hydrogen gas and water, the water being returned to the electrolytic cascade.

The modern method involves the production of water containing 15% heavy water by electrolysis and exchange, the final concentration up to 99.8% being obtained by distillation. The top fraction from the distillation column, containing 2–3% D_2O, is returned to the electrolytic cascade.

Production is 25 tonne/year which equals 30–35% recovery of the heavy water in the feed water to the electrolysers. Operating costs have not been published but the selling price for heavy water is currently U.S.$450/kg. Deuterium, obtained by the electrolysis of heavy water at Rjukan, sells at U.S.$22/kg.

3.2 Fertilizer Corporation of India Plant at Nangal[15-17]

The plant built for the Fertilizer Corporation of India at Nangal is a classic example of the cost of heavy water production being borne by another process, in this case an ammonia unit. Sited near the Bhakra Dam, the cost of power is low and the plant combines a three stage electrolytic primary unit for pre-enrichment and low temperature distillation for final concentration.

The primary unit is the biggest water electrolysis unit in the world and comprises twenty groups of De Nora bipolar filterpress electrolysers, three in each group, arranged in a three stage cascade system. The first stage contains ten groups, the second six and the third four. The hydrogen from the first and second stages (12500 Nm^3/h and 7500 Nm^3/h respectively) is passed through scrubbers to the ammonia plant. The hydrogen from the third stage (5000 Nm^3/h) passes to the water distillation unit to be stripped of its deuterium content before being fed to the ammonia plant.

Feed water is fed to the first stage electrolysers at a rate of 600 l/h/electrolyser. The condensate from this stage (109 l/min) becomes the feed for the second, and similarly the condensate from the second (67 l/min) becomes the feed for the third stage. The condensate from stage three is recycled to the same stage to help to increase the D_2O concentration of the electrolyte to 2810 ppm.

The separation factor for the De Nora electrolysers is 6.4, the output of the plant is 14.2 tonnes/year and the total recovery of the process is 55%. A useful indication of the costs borne by the ammonia plant is given by the fact that the cost of heavy water production would be U.S.$ 211.2/kg with fertilizer production and U.S. $ 660/kg without[17].

3.3 Emser Werke Plant, Domat Ems, Switzerland[18, 19]

A similar combination of electrolytic pre-enrichment and low temperature distillation is used in the 4 tonne/year plant built for Emser Werke A.G. in 1958.

Electrolysis occurs in a three stage cascade, employing four, two and one electrolysers respectively, the final stage carrying out complete dissociation. The resulting hydrogen stream (400 m^3/h) containing 0.15% HD is fed to the distillation unit for removal of deuterium and thence into an ammonia plant. The electrolytic plant is only operated during the summer when power is cheap and at 2 kWh/g energy consumption, the total operating costs are said to be competitive with the large American plants[18].

3.4 JUNTA ENERGIA NUCLEAR PLANT, MADRID[10, 20, 21]

J.E.N. utilise the facilities of a water electrolysis plant owned by the E.I.A.S.A. Fertilizer Company, together with steam/hydrogen catalytic exchange reactors to produce 500 kg of 2% heavy water/year. Final concentration to 99.8% occurs in a pilot plant which combines electrolysis, combustion and catalytic exchange.

The electrolytic installation of E.I.A.S.A. at Sabiñánigo consists of (a) six Oerlikon[22] filterpress electrolysers each containing 120 cells and rated at 2160 kW, (b) two Oerlikon electrolysers each containing 104 cells and rated at 1870 kW and (c) one Oerlikon electrolyser containing thirty-two cells and rated at 580 kW. The electrolysers are arranged in a six stage cascade, the stages comprising 532, 240, 140, 32, 23 and 9 cells respectively.

The hydrogen from the first two stages goes directly to the ammonia plant whilst enriched hydrogen from other subsequent stages is passed through a catalytic reactor together with steam for stripping of the deuterium. The resulting enriched water is mixed with condensate from the hydrogen and forms the feed to the next electrolyser.

The 2% heavy water from the Sabiñánigo plant is then fed into an electrolytic pilot unit for concentration to 20–30%. The pilot plant[23] consists of thirty monopolar diaphragm cells arranged in a five stage cascade of six, ten, nine, four and one cells, respectively. Each cell operates at 2.0 V and 700 A with a separation factor of 7.5–8 at 50°C. The yield is 97%. Recirculation of electrolyte occurs to the cells and the product gases are burned to give water which is recycled to the previous stage.

4. Final enrichment by electrolysis

Electrolytic finishing stages were used in the United States on the Manhattan Project and the Savannah River Project because of the high separation factor and the ease with which electrolysis can be operated on the small scale, especially on heavy water pre-enriched by another process.

4.1 MANHATTAN DISTRICT PROJECT, MORGANTOWN, W.VA.[1, 3]

With U.S. nuclear weapons development precipitated by World War II and the knowledge that Germany was working on heavy water processes, the Manhattan

Project was launched. Plants were not constructed with economy as a primary design factor, rather they were intended to fulfil an urgent need.

Electrolysis was used as a finishing process at DuPont's Morgantown Ordnance Works to process 90% heavy water produced by water distillation techniques to 99.7% purity, because of uncertainties in the design of the distillation process. The three distillation plants built at Morgantown Ordnance Works W.Va., Alabama Ordnance Works and Wabash River Ordnance Works, were about 50% efficient, feeding a total of 1.1 tonne/month of 90% heavy water to the electrolytic plant. The electrolytic cells, which were operated batchwise, were concentration cells in which the gaseous products of electrolysis were combined *i.e.* no diaphragms were used. Each cell was about 1.5 m high with a diameter of 40.6 cm, had a charge capacity of about 45 kg of solution and was equipped with water cooled (to maintain 40°C) inner and outer steel cathodes and concentric nickel anodes. A current of 1000 A, passing through the cells in series electrolysed about 330 g H_2O/h/cell, the cell voltage ranging from 2.6 to 3.4 V (at voltages greater than 3.5 V the electrodes tended to decompose). The power consumption of each cell was 2.5 to 3.0 kW and the separation factor from 6 to 8.2.

The mixture of hydrogen and oxygen produced in the cells was burned, the resultant water being recycled to a suitable stage in the process. The risk of flashback from the burners to the cells was a constant hazard in the operation of the plant.

The plant was constructed between January and November 1943 and steady state conditions were reached in May 1944. All operations were closed down in October 1945 because of reduced demand for heavy water and unfavourable process costs as compared to the Trail unit. The total constructional cost of the three units was U.S.$14.5m which included U.S.$215,000 for the electrolytic finishing plant, and total operating costs were U.S.$387/kg.

The Trail Plant, which also came under the project has been described earlier.

4.2 SAVANNAH RIVER PLANT[24-26]

In order to increase the output of heavy water, the U.S. Atomic Energy Commission instigated the construction of two new plants. The Dana Plant was built for DuPont on the Wabash River site by the Girdler Corp. in 1951–2, the existing power and water facilities and part of the wartime heavy water unit (described above) being incorporated into the new plant. The second plant built by DuPont on the Savannah River, S.C. site, was also commenced in 1951 but with experience of the Dana unit. Total output was about 780 tonne/year, of which some 454 tonne/year were produced in the Savannah River Unit.

The plants extracted heavy water from natural water using a combination of three processes, namely dual temperature H_2S/H_2O exchange to give 15% heavy water, vacuum distillation to increase the concentration to 90% followed by electrolysis to raise the concentration to 99.75%. The latter two processes, though less efficient in the use of energy, operated on such small volumes that they contributed only 4.2% and 2.6% respectively to the operating costs. Because of the design of any cascade system, the larger first stage will entail probably 50–75% of the capital and operating costs of the entire plant. In order that the first stage should operate at its full production rate the subsequent stages should be designed to avoid capacity limitations and for this reason, extra capacity was installed in the S.R.P. electrolytic finishing plant.

The plant contained 150 cells each of 30.5 cm diameter operating at 1000 A/cell. The anodic current density was 2.58 kA/m² and cell voltage 3.0 V. Cell operation, with burning of product gases, was similar to the Morgantown unit described above.

The S.R. plant cost U.S.$164m in 1951 including all facilities. Of this total, U.S.$2.5m represent the cost of the distillation unit and U.S.$1.5m the cost of the electrolytic plant. The operating costs in 1964 were U.S.$30.8/kg D_2O whilst the A.E.C. selling price was U.S.$53.9/kg.

The Dana Plant was shut down during the period 1957–8 since the continued operation of both the Dana and Savannah plants could not be justified on the basis of the demands for heavy water that were foreseen in the then near future. Of the two, the Dana plant was selected because it was older and in more imminent need of repair.

Two-thirds of the Savannah River Plant were also subsequently shut down, the remaining one-third operating at full capacity of 163.3 tonne/year. With Primary recovery cut by two-thirds it was also possible to shut down the electrolytic plant, the final concentration being obtained by distillation.

5. Electrolytic upgrading of diluted heavy water

Electrolysis, because of its versatility, ease of operation, and reliability, is used on several plants to upgrade diluted heavy water which has been downgraded in reactor use.

5.1 Atomic Energy of Canada Ltd. Unit, Chalk River, Ontario[27–29]

The A.E.C.L. unit at Chalk River, a plant of four electrolytic cells, was put into operation in 1953 to upgrade diluted heavy water. The plant was extended by a further four cells in which the separation factor was 11–12 and production was increased fivefold in 1963. Each batch of electrolyte (56 l) containing 8% K_2CO_3 and 0.2% KOH is electrolysed at a current density of 0.07 A/cm² (1000 A/cell). The product gases are burned, the resultant water being recycled to the cells.

5.2 UNITED KINGDOM ATOMIC ENERGY AUTHORITY UNIT AT A. E. A. WINFRITH[30]

In the heavy water reconcentration plant of the U.K. Atomic Energy Authority at Winfrith, cascades of cells are operated which take 1000 A at 4 V and a further 2 kW/cell which is required for the distillation of the concentrated water from each cell to the next in the cascade.

5.3 SHOWA DENKO K.K. UNIT, JAPAN[31]

Showa Denko K.K. built a 5 tonne/year plant in June 1963 consisting of cylindrical non-diaphragm cells, 0.4 m diameter and 2.0 m high which operated at 1 kA. The separation factor was 8–10 and the hydrogen and oxygen were burned.

5.4 ATOMIC ENERGY ESTABLISHMENT, INDIA[32]

The Atomic Energy Establishment, Trombay has considered electrolytic upgrading in De Nora electrolysers and has evolved a Process Flowsheet presumably for application at Nangal.

6. Other electrolysis methods

6.1 REVERSIBLE ELECTROLYSIS

Reversible electrolysis has been investigated as a means of making use of the large separation factor without paying the large power consumption penalty associated with complete electrolysis of water. The basic idea is the anodic oxidation of the hydrogen produced at the cathode. A suggested technique[37] was to make use of an additive which would be continuously oxidised and reduced in the cell. The cell would have a reversible voltage of zero but would still involve the discharge of hydrogen ions, the step thought to give separation.

It was expected that the separation would occur when the hydrogen ions were discharged on the surfaces of diaphragms interposed in the direction of current flow. The diaphragms would be palladium or other material permeable to atomic hydrogen[38]. The atoms would diffuse through the diaphragm and be reionised by anodic action on the other side. Deuterium would lag in the diffusion step and hence enrichment would occur. Although separation factors in the region of thirty might be expected, because of the high capital costs involved the process would not have an economic advantage over the H_2S process.

The use of gas-diffusion electrodes with which to oxidise the hydrogen produced has aroused considerable interest[39-44]. Although separation factors in the range 6–8 are obtained with a voltage of 50–100 mV, the work has not been commercially exploited[45] and it is doubtful whether reversible electrolysis will ever compete with heavy water pre-enrichment techniques employed at almost no cost in water electrolysis since there would be no hydrogen available for another process and consequently no absorption of costs by this process.

6.2 Direct electrolysis for deuterium

The possibility of deuterium selective electrolysis is intriguing and might be achieved by using pulsating direct current at a deuterium ion resonance frequency[37]. Because of the difference in mass of the two species, deuterium would excite at a different frequency from hydrogen and the technique would be to apply the required frequency to excite deuterium, but not hydrogen, to the discharge point.

If the process could be developed to give deuterium of 10% purity or better in a single stage (which would require a separation factor of 700 or more) the costs might be less than half of those for existing processes. Even if the separation factor were as low as 10, a deuterium selective electrolysis process could be competitive with existing processes, provided that power costs could be reduced by some technique like reversible electrolysis.

7. Trends

Electrolytic heavy water processes are not obsolete but their economic application depends on the right combination of circumstances, for example, low cost power and a market for the hydrogen. Where these circumstances arise, for example electrolytic hydrogen plants for ammonia synthesis, electrolytic heavy water production is economically employed. Except for upgrading, there do not appear to be any plants currently using electrolysis as a finishing stage, water distillation being preferred since this process more closely approaches reversibility.

Heavy water is an excellent moderator for nuclear reactors[33] but its use has been limited because of its cost and availability. For example, a 300 MW reactor would require about 272 tonnes of heavy water[34] and at a price of $61.6/kg this represents a large proportion of the installed capital, as well as a large proportion of the world supply.

World production was rapidly increased during the years 1953–57 owing to the Dana and Savannah River plants coming on stream, but since the closing of the Dana Plant the reduced output from the Savannah River Plant has been little more than sufficient to satisfy the needs of nuclear research programmes. The

quantity of heavy water exported by the U.S.A. up to December 1963 was 732,340 kg, the main customers having been Canada and Germany[25].
World production (excluding the Soviet Bloc and China) is as follows[27]

U.S.A.	H_2S–H_2O exchange, distillation,	163 tonnes/year
Norway	Electrolysis, H_2–H_2O exchange, distillation	25 ,,
France	NH_3–H_2 exchange	20 ,,
India	Electrolysis, H_2 distillation	14 ,,
Switzerland	Electrolysis, H_2 distillation	3 ,,

The Canadian production was expected to be about 800 tonne/year by 1970 with the coming on stream of the A.E.C.L. Glace Bay Plant in May 1967. This unit will use the H_2S/H_2O process but constructional difficulties have been met with and no heavy water has so far been produced[35].

In 1967 the Canadian nuclear association estimated that a world wide surplus of 132 tonnes of heavy water in 1970 would become a shortage of 4,570 tonnes by 1975 and of 20,250 tonnes by 1980. The U.S. is expected to produce 508 tonnes for export by 1972 by which time India will be producing 25 tonne/year.

8. Economic assessment

The cost comparison between the various processes is very difficult since capital and running costs depend so much on plant size and location. Unless at least several tons per year of heavy water are to be produced for a period of ten years or more, it would not pay to build and operate a heavy water plant; the cost of producing on a small scale or for a short term is much higher than for the long-term, large scale producer[1] unless the production can be linked with another process (which usually places a constraint on the heavy water capacity). A heavy water plant is only viable when it affords a price of $56/kg reckoned on a 10% return on capital and amortisation[36]. In water electrolysis, this limit is reached with a relatively small yield of heavy water, whereas the building in of a stripping column or exchange tower usually sends the price up steeply. The capital charges on the Canadian exchange towers were $3m in 1943, a price which would be roughly 3–4 times as much today. At $130/kg production costs, with a 6 tonne/year output, we reach a price of $230/kg heavy water, which lies outside the above index. In a newly built plant, minor improvements would only slightly improve the economics of the situation. (Chlor–alkali electrolysis cannot be used in conjunction with electrolytic heavy water production since the electrolyte is also itself a product—caustic soda.)

The only published information on the costs of water distillation as a pre-enrichment technique comes from the Manhattan Project[3] described earlier, but it must be remembered that this plant was not designed with economy in mind. In practice the balance of costs is such that electrolysis is more economic than distillation when the isotopic enrichment is above 10%[30]. The fundamental drawback with water distillation is the high steam consumption involved—about 200 tonne/kg D_2O as compared to 15 tonne/kg in the H_2S process[12]. Couple this to the unit investment of the plants involved, U.S.$560,000 per ton/year (1944) in the case of the largest Manhattan unit and U.S.$360,000/ton/year for the H_2S plants built in 1952, and it is difficult to see how distillation can ever compete with the H_2S process. Even distillation plants in conjunction with thermal power stations are problematical, since operation requires higher condenser temperatures and thus leads to loss of efficiency in the conversion of coal to steam.

Where electric power is cheap and hydrogen is produced electrolytically for ammonia synthesis for example, pre-enrichment by electrolysis followed by low temperature hydrogen distillation could be more economic than the H_2S process simply on the cost of power, even though the unit consumption of electricity is smaller in the latter case[12].

9. Conclusion

On the whole question of heavy water manufacturing costs, Barr and Drews[37] conclude that there is not a great deal of hope for a significant reduction in the cost of heavy water manufacture which was U.S.$58/kg in 1960 (A.E.C. selling price was U.S. $61.6/kg at the same period). Although significant savings may possibly still be realised, either by revision of existing processes or by development of new processes, they are likely to be only moderate and their realisation would require considerable development work.

However, improvements to the hydrogen sulphide process at Savannah River would indicate that the costs may be reduced to U.S.$37.4 kg[34] and with continued interest in nuclear reactors, the stimulus is present for even cheaper processes to be developed.

10. The literature of electrolytic heavy water production

There are numerous patents, original articles in both scientific and engineering journals, and many review articles. *Chemical Abstracts* and *Nuclear Science Abstracts* and *Nuclear Science Abstracts* are two convenient search sources.

Reviews: A Brief Survey of Processes for D_2O Production by Ryan[46], an annotated bibliography of *Heavy Water Production* by Jacobs[47], *Production of Heavy Water* by Ruzicka[48], annotated bibliography on *Heavy Water and Heavy Water Reactors*[49] and a similar work[50], *Heavy Water*[51]. The years 1934–7 (incl.) are rich in reviews. Since these can be reached through the above references they will not be quoted individually. Moreover, they have been overtaken by scientific, technological and economic trends and their interest is now purely historical. More modern reviews specialising in national aspects are not quoted when electrolytic methods are not mentioned. Thus Law[52] surveys the Canadian scene, while Shimiyu[53] reviews Japanese thought. Reference 54 surveys the operations of Norsk Hydro. Rae's survey[55] is a more general one as is that of Malkov[56], while Alvarez[21] is based on Spanish work. Bebbington and Thayer[24] and Gami[17] relate to the U.S.A. and India, respectively.

Patents: Abstracts of Canadian heavy water patents[58] from 1937–65, Japanese patents[59], use of electrolysis as a final concentration stage[60], cascaded cells (Uhde)[61] multistage cells (U.K.A.E.A.)[62]. Norwegian advances are covered in ref. 63 and relate to cell operating conditions as well as flow paths and arrangements. Reference 64 relates to isotopic exchange towers run in conjunction with cells, ref. 65 covers much the same ground as ref. 43 while a Danish patent (ref. 66) is based on the same principles. Two Swiss patents[67] relate to economics possible by heat exchange and other improvements in cell design. Spanish practice is described in ref. 68.

Other references. Work from Taiwan is reported in ref. 69, from Japan in ref. 70, 75, some earlier Spanish work in ref. 71, a laboratory study on impurity aspects of separation factors in ref. 72, while a laboratory study coupled with industrial results in Italy are quoted in ref. 73. A Swiss plant is described in ref. 74.

Acknowledgement

The author would like to thank Dr. D. W. Fry of U.K.A.E.A. Winfrith, Professor J. L. Otero of J. E. N. Madrid, Dr. H. K. Rae of A. E. C. L. Ontario, Dr. H. Smedsrud of Norsk Hydro and Professor A. Winsel, Kelkheim T.S. for their help in the preparation of this work and also I.C.I. Limited for permission to prepare and publish it.

11. References

1 M. BENEDICT, *Proc. Intern. Conf. Peaceful Uses At. Energy Geneva, 1955*, 8 (1956) 377.
2 B. TOPLEY and H. EYRING, *J. Chem. Phys.*, 2 (1934) 217.
3 P. SELAK and J. FINKE, *Chem. Eng. Progr.*, 50 (1954) 221.
4 P. R. ROWLAND, *Nature*, 218 (1968) 945.
5 D. R. AUGOOD, *Ind. Chemist*, 30 (1954) 585.
6 A. APELBLAT and Y. ILAMED-LEHRER, *J. Nucl. Energy*, 22 (1968) 1.
7 Y. ILAMED-LEHRER, Paper Presented at Torino Nuclear Symposium October 1–2 1968.
8 B. ORSONI, *Proc. Intern. Congr. Pure Appl. Chem. 11th (London)*, 5 (1947) 1057.
9 T. ECHEVARRI and J. L. OTERO, *Anales Real Soc. Esp. Fis. Quim.*, Madrid, 64 (1968) 879.
10 J. L. OTERO and A. M. ARCOCHA, Paper presented at the Colloque de Eau Lourde Societe Europeene Del'Energie Atomique Madrid, Dec. 13–14 1962.
11 J. L. OTERO and M. GISPERT, *Proc. Intern. Conf. Peaceful Uses At. Energy Geneva, 1955*, 8 (1956) 418.
12 E. W. BECKER, *Heavy Water Production, Rev. Ser., Intern. At. Energy Agency*, 21, (1962).

13 K. MYHRE, *Tidsskr. Kjemi, Bergvesen, Met.*, 21 (9) (1961) 204.
14 H. SMEDSRUD, Norsk Hydro, private communication, 4th March, 1969.
15 S. RAMANNA, *Chem. Age India*, 15 (1964) 497.
16 Anon., *Chem. Eng.*, 66 (Feb. 23rd) (1959) 68.
17 D. C. GAMI, D. GUPTA, N. B. PRASAD and K. C. SHARMA, *Proc. Intern. Conf. Peaceful Uses At. Energy 2nd, Geneva*, 4 (1958) 534.
18 N. P. CHOPEY, *Chem. Eng.*, 68 (Feb. 20th) (1961) 118.
19 L. KÜCHLER, *Chem. Ing.-Tech.*, 32 (1960) 773.
20 J. L. OTERO and M. GISPERT, *Energia Nucl. (Madrid)*, 8 (32) (1964) 4.
21 J. ALVAREZ, M. GISPERT, A. MARIA, J. L. OTERO and J. L. ROJAS, *Proc. Intern. Conf. Peaceful Uses At. Energy 3rd, Geneva*, 12 (1964) 406.
22 B. STORSAND, *Trans. 2nd Int. Congr. Electroheat and Electrochem. The Hague*, 3–4 Sept., (1947) 488.
23 J. L. OTERO, M. GISPERT and J. L. ROJAS, *Energia Nucl. Madrid*, 9 (34) (1965) 90.
24 W. P. BEBBINGTON and V. R. THAYER, *Proc. Intern. Conf. Peaceful Uses At. Energy 2nd, Geneva*, 4 (1958) 527.
25 W. P. BEBBINGTON, J. P. PROCTOR, W. C. SCOTTEN and V. R. THAYER, *Proc. Intern. Conf. Peaceful Uses At. Energy 3rd, Geneva*, 12 (1964) 334.
26 W. P. BEBBINGTON and V. R. THAYER, *Chem. Eng. Progr.*, 55 (9) (1959) 70.
27 H. K. RAE, Atomic Energy of Canada Ltd., private communication, Dec. 18th, 1968.
28 J. A. MORRISON et al., *Proc. Intern. Conf. Peaceful Uses At. Energy 3rd, Geneva*, 12 (1964) 375.
29 Anon., *Can. Chem. Process*, 53 (2) (1969) 29.
30 D. W. FRY, U.K.A.E.A. Winfrith, private communication, Jan. 22nd, 1969.
31 T. SAITO, S. SONODA, Y. KURIHARA, T. TAKAMATSU and T. MORITA, *Proc. Intern. Conf. Peaceful Uses At. Energy 3rd, Geneva*, 12 (1964) 389.
32 P. G. DESHPANDE and D. C. GAMI, *Proc. Intern. Conf. Peaceful Uses At. Energy 2nd, Geneva*, 4 (1958) 568.
33 F. BOLHAR-NORDERKAMPF, *Elektrotechn. Maschinenbau*, 77 (1960) 387.
34 J. F. PROCTER and V. R. THAYER, *Chem. Eng. Progr.*, 58 (4) (1962) 53.
35 The Sunday Times, 26th Jan., 1969.
36 E. W. BECKER, *Angew. Chem.*, 68 (1) (1956) 6.
37 F. T. BARR and W. P. DREWS, *Chem. Eng. Progr.*, 56 (3) (1960) 49.
38 J. C. SEAILLES, Brit. pat., 882,607 Nov. 15th, 1961.
39 W. VIELSTICH, *Chem. Ing.-Techn.*, 33 (2) (1961) 79.
40 R. WENDTLAND and A. WINSEL, *Chem. Ing.-Tech.*, 39 (12) (1967) 756.
41 A. WINSEL, *Chem. Ing.-Tech.*, 67 (7) (1963) 639.
42 A. WINSEL, *Z. Elektrochem.*, 65 (2) (1961) 168.
43 A. WINSEL (for Siemens A.G.) D.R. pat., 1,245,337, 27th July, 1967.
44 G. P. LEWIS and P. RUETSCHI (for Electric Storage Battery Co.), U.S. pat. 3,306,832, 28th Feb. 1967.
45 A. WINSEL, private communication, 11th Feb. 1969.
46 R. K. RYAN, *Brief Survey of Processes for D_2O Production*, AAEC/TM-368 (Feb., 1967) Australian At. Energy Comm.
47 J. M. JACOBS, *Heavy Water Production*, June, 1964 TID-3091 (Suppl. 1 from Divn. of Tech. Inf. Extn. A.E.C.).
48 J. RUZICKA, *Production of D_2O*, Technika Hlubokych Teplot, Czech. Acad. Sci., 1961, pp. 43–108.
49 A. T. CURRAN, BAW-47. *Annotated Bibliography on Heavy Water and Heavy Water Reactors*, Babcock and Wilcox Co., At. Energy Divn., Feb., 1958.
50 D. M. DUKE, BAW-1209, *Heavy Water Reactors and the Properties, Analysis and Handling of Heavy Water*, Dec., 1960, Babcock and Wilcox.
51 R. CLARKE, *Heavy Water*, AERE-Inf/Bib-87, June, 1957, Harwell, Gt. Britain.
52 C. A. LAW, *Nucl. Eng.*, June, 1968.
53 M. SHIMIYU, *Nippon Genshiryoku, Gakkaishi*, 4 (1962) 180.
54 L. MYHRE, *Tidsskr. Kjemi, Bergvesen Met.*, 21 (1961) 204.
55 H. K. RAE, Atomic Energy of Canada, Chalk River, AECL 2503, 1965.

56 I. Malkov, *Soviet J. At. Energy*, 7 (2) (1959) 101.
57 A. P. Murrenhoff, *Kerntechnik*, 6 (1964) 558.
58 F. S. Armstrong, *Nucl. Sci. Abstr.*, 20 (1966) 260.
59 Jap. pat., 7966 (1958); 1116 (1958); 525 (1956).
60 Ger. pat., 1,057,585.
61 Ger. pat., 1,058,479 (to F. Uhde).
62 Brit. pat., 726,532; 726,771.
63 Norweg. pat., 77,176 (to Norsk Hydro); 84,254 (Norsk). Brit. pat., 581,908; 580,791 (Norsk).
64 Ger. pat., 1,147,203 (to F. Uhde).
65 Can. pat., 641,333 (to Siemens, equiv. to ref. 43).
66 Danish pat., 86,137.
67 Swiss pat., 332,467; 332,465.
68 Brit pat., 837,894.
69 Yuen-Kwei Tai and Yuen-chi Tsu, (Natl. Taiwan Univ., Taipei) *Ho Tsu K'o Hsueh*, 2 (1960) 10
70 M. Shimizu, S. Nitō and A. Harada, *Nippon Genshiryoku Gakkaishi*, 4 (1962) 211.
71 M. Gispert and J. O. de la Gandera, *Anales Real Soc. Espan. Fis. Quim. (Madrid) Ser. B*, 56 (1960) 559.
72 J. Brun, W. Gundersen and Th Varberg, *Kgl. Norske Videnskab. Selskabs Forh.*, 30 (1957) 29.
73 S. Finzi, R. Renzoni and M. Silvestri, *Energia Nucl. (Milan)*, 3 (1956) 294; *ibid.* Proc. Sci. Congr., Nuclear Sessions July 2–6, 1956, 3rd Electronic-Nuclear Intern. Review, Rome.
74 L. Hänny, *Schweiz. Arch. Angew. Wiss. Tech.*, 26 (1960) 115.
75 K. Kawashima, *Kogyo Kagaku Zasshi*, 62 (1959) 487.

Chapter 6

The Electrowinning of Metals

D. GILROY

Electricity Council Research Centre, Capenhurst, Chester

1. Introduction

The metallic elements may be extracted from their naturally occurring ores by a reduction which could in every case be electrochemical. In practice, chemical reagents such as hydrogen, carbon or reactive metals may be preferred and the choice is obviously made on economic grounds by considering the required purity of the product, its physical form, and the other stages of the overall process. These include ore treatment and concentration, separation after reduction, further purification or refining, and fabrication. In this chapter we are concerned with those elements which are prepared commercially in such a way that they are first isolated to some degree of purity by electrolysis. Electrorefining is thus excluded and is treated elsewhere. However, the few metals which are initially electrodeposited as amalgams or alloys are included. In fact only a small number of the metals are manufactured solely by electrowinning and we have attempted to assess the relative importance of competing alternatives and to indicate current trends in the methods used. This is not always straightforward since published statistics normally make no differentiation. However, information has been obtained by communication with the industries involved. At the same time attempts have been made to give some idea of the scale and location of production although this cannot be regarded as being fully comprehensive since the required information is company confidential in many instances.

Electrodeposition occurs from a solution containing the metal ion in an appropriate solvent. It is necessary that cathodic reactions of the solvent play only a minor role and for this reason the use of water, for example, is restricted to elements less active than zinc and manganese. In the laboratory, organic media are employed in investigations on other more reactive metals but in practice only molten salts have found any application. The location of the electrowon elements considered in this chapter in the aqueous or molten salt groups is indicated in the Periodic Table overleaf.

In addition, several of the adjacent metals have been electrowon previously, *e.g.* K, Rb, Cs, Ca, Sr, Ba, Th, U, Sb and Bi; others have been subjected to extensive investigation with this object in mind, *e.g.* Zr, Hf, V, Mo, W, Re and Ge. Sodium has been treated elsewhere (Chapter 3).

Li	Be	B	Molten			Aqueous solution						C N O F	
Na	Mg	Al	salt group			group						Si P S Cl	
K	Ca	Sc	Ti	V	Cr	Mn	Fe	Co	Ni	Cu	Zn	Ga	Ge As Se Br
Rb	Sr	Y	Zr	Nb	Mo	Tc	Ru	Rh	Pd	Ag	Cd	In	Sn Sb Te I
Cs	Ba	Rare earths	Hf	Ta	W	Re	Os	Ir	Pt	Au	Hg	Tl	Pb Bi Po At
		Actinides											

1.1 ELECTROWINNING IN AQUEOUS MEDIA

The favourable properties of water as a solvent are well known and include the ability to dissolve salts to relatively high concentrations, and to produce solutions of high conductivity at ambient temperatures, especially when acidic. These factors are used to advantage when the overall process is considered since in most cases the following scheme is applicable:

(a) The ore is converted to an acid soluble form, usually an oxide, by roasting. The product is then leached:

$$MO_x + 2x\,H^+_{aq} \rightarrow M^{2x+}_{aq} + x\,H_2O$$

(b) The solution is purified and perhaps concentrated.

(c) The anodic and cathodic reactions during electrolysis are

$$M^{2x+}_{aq} + 2\,xe \rightarrow M$$

$$x\,H_2O \rightarrow \tfrac{1}{2}x\,O_2 + 2x\,H^+_{aq} + 2x\,e$$

(d) The acid liberated at the anode is recirculated to the leach.

The nett reaction can then be written simply as

$$MO_x \rightarrow M + \tfrac{1}{2}x\,O_2$$

and there should be no overall consumption of water or acid. The use of sulphuric acid as the leaching agent favours the anodic evolution of oxygen, whereas in HCl solution chlorine would probably be evolved. (Similarly hydrogen evolution besides giving rise to current inefficiency also leads to acid consumption.) The high conductivity of the solvated proton reduces the power losses caused by ohmic drop during electrolysis. No other solvent system can combine all these advantages at room temperatures. Some of the above stages might be briefly commented upon here as well as in the sections on the individual metals.

(a) Leaching of oxide bearing ores extracts the metal directly without a smelting process. This becomes attractive at low concentrations when much unwanted material would have to be submitted to smelting with resulting wastage of thermal energy. In many cases concentration by flotation followed by smelting is also inefficient and leaching is more satisfactory.

(b) Before electrolysis, more noble impurity metals must be removed or these will be preferentially deposited. Hence the more reactive elements, *e.g.* Zn and Mn require more extensive purification. At the same time salts of more active metals such as Mg may have to be removed simply to prevent accumulation and precipitation at inconvenient sites. It may also be found that some impurities reduce the hydrogen evolution overvoltage on active metals thus reducing the current efficiency.

(c) Since most operations involve electrolysis of sulphuric acid solutions it is to be expected that the equipment utilised would be fairly uniform. Thus lead is often used as a corrosion resistant material for lining the tanks and for the anodes. Descriptions of many plants are available in the literature and are indicated under the appropriate metal without going into much detail here. One of the major advantages of electrowinning should be the production of a high purity metal thus eliminating the need for further refining. Normally, under cathodic conditions the electrodeposit is not attacked by the solvent in aqueous systems and the only impurity is likely to be hydrogen provided the electrolyte is suitably processed. This is not always the case in molten salt regimes as will be seen later.

1.2 ELECTROWINNING IN MOLTEN SALTS

Although attempts have been made to electrowin the refractory metals from aqueous and organic media it is probably true to say that in most cases the cathodic product is grossly contaminated and it is necessary to resort to fused salt systems. In theory, any metal could be claimed by direct electrolysis of one of its salts; halides are usually considered, but in practice it is often preferable to dissolve the compound in salts of the more active metals sodium and potassium. This procedure improves the conductivity and lowers the melting point of the resultant mixture and may also permit continuous operation by replenishing the metal salt feedstock. Since alternative media are not possible for several metals it is perhaps appropriate to stress the disadvantages rather than the advantages of molten salt winning. The high temperatures of operation result in increased contamination through corrosion and reaction with the anodic products. For example, carbon monoxide may diffuse through the melt and react with the product to increase the carbon and oxygen content of the metal with deleterious effects in metallurgical properties. For similar reasons inert atmospheres and expensive cell linings may be required and the feedstock may need careful dehydration and other purification. Unless the metal is formed as a liquid it is often quite difficult to separate from the melt since deposition in powder or dendritic form is frequently encountered so that the number of operations may have to be increased. At the same time solubility in the melt becomes an important factor at the elevated temperatures considered and is often a source of current inefficiency. Continuous processes have not always been developed but where available these are based on oxide or chloride feedstocks with oxygen and chlorine evolved at the anode, respectively. In the former instance the oxygen combines with the carbon to liberate CO or CO_2. Carbon and graphite are the only suitable anode materials. The cell electrolyte is never cycled for leaching purposes, as in aqueous electrowinning, for obvious reasons and indeed the molten salt electrolysis may represent only a relatively small fraction of the total production cost since complicated concentration and separation techniques are necessary for many of the elements treated. The reduction may therefore be replaced by alternative methods without a major effect on the overall scheme. Temperature is usually maintained by resistive heating. This may imply an increase in the electrode spacing; schemes incorporating external fossil fuel heating are normally unsuccessful.

2. Aqueous processes

2.1 COPPER

Copper electrowinning is usually the final step in the production of the metal from oxide ores which have been treated by sulphuric acid leaching. The more common sulphide ores are smelted and electrorefined. The choice on which route to adopt depends on numerous factors and especially on the type and copper content of the ore. Thus low quality oxide-containing materials can be leached economically without flotation concentration, and mixed oxide sulphide bodies may be separated by dissolution of the oxide either before or after flotation. Smelting a low grade ore results in an increase of slag volume with corresponding increases in fuel consumption. The purity of the electrowon cathodes may be at least as good as that of the electrorefined copper so that this stage is eliminated, although power consumption is greater in electrowinning and the proximity of cheap hydro-electric generation as in Katanga may then become a determining factor. The sulphuric acid necessary for leaching is regenerated by the electrolysis and a recycle process is therefore possible, which means that the sulphuric acid producing plant which accompanies sulphide ore treatment is often not necessary although some "topping up" may be required.

McMahon[1] has made an extensive survey of the world copper industry and has indicated that the leaching–electrowinning method is used by Inspiration Consolidated Copper Co. Arizona; Chile Exploration Co., at Chuquicamata, Chile; Andes Copper Mining Co. at Potrerillos, Chile; Gécomin in Congo-Kinshasa (La Générale Congolaise des Minerais has now assumed the operation of plants previously operated by Union Minière) and Nchanga Consolidated Copper Mines Ltd., Zambia. The Chilean ore body has an unusually high oxide content, probably because the arid conditions have retarded the dispersion of the soluble sulphates. It is likely that some of the much smaller Chilean mines recover copper by leaching.

There are indications, however, that the oxide reserves are nearing depletion as evidenced by a recent communication to us from the Andes Copper Mining Co., which indicates that only electrorefining is now operated. Similarly, McArthur and Ledeboer[2] have suggested that the winning cells at Chuquicamata will be converted to refining as oxides become exhausted. The relative importance of leach-winning has been estimated as about 10% of the total copper production by Butts[3], (this probably refers to the U.S.A.), and as 10% of U.S. production and 3% (150,000 t) of the free world including U.S. production by Cooper[4]. However, these latter figures are not self consistent. Copper recovery from the leach solutions in the U.S.A. is more often achieved by cementation onto iron rather than by electrowinning and the Inspiration plant is probably the only major producer by this method. On the other hand the Gécomin plants at Luilu and Shituru produce about 70% of Congo-Kinshasa copper by electrowinning. The company has kindly supplied the recent data shown in Table 1. Similarly, Anglo American Corporation

TABLE 1

Company	Plant	Production (tons)	Year	Production (tons)	Year
Gécomin	Luilu	118,000	(1967)	127,000	(1968)
Gécomin	Shituru	132,000	(1967)	141,000	(1968)
Anglo American	Nchanga	104,900	(1967)	109,500	(1968)
Mufulira Copper Mines	Chambishi	13,700	(1967/68)	15,400	(1968/69)
Chile Exploration	Chuquicamata*	94,000	(1967)	95,000	(1968)
Inspiration Consolidated	Inspiration**	35,000	(1966)	20,000	(1967)

* Estimated from published production figures and tonnage of oxide and sulphide ores mined[5].
** Estimated from published production figures[6] and oxide–sulphide content of ore body[1]. Note that 1967 was a poor year for U.S. copper because of labour disputes.

have provided the statistics for Nchanga, and Roan Selection Trust those for Chambishi. A certain amount of material from the Chibuluma mine is treated at the Chambishi electrowinning plant. The total quantities involved are thus over 500,000 t/annum representing about 10% of the total copper production.

The flowsheets and details of the cells and tankhouses of the major plants are available elsewhere and only a brief description is given here. Chuquicamata practice has been described by McArthur and Leaphart[7] and by McArthur and Ledeboer[2], that at Nchanga by Chapman and Page[8] and a review of Gécomin/Union Minière experience has been published recently by Theys[9]. The development and operation of roasting, leaching and electrowinning at Chambishi Mines in Zambia has been described by Harper et al.[10]. Detailed tables are also given by McMahon[1].

Obviously the make-up of electrolyte solution entering the cells depends largely on the ore type to be leached; generally iron is the only large scale impurity to be removed since reduction from the ferric to the ferrous state reduces current efficiency of copper deposition. However, leaching of sulphides is promoted by ferric ions and at Inspiration in particular a fairly high ferric content in the electrolyte is kept for this purpose. At Nchanga, ferric hydroxide is precipitated at pH 2.2 by oxidation with air. At Chuquicamata, chloride must also be removed and nitric acid and molybdenum content controlled.

Electrolyte entering the cells may contain from 20–70 g/l Cu and similar amounts of free acid. Cell voltages are 2.2 V, current densities typically 10 mA cm^{-2} and cell temperature 30–35°C. The anodes are a lead–antimony (6–15%)–silver (0–1%) alloy. Lead shows good corrosion resistance and low oxygen overvoltage in sulphuric acid solutions; the antimony improves the mechanical properties. The cells are usually lined with lead–antimony alloy. The cathodes are deposited onto copper starting sheets at 80–90% current efficiency with a purity of over 99.5%. Power consumption is about 2.2 kWh/kg except at Inspiration where it is increased by the high iron content of the electrolyte. The solution is covered with an oil film or with plastic balls to prevent excess acid spraying.

Recently much research has been concentrated on the hydro-metallurgy of copper and the treatment of low grade ores by "wet" methods. Cooper[4] and Rosenzweig[11] have produced useful reviews especially of work in the U.S.A. Thus copper may be extracted from a leach liquor using a selective organic reagent in a kerosene solvent. The copper ions are liberated by stripping with a sulphuric acid solution and the reagent and solvent recovered for recycling. This type of operation coupled with an electrowinning cell has been initiated at the Bluebird Mine of the Ranchers Development Corporation. A new cell has been developed by the Continental Copper and Steel Industries Inc. (C.C.S.) and the Colorado School of Mines Foundation[12, 13] which operates at increased current density with a more efficient copper recovery from solution. Some of the pilot plants engaged on these new methods are mentioned in the articles cited above[4, 11]. Pressure leaching of chalcocite concentrates by the following reaction:

$$CuS + H_2SO_4 + \tfrac{1}{2}O_2 \rightarrow CuSO_4 + S + H_2O$$

has been described by Veltman et al.[14]. This process leaves the sulphur in elemental form rather than as gaseous SO_2 which usually has to be converted directly to H_2SO_4 following its formation in a conventional smelter. In many cases this acid production which must be carried out to prevent air pollution is inconvenient and the new process will enable the sulphur to be stored if necessary so that acid production may be performed elsewhere. It is apparent that the combination of such researches establishes a trend towards completely aqueous chemical smelting so that recovery by electrowinning can be expected to assume increasing importance. At the same time recovery by hydrogen reduction[4] and cementation (see for example ref. 15) will also increase, although electrowinning should gain at the expense of the latter process which gives a relatively impure product.

2.2 COBALT

World cobalt production in 1967 and 1968[5] amounted to about 20,000 t/annum of which just over 60% was produced in Congo-Kinshasa by Gécomin. This company has kindly supplied recent data for the outputs from the Shituru and Luilu electrowinning plants.

Luilu	1967	4,300 t	1968	4,400 t
Shituru	1967	6,550 t	1968	5,500 t

The fraction of cobalt electrowon is evidently at least 50% of the total quantity and is produced almost entirely at these two locations. The cobalt source is obtained by diverting liquor from the copper leaching–electrowinning circuit referred to earlier. However, the two plants use different solution treatment techniques and the cobalt from Shituru is relatively impure and is further fire refined. Bouchat and Saquet[16] have described the latter system. Here the solution is submitted to a primary electrolytic removal of copper followed by successive increases of pH by lime additions which precipitate iron and aluminium and copper in turn. Remaining copper is cemented with cobalt granules and the cobalt is finally precipitated at pH 8.3 accompanied by zinc and nickel. Most manganese and magnesium are left in solution at this stage. The cobalt is redissolved at pH 7.0–7.2 to give a pulp, containing solid calcium sulphate and some cobalt hydrates, which is delivered to the electrolysis cells. A further stage of nickel removal is effected later. As Thoumsin has indicated[17] the Luilu plant has a more sophisticated treatment sequence to produce a final solution which is clear and much less impure. The operations are performed in a somewhat different order and in addition, zinc and solid calcium sulphate are removed.

The electrolysis conditions at the two plants are summarised in Table 2.

TABLE 2

Plant	Co (g/l)	pH inlet	Temp.	Cathodes	Anodes	Cell volts	Current density ($mA\ cm^{-2}$)	Current eff.	Cathode purity
Shituru	15–20	7.0	50°C	mild steel	Pb–Sb–Ag	5.0	40	85–88%	93–95%
Luilu	40–50	1.5	65°C	stainless steel	—	—	35	75%	99.9%

Thoumsin[17] has described research into different anode materials. During electrolysis cobalt granules which fall from the electrodes are of sufficient purity to be collected and melted down.

The only other electrowinning plant of any importance is that of the Rhokana Corporation in Zambia. Total metal production in that country amounts to about 1500 t. Alternative production methods[18] as used in North America, Germany and elsewhere usually involve the leaching and precipitation of cobalt from nickel or copper ores at some stage. However, the hydrated oxide may then be calcined and reduced to the metal with carbon or hydrogen. A hydrogen reduction from solution may also be practised[19]. For recent work on electrodeposition the papers of Thoumsin[17] Löwe et al.[20] and Russian work by Lekht and Ivanova[21] may be consulted.

During the 1960's the industry as a whole has shown steady expansion, except for 1967, but the relative role of electrowinning will probably not change to any marked extent in the future.

2.3 Nickel

Direct electrowinning of nickel is of minor importance only for this metal and is practised by the International Nickel Company at the Thompson and Port Colbourne refineries in Canada and by the Outokumpu Company at Harjavalta in Finland. The quantities currently produced per annum are approximately 45,000 t in Canada[22] and 3,000 t in Finland[23], whereas total world production of metallic nickel is of the order of 450,000 t. The operations of both organisations have been described by Boldt[24], and the original papers by Renzoni et al.[25] and by Toivanen and Grönqvist[26] may also be consulted. In short, two processes are possible, either using soluble sulphide anodes or electrolysing a leach solution with insoluble anodes; the former is practised by INCO and the latter by Outokumpu.

Depending on the copper content of the ore, nickel sulphide anodes may be cast directly from the converter or following a selective flotation of converter matte. Thus at Thompson the ore bodies are low in copper which is mostly removed in the concentrator and the whole of the nickel sulphide product from the converter is cast into anodes. At Port Colbourne, copper is separated from converter matte[27] and the nickel sulphide may then be cast into anodes or roasted and reduced to metal for electrorefining. The latter route is the major one at this refinery. During electrolysis at 55°C a voluminous sulphur sludge develops at the anodes which are consequently bagged. The electrolyte contains 60 g/l Ni^{2+}, 90 g/l SO_4^{2-}, 35 g/l Na^+, 60 g/l Cl^-, and 16 g/l H_3BO_3. As the dissolution at about 20 mA cm^{-2} proceeds, the sulphur produced has the effect of increasing ohmic drop and cell overvoltage and temperature. The tank voltage must be increased from 2.8 V to about 4 V. Also some oxygen evolution commences and nickel must therefore be added to the electrolyte. This is done by leaching ground anode scrap with removal of copper and arsenic by H_2S. Further purification is effected to remove iron and cobalt by oxidation and hydrolysis. The cathodes are collected after 10 days and contain more than 99.95% nickel; the anode sludge is also collected and treated to obtain precious metals and other materials.

At Outokumpu high grade nickel matte is leached and at the same time copper iron and lead are removed by cementation and partial electrolysis (Cu), hydroxide precipitation (Fe) and sulphate precipitation (Pb). Cobalt is then removed by precipitation with anodically prepared $Ni(OH)_3$.

The conditions of electrolyte and electrolysis may be summarised as: temperature 62°C; catholyte contains Ni^{2+} 70 g/l, also boric acid and sodium sulphate; pH 3.0–3.5; anolyte contains 40 g/l free acid; current density 18 mA cm^{-2}; cell voltage 3.4 V; current efficiency 91–96%; anodes pure lead; cathode quality better than 99.9%; the cathodes are bagged in terylene cloth to maintain pH control.

Boldt[24] describes treatments of oxide ores where the nickel is recovered from solution after ammonia leaching as a hydroxide/carbonate which is then calcined to NiO. Also metal has been recovered by hydrogen reduction from a sulphuric acid leach. Possibly similar leach solutions could in future be treated by electrolysis to recover nickel, although hydrogen reduction would seem to be the preferred method. Recent developments in pressure leaching operations of this type have been reviewed by Mackiw et al.[28] Apparently it would be uneconomic to convert the Port Colbourne refinery completely to electrowinning[22] so that in general there seems little prospect of further operations by this method.

2.4 ZINC

On a tonnage basis zinc is the second most important metal after aluminium to be claimed by electrowinning methods. Statistics on all aspects of the industry are readily available and we reproduce in Table 3 a list of electrolytic zinc plants based on that given in the 1967 year book of the American Bureau of Metal Statistics[29] with data on more recently opened plants supplied by the Zinc Development Association[30].

For the year 1967 the free world production by the electrolytic method was 1,750,000 t and thus represented 53% of the total quantity of manufactured zinc. Total Soviet and other communist block production by all methods was estimated at 920,000 t. In 1968 total free world production was 3,650,000 t[5]. The industry is well established and recent trends have been the emergence of Japan as a major producer and the development of the Canadian Mining industry which is by far the largest source of ore.

Competing smelting processes are based on carbon reduction of the oxide followed by distillation of the zinc from the mixture. Quite recently a new method, the I.S.F. process, has superseded all other types and new plants commissioned or

TABLE 3

Country	Location	Company	Estd. annual capacity
U.S.A.	East St. Louis, Ill.	American Zinc Co.	75,000
	Great Falls, Mont.	Anaconda Co.	162,000
	Anaconda, Mont.	Anaconda Co.	90,000
	Corpus Christi, Tex.	Asarco	108,000
	Silver King, Ida.	Bunker Hill Electrolytic Zinc Plant	109,500
Canada	Valleyfield, P.Q.	Canadian Electrolytic Zinc.	146,200
	Trail, B.C.	Cominco Ltd.	232,200
	Flin Flon, Man.	Hudson Bay Mining and Smelting Co.	80,000
Argentina	Zarate	"Meteor" Est. Met. S.A.I.yC	7,700
	Borghi	Sulfacid S.A.	13,200
Peru	La Oraya	Cerro de Pasco Corp.	68,000
Austria	Gailitz	Bleiberger Bergwerke Union	15,400
Belgium	Balen	Soc. Anon. de la Vieille Montagne	121,300
France	Viviez	Soc. Anon. de la Vieille Montagne	99,200
Finland		Outokumpu	90,000
Germany	Datteln, NRh/W	Metallgesellschaft Ab.	100,000
Italy	Crotone	Soc. Min. and Met. di Pertusola	69,400
	Ponte Nossa, Bergamo	A.M.M.I. spa.	25,400
	Monteponi, Sardinia	Monteponi and Montevecchio S.p.A.	33,000
	Porto Marghera	Monteponi and Montevecchio S.p.A.	44,000
Norway	Eitrheim, near Odda	Det Norske Zinkkompani A.S.	70,000
Yugoslavia	Sabac	Hemijska Industrija "Zorka"	20,000
Spain	S. Juan de Nieva	Asturiana de Zinc S.A.	71,600
	Cartagena	Espanola del Zinc S.A.	27,600
India	Kerala	Cominco/Binani	18,000
	Udaipur	Debari	16,000
Japan	Kosaka, Akita-ken	Dowa Mining Co. Ltd.	19,800
	Akita-shi, Akita-ken	Mitsubishi Metal Mining Co. Ltd.	68,800
	Uguisuzawa-mura, Mikagi-ken	Mitsubishi Metal Mining Co. Ltd.	21,200
	Kamioka, Yoshiki-gun Gifu-ken	Mitsui Mining and Smelting Co. Ltd.	67,500
	Omuta, Fukuoka-ken	Mitsui Mining and Smelting Co. Ltd.	22,500
	Bandai-mura, Fukushimaken	Nisso Smelting Co. Ltd.	29,100
	Annaka-shi, Gunma-ken	Toho Zinc Co. Ltd.	179,900
Australasia	Risdon, Tasmania	Electrolytic Zinc Co. of Australasia	163,500
	Port Pirie, S.A.	Broken Hill Ass. Smelters Ltd.	30,000
Congo (Dem. Rep.)	Kolwezi	Soc. Métallurgique Katangaise	66,100
Zambia	Kabwe	Zambia Broken Hill Dev. Co.	33,600
South Africa	Johannesburg	Zincor	33,000

under construction are divided approximately equally between this and the electrolytic reduction. The chief advantage of the Imperial Smelting Furnace is its ability to handle lead-containing ores to give both lead and zinc as products and also low grade ores which are not easily treated by other methods. Additionally, zinc containing a small percentage of lead is often necessary for galvanising and this requirement is met by the blast furnace. However, electrolytic zinc is more easily made to a higher purity.

Several descriptions of the procedures adopted at individual plants, especially those in Australia and North America are available. Hampel[31] in his summary article gives references to papers on four of the North American plants and also to the Risdon plant. More recent modifications to the latter have, however, been given by Ashdown[32] and research on the solution purification described by Bratt and Gordon[33]. Italian work on the development of a continuous flow system of leaching and purification has been published by Benvenuti et al.[34]. Operations at the Valleyfield plant of the Canadian Electrolytic Zinc Company which was completed in 1963 were given by Jephson et al.[35]. Modernisation of the Bunker Hill Co. works has been described by Bird[36] and a recent paper summarises the new electrolytic installation at Port Pirie which processes zinc oxide from a slag fuming plant[37].

The starting material in most cases is roasted zinc concentrate which is subjected to a sulphuric acid leach using spent electrolyte containing approximately 10% sulphuric acid and usually an appreciable zinc ion content. An extensive purification procedure then follows necessitated by the fact that the high hydrogen overvoltage on pure zinc is easily reduced by impurities at quite low concentrations with consequent inefficiency in the cathodic reduction. Thus at Valleyfield tellurium contamination was suspected and a sensitive analytical method capable of detecting 1 μg/l was developed to test this hypothesis. The concentrations of most metallic elements more noble than zinc are reduced to below the level of 1 mg/l. The first element to be removed is iron by oxidation to the ferric state with air or manganese dioxide. The precipitate is used to carry down other impurities, and iron in scrap form may be added to the leach to facilitate this aspect of purification. Other elements are then precipitated by addition of zinc dust usually in two stages so that the cadmium may be collected separately in the second stage. The single stage precipitation practised at the Great Falls plant[38] is unusual. Analyses for cobalt, nickel, arsenic, antimony, copper and other impurities may be performed. A summary of the analytical methods used and the development of techniques of process control has been given by Krupka and Krupka[39].

In zinc ores cadmium occurs on average at a level of about 0.2% and is recovered at many plants. Other elements which are worth recovery, depending obviously on their abundance in the treated ore, include copper and cobalt as at Risdon[32] and at many locations gold and silver. A cobalt recovery scheme operated at a Polish works has been recently described[40]. Lead may be recovered but most other trace impurities such as arsenic, antimony and tellurium are rejected.

The electrolysis of the solution is carried out in large lead lined concrete tanks between lead–silver anodes and aluminium sheet cathodes at current densities of 30 mA cm^{-2} and upwards, according to the zinc content of the feed. This may be in the range from 100 to 220 g/l depending on the plant and is usually depleted to about one-third of this value on exit. At the same time the sulphuric acid content after electrolysis has built up correspondingly to a level which may be between 100 and 200 g/l. Voltage drop across a unit cell is about 3.5 V and the operation is conducted at about 35°C. Cooling is necessary to maintain this condition and this may be done externally by evaporation or in cooling towers or by inserting cooling coils in the cells. The last method probably leads to electrolyte contamination through corrosion. The use of carbon cooling elements in this context has been discussed by Alexandrov and Pamuchyski[41] and improved performance over the more usual lead and aluminium coils was claimed. Corrosion of the lead electrodes and tank linings is usually compensated for by addition of strontium carbonate, and other addition agents are present in small quantities to prevent acid spray, which may bring about copper contamination from the connectors, and to make a smoother cathode deposit. The presence of magnesium, sodium and manganese ions dissolved in the electrolyte in fairly high concentrations is tolerated because they are not co-deposited on the cathodes. However, manganese dioxide sludge may form at the anodes and must then be removed periodically. Cathodes of over 99.9% purity are deposited at current efficiencies of approximately 90%, power consumption is about 1.5 kWh/lb of zinc. It should be emphasised that the above details are intended to be representative only and there may be quite wide variations in electrolyte composition and in other details between individual plants so that specific descriptions should perhaps be consulted in the cases where these are available.

At one time zinc recovery using an amalgam method was practised[42] but this has now been abandoned because of the high capital cost of mercury.

2.5 Cadmium

Although cadmium is a relatively minor constituent of the earth's crust, it is always found in association with zinc ores from which it may be economically recovered, and production is therefore almost entirely as a byproduct from zinc plants. Normally an electrolytic plant extracts cadmium also by electrowinning whereas at a smelting plant separation is by pyrometallurgical methods in the final stages. Current world production is about 12,000 t/annum[29, 43]. The major producers are U.S.A., U.S.S.R. and Japan. Details of outputs from electrolytic zinc plants in the U.S.A. are given separately[29] thus enabling one to estimate the percentage of the metal made by electrowinning in that country. The figure turns out to be 41% averaged over the years 1960–1967 which compares with the 39% of zinc prepared electrolytically in the U.S.A. Extending this analysis to include other countries one deduces that about 50% of cadmium, i.e. 5900 t/year is made electrolytically, assuming that the cadmium content of ores treated at the various locations does not vary a great deal. This is not entirely true but the factor might be expected to average out when considering the large number of zinc plants.

During zinc smelting most of the cadmium is transferred eventually to the fume which is treated by wet methods, including ion exchange, until cadmium is finally brought down in sponge form by cementation onto metallic zinc. This impure material is melted under a caustic flux and the cadmium distilled. Further distillations may then be effected to increase the purity; Nauert[44] has recently described operations of this type at the Blackwell Okla. plant.

The cadmium source at an electrolytic plant is the purification residue resulting from the zinc addition stage during preparation of zinc sulphate solutions as described previously. The copper–cadmium–zinc residue is treated with spent electrolyte from the zinc winning cells and copper is then precipitated with zinc dust. A further stage of zinc dust precipitation brings down the cadmium and the filtrate is returned to the zinc plant. After being allowed to oxidise in the air the cadmium is leached in spent cadmium circuit solution and electrolysed.

The electrolysis is quite similar to that of the zinc in many respects. Typical data are summarised:

Anodes	Cathodes	Cell volts	Temp.	Current density	Current efficiency	Power consumption
Pb–Ag	Al	2.5–2.7	20–35°C	ca. 8 mA cm^{-2}	90%	1.5 kWh/kg

The analysis of the cadmium electrolyte at Risdon[45] is typically (in g/l):

Cd	Zn	Fe	Cu	Cl
166	22	0.5	0.00047	0.013

The spent electrolyte, generally speaking[31], contains from 60–90 g/l H_2SO_4 and 20–70 g/l Cd with a zinc content of the same order. A more detailed description of the procedures at the Risdon plant has been given by Johnstone[46] with recent changes outlined by Hopkins and Nixon[45]. An important innovation is the use of polyacrylamide flocculating agent instead of glue as a cell addition. This gives a better deposit, increases the current efficiency and the melting efficiency of the cadmium product and improves the cleaning of the cells and pipelines. The reagent also facilitates the filtration and settling in earlier stages of the treatment of the zinc dust precipitate. In the cell feed it is present at a concentration of 25 ppm by weight. Since the leaching solutions are derived from the purified zinc circuit they are very pure and zinc is the only other metal present in any quantity (apart from the more reactive elements, magnesium and sodium). The zinc is not deposited on electrolysis but is removed periodically to prevent build up by replacing portions of electrolyte with fresh acid or with low zinc content solution from the zinc circuit.

2.6 Chromium

Most chromium ore is reduced to ferrochromium products of various carbon contents and only a relatively small percentage is converted to metal. The Bureau of Mines[47] lists only two U.S. manufacturers of chromium metal namely Union Carbide Corporation Metals Division and Shieldalloy Corporation. Apparently three quarters of this output is obtained by electrolysis of chrome alum solution using ferrochromium as source material. Alternative processes are aluminothermic reduction of chromite and electrolysis of chromic acid solution, although the latter is probably used more for plating purposes. In 1963 the U.S. consumption of pure metal was quite small as can be deduced from tables given in ref.47 and probably less than 5,000 t. In that year U.S. consumption of chromite ore represented about 25% of world output.

The chrome alum electrolyte process was developed by the Bureau of Mines[48] and was first used commercially in 1954. Details of the flow sheet of the Marietta plant of the Union Carbide Corporation have been described by Bacon[49] who discusses the advantages gained over the chromic acid process. The same author also has published elsewhere[50] details of the cell construction. Ferrochromium has been found to be the cheapest starting material. Use of chromite ore is not practicable because of the number of steps involved. The crushed metal is leached in spent anolyte, chrome alum mother liquor and make up sulphuric acid. Ferrous ammonium sulphate mother liquor is then added and after removal of undissolved solids (*e.g.* silica) and conditioning, precipitation of iron sulphate is induced by cooling to about 5°C. Chrome alum is then precipitated from the remaining liquor after ageing. The cell feed is prepared by redissolving this material.

The problems encountered on electrolysis are similar to those existing in manganese electrowinning and arise because of the multivalent nature of chromium. At the anodes, oxidation to the hexavalent state occurs and after exit from the cell the excess anolyte is reduced with SO_2 before recycling to the initial ferrochromium leach. The diffusion of Cr^{6+} and H_2SO_4 into the cathode department is prevented by a cloth diaphragm. Feed solution enters at the cathode, passes through the diaphragm and leaves at the anode. Some excess catholyte is also recirculated to the chrome alum ageing and filtering circuit. Compositions and acidities of the various circuits must be closely controlled; Bacon[49] gives the typical analysis shown in Table 4 (in g/l).

TABLE 4

	Cr^{VI}	Cr^{III}	Cr^{II}	Fe	NH_3	H_2SO_4
Cell feed	0	130	0	0.2	43	3
Catholyte	0	11.5	12.5	0.035	84	pH 2.1–2.4
Anolyte	13	2	0	0.023	24	280

Other cell details are:

Anodes	Cathodes	Temp.	Cell volts	Cathode current density	Current efficiency	Power consumption	Cr purity
Pb–Ag	316 stainless steel	53°C	4.2	70 mA cm^{-2}	45%	18 kWh/kg	99.8%

2.7 MANGANESE

The free world output of electrolytic manganese has been estimated at about 34,000 t per annum[6]. All plants are relatively new; the largest has been constructed by the Foote Mineral Co. at New Johnsonville Tennessee. The new facilities will produce more than 10,000 t/annum and bring the total capacity of the company's three manganese units to over 23,000 t/annum. Other plants are those of Union Carbide Corporation at Marietta, Ohio of over 10,000 t/annum capacity and of Geo-met Reactors Ltd. at Chicoutimi, Quebec of 5400 t annual capacity[6]. Suggested flow sheets for the Knoxville plant of the Electro Manganese Corporation and the Boulder City plant of the Bureau of Mines have also been described by Bacon[49]. In addition, the Bureau of Mines[47] lists the American Potash and Chemical Corp. and refers to South Africa and Japan as exporters of electrolytic manganese.

Manganese is the most electroactive metal won from aqueous solutions and as might be expected control of the process variables is critical. Thus all the more electropositive elements must be removed from the cell feed, oxidation of manganous ion at the anode must be avoided and excessive hydrogen evolution at the cathode prevented by pH control. The electrolyte is prepared by leaching at pH 2.5 an oxide ore with ammonium sulphate–sulphuric acid solution in the cyclic process, followed by neutralisation to pH 6.5 to precipitate iron and aluminium. Molybdenum, arsenic and silica are also carried down. Copper, zinc, nickel and cobalt are then removed by sulphide precipitation. Ferrous sulphate is finally added and oxidised to precipitate ferric hydroxide and bring down other metallic impurities and sulphides. Magnesium builds up during the process and is usually removed by crystallisation at low temperatures of Mn–Mg–NH$_4$-sulphate. Manganese dioxide formation is prevented by addition of 0.10 g/l SO$_2$ to the anolyte and by keeping the manganese ion concentration below about 15 g/l. Catholyte pH is maintained at about pH 7 but hydrogen evolution still occurs and the process has a current efficiency of about 60%. As can be deduced from the above the cells must be designed to control both anolyte and catholyte parameters adequately and also to deal with sludge formation at the anodes. Designs have been proposed for bagging the anodes or the cathodes[51]. In summary, Bacon[49] describes the optimum cell conditions as:

1. Feed solution: Mn as MnSO$_4$ 30–40 g/l, (NH$_4$)$_2$SO$_4$ 125–150 g/l, SO$_2$ 0.10 g/l, glue 0.008–0.016 g/l

2. Anolyte composition: Mn as MnSO$_4$ 10–20 g/l, H$_2$SO$_4$ 25–40 g/l, (NH$_4$)$_2$SO$_4$ 125–150 g/l

3. Catholyte pH 6–7.2

4. Current density 40–60 mA cm^{-2}, cell volts 5.1; power used 8–9 kWh/kg Mn

5. Anodes Pb–Ag(1%): cathode starting sheet "Hastelloy", 316 stainless steel or Ti.

More complete descriptions of the flowsheet for solution purification are given by Bacon[49] and of the cells by Bacon[50] and by Schlain[51]. Schlain also gives references to the development work on the process by the Bureau of Mines. Further details on this work and on other aspects of the manganese research and development programme instigated by the Bureau may be found in ref. 47.

Recently laboratory studies have been carried out with the object of recovering manganese from products of the ferromanganese industry. Methods used include fused salt electrolysis in chloride melts with soluble anodes prepared from scrap manganese, ferromanganese or electrothermic metal[52]; chlorination of ores or slags followed by electrolysis of MnCl$_2$[53]; and leaching of slags followed by aqueous electrolysis[54].

2.8 Gallium

Gallium is produced at only a few locations throughout the world. Some of the companies involved are the Aluminium Company of America and the Eagle-Picher Company (U.S.A.), Johnson Matthey and Co. (United Kingdom), the Spolana plant in Kaznejov near Plzen (Czechoslovakia) and the Nippon Light Metal Company (Japan). Although the metal is as abundant as lead it occurs only as small traces in aluminous minerals, zinc and lead ores and in coal. Present demand is quite small. Hudson[55] estimates that the U.S. market is still under 500 kg/year, and existing facilities could cope with a four or five fold increase in output[47].

The extraction from Bayer process liquor has been described by Hudson[55]. Normally, gallium concentration builds up until the rate of removal as impurity in the alumina is equal to the extraction rate. However, the concentration is still quite low (about 0.2 g/l). Direct electrolysis of this solution is, however, practised in France by the Breteque method[56] using a mercury cathode. The resulting dilute amalgam is then redissolved to make a more concentrated solution for further electrolysis. In the U.S.A. a high organic content precludes this process and alumina is preferentially precipitated so that the relative gallium content may be increased. The final electrolyte is prepared by caustic leaching of a concentrate containing about 0.45% Ga_2O_3. Heavy metals and organic material may be removed from the solution before electrolysis in stainless steel cells at about 90°C. Electrodes are stainless steel with anode area several times larger than cathode area. The cathode current density is about 600 mA cm^{-2} and cell voltage 7-9 V. Current efficiency is low but electricity costs are a relatively unimportant factor in the overall process. Since gallium melts at 30°C the product may easily be collected and refined electrolytically to high purity.

The extraction from flue dust of producer gas plants as practised by Johnson Matthey is indicated by Ryan[57]. The dusts are smelted to remove alumina and silica and the metallic regulus subjected to distillation from strong HCl solution to recover germanium. The residual solution is treated with aluminium scrap to cement heavy metals and gallium is then extracted using isopropyl ether. Certain metals such as arsenic are finally removed by sulphide precipitation and neutralisation brings down iron as the hydroxide. The alkaline solution is electrolysed at platinum electrodes for an initial period then the gallium pool may be used as cathode to facilitate the deposition. The resulting product is a sodium–gallium amalgam which is subjected to further purification.

2.9 Thallium and indium

Both thallium and indium are trace metals and are normally extracted commercially from zinc or lead ores. As with gallium, production capacity can easily satisfy present demands and total annual outputs amount to a few tonnes only. Statistics of this type are usually company confidential but may be published in a few instances. References 6, 47 and 57 may be consulted for lists of manufacturers and related data. Treatment of flue dusts to obtain thallium has been described by Ryan[57] and the Bureau of Mines [47] and takes the form of sulphate leaching followed by removal of heavy metals as sulphides or hydroxides. The electrolyte solution is saturated thallous sulphate at 30°C. Cathodes are of platinum, nickel or stainless steel and anodes of platinum if a high purity product is required. Providing that the solution is replenished with electrolyte and is not allowed to become alkaline, current efficiency is high. The preferred alternative method of deposition is by cementation onto zinc or aluminium. Indium also is generally produced by this method although some electrolytic recovery from plating solutions may be practised. Statistical information where available regarding manufacturers and quantities involved may be found in refs. 6, 47 and 57. A recent paper by Bianchini and Fidani[58] gives flow sheets and process details of indium recovery from zinc. Here cementation is used as the deposition step.

Both metals may be further refined electrolytically.

3. Molten salt processes

3.1 Aluminium

Aluminium is now produced in quantities which make it the major non-ferrous metal and since it is reduced from the oxide entirely by the electrochemical process patented independently by Hall and Heroult in 1888 and 1886, its importance to the topic of electrowinning is apparent. The relevant data for 1968 are now provisionally available and are shown in Table 5 where a comparison is made with the 1967 results. The units are 1000 m tonnes.

This Table is that given in a recent paper by Ernst[59]. The 1968 total for the Soviet sphere is estimated. Similar figures appear in the Mining Annual Review[5]. The percentage increase in output over the previous year of 5.7% is below the ten year average of 9.0% since 1958. However, this may be largely explained by the drop in U.S. production caused by strikes. Since demand in 1968 showed a substantial increase, accumulated stocks were run down and some price increase was necessary. It is anticipated that consumption will continue to rise and accordingly many new plants are to be built to adjust the balance. Summaries of current producers and planned future expansion including the locations of smelters on a country by country basis may be found in refs. 5 and 59. A significant trend is the rapid growth of both output and demand in Asia and Africa.

Three steps are essential in the production of aluminium metal (a) mining of bauxite (b) production of alumina Al_2O_3 and (c) reduction to aluminium. Almost all the metal is made from bauxite ore which may contain very approximately 50% Al_2O_3. The other constituents and typical analyses are given in Pearson's review of the chemical background to the industry[60]. Jamaica is presently the largest producer of bauxite followed by Surinam, Australia and Guyana. However, many other countries have significant outputs and reserves. The tendency is to treat the ore to obtain Al_2O_3 in the country of origin thus saving transport costs and providing important employment especially in underdeveloped countries. In all but a few cases this is achieved by the Bayer process which was patented in 1888 and consists of extraction, decomposition and calcination stages. In brief, the ore is digested with caustic soda solution under pressure to dissolve the alumina as

TABLE 5

Country	1967	1968	% Increase
U.S.	2 965.8	2 952.9	— 0.4
Canada	884.9	891.8	+ 0.8
Japan	382.1	481.9	+ 26.1
Norway	362.2	473.6	+ 30.8
France	361.2	365.5	+ 1.2
W. Germany	252.9	257.4	+ 1.8
Italy	127.7	142.2	+ 11.4
India	96.4	120.1	+ 24.6
Ghana	49.0	109.0	+ 122.4
Australia	92.4	95.0	+ 2.8
Spain	78.5	88.7	+ 13.0
Austria	78.7	85.8	+ 9.0
Greece	71.6	76.3	+ 6.6
Switzerland	72.3	75.9	+ 5.0
Sweden	33.4	57.0	+ 70.7
Jugoslavia	44.6	48.1	+ 7.8
Netherlands	32.1	47.2	+ 47.0
Cameroon	48.3	45.4	— 6.0
Surinam	31.1	43.6	+ 40.2
Brazil	37.0	38.0	+ 2.7
U.K.	39.0	38.2	— 2.1
Mexico	21.5	22.5	+ 4.7
Formosa	15.4	20.0	+ 29.9
Venezuela	2.0	10.0	+ 400.0
Free World Total	6 180.1	6 586.2	+ 6.6
U.S.S.R.	1 350.0		
Hungary	61.8		
Czechoslovakia	55.0		
Poland	92.3		
E. Germany	45.0		
Rumania	52.8		
China	95.0		
Soviet Sphere Total	1 751.9	1 800	+ 2.7
World Total	7 932.0	8 386.2	+ 5.7

aluminate leaving behind a red mud containing iron oxides and other major impurities. At the same time sodium aluminium silicate should also be brought down onto the mud to minimise the silica content of the liquor otherwise the final alumina hydrate is contaminated. It can be seen that this involves some aluminium loss and for this reason ores with high silica percentages ($> 7\%$) are not economically treated by the Bayer process. The aluminate solution is seeded with crystals of hydrate to deposit this material in a form that is easily washed and handled rather than as a gelatinous mass.

$$Al_2O_3 \cdot 3H_2O + 2\ NaOH \underset{}{\overset{\text{extraction}}{\rightleftharpoons}} 2NaAlO_2 + 4H_2O$$

The precipitation is performed in cycles leaving behind a large quantity of crystals to seed the next cycle. Finally, the hydrate is calcined to alumina at temperatures above 1200°C to give product purities greater than 99%.

The oxide is reduced electrolytically to aluminium in a molten cryolite bath. In this "smelting" operation steel container tanks are lined with insulating material followed by a carbon lining which supports and makes electrical connection to the molten aluminium pool cathode. Steel current supplying bars are embedded in the carbon at the bottom of the cell. Carbon anodes dip into the bath from above and a crust of alumina and electrolyte is allowed to freeze around them to contain the evolved gases—carbon monoxide and dioxide. The electrolyte is also frozen around the side linings of the bath. The theoretical decomposition voltage of the reaction is 1.7 V; however, in practice a cell operates at about 4.5 V mainly because of ohmic losses throughout the components. There is also current inefficiency of about 15% magnitude so that the overall energy efficiency is about 35%, the remainder is taken up in maintaining the temperature around 950°C. Current densities are in the range 0.5–1.2 A cm^{-2} with modern cells in the higher ranges passing total currents of perhaps 100,000 A or more. The alumina content of the melt is initially about 5% or higher but this falls as electrolysis proceeds to less than 1%. When this occurs the cell voltage rises rapidly to over 30 V—a phenomenon known as the anode effect—and the frozen salt crust is broken to make further additions of alumina. The metal pool is tapped every day or two always maintaining a residual pad on top of the carbon lining.

Some aspects of cell operation will now be discussed in more detail with emphasis on recent research findings.

The use of carbon is of crucial importance to the operation of aluminium cells and nowadays most plants have integrated the manufacture of anodes and cell linings. A recent review of the topic has appeared in *Light Metal Age*[61]. Apparently the major tonnage components of the carbon industry are the anodes of aluminium cells, which are of two types, prebaked carbon blocks and self-baking Söderberg anodes. In the former the raw materials, coke or anthracite, are heated to induce distillation of volatiles and to promote binding, followed by pressing into the required block shape before a final high temperature baking. The blocks hang from metal rods which are set into them with a poured cast-iron joint and are connected to the bus bars with adjustable clamps. As the carbon is consumed the anodes are lowered further into the melt. The Söderberg anode consists of a steel shell into which the "green" carbon mixture before calcining is introduced from the top in the form of small briquettes. The heat at the bottom of the anode bakes and binds the mixture in this region. As carbon consumption occurs the positions of steel pins which are used as electrical connectors are adjusted to keep them away from the bottom of the anode. The arrangement of the pins may be either horizontal or vertical. For both types of anode the carbon used must be of good quality since most impurities pass eventually into the cathode metal. Indeed, the carbon rather than the alumina is usually the major source of contaminants. Further details of manufacture and carbon quality requirements may be found in refs. 60 and 61. Flowsheets of anode manufacture are also given in ref. 61 and recently a description of the Anaconda Aluminium carbon paste plant which has been redesigned has been published by Martin et al.[62]. In modern practice anode consumption is about 0.5 to 0.55 kg per kg of aluminium. Cell performances using the different anode types are compared briefly in ref. 61 and more critically by Molnar[63]. Nikitin[64] has assessed the various factors influencing consumption including the anode construction and composition and the height above the cathode, and Weiler[65] has discussed performance from the point of view of the properties of coal tar pitch used as raw material. Robozerov and Vetyukov[66] have suggested that the addition of 1 % B_2O_3 would improve the overall efficiency.

The other important usage of carbon is in the cell lining. This may be constructed in two ways either with prebaked blocks cemented with a carbon pitch or alternatively with a "green" monolithic carbon which is rammed into the cell. Many aspects of the choice and installation of the lining were discussed at a meeting on "Pot lining problems in the Aluminium Industry" and the condensed articles have been published with Traylor as editor[67]. The topics treated include economics, installation, properties of carbon materials, insulation, voltage losses and lining failure. The last aspect is important since it determines the lifetime of the cell as a whole. Once the carbon has ruptured the molten aluminium attacks the current collector bars, the insulation and finally the steel shell. Carbon pieces may also short out the electrodes. Towards the end of the cell life iron contamination becomes excessive and voltage drops increase. The heaving of the lining has been treated by Watanabe et al.[68] by considering the expansions of the carbon due to thermal gradient and sodium take up and the forces exerted by the steel shell. Several research groups have measured the penetration of sodium after starting up a new cell and their work suggests that it takes only three months for the sodium to reach the bottom surface. The resultant swelling is further exaggerated by the higher temperature at the top surface. Under the influence of these factors the lining arches and eventually fails when the restraining forces exerted by the shell and the carbon side panels prevent smooth heaving. Quandt and Begany[69] have suggested that iron sulphide impurities in the lining may contribute to the swelling process by formation of sodium sulphide. Panebianco and Bacchiega[70] recommend the use of carbon with largest crystalline form, replacement of sodium salts in the cryolite as far as possible by magnesium and lithium, and correct positioning of the steel tie bars to maximise the failure time. A different type of failure caused by potholes has been investigated by Dell and co-workers[71]. Circulation of the metal pool in combination with Al_2O_3 powder which has fallen to the cell bottom produces a scouring action which enlarges any cracks in the lining. The practical result of the various types of failure is to produce lifetimes of the order of 3 to 5 years.

The thermal aspects of cell construction have also been the subjects of recent investigation as typified by the papers of Panebianco and Bacchiega[72] and Korobov and co-workers[73-5]. It is important to maintain the correct balance between electrolyte temperature, as determined by the insulating properties of the cell construction materials, and the power efficiency. The optimum current density depends on these parameters as well as those of a purely electrochemical nature.

Although the Hall–Heroult process has a long history, detailed mechanistic explanations of the electrode reactions are still a source of controversy, probably because of the complex nature of the melt and the number of possible co-existing species. Further clarification of the reaction between aluminium and the melt, of the anode effect, and of the effect of the various additives should be attempted. At 100% current efficiency one would have a situation in which an oxygen-containing species discharged at the anode to produce CO_2 only, but in fact some percentage of CO is also formed and an equivalent redissolution of aluminium metal occurs. It is generally accepted that in fact the CO production is due to a subsequent reduction of CO_2, for the reasons given by Pearson[60] for example, and Thonstad[76] has shown that this recombination is primarily due to the universally assumed overall reaction:

$$2Al + 3CO_2 \rightarrow Al_2O_3 + 3CO$$

Further, Gjerstad and Welch[77] and also Bersimenko and Vetyukov[78] showed that the rate determining step of the oxidation by CO_2 is the transport of the reductant from the Al/cryolite interface to the gas/cryolite interface. However, the actual reducing species might be Al dissolved in the melt, Na, Na_2F, AlF or even metallic globules, all of which have been postulated at one time or another. Recently Snow and Welch[79] have put forward evidence suggesting that a combination of both Na_2F and AlF gives the best fit to the experimental data. The overall reactions of aluminium with the melt would then be written:

$$2Al(l) + AlF_3 \text{ (melt)} = 3AlF(g)$$

$$Al(l) + 6NaF \text{ (melt)} = 3Na_2F(g) + AlF_3 \text{ (melt)}$$

In practice the phenomenon is readily observed as a "metallic fog" which diffuses away from the liquid aluminium/electrolyte interface. Gerlach et al.[80] measured the oxidation of dissolved aluminium in the melt by CO_2 and indicated that the metal was also present to a large extent as globules.

The effect of various additions to the melt has been reviewed by Welch[81] especially in terms of their influence on the aluminium/cryolite interaction. Calcium fluoride is added to improve conductivity and to lower the liquidus temperature. The operative concentration range is governed by the latter factor at the lower end and by the increasing melt density at the higher end which can lead to inversion with the metallic pad. AlF_3 is usually added to reduce fog formation and hence to increase current efficiency. However, at high concentrations volatilisation becomes a problem. The best method of adding the material has been investigated by Botor and Kalinowski[82]. Other additives which have been accepted into the industry include MgF_2[83] and LiF[84] for reasons similar to those given above for CaF_2. Much of the data on this aspect of the electrolysis has been obtained by Belyaev and co-workers. This author has assessed the contributions of Soviet scientists to the industry as a whole in a recent review[85]. Apart from the additions described above the current efficiency may also be increased by lowering the temperature as reported for example by Saakyan[86] and by Welch[81] among others (previous work is referred to by Welch). This presumably is a reflection of the fact that the rate of the aluminium redissolution reaction is increased by temperature. Whilst most of the additives lower the liquid temperature the lower operating limit is set by the solubility of alumina. In practice the actual current efficiency obtained from a given cell is measured by anode gas analysis. However, Richards and Russell[87] have recently proposed a method, based on use of tracer metals such as copper and titanium, which can determine short term efficiencies as well as the weight of aluminium at the cathode. This paper also gives references to previously published analytical techniques of efficiency measurement.

When alumina is dissolved in cryolite the possibility of the existence of many complex oxy and oxyfluoride ions arises: Foster and Frank[88] have selected AlO_2^- and $AlOF_2^-$ as being the most likely and their conclusions find acceptance with other workers. The ion discharged at the anode to liberate CO_2 is presumably one or other of these two. However, a precise mechanism cannot be written down at this time and a similar situation exists at the cathode where the role of sodium ion discharge, for example, is not clear. Frank and Foster[89] showed that this ion (i.e. Na^+) carries most of the current across the cell. Thonstad[90] established a linear relationship between critical current density of the anode effect and alumina concentration and later suggested that a slow dissociation of the dissolved alumina preceded the discharge step[91]. These observations indicate diffusion control in the transfer of oxygen species to the electrode and that the anode effect occurs when their concentration at the surface approaches zero. The current is then maintained by the discharge of fluoride ions and the product gas contains CF_4. However, the large rise in potential has not been explained satisfactorily.

One of the most important operational variables is the anode–cathode separation, since this parameter influences the diffusion of the redissolved aluminium or reducing agent towards the gaseous anodic products; (for a summary of results on this aspect of current inefficiency, see Welch[81]). It also determines the power dissipation due to ohmic drop in the electrolyte. As shown in Table 6 this part of the voltage loss is quite significant in comparison with busbar and electrode losses[61].

The total contribution of voltage drop at the cathode has been discussed by

TABLE 6

	Voltage drop (V)
Electrolysis–decomposition and polarisation	1.8
Anode effect	0.1
Pot lining	0.4
Carbon electrode and clamp	0.35
Busbars and joints	0.15
Electrolyte	1.9
Total	4.7

White[92] who has presented typical potential profiles and suggested how the various component losses may be minimised. However, these are generally time invariant as distinct from the electrolyte resistance which increases as the anode is consumed. It is important to correct for this so that cell temperature and efficiency may be controlled. The modern trend is to do this automatically using computer scanning techniques. A typical description of the methods of measuring the potentials and controlling the cell resistance has been given by Derkach et al.[93]. The potential probes can also give warning of the onset of the anode effect. The importance of anode gas control has also been stressed, for example by Orman and Lach[94] since as shown previously this can give a good indication of cell current efficiency. McMinn[95] has reviewed this topic and suggests that Al_2O_3 concentration must be stabilised for optimum results. This involves continuous alumina feeding which could also be computer controlled. The combination of controlling both Al_2O_3 supply and electrode separation has been discussed by McMahon and Dirth[96] and apparently practised at the Nippon Light Metal Company[97]. Putman and Carlson[98] have indicated how the computer may be applied as a controller throughout the industry in the Bayer process, as well as on the pot line. In large modern cells operating in the 100,000 A range the fields set up by the high currents produce undulations in the molten metal and the anode–cathode distance may then not be constant. For adequate process control these disturbances must be minimised by suitable arrangements of the current leads and by sectioning the cathode as discussed by Gefter et al.[99], Rudakov et al.[100] and Kostin[101]. This is one aspect of the trend in the industry towards larger cells and higher efficiency. The developments occurring since 1910 have been reviewed by Schmidt–Hatting[102] and the principles underlying the modern advances in design of plant have been discussed by Bobkov[103]. The factors that determine the location of facilities and the structure and economic aspects of the world industry have been detailed by Brubaker[104].

Numerous attempts have been made to devise other reduction methods for aluminium, none of which has been adopted commercially. This topic has been treated by Pearson[60].

3.2 Magnesium

The electrolytic manufacture of magnesium is well established and the historical developments are described by Eger[105], Kirk–Othmer[18] and Emley[106]. Accordingly there have been few recent fundamental developments and the standard sources may be consulted for more detailed coverage than will be attempted here. General considerations to be kept in mind when discussing the cell types now used evolve around the fact that magnesium is lighter than the melt, and that the melts themselves tend to be hygroscopic and sometimes of relatively low conductivity. The two processes now used were developed by Dow Chemical Company and by I.G. Farbenindustrie; the main difference between them is that the former accepts a magnesium chloride feed prepared from sea water which contains about 25% water, whereas the latter operates on dehydrated molten carnalite. The cells are compared in Table 7 which is based on that given by Krenzke[107].

TABLE 7

Process	Cell body	Volts/cell	Cath. eff.	kWh/kg Mg	Anode–cathode spacing	Temp. (°C)	Graphite kg/t Mg	Mg purity (%)
Dow	Steel	6.5	78	18.5	4 cm	700	100	99.8
I.G.F.	Ceramic	7.5	90	18.5	13 cm	790	15	99.8

ANALYSES OF FEED (a) AND ELECTROLYTE (b)

Process	$MgCl_2$ (%)	$CaCl_2$ (%)	NaCl (%)	KCl (%)	CaF_2 (%)	MgO (%)	H_2O (%)
Dow (a)	72	0.5	1.0			1.5	25
Dow (b)	20	20	57	2	1		
I.G.F. (a)	96.3	1.4	2.2			0.1	
I.G.F. (b)	12	35	28	24	1		

In both cells graphite anodes dip into the electrolyte from above; the molten magnesium is diverted upwards by the cathodes in the Dow cell which are very close to the anodes and conically shaped. On the other hand in the I.G.F. design ceramic spacers between the electrodes separate the products. Some of the high water content of the sea water feed flashes off on addition to the melt and reacts with chlorine to give high HCl impurity; at the same time a pronounced graphite consumption and reduced cathodic current efficiency are inevitable. However, these factors are balanced by the cheapness of the source material and the lower power consumption and higher current density arising from the closely spaced electrodes. Other variables which arise in a detailed comparison of the two cells are more fully treated by Krenzke[107]. Present practice is to heat the cell by ohmic drop rather than by a fossil fuel as originally designed.

Several variants on the present method have been proposed, although none has achieved commercial exploitation. These include the use of tin plated or zinc plated cathodes[108], magnesium oxide feed[109], electrolytes lighter than magnesium[110] and liquid cathodes[111]. The latter two are attempts to contain the magnesium at the bottom of the cell rather than at the top where it can all too easily react with the evolved chlorine. Unfortunately, separation from a liquid amalgam cathode increases the overall operational cost although high purity metal can be obtained.

The competing non-electrochemical reaction for magnesium winning is that of reduction of dolomite (MgO CaO) by ferrosilicon. At present most magnesium is made electrolytically but a recent review[6] has suggested that many new plants might use the thermal process because of ready availability of dolomite and improvement in materials of construction. Power requirements of about 11 kWh/kg are also lower than the usual electrolytic plant which may consume about 22 kWh/kg. Additionally, the thermic plants can be erected rapidly, operations may be on a small scale and the lower purity may not be important for some applications. Disadvantages include the high cost of ferrosilicon and the batch mode of operation. The present demand for magnesium is such that new plants are becoming necessary; most of the expansion scheduled for starting before 1971 is for electrolytic plants[5]. Continued growth can be expected as new markets are found for magnesium products and in particular, as the aluminium industry expands, since Mg–Al alloys represent the largest use of the metal. In fact the present (1968) capacity of the U.S.A., *i.e.* about 120,000 t, will be increased to 220,000 t by 1972. These data are included in Table 8 which lists the present and planned centres of production. References 5, 6 and 112 may be consulted to obtain the actual output statistics which amounted to 200,000 t in 1967 and 206,000t in 1968. The U.S.A. accounts for 50% of these quantities. Some details of the announced expansion and reviews of the industry are given in refs. 5, 112 and 113.

Most of the recently published research on the present electrolysers is of Russian origin and emphasises the effects of operating conditions on product purity[114, 115]. However, Höy-Petersen[116] in a useful paper has discussed various aspects of the operation of I-G cells including materials of construction, structure and positioning of electrodes, composition of electrolyte, effect of impurities, heat and energy balance, chlorine recovery, and costing. References to recent Russian patents are also given. The fact that high chlorine recovery is possible may be utilised when the I-G cell is coupled with a titanium or zirconium producing unit since chlorination of the oxides from the ores produces a metal chloride which is then reduced with magnesium. The magnesium chloride is recycled to the electrolyser.

TABLE 8

Country	Company	Raw material	Process	Estimated capacity (1966)	Planned capacity (1971–2)
U.S.A.	Alamet Div. of Calumet and Hecla Inc.	Dolomite	E	7,000	9,000
	Dow Chemical Co. (Freeport, Tex.)	Sea water	E	100,000	125,000
	Dow Chemical Co. (Dalesport, Wash.)	Brine	E	—	25,000
	Titanium Metals Corp.	Recycled $MgCl_2$	E	12,000	15,000
	American Magnesium Co.	Brine	E	—	30,000
	National Lead Co.	Brine	E	—	45,000
	Oregon Metallurgical	—	—	—	10,000
	Nalco Chemical	—	—	—	10,000
	Nelco Div. of Charles Pfizer Co.	Dolomite	F	5,000	—
Canada	Dominion Magnesium Ltd.	Dolomite	F	11,300	11,300
France	Societe Magnetherme	Dolomite	F	3,900	7,700
W. Germany	Norsk-Hydro and Salzdetfurth	—	—	—	26,000
Italy	Soc. Italiano per il Magnesio	Dolomite	F	7,000	9,000
	Soc. Italiano per il Magnesio	—	—	—	13,000
Norway	Norsk-Hydro		E	29,700	48,000
U.K.	Magnesium Elektron Ltd.	Dolomite	F	5,000	closed
Japan	Furukawa Magnesium Co.	Dolomite	F	5,000	6,600
	Ube Kosan K.K.		F	2,000	2,000
U.S.S.R.			E	50,000	—
China				1,100	—

E = electrolytic   F = ferrosilicon

3.3 TITANIUM

In view of the interesting problems which arise in the electrolytic production of titanium, more space will be devoted to this metal than is perhaps deserved on the basis of production quantities alone. Since 1940 much effort has been expended in the U.S.A. on this work and earlier results might have been expected if it were not for the recession in demand in the late 1950's. At the present time there are probably only two pilot scale plants in operation. The various difficulties encountered have been reviewed by Kroll[117] and by Senderoff[118].

Initially it should be stressed that all attempts to deposit titanium in aqueous or organic media have been failures; usually a grossly contaminated product results. In molten salt electrolysis the feed material is the tetrachloride, a liquid which is not particularly soluble in molten chlorides and volatilises rapidly above 750°C. At the same time a reaction with titanium metal to give the di and tri chlorides occurs so that it is necessary to regulate the bath valence to preclude this possibility. Most investigators find that this factor is the critical one as regards the efficiency and quality of electrodeposition, thus Leone et al.[119] recommend an average valence of 2.0 to 2.1 and Rand and Reimert[120] suggest that in operation an equilibrium value of about 2.10 is attained. Working at higher valencies simply results in some fraction of the current being spent on reduction of ions to the divalent state. Given this valency requirement it is then unfortunately the case that $TiCl_2$ oxidises at the anode to $TiCl_4$ which may be lost as evolved gas.

It becomes necessary, therefore, to separate the anode and cathode compartments and choice of a suitable membrane is probably the most difficult aspect of the research, since $TiCl_2$ attacks most ceramics and metals. Leone et al.[119] used an alumina diaphragm and experienced failure due to cracking and penetration after about 30 cycles which they associated with the actual passage of current through the ceramic. Material simply immersed in the electrolyte for long periods showed little sign of attack. The solution adopted by the New Jersey Zinc Company as described by Myhren et al.[121] was to use a titanium metal diaphragm formed by deposition onto a nickel mesh screen. Eventually it was decided that the quality of the metal on the screen was of sufficient quality as to eliminate the need for a separate cathode and this system was adopted in the pilot plant experiments.

All of the fairly successful investigations so far mentioned have used low melting chloride eutectics and operating temperatures of about 550°C, simply to prevent $TiCl_4$ losses. Other investigators including Brenner and Senderoff[122] and Steinberg et al.[123] have used K_2TiF_6 in chloride melts. Whilst eliminating $TiCl_4$ evolution this type of bath gradually becomes enriched in fluoride as chlorine is evolved at the anode and this makes the leaching of the titanium crystals more difficult. Also the tetravalent salt tends to attack the metal. According to Smirnov et al.[124] K_2TiF_4 deposition may occur and lower the product purity and increase polarisation. Baitenev et al.[125] have examined the possibility of using oxide feed by studying the system K_2TiF_6–$NaCl$–TiO_2. It is difficult to see how metal contamination by oxygen can be avoided in this situation. A similar situation exists in the use of soluble titanium-containing anodes, e.g. TiO and TiC although the valency problem is avoided. This work probably comes within the realm of refining and Senderoff's review[118] may be consulted for references to the extensive work by the U.S. Bureau of Mines.

Most Russian work also deals with refining. Anufrieva and co-workers have published several papers in recent years on this aspect[126-131]. However, other groups have obtained results more related to electrowinning. Thus Gitman and Sheiko[132] investigated the cathodic deposits on electrolysis of Na_2TiF_6 in alkali chloride melts and Gopienko and Timofeev[133] studied the reaction of $TiCl_4$ with fused halides. Earlier, Ivanov and others[134] had electrolysed $TiCl_4$ and TiO_2 in molten chlorides.

Production figures for titanium metal are now withheld by the Bureau of Mines and are regarded as company confidential. However, estimates for the U.S. industry are of the order of 15,000–25,000 t/annum[5, 6, 135, 136]. The three major producers are Titanium Metals Corporation, Reactive Metals Inc., and Oregon Metallurgical Corporation. Titanium tetrachloride made from rutile is reduced with magnesium or sodium. The U.S. which is the world's major consumer and producer, also imports titanium metal from Japan, and recently from Russia.

Electrolytic production is as yet in the pilot plant stage. The New Jersey Zinc process has been mentioned above. Previous patents of this Company may be found in ref. 121. The cell of 3.4 m^3 volume using Li–Na–K–Cl eutectic operates at 500°C with a graphite anode and the metal screen cathode already discussed. Argon purging is required. Initially TiCl$_4$ is fed to the cell to a solution composition of about 3% Ti and valency adjusted with auxiliary cathodes. Then current is switched entirely to the primary cathode and deposit build up commences with a current density of about 0.5 A cm^{-2} and efficiency of about 90%. This cycle is repeated from 3 to 6 times and the cathode is removed from the melt after some cooling to 400°C and allowed to further cool in the atmosphere, surprisingly without much contamination. The deposit is crushed and leached and the dragged out salt recovered. A scale up to a 20 t/day plant with 62 cells is envisaged.

The Titanium Metals Corporation cell has recently been described by Priscu et al.[137] (see also ref. 136) although in less detail. Rand and Reimert[120] refer to Russian investigations on a pilot plant scale.

Obviously once the rather special difficulties associated with the titanium system are overcome electrolytic processes will become increasingly important since the product is of much higher purity than that obtained in Kroll reduction. The industry as a whole does depend, however, to a marked extent on the expansion of the space and aircraft programmes although new uses for the metal are continually developed.

3.4 Niobium and tantalum

The major ore sources for niobium and tantalum are the minerals colombite and tantalite (Fe, Mn)(Ta, Nb)$_2$O$_6$ and pyrochlore (Na, Ca)Nb$_2$O$_6$F. Because of their occurrence together and their similar chemical properties the separation of the metals represents an important part in the overall preparation. Apparently it is necessary to reduce the fraction of one in the other to less than 0.1% and amounts of other impurities must also be minimised to obtain the required metallurgical characteristics. Electrochemical reduction should be attractive from this point of view. Production statistics of niobium and tantalum are difficult to estimate; most data relate to shipments of ore concentrates and in any case the alloys ferro-colombium and ferro-tantalum may be produced directly without isolating the pure metals. However, in 1960 about 150 t of each metal were made in the U.S.A.[138] and installed capacity in 1965 in the U.S.A. was about 300 t/annum[47], several times larger than the rest of the free world. Known producers of the metals and the

TABLE 9

Company	Country	Reduction method	
Fansteel Metallurgical Corp.	U.S.A.	E*	Na
Electro Metallurgical Co. (Union Carbide Corp.)	U.S.A.	E	C
Hermann C. Starck, Berlin	W. Germany	E	Na
Kawecki Chemical Co.	U.S.A.	Na	
National Research Corp.	U.S.A.	Na	
Wah Chang Corp.	U.S.A.	Na	C
Murex Ltd.	U.K.	Na	
Kennametal Inc.	U.S.A.	C	
Du Pont	U.S.A.	H_2	
Stauffer Chemical Co.	U.S.A.	H_2	
Metallurgie–Hoboken	Belgium	—	

E = electrolysis; Na = sodium reduction of halide; C = carbon reduction of oxide; H_2 = hydrogen reduction of halide; * no longer used[139].

reduction process used are given in Table 9. This information was obtained by communication with Fansteel, Murex, Hermann C. Starck (who recently took over the activities of CIBA in this field) and Union Carbide.

Although three companies are listed as using an electrochemical preparation there seems to have been a recent trend away from this method towards that of sodium reduction. In any case niobium is difficult to prepare electrolytically because of the greater ease with which intermediate valency states are produced, and is made almost entirely by thermochemical reactions.

Three systems have been studied. Initially tantalum was deposited by Balke[140] and later by Driggs and Lilliendahl[141] using the oxide Ta_2O_5 as feed in a fluoride or fluotantalate bath. This method is open to the criticism that the cathode deposit is contaminated with carbon and oxygen resulting from CO diffusing across from the anode. Electrolysis of fluotantalate K_2TaF_7 results in a build up of fluoride in the melt which gradually becomes more viscous and a batch process seems to be unavoidable. Use of a chloride feed into a fluoride bath seems to be the most satisfactory solution. The advantages and disadvantages of the various methods have been reviewed by Senderoff[118] and by Rockenbauer[142]. In addition, Rockenbauer describes the practices developed by CIBA now operated by Hermann Starck. The chloride feed system is favoured and has been adapted to continuous running. The cell consists of a carbon crucible and anodes in a metal oven with a central cathode. The KF/KCl/NaCl mixture is melted and gaseous $TaCl_5$ added to give the desired concentration, and electrolysed at 850–950°C at a cathodic current density of 2.7 A cm^{-2} and cell voltage of 6–8 V. The total current is 1500 A. The cathode is stripped periodically by lifting it through a concentric cylinder and catching the dendrites in a tray. The whole process can be automated and current efficiency is 85–88%. The cell may also be used for niobium electrolysis although the efficiency is reduced to about 30%. However, both metals can be obtained with high purity and may be relatively easily fabricated. Present development of the process is concerned with the corrosive nature of the melt and with further automation.

The complete operations of the Fansteel niobium–tantalum plant have been summarised by Soisson et al.[138]. K_2TaF_7 is electrolysed in batches of about 50 kg in cast iron cells with graphite anodes and the pot as cathode. The tantalum is deposited as crystalline aggregates in a fused melt and is separated by pulverising followed by various washes. Niobium is not produced electrochemically. A process flowsheet of the Electro Metallurgical Co., has been given by Chilton[143]. Fluotantalate is again the compound submitted to electrolysis.

Techniques for processing the ores and producing the cell feeds are obviously important and the books by Fairbrother[144] and by Sisco and Epremian[145] may be consulted in this context in addition to the papers describing the specific plants mentioned above. Recently Bowles[146] has attempted to separate the metals electrochemically following the dissolution of pyrochlore and colombite in fluoride melts. Some increase of the Nb/Ta ratio was achieved. The problems of carbide formation were also investigated. Decroly et al.[147] have investigated the electrodeposition of both metals from $KF-K_2MF_7$ baths and have suggested that the deposited metal may be redissolved through a chemical reaction to give species of lower valency. Some of these species are oxidised at the anode and the rest are redeposited at the cathode. The process obviously leads to current inefficiency and is more apparent in niobium melts as already indicated. Wong and Kirby[148] gave the optimum operating conditions for melts of the type $KCl-KF-K_2NbF_7-Nb_2O_5$. The highest current efficiency was only 50% and carbon and nitrogen contents were an order of magnitude higher than those reported by Rockenbauer[142] for metal deposited from chloride feedstock, although oxygen contents were comparable.

3.5 RARE EARTH METALS

The rare earths are extracted from monazite sand, bastnasite and other ores as hydrated or anhydrous chlorides and reduced to the metal by the usual methods. The mixture known as mischmetal obtained from the extract without separating into component elements is the only metallic product prepared in tonnage quantities by electrochemical means. Some of the metals themselves, especially cerium and lanthanum, and the mixture didymium (Nd + Pr) are also made commercially by electrolysis in amounts ranging from a few hundred kg/annum downwards. Exact figures are not available. Because of the fairly small quantities involved it is more economic to produce the elements by metallic reduction techniques in batch processes as typified by the description published by Lever and Payne[149]. On a larger scale electrolysis may become more suitable especially when the metal is liquid at the operating temperature. The higher melting metals and yttrium are made by calcium or lithium reduction and those with high vapour pressure—samarium, europium and ytterbium metals—by reduction–sublimation with lanthanum or mischmetal[150].

Commercial producers by electrolysis include Ronson Metals Corporation, American Metallurgical Products Company and Molybdenum Corporation of America, all in the U.S.A.; Th. Goldschmidt A-G and Prometheus Metallwerke, West Germany; and Treibacher Chemische Werke A-G, Austria. A list of suppliers of rare earth metals, alloys and compounds has been compiled by Smith and Gschneidner[151] of the Rare Earth Information Centre, Ames Laboratory. Hirschhorn[152] has summarised the methods and cells used in fused salt electrolysis, which is at present based on the use of chloride feedstock. This is normally produced by wet chemical treatment of the ore and is consequently partially hydrated so that some drying procedure is needed. A new method developed by Th. Goldschmidt A–G[153] gives an anhydrous material directly which is especially suited to electrolysis. Two types of ceramic cell are in general use; both have an iron cathode at the bottom on which the pool of molten mischmetal collects. Iron contamination is often not a serious problem since there are applications for ferrous–rare earth alloys and the pots themselves may also then be made of iron rather than graphite. The electrolyte is based on $NaCl-KCl-CaCl_2$ and feedstock, and is run at 800–900°C. Cell voltages are between 12 and 14 V and currents typically 1500–2500 A. Current efficiency is about 50%. Anodes are carbon or graphite. The metal may be deposited as a solid on a cooled cathode or as a liquid. However, it is ladled from the cell in the molten form together with some melt to prevent oxidation. The two cells are reported in refs. 154 and 155.

A report by Morrice et al.[156] summarises the work of the Bureau of Mines on the development of a new process using oxide feedstocks. The melt is based on a mixture of LiF with the rare earth fluoride run at high temperatures to produce liquid metals. The cathodes are tungsten or molybdenum rods or consumable metals when alloy formation is desired, anodes and crucibles are graphite, and an inert atmosphere is provided. The materials of construction are tested almost to the limit during yttrium production at 1620°C; other metals and alloys require gradually less severe conditions depending on their melting points. The liquid products drop from the cathodes and may be tapped from a pool or allowed to freeze by keeping the bottom of the cell relatively cool to prevent reaction with the melt. Current efficiencies depend on this dissolution and increase at the lower temperature; that for cerium may be as high as 96%. As an example of the operating conditions the data for gadolinium winning may be quoted:

Temperature of electrolyte 1370°C, cell bottom 810°C, DC current 95 A, cathode current density 31.4 A cm^{-2}, cell voltage 22 V, current efficiency 50%. Additional AC power is supplied to maintain the temperature. Two of these cells are presently being developed on a commercial basis by the Molybdenum Corporation of America.

3.6 LITHIUM

Several processes for the extraction of lithium from its ores have been described by Laidler[157]. They depend on an acid-roast to extract the metal as the sulphate. After purification the lithium may then be precipitated as the carbonate or hydroxide and used in this form and only a small percentage of material is converted to metal. This is always done by electrolysis of the chloride which is easily made from either of the above compounds. The cells are similar to those used for making sodium and consist of a carbon anode and a steel cathode separated by a perforated steel diaphragm to prevent mixing of the products. In most operations the LiCl–KCl eutectic is employed to lower the working temperature and because pure LiCl is very corrosive towards the iron container and other materials of construction. The geometrical arrangement of the components is important as it determines the way in which the lithium and the chlorine are separated and collected. Several designs have been suggested. Mantell[158] shows that of Degussa Rheinfelder in which the anodes enter horizontally from the sides of the cell and the cathode from the bottom; on the other hand American cells have vertical top entering anodes and the pot may sometimes be used as cathode. More recently Motock[159] has considered the various possible combinations and the one finally developed had a bottom entering anode (15 cm diameter, 30 cm high) and a top entering cathode (27 cm inside diameter, 30 cm high). This cell was designed for total currents of about 1000 A. Average operating details are given below:

Cell volts	Temp.	kWh/kg Li	Current eff.	Total current	Cathode current density
6.8	450°C	36	80%	1420 A	0.5 A cm^{-2}

The lithium and the chlorine both pass to the top of the cell, the lithium to a holding tank where it is kept molten. Both products have purities better than 99%. The paper describes in addition, the corrosion problems, the material and energy balances, and the preparation of raw materials, which should be very pure (LiCl > 99.6% KCl > 98.92%). A pre-electrolysis is performed and continuous feeding gives better results than the batch method.

Previously a 30,000 A cell was operated, also at Olin Mathieson Chemical Corporation, on a trial basis by Schoepfer et al.[160], the objectives being to improve product collection and purity and to work at high currents with continuous feed. In this design, four graphite anodes 40 cm in diameter entered from the bottom and were surrounded concentrically by steel ring cathodes. Perforated stainless steel diaphragms served to channel the products. Operating conditions were similar to those above and an inert gas blanket was used in both cases.

The major impurity in lithium metal is sodium which is co-deposited if present in the melt. Current production in the U.S.A. is about 250 t/annum [61].

3.7 Beryllium

Beryllium may be produced by electrolysis of the chloride although at present Pechiney is the only organisation to do this commercially. General Astrometals, New York are believed to be piloting an electrowinning process in collaboration with the French company and have conducted some development work under a NASA contract using a mercury cathode cell. Electrorefining of the metal may be more widespread although the demand for ultra high purity is probably minimal and the alternative process, that of magnesium reduction of the fluoride, may give a product suitable for most purposes. The structure of the industry including a historical survey, the uses of the metal and its alloys and the current producers is available[161]. Present production seems to be of the order of a few hundred t/annum.

The established electrolytic method has been described by Windecker[162] and consists of batch operation of a KCl–NaCl melt containing initially about 14% $BeCl_2$. The stainless steel cell acts as cathode with a central graphite anode. Maximum current densities are 1 A cm^{-2} and cell voltages between 6 and 9 V. As the beryllium content of the melt decreases, the temperature must be increased from about 760° to over 900°C. On completion the melt and the product metal in flake form are transferred to a receiver vessel by tilting the cell. Separation is achieved by aqueous washes and the salts are reclaimed. An improved method uses a $BeCl_2$–NaCl (54–46%) melt which is depleted to about 45% $BeCl_2$ before product separation. Since this mixture may be operated at a lower temperature, 370°C, corrosion problems are less severe and the metal does not oxidise rapidly when it is withdrawn. Cathodic current densities are about 0.1 A cm^{-2} and cell voltages 4 to 6 V. The metal may be deposited onto a perforated inner liner which is removed, drained above the cell and then dumped into water, or alternatively the melt may be siphoned off to leave the beryllium flakes on the cell walls. Purity greater than 99.4% may be obtained.

The various methods of reducing beryllium compounds to the metal, including electrolysis, have been reviewed by Magel[163], Jaeger[164] and Sibert and Hultquist[165], among others. Apart from the chloride process described above attempts have been made to utilise BeO usually in fluoride melts. The latest work in this direction is that of Wong *et al.*[166] who found that the limited solubility of the oxide (5% in BeF_2–LiF) might be a problem. Some contamination from the cathode was also evident. Earlier workers tried direct electrolysis of fluorides (for details see refs. 163–165), and an amalgam procedure was developed by Sylvania[167].

3.8 BORON

Much effort has been expended over a period of years in attempts to make pure boron. Most thermochemical methods involve the decomposition of halides in the presence of hydrogen onto hot wires. The only commercial electrochemical methods were developed by Cooper[168]. This work has been described as articles in several books[169]. References to other electrochemical investigations are also given by Monnier et al.[170]. Until recently U.S. demand was about 500 kg/month and was satisfied by the output from Cooper Metallurgical Associates. Probably the only other production is at Model City, New York where the ^{10}B isotope is made on a larger scale (of the order of 100 t/year) for the Atomic Energy Commission.

Cooper developed two processes, one depends on the electrolysis of potassium fluoborate KBF_4 in KCl, the second on electrolysis of B_2O_3–KBF_4–KCl. The first method has been extended by using a BCl_3 feed thus preventing fluoride build up in the bath[171]. Similar cells are used in all cases; these consist of a graphite lined pot with an iron cathode suspended centrally in the melt. It was found more suitable to use suspended graphite anodes than to use the cell lining. Chlorine is the anodic product in the KBF_4–KCl melt, and oxygen when B_2O_3 is added. Cell temperatures are about 800°C; voltages 6 to 12 V, and 3000 A are consumed at efficiencies of about 75%. The purity of the boron is well over 99%. Carbon and iron are the major impurities.

Apparently there is some discrepancy concerning the purity of boron obtainable from a given melt under defined conditions. Russell and Kelly[172] have indicated that many groups report purities much lower than those of Cooper, and have attempted to evaluate statistically some of the factors involved. In this work the highest boron percentage was 98.1% and oxygen levels were quite high, always greater than 1%, using melts containing B_2O_3.

4. Other metals

The rest of the alkali and alkaline earth metals (K, Rb, Cs, Ca, Sr, Ba) were produced previously by electrolysis but this method has now been abandoned in favour of thermochemical reduction, the main reason being the high vapour pressure of the metal at the temperature of formation. This is advantageous when the metal can be distilled from a reaction mixture but very inconvenient in an electrolysis cell. Examination of the melting and boiling points of the elements and the typical operating temperatures of molten salt baths brings home this point. The fact that the metals may be lighter than the bath is an additional complication since the apparatus must be designed to prevent air contamination and to separate the metallic and gaseous products. A paper on calcium electrolysis by Threadgill[173] illustrates these aspects. However, descriptions of the methods pertinent to the alkali metals may be found in Hampel's book[174].

Laboratory experiments on deposition of germanium from borate and silicate melts have been reported by Ruis et al.[175]. The Broken Hill Associated Smelters Proprietary carried out the Halkyn process to an advanced pilot scale before abandoning it for economic reasons. Here lead is won from galena dissolved in a lead chloride bath. Several articles refer to this method[176]. Schlain[51] has described the winning of antimony from sulphide solutions using steel cathodes and steel or lead anodes, and Russian groups have investigated the electrochemistry of the various complex sulphide ions involved[177] and the way they affect the current efficiency. Mercea et al.[178] have described a method of obtaining antimony from lead anode sludge. This material is cast into anodes containing about 65% antimony and recovered from an acidic bath at elevated temperatures. Several investigations on bismuth electrodeposition have been performed both in aqueous and in molten salt media. The papers of Morris et al.[179] and Gruzensky[180] contain further references.

Steinberg and coworkers at Horizons Incorporated have developed the reduction of K_2ZrF_6 to give zirconium. One paper[181] describes the expansion from laboratory scale upwards to a cell capable of making approx. 15 kg in each run. The metal was obtained in good yield and purity and in quantities approaching those of the original Kroll sponge producing reactors. The apparatus consisted of a graphite crucible anode contained in an insulated steel container. The working section of the cathode was stainless steel, current density 2.6–3 A cm^{-2}, operating temperature 830–850°C. The cell charge was a mixture of 76.5 kg NaCl and 22.5 kg K_2ZrF_6. Senderoff[118] has reviewed this and other work and concludes that the use of all-chloride baths is not successful perhaps because it is difficult to prepare pure $ZrCl_4$. However, Smirnov et al.[182] prefer chloride systems and showed that the reduction potential indicated the divalent species as a stable intermediate. Winand[183] on the other hand found no evidence for intermediates but rather reductions corresponding to free Zr^{4+} ions and to complex ions. At the anode, oxidation of the complex ions as well as chlorine evolution was observed. The problems of current efficiency would then be similar to those encountered in the titanium work. Most of the electrochemical experiments on hafnium and vanadium have dealt with electrorefining.

Much work has been done on the electrodeposition of molybdenum from aqueous and organic solvents but Senderoff and Brenner[184] first obtained the metal without serious contamination by reduction of K_3MoCl_6 in molten alkali chloride mixtures. The metal may also be won from melts of oxygenated salts, for example Heinen and Baker[185] have studied the system $Na_2B_4O_7$ (54%)–NaCl(23.6%)–NaF (14.9%)–MoO_3 (7.5%). MoO_3 was added to the bath at intervals during the electrolysis. It was found that a cathodic crucible and anodic central electrode gave the better result and prevented carbide formation which is often a problem in oxygen-containing melts. Similarly tungsten is usually won from this type of melt, for example Fink and Ma[186] dissolved WO_3 or wolframite in borates and produced a powder deposit. They also used phosphate baths as did Zadra and Gomes[187] Mixed phosphate–borate melts were also tried by these workers and by Newdick and Bowles[188] and scheelite ($CaWO_4$) was used as cell feed. Carbide formation was again a problem in the latter case. It is unlikely that commercial applications of electrowinning for tungsten or molybdenum will be realised since the two metals are very easily prepared by hydrogen reduction of the purified oxides.

An interesting method of recovery of rhenium from flue dusts of molybdenum sulphide roasters has been described by Churchwood and Rosenbaum[189]. The dusts are leached with water and rhenium preferentially extracted into an organic solvent phase and recovered by stripping with perchloric acid solution. After addition of a sulphate, electrolysis at 200–400 mA cm^{-2} at a copper cathode yields a deposit of rhenium containing only 0.1% Mo. This method may well be adopted commercially.

In the early days of the atomic energy programme uranium was made by electrolysis at Westinghouse but currently the metal is made by other methods. Both the U.K.A.E.A. and the U.S.A.E.C. have investigated electrowinning fairly recently and the work in the U.S.A. was terminated in 1966[190]. The problems arising from the use of UO_2 feedstock include the temperature balance of the cell, the usual difficulty of intermediate valency states, materials of construction and limited solubility of UO_2. Contamination by oxygen and carbon is also difficult to avoid, as might be expected. The Bureau of Mines has studied uranium as part of its electrowinning programme[191] and similar difficulties were encountered. A diaphragm was used to prevent oxide contamination of the metal and qualities of 99.8% were obtained with carbon as the major impurity. Similar studies were made on thorium alloy winning[192] using feeds of ThO_2. The U.K.A.E.A. also developed thorium winning on a pilot plant scale[193].

5. Summary and conclusions

Because of the wealth of detail available on electrowinning we have given fairly brief descriptions only of the salient features of the methods used to obtain the various metals. For this reason diagrams of cells and flowsheets have been omitted. However, wherever possible further sources of information have been listed and an effort has been made to ensure that these are contemporary. The references have not been presented in a particularly critical manner but rather to serve as indications of current trends in research and industrial practice. The quantities of each metal produced by electrowinning and the percentage of the total which this represents have been summarised in Table 10 (total production refers to all methods).

Current trends indicate that the aluminium industry will continue to expand and

TABLE 10

Metal	Quantity produced by electrowinning	% of total	Remarks
Al	8,400,000	100	
Zn	2,400,000	53	
Cu	540,000	10	
Na	163,000 (U.S.A.)	100	
Mg	206,000	80	
Ni	48,000	11	
Mn	37,500 (free world)	—	
Co	10,000	50	
Cd	6,500	50	
Cr	$\simeq$ 3,000 (U.S.A.)	$\simeq$ 75 (U.S.A.)	
Li	250 (U.S.A.)	100	
B	$\simeq$ 100 (U.S.A.)	—	U.S. Atomic Energy Commission—conjectural
Be	—	—	Electrolytic by Pechiney only. Total U.S. market: few hundred t.
Ta	—	—	At present only Hermann C. Starck produces by electrolysis.
Nb	pilot scale	—	Total U.S. capacity about 300 t for each metal.
Rare earths	a few tonnes	—	Only mischmetal in tonnage quantities.
Ga	0.5 (U.S.A.)	—	Mostly by electrolysis.
Ti	Pilot scale	—	Total U.S. production about 20,000 t.
Tl	—	—	
In	—	—	Total prod. probably just under 100 t.

this will be reflected in increases in magnesium production where most of the scheduled new plants are electrolytic. The application of new techniques in hydrometallurgy coupled with the need to treat low grade ore will enhance the quantity of copper electrowon, especially in the U.S.A. The well established electrolytic methods of zinc and cadmium production should maintain their position. However, there seems to be a trend towards thermochemical reduction for the reactive metals which are produced on a fairly small scale, although electrolytic titanium production would be a distinct possibility if the long awaited expansion of this industry should occur.

Acknowledgements

Thanks are due to the following individuals and organisations who have patiently answered our enquiries into many aspects of electrowinning. E. W. Page, Anglo American Corp. Ltd.; L. Theys, Union Minière; L. O. Fines, Andes Copper Mining Co.; A. H. Neal, Inspiration Consolidated Copper Co.; K. G. Robb, International Nickel; Outokumpu Oy; A. R. L. Chivers, Zinc Development Assoc.; W. Chynoweth, S. T. Wlodek, Union Carbide Corp.; G. J. Korinek, CIBA Corp.; O. N. Carlson, Iowa State University; L. M. Pidgeon, University of Toronto; R. C. Davis, American Potash and Chemical Corp.; J. J. McGlynn, British Non-Ferrous Metals Research Assn.; K. Boodson, Imperial Metal Industries Ltd.; E. H. Amstein, Associated Lead Manufacturers Ltd.; J. Funnell, Broken Hill Associated Smelters Proprietary Ltd.; C. A. Keller, U.S.A.E.C.; L. H. Ford, H. K. Hardy, T. J. Heal, C. T. John, U.K.A.E.A.; R. A. Foos, Brush Beryllium Co.; H. S. Cooper; W. H. Gross, Dow Chemical Co.; F. Hock, Magnesium Elektron Ltd.; P. A. Butters, H. H. Varrier, Murex Ltd.; C. J. Bradford, Fansteel Inc.; Hermann C. Starck, Berlin; K. A. Gschneidner Jr., Ames Laboratory; Th-Goldschmidt, A-G; T. J. McLeer, Foote Mineral Co.; E. Morrice, U.S. Bureau of Mines; J. G. Cannon, Molybdenum Corp. of America; P. J. Bowles, Warren Springs Laboratory; Johnson Matthey Chemicals Ltd.; A. W. Smith, Roan Selection Trust Ltd.

The work is published by permission of the Electricity Council.

6. References

1 A. D. McMahon, *Copper, a Materials Survey, U.S.Bur. Mines Inform. Circ.*, 8225 (1965).
2 J. A. McArthur and B. J. Ledeboer in P. Queneau (Ed.), *Extractive Metallurgy of Copper Nickel and Cobalt*, Interscience, New York, 1961.
3 A. Butts in C. A. Hampel (Ed.), *The Encyclopaedia of Electrochemistry*, Reinhold, New York, 1964.
4 F. D. Cooper, *U.S. Bur. Mines Inform. Circ.*, 8394 (1968).
5 *Mining Annual Review, Mining Journal*, London (June 1969).
6 *Mining Annual Review, Mining Journal*, London (May 1968).
7 J. A. McArthur and C. Leaphart in P. Queneau (Ed.), *Extractive Metallurgy of Copper, Nickel and Cobalt*, Interscience, New York, 1961.
8 F. H. Chapman and E. W. Page, *ibid*.
9 L. Theys, *Rev. Met. Paris*, 8 (1968) 113.
10 J. E. Harper, P. N. Vernon and L. R. Verney, *Extractive Met. Div. Sumposium on Electrometallurgy, Cleveland, 1968*, Met. Soc. AIME, New York, in press.
11 M. D. Rosenzweig, *Chem. Eng.*, 74 (25) (1967) 88.
12 Anon., *Eng. Mining J.*, 168 (12) (1967) 88.
13 A. K. Anderson and T. Balberyszski, *TMS Paper No. A68-17*, Met. Soc. AIME, New York, (1968).
14 H. Veltman, S. Pellegrini and V. N. Mackiw, *J. Metals*, 19 (2) (1967) 21.
15 H. W. Jacky, *J. Metals*, 19 (4) (1967) 22.
16 M. A. Bouchat and J. J. Saquet in P. Queneau (Ed.) *Extractive Metallurgy of Copper, Nickel and Cobalt*, Interscience, New York, 1961.
17 Fr. Thoumsin, *Acta Techn. Belgica*, 9 (1969) 33.
18 *Encyclopedia of Chemical Technology*, Kirk–Othmer (Ed.), Interscience, New York, 2nd ed., 1964.
19 V. N. Mackiw and T. W. Benz in P. Queneau (Ed.), *Extractive Metallurgy of Copper, Nickel and Cobalt*, Interscience, New York, 1961.
20 D. Löwe, L. Müller and H. Ufer, *Neue Hütte*, 13 (1968) 281.
21 E. S. Lekht and L. S. Ivanova, *Tsvetn. Metal.*, 39 (5) (1966) 37.

22 International Nickel Company of Canada Ltd., private communication, 1969.
23 Outokumpu Oy, private communication, 1969.
24 J. S. BOLDT JR., *The Winning of Nickel*, Methuen, London, 1967.
25 L. S. RENZONI, R. C. MCQUIRE and W. V. BARKER, *J. Metals*, 10 (1958) 414.
26 T. TOIVANEN and P. O. GRÖNQVIST, *Can. Mining Met. Bull.*, 57 (1964) 653.
27 K. SPROULE, G. A. HARCOURT and L. S. RENZONI in P. QUENEAU (Ed.), *Extractive Metallurgy of Copper, Nickel and Cobalt*, Interscience, New York, 1961.
28 V. N. MACKIW, T. W. BENZ and D. J. I. EVANS, *Met. Rev.*, 11 (1966) 109.
29 *Year Book of American Bureau of Metal Statistics 1967*, American Bureau of Metal Statistics, New York, issued June, 1968.
30 Zinc Development Association, private communication, 1969.
31 C. A. HAMPEL in C. A. HAMPEL (Ed.), *The Encyclopedia of Electrochemistry*, Reinhold New York, 1964.
32 N. C. ASHDOWN, *Commonwealth Mining Met. Congr. 8th*, Vol. 3, Australasian Inst. Mining and Met., 1965, p. 95.
33 G. C. BRATT and A. R. GORDON in *Research in Chemical and Extraction Metallurgy (Sydney 1965)*, Australasian Inst. Mining and Metal., 1967, p. 197.
34 P. BENVENUTI, T. DEMICHELIS, A. D'ESTE and F. GRESOTTO, *Met. Ital.*, 60 (1968) 417.
35 A. C. JEPHSON, A. Y. BETHUNE and R. C. KELAHAN, *J. Metals* 18 (1966) 947.
36 R. L. BIRD, *Eng. Mining J.*, 170 (5) (1969) 83.
37 Anon., *Australian Mining*, 61 (3) (1969) 48.
38 H. BARDWELL, *Eng. Mining J.*, 170 (5) (1969) 102.
39 W. KRUPKA and D. KRUPKA, *Rudy Metale Niezelazne*, 13 (1968) 137.
40 J. WLODYKA and J. WOJTOWICZ, *Prace Inst. Hutniczych*, 18 (1966) 243.
41 A. ALEKSANDROV and K. PAMUCHYSKI, *Rudodobiv. Met.*, 22 (11) (1967) 55.
42 E. KUSS, *Z. Angew. Chem.*, 22 (1950) 519.
43 *World Metal Statistics*, September, 1968.
44 R. L. NAUERT, *J. Metals*, 18 (1) (1966) 15.
45 R. J. HOPKINS and J. C. NIXON in *Commonwealth Mining Metal. Congr. 8th*, Vol. 3, Australasian Inst. Mining and Metal., 1965.
46 D. H. JOHNSTONE in F. A. GREEN (Ed.), *Extraction Metallurgy in Australia: Non Ferrous Metallurgy, 1953*, Australasian Inst. Mining and Metal., 1953, p. 180.
47 *Mineral Facts and Problems*, 1965 ed., *U.S. Bur. Mines Bull.* 630.
48 J. B. ROSENBAUM, R. R. LLOYD and C. C. MERRILL, *U.S. Bur. Mines Rept. Invest.*, 5322 (1957).
49 F. E. BACON in C. A. HAMPEL (Ed.), *The Encyclopedia of Electrochemistry*, Reinhold, New York, 1964.
50 F. E. BACON in W. J. MEAD (Ed.), *Encyclopedia of Chemical Process Equipment*, Reinhold, New York, 1964.
51 D. SCHLAIN in R. F. BUNSHAH (Ed.), *Techniques of Metals Research*, Vol. 1, Pt. 2, Interscience, New York, 1968.
52 A. B. SUCHKOV, V. N. MIKHINA and A. S. VOROB'EVA, *Zh. Prikl. Khim.*, 39 (1966) 2157.
53 A. A. COCHRAN and W. L. FALKE, *J. Metals*, 19 (4) (1967) 28.
54 E. G. DAVIS, F. C. BRANTLEY and E. C. WRIGHT, *U.S. Bur. Mines Rept. Invest.*, 6728 (1966).
55 L. K. HUDSON, *J. Metals*, 17 (1965) 948.
56 P. BRETEQUE, *Etudes sur le Gallium*, Doctoral Thesis, University Aix-Marseille, 1955; Imprimerie Vaudoise, Lausanne, Switz., 1956; *J. Metals*, 8 (1956) 1528.
57 W. RYAN, *Non-ferrous Extractive Metallurgy in the United Kingdom*, Inst. Mining and Metal. London, 1968.
58 A. BIANCHINI and A. FIDANI, *Met. Ital.*, 60 (1968) 395.
59 L. ERNST, *Aluminium*, 45 (1969) 253.
60 T. G. PEARSON, *Roy. Inst. Chem. London, Lectures, Monographs Repts.*, 3 (1955).
61 Anon., *Light Metal Age*, 26 (9/10) (1968) 21.
62 J. R. MARTIN, L. L. PORTER and L. H. ROSE, *Eng. Mining J.*, 170 (5) (1969) 88.
63 I. MOLNAR, *Kohasz. Lapok*, 101 (3) (1968) 97.
64 V. YA NIKITIN, *Tsvetn. Metal.*, 39 (5) (1966) 61.
65 J. F. WEILER, *T.M.S. Paper No. A68–18*, Met. Soc. AIME New York, 1968.

66 V. V. ROBOZEROV and M. M. VETYUKOV, *Tsvetn. Metal.*, 41 (6) (1968) 55.
67 G. H. TRAYLOR, *J. Metals*, 20 (7) (1968) 63.
68 T. WATANABE, H. HAYASHI and K. MOCHIZUKI, *J. Electrochem. Soc. Japan*, 36 (1968) 123.
69 H. C. QUANDT and S. P. BEGANY, *T.M.S.Paper No. A68-35*, Met. Soc. AIME, New York, 1968.
70 B. PANEBIANCO and R. BACCHIEGA, *Alluminio*, 35 (1966) 69.
71 M. B. DELL, R. W. PETERSON and J. N. RUMBLE, *J. Metals*, 20 (9) (1968) 55.
72 B. PANEBIANCO and R. BACCHIEGA, *T.M.S. Paper No. A68-22*, Met. Soc. AIME, New York, 1968.
73 M. A. KOROBOV and E. A. YANKO, *Tsvetn. Metal.*, 39 (11) (1966) 59.
74 M. A. KOROBOV, E. A. YANKO and V. N. DERYAGIN, *Tsvetn. Metal.*, 41 (1) (1968) 49.
75 M. A. KOROBOV and E. A. YANKO, *Tsvetn. Metal.*, 41 (2) (1968) 71.
76 J. THONSTAD, *J. Electrochem. Soc.*, 111 (1964) 955.
77 S. GJERSTAD and B. J. WELCH, *J. Electrochem. Soc.*, 111 (1964) 976.
78 O. BERSIMENKO and M. M. VETYUKOV, *Zh. Prikl. Khim.*, 39 (1966) 1696.
79 R. J. SNOW and B. J. WELCH, *J. Electrochem. Soc.*, 115 (1968) 1170.
80 J. GERLACH, W. SCHMIDT and H. SCHMITT, *Z. Erzbergbau Metallhüttenw.*, 20 (1967) 111.
81 B. J. WELCH, in *Research in Chemical and Extraction Metallurgy*, Australasian Inst. Mining and Metall., 1967.
82 J. BOTOR and J. KALINOWSKI, *Rudy Metale Niezelazne*, 12 (1967) 563.
83 A. I. BELYAEV, *The Electrolyte of Aluminium Cells*, Metallurgizdat, Moscow, 1961, Chap. 3.
84 R. A. LEWIS, *J. Metals*, 19 (5) (1967) 30.
85 A. I. BELYAEV, *Izv. Akad. Nauk SSSR, Metally*, (Sept./Oct. 1967) 26.
86 P. S. SAAKYAN, *Zhr. Prikl. Khim.*, 39 (1966) 347.
87 N. E. RICHARDS and E. R. RUSSELL, *Trans. AIME*, 242 (1968) 2495.
88 P. A. FOSTER JR. and W. B. FRANK, *J. Electrochem. Soc.*, 107 (1960) 998.
89 W. B. FRANK and L. M. FOSTER, *J. Phys. Chem.*, 61 (1957) 1531.
90 J. THONSTAD, *Electrochim. Acta*, 12 (1967) 1219.
91 J. THONSTAD, *Electrochim. Acta*, 14 (1969) 127.
92 G. E. WHITE, *T.M.S. Paper No. A68-27*, Met. Soc. AIME, New York, 1968.
93 A. S. DERKACH et al., *Tsvetn. Metal.*, 39 (2) (1966) 39, 44. *Soviet J. Non-Ferrous Metals, English Transl.*, 7 (2) (1966) 40, 45.
94 Z. ORMAN and T. LACH, *Prace Inst. Hutniczych*, 18 (1966) 301.
95 C. J. MCMINN, *J. Metals*, 18 (1966) 308.
96 T. K. MCMAHON and G. P. DIRTH, *J. Metals*, 18 (1966) 317.
97 T. KOJIMA and S. ICHIHARA, *J. Metals*, 18 (1966) 312.
98 R. E. J. PUTMAN and N. R. CARLSON, *T.M.S. Paper No. A68-16*, Met. Soc. AIME, New York, 1968.
99 S. E. GEFTER et al., *Tsvetn. Metal.*, 39 (11) (1966) 54.
100 V. N. RUDAKOV et al., *Tsvetn. Metal.*, 41 (1) (1968) 54.
101 A. A. KOSTIN, *Tsvetn. Metal.*, 41 (1) (1968) 57.
102 W. SCHMIDT-HATTING, *Chem. Ing.-Tech.*, 41 (1969) 163.
103 L. N. BOBKOV, *Tsvetn. Metal.*, 41 (3) (1968) 1.
104 S. BRUBAKER, *Trends in the World Aluminium Industry*, Johns Hopkins Press, Baltimore, 1967.
105 S. EGER, *Handbuch der Technischen Elektrochemie*, Band III, 2nd ed., 1961, Akad. Verlag, Leipzig.
106 A. S. EMLEY, *Principles of Magnesium Technology*, Pergamon, Oxford, 1966.
107 F. J. KRENZKE in C. A. HAMPEL (Ed.), *The Encyclopedia of Electrochemistry*, Reinhold, New York, (1964).
108 U.S. Pat., 1,190,122.
109 R. R. LLOYD et al., *Metals Technology*, (1945) T.P. 1848.
110 U.S. Patents Nos. 2,880,151; 2,950,236; 2,888,389.
111 B. CARTWRIGHT, L. R. MICHELS and S. R. RAVITZ, *U.S. Bur. Mines Rept. Invest.*, 3805 (1945).
112 Anon., *American Metal Market*, Vol. LXXVI, No. 107, Section 2, *Magnesium* (June 9th, 1969).
113 W. H. JACKSON, *Can. Mining J.*, 90 (2) (1969) 118.

114 N. M. ZUEV et al., *Tsvetn. Metal.*, 41 (3) (1968) 50; 41 (7) (1968) 54; KH. L. STRELETS et al., *Tsevtn. Metal.*, 41 (6) (1968) 59.
115 K. D. MUZHZHAVLEV et al., *Tsvetn. Metal.*, 39 (7) (1966) 62.
116 N. HÖY-PETERSEN, *J. Metals*, 21 (4) (1969) 43.
117 W. J. KROLL, *Chem. Ind.*, (1960) 1314.
118 S. SENDEROFF, *Met. Rev.*, 11 (1966) 97.
119 O. Q. LEONE, H. KNUDSEN and D. COUCH, *J. Metals*, 19 (3) (1967) 18.
120 M. J. RAND and L. J. REIMERT, *J. Electrochem. Soc.*, 111 (1964) 429, 434.
121 A. J. MYHREN et al., *J. Metals*, 20 (5) (1968) 38.
122 A. BRENNER and S. SENDEROFF, *J. Electrochem. Soc.*, 99 (1952) 223 C.
123 M. A. STEINBERG, S. S. CARLTON, M. E. SIBERT and E. WAINER, *J. Electrochem. Soc.*, 102 (1955) 332.
124 M. V. SMIRNOV, W. A. LOGINOV and L. A. TSIOVKINA, *Proceedings of the All-Union Conference on Physical Chemistry of Molten Salts and Slags, 1960*, Acad. Sci. U.S.S.R., Moscow, (1962), p. 337.
125 N. A. BAITENEV, A. J. MILOV and V. D. PONOMAREV, *Tr. Inst. Met. i Obogashch. Akad. Nauk Kaz. SSR*, 22 (1967) 75.
126 N. I. ANUFRIEVA and Z. N. BALASHOVA, *Izv. Akad. Nauk SSSR, Metally.*, No. 2, (March-April 1969) 69.
127 N. I. ANUFRIEVA, *Zh. Prikl. Khim.*, 40 (1967) 1288.
128 N. I. ANUFRIEVA, *Izv. Akad. Nauk SSSR Metally.*, No. 2, (March-April 1968) 89.
129 N. I. ANUFRIEVA and Z. N. BALASHOVA, *Tsvetn. Metal.*, 39 (9) (1966) 75.
130 N. I. ANUFRIEVA and Z. N. BALASHOVA, *Russ. Met.*, (Feb. 1967) 29.
131 N. I. ANUFRIEVA, *Izv. Akad. Nauk SSSR, Metally.*, No. 3, (1967) 81.
132 E. B. GITMAN and I. N. SHEIKO, *Ukr. Khim. Zh.*, 34 (1968) 129.
133 G. N. GOPIENKO and V. V. TIMOFEEV, *Zh. Prikl. Khim.*, 40 (1967) 1610.
134 A. I. IVANOV et al., *Titan i Fgo Splavy, Akad. Nauk SSSR, Inst. Met.*, (1961) 131, 136, 153.
135 J. B. ROSENBAUM, *J. Metals*, 21 (1969) 18.
136 Anon., *Chem. Eng. News*, (Dec 23rd 1968) 32.
137 J. C. PRISCU, E. R. POULSEN and L. C. SNYDER, *Extractive Metallurgy Division Symposium on Electrometallurgy, Cleveland, Dec. 1968*, Met. Soc. AIME, New York, in press.
138 D. J. SOISSON, J. J. MCLAFFERTY and J. A. PIERRET, *Ind. Eng. Chem.*, 53 (1961) 861.
139 Fansteel Metallurgical Corp., private communication, 1969.
140 C. W. BALKE, *Ind. Eng. Chem.*, 27 (1935) 1166.
141 F. H. DRIGGS and W. C. LILLIENDAHL, *Ind. Eng. Chem.*, 23 (1931) 634.
142 W. ROCKENBAUER, *Chem. Ing.-Tech.*, 41 (1969) 159.
143 C. H. CHILTON, *Chem. Eng.*, 65 (22) (1958) 104.
144 F. FAIRBROTHER, *The Chemistry of Niobium and Tantalum*, Elsevier, Amsterdam, 1967.
145 F. T. SISCO and E. EPREMIAN, *Columbium and Tantalum*, J. Wiley and Sons Inc., New York, 1963.
146 P. J. BOWLES in *Advances in Extractive Metallurgy*, Inst. Mining and Metal., London, 1968.
147 CL. DECROLY, A. MUKHTAR and R. WINAND, *J. Electrochem. Soc.*, 115 (1968) 905.
148 M. M. WONG and D. E. KIRBY, *Electrochem. Tech.*, 6 (1968) 119.
149 F. M. LEVER and J. B. PAYNE in *Advances in Extractive Metallurgy*, Inst. Mining and Metal., London, 1968.
150 E. MORRICE, private communication, 1969.
151 J. E. SMITH and K. A. GSCHNEIDNER JR., *Compilation of Rare-Earth Products Available from Commercial Suppliers*, I.S.-R.I.C.–1, Ames Laboratory, Iowa State University of Science and Technology, 1967.
152 I. S. HIRSCHHORN, *J. Metals*, 20 (3) (1968) 19.
153 W. BRUGGER and E. GREINACHER, *J. Metals*, 19 (12) (1967) 32.
154 J. A. LIVINGSTON, FIAT Final Report No. 909 (1946).
155 R. SINGER et al., B.I.O.S. Final Report No. 400 (1945).
156 E. MORRICE, E. S. SHEDD and T. A. HENRIE, *U.S. Bur. Mines Rept. Invest.*, 7146 (1968).
157 D. S. LAIDLER, *Roy. Inst. Chem. London, Lectures, Monographs and Repts.*, 16 (1957).
158 C. L. MANTELL, *Electrochemical Engineering*, McGraw-Hill, New York, 4th ed., 1960.

159 G. T. Motock, *Electrochem. Tech.*, 1 (1963) 122.
160 A. Schoepfer, J. G. Cecala and R. I. Elliot, *U.S. Dept. Comm. Office, Tech. Serv. PB Rept.*, 157, 068. Contract 33 (600 - 33920), 1959.
161 M. P. Margolis (Ed.), *Market Guide, Beryllium, Eng. Mining J.*, Metal and Mineral Markets, New York, (Dec. 5th 1966).
162 C. E. Windecker in D. W. White Jr. and J. E. Burke (Eds.), *The Metal Beryllium*, A.S.M., Cleveland, 1955.
163 T. T. Magel in D. W. White Jr. and J. E. Burke (Eds.), *The Metal Beryllium*, A.S.M., Cleveland, 1955.
164 G. Jaeger, *Metallwissenschaft und Technik*, 4 (9/10) (1950) 183.
165 M. E. Sibert and A. E. Hultquist, *J. Metals*, 18 (1966) 1215.
166 M. M. Wong, D. E. Couch and D. A. O'Keefe, *J. Metals*, 21 (1) (1963) 43.
167 M. C. Kells, R. B. Holden and C. Whitman, *2nd Intern. Conf. on Peaceful Uses of Atomic Energy*, A/Conf 15/P/717 (1958).
168 H. S. Cooper, U.S. Patents Nos. 2,572,248; 2,572,249.
169 H. S. Cooper and J. C. Schaeffer in C. A. Hampel (Ed.), *The Encyclopedia of Electrochemistry*, Reinhold, New York, 1964; H. S. Cooper in G. L. Clark (Ed.), *The Encyclopedia of Chemistry*, Reinhold, New York, 1966; J. C. Schaeffer in C. A. Hampel (Ed.), *The Encyclopedia of the Chemical Elements*, Reinhold, New York, 1968.
170 R. Monnier, P. Tissot and P. Pearson, *Helv. Chim. Acta*, 49 (11) (1968) 67.
171 H. S. Cooper, and J. C. Schaeffer, U.S. Pat., 2,918,417.
172 J. H. Russell and H. J. Kelly, *U.S. Bur. Mines Rept. Invest.* 7028 (1967).
173 W. D. Threadgill, *J. Electrochem. Soc.*, 111 (1964) 1408.
174 C. A. Hampel, *The Encyclopedia of Electrochemistry*, Reinhold, New York, 1964.
175 A. Ruis, F. Colom and A. Artacho, *Electrochim. Acta*, 11 (1966) 1497.
176 H. Winterhager and B. Kammel, *Erzmetall*, 9 (March 1956) 97; H. Sawamoto and T. Saito, *J. Min. Inst. Japan*, 68 (1953) 555; G. Angel and E. Garnum, *Tek. Tidsk.*, 75 (1945) 297; A. P. Newall, *Ind. Chem.*, 14 (1938) 184; J. B. Richardson, *Trans. Inst. Mining Met.*, 46 (1936-7) 383, 405.
177 N. N. Sevryukov and R. Murti, *Tsvetn. Metal.*, 41 (8) (1968) 50; O. M. Tleukulov and I. R. Polyvyanny, *Tsvetn. Metal.*, 40 (3) (1967) 58.
178 V. Mercea et al., *Revista de Chemie*, 16 (1965) 352.
179 K. B. Morris, D. Z. Douglass and C. B. Vaughn, *J. Electrochem. Soc.*, 101 (1954) 343.
180 P. M. Gruzensky, *J. Electrochem. Soc.*, 103 (1956) 171.
181 B. C. Raynes, E. L. Thellmann, M. A. Steinberg and E. Wainer, *J. Electrochem. Soc.*, 102 (1955) 137.
182 M. V. Smirnov, A. N. Baraboshkin and V. E. Komarov, *Zh. Fiz. Khim.*, 37 (1963) 1669, 1677.
183 R. Winand, *Electrochim. Acta*, 7 (1962) 475.
184 S. Senderoff and A. Brenner, *J. Electrochem. Soc.*, 101 (1954) 16.
185 H. J. Heinen and D. H. Baker Jr., *U.S. Bur. Mines Rept. Invest.* 6834 (1966).
186 C. E. Fink and C. C. Ma, *Trans. Electrochem. Soc.*, 84 (1943) 33.
187 J. B. Zadra and J. M. Gomes, *U.S. Bur. Mines Rept. Invest.*, (1959) 5554.
188 P. C. Newdick and P. J. Bowles, *Warren Springs Lab. Rept.* No. LR 92 (MST) (Dec. 1968).
189 P. E. Churchwood and J. B. Rosenbaum, *J. Metals*, 15 (1963) 648.
190 J. D. Vie, J. W. Stevenson et al., *Electrolytic Uranium Project Terminal Progress Report—Cell Design/Cell Operation*, MCW-1513 and MCW-1514. U.S.A.E.C. Contract No. W-14-108-Eng.-8, Mallinckroft Chemical Works, Uranium Division (Sept.–Oct. 1966).
191 D. G. Kesterke, D. C. Fleck and T. A. Henrie, *U.S. Bur. Mines Rept. Invest.*, 6436 (1964).
192 D. G. Kesterke, D. C. Fleck and T. A. Henrie, *U.S. Bur. Mines Rept. Invest.*, 6789 (1966).
193 A. R. Gibson and J. R. Chalkley, *Trans. Inst. Min. Met.*, 69 (1966) 281.

Chapter 7 (Part 1)

Electrorefining in Aqueous Electrolytes

J. M. STEELE

Capper Pass and Son Ltd., Melton Works, N. Ferriby, Yorks.

1. Introduction

1.1 HISTORY OF ELECTROREFINING

The first experiments in this field were carried out by chemists in the early 19th century. The relationships between current used and quantity of metal deposited were worked out and published by Faraday in the early 1830's.

Large scale production did not take place until the 1860's after the development of the dynamo. The first patents for copper refining were published by James Elkington[1] in 1865 and the first copper refinery built in America at Newark by the Balbach Smelting and Refining Company in 1883. The industry quickly developed and electrolytic copper became the official basis for price quotation in 1914.

During the same period silver was produced with the development of the Balbach–Thum cells. In 1878 the first gold refinery was introduced by the Norddeutsche Affinerie in Hamburg, using the Wohlwill process. This was dominant until the 1920's when it went into decline as the Miller process was established for large scale gold refining.

The patent for the Betts[2] method of refining was granted in 1901 and the first Betts[3] refinery for the production of lead was established at Trail in Canada in 1902 with a production of 50 tonnes/day. In 1964 this had reached 450 tonnes/day[4]. The same basic process is used on a limited scale for the production of electrolytic solder.

In 1905 a patent was granted to Hybinette[5] for a method of refining which used a divided cell. This development enabled nickel to be refined electrolytically for the first time and one of the first refineries was set up in Kristiansand, Norway in 1910.

The electrolytic refining of tin using acid electrolyte was developed in the United States at Perth Amboy during the first world war. The initial electrolyte was a mixture of fluosilicic acid and sulphuric acid but was later changed to sulphuric acid/cresyl sulphonic acid. Production was discontinued in 1923 and many other ventures using this method have been tried and abandoned since. The method is at present carried out only at the Estanifera do Brazil Refinery in Brazil.

The electrolytic refining of tin using an alkaline electrolyte was established in this country in 1936, by Capper Pass and Sons Ltd. to treat complex ores and secondary materials. In Germany, tin is refined using an alkaline sodium sulphide electrolyte by Berzelius Metallhütten Gesellschaft at Duisburg, West Germany.

The early history of electrorefining is discussed in detail by Bouchers and Macmillan[6].

1.2 Basic principles

In the electrorefining of metals an impure anode is corroded while purified metal is deposited at the cathode. The cell voltage for such a system under reversible conditions is, of course, zero and even under normal operating conditions, electrorefining cells have a smaller cell voltage, and hence power consumption, than any other electrochemical process handling comparable tonnages. The result is that as a technology electrorefining is much less sensitive to the cost of electric power than most other processes described in this book. Electrorefining is carried out in most industrialised countries of the world, including those where power costs are high. Its relationships to electrowinning on the one hand, and electroplating on the other, will not have escaped the reader.

The reason why the process of electrorefining is effective, would, in many cases, not be obvious from a simple consideration of the electrochemical series of the metals, for under "irreversible" electrode conditions, the various rates of anodic dissolution or cathode deposition, frequently deviate from the order of the electrochemical series. For example, lead may be deposited from an electrolyte containing both lead and hydrogen ions. This is because the hydrogen over-potential on lead exceeds the difference in the standard electrode potentials. Similarly, nickel may be deposited from an electrolyte containing both nickel and hydrogen ions, even though the electrode potentials are well separated. In this case the hydrogen ion content must be carefully controlled.

The more modest power consumption referred to above, has meant that the competition between electrochemical and non-electrochemical processes has been less marked in this, than in many other fields. However, new processes are considered from time to time, and some of these are mentioned. One of these, the use of high pressure hydrogen, is of interest in that it is, in all probability, an electroless electrochemical process, like wet corrosion, electroless metal plating, and cementation.

This chapter deals with straightforward electrorefining, and the manufacture of metal powders. The principles involved in scrap-recovery are little different. The metal is often shredded and refined from baskets. References 41–5 kindly supplied by K. Boodson of Imperial Metal Industries give a typical picture of this activity, describing nickel-rich scrap recovery[41] and Co–Ni alloy from superalloy processing[42], copper recovery[43] and tin-rich copper scrap[44] with electrolytic de-tinning[45, 24]. The latter activity has undergone a sharp decline, resulting mainly from the reduction in the amount of tin used in the protection of modern tin cans.

TABLE 1

	Copper	Nickel	Cobalt	Lead	Tin	Silver	Gold
Electrolyte concn. (g/l)	Cu 40–50 H_2SO_4 175–225	Ni 50–60 SO_4 90–95 Cl 55–60 H_3BO_3 10–20 Na 35	Co 50–60 SO_4 150–200 Cl 10 H_3BO_3 10–20 Na 40	Pb 60–90 H_2SiF_6 50–100	Sn 30–40 H_2SO_4 75	Ag 30–150 HNO_3 0–10 Cu 5–80	Au 100 HCl 100
Slime impurity	Ag Au Ni Sb Pb	Ag Au Pt metals		Bi Ag Au Sb	Pb Sb	Au	Ag
Major electrolyte impurity	Ni As Fe	Cu Co	Ni Cu			Cu	Pt metals
Current density (A/m²)	100–200	150–200	150–200	150–250	100	200–500	600–1500
Cells	Simple	Diaphragm	Diaphragm	Simple	Simple	Diaphragm	Simple
Temp. (°C)	60	60	60	30–40	20–30	25–45	60
Cell voltage	0.15–0.3	1.5–3.0	1.5–3.0	0.3–0.6	0.3–0.6	1.5–5.0	0.5–2.0

2. Process summary

The metals produced by electrorefining which are discussed here are copper, nickel, cobalt, lead, tin, silver and gold.

Each metal is produced in different parts of the world, often by the same basic system. A general summary is given in Table 1 (p. 221).

TABLE 2

LIST OF MAJOR REFINERIES (INCLUDING BRITISH REFINERIES)

Metal	Company	Location
Copper	Details of the thirty leading Copper Refineries are given in the first reference of this section.	
	British Copper Refineries Ltd.	Prescot, Lancs.
	Elkington Copper Refineries Ltd.	Walsall, Staffs.
	Wolverhampton Metal Company Ltd.	Walsall, Staffs.
	McKechnie Chemicals Ltd.	Widnes, Lancs.
Nickel	International Nickel Company	Port Colborne, Ontario
	International Nickel Company	Thompson, Manitoba
	Falconbridge Nikkelwerk Aktieselskap	Kristiansand, Norway.
Cobalt	International Nickel Company	Port Colborne, Ontario
	International Nickel Company	Thompson, Manitoba
	Falconbridge Nikkelwerk Aktieselskap	Kristiansand, Norway
Lead	Consolidated Mining and Smelting Company of Canada	Trail, British Columbia
	Cerro de Pasco Copper Company	Oroya, Peru.
	U.S. Smelting, Refining and Mining Company (U.S.S. Lead Refinery)	East Chicago, U.S.A.
	Monteponi and Monteveccio Company	Sardinia
	Capper Pass and Sons Ltd.	North Ferriby, Yorks.
Tin	Estanifera do Brazil	Volta Redmonda, Brazil
	Berzelius Metallhütten Gesellschaft	Duisburg, West Germany
	Capper Pass and Sons Ltd.	North Ferriby, Yorks.
Silver	The Anaconda Company (International Smelting and Refining Company)	Perth Amboy, New Jersey
	American Metal Climax Inc. (U.S. Metals Refining Company)	Carteret, New Jersey
	International Nickel Company	Copper Cliff, Ontario
	Johnson Matthey Chemicals Ltd.	Royston, Herts.
	Sheffield Smelting Company Ltd.	Sheffield, Yorks
Silver	Electrolytic Refining and Smelting Company of Australia	Port Kembla, New South Wales
	Eastman Kodak Company	Rochester, New York
	Canadian Copper Refineries Ltd.	Montreal, Quebec
	American Smelting and Refining Company	Barber, New Jersey
Gold	American Smelting and Refining Company	Barber, New Jersey
	International Nickel Company	Copper Cliff, Ontario
	The Anaconda Company (International Smelting and Refining Company)	Perth Amboy, New Jersey
	American Metal Climax (U.S. Metals Refining Company)	Carteret, New Jersey
	Nord-Deutsche Affinerie	Hamburg, Germany
	Cerro de Pasco Copper Company	Oroya, Peru

TABLE 3
REFERENCES DESCRIBING OPERATIONS AND DETAILS OF SOME MAJOR REFINERIES

Metal	Title	Publisher	Author
Copper	Extractive Metallurgy of Copper Nickel and Cobalt. Industrial Report on Modern Tankhouse Practice	AIME Interscience, New York, 1960	J. H. Schloer and S. S. Forbes
Copper, nickel, cobalt, lead, tin, silver, gold	Electrochemical Engineering	McGraw-Hill, New York, 1960	C. L. Mantell
Copper	Copper, the Metal, its Alloys and Compounds	Reinhold, New York, 1954	A. Butts
	Copper Refining in the Nord-Deutsche Affinerie	Chem. Ing.-Tech. 37 (1965) 590	E. Süssmuth
Nickel	The Falconbridge Story	Can. Mining J., 81 (1959)	
Nickel	The Kristiansand Nickel Refinery	J. Metals, September, 1952	F. R. Archibald
Nickel	The Winning of Nickel	Longmans, Canada, 1967	J. R. Boldt and P. Queneau
Nickel	The Thompson Smelter	Trans. C.I.M.M., 67 (1964)	Spence and Cooke
Nickel	Direct Electrorefining of Nickel Matte	J. Metals, 10 (1958)	L. S. Renzoni, R. C. McQuire and M. V. Barker
Cobalt	Extractive Metallurgy of Copper Nickel and Cobalt. Production of Cobalt at Inco's Port Colborne Nickel Refinery	AIME, Interscience, New York, 1960	L. S. Renzoni and M. V. Barker
Silver	Silver, Economics, Metallurgy and Use	Van Nostrand, Canada, 1967	
	1. Refining Processes		O. C. Johnson
	2. Silver Refining, Operation at Cerro de Pasco		I. L. Barker
Silver	Refining of Non-Ferrous Metals	Institute of Mining and Metallurgy, Can. Mining J., 95 (1954)	A. E. Richards
Gold	The Refining of Silver and Gold		
Lead	The Story of COMINCO		
Lead	Electrolytic Refining in Sardinia	J. Metals, 17 (1965)	E. R. Freni
Tin	Electrolytic Tin Refining	Electrochemical Technology, Jan./Feb. (1966) 42	C. L. Mantell

Electrolyte concentrations and current densities vary widely and typical ranges or approximate values are given for some major refineries.

3. Electrorefining compared with other methods of refining

There are a number of ways in which metals may be processed from ores and residues. They may be broadly grouped as follows:

(i) Smelting and firerefining
(ii) Calcination, leaching and purification by precipitation or electrowinning
(iii) Smelting and electrorefining.

The quantity of metal produced by electrorefining changes as new processes are developed and existing ones are improved. The choice of electrorefining may be for one of the following reasons:

(1) A premium for high purity
(2) Concentration of other valuable metals or their co-production
(3) It is the only commercially acceptable method.

Of the metals considered, lead, tin, gold, and cobalt are refined mostly by methods other than electrorefining, whereas copper, nickel and silver are produced in the main by electrorefining.

3.1 COPPER

Much of the copper for industry is required to be of the highest purity. The British Standard[7] and American Standard for Testing Materials[8] impose a limit on chemical purity and a minimum standard for resistivity; for example, bismuth must not exceed 0.001% and lead 0.005%.

The bulk of the world's copper ores contain significant amounts of silver, gold, nickel, selenium and tellurium. The silver and gold cannot be separated by fire-refining processes. Nickel, selenium and tellurium are also difficult to separate by this method. The impurities are all easily separated in the electrolytic process. A detailed description of operation of a large copper refinery is given in ref. 46.

3.2 NICKEL

Nickel was originally produced by the carbonyl[9] process which is still in operation today. It was later found possible to refine nickel electrolytically by the use of diaphragm cells (The Hybinette Process).

The latest development has been to produce nickel by direct hydrogen reduction of alkaline solutions and slurries at elevated pressures[10].

3.3 COBALT

The bulk of the world's cobalt is refined by pyrometallurgical or electrowinning techniques. Cobalt separated from nickel refinery electrolytes is usually purified by electrorefining, although in recent years it has also been produced by hydrogen reduction of alkaline solutions at high pressure.

3.4 LEAD

Lead is produced by firerefining and electrorefining processes. The electrorefined product is normally of very high purity as a good separation of bismuth and lead is readily achieved. For example, lead produced at the Consolidated Smelting and Refining Company of Canada[11] assays 99.998% lead with typical impurities in parts per million as follows:

Sb	Bi	Fe	Ag	As	Cu	Zn	Ni + Cu
2	5	1	4	<1	4	1	2

In the firerefined process bismuth accompanies lead through all stages, and if present in significant amounts is removed by the Kroll Betterton Process[11] using calcium metal.

3.5 SILVER AND GOLD

One of the disadvantages of electrorefining is that there is a large quantity of metal "locked up" in the process. This is particularly significant in the refining of silver. Successful refining processes must therefore be rapid and not leave residues which are difficult to treat.

The first attempts to purify these metals were simple smelting operations using clean ore or native metal with subsequent oxidation of the impurities from the molten metal. This was replaced by leaching processes using nitric or sulphuric acid. The silver dissolved leaving a residue of gold which was refined by the Wohlwill[13] electrolytic process. Both processes had serious disadvantages: the loss of acid and treatment of nitrous fumes in the nitric acid process, the disposal of sulphur dioxide and the contamination of the gold with lead sulphate in the sulphuric acid process. These processes have been superseded by the electrorefining of silver/gold bullion using a nitric acid electrolyte where pure silver is produced directly and the acid regenerated. The slimed gold is then melted and impurities removed by treatment with chlorine gas (Miller Process)[14]. The Miller Process is superior to the electrolytic process in that the "lock up" of gold is considerably reduced and the reagent cost is low.

4. Electrorefining circuits

4.1 SIMPLE LAYOUT

A simple electrolytic refinery is shown in Fig. 1 and is composed of:
(a) A batch of cells where metal transfer takes place

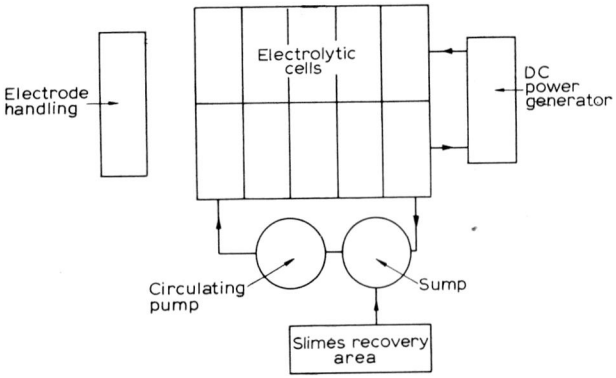

Fig. 1.
General arrangement of a simple electrorefinery.

(b) A liquor storage and transfer system to provide adequate flow of liquor through the cells
(c) Handling equipment for cathodes and anodes
(d) Rectification equipment for production of direct current
(e) Equipment for slimes recovery.

Wide variations and types are encountered in all five components; in many refineries an electrolyte purification system may be required.

4.2 Electrolytic cells

Electrolytic cells are rectangular tanks in which the electrodes are placed alternately along the length of the tank. Flow of liquor is normally at right angles to the face of the electrodes with the feed at the bottom of the cell and discharge at the top.

The materials of construction vary with the electrolyte used. With alkaline electrolytes cells constructed from mild steel are satisfactory. Acid electrolytes present severe corrosion problems. The base materials used are wood, concrete or steel, with acid resistant linings of rubber, asphalt, lead or plastic.

In copper, nickel and cobalt refineries it is necessary to construct specially deep vats in which the starting sheet cathodes for the normal vats are grown by deposition on copper or stainless steel blanks. To ensure that a good deposit is obtained the liquor is often filtered. In tin and lead refineries the starting sheets are prepared either by pouring metal over inclined steel plates or by revolving cooled drums in which a sector is immersed in molten cathode metal.

In some cases specialised cells are required as in the refining of silver and gold. The high price of the metals demands that the refining be carried out as quickly as possible and consequently the cells run with the very high current densities.

Two cells are currently in use in the refining of silver.

4.2.1 Thum cell[15]

In this cell the bottom is lined with graphite and constitutes the cathode. The anode is laid horizontally above the cathode in a canvas diaphragm. Due to the high current density the silver is deposited in crystalline form which is easily detached from the graphite cathode.

4.2.2 Moebius cell[15]

The anode and cathode are arranged vertically in the conventional manner, the anodes enclosed in canvas bags. The cathodes are constructed of stainless steel fitted with mechanical scrapers which continuously knock off the silver crystals formed.

4.3 ELECTROLYTE CIRCULATION

Electrolyte must be circulated through cells at a constant rate and at constant temperature. This is most easily achieved by pumping electrolyte from a sump tank to a header tank and then by gravity to the cells, the liquor overflowing the cells returning by gravity to the sump. Electrolyte may be fed to each individual vat or by a cascade system where it flows through several vats before returning to the sump. The advantage of simplicity in the cascade system may be outweighed by difficulties encountered in carry over of solids from one vat to another, build up of impurities in the electrolyte as it passes through the cascade and lack of control of temperature. The flow pattern of electrolyte through a cell is normally from bottom inlet to top outlet, as shown in Fig. 2. The reverse flow pattern has been tried in systems where the anode slime falls to the bottom of the vat[16]. The advantage is claimed that there is less carry over of slimes onto the cathode. The system has not been widely adopted, since in a cascade system the solids may collect at the outlet and carry over into the next vat.

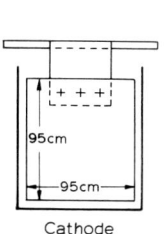
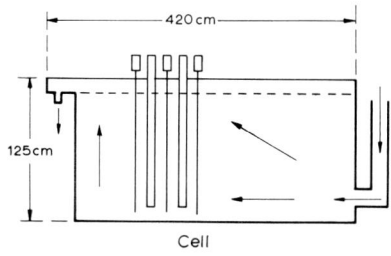
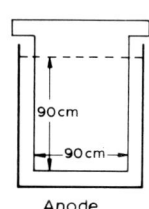

Fig. 2.
Typical simple cell (approximate dimensions).

The electrolyte in the cell must be continuously replaced or stratification occurs. Dense liquor formed at the bottom of the cell is denuded in essential addition agents and poor cathode condition ensues in this area of the cell.

In some cases simple cells are not satisfactory for the production of high purity metal as in the refining of nickel and tin. In both these cases impurities would be deposited on the cathode in simple cells. The problem is overcome by the use of the Hybinette divided cell[5] (Fig. 3) where a diaphragm is used to separate anolyte and

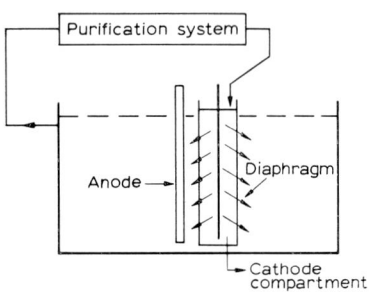

Fig. 3.
Hybinette divided cell.

catholyte. Electrolyte from the anode compartment overflows from the cells and passes through a purification system where the impurities are removed. The clean liquor is then fed back to the cathode compartment where pure metal is deposited. A slight head of liquor is often maintained in the cathode compartment and the circuit is completed by electrolyte diffusing through the diaphragm to the anode compartment. There is, however, a large increase in power cost, due to the voltage drop across the diaphragm. In copper and lead refineries the cell voltage is usually less than 0.5 V whereas in nickel refineries the voltage may be as high as 2.0 V.

4.4 Electrodes

The normal arrangement of electrodes in a cell is shown in Fig. 2. Both anode and cathode must hang vertically in the cell and make good contact with the bus bar system carrying the current. They must be well clear of the sides and bottom of the cell. Anodes are cast so that they are self supporting in the cell. The cathodes are made by hanging the starting sheets from a carrier bar which makes electrical contact with the bus bars.

In some refining operations the cathodes are subject to corrosion at the liquor line due to oxidation reactions. Copper refining electrolytes often contain significant quantities of iron which may cause re-solution of copper as follows:

$$2FeSO_4 + (O) + H_2SO_4 = Fe_2(SO_4)_3 + H_2O$$

$$Fe_2(SO_4)_3 + Cu = CuSO_4 + 2FeSO_4$$

High acid concentrations may also occur at the liquor line due to poor circulation of electrolyte. The life of the cathode may be extended by raising the liquor level in the cell during the first few days of life, providing a thickened strip at the liquor line throughout the remainder of the life.

In all cases, cathodes, anodes and starting sheets are handled by overhead cranes. The development of cranes capable of lifting complete sets of anodes or cathodes from one vat has significantly reduced the refining dead time.

4.5 ELECTRICAL CONNECTION

There are several systems for the transport of current through the refinery. The most widely used are shown in Fig. 4.

The individual cells are connected in the series where the anodes in one tank make contact with the cathodes in the adjacent tank either directly or by use of a

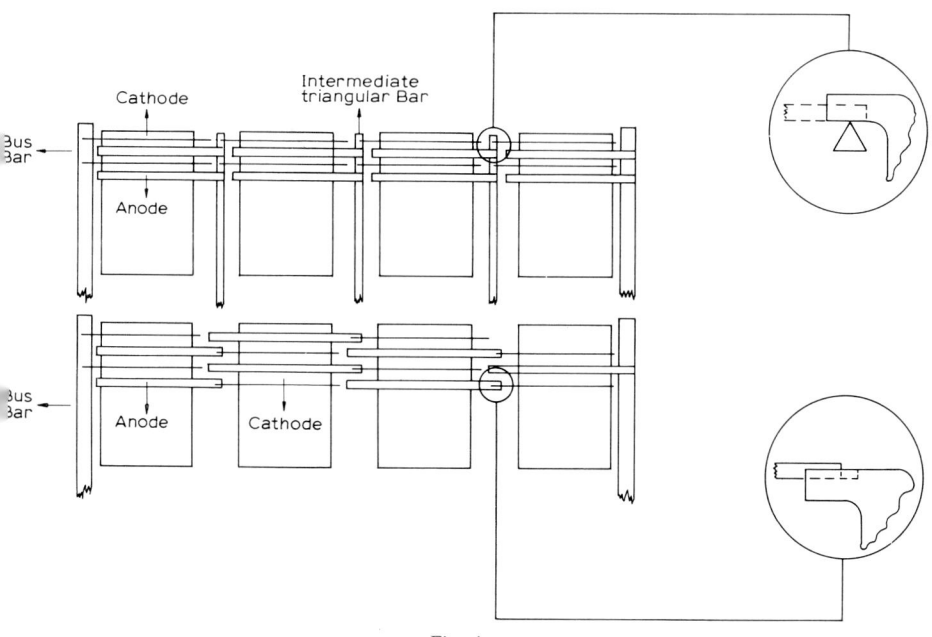

Fig. 4.
Top, Walker system;
bottom, Whitehead system.

small triangular copper bar, the end vats carrying the main bus bars to and from the next series of vats. In smaller refineries the vats may be arranged in long lines; both bus bars are then carried on one side of the vat as shown in Fig. 5.

A second method of operation has been used for copper electrolysis where the two end electrolytes are connected externally. In operation the intermediate electrodes are then bipolar, one face acting as anode, the other as cathode (see Fig. 6).

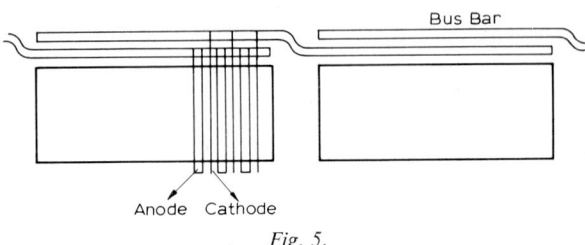

Fig. 5.
Alternative electrical connection.

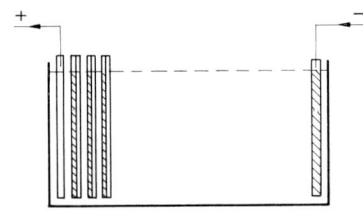

Fig. 6.
The Nicols series system (electrolytic cells).

The anodes are prepared by coating one side with a resin film. As electrolysis proceeds, copper is built up on one face and dissolved from the other. At the end of the process, the refined copper is removed from the anode residue.

The system has not been widely adopted, although starting sheets are not required and the cell voltage is 33% lower than that in normal refining operations. These advantages are offset by the need to make and align electrodes carefully and the loss of current efficiency.

The difficulty in separating anode remnant from cathode may be reduced by the development of the Hazelet continually cast anode.

4.6 Slimes Recovery

Slimes are recovered in two ways, the method depending on whether they adhere to the anode or not. Where the slimes are adherent, the anodes are withdrawn from the cell and transported to the treatment area where they are removed by scrubbing or scraping for electrolyte recovery. Where the slime is non-adherent it falls to the bottom of the cell and is sluiced out into launders which carry it to a recovery section.

5. Anodes

5.1 Production of Anodes

Ideally anodes should be:
(a) Of good geometrical shape so they will hang vertically in the vats and particular attention is paid to the parts which are in contact with the bus bars and adjacent electrodes.
(b) Of equal thickness within fairly close limits.
(c) Homogeneous. Most anodes are cast in horizontal water-cooled moulds. Under these circumstances there is a tendency for liquation of impurities to occur and for the two faces of the anodes to corrode at different rates. This has been observed with horizontally cast solder anodes where one of the faces has become passive whilst the other face corrodes normally. In general, finer grain-sized anodes with impurities in finer dispersions (supersaturated solid solutions, etc.) tend to dissolve more readily.

Lead anodes have been cast very successfully in vertical water-cooled moulds. In this case the thickness is very reproducible and both faces have identical corrosion characteristics.

5.2 Classification of Anodes

Anodes may be classified according to the type of slime they produce.

5.2.1 Adherent slimes

In the Betts process for the production of electrolytic lead, adherent slimes are formed.

As electrolysis proceeds metal dissolves from the anode and the slimes remain as a shell on the anode face. The anode voltage rises throughout the life and may rise to a point where the impurities dissolve. The anodes are removed before impurities dissolve and the slime is removed by brushing or scraping. The anode remnant is sent back for re-use or to be re-melted as appropriate, and the slime is washed to recover electrolyte. Although adherent slimes are easily handled and do not become mechanically entrained on the cathode they significantly increase the power cost of the operation. An increase in cell voltage of 50–75% is commonly encountered as electrolysis proceeds.

5.2.2 Falling slimes

Non-adherent slimes are encountered in copper refining. As electrolysis proceeds the slimes normally fall to the bottom of the cell and there is, therefore, no significant increase in cell voltage. However, some of the falling slimes may be carried by the electrolyte on to the cathode surface. The rate and method of circulation are therefore more critical than cells with adherent slimes.

In nickel refining operations using metal anodes, the slime falls to the bottom of the vats and is sluiced out to the recovery section. No contamination of cathode occurs as they are protected by the diaphragm.

5.2.3 Floating slimes

Nickel sulphide anodes are also used for the production of cathode nickel; in this case due to the large quantity of elemental sulphur formed the anodes are surrounded by bags. This causes a further rise in cell voltage but this is offset by the reduction in the cost of preparing the anode since the high temperature converter operation is not required. The elemental sulphur is easily recovered from the slime.

In copper electrorefining floating slimes are frequently encountered and give considerable trouble by impinging on the cathode. This causes irregular growth and increases cathode impurity levels.

Both Mantell[18] and Butts[19] report the formation of a floating slime which contains antimony and arsenic, where it is thought that the two elements at first dissolve and are later precipitated in the liquor. There is also evidence to show that these elements may be precipitated by glue or its decomposition products.

5.3 SLIME CHARACTERISTICS

In an ideal system only the metal to be deposited should dissolve in the electrolyte leaving the slimes as a concentrate of the impurities. In practice this does not happen. Impurities tend to dissolve and the metal to be deposited often remains as a constituent of the anode.

Slimes are formed in one of four ways:

(a) The impurity metals may be more electropositive than the refined element. For example, platinum, palladium and gold in the refining of copper, and bismuth and silver in the refining of lead.

(b) Compounds may be formed in the metal which are not dissolved. Silver and copper form sulphides, tellurides and selenides in copper anodes. It has been shown that silver compounds are first formed and if there is an excess of selenium and tellurium, copper compounds are formed, thus reducing the efficiency of the operation.

In the same system nickel may be fixed in the slime as nickel oxide by controlling the oxygen content of the anode. This is undesirable as modern secondary copper anodes may be high in nickel and the increase in the nickel content of the slimes increases the slimes' treatment costs[20]. Modern practice is therefore to reduce the oxygen content and recover the nickel from the electrolyte.

(c) Some of the impurities may dissolve but become part of the slime by reaction with the electrolyte.

Any silver which dissolves in copper refining electrolyte is precipitated as silver chloride by maintaining a low level concentration of chloride in the electrolyte. Lead dissolving in copper electrolyte is immediately precipitated as sulphate.

In the refining of tin using sulphuric acid electrolyte, lead sulphate slimes are produced. The slimes produced are dense and have a high resistance, consequently anode composition has to be closely controlled and intermediate scrubbing of the anode may be required.

(d) Intermetallic compounds are often formed between impurities and the metal to be deposited. In the refining of solder by the Betts Process a peritectic compound of tin and antimony is formed.

5.4 Anode life

In practice the most significant refinery variable is the change in the properties of the anode. The points of control in a refinery are mainly concerned with the effect of these changes which may lead to increased operating costs or to reduction in the purity of the product metal. In refineries receiving feed from a single source, changes in anode behaviour are likely to be small from day to day, but where the feed is from a number of sources significant changes in behaviour may occur very quickly. In this case the refinery task is much more difficult and a comprehensive system of records is required so that conditions may be adjusted to meet a given anode feed.

The anode life (conveniently measured in ampere days) is deduced using information from the anode casting procedure and anode impurity content. Ideally the major portion of the anode is consumed leaving a thin profile of the original anode as a residue. In normal circumstances the life is set for a given thickness of anode. The life, of course, must never be exceeded but may be shortened to overcome difficulties associated with change in the anode behaviour.

6. Electrolyte

6.1 Function

The function of the electrolyte is to provide a base for the transfer of metal ions produced at the anode to the cathode. In all refining operations impurities are dissolved at the anode and must be removed. In some cases the production of commercially acceptable cathodes is only brought about by addition of reagents which modify the deposition process.

The electrolyte condition therefore, must be closely controlled. The important factors are:
(a) Impurity content
(b) Addition agents
(c) Temperature
(d) Balance of metal ion concentration.

6.2 Purification of electrolyte

Impurities in the anode which dissolve will either be deposited on the cathode or build up in the electrolyte. Small amounts of impurities may often be tolerated in the cathode and are consequently not removed. Electrolytically refined cobalt may contain up to 0.4% nickel and refined nickel up to 0.5% cobalt[21].

In most cases impurities cannot be tolerated and are removed by a variety of purification procedures.

The procedures used are:
(i) Removal of the metal to be refined and rejection of the stripped electrolyte
(ii) Precipitation of the impurity by:
 (a) Cementation
 (b) Reduction to a lower valency
 (c) Oxidation to a higher valency
 (d) Adjustment of pH.

6.2.1 Stripping of electrolyte

Copper refinery electrolytes may contain significant amounts of iron, nickel and arsenic and they are removed by electrochemically stripping copper from a portion of the electrolyte using insoluble anodes. As the copper content is reduced to a few g/l, arsenic is then deposited with the copper. At very low levels of copper, arsine may be produced at the cathode and because of its toxicity the last few cells are either hooded or located in the open air. The electrolyte now denuded in copper is then evaporated for recovery of nickel and acid.

An interesting separation can be achieved when copper is to be stripped from the electrolyte by the use of Pyne–Green Cells[22]. The electrolyte is withdrawn from the cells at two points simultaneously. The electrolyte from the top overflow is higher in acid and impurities but lower in copper than that which is drawn from the bottom. The electrolyte low in copper is then fed to the cells with insoluble anodes (Liberator Cells).

	Electrolyte concentration (g/l)		
	Cu	H_2SO_4	Ni
Top overflow	27	212	12.05
Bottom overflow	49.5	197	11.96

At the International Nickel Company Refinery[23] the corresponding figures for copper are 38 and 14 g/l.

(a) Cementation is used in lead refining to remove bismuth which dissolves in the electrolyte in the slimes treatment operation. The bismuth is simply removed by passing the electrolyte through a tank containing lead with a large surface area. In nickel refining, copper is removed by cementation on nickel powder.

(b) Reduction to a lower valency. In the alkaline tin electrolysis the anode reaction produces stannite, stannate, plumbite and antimonite[24]. The stannite in the liquor subsequently reduces the plumbite and antimonite to metal

$$SnO + PbO = SnO_2 + Pb$$
$$3SnO + Sb_2O_3 = 3SnO_2 + 2Sb$$

(c) Oxidation to a higher valency. One of the methods used to remove cobalt from nickel electrolyte depends upon the precipitation of oxidised cobalt compounds.

At pH 5 cobalt is oxidised from cobaltous to cobaltic by treatment with either chlorine or nickelic hydroxide.

$$Ni(OH)_3 + CoSO_4 = NiSO_4 + Co(OH)_3$$
$$Cl_2 + 3H_2O + CoSO_4 = H_2SO_4 + HCl + Co(OH_3)$$

(d) Control of pH. Iron, arsenic, lead and tin are common impurities in nickel anodes and if the iron in the electrolyte is oxidised and the pH adjusted to 5, ferric hydroxide is precipitated, arsenic, lead and tin being absorbed on the precipitated hydroxide and removed.

6.3 ADDITION AGENTS

Reference has already been made to the role played by the chloride ion in the control of silver in copper electrorefining. In all refineries additions are made to the electrolyte to
(a) Reduce anode polarisation
(b) Improve cathode condition.

Originally nickel refiners used a simple sulphate electrolyte. It was found necessary to control the pH of the cathode compartment to produce good quality nickel cathodes. This was achieved by addition of boric acid which, acting as a buffer, effectively controlled the pH of the compartment and enabled higher current densities to be employed.

Further development has led to the addition of sodium chloride to the electrolyte which markedly reduces anode polarisation and power cost. In addition it opened up a new route for the separation of cobalt by selective oxidation with chlorine.

It has been found necessary in most refineries to add substances which modify the grain size of the deposit. In copper refining, if nothing is added to the electrolyte, deposits are irregular with nodule formation. This leads to slimes pick up on the cathode and loss of efficiency due to short circuiting. Although chloride ion is said to be useful in modifying grain size[25] the reagent used by most copper refineries is bone glue. Quantities added are very small ranging from as little as 8 g/tonne to over 230 g/tonne of cathode deposited. Addition agents of this type increase polarisation of the cathode and therefore power cost; this has to be accepted to achieve the required cathode condition. Other addition agents have been tried for reducing polarisation whilst maintaining cathode condition, but to date no satisfactory substitute for glue has been found. Among those tried are, Goulac, thiourea, casein, Seperan and Avitone A. For the same reasons additions are made in the Betts process. In this case it has been found possible to reduce the amount of glue used by a mixture of glue and calcium lignosulphonate (Goulac). Some systems are specially sensitive to the addition agent content. The most successful agent used in the sulphuric acid electrolyte for the refining of tin has been a mixture of glue and cresol sulphonic acid. The cathodes often develop tree-like growths which form in a period of a few hours and are a serious limitation on the refining method, so current densities do not exceed 10 A/dm^2.

In all cases the addition agents are consumed by degradation and by losses in slimes and have to be replaced continuously.

6.4 Temperature

Control of electrolyte temperature is very important in electrorefining in that it affects
(a) Cathode polarisation
(b) Anode polarisation
(c) Rate of evaporation of electrolyte.

Generally the lower the temperature the better the appearance of the metal although the effect is not very marked in most cases over a range of 5–10°C.

The effect of temperature on the anode dissolution is much more critical. In solder refining by the Betts process a change of 5°C has a marked effect on the anode polarisation, an increase in temperature reducing the polarisation. At the Noranda Copper Refinery[26] anode dissolution characteristics were considerably changed by an increase in electrolyte temperature from 61.5° to 66.5°C. As temperature increases in the electrolytic cell, voltage is reduced significantly and a cost saving is possible if power provided by steam is cheaper than electrical power. The operating temperature is important in maintenance of the water balance in the circuit. In some cases in the Betts refining process, where there is no electrolyte outlet from a stripping process, at 40°C enough water is evaporated to balance the water introduced by washing the slimes, whilst at 30°C evaporation has to be used to maintain the balance.

6.5 Metal ion balance

In electrorefining, anode and cathode efficiency are rarely the same resulting in the metal ion building or falling in the electrolyte. Where the metal ion content builds as electrolysis proceeds, the balance is maintained by the use of cells with insoluble anodes (copper and alkaline tin electrolysis). Where the metal ion content falls, restoration is made by chemical dissolution of the metal or one of its compounds. The balance may be affected by the method of anode purification. For example, it has been noted that when lead anodes are treated by the addition of sodium metal, there is a marked drop in anode polarisation with a tendency for lead ions to build up in the electrolyte.

7. Cathode

The production of smooth dense cathodes may be essential in some electrorefining processes since the cathodes themselves are sold on the metal market. The British Standard for cathode copper[7] states "Cathodes should be tough, dense and free from loose or brittle lumps, nodules and other excrescences. The surface should be free from slime and from copper sulphate".

In other refineries (tin, lead solder) whilst it is not essential to produce smooth dense cathodes, it is advisable since impurities may be trapped in the cathode cavities, and will dissolve when the cathodes are melted.

The important factors in the production of good cathodes are as follows.

7.1 Positioning of electrodes in cells

The edges of the cathodes must always be outside the edges of the corresponding anodes. If the electrodes are badly aligned so that one edge of the cathode is inside of the anode, a heavy deposit grows on the edge of the cathode often with distinct breaks in the surface.

7.2 Flow of electrolyte through the cell

As already mentioned the flow of electrolyte through the cell is important. The flow should be such that electrolyte is transported continuously past the entire cathode face. Particular attention is paid to the surface flow of electrolyte. In a particular case in a Betts lead refinery it was found that at high current densities irregular growth on the cathode appeared at the electrolyte surface. Examination of the system showed that whilst the bulk of the electrolyte contained 40 g/l lead, the electrolyte at the surface contained only 5 g/l lead, and in fact there was very little flow in this region.

7.3 Maintenance of an adequate supply of addition agent

This has already been discussed in Section 6. It has been impossible in the past to find a reliable rapid method for the determination of the glue content in refinery electrolytes. The British Non-Ferrous Metals Research Association have published a method which is reliable but unfortunately results cannot be obtained on the same day as the sample is drawn[27]. The control of the amount required is normally by a day to day examination of the cathode condition.

7.4 Absence of solids in the electrolyte

High concentrations of solids in the electrolyte are undesirable since they may impinge on the cathode surface causing irregular growth and increase the impurity content of the cathode.

7.5 Elimination of short circuits between anode and cathode

In all refineries short circuits between anode and cathode occur and can only be controlled by a rigorous system of inspection. When a short circuit occurs in a cell the current passing through the other electrodes is greatly reduced and thin cathodes are produced. The vat voltage is significantly reduced if a bad short circuit has been formed.

Two methods for the identification of short circuits have been used. The first is based on the heating effect of the current. If a cathode is in contact with the anode, the cathode carrier bar becomes hot and may be easily identified. The second is based on the increase in magnetic flux created by an excess of current carried by the shorting cathode.

8. Refinery costs

The major refinery costs may be broken down as follows.

8.1 POWER COSTS

The power cost is derived from the product of the current required to deposit metal in a given time and the voltage required to drive the current through the circuit. In theory 96,500 C (1 A passed for 1 sec) will dissolve or deposit 1 g equiv. of any metal. There will, therefore, be a considerable difference in the quantity of metals deposited for a given current. For example, 96,500 C will deposit as much as 108 g silver compared with only 9 g of aluminium. In practice the current efficiency is never 100% but usually well above 90%. The current is lost by:

(a) Leakage to earth due to faulty insulation.

(b) Other possible electrochemical reactions.

In nickel refining, hydrogen is produced simultaneously with nickel thus producing a direct reduction in current efficiency. Indirect reductions in current efficiency occur in copper refining where the electrolyte contains iron. The electrodeposited copper may be redissolved by ferric ions in the electrolyte. The mechanism is discussed in Section 4.

(c) Short circuiting between anode and cathode.

The voltage required to drive current through the system may be considered in two parts. In the first place there is a voltage drop between the rectifier and the carrier bars of the electrodes. High resistance points occur at each joint in the circuit. They cannot be eliminated but particular attention is paid to maintaining good clean connections throughout the system. Normally the voltage drop attributed to poor connections is less than 5% of the total.

The second component is the voltage drop across the cell itself; this may be subdivided as follows:

(i) Anode voltage
(ii) Cathode voltage
(iii) Electrolytic voltage.

The anode voltage may be affected by changes in temperature, impurity content, the nature of slimes produced, concentration of the metal ions in solution and current density.

Cathode potential is only significantly affected by addition agents and the concentration of metal ions in solution.

The voltage drop between electrodes is mainly affected by temperature, electrolyte composition and distance between electrodes; the distance between electrodes being particularly important as shown by the results of an experiment in a Betts solder refinery where the electrode spacing was varied and the cell voltage measured.

Anode/cathode spacing (cm)	Cell voltage (V)
11.4	0.62
8.9	0.50
6.2	0.36
5.1	0.33

A detailed discussion of power costs is given by Addicks in a book published in 1921. The points are still relevant today.

8.2 Labour costs

Labour costs are one of the highest in the electrorefining operation. Labour is required for:
 (a) Systematic inspection of plant and cells
 (b) Casting of anodes
 (c) Treatment of slimes
 (d) Changing of anodes and cathodes and preparation of starting sheets.

Improvements in mechanical handling over the past few years have brought significant reduction in labour costs.

8.3 Fuel and reagent costs

These must vary widely according to the metal refined and the complexity of the refinery. Fuel costs are incurred in the melting of anodes and cathodes and in the heating of electrolyte. The costs of reagents may be considerable since expensive chemicals are continuously being used in refining operations. In some cases the cost of transport may be significant. There may be a significant loss of expensive electrolyte in the slimes treatment procedure, *e.g.* loss of fluosilicic acid in the Betts refining processes.

8.4 Maintenance

In the efficient running of a refinery it is essential for the plant to be kept running continuously as breakdown in the circulation system will result in stripping of the cations from the electrolyte with subsequent deterioration of the cathode condition. The breakdown of the power supply may result in chemical re-solution of the cathode metal because the cells may be reversible.

In general the following are the most difficult problems encountered.

8.4.1 Corrosion

Corrosion problems are varied and numerous as refineries run using strong concentrations of acids and alkalies. Faulty insulation may also result in electrochemical corrosion. The advance of the plastics industry has solved many of the problems and plastics are gradually replacing the wood, metal and rubber formerly used. Lead covered vats are still widely used in the copper refining industry.

8.4.2 Breakdown in pumping and circulating equipment

Duplication of key pieces of equipment is necessary for the maintenance of continuous flow of electrolyte. The principle of using three circulating pumps (two in circuit and one stand-by) ensures that at least one pump is in circuit the whole time.

8.4.3 Breakdown in electrode handling and slimes treatment equipment

Overhead cranes for the handling of electrodes have become an integral part of the process and because duplication is not normally desirable they must be very reliable. Maintenance work is best carried out on planned stops.

Maintenance thus becomes a significant part of refinery cost and is often the factor where the largest saving can be made.

9. Future developments

Attempts have been made in the past few years to cut the costs in electrorefining by one of the following methods:

(i) Increasing the current density applied

The major barrier to the increase in current density is that as current density increases so anode and cathode polarisation increase. The rate of removal of metal ions at the cathode surface exceeds the rate of transfer from the bulk of the solution to the cathode surface. Recently attempts have been made to overcome this by introducing the electrolyte parallel to the face of the cathode at a higher flow rate. Several investigations have shown that the current density can be significantly increased by this method without affecting cathode condition[28].

(ii) Reducing anode preparation costs

The direct refining of nickel sulphide anodes provides an instance where anode preparation costs have been reduced. Several systems have been proposed where copper alloys may be used directly as anodes[29]. There are several difficulties to overcome, notably the imbalance in the electrolyte created by the high impurity content of the anode and reproducibility of the anode if perforated titanium baskets are used as scrap carriers.

(iii) Direct production of salable products

An improvement in cost would be achieved if salable products could be made directly. Copper sheet is already made in a limited scale using a revolving drum as cathode. Experiments at British Non-Ferrous Metals Research Association have been carried out to produce copper wire, tube and strip directly[30, 31].

(iv) Improvements in materials of construction

The high cost of corrosion in electrorefining plants is gradually being reduced by the introduction of new materials of construction. The increased use of plastics may prolong the life of some of the major plant components.

(v) Improvements to cathode purity

Attempts are being made to increase cathode purity. In some instances this may be accomplished by the development of new techniques for electrolyte purification. Solvent extraction previously used for the separation of rare metals has been developed to the point where it may be useful for the purification of refinery electrolytes. For example, separation of cobalt and zinc from nickel.

10. Electrolytic production of metal powders

The electrolytic manufacture of metal powders is based on the same basic processes as those found in the electrowinning and electrorefining of metals, and these subjects should be studied for further details. Just as in electroplating, the aim is to obtain a smooth, coherent deposit; by suitably altering the conditions it is possible to obtain very poorly adherent deposits. In certain cases such as copper powder they may be removed by a simple wiping action, while in the case of iron powder a more severe mechanical treatment such as percussion is needed to remove the deposit, which often needs crushing before attaining the desirable physical form.

In electrowinning and refining from molten salt electrolytes, it is all too easy to obtain a cathodic deposit in powder form; indeed major effort has gone into obtaining a continuous deposit from these melts. Thus tantalum, for example, is obtained both by electrowinning and by sodium reduction. The former is often used when the metal is destined for use as a powder, the latter is preferable when remelting or other treatment is demanded.

In aqueous electrolytes, copper, iron and zinc are made as powders by electrolytic methods, though with iron especially there is severe competition from other purely chemical routes to the powder form. Different uses of the powders favour one or other method of preparation, and the more random size distribution found in electrolytic powders is often a factor in their favour. On the other hand, the metals are often deposited in a very active state which causes problems with washing/drying and air oxidation. An excellent review of the subject of powder metallurgy is found in ref. 32 where the author devotes some 25 pages to the subject of electrolytic powder manufacture. He indicates that among the 30 or more metals which may be made electrolytically, not only copper and iron but also zinc, nickel and tin have commercial significance. However we find it hard to locate references to the latter two metals, in a commercial context. Jones deals with the properties and manufacture from aqueous, molten salt and mercury cathode cells. In ref. 33 several papers on industrial practice are given. Ljungberg of Husqvarna, discusses iron and nickel and also the manufacture of ferro-alloys by electrolysis, without explicitly stating the commercial importance of the latter. Mehl discusses the manufacture of copper powders while Miller discusses molten salt electrolyses such as those of Be, Ta, Nb, Ti and Zr. Further points on all the above contributions are made in the discussions reported. The manufacture of Ti powders of different grain sizes is dealt with in ref. 34.

An up to date review of the overall situation in iron powder manufacture, with economics of the main competitive processes is found in ref. 35, while Ibl[36, 37] has made two very comprehensive surveys of the entire field of electrowinning of metal powders containing over 350 references.

As in the other branches of metal winning, methods based on reduction with hydrogen compete strongly with electrochemical methods—indeed, many would say they are themselves electrochemical in their *modus operandi*. Details of the Sherritt Gordon Mines process are found in ref. 38 while ref. 39 gives an overall view of methods of powder manufacture, including work on iron powders with mercury cathodes. Schulz[40] refers to the commercial importance of zinc, but gives no references other than those already quoted here.

11. References

1 J. ELKINGTON, English pat., 2838, Nov. 1865; 3120, Oct. 1869.
2 A. G. BETTS, U.S. pat., 713, 277; 713, 278 (1902); 891, 395; 891, 396 (1908); 918, 647 (1909).
3 R. C. ROWE, *Can. Mining J.*, 75 (1954) 249.
4 H. G. CORDERO and T. J. TARRING, *Non-Ferrous Metal Works of the World*, 1965, p. 549.
5 N. V. HYBINETTE, U.S. pat., 805, 555; 805, 969, Nov. 1905.
6 W. BORCHERS and W. G. MACMILLAN, *Electric Smelting and Refining*, C. Griffiths, 1897.
7 British Standard 1035-40, 1964.
8 American Standard for Testing Materials, 1955.
9 *Can. Mining J.*, 67 (1946) 535.
10 F. A. FORWARD and V. N. MACKIW, *J. Metals*, (1955) 457; V. N. MACKIW, W. C. LIN and W. KUNDA, *J. Metals*, (1957) 786.
11 R. C. ROWE, *Can. Mining J.*, 75 (1954) 316.
12 J. O. BETTERTON, Y. E. LEBEDEFF, *Metallurgy of Lead and Zinc*, AIMME, 1936, p. 205.
13 C. L. MANTELL, *Electrochemical Engineering*, McGraw-Hill, New York, 1960, p. 171.
14 A. E. RICHARDS, *The Refining of Non-Ferrous Metals*, I.M.M., London, 1950, p. 97.

15 A. E. RICHARDS, *The Refining of Non-Ferrous Metals*, I.M.M., London, 1950, p. 107.
16 A. BUTTS, *Copper, the Metal, its Alloys and Compounds*, Reinhold, New York, 1954, p. 177.
17 C. S. HARLOFF and H. F. JOHNSON, *Trans. Am. Inst. Mining Met. Engrs.*, 106 (1933) 398.
18 C. L. MANTELL, *Electrochemical Engineering*, McGraw-Hill, New York 1960, p. 163.
19 A. BUTTS, *Copper, the Metal, its Alloys and Compounds*, Reinhold, New York, 1954, p. 173.
20 J. S. JACOBI, *Advances in Extractive Metallurgy*, I.M.M., London, 1967, p. 395.
21 British Standard, 375, 1930.
22 F. R. PYNE, *Trans. Am. Electrochem. Soc.* 28 (1915) 111.
23 *Can. Mining J.*, 67 (1946) 451.
24 P. A. WRIGHT, *Extractive Metallurgy of Tin*, Elsevier, Amsterdam, 1966, p. 157.
25 YU-LIN YAO, *Trans. Electrochem. Soc.*, 86 (1944) 371.
26 A. BUTTS, *Copper, the Metal, its Alloys and Compounds*, Reinhold, New York, 1954, p. 175.
27 R. A. WHITE and A. M. SZOKOLAY, *B.N.F. Research Report*, A. 1605, 1966.
28 A. K. ANDERSON and T. BALBERYSZSKI, *T.M.S. Paper*, A.68-17, 1968.
29 F. JOHANNSEN and W. LIEBE, *Z.Erzbergbau Metallhuettenw.*, 16 (1963) 558.
30 N. J. FINCH, *B.N.F. Research Report*, A.1676.
31 A. CIBULA and N. J. FINCH, *B.N.F. Research Report*, A.1696.
32 W. D. JONES, *Principles of Powder Metallurgy*, Arnold, New York, 1960.
33 *Powder Met.*, No. 1/2 (1958).
34 A. B. SUCHOV et al., *Porosh Kovaya Met. Akad. Nauk Ukr. SSR*, (1966) 46.
35 P. C. FINLAYSON and A. P. MORRELL, *Commonwealth Mining Met. Congr. 9th, 1969*, IMM.
36 N. IBL in P. DELAHAY and C. W. TOBIAS (Eds.), *Advances in Electrochemistry and Electrochemical Engineering*, Vol. 2, 1962, pp. 49–143.
37 N. IBL, *Chem. Ing.-Tech.*, 36 (1964) 601.
38 C. F. BRENTHEL, *Erzmetall*, 8 (1955) 422.
39 R. KIEFFER and G. JANGG, *Chem. Ing.-Tech.*, 39 (1967) 43.
40 W. SCHULZ, *Chem. Ing.-Tech.*, 40 (1968) 570.
41 *Foundry Trade J.*, 123 (1967) 47.
42 M. HAYASHI et al., *U.S. Bur. Mines. Rept. Invest.* No. 6445.
43 F. JOHANNSEN and H. STOCKHANS, *Erzmetall*, 16 (1963) 227.
44 F. JOHANNSEN and W. LIEBE, *Erzmetall*, 16 (1963) 558.
45 B. N. SINGH and H. K. CHAKRAVARTI, *Natl. Metal. Lab. J. (India)*, 7 (1965).
46 E. SÜSSMUTH, *Chem. Ing.-Tech.*, 37 (1965) 590.

Chapter 7 (Part 2)

Electrorefining in Molten Salts

P. F. HART

G.E.C., Hurst Research Centre, Wembley, Middlesex

A. W. D. HILLS

Metallurgy Department, Sheffield Polytechnic, Sheffield S1 1WB, Yorkshire

1. Introduction

Most electrorefining processes take place with an aqueous electrolyte, and these have been discussed elsewhere, but it is also possible to use molten salt electrolytes during metal refining processes. These processes naturally operate at a higher temperature than aqueous refining processes and are therefore basically more expensive to run. For them to be economic, then, the use of a molten salt electrolyte must offer certain specific advantages over an aqueous electrolyte. One obvious advantage arises from the absence of hydrogen ions from the electrolyte thus enabling metals such as aluminium and magnesium, high in the electrochemical series, to be refined. Another possible advantage arises because the metal electrodes are frequently molten at the temperatures of operation. Overvoltages at molten metal electrodes are negligibly small[1], so that such cells can be operated at high current densities and, of course, without any problems resulting from dendritic growth at the cathode or inter-granular attack in the anode. Moreover, diffusion rates are very much more rapid in liquid electrodes, and it becomes possible to refine the anode by electrolysing impurities into the cathode. Processes of this type offer obvious cost advantages over the more normal refining processes since very much less metal needs to be transferred across the refining cell. Molten salt electrorefining has also been used for refractory metals which are solid at the temperatures of operation. Such processes are only economic if the molten salt electrolyte offers definite thermodynamic advantage over an aqueous electrolyte, or if the high temperatures involved have a beneficial effect on the kinetics of the electrode reactions.

In spite of its possible advantages, and in spite of the large amount of experimental work that has been done on molten salt refining, the process has found relatively little commercial use. In the main, then, this review is concerned with preliminary laboratory investigations into various molten salt refining processes, with some pilot plant developments, but with very few commercial industrial processes.

Two high temperature refining processes apparently related to molten salt electrorefining, electrotransport and electroslag refining, will not be dealt with here in any great detail since they are not truly electrochemical in nature. Electrotransport is the movement of solute atoms in metals under the action of an electric field[2].

Although it represents a possible tool for removing interstitial solutes from metals, especially where super purity metals are required, it does have marked disadvantages[3]. It is extremely inefficient since very large electric currents are produced by fields that have a moderate effect on the movement of solutes, so that large amounts of energy are consumed for relatively little refining effect. Moreover, the migration effects are greatest in liquid metals but the Joule heating effect resulting from the large electric currents sets up intense natural convection currents which stir the liquid metal and greatly reduce the refining effect[4]. The intensity of these convection currents increases with the diameter of the liquid metal channel so that virtually no refining effect occurs in channels greater than 1 mm in diameter[4]. Thus very small amounts of metal could be treated in this way. The charges of transport deduced from experimental results are positive, slightly dependent on temperature, and markedly larger for the heavier solute species[5]. This suggests that the migration process is due to the transfer of momentum from conductivity electrons to impurity ions[5], and the process is not truly electrochemical in nature.

Another high temperature refining process involving the passage of electric current is the electroslag refining of steel[6]. Electric current is passed through a slag bath of calcium fluoride, calcium oxide and alumina, from an impure metal electrode to a water cooled purified electrode below the bath. The heat generated in the slag melts the tip of the impure electrode which takes on the shape of an inverted cone. Impurities are removed from the thin molten film as it runs down the cone to form a large drop at the tip. Once this drop reaches a critical size it breaks free and falls through the slag into a molten metal pool on top of the water cooled electrode, eventually solidifying at the bottom of this pool. The process produces pure steels containing a very low number of non-metallic inclusions. Inclusions in the impure electrode dissolve in the slag as the metal around them melts away, and further refining results from the chemical action of the slag on the thin film of molten metal at the electrode tip[7]. Once again, it would appear that truly electrochemical effects play a negligible part in the refining action.

The detailed information that follows in this review is concerned with commercial molten salt electrorefining processes where they exist, and with laboratory and pilot plant feasibility studies on other possible molten salt refining processes. The theory of the process will not be discussed here, the interested reader is referred to a previous paper[8] to which the authors contributed.

2. Aluminium

Aluminium is too reactive a metal to be refined by controlled oxidation of impurities, and has too high a boiling point to be refined by distillation. It is refined industrially by the most successful molten salt electrorefining process, often described as the "three-layer" process[9]. The process depends on adjusting the densities of the molten impure aluminium and of the electrolyte so that a pure molten aluminium cathode will float on the electrolyte, which in turn will float on the impure anode. The density of the impure aluminium is increased by adding copper, the melting point of the resulting alloy being lowered by additions of small amounts of silicon, and the electrolyte is a mixture of barium, sodium and aluminium fluorides with some alumina. A typical aluminium refining cell is shown in Fig. 1.

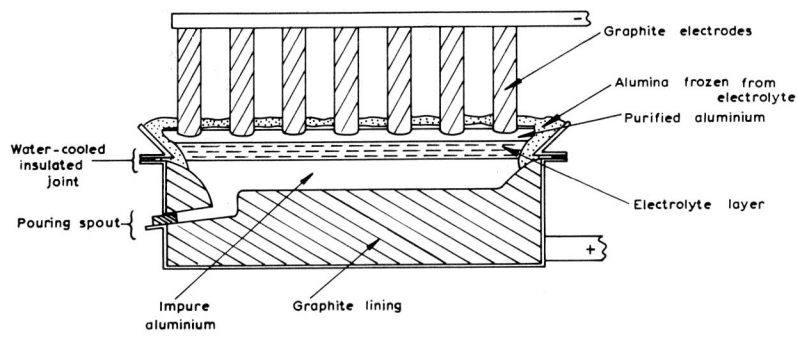

Fig. 1.
Typical cell for the refining of aluminium.

It consists of a steel shell made in two sections, a lower section lined with graphite, and an upper section lined with alumina frozen from the molten electrolyte. An insulated water cooled joint is maintained between the two sections of the shell. Graphite electrodes dip into the floating molten cathode and the impure anode is connected to the power supply *via* the graphite lining of the lower section. A thin solid crust of frozen alumina is maintained on top of the molten cathode protecting it from oxidation.

When the cell is in operation, the cathode is tapped through a pouring trough, not shown in the diagram, and the anode is made up by pouring premelted aluminium–copper alloy through a graphite funnel directly into the anode layer. The depth of the molten cathode must not be reduced too much when tapping the cell, since highly turbulent conditions can be set up in the resulting thin electrode layer by electromagnetic forces. This turbulence can so distort the electrolyte layer that the anode and cathode come into contact, thus contaminating the cathode. This turbulence is avoided if a relatively deep cathode pool is maintained on top of the electrolyte.

The process is used to manufacture high purity aluminium from aluminium produced by electrolysing bauxite by the Hall–Heroult process. Aluminium of 99.9% purity and sometimes even 99.99% can be produced. Akerman *et al.*[10] have used radioactive traces to study the movement of impurities during the molten salt electrorefining of aluminium and they suggest that a double electrolysis process could produce aluminium of 99.999% purity.

Other work on theoretical and scientific aspects of aluminium electrorefining is found in refs. 96, 97 and 98 with physico-chemical electrolyte properties described in ref. 99 and electrode processes described in ref. 100.

Commercial data on electro-refined ("S.P." Superpurity) aluminium is found in refs. 94 and 95. In addition, further information has been kindly supplied by British Aluminium (J. Waddington) and Alcan (I. H. Jenks). Table 1 gives capacity and production.

TABLE 1

SUPERPURITY ALUMINIUM

Alcan, Arvida, Quebec	Capacity:	780 tonnes/annum
Aluminium Co. of America, Badin, N.C.		3,200
Consol Alumin. Corp., N. Johnsonville, Tenn.		1,500
Kaiser Al Corp., Spokane, Wash.		2,650
Reynolds Metals, Listerhill, Ala.		600
Vereinigte Metall Werke, Ranshofen, Austria		3,000
Pechiney, Ariege, France		3,300
Aluminium Hütte, Rheinfelden, Germany		1,650
Ver. Aluminium Werke, Erftwerk, Germany		3,000
S.A.V.A. Porto Marghera, Italy		400
Nippon Light Metals, Kambara, Japan		1,000
Sumitomo Chemical, Kikumoto, Japan		2,500
A/S Vigelands Brug, Vigeland, Norway		3,000
Alusuisse, Chippis, Switzerland		3,200
Taiwan Aluminium Corp., Takao, Taiwan		770
Soiusalumini, Volgograd, USSR		11,500
Volkov, USSR		3,300
Tatabanya, Hungary		650
Loch Foyers, British Aluminium, U.K.		740

It must be stressed that many of the above mentioned plants have made no product for several years, and may be mothballed or closed down.

3. Lead

A considerable amount of work has been done on the electrorefining of lead—mainly in Russia. Generally, crude lead is purified by a complex metallurgical process finishing with electrodeposition from aqueous electrolytes. It has been suggested that molten salt electrorefining might offer a cheaper refining method, and such a process was developed by Borchers at the turn of the century[11]. The cell feed for this process was molten crude lead and the electrolyte was a mixture of lead oxychloride with potassium and sodium chlorides. The cell operated at 550°C and at a cathodic current density of 1 A/cm². The high density of lead makes it impossible to use a three-layer process, and so a different cell design was developed (see Fig. 2). The molten electrodes were separated by a water-cooled insulating joint

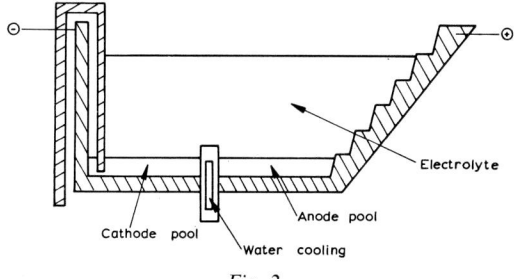

Fig. 2.
Borchers' cell for refining lead in fused salts.

which was surrounded and protected by a coating of solidified salt. The anode side of the electrolyte cell was sloped and had a series of deep horizontal platforms which served to retain some of the crude molten lead feed. Purified lead was siphoned off from the cathode pool. Borchers' process received no encouragement and is no longer used, but Delimarskii and his co-workers[12] in Russia have carried out extensive investigations into the transfer of impurities from crude lead anodes during molten salt electrolysis. Electrolytes used have included chloride and silicate melts, silicate–chloride mixtures and fused sodium hydroxide. Using lead chloride and sodium–potassium chloride eutectic mixtures, refining has been carried out at 500°C with high current efficiency (97–98%)[12-15]. Only 0.7% of the total impurities in the original anode feed was transferred to the cathode. Certain impurities—bismuth, silver, antimony and arsenic—were retained by the anode, whereas copper, silver, gold and zinc accumulated in the electrolyte, and tin was lost by evaporation.

Roms *et al.* have recently described a pilot plant cell for refining lead using this

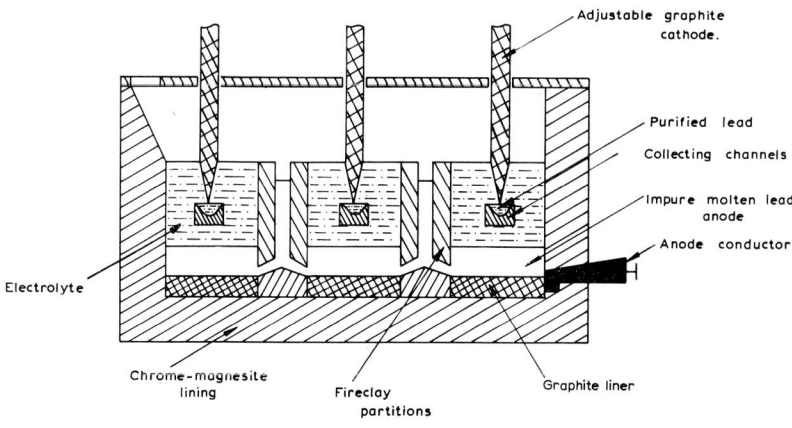

Fig. 3.
Cell developed by Roms, Belen'kii and Delimarskii for the molten salt electrorefining of lead, and other dense metals.

chloride electrolyte[16], and a sketch of this cell is shown in Fig. 3. It consists of a steel jacket lined with chrome–magnesite refractory set in powdered magnesite. Two fire clay block partitions divide the cell into three compartments, but contain passages to connect the resulting three molten lead anodes. Current is supplied to the anodes through shaped iron conductors in the cell walls and these are protected from corrosion by a coating of the crude metal. Each compartment contains an adjustable graphite cathode upon which the purified liquid lead is deposited. The lead runs down these cathodes into special collecting channels immediately below them from which it is withdrawn from the cell. The cell operates at 480/520°C at a current density of 0.5 to 0.6 A/cm², the total current being 1100 A. The authors report the results of one 80 hour test in which anode metal containing 85.3% lead and 14.7% bismuth was electrolysed to give the cathode containing 99.4% lead and an anode containing 48.4% bismuth.

In their experiments with silicate melts Delimarskii *et al.* found that the refining effect could be varied by changing the amounts of sodium and lead oxides and lead chloride in the electrolyte[17]. These vary the electrochemical relations between the lead and the impurity metals. The refining of lead has also been studied in mixed silicate–chloride electrolytes[18], and the removal of antimony, silver, tin and bismuth found to be satisfactory. However, copper present in the impure anode is deposited at the cathode together with the lead.

Delimarskii *et al.* also describe a process for the separation of bismuth from lead using a molten sodium hydroxide electrolyte in which a complex intermetallic compound of the form $Na_xPb_yBi_z$ is formed at the cathode[19]. This dissolves in the electrolyte and is oxidized at the surface of the liquid lead anode forming a lead–bismuth alloy. When the temperature of operation is 350°C the bismuth content of the lead cathode can be lowered to 0.0005%. At higher temperatures (400°C) the bismuth concentration in the anode can be raised to 70%[20]. Obviously, such a process could be regarded as an anode-purification process for bismuth, as well as a cathode purification process for lead. Delimarskii and co-workers also studied the purification of bismuth–lead alloys using molten chloride electrolytes[21]; and the electrolytic separation of lead and tin has also been reported[22].

4. Plutonium

Bjorklund *et al.*[23] studied the refining of plutonium using a molten electrolyte consisting of eutectic mixtures of plutonium chloride with alkali chlorides. The anode is a solid rod of reactor plutonium, and solid purified metal is deposited on an iron cathode. In the method developed by Mullins *et al.*[24-26], which operates at a higher temperature, a perforated tungsten cathode is used, and the electrolyte is a mixture of potassium and sodium chlorides with plutonium fluorides and chlorides.

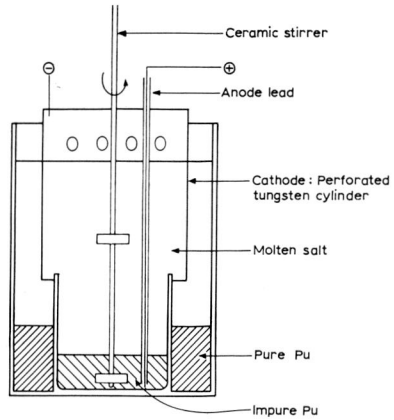

Fig. 4.
Lamex process for electrorefining plutonium.

The cell design is illustrated in Fig. 4. The impure molten anode is held in a central cylindrical container, and the cathode is a perforated tungsten tube placed above the anode compartment but concentric with it. Its diameter is larger than that of the anode compartment, and the purified plutonium flows down the cylinder, being collected in the annular space between the anode compartment and the outer electrolyte container. The cell has been operated at 740°C under an argon atmosphere, producing 250 g of pure plutonium in one hour.

Other cells have been reported[27] with molybdenum or tantalum cathodes, and Benedict and Pigford[28] developed an ingenious method of separating plutonium and other fission products from irradiated uranium by anodic dissolution in a molten mixture of calcium and uranium chlorides. In this case uranium metal is deposited on a manganese cathode and the manganese is subsequently removed by a vacuum distillation. Fission products—mainly Zr, Nb, Mo, Ru and Rh—pass into the anodic sludge and plutonium remains dissolved in the melt, from which it is later distilled as a volatile chloride.

5. Beryllium

Several different electrolyte mixtures have been used in the electrorefining of beryllium. Menzies *et al.*[29], for example, used an electrolyte containing lithium, calcium and beryllium chlorides or sodium and beryllium chlorides. Wong *et al.*[30-34] also used mixtures of lithium chloride, calcium chloride and beryllium chloride, finding that the impurity of the cathode metal increased with the concentration of beryllium chloride in the electrolyte, but that the cathode current efficiency fell. Between 450 and 550°C they obtained maximum current efficiencies of about 92%.

Bilard et al.[35] have used a lithium chloride–potassium chloride eutectic mixture containing 10–20% of beryllium chloride and they describe a pilot plant cell which has produced 100 g of beryllium in 24 hours. The cell operated at about 500°C under an atmosphere of argon and contained a central anode in the form of a rod of impure beryllium. The cathode was a perforated nickel cylinder surrounding this rod and crystals of pure beryllium were formed on the cylinder. The authors report that the impurity levels of aluminium, chromium, copper, iron, magnesium, manganese, and nickel could be reduced from up to 4000 ppm in the impure anode down to about 35 ppm in the cathode. However, oxygen impurity levels could only be reduced from up to 5000 ppm to about 300 ppm.

Wong and O'Keefe[34] have developed a two-stage refining process for the electrorefining of beryllium in which flakes from the first stage cathode were used as feed for the second cell. The resulting beryllium was spectrographically pure, except for 35 ppm of calcium.

6. Niobium

Sibert[36] describes a method for the production of niobium in which Nb_2O_5 is reduced with carbon at high temperature, and the product is electrolytically refined using a molten mixture of K_2NbF_7 and alkali chlorides. Steinberg[37] has used a similar method starting with mixtures of Ta_2O_5 and Nb_2O_5, in which a molten salt refining process served to separate the niobium from the tantalum. A similar process has been developed by Mitsuhiro Sakawa and Tadashi Kuroda for the production of metallic niobium from niobium carbide[38]. This is also a two-stage process, the niobium carbide being treated as a soluble anode for the first electrolytic step and the cathode product then refined in a molten salt refining cell. The electrolyte in this cell contains 5% by weight of K_2NbF_7 in a sodium chloride–potassium chloride eutectic mixture. The cell operates at 800°C with a rather low current density of 0.05 A/cm² and a current efficiency of 90% is achieved.

Suchkov et al.[39] report a process for the refining of niobium in which the electrolyte contains $NbCl_3$ dissolved in sodium and potassium chlorides. The cell operates at temperatures between 680 and 720°C at current densities up to 0.9 A/cm², under an argon atmosphere. The central cathode is a niobium tube inserted into the melt and surrounded by a molybdenum diaphragm. Impure granules of niobium are contained between this diaphragm and the graphite crucible which acts as the anode contact. Niobium is transferred through the molybdenum diaphragm to form a granular deposit on the cathode. This can then be withdrawn into an upper chamber where it is cooled in the argon atmosphere. The cathode product prepared from a niobium anode of 97% purity contained 99.2% niobium; tantalum, titanium, iron, silicon and oxygen being removed quite satisfactorily.

In a series of papers Grinevich and Reznichenko[40-42] describe another process for the electrolytic refining of niobium in which they use a sodium chloride–potassium chloride eutectic electrolyte containing K_2NbF_7. They report that tantalum, titanium, aluminium, zirconium, silicon, tungsten and molybdenum are satisfactorily removed from the impure niobium, although iron is not. The electrochemical potentials of iron and niobium are very similar in this electrolyte and the iron concentration of the impure anode must therefore be kept low. Furthermore, the iron parts of the apparatus had to be protected from the melt to avoid the dissolution of iron in the melt and subsequent contamination of the cathode. However, they found that molybdenum, tungsten and graphite were not attacked by the electrolyte and could be used as container materials. They also found that the niobium cathode was only free of oxygen if the apparatus was hermetically sealed.

7. Titanium

A series of Russian papers[43-51] has been published on the electrolytic refining of impure titanium–aluminium alloys using fused chloride electrolytes. In his review on the electrolysis of titanium, Kroll[52] reported an experimental process for titanium purification in which the soluble anode can be made from a range of titanium-containing materials, including 50–50 ferrotitanium. Several papers[53-58] have also been published on the refining of a series of complex titanium alloys. The work carried out at the U.S. Bureau of Mines[56-58] has shown that the most suitable electrolyte for refining titanium is a sodium chloride–titanium chloride mixture. This work[59] has also shown that nitrogen, the major contributor to the brittleness of impure titanium, can be removed by molten salt refining in this electrolyte. Scrap titanium alloy is suspended in the electrolyte in a perforated iron basket anode held at 830°C under a helium atmosphere. Refining can also be conducted at 480°C in a lithium chloride–potassium chloride–titanium chloride electrolyte under an atmosphere of nitrogen. No nitrogen is picked up by the electrolyte and the nitrogen in the original feed material is retained as an anode scale, whose composition can approach that of titanium nitride. Further work is listed in ref. 4.

8. Vanadium

Impure metallic vanadium has been refined using molten electrolytes of sodium and vanadium chlorides[60], a laboratory process having been operated[61] at 800°C under an atmosphere of helium, and with a cathode current efficiency of about 91%. Similar processes have also been operated at about 650°C with electrolytes of potassium and sodium chlorides[62], and potassium and lithium chlorides[63] both containing vanadium chloride. Gaseous impurities were satisfactorily removed by these processes, but the purified metal contained 0.1% of iron. When a bromide electrolyte was used[64], the iron level was reduced to 50 ppm, and ductile vanadium could be produced by vacuum melting the cathode deposit, even though it contained 250 ppm of oxygen. Lei and Sullivan[65] have also refined scrap vanadium using a potassium chloride–lithium chloride electrolyte containing vanadium chloride. Their cell was operated at 615°C under a helium atmosphere. Oxygen, nitrogen and most of the metallic impurities were removed satisfactorily and ductile metal could be produced from the dendritic cathode deposit.

9. Zirconium, tungsten and molybdenum

Zirconium has been refined in a sodium chloride melt containing K_2ZrF_6[66] and in chloride and chloride–fluoride melts[67]. Suchkov et al.[68] have reported a process for the electrolytic refining of zirconium carbide, in which it is first treated as a soluble anode, and the zirconium refined.

Tungsten has been refined by Cattoir[69] using melts of lithium and potassium chlorides containing $Na_2B_4O_7$, and by Balikhin and Reznichenko[70] using chloride/fluoride or chloride/phosphate melts containing WO_3. Cattoir et al.[71] have also prepared high purity molybdenum using molten salt refining. They used many different electrolytes but found that a mixture of potassium chloride and potassium hexachloromolybdate was best because of its easy preparation, the nature of the deposits obtained, and the low vapour pressure of volatile molybdenum species above it. The cathode deposit was plate-like at the lower operating temperatures, about 770°C, but dendritic at about 900°C. Increasing the current density had little effect on the nature of the deposit; it merely increased the crystal size.

10. Uranium

Commercial uranium is produced by oxide reduction with magnesium. Sullivan et al.[72, 73] develop a method whereby it is subsequently electrolytically refined using molten sodium chloride–uranium chloride melts, solid uranium being deposited on a molybdenum cathode. Chauvin and his co-workers[74–77] refined uranium in a potassium chloride–lithium chloride melt containing up to 30% by weight of UCl_3. The cell operates at 450–500°C under an argon atmosphere and the metal is deposited on a molybdenum cathode. A similar process has been developed by Ford[78].

11. Tin and antimony

Two interesting processes have been developed for refining tin and antimony, both depending on the removal of impurities from liquid electrodes so that the purified metal itself does not migrate across the cell. In the process developed by Samodelov and Delimarskii[79-81], impure tin forms a molten cathode and the electrolyte contains sodium, potassium and calcium chlorides. An inert graphite anode is suspended in the electrode and the alkali and alkaline earth metal ions are discharged at the cathode to form inter-metallic compounds with the impurities in the tin. These compounds then diffuse, or migrate, to the anode where they are oxidised. The authors report that iron, arsenic, antimony and bismuth can be removed quite satisfactorily using this cell but not lead or copper.

On the other hand, Morachevskii has reported that Sazhin and his co-workers have studied the purification of antimony by the dissolution of impurities from an impure molten anode into a molten sodium–potassium chloride eutectic[82]. The lead concentration in the anode can be reduced from 10 to 0.1% and tin, copper, silver, nickel, iron, zinc and bismuth are similarly reduced. Rozalovskii and co-workers[83] have reported a similar process which operates between 750 and 850°C at a current density of 0.22 A/cm².

12. Other metals

Many attempts have been made to use the three layer process for the refining of magnesium and its alloys[84-87] and experiments have been carried out on a semi-industrial scale. The density of the impure anode has been increased by additions of copper or zinc or zinc and aluminium together, and the process has been shown to work satisfactorily. It does not, however, appear to offer cost advantages over other refining methods, and is not used commercially.

Molten salt electrorefining process for the refining of gold and silver have been investigated[88]. Base impurities such as zinc, lead and copper are removed from an impure anode by electrolysis into a molten chloride eutectic. The electrorefining of thorium has been investigated by Noland[89] using lithium and potassium chloride mixtures containing $ThCl_4$ or ThF_4. The cells were operated under inert atmospheres, and used molybdenum cathodes. Nettle and others[90] reported experiments on the refining of hafnium using various molten electrolytes containing $HfCl_4$, K_3HfCl_7 or K_2HfCl_6. Merrill and Wong[91] have investigated the electrorefining of yttrium in a lithium chloride eutectic containing 1.7 to 13.4% of yttrium chloride, at 710°C. Metallic impurities with the exception of lithium could be removed satisfactorily, but vacuum distillation had to be employed to remove salt from the cathode deposit. Wong et al.[92] have also studied the refining of hafnium using electrolytes of potassium and barium chloride containing rubidium or caesium chloride and hafnium tetrachloride. They found that the presence of rubidium or caesium chloride suppressed the volatility of the hafnium tetrachloride, and that all metallic impurities were successfully removed, except titanium, manganese and tin. Suchkov et al.[93] have also successfully refined the electrorefining of manganese, in a sodium–potassium chloride melt containing manganese chloride, the cell operating at 470–550°C.

13. Future possibilities

It is apparent from the large number of successful refining experiments reported here that molten salt electrorefining is an extremely versatile refining method, and, moreover, one that is very precise in its effect. It is equally apparent from the paucity of actual industrial refining cells that the process is expensive to operate on a commercial scale. It is possible, however, that the future economics of metal production will turn increasingly in its favour. From both an economic and an environmental control point of view, it is becoming increasingly important for society to re-use metallic scrap. As alloys become more and more complex, the refining problem involved in unscrambling alloy scrap will increase in magnitude. Thus we can expect that the extra expense involved in molten salt electrorefining will be outweighed by its ability to separate the several constituents of complex metal alloys.

14. References

1 R. PIONTELLI and G. STERNHEIM, *Overvoltages at the electrode: melted Pb/PbCl$_2$*, J. Chem. Phys., 23 (1955) 1358.
2 J. D. VERHOEVEN, *Electrotransport as a means of purifying metals*, J. Metals, 18 (1966) 26.
3 D. T. PETERSON, F. A. SCHMIDT and J. D. VERHOEVEN, *Electrotransport of carbon, nitrogen and oxygen in thorium*, Trans. AIME, 236 (1966) 1311.
4 J. D. VERHOEVEN, *Effect of electric field on solute distribution during solidification of bismuth–tin alloys*, Trans. AIME, 239 (1967) 694.
5 R. A. ORIANI and O. D. GONZALEZ, *Electro-migration of hydrogen isotopes dissolved in alpha iron*, Trans. AIME, 239 (1967) 1041.
6 B. I. MEDOVAR, YU. V. LATASH, B. I. MAKSIMOVICH and L. M. STUPAK. *Electroslag Remelting*, State Scientific and Technical Publishing House of Literature on Ferrous and Non-ferrous Metallurgy, Moscow, U.S.S.R., 1963, (Russian text).
7 M. W. DAVIES, R. J. HAWKINS and P. N. SMITH, *The application of thermodynamics to the behaviour of slags in electroslag refining*, BISRA Open Report C/36/68, 1968, p. 13.
8 P. F. HART, A. W. D. HILLS and J. W. TOMLINSON, *Electrorefining using molten salt electrolysis*, Advances in Extractive Metallurgy, Inst. of Min. Met., 1967, p. 624.
9 J. D. EDWARDS, F. C. FRARY and Z. JEFFRIES, *The Aluminium Industry. Aluminium and its Production*, McGraw-Hill, (New York), 1930, pp. 321–327.
10 K. AKERMAN, M. BRAFMAN, D. KOZMINSKA, N. BIALA, J. KIERZEK, W. TABOR and T. PIECHOTA, *Use of radioactive tracers for the study of industrial electrolytic refining of Al*, Rev. Met. (Paris), 64 (1967) 1013 (French text). Chem. Abstr., 68 (1968) 80648.
11 E. K. RIDEAL, *Industrial Electrometallurgy including Electrolytic and Electrothermal processes*, Balière, Tindall and Cox, London, (1918) p. 247.
12 YU. K. DELIMARSKII, P. P. TUROV and E. B. GITMAN, *Electrochemical separation of lead from the binary lead alloys with bismuth, antimony, arsenic, and tin in molten electrolyte*, Ukr. Khim. Zh., 21 (1955) 687 (Russian text).

13 E. B. GITMAN and YU. K. DELIMARSKII, *Electrochemical separation of lead–gold alloys*, *Ukr. Khim. Zh.*, 22 (1956) 731 (Russian text).
14 E. B. GITMAN and YU. K. DELIMARSKII, *Electrolytic separation of binary alloys of lead with silver and arsenic*, *Zh. Prikl. Khim.*, 32 (1959) 578; *J. Appl. Chem. USSR*, 32 (1959) 607.
15 I. D. PANCHENKO and YU. K. DELIMARSKII, *The electrolytic recovery of lead from factory crudes and bismuth-containing drosses by the use of fused electrolytes*, *Zh. Prikl. Khim.*, 33 (1960) 153; *J. Appl. Chem. USSR*, 33 (1960) 147.
16 Y. G. ROMS, B. S. BELEN'KII and YU. K. DELIMARSKII, *Pilot plant electrolytic cell for producing and refining heavy non-ferrous metals in salt melts*, *Tsvetn. Metal.*, 41 (1968) 34 (Russian text). *Chem. Abstr.*, 69 (1968) 82816.
17 YU. K. DELIMARSKII, I. G. PAVLENKO and YU. E. KOSMATYI, *Electrolytic refining of lead in molten silica electrolytes*, *Zh. Prikl. Khim.*, 33 (1960) 1840; *J. Appl. Chem. USSR*, 33 (1960) 1819.
18 YU. K. DELIMARSKII and Y. E. KOSMATYI, *Electrolytic refining of crude Pb from silicate-chloride electrolytes*, *Ukr. Khim. Zh.*, 34 (1968) 767 (Russian text). *Chem. Abstr.*, 69 (1968) 102392.
19 YU. K. DELIMARSKII, I. G. PAVLENKO and O. G. ZARUBITSKII, *Electrolytic removal of bismuth from lead*, *Zh. Prikl. Khim.*, 35 (1962) 322; *J. Appl. Chem. USSR*, 35 (1962) 302.
20 V. D. PONOMARYOV, I. Z. SLUTSKY and G. A. KONONENKO, *Lead and bismuth electrolytic separation in leaching melts*, *Izv. Vysshikh Uchebn. Zavedenu Tsvetn. Met.*, 7 (1964) 67 (Russian text).
21 YU. K. DELIMARSKII, et. al, *Electrolytic production and refining of bismuth in melts*, *Zh. Prikl. Khim.*, 35 (1962) 317; *J. Appl. Chem. USSR*, 35 (1962) 298.
22 G. B. IRGALIEV et al., *Anode refining of crude lead in a fused alkali*, *Met. i. Khim. Prom. Kazakkstava, Nauchno-Tekhn. Sb.*, no. 5 (1962) 48 (Russian text). *Chem. Abstr.*, 61 (1964) 348.
23 C. W. BJORKLUND et al., U.S. pat., 2923, 670/1960.
24 L. J. MULLINS et al., *Plutonium electrorefining*, *I & EC Process Design Develop.*, 2 (1963) 20.
25 L. J. MULLINS and J. A. LEARY, *Fused-salt electrorefining of molten plutonium and its alloys by the LAMEX process*, *I & EC Process Design Develop.*, 4 (1965) 394.
26 J. A. LEARY, L. J. MULLINS JR. and J. F. BUCHEN, *Electrolytic refining of Pu alloys containing Ga to recover Pu metal*, U.S. pat.,3, 417,002/1968.
27 B. BLUMENTHAL and M. B. BRODSKY, *The Preparation of High Purity Plutonium*, in *Plutonium 1960*, Cleaver–Hume Press, London, (1961) pp. 171–86.
28 M. BENEDICT and T. PIGFORD, *Nuclear Chemical Engineering*, McGraw-Hill, New York, 1957, p. 572.
29 I. A. MENZIES, D. L. HILL and L. W. OWEN, *Electrolytic deposition of beryllium*, *Nature*, 183 (1959) 816.
30 M. M. WONG, R. E. CAMPBELL and D. H. BAKER JR., *Fused salt electrorefining of beryllium*, *J. Metals*, 12 (1960) 786.
31 M. M. WONG, F. R. CATTOIR and D. H. BAKER JR., *Electrorefining beryllium; preliminary studies*, *U.S. Bur. Mines Rept. Invest.*, 5581 (1960) 9.
32 M. M. WONG, F. R. CATTOIR and D. H. BAKER JR., *Min. J.*, *Lond.*, 255 (1960) 104 (abstract of ref. 31).
33 M. M. WONG and J. E. KLOSTERMAN, *Electrorefining beryllium, operation of a prototype cell*, *U.S. Bur. Mines Rept. Invest.* 6489 (1964) 17.
34 M. M. WONG and D. A. O'KEEFE. *Electrorefining beryllium, two-cycle electrolysis*, *U.S. Bur. Mines Rept. Invest.*, 6570 (1964) 8.
35 J. BILARD, G. BOISDE, M. BROC, G. CHAUVIN, H. CORIOU and J. HARDY, *Production of high purity beryllium by electrorefining in molten salt baths*, *Metaux Corrosion Ind.*, 42 (1967) 259 (French text). *Chem. Abstr.*, 68 (1969) 5160.
36 M. E. SIBERT, *Investigation of processes for preparation of high-purity niobium metal*, *U.S. At. Energy. Comm.* AECU-3798 (1958) 79. *Chem. Abstr.*, 55 (1961) 19558.
37 M. A. STEINBERG, U.S. pat., 2913 379/1959.
38 MITSUHIRO SAKAWA and TADASHI KURODA, *Production of metallic Nb from niobium carbide by molten salt electrolysis. IV. Metallic niobium and its electrorefining*, *Denki Kagaku*, 36 (1968) 792 (Japanese text). *Chem. Abstr.*, 70 (1969) 73488.

39 A. B. SUCHKOV, T. A. LABOVA, G. A. MEERSON, V. N. MIKHINA, V. V. SAFONOV and B. G. KORSHUNOV, *Electrochemical refining of niobium in chloride melts*, Izv. Akad. Nauk SSSR, Metall, (1968) 52 (Russian text). Chem. Abstr., 69 (1968) 82818.
40 V. V. GRINEVICH and V. A. REZNICHENKO, *Apparatus for electrolytic refining of niobium in a fluoride–chloride melt*, Met. Vol'frama, Molibdena, Niobiya, (1967) 177 (Russian text); Ref. Zh. Met., (1968) No. 4G285. Chem. Abstr., 69 (1968) 98640.
41 V. V. GRINEVICH and V. A. REZNICHENKO, *Electrolyte for production of high-purity niobium by electrolytic refining*, Met. Vol'frama, Molibdena, Niobiya, (1967) 172, (Russian text); Ref. Zh. Met., (1968) No. 4G284. Chem. Abstr., 69 (1968) 98641.
42 V. V. GRINEVICH and V. A. REZNICHENKO, *Behaviour of impurities during electrolytic refining of niobium*, Met. Vol'frama, Molibdena, Niobiya, (1967) 184 (Russian text). Ref. Zh. Met., (1968) No. 4G283; Chem. Abstr., 69 (1968) 98642.
43 M. V. SMIRNOV and N. A. LONGINOV, *Current efficiency for anodic dissolution of titanium in chloride and mixed chloride–fluoride melts*, Tr. Inst. Elektrokhim. Akad. Nauk SSSR, Ural.'Sk Filial, 4 (1963) 29 (Russian text); Chem. Abstr., 60 (1964) 15426.
44 B. F. MARKOV and T. A. TISHURA, *Anode polarization on dissolving titanium in fused chlorides; on the form of cathodic deposits of titanium obtained by the electrolysis of fused chlorides*, Ukr. Khim. Zh., 29 (1963) 1043, 1155 (Russian text).
45 V. G. GOPIENKO and G. P. KHRISTYUK, *Pilot-plant production of melts containing lower titanium chlorides*, Tr. Vses. Alyumin-Magnievyi Inst., 50 (1963) 135 (Russian text); Chem. Abstr., 59 (1963) 12428.
46 N. I. ANUFRIEVA, *Effect of current density on refining of titanium*, ibid., 50 (1963) 177 (Russian text). Chem. Abstr. 59 (1963) 9591.
47 A. I. IVANOV and V. G. GOBIENKO, *Electrolytic refining of Ti in a cell with lined bath and internal heating*, ibid., 50 (1963) 182 (Russian text). Chem. Abstr., 60 (1964) 1333.
48 N. I. ANUFRIEVA, et al., *Electrolytic refining of titanium sponge wastes*, ibid., 50 (1963) 167 (Russian text). Chem. Abstr., 59 (1963) 9591.
49 V. A. SUKHODSKII and N. I. ANUFRIEVA, *Electrolytic refining of some Ti alloys*, ref. 84, p. 188 (Russian text). Chem. Abstr., 60 (1964) 3718.
50 E. K. KLEESPIES and T. A. HENRIE, *Transfer of selected metals in titanium electrorefining*, U.S. Bur. Mines Rept. Invest. 6437 (1964) 9.
51 V. S. BALIKHIN and V. A. REZNICHENKO, *Electrolytic refining of alumino-thermic titanium*, Probl. Met. Titana, (1967) 175 (Russian text); Ref. Zh. Met., (1968) No. 4G231. Chem. Abstr., 69 (1968) 108894.
52 W. J. KROLL, *The fusion electrolysis of titanium*, Chem. Ind. London, Oct. 22 (1960) 1314.
53 N. I. ANUFRIEVA and Z. N. BALASHOVA, *Electrolytic refining of shavings of complex titanium alloys*, Izv. Akad. Nauk SSSR, Metal, (1969) 69 (Russian text). Chem. Abstr., 71 (1969) 8987.
54 *Electrolysis cell for refining titanium and other metals*, U.S.S.R. pat., 191,132/1967.
55 SAKAE TAKEUCHI and OSAMU WATANABE, *Electrorefining of titanium from molten salts*. Part I. Influence of various salt baths on the electrorefining, Nippon Kinzoku Gakkaishi, 28 (1964) 627 (Japanese text). Chem. Abstr., 66 (1967) 253642.
56 W. W. GULLETT, *Refining titanium–vanadium alloys*, U.S. pat., 2,913, 380/1959.
57 K. KOMAREK and P. HERASYMENKO, *Equilibrium between titanium metal and solutions of titanium dichloride in fused sodium chloride*, J. Electrochem. Soc., 105 (1958) 216.
58 F. J. SCHULTZ and T. M. BUCK, *Electrolytic method for refining titanium metal*, U.S. pat., 2,734, 856/1956.
59 O. Q. LEONE and D. E. CUCH, *Electrorefining of titanium–nitrogen alloys*, U.S. Bur. Mines Rept. Invest., 6878 (1968).
60 *Treatment processes investigated in the United States*, Mining J., 255 (1960) 286.
61 D. H. BAKER JR. and J. D. RAMSDELL, *Electrolytic vanadium and its properties*, J. Electrochem. Soc., 107 (1960) 985.
62 V. V. VOLEINIK and A. M. KUNAEV, *Electrolytic refining of crude vanadium in salt melts*, Tr. Inst. Met. Oibogashch., Akad. Nauk Kaz. SSR, 21 (1968) 29 (Russian text). Chem. Abstr., 71 (1969) 8988.
63 K. P. V. LEI and T. A. SULLIVAN, *High purity vanadium*, J. Less-Common Metals, 14 (1968) 145; Chem. Abstr., 68 (1968) 23830.

64 T. A. SULLIVAN and F. R. CATTOIR, *Electrorefining vanadium in a molten bromide electrolyte*, U.S. Bur. Mines Rept. Invest., 6631 (1965) 12.
65 K. P. V. LEI and T. A. SULLIVAN, *Fused salt electrorefining vanadium scrap*, U.S. Bur. Mines Rept. Invest., 7036 (1968).
66 D. H. BAKER JR., J. R. NETTLE and H. KNUDSEN, *Electrorefining zirconium*, U.S. Bur. Mines Rept. Invest., 5758 (1961) 12.
67 M. V. SMIRNOV, A. N. BARABOSHKIN and V. E. KOMAROV, *Cathodic processes during deposition of zirconium from chloride melts; cathodic processes in the deposition of zirconium from mixed chloride–fluoride melts*, Zh. Fiz. Khim., 37 (1969) 1677; Russ. J. Phys. Chem., 37 (1963) 901, 905.
68 A. B. SUCHKOV, G. A. MEERSON and Y. G. OLESOV, *Electrolytic refining of zirconium carbide*, Izv. Akad. Nauk SSSR, Metal, (1968) 106 (Russian text). Chem. Abstr., 69 (1968) 45416.
69 F. R. CATTOIR, *Experiments in fused-salt electrolysis of tungsten*, U.S. Bur. Mines Rept. Invest., 6154 (1963) 10.
70 V. S. BALIKHIN and V. A. REZNICHENKO, *Electrolytic refining of tungsten*. Met. Vol'frama, Molibdena Niobiya, (1967) 166. (Russian text); Ref. Zh. Met., (1968) No. 4G270. Chem. Abstr., 69 (1968) 98639.
71 R. CUMMINGS, F. CATTOIR and T. SULLIVAN, *Preparation of high purity molybdenum by molten salt electrorefining*, U.S. Bur. Mines Rept. Invest., 6850 (1966) 24.
72 R. E. CAMPBELL and T. A. SULLIVAN, *Electrorefining uranium in a chloride electrolyte*, U.S. Bur. Mines Rept. Invest., 6624 (1965) 14.
73 F. R. CATTOIR and T. A. SULLIVAN, *Molten-salt electrorefining of uranium*, U.S. Bur. Mines Rept. Invest., 6507 (1964) 24.
74 G. CHAUVIN, et al., *Production d'uranium de haute pureté par électroraffinage en bains de sels fondus*, J. Nucl. Mater., 11 (1964) 183.
75 G. CHAUVIN, H. CORIOU and A. SIMENAUER, *Phénomène de concentration de fer au voisinage de la cathode au cours de l'électroraffinage de l'uranium en bains de sel fondus*, Electrochim. Acta, 8 (1963) 323.
76 G. CHAUVIN, H. CORIOU and J. HURE, *Electroraffinage de certains métaux nucléaires en bain de sels fondus*, Métaux Corrosion-Ind., 37 (1962) 112.
77 G. BOISDE, et al., *Contribution à la connaissance du mécanisme de l'électroraffinage de l'uranium en bains de sel fondus*, Electrochim. Acta, 5 (1961) 54.
78 L. H. FORD, Personal communication, Fuel Element Development Branch, UKAEA, Springfields.
79 A. P. SAMODELOV and YU. K. DELIMARSKII, *Electrochemical refining of black tin in fused salts*, Tsvetn. Metal, 41 (1968) 44 (Russian text). Chem. Abstr., 69 (1968) 45414.
80 YU. K. DELIMARSKII and A. P. SAMODELOV, *Electrolytic refining of Sn in fused chlorides*, Ukr. Khim. Zh., 34 (1968) 522 (Russian text). Chem. Abstr., 69 (1968) 45415.
81 YU. K. DELIMARSKII and A. P. SAMODELOV, *Electrode processes during cathodic–anodic refining of tin in fused salts*, Ukr. Khim. Zh, 35 (1969) 245 (Russian text); Chem. Abstr., 70 (1969) 120487.
82 A. G. MORACHEVSKII, *Review of electrochemistry*, J. Appl. Chem. USSR, 33 (1960) 1422.
83 A. A. ROZLOVSKII, A. A. BULDAKOV, I. V. DEMINA and G. N. EFIMOV, *Electrolytic refining of antimony in fused salts*, Tsvetn. Metal., 41 (1968) 36; Chem. Abstr., 70 (1969) 63445.
84 O. A. LEBEDEV, *Electrolytic refining of magnesium alloys by the three-layer method under laboratory conditions*, Tr., Vses. Alyumin.-Magnievyi Inst., 48 (1962) 102 (Russian text). Chem. Abstr., 60 (1964) 1333.
85 YU. V. BAIMAKOV and O. A. LEBEDEV, *Polarization E.M.F. triple layer electrolytic refining of magnesium and its alloys: Measuring E.M.F. amalgam type circuits at three-layered electrolytic magnesium refining*, Izv. Vysshikh Uchebn. Zavedenii Tsvetn. Met.,7 (1964) 99–105; no 1, 112–7 (Russian text).
86 YU. V. BAIMAKOV, *Theory of electrorefining of metals in melts with liquid electrodes*, Tr. Leningr. Politekhn. Inst., 239 (1964) 70 (Russian text). Chem. Abstr., 63 (1965) 1477.
87 U. DEITER and A. I. BELAYEV, *To the question of obtaining pure magnesium by electrolytic refinement*, Izv. Vysshikh Uchebn. Zavedenii Tsvetn. Met. 6 (1963) 94 (Russian text).
88 S. V. LIPKIN, *Direct production of DORE alloy, zinc and lead electrolysis of silver–zinc alloy*, Sb. Nauchn. Tr. Mosk. Inst. Tsvetn. Metal. Zolota, 6 (1960) 323 (Russian text). Chem. Abstr., 56 (1962) 4510.

89 R. A. NOLAND, *Electrolytic refining of thorium*, in *Proc. Conf. Metal Thorium*, Cleveland, 1956 Am. Soc. Metals, Cleveland, Ohio, 1958, pp. 124–32.
90 J. R. NETTLE, J. M. HIEGE and D. H. BAKER JR., *Hafnium electrorefining, U.S. Bur. Mines Rept. Invest.*, 5851 (1961) 18.
91 C. C. MERRILL and M. M. WONG, *Electrorefining yttrium, U.S. Bur. Mines. Rept. Invest.*, No 7018 (1967) 10.
92 M. WONG, J. HIEGEL and G. MARTINEZ, *Electrolytes for electrorefining hafnium, U.S. Bur. Mines Rept. Invest.*, 6818 (1966) 9.
93 V. V. SAFONOV, V. N. MIKHINA, A. S. VOROB'EVA, B. G. KOROHUNOV and A. B. SUCHKOV, *Electrochemical refining of Mn in chloride melts, Izv. Vysshikh Uchebn. Zavedenii Tsvetn. Met.* 11 (1968) 57; *Chem. Abstr.*, 6 (1968) 82817.
94 *Alluminio*, 37 Feb. (1968) (Italian text).
95 *Rev. Aluminium*, March (1967) 287 (French text).
96 YU BAIMAKOV, *Trans. Leningrad Polytechn. Inst.*, 239 (1964) 70; English Transl., FTO-MT-65-193, USAF, Wright-Patterson Base.
97 V. A. GARMATA and A. I. BELYAEV, *Tsvetn. Metal.*, 30 No 9 58 (Russian text).
98 J. ISHIHARA, *J. Metals*, 17 (1965) 944.
99 A. I. BELYAEV et al., *Sb. Mosk. Inst. Stali i Splavov*, 41 (1966) 273 (Russian text).
100 M. I. LAVRENT'EV, L. A. FIRSANOVA et al., *Izv. Tsevtn. Metal.*, 10 (1967) 52 (Russian text).

15. Bibliography

The following additional bibliography on titanium electrorefining was kindly provided by K. Boodson, Librarian, Imperial Metal Industries, Witton.

P. EHRLICH, W. GUTSCHE and H. KUEHNL, *Electrolytic refining of titanium in fused alkali bromides*, Z. Anorg. Allgem. Chem., 312 (1961) 70.
A. I. IVANOV and A. B. PICHUKOV, *Large scale laboratory investigation to refine tails from titanium sponge*, Titan. i Ego Splavy Akad. Nauk SSSR., Inst. Met., 8 (1962) 227.
V. A. SUKHODSKII and M. M. TSYPLAKOVA, *Effect of the middle layer in electrolyte on the process index of electrorefinement of titanium*, Titan. i Ego Splavy Akad. Nauk SSSR, Inst. Met., 8 (1962) 237.
E. B. GITMAN, *Potential/current characteristics of electrolytic dissolution of titanium in molten salts*, Ukr. Khim. Zh., 28 (1962) 1116.
G. CHAUVIN, H. CORIOU and J. HURÉ, *Electrorefining of some nuclear metals in a fused-salt bath*, Métaux Corrosion-Ind., 37 (1962) 112.
A. I. IVANOV and V. G. GOPIENKO, *Electrolytic refining of titanium in a cell with lined bath and internal heating*, Tr. Vse. Alyumin. Magnievyi Inst., 48 (1962) 182.
J. R. NETTLE et al., *Electrorefining of selected titanium ternary alloys*, U.S. Bur. Mines Rept. Invest., 5998 (1962) 14.
N. I. ANUFRIEVA, *Electrode processes in titanium refining*, Titan. i Ego Splavy Akad. Nauk SSSR, Inst. Met., 9 (1963) 236.
N. I. ANUFRIEVA, *Studies on some technological problems of titanium electrorefining*, Titan. i Ego Splavy Akad. Nauk SSSR, Inst. Met., 9 (1963) 242.
P. F. MARKOV and T. A. TISHURA, *The form of the cathode deposits of titanium obtained by electrolysis of the molten chlorides*, Ukr. Khim. Zh., 29 (1963) 1155.
A. I. IVANOV et al., *Electrode processes in electrolytic production and refining of titanium in molten salts*, Tr., Vses. Soveshch. po Fiz. Khim. Rasplavlen Solei, 2nd Kiev., (1963) (pub. 1965) 139.
O. Q. LEONE and D. E. CUCH, *Electrorefining of titanium–nitrogen alloys*, U.S. Bur. Mines Rept. Invest., 6878 (1966) 11.
N. I. ANUFRIEVA and N. BALASHOVA, *Influence of operating conditions on the electrolytic refining of titanium and titanium alloy scrap*, Russ. Met., 2 (1967) 29.
N. I. ANUFRIEVA, *Influence of certain factors on the electrodeposition of titanium from salt melts (i.e. the refining of titanium from waste materials)*, Izv. Akad. Nauk SSSR, Metally, 3 (1967) 81.

V. T. MUSIENKO and V. M. ORLOV, *Electrolytic refining of titanium alloys*, Izv. Akad. Nauk SSSR, Metally, 3 (1967) 88.

N. I. ANUFRIEVA, *The effect of some impurities on the electrocrystallization of titanium*, Zh. Prikl. Khim., 40 (1967) 1288.

E. B. GITMAN and I. N. SHEIKO, *Nature of the cathode deposits in the electrolysis of sodium fluorotitanate in alkali chloride metals (production of metallic titanium)*, Ukr. Khim. Zh., 34 (1968) 129.

N. I. ANUFRIEVA, *Behaviour of the anode material in the electrolytic refining of titanium and its alloys*, Izv. Akad. Nauk SSSR, Metally, Mar.-Apl. (1968) (2), 89.

YASUHIKO HASHIMOTO, *Fused salt electrolysis of titanium metal from titanium–carbon–oxygen alloys or TiC*, Nippon Kinzoku Gakkaishi, 32 (1968) 1327.

D. H. BAKER JR., *Titanium electrorefining*, High-Temperature Refractory Metals, New York, 1965, Met. Soc. Conf. 34 (1969) 223.

YU. G. OLESOV et al., *Hydrogenation of titanium alloys waste with subsequent electrorefining*, Tsvetn. Metal., 3 (1969) 79.

V. G. GOPIENKO and I. B. FIRFAROVA, *Phase composition of anodic residues in the refining of titanium alloy waste*, Izv. Akad. Nauk SSSR, Metally, 3 (1969) 64.

M. M. WONG, *Electrolytic methods of preparing cell feed for electrorefining titanium*, U.S. Bur. Mines Rept. Invest., 6161 (1963) 22.

S. TAKEUCHI and O. WATANABE, *Electrolytic extraction of titanium from a soluble anode of titanium alloys*, Denki Kagaku, 32 (1964) 122.

H. KEUHNL and G. FLEISCHHAUER, *Melt-electrolytic refining of titanium and titanium alloys*, Chem. Ing.-Tech., 36 (1964) 729.

O. Q. LEONE, *The development of titanium-refining cells*, U.S. Bur. Mines Rept. Invest., 6432 (1964) 27.

V. S. BALIKHIN, *Electrorefining of titanium from molten salts*, Izv. Akad. Nauk SSSR, Metally, 2 (1965) 77.

V. T. MUSIENKO, *The shape and size of crystals of electrorefined titanium and its alloys*, Izv. Akad. Nauk SSSR, Metally, 3 (1965) 34.

SAKAE TAKEUCHI and OSAMU WATANABE, *The electrorefining of crude titanium in the fused-salt bath, and mechanical proporties of the refined titanium*, Nippon Kinzoku Gakkaishi, 29 (1965) 263.

E. B. GITMAN, *Current efficiency and nature of the cathode deposit on electrolysis of the lower chlorides of titanium with a soluble anode*, Ukr. Khim. Zh., 31 (1965) 1275.

V. G. GOPIENKO, N. U. ANUFRIEVA and E. F. KLYUCHNIKOVA, *Cathodic crystallization during the refining of titanium in fused salts*, Zh. Prikl. Kh., 39 (1966) 577.

YU. G. OLESOV et al., *Electrolytic refining of wastes of industrial titanium alloys*, Soviet J. Non-Ferrous Metals, English Transl., May (1966) 79.

O. Q. LEONE and F. S. WARTMAN, *Electrorefining of titanium–oxygen alloys*, U.S. Bur. Mines, Rept. Invest., 6588 (1965) 20.

N. I. ANUFRIEVA and Z. N. BALASHOVA, *Characteristics of products obtained after electrolytic refinement of certain titanium wastes*, Soviet J. Non-Ferrous Metals, English Transl. 7 (1966) 87.

N. I. ANUFRIEVA and Z. N. BALASHOVA, *Products obtained by the electrolytic refining of titanium waste*, Tsvetn. Metal., 9 (1966) 75.

Chapter 8

Electrochemical Machining

E. V. TUCK

Manager, Machining Development, Manufacturing Development Dept., Rolls-Royce Ltd., Bristol Engine Division, Filton, Bristol

1. Introduction

The first application of electrolysis to metal removal machining was described in a patent by W. Gusseff in 1929. This patent describes very closely the process as it is practised today except that the voltage now used is lower, 6–24 V as opposed to the 110 V used by Gusseff. Little interest was shown in the process at that time or for many years, mainly because the materials then being used could be readily and economically machined by conventional metal cutting processes. Other reasons were the difficulty and cost of making machines and equipment to withstand the corrosive effect of the electrolyte.

In the late 1950's work in the United States carried out at the Battelle Memorial Institute resulted in the development of a machine for forming aerofoil surfaces of turbine blades for gas turbine engines. One of these machines was imported into the UK by Metachemical Machines Limited and in 1960 was obtained by Rolls-Royce Limited, Derby. This machine, although extremely crude by present day standards, did show the advantages of the process and stimulated considerable interest. As a result of this, improved machines for the shaping of blade aerofoils were manufactured in the UK.

Also about this time the Anocut Company in the USA headed by W. L. A. Williams was developing an electrochemical machine to enable the electrochemical process to be applied to a wide range of applications. This company is now probably the largest supplier of electrochemical machining equipment in the world. Machines are being manufactured in both the UK and Europe under agreements from Anocut. Other machine tool manufacturers also making equipment for electrochemical machining are Cincinnati and Ex-Cell-O in the U.S.A. and B.S.A., Associated Engineering, Alfred Herbert, H.P.E. of Bletchley and Shands of Axbridge in the U.K. Rolls-Royce have continued to co-operate with machine tool manufacturers to develop the blade aerofoil forming machines.

Surprising as it may seem at first sight the electrochemical machine tool should be as rigid, and in many cases even more rigid than its conventional counterpart. The control systems and ancillary equipment required to go with the machine tool are now very sophisticated. These machines are capable of forming complicated contoured shapes quickly, to close tolerances, and with accurate repeatability.

One reason for the rapid growth in electrochemical machining in the last few years has been the need to machine the newly developed, tough, heat resisting materials demanded in the latest generation of gas turbine engines. These materials, the Nimonics and the cast nickel–cobalt alloys are extremely difficult and costly—in some cases virtually impossible—to machine by conventional metal cutting methods.

Another reason is the need to remove every ounce of surplus material from engine components to reduce their weight. This entails machining very complicated contoured shapes, at times combined with very thin sections. These shapes are very difficult and costly to machine by conventional methods. Using electrochemical machining these shapes can be formed quickly, accurately and in one plunge forming operation. The surfaces produced are free from residual stresses—this aspect makes the process ideally suitable for producing thin sections.

The need to drill cooling holes in turbine blades has given considerable impetus to the development of the electrochemical drilling process, and during the last year this technique has been developed into a production process. Shaped holes, small deep holes, impossible to drill by conventional methods, can be produced with relative ease. Holes from 1.016 mm to 3.81 mm are being drilled to depths of up to 304.8 mm with good directional accuracy and maintaining good tolerances on the hole diameter.

ECM is also being introduced to the less spectacular field of production, *i.e.* the removal of burrs and sharp edges and both the aeroengine and automobile industries are using the process for these applications.

2. Plant

The plant required to successfully perform the electrochemical machining process can be divided into four basic elements, which have to be integrated into one unit.

(1) The machine tool which will hold the workpiece, and locate the tool accurately in relation to the workpiece with the means to advance the tool towards the workpiece at a steady feed rate with the minimum deflection.

(2) An electrolyte supply and clarifying system which will supply clean electrolyte to the machining gap at the required pressure, flow and temperature.

(3) An extraction system to carry away the hydrogen gas generated in large quantities during the machining process.

(4) The means to supply the required capacity of DC current to the machining gap, with the necessary control to give accurate machining and means of shut down when detecting conditions which could lead to arcing or sparking causing damage to tool and workpiece.

2.1 THE MACHINE TOOL

The main failings of the early electrochemical machines were their lack of rigidity and poor feed systems. There is now a much better understanding of the electromagnetic, hydrodynamic and hydrostatic forces, which on the larger machines can be considerable. These are discussed in detail in De Barr and Oliver[1].

The major forces acting upon the machine structure are the hydrodynamic and hydrostatic forces produced by the electrolyte pressure between the tool and workpiece. Considering a large machine with 20,000 A power supply available, there is very often in the region of 645 mm² of tool area. With electrolyte supply pressures of 20 kg/cm² available the hydrostatic forces acting on the tool and workpiece could be up to 2764 N. It would of course be desirable to have no deflection in the machine tool, but in order to be realistic and keep the machine to a reasonable size and price, deflection figures of the order of 0.1778 mm per 9171.8 kg load axial to the tool feed ram and deflections in the transverse plane of 0.0127 mm per 453.59 kg load with the ram fully extended are acceptable.

Deflections of this order can be tolerated on the gap sizes used when machining with large area tools. Approximately half the deflection can be attributed to the feed system. The remainder is accounted for in the main machine structure and unclamped slide ways.

In addition to good rigidity of the machine frame it has to have a high natural vibration frequency, higher than the vibration frequencies encountered during the machining process. One source of vibration which can be set up during machining can be from the electrolyte supply pump. If this pulsing is not damped before the electrolyte reaches the machining gap it will be transmitted through the ram and feed system. Another, and probably the greater, source is from the pressure hammer effect. This can occur if the pump is switched on when the electrolyte is fed into a small machining gap with a high flow resistance. In this instance vibration can be introduced into the machine frame and workpiece and could continue in sympathy with either or both.

The main structure of the machine is usually of fabricated steel plate construction, in the 20,000 A machine the steel plate is 38.1 mm thick. The whole of the structure must be given a protective coating of corrosion resistant paint. A list of protective coatings are given in De Barr and Oliver[2] and the effect of cathodic protection is discussed below. Particular attention is required where foot traffic is likely—on large machines the operator has to walk into the work enclosure to load the workpiece and tools—and in the work enclosure where electrolyte could be jetting on the surfaces. In these areas the protective coating has to withstand the impact of tools and equipment that may be accidentally dropped. In these critical areas it is advisable to radius all corners before applying the protective coating. In such areas it may be advisable to use stainless steel.

Where bellows are used to protect slideways these should be preferably of the one-piece moulded or welded type. For the moulded type, polyurethane has been found the most suitable material. Bellows with stitched seams have been found to give trouble due to the salt crystals wicking through the stitching and getting inside the bellows. Slideways can also be protected by labyrinth covers, or pressurisation. Again, where the areas are subjected to jetting electrolyte, particular attention should be paid to sealing, as salt crystals will wick through very small apertures even against air pressurisation. For sealing rotating parts, for example a rotating work table on an electrochemical turning machine, a labyrinth fed with water to the inner groove and spilling outward across the labyrinth provides excellent protection.

The machine work table should be made of a good grade of 18/8 stainless steel (EN 58J) which has been carefully stress relieved during manufacture, and it should be mounted on a substantial steel base before final machining. The ram tool mounting platen should also be made of a good grade stainless steel (EN 58J). Where this tool mounting platen is also an electrolyte delivery manifold, frequent inspection should be made as it has been found that corrosion can occur, particularly at joint faces. Consideration should be given to using Monel metal for this part of the machine.

The present practice is to make the whole machine structure at a negative potential, the work table being insulated from the machine and being at a positive potential. Cathodic protection can be given to the machine table by mounting a copper bar in close proximity around the table.

For the ram feed the present practice is to use preloaded heavy duty recirculating ball screws which are mounted along the ram centre line. The lead screws are mounted in preloaded taper roller bearings which take the axial and transverse loads. Feeds are generally provided in both directions. These are usually constant line feeds and are accurate to $\pm 1\%$. The drive of the lead screw is normally provided by either hydraulic motors or DC fractional horse power electric motors operating through a gear box and electromagnetic clutches. Other types of feed systems are described in ref. 3.

The ram is usually mounted on recirculating roller bearings to give a smooth action with no stick-slip. As this slide is always unclamped, further rigidity could be gained by using hydrostatic slides, as this type of slideway is of the order of two to three times stiffer than a roller slide.

2.2 Electrolyte supply system

The electrolyte supply system consists of a tank to hold the electrolyte with the appropriate heat exchangers to maintain the electrolyte at the correct working temperature; a pump to supply the electrolyte to the working gap at the required pressure and flow; and means to separate the products of machining (metal hydroxides) from the electrolyte. Also a filter system to remove any foreign material such as particles of cloth, paper, hair, steel etc. which can get into the system from the workpiece, operator's clothing, shoes, salt, etc.

The electrolyte tanks can be made of epoxy resin reinforced with glass fibre, or steel lined with PVC or rubber. Consider first the case where one tank acts as the sole source of supply to a single machine. The majority of these are the rectangular lined steel type, divided into clean and dirty compartments by a weir, the capacity of the dirty side being approximately one fifth that of the whole tank. The supply pump sucks electrolyte from the clean side of the tank and delivers it to the machining gap where, after fulfilling its function in the machining action, it is returned, with the products of machining, to the dirty side of the tank.

2.3 CLARIFICATION OF ELECTROLYTE

Clarification in this type of installation is usually by centrifuge, the supply to which is pumped from the dirty side of the tank into the centrifuge where the hydroxides are separated from the electrolyte and the clarified electrolyte returned to the clean side of the tank. During the machining cycle, agitation in the tank is good and hydroxide contamination of the electrolyte is reasonably uniform in each section of the tank. Difficulties do arise, however, during down time when the centrifuge can still be operating from the dirty side of the tank. The electrolyte in this section of the tank is rapidly clarified but there is not sufficient agitation to carry the settling hydroxides back over the weir and these then settle to the bottom of the clean side of the tank. This can cause undesirable conditions on restarting, as thick sludge can be sucked in by the electrolyte supply pump and block the filters which are between the pump and tool, or can be pumped *via* the tool to the machining gap giving conditions in which sparking or arcing could occur. Other disadvantages with centrifuges are that they are expensive to purchase and the maintenance cost can be high. Probably the next most used method is the settling tank or swimming pool. With this system the hydroxides settle out from the electrolyte in its passage from one end of the tank to the other. This is most suited to large installations but requires quite a large space for the tank or tanks. These settling tanks work very well and require little maintenance but the capital cost is probably equal to a centrifuge system. The large volume of electrolyte held makes a good heat sink and eases the difficulties of temperature control. A small tank by the machine is still required to supply the high pressure pump and in which the electrolyte conditions can be monitored.

A third system is the Anocut clarifier. This is an accelerated settling tank system which requires approximately twice the space required for a centrifuge clarifying system but much less than the "swimming pool" system.

2.4 Hydroxide sludge disposal

When the hydroxides have been separated from the electrolyte the sludge discharged from the centrifuge or settling tank—which can vary in constituency from a thin mud to a very stiff mud—has to be transferred to a disposal point, transported away and dumped. This can become quite a major function in a large electrochemical machining facility. The following figures give some idea of the amount of sludge which results from the electrochemical machining process: 0.5 kg of stainless steel machined away will produce approximately 20 litres of quite thick sludge.

Taking for example a workpiece having between 385 and 431 kg of material removed by electrochemical machining, resulting in between 15,400 and 17,240 litres of sludge, some idea of the size of this function can be realised. Another important point to be considered is that the sludge consists of approximately 25% metal hydroxides and 75% electrolyte and this is a good reason why the electrolyte should be inexpensive.

2.5 Hydrogen extraction

During the electrochemical machining process large quantities of hydrogen are produced. If this is not removed it will gather in the work enclosure and constitute a very serious safety hazard as any sparking or arcing between tool and workpiece will result in an explosion. An extraction system, then, is an essential part of the unit. Extraction must be provided from the work enclosure and electrolyte tank. Air is drawn in through the gaps in the work enclosure and tank cover, this mixes with the hydrogen gas and is extracted to atmosphere. On a 10,000 A installation the extraction fan should deliver 28 m^3 of air per minute in order to reduce the hydrogen to a level below its lower explosive level. It is clear, then, that the extractor fan must be working efficiently while machining is taking place, and safety interlocks must be fitted to the system to ensure that machining cannot take place unless this is so.

Air flow detectors should be positioned in the outlet from the work enclosure and be interlocked with the machine operating control. The system whereby the fan motor is interlocked to the machine operating control can be dangerous, as there have been instances where the fan has been spinning on the motor shaft or has fallen off. In another instance the fan blades were badly corroded, causing poor air flow. Whatever system is used it is essential that machining does not take place unless the extraction system is working efficiently.

2.6 Pumps

The pump which supplies electrolyte to the working gap is an important part of the equipment. It is required to supply electrolyte at 14 to 21 kg/cm^2 and this delivery has to be smooth and without pulsation. The detrimental effects of pressure pulsing on the machine tool and workpiece have been explained earlier. This requirement would appear to eliminate most positive displacement pumps. There is however one kind of positive displacement type pump which has given good service, this being the Archimedian screw type marketed by Mono Pumps. The pump which gives the best results is the centrifugal type, either single or multi-stage.

In view of the corrosive liquids being pumped the pump must be manufactured from corrosion resistant material. Where neutral salts are being used, stainless steel (EN 58J), Monel or gun metal are suitable. Where the more inhibiting salts—nitrites and nitrates—are used, cast iron pumps have been found to be quite serviceable. In practice, cast iron pumps have given surprisingly good service even on sodium chloride solutions, in spite of the fact that the corrosion rate in 20% sodium chloride is between 10 and 50 times that of austenitic stainless steels. Lined pumps are not recommended due to the danger of particles of the lining getting into the electrolyte circuit.

When determining the size of pump required it has been found in practice that a 3.5 kg/cm^2 rating for every 2.6 cm of flow across the tool face is sufficient for a 0.4 mm to 0.5 mm machining gap, and an allowance of not less than 1.1 litres minimum per minute for every cm^2 of tool face. Pumps delivering 1363 litres per minute at 21 kg/cm^2 are in use on 20,000 A units. These pumps, usually driven through V belts and pulleys or a gear box, rotate at high speeds and require 100 to 150 hp motors for drive. Most of these pumps are very noisy and to make working conditions bearable for personnel they have had to be situated in a sound insulated compartment.

For acid electrolytes pumps should be made of Langalloy 5R or titanium. Experience with a pump constructed of titanium showed that anodic polarisation reduced the corrosion rate from 95.4 mm/yr for the unprotected case to only 0.0025 mm/yr for the protected case when handling hot 50% sulphuric acid. The power consumption was only 3 W per 93 m^2 of immersed titanium[4].

3. Power supplies

The power unit required for electrochemical machining needs to be specifically designed for the process. The DC output usually ranges from 6–18 V or 8–24 V, and if titanium alloys are being considered it may be advisable to have 30 V available.

As the power supply available is usually three phase AC a step-down transformer and rectifier are required. A typical power circuit consists of a star wound transformer with voltage controlled to the primary input by a low gain three phase transductor. Output from the secondary windings is to silicon diode rectifiers. From the rectifier the power is fed *via* tool and workpiece to the machining gap. Fig. 1 shows a typical layout of the electrical unit.

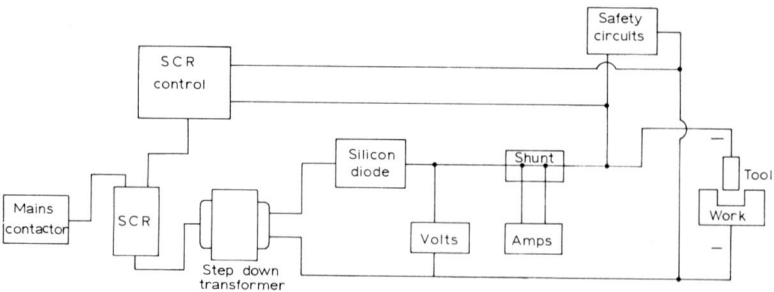

Fig. 1.
Typical layout of ECM electrical unit.

Existing power supplies are mainly transductor controlled but recently silicon controlled rectifiers have been used for voltage control. As a means of safeguarding and controlling, large banks of silicon control rectifiers have been perfected, and it is expected that this system will replace the transductor. Silicon control rectifiers are not so costly as the transductor and a faster response can be attained—within half a cycle, which would be 10 msec as against 100–500 msec for the transductor. M. and T. Chemicals Inc., who manufacture power supplies for electrochemical machining units, claim up to 90% efficiency for the silicon control rectifier as against 80% for the transductor. However, Anocut note the improving quality of silicon control rectifiers[5] but sound a note of caution. With the wide voltage and current ranges and also the low ripple consideration required for ECM, the reliability of silicon control rectifiers is not yet considered to be as good as that of the transductor. The accurate control needed on firing angle and timing is critical if the load is to be evenly shared on a bank of silicon controlled rectifiers arranged in parallel. Also the sensitivity of silicon controlled rectifiers to overload makes good protection essential against transients and surges resulting from short circuit conditions, as any fluctuations in voltage can alter the machining gap and affect the accuracy of the process. Good voltage control over the full rated range is therefore essential.

Most electrochemical machining power supplies are specified to ± 0.5% of maximum rated output. This would be ± 0.120 mV at 24 V. Voltage regulation is achieved by continuous comparison of the machining gap volts against a preset reference voltage on the control console. Voltage variations are signalled through the control circuit and are corrected. An improvement in voltage control is claimed in ref. 6 which describes a new voltage regulation system that permits holding the DC output voltage of Anocut power supplies to a virtually constant level. Voltage can be held to 0.05% of the full scale voltage regardless of load or line voltage variations. This contains voltage fluctuations to within ± 12 mV on an 8–24 V supply and will greatly facilitate the production of close tolerance work.

The maximum acceptable ripple which is superimposed on the output voltage is usually specified as 8% to 12% at full load conditions. The amount of ripple depends on the form of the connection in the rectifier bank, the most common being a six phase connection. Despite the theoretical effects on efficiency, there is no practical evidence to indicate that large ripple has any noticeable detrimental effect on the process. One detrimental effect of excessive smoothing is that the greater the smoothing, the greater the storage energy, with adverse effects on response time and spark damage.

Finally, one of the most vital features of an electrochemical machining power supply, namely, safety circuits. Sparking or arcing between the tool and workpiece should not occur if the electrochemical machining operation is well controlled; however, there are conditions when sparking or arcing can occur, and the more common causes of these conditions are listed below:

(1) Cavitation of the electrolyte flow.
(2) Boiling of the electrolyte in the machining gap.
(3) Electrode vibration.
(4) Feed rates too high.
(5) Sharp projections on the workpiece.
(6) Failure of the electrolyte filter system allowing conductive or non-conductive particles to find their way into the machining gap.

Safeguards directly affecting the machining gap are supplied with modern electrical supply units and will shut the power supply off in the event of:

(a) Voltage transients in the spark frequency range 700 Hz to 1,400 Hz.

(b) Exceeding the allowable rate of current increase $+ di/dt$. This would probably be adjustable in the range 1% change per second to 20% change per second, but is unlikely to exceed the latter figure.

(c) Exceeding the allowable rate of current decrease $- di/dt$. This would probably be adjustable in the range 4% per second to 30% per second, but is unlikely to exceed the latter figure.

(d) Short circuit before start up and overcurrent.

The damage caused by sparking increases with its duration and when, for example, a sharp projection on the workpiece or a conductive particle in the machining gap causes sparking and this condition is allowed to continue unchecked, sparks will occur at a frequency of several hundred times a second, causing considerable damage to the tool and workpiece. Detection of this type of fault is by the spark frequency pick up[7]. Where a non-conductive particle is jammed in the machining gap, this could cause cavitation of the electrolyte flow. When the area affected is reasonably large, the effect gives a detectable current change and this is detected by the $- di/dt$ circuit. This circuit will also pick up conditions when boiling of the electrolyte in the machining gap occurs.

The most difficult fault to pick up is when a small area of the workpiece passivates, for example, when a small area of the tool is affected by flow cavitation or inadequate flow. Then the resulting small current fall can pass undetected. The tool feeds in until it touches the workpiece but as the latter will have acquired an insulating oxide film there is no immediate short circuit. If the tool continues to feed towards the raised portion of the workpiece they will eventually touch and the insulating oxide film on the unmachined area becomes progressively more conductive as pressure of the tool on the work increases. Eventually the current increases sufficiently to make the tool and workpiece become molten, and finally a severe spark is passed. If the fault is detected before sparking occurs it will be by means of the $+ di/dt$ circuit, but if sparking occurs detection would be *via* the over current trip.

The damage caused by sparking or arcing depends mainly on the duration of the fault. The speed with which the fault is detected and the power cut off is then very important. A comprehensive list of detection devices and circuits is described in ref. 7. On most existing machines the shut-off times using the relay cut off in the transformer primary is 25 msec. Silicon control and rectifier control supplies are shut off by using the voltage transient signal to prevent firing of the next silicon control rectifier phase. This silicon control rectifier primary shut off can be effected within 8–9 msec. The new Anocut Anoguard protective systems contain safety circuits to detect the first spike transient of the spark, in addition to the rising and falling current and over current detection systems[5]. Ref. 5 also describes the Anocut micro bar shut-off system. This is probably the fastest shut-off system in use, and

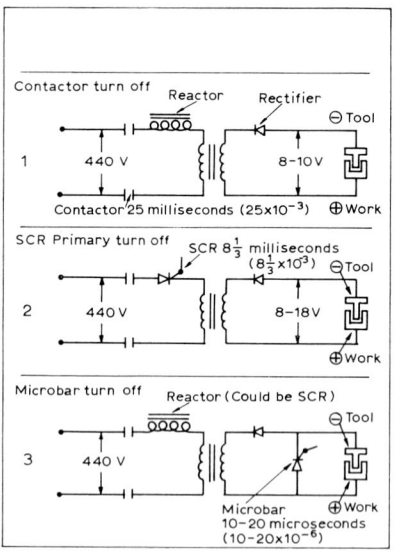

Fig. 2.
Power shut-off systems used in ECM control systems. Courtesy Anocut Engineering Co.

comprises a bank of silicon control rectifiers connected across the machining gap. When a spark is detected, the silicon control rectifiers are fired and become fully conductive, short circuiting the output voltage. This drops the voltage at the machining gap to nearly zero in a total time of 10–20 μsec. At the same time a shut down signal is transmitted to the control switch of the power supply which cuts off the input power within a period of approximately 40 msec. Figure 2 shows the three systems.

4. Dimensional accuracy and surface finish

Many claims have been made regarding the accuracy attainable by electrochemical machining, and without doubt in many small-scale applications with the machining parameter closely controlled, accuracies of $\pm$ 0.0254 mm can be achieved. However, when working under production conditions, variations in the process parameters do occur which makes the holding of tolerances of 0.0254 mm extremely difficult, if not highly improbable, for how much time can be spent on correcting a complicated form tool to achieve a 0.0254 mm tolerance on the shape? Having reached an acceptable form on the tool, it is then used on a production run. Other conditions which can alter to give variations in the final accuracy of the workpiece are described below.

Firstly there are losses of water due to evaporation and electrolysis. As water is lost the concentration of salt in the electrolyte increases, thus making the electrolyte more conductive. Then there is the water added from the water wash spray—on most machines there is a water spray to wash the workpiece and working area before the work area cover is opened—when this water runs into the electrolyte tank this can off-set the evaporation loss to some degree, but again some variation in electrolyte conductance can occur.

Secondly there is a pressure drop and reduction in flow through the filters as they tend to become blocked with hydroxide. This can cause variations in the flow rates from day to day, again affecting the dimensional accuracy.

Thirdly there are the voltage variations, and finally there are the small variations in the machine performance due to variations of operator control and to machine temperature change throughout the day. All these small variations can and do occur, affecting the accuracy of the process attainable under day-to-day production conditions.

4.1 DIMENSIONAL ACCURACY

The tolerances listed below have been found to be repeatedly achieved under day-to-day production conditions.

4.1.1 For forming three dimensional shapes and general formed surfaces where no vertical or near vertical walls are being formed

General tolerance ± 0.127 mm
Minimum tolerance ± 0.051 mm

4.1.2 For hole drilling

General tolerance + 0.127 mm
Minimum tolerance + 0.076 mm

When holes of great depth-to-diameter ratio are being drilled, the back pressure formed as the drill feeds deep into the workpiece tends to make the hole larger due to the increased efficiency of the electrolytic cell.

4.1.3 For generated forms

Generated forms where vertical walls are being formed and the surface is subject to stray attack and poor voltage gradient conditions.

General tolerances ± 0.254 mm
Minimum tolerances ± 0.127 mm

4.2 Surface finish

The surface finish produced by electrochemical machining varies according to the application and the material being machined. The best finishes are obtained when the surface being machined is formed by the tool in one operation, and the whole surface is subject to small gap and high voltage gradient conditions.

The figures given below are in the writer's experience readily attainable, and are repeatable.

For forged or rolled steels and nickel chrome alloys

600 to 2,400 $\sqrt{}$ (15–60 μ'')

For cast irons and cast nickel chrome alloys

1,600 to 3,200 $\sqrt{}$ (40–80 μ'')

5. Operating conditions

ECM can most simply be described as an electroplating process in reverse, the tool being the cathode and the workpiece being the anode. The difference between ECM and other electrolytic processes being used in industry is that with ECM the workpiece can be shaped, while there is also a difference in the speed at which the metal is removed. The metal removal rate is not influenced to any great degree by the mechanical properties of the metal. Another important difference is in the high current densities employed. For example, where the electrolyte flow path is very short, as in the case of drilling or trepanning operations, current densities can be as high as 465 A/cm². In general forming operations where the flow path is longer, 233 A/cm² would probably be the maximum. The current densities used depend to a great extent on the application, and where large area tools are used can be as low as 5 A/cm², but for most general plunge-forming applications the aim is to work with current densities in the region of 45 to 110 A/cm².

In a typical ECM operation, the cathode or tool, which for this purpose can be described as a mirror image of the desired shape, is placed in close proximity to the workpiece to be machined. The gap, normally between 0.127 mm and 1.016 mm, is flooded with a conducting electrolyte, and the potential, which could be from 3 to 30 V, is applied. A large current then flows causing metal to be anodically dissolved from the workpiece. If at the commencement of the operation the tool shape does not resemble the workpiece, the gap size will vary over the area between the tool and the workpiece. Where the gap is narrow the current densities will be higher than in the wider regions (narrow gap, lower resistivity; wider gap, higher resistivity.) The metal will thus be dissolved away at a high rate in the high current density regions. As the tool is fed towards the workpiece, the surface of the latter will eventually become an imprint of the tool shape (see Fig. 3).

The electrolyte, normally a dilute solution of sodium chloride or sodium nitrate

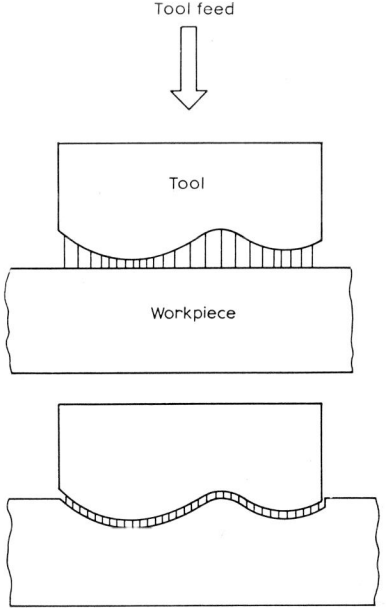

Fig. 3.
Showing current density as tool forms imprint of the tool in the workpiece.

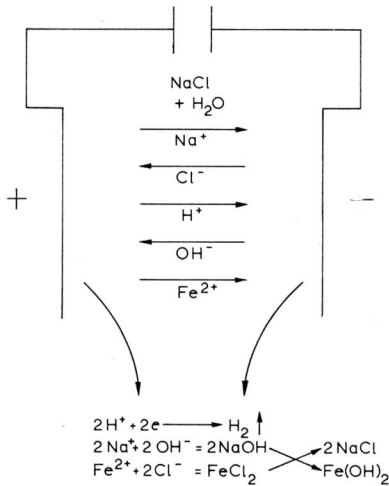

Fig. 4.
Possible chemical reactions in the machine gap.

or a mixture of both, is pumped through the gap at speeds which can be from 9 to 60 m/sec at pressures varying from 1.4 kg/cm² to 35 kg/cm². This is required to make possible a continuous passage of high current to achieve the high metal removal rates, and to wash away the products of the anodic dissolution.

The metal is removed from the workpiece atom by atom, in contrast to the shearing action brought about by conventional machining. When machining multi-phase alloys with salt electrolyte the chemical reactions which occur at the cathode face and in the electrolyte are numerous and complex, but taking a simple example of iron being machined with a sodium chloride electrolyte, Fig. 4 shows the possible chemical reactions which occur.

Assuming the process is proceeding using the electric current at 100% efficiency the metal removal rates in a given time are determined by Faraday's Laws of electrolysis, and are proportional to the applied current. The formula for computing metal removal rate is given below.

$$\text{Metal removal rate } u = \frac{A_t \, J \, t}{F} \cdot \frac{A}{(Z/\varrho_m)} \text{ cm}^3/\text{sec}$$

where A_t = Tool area
J = Current density, A/cm²
t = Time in seconds
F = Faraday, coulombs per gram equivalent (96,500)
Z = Valency
A = Atomic weight
ϱ_m = Density of workpiece g/cm³

Tool feed rates can also be calculated from this formula.

In practice, metal removal rates have been found to be between 75% and 100% efficient, depending upon the material and the electrolyte.

When discussing feed rates one of the interesting and fortunate features of ECM is the way in which the parameters adjust themselves to suit the feed rate. If the tool is feeding towards the workpiece at a faster rate than the metal is being removed, the gap will become progressively smaller. As the gap becomes smaller the current will rise due to the lower resistivity of the electrolyte in the gap, and the metal removal rate will increase until the equilibrium point is reached. The metal removal rate will then equal the feed rate of the advancing tool. Conversely if the feed rate is reduced the gap will increase and due to the increase in the resistivity of the electrolyte the current will decrease until the equilibrium point is reached and again the metal removal rate will match the tool feed rate.

The equipment required for electrochemical machining is shown in Fig. 5. This shows the machine, which has a ram which can be fed towards the workpiece at a steady constant feed rate and with no stick slip, this ram having a tool mounting platen on which can be mounted the forming tools. The machine must be of a good design, be very rigid and free from vibration. Also shown is the electrolyte system consisting of a tank to hold the electrolyte, a pump to deliver the electrolyte to the tool at the required pressure and flow, a means of filtering the electrolyte, a centrifuge or settling tank to remove the hydroxides formed during the machining process, a heat exchanger for maintaining the electrolyte at a constant temperature, an extraction system to remove the hydrogen evolved during the process, a DC power

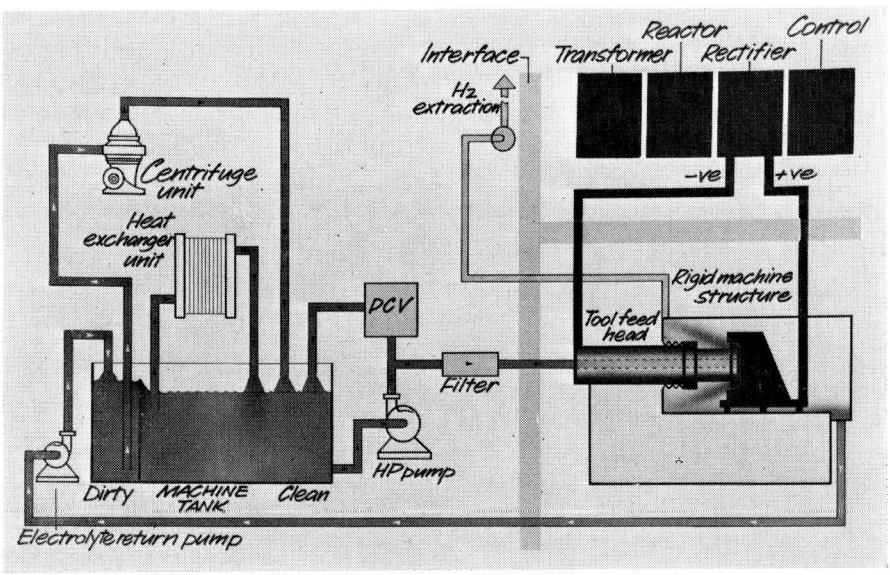

Fig. 5.
Typical ECM plant layout.

supply capable of a continuous output at the required power level with voltage adjustments usually between 6 and 24 V, spark detection to monitor both rising and falling current, and a high sensitivity frequency detector circuit. If a spark condition arises, shut down must be effected within 15 to 20 μsec if serious damage to the tool is to be avoided.

There are other types of equipment in which no feed mechanism is required. These use a static tool and either reduce the external shape or increase internal bore by increasing the gap size. Operations for removing burrs or sharp corners are carried out by this method.

Where metal removal rates are low the equipment can be relatively simple and the electrolyte can be run to waste. These units, due to low machine tool costs and simple ancillary equipment, can be obtained for a much lower cost than the forming machines.

5.1 Factors affecting the efficiency

The gap size is very important as any variation in it will affect the shape being produced. The factors which can affect the efficiency in the gap between tool and workpiece are numerous, the main ones being as follows: electrolyte conductivity, electrolyte temperature, concentration of salts in the solution, gas evolution, cavitation of flow, length of gap, flow rate, composition of the material and voltage. Electrolyte conductivity is one of the main factors influencing the gap size. Again the conductivity of the electrolyte can be affected by various factors, for example the amount of salts in the solution, and the type of salts used; in industry these are mainly NaCl or $NaNO_3$, or a mixture of both. The temperature of the electrolyte also affects the conductivity, for example with a 20% v/v mix of sodium chloride with water, a 5°C variation in temperature can give a 0.0305 mm variation in a 0.508 mm gap. Gas evolution will also reduce the conductivity and if there is not sufficient electrolyte flow to remove the gases formed, machining can stop completely. Back pressure can be applied at the gap outlet to overcome troubles encountered due to gas generation. This will reduce the gas bubble size and enable machining to continue in conditions where normal machining would not be supported.

Tool design must be such that all parts of the gap are supplied with an even flow of electrolyte. Cavitation of the flow will leave a part of the workpiece unmachined, or, being machined more slowly, will lead eventually to contact with the tool and arcing will occur.

The applied voltage will also affect the gap, the higher the voltage the wider the gap between the workpiece and tool. An increase of 1 V can give an increase in the gap of 0.051 mm. The length of the flow path the electrolyte has to travel is important, as products of machining, *i.e.* gas and hydroxides, will progressively affect the conductivity the further the electrolyte has to flow. If the flow path is too long the electrolyte will become saturated with the products of machining and will not support any further machining. There is, however, a factor which tends to offset this to a remarkable degree, and this is the rise in conductivity due to the temperature increase on its passage through the gap. It has been found that one offsets the other to a remarkable degree. Tools in which there are flow paths of 127 mm to 152.4 mm are being used with little or no correction being made to offset the taper one would expect to get towards the electrolyte outlet. It must be stressed, however, that high electrolyte flow rates must be maintained to reduce the build-up of hydrogen gas, and to prevent the electrolyte boiling in the gap, another condition which can cause a breakdown of the process.

The gap between the tool and the workpiece, and the current density, also affect the surface finish obtained by ECM. In general the best surface finish is obtained by using high current densities with relatively small gaps. In areas which are subjected to low voltage gradient conditions, or stray attack, the surface finish deteriorates. This shows itself particularly when generating vertical walls. Fig. 6 illustrates this.

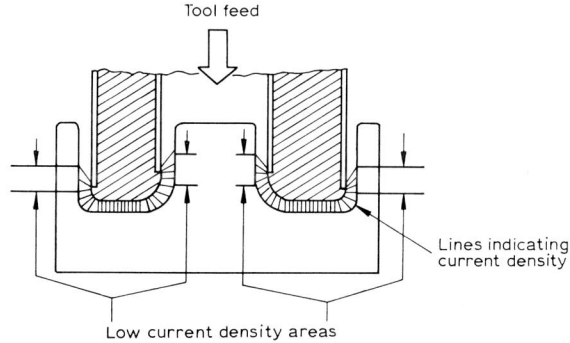

Fig. 6.
Showing areas which will be subject to low current density conditions or stray attack.

The majority of metals on which ECM operations are being carried out are not homogeneous, so that the surfaces of the workpiece being machined have a granular structure. The adjacent grains of metal, due to their differing constituents, will have varying potentials. The grain which has the higher positive potential will be machined away faster than the adjacent grain which has a lower positive potential. This will leave some grains standing proud. These protruding grains will start to be machined when due to the voltage gradient in the machining gap the electrode potential reaches a value equal to that of the material of the grain.

The effect of stray attack can be minimised by the careful selection of the electrolyte, the most suitable electrolyte being that which reduces the difference between the dissolution voltages of the varying constituents of the material. In many cases the only way to eliminate entirely the stray attack is to subject the whole surface to high current density conditions. This can be done by leaving material for a final finishing operation. Assuming the form shown in Fig. 7 has been produced leaving 0.254 mm on the surface for a finishing operation, then an uninsulated tool having a 0.254 mm smaller envelope than the hole produced is held static in the cavity. The electrolyte and the power are switched on for a sufficient period to remove 0.254 mm of material, bringing the cavity to the required size and depth and also removing material from the whole surface under high current density conditions. The restrictor, as shown in Fig. 7, is to provide full flow conditions and to restrict the formation of hydrogen bubbles.

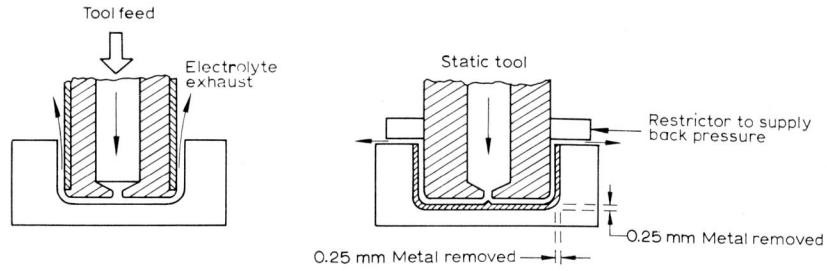

Fig. 7.
Finishing a cavity using static tool.

6. Design of tools

6.1 Materials for tools and fixtures

The tools and fixtures used for electrochemical machining are required to withstand for long periods the corrosive effect of hot electrolyte, possibly in the presence of stray electric currents. The choice of materials used for the tools and fixtures is very important if erosion and failure is to be avoided. The materials being used for the work holding fixtures are generally stainless steel, bronze, brass, copper, resin bonded glass fibre based laminates and Tufnol. The use of granite and special cement mixes is also being assessed.

For the cathode tools, copper, copper–tungsten, or stainless steel are used, but copper is the most commonly used material due to its high electrical conductivity and ease of manufacture and repair. In the event of spark or handling damage the tool can easily be repaired by electroplating or plugging. One disadvantage of copper is that it is difficult to maintain a good bond between tool and insulating material. Copper–tungsten (Elkonite 10W 3 grade) is recommended for close gap work where sparking is likely[8]. Stainless steel does not have the thermal or electrical conductivity of copper or copper–tungsten, but it has good resistance to corrosion and handling damage. For the drilling of small diameter holes thin wall stainless steel or titanium tube are used.

Both tools and fixtures should be constructed to withstand the high deflection forces encountered during electrochemical machining. With this in mind, care should be exercised when using non-metallic materials in fixtures. Any materials used should be stable enough to maintain the accuracy required from the machining operation.

Corrosion of normally non-corrosive metals can occur when two differing materials are used; this is due to the galvanic effect between the materials in the presence of the conducting electrolyte. Instances of severe damage to fixtures have been experienced due to this effect. Also, there have been instances of stainless steel fixtures which are normally in the electrolyte environment which have been in the presence of quite low stray currents resulting in an etching away of the grain boundaries, leaving the metal in a crumbling condition. This condition can give rise to very undesirable results where it could be possible for metal grains to get into the electrolyte circuit.

Any metal part of any fixture which is in the path of electrolyte exhausting from the tool should be protected by a PVC or similar protective coating. This is particularly so if there are joints in the area, experience having shown that quite severe corrosion can take place in these areas.

6.2 Power connections

Electrical contacts should be reduced to a minimum and electrical power connections should be as close to the component and tool as possible, as each joint is a potential source of power loss. It is important that the power connections to the fixture or component and the tool be of sufficient area and that good joints are established in view of the high currents being carried. Copper cables, copper braid or copper buss bars are generally used to bring the electrical power to the fixture and tool. In the event of too small an area or too thin a section being used, overheating will occur.

As important, if not more important, than area is the flatness of the contact faces and the pressure applied to the joint faces. Badly fitting joints can cause losses in power, burning and corrosion. The use of good conductive contact grease can to a great extent reduce the chances of these undesirable effects by keeping out air and electrolyte. Where possible the connection points should be near the heavier sections of the fixture or tool in order to dissipate quickly any heat generated at the joints.

It is sometimes difficult to make a good electrical contact with the workpiece due to irregular and varying shapes or small contact faces. Small contact faces between component and fixture can give rise to local hot spots which could cause metallurgical damage and/or distortion. It is usually found that supplementary contact areas can be made. Areas between locating faces can be reached by using copper braid on a soft rubber pad. This copper bound pad takes up the irregularities in other faces or shapes even if these are a forging or casting and provides a large area of contact. This method of contact is especially suitable for supplying current when machining thin sections.

Before discussing the design of tooling, the design of the component to be machined should be considered. Tooling for electrochemical machining is expensive, but tools can be made simpler and less costly if at the design stage consideration is given to the various factors which will be outlined in this section. For example, a true radius on the tool does not produce a true radius on the component, but a radius which is slightly malformed. This shape will most probably be acceptable to the designer if the situation can be discussed prior to detailed drawings being issued. Each component has to be considered on its own merits, but usually some compromise can be reached between the component and tool designers so that the component is produced to the dimensions required with the least expensive tooling.

Another point to be considered is that when ordering raw material it must be borne in mind that unlike conventional machining, electrochemical machining requires a depth of metal to be removed in order to produce the final form required. It must be understood that whilst the tool is being fed in, metal is being removed from all the surface of the workpiece facing the tool, but at differing rates. Where there is a wide gap some machining will take place, but at a much slower rate than where the gap is small. There must therefore be sufficient material on the parts of the component where large gap conditions prevail at the start of the machining operation to allow for some metal removal whilst the tool is catching up due to preferential rates of machining.

6.3 Tool insulation

The parts of the tool which are not required to remove metal must be insulated to prevent machining taking place. It is essential that the insulation remains secure and intact; it must, therefore, be securely bonded to the tool. Insulation is primarily used on tools used for generating operations, but at times it is required on die sinking tools to prevent stray cutting from the sides of the tool. The insulating coating has to be sufficiently flexible to accommodate the dimensional changes due to temperature variations during the machining operation or, if the tool has thin walls, the flexing of the walls due to electrolyte pressure in the body of the tool. There are many materials that can be used for insulating, PVC, resin bonded glass fibre, rubber-based coating and epoxy resins. The rubber based coatings have been found to be very serviceable, but the most recent and best is an epoxy resin mix with mineral fillers.

When applying insulating coatings to copper tools the surface to be coated should be grit blasted to roughen the surface and ensure complete cleanliness.

6.4 Tool design

The electrochemical machining process requires a sufficient flow of electrolyte between the tool and workpiece to carry away the heat and products of machining, support machining at the required feed rate and produce a satisfactory surface finish. Any features which give rise to starvation of flow in certain areas will cause cavitation or streaming and will give undesirable features from poor surface finish to a complete breakdown in the process. The observance of several basic rules can do a great deal to reduce the problems that can be encountered in maintaining a full flow of electrolyte over the whole face of the tool. There should be no sharp corners in the flow path; sharp corners cause cavitation of the flow which will at best leave an uneven surface on the workpiece. All corners in the flow path should have at least a 0.762 mm radius (see Fig. 8).

Each tool and component should be considered on its merits and full use made of any natural features to achieve full flow conditions. The electrolyte can be made to flow from one side of the tool across the face and exhaust from the opposite side, flow outwards from slots or ports in the tool face or fed from the outside of the tool exhausting through slots or ports in the tool face. A shape which is very common is a projecting boss. This can be formed with a very simple tool. Whatever the shape of the boss a corresponding but slightly larger shape will be cut into the tool,

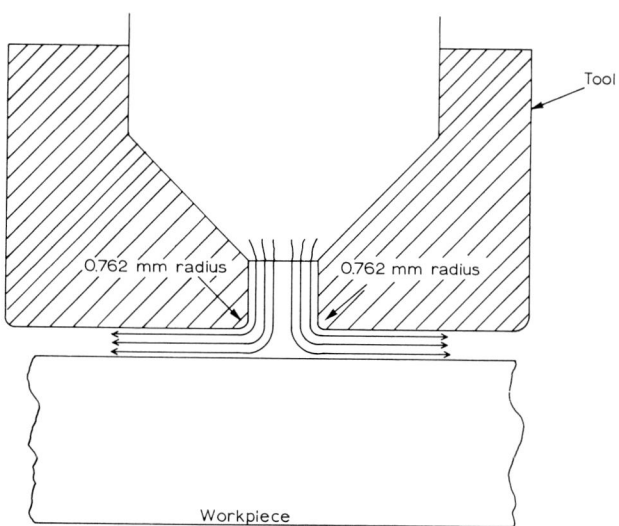

Fig. 8.
Showing feed slot correctly radiused.

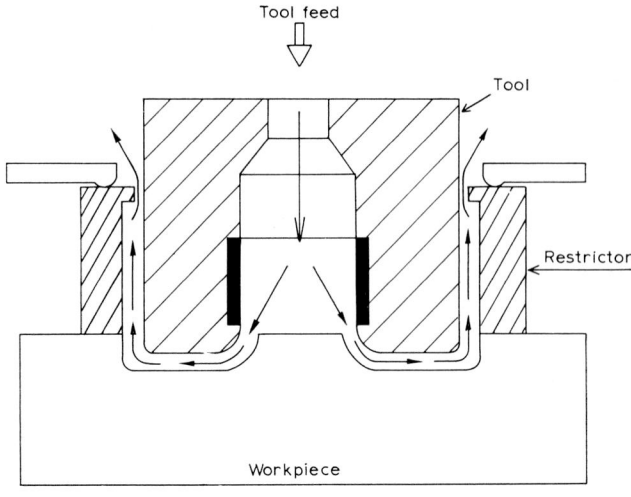

Fig. 9.
Boss forming tool

electrolyte then being fed down through the aperture and flowing out across the face of the tool (see Fig. 9).

When sinking cavities or forms, tools with electrolyte supply slots in them are the simplest to manufacture, these slots then leaving small ridges on the workpiece. In many instances these ridges can be left, but if they have to be removed a simple polishing operation is all that is required.

To supply an adequate flow, these electrolyte feed slots should be approximately twice the estimated operating gap. For example, if the operating gap is estimated to be 0.508 mm the feed slots should be 1.016 mm wide. The flow from these slots is orthogonal to the slot so that the flow from the ends of the slot is poor. If the profile of the tool has sharp corners or corners with less than a 2.54 mm radius then the slot needs to terminate as near to the corner as possible (see Fig. 10). When the

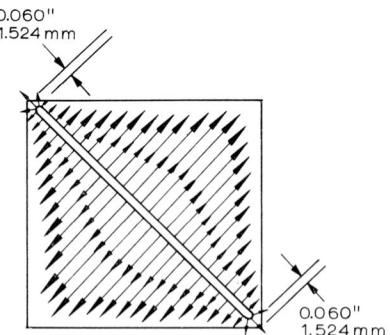

Fig. 10.
Showing electrolyte feed slot in tool with sharp corner profile.

tool has a profile with radii greater than 2.54 mm then the slot should terminate in a radius larger than the slot (see Fig. 11). Straight slots ease tool manufacture but where profiles are irregular shapes, curved flow slots are usually a necessity. Sharp changes in the direction of the slots should be avoided as this will give areas which are starved of flow. Figs. 12 and 13 show typical slot patterns for several shapes.

Where large areas are being machined, the flow path should not exceed 15 cm. If the area being machined necessitates a flow path longer than this, then exhaust slots as well as feed slots will be required to ensure good flow conditions. Examples of this are shown in Fig. 14.

Where the electrolyte has to flow across the face of the tool the use of chambers or flow restrictors is required to direct the flow in the required direction. These chambers or restrictors should be made of insulating material, locate on the tool

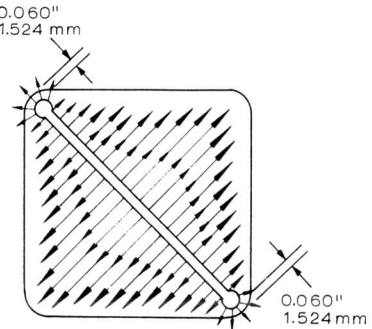

Fig. 11.
Showing electrolyte feed slot in tool with radiused profile.

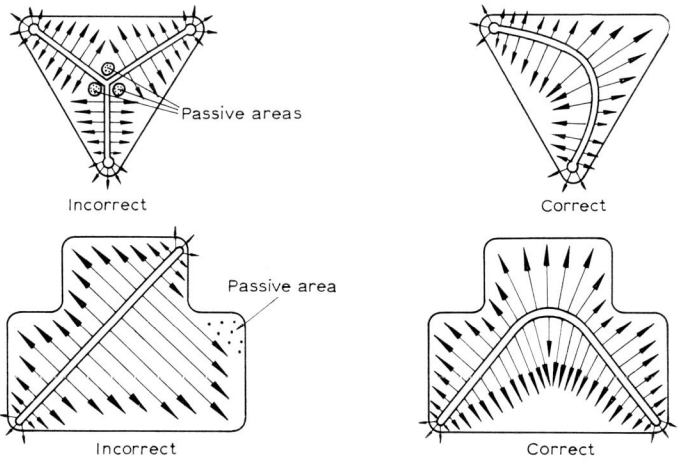

Figs. 12, 13
Showing incorrect and correct slot patterns in tools.

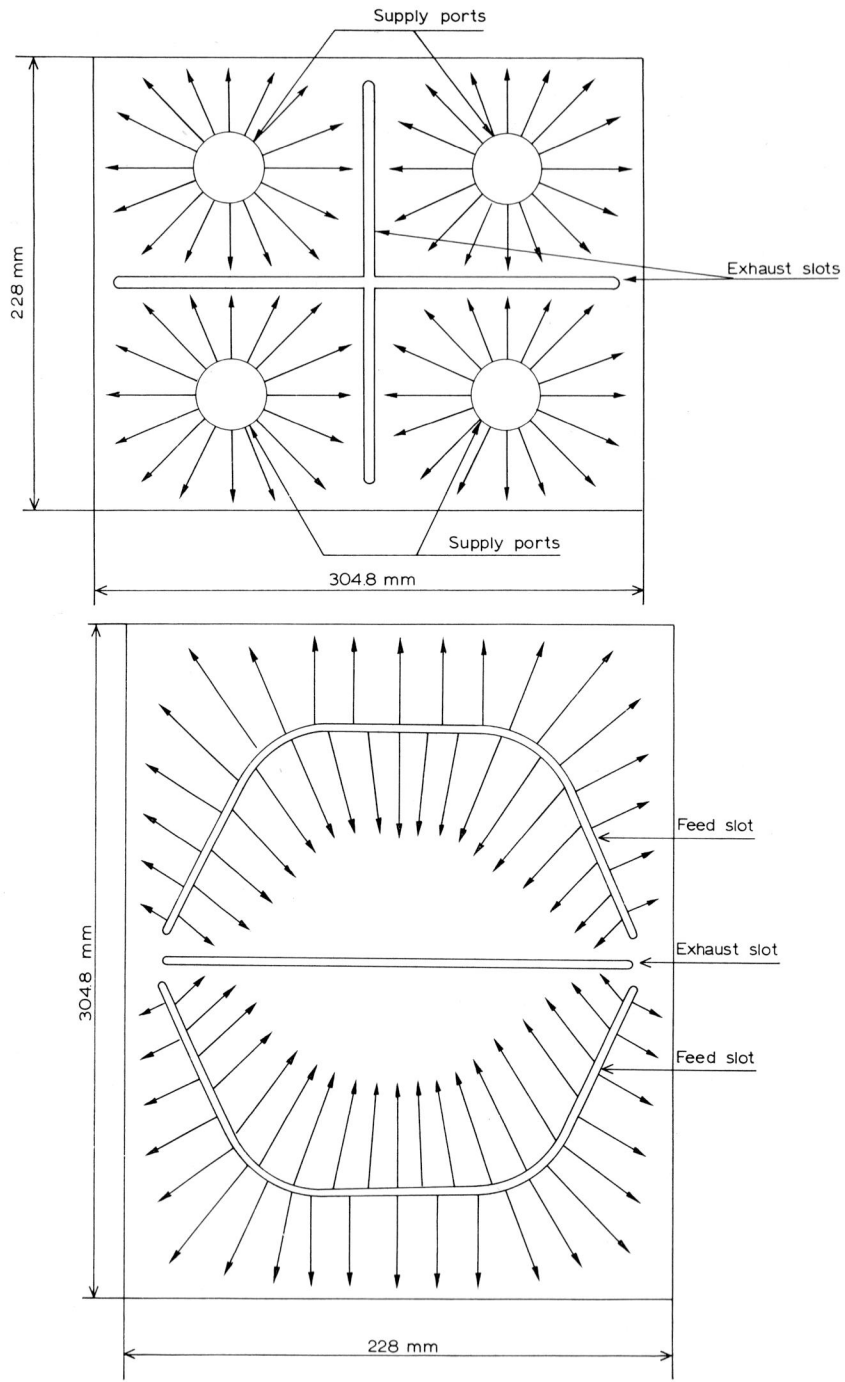

Fig. 14.
Electrolyte feed slot patterns for large area tools.

and seal onto the workpiece (see Fig. 15). These chambers and restrictors give good control of flow conditions, but can be expensive, so should only be used when flow conditions cannot be achieved by simple tooling.

The area to be machined in a single plunge forming operation may contain several features each of which requires an electrolyte supply port. Such an example is shown in Fig. 16 where several bosses are required to be formed in a single operation. If two or more adjacent sources are used for electrolyte supply the opposing flows result in static or slow moving areas (see Fig. 17). To prevent this

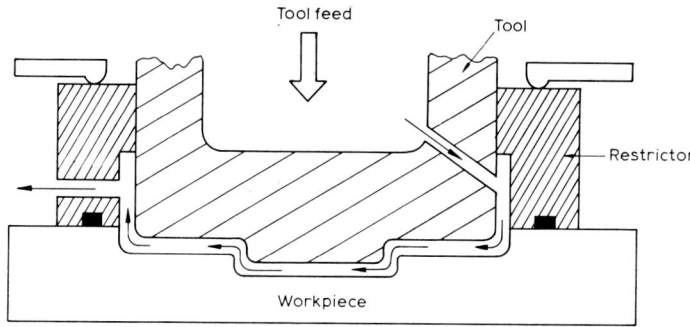

Fig. 15.
Showing restrictor to ensure flow over the whole tool face.

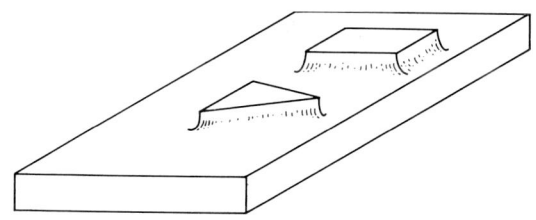

Fig. 16.
Showing area to be EC machined which contains two bosses.

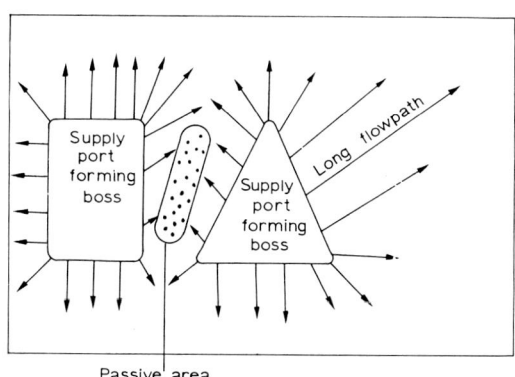

Fig. 17.
Showing tool with poor electrolyte supply.

condition the area should be sub-divided by exhaust ports, usually in the form of slots as shown in Fig. 18. Each supply port is then isolated from the other by exhaust slots. In this way a continuous supply of electrolyte is supplied over the whole area. When forming round bosses or circular cavities using forward flow, *i.e.* from the centre of the tool to the outside diameter, the maximum flow path which should be attempted without restriction should not be more than twice the distance from r to R (see Fig. 19). Where the flow path on this type of tool is long, some

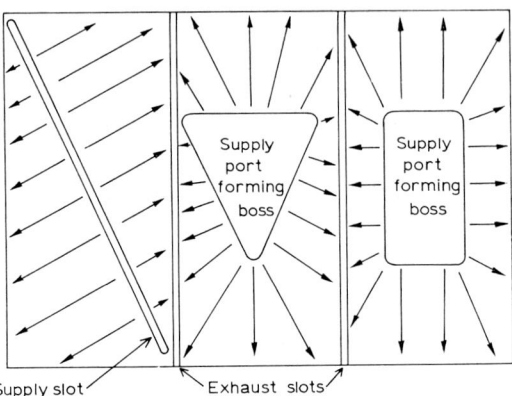

Fig. 18.
Tool as in Fig. 17 with correct electrolyte supply ports and slots, also exhaust slots.

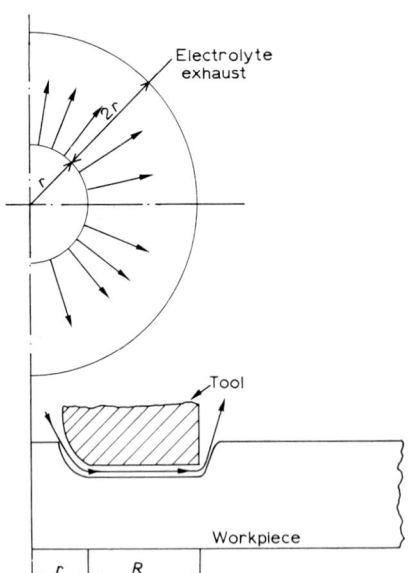

Fig. 19.
Showing maximum flow path for tools forming round bosses or circular cavities. Maximum flow path for divergent flow should not exceed $2R$ or $6'$.

form of restriction or reverse flow is advised as forward flow is very prone to streaming.

Where several electrochemical machining operations are being carried out on a workpiece, care should be taken to prevent electrolyte exhausting from the tool impinging on to a finished surface. If this does occur the finished surfaces will be affected by stray attack. The tool should be designed to direct the exhausting electrolyte away from any finished machined surfaces. Fig. 20 shows an example of this. This will leave a small cusp on the workpiece which according to requirements can either be left or polished away. It is also not advisable to have electrolyte collecting in pools on finished surfaces and where this cannot be prevented some form of protective coating should be applied to the finished surfaces.

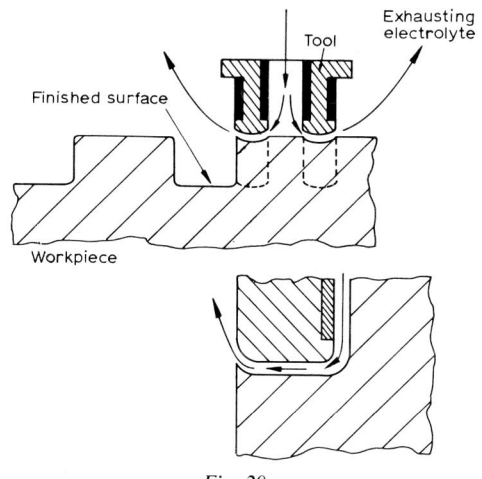

Fig. 20.
Showing method for directing electrolyte away from finished machined surfaces.

6.5 STARTING ON SURFACES WHICH DO NOT CONFORM TO THE TOOL SHAPE

The previous examples describe the tool design for machining a workpiece which conforms closely to the tool shape. Where, however, the workpiece shape is completely different from the shape to be formed, consideration must be given to maintaining an adequate electrolyte flow between the highest point of the tool and the workpiece. An example of this situation is shown in Fig. 21. With unrestricted flow the electrolyte will flow freely from the tool as it approaches the work, but when point A reaches the normal machining gap there will not be sufficient flow to support machining at the normal feed rate. A dam positioned around the tool will raise the pressure under the tool and sufficient electrolyte will be present at point A to support machining.

290   ELECTROCHEMICAL MACHINING

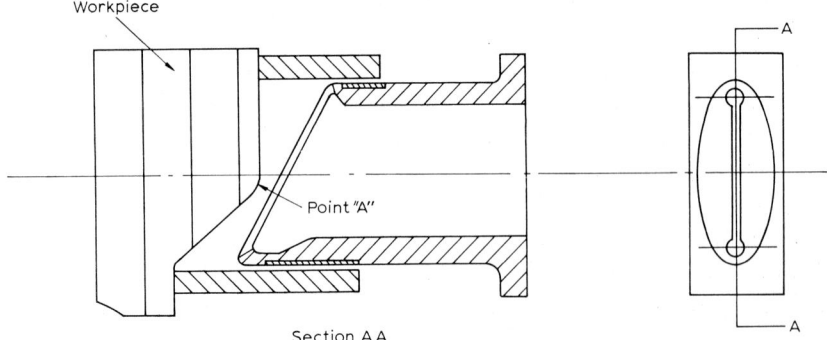

Fig. 21.
Showing position where workpiece shape is different from tool shape.

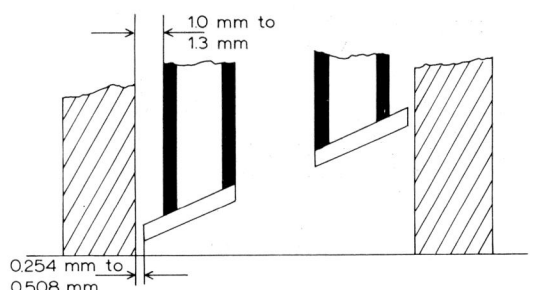

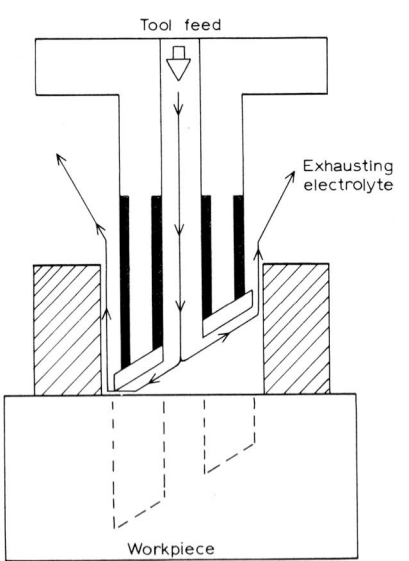

Fig. 22.
Showing angular tool starting on a flat workpiece surface.

A further example is shown in Fig. 22. This shows the forming of a boss on an angular face with the tool starting on a flat face. The tool should have a restrictor with a small annular clearance of 0.254 to 0.508 mm around the tool lip, and above the tool lip an annular clearance for the escaping electrolyte. This will provide a low pressure exhausting condition and give a good pressure differential in the critical area.

Another method which dispenses with restrictors is to cast a conductive low temperature alloy onto the workpiece to build up the surface to conform to the tool shape. Any alloy which is left on the workpiece after the operation can be melted off.

6.6 Tool correction

An electrochemical machining tool does not produce in the workpiece the reverse image of its own shape. In practice, the geometry of the tool, the feed direction, the length of the flow path and the variables of the process all contribute to the final shape produced. The complex nature of the problem indicates the need for an empirical method of determining tool shapes. Some knowledge has been gained by this method and information for correction on simple tool shapes is available from users of the process and the machine tool manufacturers. Application of the empirical approach to other than simple tool shapes involves a considerable amount of experimental work and adjustment of tool shape on the cut and try technique. The complexity of the problem can be reduced if the process variables—hydrogen evolution, voltage variation etc.—are reduced to a minimum by careful selection of the process conditions. Theoretical methods can then be used to predict tool corrections.

The simplest form of tool correction is that used for generating tools (see Fig. 23). For a given tool lip size, the overcut will be approximately one to one and a half times the end gap when the electrolyte is exhausting to atmosphere. For the average size tool the end gap will be in the region of 0.254 mm but generally for smaller, more intricate tools the gap will be smaller, and for larger tools with longer flow paths the operating gap will be larger. The operating gap and the side gap can be adjusted to some extent by varying the electrolyte, the electrolyte conductivity or the voltage. In a generating operation the tool overcut increases with the magnitude of the radius. The size of the overcut does not bear a linear relationship to the magnitude of the radius, since the effect of hydrogen evolution over longer flow paths, and divergence of flow, tend to minimise the degree of overcut on a tool with a large radius. Providing the hydrogen evolution in the gap is reduced, and electric field concentration is minimised, theoretical methods of tool design are available which will allow a first approximation of the tool shape to be calculated. Concentration of the machining current at boundaries of the areas being machined, and on sharp features of the workpiece whose radii are smaller than 1.27 mm, will

292 ELECTROCHEMICAL MACHINING

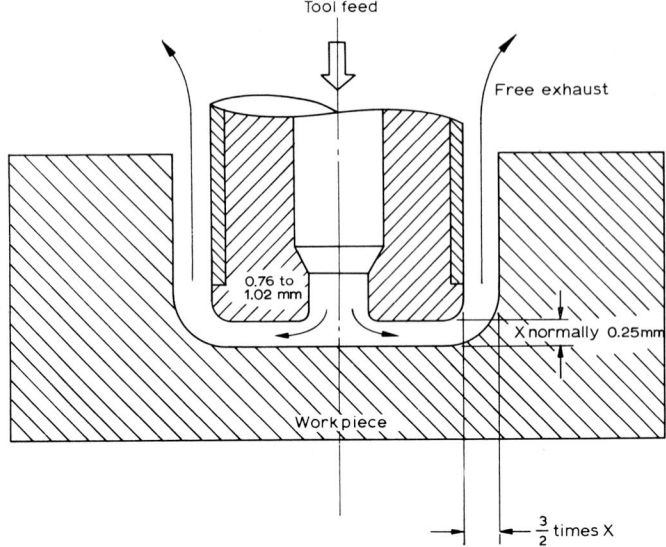

Fig. 23.
Simple form of tool correction.

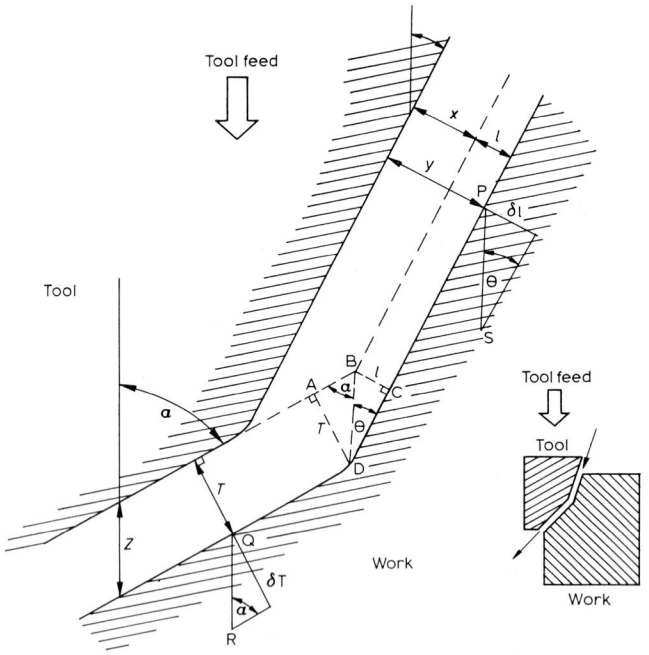

Fig. 24.
Corrected tool profile and work profile. Courtesy Rolls-Royce Ltd.

produce excess machining. Caution should be observed in applying the theory to these areas.

The general relationship between tool and workpiece is shown in Fig. 24. The work profile and corrected tool profile are shown by the heavy lines and the dotted lines indicate the shape of the tool if it were an exact replica of the work. The difference between workpiece and tool shape at any angle θ on the surface of the workpiece is x. This is the value required in order that the tool shape can be designed straight from the component drawing. The minimum operating tool end gap is T, which occurs at the maximum angle x to the work surface ($x = 90°$) in most die sinking operations.

Tool correction: $x = y - l$ (1)

From triangles ABD and BCD (see Fig. 25):

$$\frac{T}{\sin \alpha} = \frac{l}{\sin \theta}$$

$l = T \sin \theta \operatorname{cosec} \alpha$ (2)

Under the conditions previously given as applicable for this analysis the potential gradient across the gap at any point is uniform. Metal removal ratio is therefore inversely proportional to the gap size.

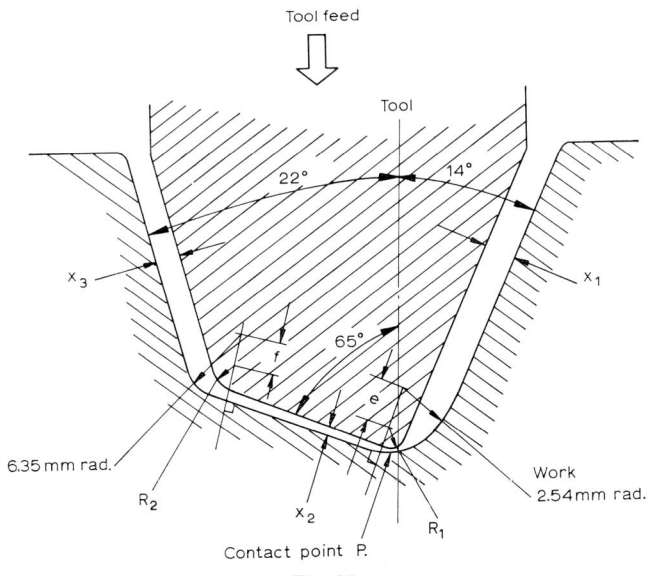

Fig. 25.
Workpiece shape and final corrected tool shape. Courtesy Rolls-Royce Ltd.

The incremental metal removals at P and Q are related as follows:

$$\frac{\delta l}{\delta T} = \frac{T}{y} \tag{3}$$

As the tool moves forward, metal removal in the direction of the tool feed at any point on the work surface is the same.
Therefore

$$PS = QR$$

and so

$$\frac{\delta l}{\sin \theta} = \frac{\delta T}{\sin \alpha}$$

$$\frac{\delta l}{\delta T} = \frac{\sin \theta}{\sin \alpha}$$

Substituting from (3)

$$\frac{T}{y} = \frac{\sin \theta}{\sin \alpha}$$

$$y = T \sin \alpha \operatorname{cosec} \theta \tag{4}$$

Substituting (4) and (2) in (1)

$$x = T (\sin \alpha \operatorname{cosec} \theta - \sin \theta \operatorname{cosec} \alpha)$$

When $\alpha = 90°$

$$x = T (\operatorname{cosec} \theta - \sin \theta)$$

Figure 25 shows a component shape and the final corrected tool shape. The tool shape will be the same as the component shape less amount x_1, x_2 and x_3 on its various sloping faces. Small radii may be blended while on large radii, values of the corrections x can be found for individual points from which the tool shape is plotted.

The minimum gap T will occur at P and is estimated to be 0.203 mm for this tool. At P, $\alpha = 90°$, so that:

$$x = T (\cos \theta - \sin \theta)$$
$$x_1 = 0.203 \text{ mm} (\operatorname{cosec} 14° - \sin 14°) = 0.787 \text{ mm}$$
$$x_2 = 0.203 \text{ mm} (\operatorname{cosec} 65° - \sin 65°) = 0.0508 \text{ mm}$$
$$x_3 = 0.203 \text{ mm} (\operatorname{cosec} 22° - \sin 22°) = 0.457 \text{ mm}$$

The tool radius R_1 to give 2.54 mm radius on the work becomes:

$$R_1 = 2.54 \text{ mm} - 0.787 \text{ mm} = 1.753 \text{ mm}$$

The new centre of curvature will be as shown on the diagram where

$$e = (2.54 \text{ mm} - 1.702 \text{ mm} - 0.0508 \text{ mm}) = 0.7366 \text{ mm}$$

7. Electrolytes

As mentioned previously, electrolyte is pumped to the machining gap completing the electrical circuit and allowing the machining reaction to take place. In addition it carries away the heat and products of the machining process.

A good electrolyte should have good electrical conductivity, leave a good surface finish on the workpiece, be inexpensive, readily available, safe to handle, and as non-corrosive as possible. Salt solutions generally have a lower electrical conductivity than acids, but they are safer to handle and are much less corrosive. Except for a few special applications, salt solutions are therefore normally used as electrolytes.

The three most favoured electrolytes are sodium chloride, sodium nitrate, and mixtures of the two. Of these, the most widely used, without doubt, is 15–20% w/w sodium chloride in water. This is inexpensive, readily available in bulk or small quantities, and has good electrical conductivity (see Fig. 26). It is corrosive, but stainless steel and the modern epoxy based and rubber based protective coatings withstand its corrosive action. This electrolyte does have some undesirable effects when machining certain materials, these being:

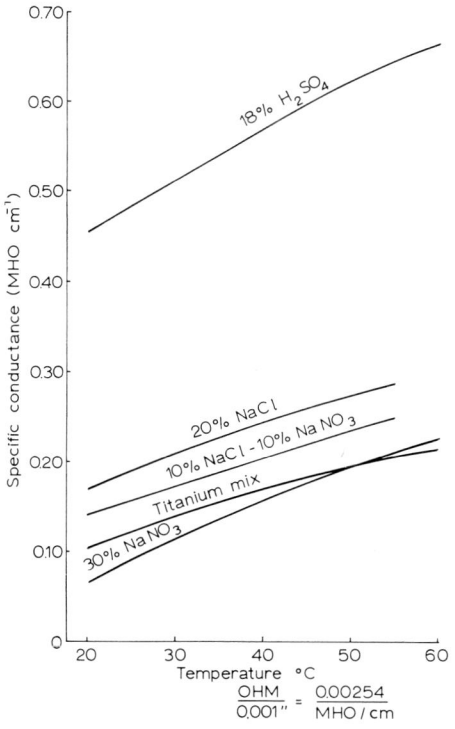

Fig. 26.
Conductivity of nominal electrolytes *vs.* temperature (concentration by weight).

(a) Overcuts can be a little unpredictable and stray attack can be a problem, particularly when generating surfaces.

(b) Fatigue properties can be badly affected due to selective etching at the grain boundaries.

The second most commonly used electrolyte is sodium nitrate, 20–40% w/w aqueous solution. Such electrolytes are less conductive (see Fig. 26) and more expensive than sodium chloride, but overcuts are normally smaller and more controlled, and where there are problems with surface finish this electrolyte will usually give improved results. This is due to a tendency of the electrolyte to passify the metal surface. This electrolyte is less corrosive than the sodium chloride mix and is kinder to the machine and the electrolyte system. Thirdly, combinations of sodium chloride and sodium nitrate are used, the percentage of each salt being chosen to give the best results on the material being machined. A good compromise can usually be reached between high metal removal rates and the surface finish required. Most of the stainless and heat resisting steels are machined with such mixed electrolytes.

Very weak electrolytes can be used to advantage in applications where large area tools are required and the power supply is limited. Taking for example a tool having a working area of 1290 cm² being used on a 10,000 A machine, this would allow a maximum current of 8 A/cm² to be drawn. The feed rate—which is proportional to the current being drawn—would be approximately 0.127 mm per minute. Using the normal voltage and 20% sodium chloride mix under these conditions the machining gap would be large, giving poor finish due to selective etching. Using a 2% to 3% sodium chloride mix the machining gap would be reduced to approximately 0.245 mm, giving good surface finish as there would be a high voltage gradient across the gap.

These electrolytes are normally used at a temperature of 40°C. This temperature gives good conductivity yet still leaves capacity to absorb the heat generated in the machining process. The electrolyte temperature is controlled to ± 2°C in order to maintain constant conductivity and workpiece dimensions. As stated on page 278 a change of 5°C in the electrolyte temperature can give a change of working gap of 0.0305 mm when working on a gap of 0.508 mm.

These salt electrolytes are neutral solutions and have a pH of 7. However, during the machining process, particularly if machining conditions are inefficient and a certain amount of electropolishing is taking place, oxygen and possibly other gases (for example chlorine when sodium chloride is being used) will be evolved at the anode surface. This will cause an imbalance in the chemical composition of the electrolyte, and the solution tends to become alkaline. This condition is more prevalent when machining nickel based alloys than when machining iron based alloys. The pH will rise to a figure of 9 or 10 or even higher if not checked, and if it cannot be reduced by normal topping-up then the solution should be replaced. When the pH of the electrolyte reaches 9 or 10 trouble can arise due to passifying of the workpiece, with the undesirable results outlined on page 271. It has been found in practice that the normal topping-up to replace losses due to evaporation and electrolyte in the sludge keeps the pH at or below the figure of 9.

Titanium is a difficult material to machine electrochemically due to its natural tendency to form a very tenacious and non-conductive oxide film on the surface. Small quantities of very active salts such as potassium fluoride are added to the electrolyte to break through this film. It is reported that some titanium alloys are being machined with a sodium chloride electrolyte but when certain other metals are present in the alloy, copper in particular, the more active salts have been found to be essential. Another electrolyte used to machine titanium is a sodium chloride mix with a small addition of hydrochloric acid. Corrosion and handling could be a problem with this electrolyte. Higher voltages than those usually used on steel alloys are required to machine the titanium alloys.

For other materials and for special applications, a wide variety of salts or acids are used as electrolytes or as additives to the more standard electrolytes[9]. These include: sodium chlorate, potassium chloride, sodium hydroxide, sodium citrate, Rochelle salt, potassium nitrate, sulphuric acid, nitric acid, hydrochloric acid.

It is not difficult to select an electrolyte which will give rapid metal removal rates for a particular material. It can, however, be more difficult to find the electrolyte which gives a good surface finish, and which gives the minimum worsening in fatigue properties. On the majority of the materials being processed by electrochemical machining it is found that prior to any post-ECM treatment the fatigue properties are from 10% to 40% lower than for the same material machined by conventional metal cutting processes. The nature of the effect is related to the material and there are recorded cases where there has been no worsening of fatigue properties after electrochemical machining[10].

It is generally agreed that the surface left by electrochemical machining is in a virtually stress-free condition. Now it is contended that the fatigue properties recorded after electrochemical machining are the true properties, whereas the fatigue properties recorded after conventional machining display enhancement due to the compressive stress left in the surface by the machining process. In many cases this is probably a correct assumption, although particularly in the case of the multi-phase heat resisting alloys, non-uniform metal dissolution can occur, giving pitting or preferential etching at the grain boundaries up to 0.02 mm deep even when the surface finish is good. This does of course have an adverse effect on the fatigue properties.

It has been found that preferential etching at the grain boundaries can be minimised or eliminated by developing a more suitable electrolyte and/or altering the machining conditions. Heat treating the material to change the grain structure to give more uniform metal dissolution can also be a remedy.

Preferential etching at the grain boundaries appears to be more prevalent when using sodium chloride electrolyte; in some cases, changing to a sodium nitrate, or sodium chloride–sodium nitrate mix, has reduced or eliminated this undesirable characteristic.

7.1 POST-ECM TREATMENT

When the fatigue properties of metal processed by electrochemical machining are required to be improved, this can be achieved quite readily by various treatments which impart residual stresses in the machined surfaces. These include vibro polishing, grit blast, glass bead blasting, vapour blast or hand polish. Care is required when hand polishing is used and steps should be taken to ensure that the whole surface has been polished. Improved finish and a very sharp cut-off in throwing power—the distance at which metal dissolution ceases—has been claimed for sodium chlorate[11]. Due to the fire hazard associated with this electrolyte it is not considered advisable to have it in general use in production shops. The salt itself will not ignite and organic combustible materials such as paper, wood, cloth, etc. will not burn if soaked in the electrolyte and kept wet. The danger occurs when organic material soaked in the electrolyte is allowed to dry. Material in this condition will ignite very easily on heavy impact, and once ignited burns at the same rate as if pure oxygen were being supplied. However, for special applications where the electrolyte is particularly suitable it could be considered, although the necessary safety precautions must then be strictly adhered to.

The other group of electrolytes are the acids. These are more conductive than the salts but are difficult to handle and are very corrosive to the machine and tooling. There is also the adverse effect of the metal in solution plating out on the tool. Acid electrolytes are nevertheless particularly suited to some applications, for example the drilling of small deep holes where, due to the high conductivity of the electrolyte, flow rates and pressures can be kept low, thereby helping to reduce vibration at the drill tip and improve directional stability. Filtering problems are also eased as the metal removed is taken into solution. The electrolyte gradually becomes contaminated with metal salts and has to be replaced when their concentration reaches an unacceptable level[12]. Titanium tools and fixtures have to be used to withstand the corrosive action of the acid. The problem of metal plating out on the tool is overcome by periodically reversing the polarity of the current, which deplates the tool.

8. Electrochemical grinding

Electrochemical grinding was probably used in industry before electrochemical forming, but until recently its development lagged behind that of the latter. The development in recent years of extremely tough heat-resisting materials—for example, the Nimonics and cast nickel alloys—which are difficult to grind by conventional methods, has given considerable impetus to the development of electrochemical grinding. The economic advantages of electrochemical grinding have been demonstrated, and its application is encountering expanding scope and stimulating fresh research to find a more complete understanding of the process. From a study of the current literature it is evident that there is still much to be learnt before there is a complete understanding of the process, its application, the composition of the grinding wheel and the methods of forming these wheels.

Electrochemical grinding is very similar to electrochemical machining in that the removal of the majority of the metal is achieved by electrolysis. The amount is generally put at 90% removed by electrolysis and the remaining 10% removed by the abrasive action of the grinding wheel, but these figures can vary considerably according to the power and feed rates being used. As in electrochemical machining, the workpiece is connected to the positive side of the DC power supply and therefore forms the anode, but whereas in machining the tool is shaped to produce the form required, in electrochemical grinding the cathode is a rotating grinding wheel consisting of an insulating abrasive set in a conductive bonded material. The electrolyte is injected between the workpiece and the grinding wheel which completes the circuit, and allows the electrochemical action to take place. The metal hydroxides formed by the electrolysis and any metal removed by the abrasive action of the grinding wheel are washed away in the electrolyte. Fig. 27 shows a schematic layout of the equipment required.

One of the most important elements in the electrochemical grinding process is the conductive grinding wheel. Electrochemical grinding wheels are now commer-

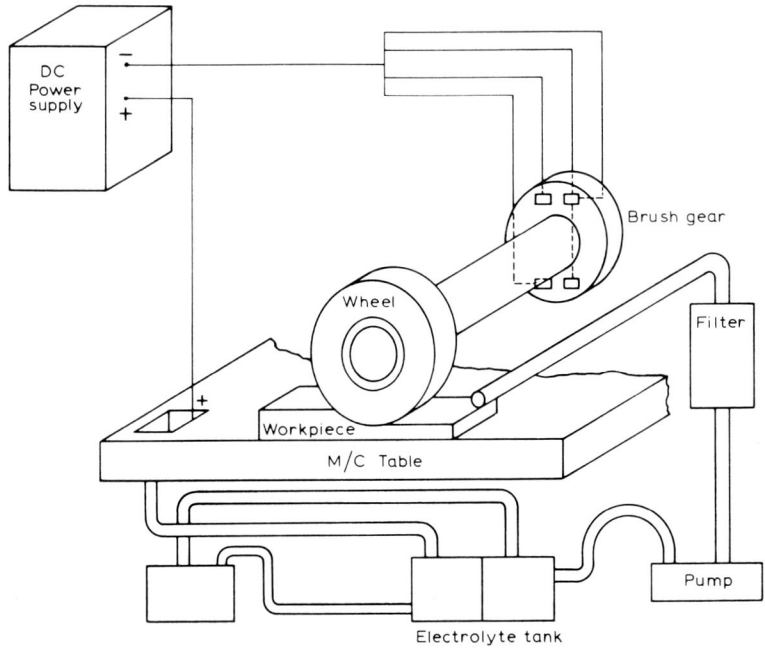

Fig. 27.
Schematic layout of electrochemical grinding machine.

cially available from most suppliers of conventional grinding wheels, and usually consist of an aluminium oxide or diamond abrasive bonded with metal, usually copper. Other types of wheel consist of an aluminium oxide abrasive bonded with graphite or diamond plated metal wheels. The abrasive in the wheel has two functions: firstly to act as a spacer between the conductive bond of the wheel and the workpiece, leaving a gap which is filled with electrolyte; secondly to remove the oxide layer formed by the electrochemical action, leaving a fresh clean surface on which the electrochemical action can take place. The power is carried to the wheel by brushes which run on the spindle or on a disc mounted on the spindle. The current then flows through the spindle to the grinding wheel. Usually the wheel head is insulated from the remainder of the machine, but in some cases the whole machine is cathodic with an insulating plate beneath a false table, or an insulating plate is in itself the fixture mounting plates.

The DC voltages required are less than those normally used for electrochemical forming, usually ranging from 4–8 V. Lower voltages are required due to the smaller working gaps, i.e. 0.0254 mm or less. The current rating of the power supplies ranges from 50 to 3000 A.

The electrolytes used for electrochemical grinding are usually less corrosive mixes such as sodium nitrate or sodium nitrite, but as in electrochemical forming all types of electrolyte mixes are in use. There are several commercial mixes being marketed; these appear to function quite well and are less corrosive than sodium chloride or sodium chloride–sodium nitrate mixes. The supply of electrolyte is less than is required for electrochemical forming but needs to be a steady even flow over the whole surface of the wheel which is removing metal. A poor supply of electrolyte will result in poor metal removal and excessive wheel wear. An excessive supply of electrolyte could result in stray etching, as in electrochemical forming. A system is required for separating the metal hydroxides from the electrolyte, and also a filter to prevent other foreign matter getting into the working gap. The units are much smaller than those required for electrochemical forming as the electrolyte flow rates are much lower.

The accuracy obtained by electrochemical grinding should be better than that obtained by electrochemical forming, and is generally in the region of 0.0254 mm to 0.0762 mm, with surface finishes from 320 to 1,200 $\sqrt{}$ (8 to 30 μ''). If closer tolerances are required, a final pass of 0.0127 mm to 0.0254 mm can be taken with the DC power switched off, allowing the grinding wheel to cut in the conventional manner. If this is done then the tolerances normally obtained by conventional grinding can be achieved. It should be borne in mind, however, that the more conventional grinding the wheel is called upon to do, the greater will be the wheel wear.

Metal removal rates vary widely according to the application, and figures of up to 0.443 mm^3 per minute per hundred amperes have been quoted. A general figure of 0.0164 mm^3 per minute per hundred amperes is suggested as a good approximation[13].

Until recently the most widely used application of electrochemical grinding has been the grinding of carbide tools. The advantages of electrochemical grinding for these tools are that no burrs are produced, surface finish is improved and there are no cracking problems, the latter condition often being present when carbide tools are ground conventionally. Economically, electrochemical grinding compares very favourably with the conventional grinding of carbide tools, savings of up to 80% on wheel costs and 50% on labour costs are quoted as being common[13]. Other applications are the cutting of stainless steel and titanium tube and bar, and the machining of stainless steel and titanium honeycomb materials. Here again the advantages of electrochemical grinding are that no burrs are formed and that there is no destruction of the fragile metal sections, this being coupled with high production rates and low wheel wear. Another application now rapidly being developed is the grinding of the difficult-to-machine materials mentioned previously as being demanded by the present generation of gas turbine engines. The rotor and stator blades of the turbine are manufactured from materials which are extremely difficult to machine by conventional grinding and milling techniques. Electrochemical grinding is being developed to perform many of the difficult operations on these components and is showing good economies in wheel life and machining times.

9. Economics

The cost of the plant for electrochemical machining is high; for example a complete unit suitable for use with a 10,000 A power unit will cost £80,000 and a 20,000 A unit will cost £100,000. Comparing this with £25,000 for a three dimensional copy milling machine, then it is clear that an economic advantage must be shown before embarking upon electrochemical machining for any particular application. There are of course cases where electrochemical machining is the only method by which the job can be done, an example of this being the drilling of the cooling holes in turbine rotor blades. In this case electrochemical drilling is the only method by which these holes can be produced, and the economic advantage is in the considerably increased life of the blade.

The operating costs of electrochemical machining differ somewhat from those of a conventional machine tool. Power consumption per cm^3 of metal removal has been found to be very similar, but in some cases is a little higher. Operator costs for each component are lower due to the high metal removal rate. Amortisation rate per hour is higher due to the high initial cost. Shop overheads are slightly higher due to the greater shop floor space required, and the additional services required. For example, electrolyte mixing and supply equipment will be required. The costs which are additional to those associated with conventional machining are the cost of the electrolyte used and the high cost of tooling.

The high capital cost of an electrochemical machining unit could not be justified for carrying out simple operations on easy-to-machine materials such as aluminium

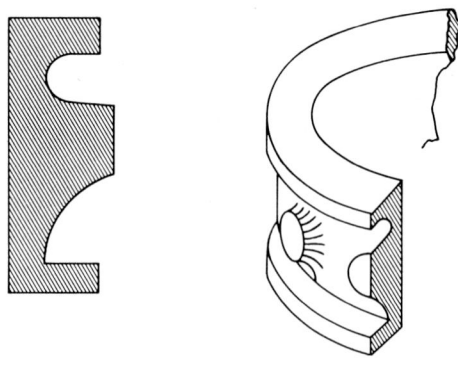

Fig. 28.
A contoured boss shape.

or plain carbon steels. The applications which show the best economic advantage are the contouring of two- or three-dimensional shapes in the difficult-to-machine materials, for example the Nimonics, the cast nickel–cobalt alloys and titanium alloys.

Taking an example of a three-dimensional contouring operation (similar to that shown in Fig. 28) in a difficult-to-machine material, it can be shown that a considerable economic advantage can be derived by using electrochemical machining. We can take for example an application where sixteen bosses are formed on a component. In view of the complex shape of the bosses, copy milling would be the only conventional method of producing them. Assuming that the end mill size would be restricted to 8 mm then only a limited depth of cut could be taken. Constant cutter changes would become necessary because of the rapid deterioration of the cutting edge. Let us then assume that the time taken to produce the sixteen bosses on this component would be 250 hours.

We can now look at the cost of producing the sixteen bosses by copy milling.

	£
Capital cost of copy milling machine, £25,000: amortised over 10 years, £2,500 per annum. Assuming 3,000 working hours per year (two shift working), the cost per hour is £0.833.	
Then the cost per component (250 hours) is:	208.25
Allowing skilled labour cost of £0.5 per hour:	125.00
Fixtures, models and cutters cost, say, £900 spread over a production quantity of 1,500 components. Cost per component:	0.6
Electrical power cost per component:	5.6
Then the machining cost per component, ignoring cutter grinding coolant and overheads, is:	£339.45

Now take the cost of the same operation carried out on a 10,000 A electrochemical machine.

The total time to produce the sixteen bosses on the component would be 12 hours.
The capital cost of a 10,000 A electrochemical forming machine, £90,000, amortised over 10 years is £9,000 per annum. Assuming 3,000 working hours per year (two shift working), the cost per hour is £3.

Then the cost per component (12 hours) is:	36.0
Allowing skilled labour at £0.5 per hour:	6.0
Tools and fixtures cost, say, £4,000 spread over 1,500 components:	2.67
Electrical power costs:	7.2
Electrolyte cost:	5.0
Then the machining costs per component, ignoring sludge disposal and overheads, is:	£56.87

These figures are calculated on a production run of 1,500 components. The cost of machinery, considering the cost of machining one component, will be £338.85 plus tooling costs, and £54.2 plus tooling costs for copy milling and for electrochemical machining respectively. For one component, then, the machining cost would be:

milling, £338.85 plus tooling costs of £900 which equals £1238.85;
electrochemical machining, £54.2 plus tooling costs of £4000 which equals £4054.2.

Considering now a production run of 10 components, the machining cost would be:

milling, £3388.5 plus tooling costs of £900 which equals £4288.5;
electrochemical machining, £542 plus tooling costs of £4000 which equals £4542.

It can be seen from these figures that at a production run of 10 components the break-even point is very close. There is, then, a point where electrochemical machining becomes an economic proposition. This is shown graphically in Fig. 29, where it can be seen that the break-even point is 11 components; thereafter electrochemical machining is progressively more economic as production quantities increase.

A point to be considered is the utilisation figures for electrochemical machines which tend to be lower than those for conventional machines. This is due to the sophisticated nature of the machine and ancillary equipment which need more maintenance. The three most time-consuming factors which make the machine utilisation low are the cleaning of filters, tool try-outs and tool fixture set-up times. As the technique and the ancillary equipment improve through feedback of infor-

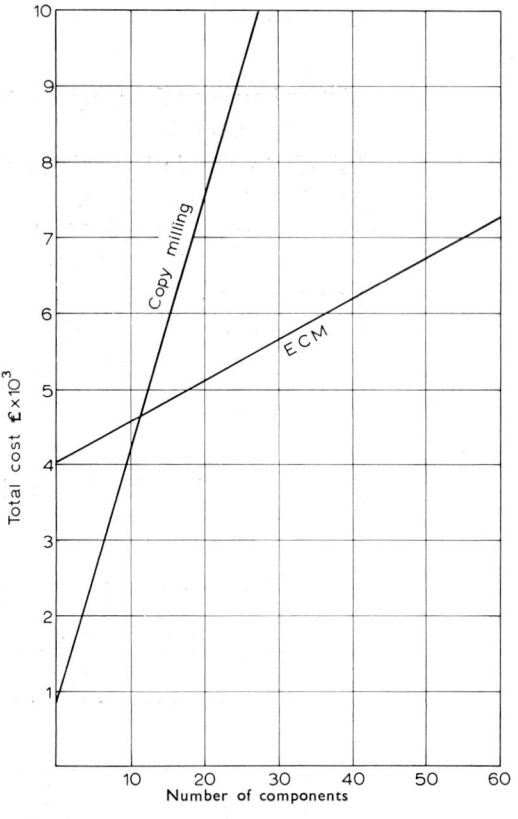

Fig. 29.
Cost comparison of conventional and EC machining.

mation to the machine tool manufacturers, and from improvements by the machine tool manufacturers themselves, then the utilisation of electrochemical machines should improve and become at least equal to that of conventional machines.

10. Applications

Electrochemical forming is probably the application most widely practised. These are the types of operation which would, if carried out by conventional machining methods, be done by two- or three-dimensional milling. Typical operations of this type are the forming of bosses, contoured surfaces, sunken cavities and blade aerofoil shapes. Another type of application is the drilling of holes having a depth-to-diameter ratio unobtainable by conventional drilling. With electrochemical drilling large numbers of holes can be drilled in one pass. Some broaching applications are being carried out to form slots in discs and to finish splined bores in hardened gears. When finishing splines in bores, the design of tools and fixtures has to be good in order to produce the fine tolerances required. Another application which is increasing in importance is electrochemical turning. As no stresses are imparted by electrochemical metal removal, very thin sections can be produced in tough materials with none of the distortion encountered when using conventional turning.

10.1 ELECTROCHEMICAL FORMING

Fig. 30 shows an engine casing in a nickel iron material. This object is a forged ring, 1066 mm in diameter and 50.8 mm thick, the weight of which is between 408.4 and 430.9 kg. Fig. 31 shows the casing after electrochemical machining; the general wall section between the contoured bosses is 3.81 mm thick. It can be seen that the aerofoil shape bosses have been produced in the bore of the casing and that the metal between these bosses has been removed. On the outside diameter of the casing, bosses of various shapes have been produced and again the metal between the bosses removed. The weight of the casing after electrochemical machining is 68.03 kg.

Fig. 30.
Engine casing before ECM.

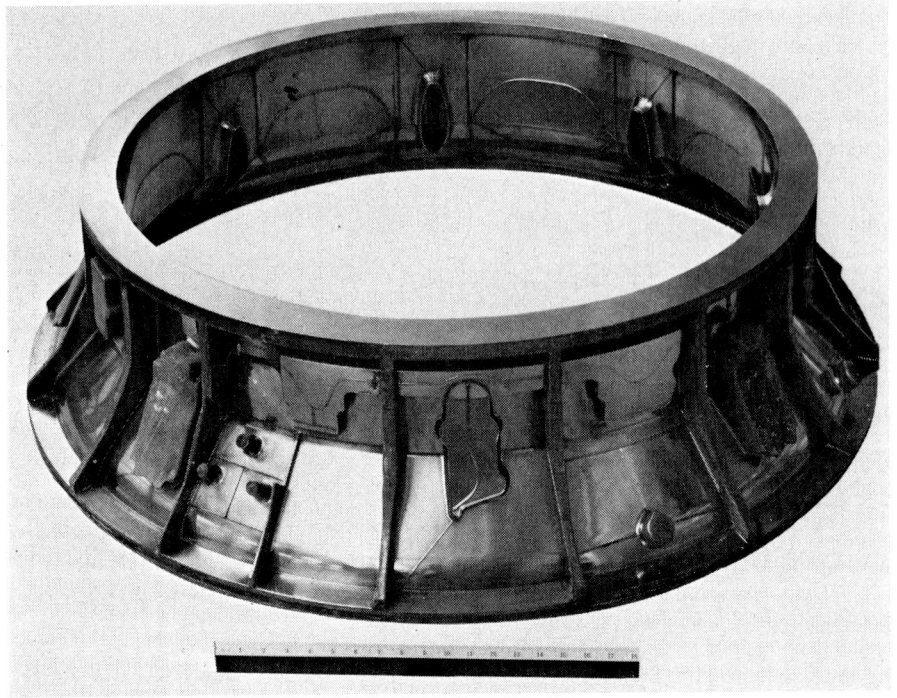

Fig. 31.
Engine casing after ECM

10.2 Electrochemical turning

Fig. 32 shows a turbine rotor disc on which the web section has been turned to 0.635 mm thick in the thinnest portion. The tool used to produce this web section is also shown, and it can readily be seen that the tool has only a fraction of the total area of the workpiece. The disc is rotated beneath the tool which is fed downwards as the metal is removed. With this method large surface areas can be machined using a reasonably small power supply, but the feed rate will of course be proportionately slower.

10.3 Electrochemical drilling

Fig. 33 shows a photograph of a turbine blade together with an X-ray photograph of the same blade, which shows the holes passing down through the aerofoil section. These holes range from 2.032 to 3.81 mm diameter and the depth of the drilled holes is 317.5 mm. The material is Nimonic 115 and holes of these diameters

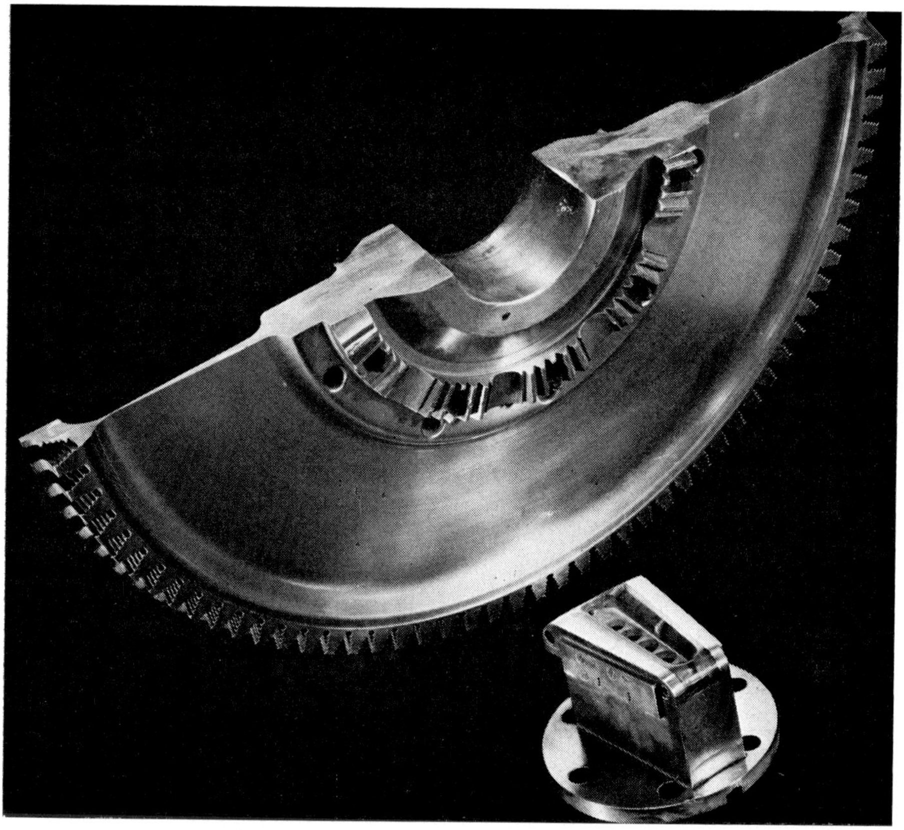

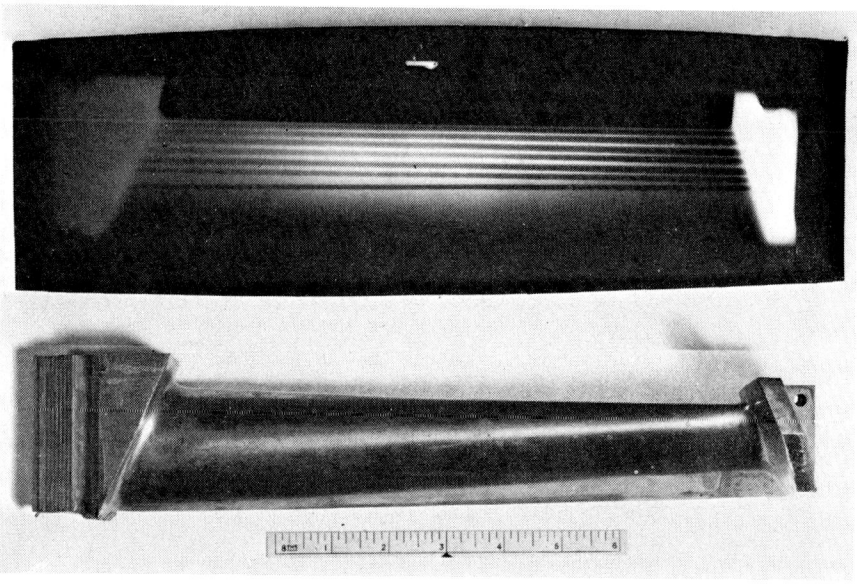

Fig. 32.
Turbine disc with web section produced by EC turning.

Fig. 33.
Turbine blade with cooling holes drilled by EC drilling.

to this depth in this material would be impossible to attain by conventional drilling methods. All the holes can be drilled simultaneously and the feed rate is 1.27 mm per minute. In similar applications, holes of 0.635 mm diameter are being drilled to a depth of 63.5 mm and holes of 3.81 mm diameter have been drilled to a depth of 609.6 mm. Shaped holes, oval, triangular, square, etc. can also be drilled as neither the drill nor the workpiece has to be rotated. The directional accuracy can be held to 0.0508 mm per 25.45 mm. Drill tip grinding, drill straightness and guide bushing must be very precise to achieve this directional accuracy.

10.4 Trepanning

If a large hole is to be produced or a slug of material removed from a large billet or sheet then a trepanning operation should be considered. It is more economical and efficient to machine away a narrow band of material than to machine away a large area. Fig. 34 shows an example where a shaped aperture has been produced in a piston engine sleeve valve. Also shown is the tool used. This operation was formerly carried out by punch and finishing milling. Electrochemical machining produced this shape in one operation in three minutes, many times faster than the punching and milling operation. The centre of the tool is a spring loaded pad made

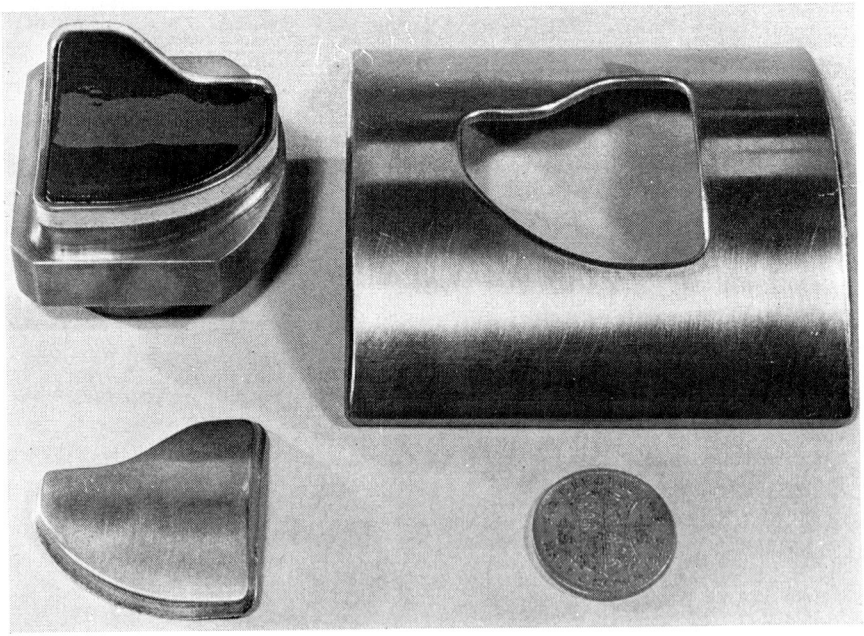

Fig. 34.
Shaped aperture produced by EC trepanning.

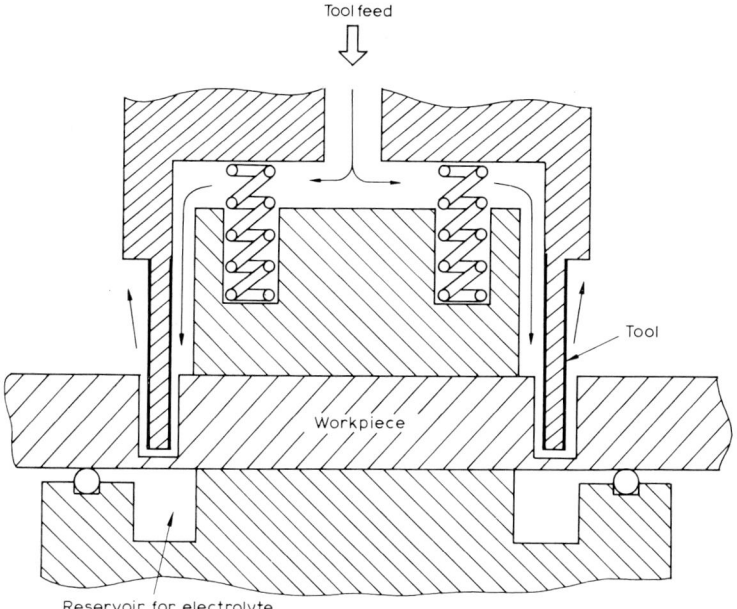

Fig. 35.
Showing tool design for trepanning operation.

of insulating material. This pad holds the material which is left in the centre to prevent movement when the tool breaks through. This is essential because if the freed centre piece were to move sideways, sparking would occur with consequent damage to the tool. Another point is that a reservoir to hold electrolyte is required where the tool breaks through. It is necessary to have this reservoir of electrolyte to allow for complete removal of excess metal (see Fig. 35). Another application where trepanning is employed is in the forming of blades on rotors where the blades are integral with a disc. In this application the centre slug is machined to the blade aerofoil shape, and of course is not removed.

11. Future developments

Electrochemical machining is rapidly becoming an established production process, but there are problems in the process which require attention. Development is required to reduce or eliminate these aspects of the process, which cause loss of machine utilisation, make plant costly and restrict applications because of lack of accuracy.

One area in which problems are encountered is in the filtration of used electrolyte. With the present filtering system there are two stages: one in which the hydroxides are separated from the electrolyte, and another when the electrolyte is filtered on its path between the high pressure pump and the tool. This second stage of filtration is required to prevent foreign matter reaching the machining gap, because if this occurs, the flow of electrolyte is interrupted and sparking will take place. These filters are usually of the mesh or cartridge type on which the metal hydroxide gradually builds up until the flow of electrolyte is restricted to an unacceptable level. In some cases, when high pressures are being used, the filters collapse, allowing unfiltered electrolyte to flow into the machining gap. At the present time these filters have to be cleaned at frequent intervals, and as this entails stopping the machine it is one of the major causes of machine down time.

Improved systems are required with facilities for quick change of filter units or a dual system in which the electrolyte can be switched when the pressure drop across the filters reaches an unacceptable level. Such a system would allow the machine to be productive whilst the filters were being cleaned or changed. The method of separating the hydroxides from the electrolyte also needs careful consideration. Centrifuges are expensive and maintenance costs can be high. The swimming pool system is costly in space. Probably the best system to develop is the accelerated settling tank combined with a filter press to give good separation and extract as much electrolyte as possible from the sludge.

Prediction of tool shape to produce the required workpiece shape is another area in which further development is required. Ideally, it should be possible to feed into a computer all the relevant details regarding the workpiece shape, material composition, electrolyte, feed rate, voltage, etc. The computer would calculate and predict the required size and shape of the tool. The state of the art at this time is such that this cannot be achieved for complicated three-dimensional shapes, although it can be done for relatively simple shapes. Many of the data for tool correction are at present based on information collected from empirical work. Much information regarding over cuts on various materials and electrolytes is being gathered together and recorded, and this will eventually make the prediction of tool shape much less of a "cut and try" solution.

The cost of tool manufacture needs to be reduced. Electroforming is one method which could help in this field, and E.H.E. Ltd. of Runcorn are active in this area (see Chapter 10, Electroforming). Forming techniques could also be investigated in an endeavour to reduce the manufacturing cost of electrochemical machining tools.

The machining parameters need to be controlled more precisely. Voltage fluctuations need to be controlled to finer limits than they are at present. Conductivity and salt concentrations in the electrolyte need to be automatically controlled. The new Anocut voltage control is a step forward in the control of the process parameters.

The existing conventional shop floor lay-out is not the best for electrochemical machining, as the floor space required is large and partitions have to be built between the machines and the high pressure pump and centrifuge to maintain an acceptable noise level. It is generally considered that a three tier system would be ideal, having on the top floor all the electrical power units, on the ground floor the machine and control console, and in the basement the electrolyte tank, pumps, filters and separators.

Production time is also lost due to tool changing. The machines being used at the present time normally have one tool mounted on the ram. It would be advantageous to have a number of tools mounted on a turret, allowing six or eight different tools to be used and thus eliminating time lost due to the changing of tools.

Efforts are required to reduce the cost of electrochemical machining equipment, but with the increasing sophistication in the control of the process parameters there appears to be little chance of the equipment becoming less expensive. The use of AC current as opposed to DC current would considerably reduce the cost of the power supply unit, but as the only practical material for tools is graphite—which does not machine away on the reverse current wave—improvements will have to be made in the material to reduce the erosion caused by the high velocity electrolyte flow.

Acknowledgements

The author wishes to thank Rolls-Royce Ltd., Bristol Engine Division, for permission to publish this information, and to express his appreciation of the assistance given by D. Glew, Senior Project Engineer ECM, Rolls-Roye Ltd., Bristol Engine Division.

12. Key patents (UK numbers)

335 003	W. Gusseff	Method and apparatus for the electrolytic treatment of metals.
563 360	British Thompson, Houston	Electrochemical machines.
704 945	Société Jacquet-Hispano, Suiza	Improvements in or relating to electrolytes and to methods for electropolishing and electrode-burring.
722 939	Toolmasters	Improvements in apparatus for producing shaped members and/or for checking the shape of members.
724 521	Sparcatron	Working metals and other materials by electrical erosion.

748 485	D. W. A. F. Rudorff	Improvements in apparatus for removing materials by electrolytic processes.
753 466	Boart Products, South Africa	Electric cutting and shaping.
753 467	Boart Products, South Africa	Method and apparatus for electrolytic cutting and shaping.
755 826	N.R.D.C.	Improvements in and relating to the removal of metal from a workpiece.
757 143	Boart Products, South Africa	Electric supply system for electrolytic grinding.
760 350	Norton Grinding Wheel Co.	Electrolytic grinding machine.
761 483	Norton Grinding Wheel Co.	Electrolytic grinding apparatus.
762 993	General Motors	Improvements in electrical cutting or grinding of electroconductive material.
765 101	Norton Grinding Wheel Co.	Machine wheel and method for the electrolytic erosion of workpieces.
770 754	Norton Grinding Wheel Co.	Electrolytic grinding apparatus.
776 046	Deutsche Edelstahlwerke	Process for electrolyte machining of cemented carbide materials.
778 053	Sparcatron	Improvements in and relating to methods and apparatus for working electrically conductive materials by electroerosion or electrolysis.
789 293	Sparcatron	Improvements in and relating to methods and means for removal of material from solid bodies.

803 887	Western Electric	Improvements in methods and apparatus for the electrolytic shaping of electrically conductive bodies.
806 805	Ernst Bisterfield	Improvements in or relating to grinding tools.
846 278	Rolls-Royce Ltd.	Manufacture of turbine and compressor blades.
846 279	Rolls-Royce Ltd.	Improvements relating to turbine and compressor blades.
846 280	Rolls-Royce Ltd.	Improvements relating to turbine and compressor blades.
854 541	Cleveland Twist Drill	Electrolytic sharpening, shaping and finishing of electrically conductive material.
861 039	Hammond Machinery Builders	Electrolyte or coolant collecting and supplying apparatus for machine tools.
892 999	Anocut	Electrolytic grinding.
927 080	Steel Improvement and Forge Co.	Electrolytic removal of material.
933 731	Rolls-Royce Ltd.	Improvements relating to the electrolytic removal of metal.
936 566	Rolls-Royce Ltd.	Electrolytic marking of metal workpieces.
936 871	Cleveland Twist Drill	Improvements in or relating to the electrolytic machining of metals.
937 263	Rolls-Royce Ltd.	Method of making an electrode for use in the electrolytic formation of a hole in a metal workpiece.

937 286	Rolls-Royce Ltd.	Improvements relating to electrolytic removal of metals.
937 287	Rolls-Royce Ltd.	Improvements relating to electrolytic removal of metals.
937 681	Rolls-Royce Ltd.	Method and apparatus for the electrolytic removal of metal.
937 737	Rolls-Royce Ltd.	Electrolytic shaping of metallic aerofoil blades.
939 402	Battelle Development	Electrolytic cutting.
943 101	Anocut	Electrolytic shaping.
944 613	Anocut	Improvements in electrolytic machining apparatus.
946 938	Anocut	Electrolysing electrode.
948 388	General Motors	Electrolytic erosion apparatus.
950 506	Rolls-Royce Ltd.	Improvements relating to electrolytic removal of metal.
952 719	General Electric	Improvements in or relating to electrode tools for electrical machinery.
954 557	Philips Electrical Industries	Improvements in or relating to devices for determining the conductivity of electrolytes.
958 401	Charmilles	Improvements in or relating to a method and an apparatus for electrolytic machining.
962 656	Anocut	Grinding and shaping internal cylindrical surfaces.

962 932	S. L. Marosi	Method and apparatus for electrolytic production of printed circuits.
967 571	H. G. Amrein & G. Wendt	Metal bonded grinding wheels.
968 239	Hammond Machinery Builders	Belt-type grinding machine.
969 175	Hammond Machinery Builders	Table support structure for grinding machine.
969 957	Anocut	Electrolytic removal of work material.
970 436	Anocut	Methods and apparatus for use in electrolytic machining.
972 620	General Electric	Improvements in or relating to electrode tools for electrical machining.
973 164	Deutsche Edelstahlwerke	A circuit arrangement for the electric generator required for electroerosive machining of metallic conducting materials.
974 338	Rolls-Royce Ltd.	Apparatus suitable for forming holes electrolytically in a metallic workpiece.
974 339	Rolls-Royce Ltd.	Electrode for forming holes electrolytically in a metallic workpiece.
976 493	General Electric	Electrode holder for replaceable electrodes for use in electrical machining.
979 968	Nippon Kogaka K.K.	Method and apparatus for electrolytically working metals and quasimetals by anodic decomposition.

981 993	Steel Improvement & Forge	Electrolytic erosion.
982 122	Ex-Cell-O	Purification of electrolyte used in electrolytic machining.
985 158	General Electric	Electrolyte for electrochemical removal of metals.
985 334	Pressed Steel	Improvements in and relating to tool for electrochemical machining.
985 676	Pressed Steel	Improvements in and relating to tool for electrochemical machining.
986 218	Anocut	Improvements in and relating to a method and apparatus for electrolytically removing material from a workpiece.
987 331	General Electric	Improvements in a cathode for electrolytic erosion of metal.
988 545	Pressed Steel	Method of machining at least one tool in a set of power press tools.
991 143	Norton Abrasives	Grinding wheels.
991 454	Anocut	Electrolytic shaping.
991 607	Anocut	Electrolytic cavity sinking apparatus and method.
991 872	Hammond Machinery Builders	Table structure for an electro-erosion apparatus.
992 392	Anocut	Electrolytic cavity sinking apparatus.
997 910	Midvale-Heppenstall	Apparatus for removing metal electrolytically from metal objects.

998 739	Anocut	Improvements in or relating to electrodes for electrolytic hole sinking.
1002 802	Soudure Electrique Languepin	Methods of and applications for electrically treating workpieces by means of an electrode.
1002 956	Anocut	Electrolytic cavity sinking apparatus and method.
1002 957	Anocut	Improvements in or relating to electrolytic machining.
1004 009	Anocut	Improvements in or relating to a process and system for electrolytically removing metal.
1004 631	Anocut	Improvements in or relating to electrolytic cavity sinking apparatus and method.
1004 829	Anocut	Improvements in or relating to a method and apparatus for use in electrolytic machining.
1005 458	Charmilles	Electrolytic removal of materials from the surface of an elongated body.
1010 640	Anocut	Improvements in or relating to electrolytic cavity sinking.
1011 147	L. V. Rozl	Device for electroerosion treatment of metals.
1012 161	Associated Engineering	Improvements in or relating to electrochemical machining.
1012 587	General Electric	Electrode for electrolytic machining.

1013 242	Ex-Cell-O	Method and apparatus for a combined machining and electrolytic grinding operation.
1014 095	Rolls-Royce Ltd.	Electrode for forming holes electrolytically in a metallic workpiece.
1014 313	B.S.A.	Electrochemical machining of metals.
1015 482	Machinenfabrik Verkheim	Contact wheel for electrolytic stock removal.
1021 447	Anocut	Electrolytic machining or hole sinking electrode.
1023 256	General Electric	Electrolytic material removal method and electrode.
1023 464	Anocut	Electrolytic machining or hole-sinking electrode.
1024 567	Cincinnati Milling Machine Co.	Electrochemical machining.
1024 699	English Electric	Formation of cavities in metals.
1025 297	Rolls-Royce Ltd.	Electrolytic machining.
1025 817	Rolls-Royce Ltd.	Electrolytic machining.
1028 247	Anocut	Electrolytic removal of metal.
1029 233	Associated Engineering	Electrochemical hole boring machine.
1029 234	Associated Engineering	Electrodes for electrochemical machining.
1031 663	Anocut	Improvements in or relating to an electrolyte for electrolytic shaping of a workpiece containing niobium, tantalum and/or vanadium.

1032 799	Beteiligungs und Patent verwaltungsgesellschaft mit Beschränkter Haftung (Essen)	Apparatus for the electrical treatment of electrically conducting workpieces.
1033 769	General Electric N.Y.	Improvements in control of electrolyte flow in electrolytic removal of metals.
1036 549	Rolls-Royce Ltd.	Electrolytic machining.
1037 661	B.S.A.	Electrochemical machining of metals.
1037 662	B.S.A.	Improvements in or relating to the electrochemical machining of metals.
1041 776	Mitsubishi Denki Kabushiki-Kaisha	Electrolytic machining process and apparatus.
1044 049	B.S.A.	Electrochemical machining of metals.
1044 224	Steel Improvement and Forge	Electrolytic removal of material.
1045 335	General Electric	Electrolytic and mechanical removal of metal.
1045 489	Pressed Steel	Improvements in and relating to methods for electrochemical machining.
1046 198	Standard Telephones and Cables	Electrolytic grinding processes.
1046 471	B.S.A.	Methods and apparatus for electrochemical machining of metals.
1050 139	M.T.I.R.A.	Electrolytic machining.
1051 590	Rolls-Royce Ltd.	An electrode for use in electrochemical machining.

1054 934	Anocut	Improvements in or relating to a workpiece-holding fixture of electrolytic shaping apparatus.
1055 170	Anocut	Improvements in or relating to a multi-phase electrolytic removal process and apparatus therefore.
1057 258	Collsman Instrument Corporation	Electrolytic machining
1058 930	Anocut	Electrolytic shaping apparatus.
1059 526	Sandvikens Jernverks	Electrolytic grinding.
1060 363	Standard Telephones and Cables Ltd.	Etching sheet metal by jet electrolyte.
1061 295	Anocut	Electrolytic polishing.
1061 442	Associated Electrical Industries	Electrolytic machining.
1062 050	Anocut	Electrolytic shaping.
1062 343	Ex-Cell-O	Electro-machining.
1062 593	Associated Electrical Industries	Improvements relating to electrolytic machining.
1063 721	Wickman Machine Tool Sales Ltd.	Electro-machining and grinding balls and like objects.
1064 508	Rolls-Royce Ltd.	Electro-machining annular slots.
1064 995	Vuma Vyskumny Ustav Mechanizacie A. Automatizacie	Electrolytic shaping and grinding.
1065 145	General Electric	Electrolytic machining.
1065 485	Rolls-Royce Ltd.	Electrolytic machining.
1066 220	Electrochemical Machines Ltd.	Electrolytically removing metal.

1066 197	Associated Electrical Industries Ltd.	Electrolytic machining.
1067 647	Sparcatron Ltd.	Electrolytic machining.
1070 912	A.G. für Industrielle Elektronik Agie Losone bei Locarno	Electrolytic working of metallic material.
1071 432	Rolls-Royce Ltd. and Ferranti Ltd.	Control of electrochemical machining.
1071 388	General Electric	Electrolytic machining.
1072 952	K. Inoue	Electrolytic machining.
1073 666	Norton Co.	Electrolytic grinding.
1074 749	Ex-Cell-O Corporation and Steel Improvement & Forge Co.	Electrolytic shaping.
1076 141	Wickman Machine Tool Sales Ltd.	Electrolytic machining of multi-toothed members.
1076 171	K. Inoue	Electrolytic machining
1076 173	Kyowa Hakko Kogyo Co.	Electrolytic etching of iron.
1076 633	K. Inoue	Electrochemical machining.
1079 982	Anocut	Electrolytic machining.
1080 065	Cincinnati Milling Machine Co.	Electro-machining holes.
1081 101	Imperial Metal Industries (Kynoch) Ltd.	Electro-machining.
1081 901	Production Engineering Research Association of G.B.	Electrolytic machining.
1081 902	Production Engineering Research Association of G.B.	Electrolytic machining.

1088 013	Sandvikens Jernverks	Electrolytic grinding.
1090 719	Siemens-Schückertwerke A.G.	Electrochemical machine tools.
1093 077	Mitsubishi Denki Kakushiki-Kaisha	Electrolytic machining, alloys.
1093 114	Mitsubishi Denki Kakushiki-Kaisha	Electrolytically machining tungsten carbide alloys.
1093 932	Associated Engineering Ltd.	Electrochemical machining.
1093 947	Associated Engineering Ltd.	Electrochemical machining.
1093 962	Associated Engineering Ltd.	Electrochemical machining.
1093 963	Associated Engineering Ltd.	Electrochemical machining.
1096 411	Wilkinson Sword Ltd.	Electrolytic forming and sharpening of razor blades.
1096 935	Deutsche Gold und Silber Scheideanstalt	Electro-machining.
1097 100	Hirst Electric Industries Ltd.	Electrochemical removal of metal.
1099 703	Bay State Abrasive Products Co.	Electrolytic grinding.
1100 928	Büro-Maschinenwerk Veb	Electro-machining.
1103 486	Associated Engineering Ltd.	Electrolytic machining.
1108 937	Associated Engineering Ltd.	Electrolytic machining tool.
1109 720	Ateliers Des Charmilles S.A.	Electrolytic machining.
1114 658	Cincinnati Milling Machine Co.	Electrolytic machining tools.
1115 816	Cincinnati Milling Machine Co.	Electrolytic machining tools.
1118 502	Phillip Morris Inc.	Electro-shaping and polishing.

1 119 881	Supfinawieck & Hentzen	Combined electrolytic and mechanical machining.
1 119 934	Norton Co.	Electrolytic grinding.
1 129 268	Morle (Kiyoshi Inoue)	Electrolytic machining.
1 129 389	Micromatic Hone Corp.	Electrolytic machining.
1 129 812	Harrison & Sons Ltd.	Electrolytic etching.
1 130 561	Rolls-Royce Ltd.	Electrolytic drilling tools.
1 143 534	G.E.C.	Electrolytic machining and metal removal process.

13. Bibliography

J. MOLLOY, *Electro-chemical machining*, Machinery, November 10th, 1965.
J. MOLLOY, *Electro-chemical machining of a turbine blade*, Machinery, December 1st, 1965.
J. MOLLOY, *Comparisons between electro-chemical and conventional methods of machining complex shapes*, Machinery, January 5th, 1966.
A study—electrolytic grinding, Grinding & Finishing, March, 1962.
J. B. POND, *Tools performance changed by grinding electrolytically*, Cutting Tool Engineering, Publication No. 2498, July, 1962.
H. R. DANIELS, *ECM A hot new cost cutter*, Metal Working, July, 1963.
R. H. ELSBELMAN, *ECM attains production status*, Iron Age, July 12th, 1962.
Publication No. M.2558, Cincinnati Milling Machine Co.
A. E. DE BARR, *Electro-chemistry in the machine shop*, The New Scientist, No. 398, July 1964.
Electro-chemical machining, paper 13, Engineering Materials and Design Conference, London, 14th November, 1963.
A. E. DE BARR, *ECM potential and problems*, Metal Working Production, November 18th, 1964.
Publication No. 2498, Cutting Tool Engineering, July, 1962 (Technifax publication).
Turning breakthrough for ECM, "*Anocut*" publication, Anocut Engineering Co., Chicago.
Elements of ECM, Cincinnati Milling Machine Co., Publication, Meta-Dynamics Division.
E. EVANS FOERTMEYER, *The versatile metal removal process*, American Society of Tool and Manufacturing Engineers, Technical Paper No. 643.
Bulletin 700, Anocut publication, Anocut Engineering Co., Chicago.
C. R. ALLISON, *How electrolytes influenced the ECM process*, American Society of Tool Manufacturing Engineers, Technical Paper SP./64/79.
D. FISHLOCK, *Shaping metals by ECM*, New Scientist, May 24th, 1962.
W. B. KLEINER, *ECM cutting fluids and "chemical chip" disposal*, Automotive Engineering Congress, Detroit, Michigan, January 14th–18th, 1963.
R. R. COLE, *Two reports shed light on electrolytic grinding*, Machinery, March, 1961 (U.S.).
R. R. COLE, *An experimental investigation of the electro-chemical machinery*, Trans. ASME, May, 1961 (U.S.).
Applying electro-chemical machinery, Production Technology, June, 1963.
Machining thermal-resistant high-strength materials, Western Machinery and Steel World, December, 1960.

Electro-chemical machining equipment results from cooperative efforts of three organisations, Machinery, September, 1963.

"*Etchineering*" *contract chemical milling service, Machinery*, August 28th, 1963.

C. L. FAUST, *Electro-chemical machining of metals*.

L. A. WILLIAMS, *Turning–breakthrough for ECM, American Machinist Metalworking Manufacturing*, July 23rd, 1962 (U.S.).

D. FISHLOCK, *What electro-chemical machining can do, Metal Working Production*, December 7th, 1960 and October 3rd, 1962 (U.S.).

W. A. HAGGERTY, *Electro-chemical machining, Metal Working Production*, October 3rd, 1962 (U.S.).

W. B. KLEINER, *Technical paper, American Society of Tool and Manufacturing Engineers*, 1962-2 (U.S.).

L. A. WILLIAMS, *Anode cutting, science, art and skill, Am. Soc. Mech. Engrs.*, 62 Prod. 12.

Anocutting, Machine Tool Review, November–December, 1962.

ECM, Aircraft Production, December, 1962 and February, 1961.

C. L. FAUST, *Electro-chemical machining of metals, Batelle Technical Reviews*, January, 1963, Vol. 12, No. 1.

L. A. WILLIAMS, *Electrolytic machining. A new competitor for numerical control, American Society of Tool and Manufacturing Engineers*, No. 44, September, 1963.

D. FISHLOCK, *ECM What it can do, Metal Working Production*, December 21st, 1960.

ECM takes big bites at low costs says firm, Steel, September 28th, 1964.

L. A. WILLIAMS, *Review of electrolytic machining*, President, Anocut Engineering, Chicago.

L. A. WILLIAMS, *How to apply electrolytic machining, Tool Engineering*, December, 1959.

F. A. PITSCHKE, *General applications of ECM* (Society of Automotive Engineers), Hanson van Winkle Co., Rockford-Beloit Section, May 8th, 1962.

Electro-chemical machining, Machinery, January 18th and 24th, 1961.

Electro shaping, Aircraft Production, March, 1961.

G. ROWDEN, *Effects of non-conventional machining processes on the mechanical properties of metals, Metallurgia*, May, 1968.

J. L. WENNBERG and W. PENTLAND, *The manufacture of turbine blade aerofoil contours by electrochemical machining, Am. Soc. Mech. Engrs. Paper*, January, 1968.

G. L. BALDWIN, D. C. BROWN and J. L. GULATI, *Electro-chemical machining, The Engineer*, February 23rd, 1968.

R. K. SPRINGBORN, *Non-traditional machining processes*, Dearborn, Mich., U.S.A.

Contents of material removal by electromachining methods, Vol. 66, Book 1.

F. HUGHES and A. NOTTER, *Evaluation of the electrolytic grinding process*, Diamond Research Lab., Information L1.

Metal cutting. Electrochemical machining. Development and application of PERA electrochemical machining theory, PERA Rept. No. 159.

C. E. GLYNN and J. BAYER, *Machinability of beryllium, final report*, G.E.C. Ohio. Contract AF33 – (615)-2241, Part II.

J. A. CROSS, *Machine tool requirements for electrochemical machining*, Automatic Engineering Congress, 618C, Detroit, Mich., January 14–18, 1963.

J. A. CROSS and R. W. DRUSHEL, *Electronic controls for use with ECM*, Ex-Cell-O Corp., Report MR. 67-550.

J. A. CROSS, *Principles of electrochemical machining*, AST Report MR 67-640.

J. A. CROSS, *A discussion of electrochemical deburring*, Ex-Cell-O Corp., Report MR. 67.234.

M. A. LABODA and M. MCMILLAN, *New electrolyte for electrochemical machining, Electrochem. Technol.*, July–August, 1967.

Electrochemical machining, American Machinist, Special Report, October 23rd, 1967.

C. A. SNAVELY, *Principles of electro machining applications, ASTME*, SP. 62-39, 1962.

C. R. ALLISON, *ECM electrolyte manual*, Cincinnati Milling Machine Company, Technical Report, Proj. No. 98000-30, June 11th, 1965.

Electrolytes for electro-chemical machining, Cincinnati Milling Machine Company, Progress Report No. 2, Contract No. DA-19-085-ORD1121.

E. E. WEISMANTEL and R. H. KUHN, *Electrochemical machining of aerospace and commercial fabrications, ASTME*, MR.68-206, 5.3.68

RICHARD C. MOVICK, *Electrochemical machining titanium alloys*, ASTME, MR. 68-207, 5.3.68
JACOB B. DARLING JR., *Electrochemical turning of Inco 713 cast turbine wheels*, ASTME, MR. 68-208, 5.3.68
PETER L. RECCHIA, *Innovations in ECM technology*, ASTME, MR. 68-709, 3.12.68
SAMUEL RATMANSKY, *What is new in ECM*, ASTME, MR 68-813, 10.9.68.
C. R. STROUPE, *ECM at the crossroads*, ASTME, MR. 68-814, 10.9.68.
JAMES W. THROOP, *Sodium chlorate for ECM cavity sinking*, ASTME, MR. 68T009, 1-68.
GUY BELLOWS, *Effect of ECM on surface integrity*, ASTME, MR. T086, 9-68.
J. P. HOARE, A. LABODA, M. L. MCMILLAN and A. J. WALLACE JR., *An investigation of the differences between NaCl and NaClO_3 as electrolytes in electrochemical machining*, Electrochemical Science, February, 1969.
R. J. HACK, *Electrochemical machining electrolytes*, ASTME, MR. 69-136.
J. WAITHBERTSON and T. S. TURNER, *EC machining further studies of process variables*, Production Engineer, January, 1967.
W. B. KLEINER, *Which cutting fluids for ECM*, Metal Working Production, May 3rd, 1963.
N. D. G. MOUNTFORD, *Electrochemical machining*, Trans. Inst. Metal Finishing, 40 (1963).
J. A. GURKLIS, *Electrochemical machining of heat resisting alloys*, Cobalt 39, June, 1968.
W. B. KLEINER, *Electrochemical machining*, Techn. Proc., Am. Electroplaters' Soc., 6 (1963) 147.
N. D. G. MOUNTFORD, *Electrochemical machining*, Trans. Inst. Metal Finishing, 40 (1963) 171.
G. BOOTHROYD, *Fundamentals of metal machining*, Edward Arnold, London, 1st ed., 1965.
T. P. HOAR and G. P. ROTHWELL, *The influence of flow on anodic polishing*, I, Electrochim. Acta, 135 (1964) 9.
T. P. HOAR and G. P. ROTHWELL, *The influence of flow on anodic polishing*, II, Electrochim. Acta, 10 (1965) 403.
J. K. HIGGINS, *The anodic dissolution and electrolytic polishing of metals*, J. Electrochem. Soc., 106 (1959) 999.
J. L. ORD and J. H. BARTLETT, *Electrical behaviour of passive iron*, J. Electrochem. Soc., 112 (1965) 160.
R. A. JACQUET, *The principles and scientific applications of the electrolytic polishing of metals*, Sheet Metal Ind., 24 (1947) 2015.
J. EDWARDS, *The mechanism of electropolishing of copper in phosphoric acid solution*, J. Electrochem. Soc., 100 (1953) 189C, 223C.
W. C. ELMORE, *Electrolytic polishing*, I and II, J. Appl. Phys., 10 (1939) 724; 11 (1940) 797.
R. R. COLE, *Basic research in electrochemical machining–present status and future directions*, Intern. J. Production Research, 4 (1965) 2, 75.
R. R. COLE, *An experimental investigation of the electrolytic grinding process*, American Soc. Mech. Engineers, J. Eng. Ind. B, 83 (1961) 194.
J. FRISCH and R. R. COLE, *Surface effects and residual stresses in electrolytically ground steel*, Am. Soc. Mech. Engrs., Paper No. 61-WA-94 (1962).
R. R. COLE and H. HOPENFELD, *An investigation of electrolytic jet polishing at high current densities*, J. Eng. Ind., B 85 (1963) 395.
J. W. CUTHBERTSON and T. S. TURNER, *Electrochemical machining–study of the effects of some variables*, Production Engineer, May 27, 1966.
H. TIPTON, *The dynamics of electrochemical machining*, International Machine Tool Design and Research Conf., University of Birmingham, 1964.
C. L. FAUST, *Electrochemical machining of metals*, Trans. Inst. Metal Finishing, 41 (1964) 1.
J. BAYER, *Final Report on electrolytic machining development*, G.E.C., Cincinnati, 1964.
C. E. FOERTMEYER, *Electrochemical machining–the versatile metal removal process*, American Society Tool and Manufacturing Engineers, Paper No. 643, 1964.
J. HOPENFELD and R. R. COLE, *Electrochemical machining–prediction and correlation of process variables*, Am. Soc. Mech. Engrs. Paper No. 66, Prod.-5.
B. H. WILKINSON and G. R. STUART, *Development of electrochemical machining techniques*, Electonics and Power, December, 1964, p. 420.

14. References

1. A. E. DE BARR and D. A. OLIVER, *Electro-Chemical Machining*, Macdonald, London, 1968, pp. 178-179.
2. *Ibid.*, p. 181.
3. J. L. WENNBERG and W. PENTLAND, *The Manufacture of Turbine Blades Aerofoil Contours by Electro-Chemical Machining*, Am. Soc. Mech. Engrs. Paper, 68-GT-30 (1968); Brit. Pat. 1050139, 1037662, 1037661, 1062593, 948388, 943101, 937681, 1070351, 1062343, 1115816, 1004829.
4. A. E. DE BARR and D. A. OLIVER, *Electro-Chemical Machining*, Macdonald, London, 1968, p. 184.
5. *Power Supply Units for ECM*, Anocut Engineering Co., Chicago, Paper JR. 7, May, 1968.
6. *Anocut News Letter*, Vol. 3, No. 6, Anocut Engineering Co., Chicago, 1969.
7. J. A. CROSS and R. W. DRUSHEL, *Electronic Controls for Use with EC Machining*, Ex-Cell-O Corp., Detroit, MR. 67-550.
8. B. WILKINSON and P. WARBURTON, *Electro-Chemical Machining*, Conference on Machineability, October, 1965, Iron and Steel Institute, London.
9. The Cincinnati Milling Machine Co., *Electrolytes for Electro-Chemical Machining*, Contract No. DA-10-058-ORD-1121; Springfield Armory Facility, Progr. Rep. No. 2, SP 64079, Part 3, p. 5.
10. G. ROWDEN, *Effects of Non-Conventional Machining Processes on the Mechanical Properties of Metals*, Metallurgia, 77 (1968) 189.
11. M. A. LA BODA and M. L. MCMILLAN, *Electrochem. Technol.*, (1967) 346.
12. W. B. KLEINER, *Metal Working Production*, May 18th, 1963.
13. *Non-Traditional Machining Processes*, American Society of Tool and Manufacturing Engineers, Dearborn, Mich., 1967, p. 55.

Chapter 9

The Electrolytic Finishing of Metals

G. ISSERLIS

Head of Department of Metal Science, Polytechnic of the South Bank, London, and Past President, Institute of Metal Finishing

1. Introduction

The field of electrolytic finishing of metals embraces the following main areas:

(a) Electrodeposition (Electroplating)
(b) Electropolishing
(c) Electrocolouring
(d) Anodising.

The electro-finishing industry is remarkable in the way it ranges from the highly sophisticated fully automated plants found in the automobile and other industries, backed by the highest degree of metallurgical and electrochemical expertise, down to the simplest tanks, manually operated in the smallest conceivable premises, often by those whose knowledge of the processes is largely a mixture of hearsay and empiricism. This chapter outlines the principles underlying the main processes and the ways in which they are executed in practice.

2. Electrodeposition of metals

2.1 INTRODUCTION

Electrodeposition, commonly known as electroplating, is the deposition of metal coatings onto conducting surfaces by making these the cathode in an electrolytic cell with a suitable electrolyte containing heavy metal ions. A large variety of items, such as pieces of machinery, engineering components, various household and other articles, are plated with metal films, with the object of

(a) protecting them against corrosion and imparting a pleasing appearance to them; or

(b) endowing them with specific properties, *e.g.* hardness, wear resistance, anti-frictional properties, electrical, magnetic properties, etc.

In order to ensure good adhesion of the deposit to the substrate, there must be an intimate linkage between the atoms of both materials, and for this reason the surface of the substrate must be free from scales, grease etc. Therefore, appropriate pre-treatment of items prior to plating is an essential operation (see pp. 339-344).

Electrodeposited metals and alloys are invariably crystalline, and their microstructure, and hence their properties, depend on the composition of the electrolyte and the operating parameters (bath temperature, cathode and anode current density, pH of electrolyte, agitation etc.). However, crystal growth can rarely proceed under conditions even approaching equilibrium, so that there is insufficient time for all the metal ions to diffuse to equilibrium positions before being reduced and accommodated in the lattice of the growing crystals, with the result that electrodeposits usually exhibit lattice distortions. These can often be very severe and influence the properties of the deposit.

2.2 Mechanism and kinetics of electrodeposition

Although the techniques of electroplating are well established and processes can run very smoothly, provided that adequate control is exercised, the mechanism of ion reduction and accommodation in the lattice is extremely complex, and it is only comparatively recently that light has been shed on this subject.

The reactional steps involved are (a) transport of hydrated ions to the cathode diffusion layer, (b) their passage across the diffusion layer and partial dehydration, (c) adsorption of the partly dehydrated ions (adions) at the cathode surface and diffusion along this surface to low energy positions, (d) complete dehydration, discharge and accommodation of these ions at appropriate lattice sites. Such sites are known as growth sites or active sites, and in ideal cases are kinks in monatomic microsteps[1,2].

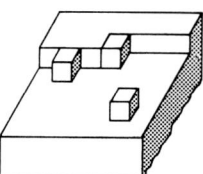

Fig. 1.
Landing and diffusion of adions along the cathode surface.

At the kink site, the ion, having completely lost its molecules of hydration, is no longer an adion, and being in contact with one half of the number of atoms it would be in contact with in the bulk of the metal, is said to be in a half-crystal position.

Two distinct reaction steps are involved in the deposition of an atom, namely (a) charge transfer across the double layer with formation of adions, and (b) surface diffusion of adions from their landing sites to monatomic steps. Each of these reaction steps may control the overall rate of reaction at a given electrode potential.

The conditions of electrodeposition determine which of these steps controls the rate of the reaction. Surface diffusion of adions is likely to be the controlling factor of the overall rate of deposition at low cathode overpotentials, whereas charge transfer should become rate-controlling with increasing current density, and hence with increased overpotential[1, 2].

The potential–log current density diagram for copper is shown in Fig. 2.

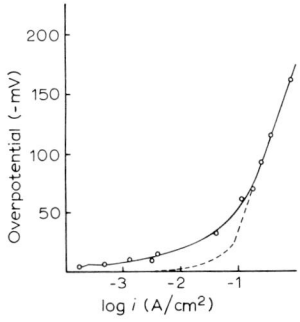

Fig. 2.
Typical Tafel line for copper electrodeposition on a polycrystalline electrode quenched in He. Broken line is that calculated for charge transfer as rate-controlling reaction step.

It can be seen that the relationship is linear at higher current densities, whereas it deviates from linearity at lower current densities. The linear relationship indicates that charge transfer is the controlling factor at higher current densities. However, where the relationship deviates from linearity, the overpotential for the same current density is higher than if charge transfer were the rate-controlling step. This is characteristic of the surface diffusion of adions being rate-controlling[3].

2.2.1 Growth along monatomic steps

Perfect monatomic steps would, during deposition, soon grow laterally out of the surface, leaving the surface without steps. Therefore, a mechanism must exist by means of which these steps are created and maintained on the surface.

There are two possibilities, (a) two-dimensional nucleation, whereby adions cluster on the surface forming "embryos", which are unstable at low overpotentials, but whose stability becomes very much greater at higher overpotentials, thereby creating a high density of monatomic steps; (b) formation and propagation of monatomic steps by incorporation of atoms at points where screw dislocations emerge on the surface[3].

Incorporation of atoms at such points will create advancing growth steps. However, the step is fixed at the point where it originates (point of emerging screw dislocation), and therefore it must rotate around that point. In this process it does not

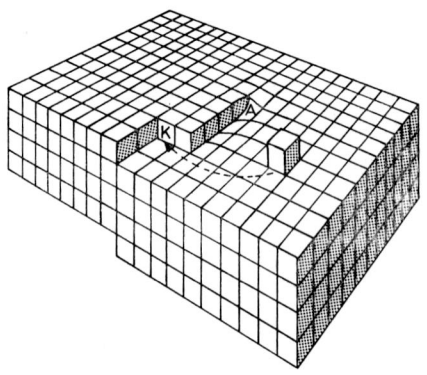

Fig. 3.
Model of a surface with a screw dislocation emerging at A. At a point where the dislocation emerges, a step originates. These steps are suitable for growth by incorporation of adions. K is a kink site.

grow out of the surface, and under favourable conditions, a step may wind itself up into a spiral which eventually, as growth proceeds, rotates as a whole around the dislocation.

The more growth steps due to screw dislocations are present on an electrode surface, the more easily will adions reach them and become incorporated into the metal lattice and the surface diffusion of adions may cease to be the rate-controlling reaction step in the overall reaction.

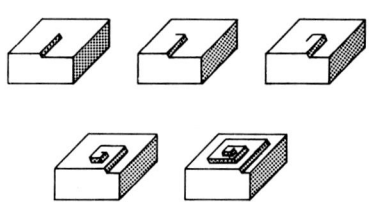

Fig. 4.
Successive stages of spiral crystal growth caused by a dislocation.

Further growth of an electrodeposit takes place by the formation of growth layers. The thickness of these layers depends on the orientation of the substrate planes and the deposition conditions. It is generally of the order of 10^{-4} to 10^{-5} cm.

Electrocrystallisation is the building-up of crystals in the deposit after the first growth layer, or "monolayer", is complete. Mechanisms and rates of electrocrystallisation on the different facets of the individual crystals composing a polycrystalline electrode may differ from one surface to another. For copper deposits the types of growth observed are pyramids, blocks, macrosteps and ridges[3].

The potential changes with time during constant current electrodeposition, particularly during the initial stages of deposition, are due to changes occurring in the morphology of the deposit and the formation of new crystal faces with different activity.

2.3 MODE OF GROWTH OF ELECTRODEPOSITS

From an overall crystallographic point of view, an electrodeposit will tend to grow on a substrate in such a manner that one of its lattice parameters will try to match that of the underlying metal. This is possible provided that the two lattices are similar in dimensions (the lattice parameter of the deposit metal must be within the limits −2.4 and + 12.5% of the substrate metal) and that the substrate surface has not suffered too great a distortion by preliminary machining or polishing.

Such a mode of growth is referred to as "epitaxy". Under favourable conditions, epitaxial growth can continue[4] up to some 30,000 Å. However, the greater the difference between the lattice parameters of deposit and substrate metal, the greater the degree of distortion of the latter and the higher the cathode current density used in deposition, the sooner will epitaxial growth cease and give way to a type of growth that is determined by operating parameters and electrolyte composition. Epitaxy is not necessarily a prerequisite for good adhesion of the deposit. Mechanical deformation of the substrate often prevents such a mode of growth. For instance, in the case of zinc deposits on heavily polished steel, epitaxial growth is impossible. Instead, alloy formation occurs at the interface between deposit and substrate by atomic diffusion of the two types of atom into each other's lattice to a depth of a few atomic diameters[5]. The cohesion between the two metals will be very strong, and in tensile testing parting will occur by rupture of the zinc, rather than detachment at the interface.

2.4 THROWING POWER OF AN ELECTROLYTE

One of the most important properties of an electroplating solution is its macro throwing power, *i.e.*, its ability to deposit a metal coating of as nearly uniform as possible thickness on a cathode surface not all areas of which are equidistant from the anode (see p. 332, the Haring Cell).

The chief factors affecting throwing power are the degree to which the cathode polarises with increase in current density, the electrical conductivity of the electrolyte and the relationship between cathode current efficiency and cathode current density. The steeper the slope of the cathode polarisation curve and the greater the electrical conductivity of the electrolyte, the more uniform will be the current distribution over the cathode surface. The uniformity of metal distribution will be the higher, the greater the decrease of cathode current efficiency with increase in current density. Any factor influencing the above conditions will therefore have a bearing on throwing power.

The temperature of the electrolyte has a pronounced effect on throwing power, since the bath conductivity increases, but the cathode polarisation decreases, with rise in temperature. Depending on which of these two factors has the greater effect, the throwing power will either increase or decrease with rise in temperature. In most cases it decreases.

Agitation of the electrolyte, by supplying metal ions to the cathode, reduces cathode polarisation and hence diminishes the throwing power.

However, both the above factors increase the conductivity of the electrolyte, and hence enable higher current densities to be used. If the cathode polarisation increases with increase in current density, the throwing power can be restored with the added advantage of a more rapid rate of deposition.

In electrolytes of simple salts, the cathode polarisation will usually be low, and such solutions can therefore only be used for plating products of simple shape, *e.g.* strip, sheet and wire. Such electrolytes will give deposits of relatively coarse crystal sizes, which are associated with a low hardness and low strength. Addition of small quantities of surface-active substances (surfactants) will often increase cathode polarisation and refine the crystal size (see pp. 333-336, 346).

Geometrical factors, such as shape and size of cell, have a bearing on current distribution.

In order to ensure a better throwing power, not only must an appropriate bath composition and operating parameters be chosen, but often resort is had to purely mechanical devices, *e.g.* the use of internal anodes for coating internal surfaces, the shape of these anodes being adjusted to those of the surfaces being plated; or the use of additional (dummy) cathodes on either side of a flat article, in order to take off excess current from the edges. Screens made of a non-conducting material are also sometimes used in order to prevent excess current from reaching specific portions of the articles being plated.

Finally, so-called "chance" factors play a part in determining throwing power. These include the type of basis metal or substrate, the condition of its surface (whether passive or active), its compositional and structural homogeneity, type of pretreatment of electrodes, etc.

2.5 Metal distribution formulae

The Haring cell is conveniently used for carrying out throwing power measurements, although other cells may also be used. It is a rectangular cell containing two sheet metal cathodes which fill the entire cross-section of the cell, and a perforated or wire mesh anode placed between the two cathodes so that the distance between it and one of the cathodes is 1/5th that between it and the other cathode. If it is assumed that the polarisation is low as compared with the potential drop in the electrolyte, and that the cathode current efficiency is 100%, the metal distribution will be determined by the inter-electrode distances, *viz.*, the weight of metal deposited on the nearer cathode (C_n) will be five times greater than that deposited on the more distant cathode (C_f).

In the Haring and Blum formula, the throwing power is evaluated from the following equation:

$$\text{T.P.} = \frac{K - C}{K} \times 100\% \qquad (1)$$

where

$$C = \frac{C_n}{C_f} = \text{metal distribution ratio},$$

and K is the ratio of distances from the more distant and the nearer cathodes to the anode. Thus, K is the current distribution ratio. In the particular arrangement described above its value is 5.

The optimum metal distribution would be that in which equal weights of metal would be deposited on the two cathodes, *i.e.* when $C = \dfrac{C_n}{C_f} = 1$.

Using Haring and Blum's formula, the throwing power will be 80% under these conditions, which represent maximum throwing power, whereas if the weight of metal deposited on the nearer cathode were to be five times that deposited on the more distant one, the throwing power would be zero.

In order to equate the optimum throwing power to 100%, Heathley proposed the following modification:

$$\text{T.P.} = \frac{K - C}{K - 1} \times 100\% \qquad (2)$$

Using this formula, the throwing power will vary between 100% and $-\infty$.

Field's formula is the one generally used in the United Kingdom since the values obtained with it range from $+100\%$ to -100%, irrespective of the value of K. According to this formula,

$$\text{T.P.} = \frac{K - C}{K + C - 2} \times 100\% \qquad (3)$$

2.6 Micro throwing power and levelling

In contrast to macro throwing power, certain electrolytes have the ability to even out the microtopography, *i.e.* to reduce the degree of roughness, of metal surfaces. This property is called the micro throwing power. In addition, a marked levelling often occurs in electrolytes which contain specific additives, for example, small quantities of coumarin in a nickel solution have a pronounced levelling effect on the microtopography. Levelling of a surface by an electrodeposit is thought to be due to different degrees of polarisation on micro excrescences as compared with micro recesses. These differences in polarisation are due either to a higher degree of adsorption of the molecules (ions) of addition agents at micro excrescences, or to their greater rate of diffusion to them.

2.7 Codeposition of foreign substances

Coatings deposited from commercial plating baths nearly always contain impurities, *e.g.* organic and inorganic colloids, anions and uncharged particles. The degree of contamination of the deposit depends on the rate of reaction of the deposit metal with the surrounding medium. On this basis, metals can be divided into three groups, namely,

(a) those which deposit with a low overvoltage (*e.g.* Ag, Sn, Zn, Cd, Th, Pb, Bi)
(b) those which deposit with a high overvoltage (*e.g.* Fe, Ni, Co, Cr, Mn, Pt), and
(c) those which cannot be deposited from aqueous solution as single metals (*e.g.* Mo, W, Zr, Nb, U, Ta).

The rate of reaction of metals with their surrounding medium is related to the surface energy of their atoms. In the body of a metal, all atoms are completely surrounded by others. At a metal surface, however, the atoms are partly surrounded by other atoms and partly exposed to the surrounding medium. Consequently, their energy is greater than the energy of the atoms inside the metal. This excess energy is termed surface energy. An indication of the magnitude of the surface energy of a metal is its melting point. The stronger the bond forces, the greater the energy required to disrupt them. The melting point of a metal is that temperature at which the disrupting forces (thermal vibrations) exceed the bond strength between the atoms. Hence, the higher the melting point of a metal, the greater its surface energy.

In an effort to satisfy its available surface energy, the metal tends to attract any substances (*e.g.* hydrogen, impurities, basic salts, etc.) available in its immediate vicinity within the surrounding medium. The greater the surface energy, the more strongly will the metal attract and bond these substances to its surface. The latter will thereby tend to be rendered passive.

Metals of group (a) have a relatively low surface energy, and there is therefore little tendency for them to become passive. Metal ions in solution have little competition from other substances for the active sites on which growth occurs. Hence, the resistance to be overcome during the discharge and accommodation process is low, and this is reflected by the low metal overvoltage accompanying deposition.

Metals of group (b) have a much higher surface energy than those of group (a) and hence have a greater tendency to passivity. This greatly reduces the number of active sites for metal atom accommodation. Nucleation will take place from a great many sites and crystal growth will be restricted. This results in a much finer crystal size than in the case of group (a) metals.

Electrodeposits of this group of metals will contain a relatively high proportion of impurities. Since their adsorptive capacity for foreign substances is high, their surface will be impure. This will result in a considerable deviation from equilibrium of the potential at the metal/electrolyte interface. The tendency of these metals to passivity will affect the rate of cathodic reduction of their ions, and is one of the main reasons for the high overvoltage. Before an ion can enter the crystal lattice it must displace any adsorbed substances from the electrode surface, and this retards its rate of reduction.

In addition, since the affinity of these metals for hydrogen is greater than that of group (a) metals, their deposition is always accompanied by a greater hydrogen evolution, which is another contributory factor to the retardation of metal ion reduction and incorporation in the lattice.

Metals of group (c) have a very high surface energy, *i.e.* they adsorb specific constituents from the surrounding medium, forming films which normally render the metal passive. Therefore, such metals cannot be deposited singly, but only in the presence of such other metals as are capable of exerting a depolarising action, so that they are deposited as alloys, and then only as very thin films.

Although foreign substances are present in many deposits, particularly in group (b) metal deposits, the lattice constant of the latter is not affected by this co-deposition, even if their content is high. The metal lattice is, however, highly stressed and this has an important bearing on the chemical and physical properties of the deposit. The potential of a coating containing impurities is usually less noble than that of the pure metal and the deposit has a lower resistance against corrosion.

Additions of various substances, particularly certain specific organic compounds, are often made intentionally to plating baths in order to modify, or completely change, the properties of the deposits. Very often these additives decompose at the cathode and their decomposition products are co-deposited.

2.8 Deposition of bright metal coatings

Deposits obtained from electrolytes free from brightening additives are normally matt. Their surface is uneven on a microscale, so that a high proportion of light reflected from them is scattered. In order to ensure maximum specular reflection of light (*i.e.* a minimum loss due to scattering), the micro roughness of a surface must be less than the shortest wavelength of visible light. Roughness on a micro scale has little effect on the brightness of a surface.

Matt deposits can be made bright by mechanical polishing provided that the substrate is smooth. However, mechanical polishing is costly in terms of labour, time and metal lost due to mechanical thinning.

Bright deposits can be obtained directly from the plating bath by incorporating specific additives into it. These are surface-active substances (surfactants) and colloids. The mechanism of their action is not yet fully understood. Among the theories proposed, two are of importance: one is based on complex formation and the other on adsorption.

According to the complex formation theory, colloid compounds form complexes with the metal cations. Owing to the strong adsorption bond between organic colloids and metal cations, discharge of the complex ions is retarded, and hence the discharge of metal ions at the cathode in the presence of colloid additives is accompanied by a high polarisation. Surfactants can be adsorbed either on the entire cathode or on preferential sites. In the former case, discharge of cations is effected across the entire film formed by the surfactant, and in the latter case, this occurs only on the free sites of the cathode surface.

The increase in cathode polarisation in the case of adsorption of surfactants occurs as a result of either a sharp reduction in size of the active cathode surface, and hence local increase in current density, or an increase in activation energy required for the penetration of cations to the cathode surface across the adsorbed layer.

The metal surface charge has an important bearing on the adsorption of both ions and neutral molecules. If the surface charge is negative, the electrostatic repulsion will oppose adsorption of anions, whilst favouring adsorption of cations. However, a positive surface charge will promote adsorption of anions.

The potential at which there is no charge on a metal surface is called the zero charge potential. In order to determine whether adsorption will occur on a metal surface at a given potential, the position of this potential relative to the zero charge potential of the metal must be ascertained. This information is very important for the selection of the appropriate additives for an electrolyte. The latter enable the range of potential over which preferential adsorption of cations, anions and molecules occurs to be widened.

Adsorption of neutral organic molecules occurs only within a specific potential range on each side of the zero charge potential of the cathode surface. Beyond that range (in either direction from the null point), desorption of these molecules occurs.

If it is known how adsorption of specific surfactants occurs on any one metal, *e.g.* mercury, the potential range over which adsorption of these substances on the surface of any other metal is likely to occur can be found with the help of the "reduced or φ-scale" proposed by Antropov[6].

The way in which electrocrystallisation proceeds is also influenced by colloidal suspensions of metal hydroxides which form in the catholyte during electrolysis in weakly acid or neutral media under those conditions when hydrogen and metal ions are reduced simultaneously. Hydrolysis is then likely to occur in the catholyte and the colloidal suspensions thus formed will possess properties characteristic of jellies. They are readily adsorbed on the metal surface and have a strong bearing on the structure and properties of the deposit.

Deposition of bright deposits requires a higher energy than that of matt ones, since the inhibition due to adsorption of foreign substances at the cathode must be overcome. Deposition of bright deposits therefore generally takes place at higher overpotentials (*i.e.* increased polarisation).

2.9 Effect of dissolved hydrogen

Dissolved hydrogen has a marked effect on the structure of the deposit. The mechanism of inclusion of this element in the deposit can be diverse. One of the methods by which hydrogen can enter the deposit is by adsorption of atomic hydrogen at the surface during metal deposition. The adsorbed hydrogen partly recombines and is converted to its molecular form, and partly it enters the crystal lattice of the metal where it will occupy atomic positions or be accommodated interstitially, forming solid solutions. Direct incorporation of hydrogen ions into the crystal lattice in the form of protons is also possible. Yet another method of hydrogen entry can be by formation of chemical compounds with the metal (hydrides).

The specific mode of hydrogen entry into a metal deposit depends on the metal involved and on the parameters of deposition. Although the total hydrogen content of an electrodeposited coating is low, the crystal lattice suffers considerable deformation. Large internal stresses are set up in the deposit, which often result in its becoming brittle.

In addition, atomic hydrogen may diffuse into the basis metal, followed by the migration of these atoms to, and their retention in, imperfections in the crystal lattice itself or in much larger mechanical discontinuities, for example those associated with non-metallic inclusions. In these "collectors" the atoms unite to form molecules, and it is the significant increase in volume associated with this which leads to the creation of very high stresses in the vicinity of the gas pockets.

In ductile metals this phenomenon can sometimes be accommodated by plastic deformation, but when, as in high-tensile steels, the ductility is low, cracking may occur. According to circumstances, cracks thus formed can range from tiny fissures to complete and often sudden cleavage through a heavy section.

Baking at the highest temperature that will not of itself injure the component, immediately after the plating operation, will cause most of the hydrogen to diffuse out. This treatment may be effective provided that cracks, however small, have not already formed during plating, previous pickling treatments, or cathodic cleaning.

2.10 Internal stresses in electrodeposits

Internal stress appears to be unrelated to any other physical property of a metal, such as strength, ductility, brightness, etc. In a deposit it manifests itself by the latter being either stretched or compressed as compared with its normal state. When stretched, the deposit is in tension and tends to contract, whilst when compressed (*i.e.* in the presence of compressive stresses), it tends to expand.

The nature of internal stresses and their magnitude depend on the particular metal being deposited, on the operating parameters and on the composition of the electrolyte.

High internal tensile stresses may markedly reduce the corrosion resistance of the deposit, and in extreme cases may even cause it to crack and flake off. Such stresses will also reduce the fatigue life of a plated component, in some instances very seriously.

Internal stresses may arise from two basic mechanisms: (1) lattice misfit between the initial atomic layers of the deposit and the substrate, and (2) the manner in which the metal is deposited from the electrolyte.

In the first mechanism, the basis metal plays a dominant role, whereas in the second mechanism one or more of a number of factors may be involved. For example, foreign matter may become occluded in the deposit, or hydrogen may be co-deposited. Hoar and Arrowsmith have explained stress formation in nickel deposits on the basis of formation and behaviour of dislocations[7].

The fundamental principles and underlying methods of measuring internal stress in electrodeposits have been reviewed by Gabe and West[8].

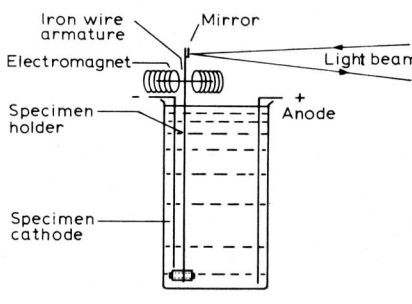

Fig. 5.
Hoar and Arrowsmith instrument (Courtesy Institute of Metal Finishing).

2.11 Measurement of internal stress

Although numerous methods have been developed for measuring internal stress in electrodeposits, only a few have found general application. Those used are based on strain measurement which is subsequently converted to stress. Practically all are modifications of Stoney's method, introduced in 1909 and itself still widely used. The latter involves plating one side only of a metal strip which is rigidly held at its upper end. The other side is stopped-off with a suitable lacquer. Tensile stresses cause the strip to curve in the direction of the plated side, while compressive stresses cause it to deflect in the opposite direction. The radius of curvature of the plated strip is a measure of the stress.

In the Spiral Contractometer, developed by Brenner and Senderoff in 1948, the strip is in the form of a helix, one end of which is fixed to the sleeve of the housing of a calibrated dial and the other end is attached to the end of a co-axial torque rod free to move within the sleeve and joined to a pointer traversing the calibrated dial. The helical form allows the use of a very long strip within a given compass, thereby greatly increasing the accuracy of measurement.

In 1957, Hoar and Arrowsmith[9] published a method which, although incorporating a completely new principle in measuring stress, is basically a variation of the bent strip test. However, instead of being allowed to deflect freely, the strip is held in its original position electromagnetically during plating. The advantage of this instrument is that stress can be recorded continuously during electroplating without allowing the strip to strain.

2.12 Hardness and wear resistance

Hardness and wear resistance are not basic physical properties, but are due to the interaction of various properties. Although the wear resistance generally increases with hardness, this is not always so, and sometimes hard deposits exhibit a lower wear resistance than softer ones.

It often happens that a metal is harder when electrodeposited than after having been severely cold-worked. Since an increase in hardness by cold work is due to lattice distortion, it is evident that the lattices of electrodeposited metals are severely distorted. Impurity incorporation strongly contributes to this lattice distortion.

2.13 Porosity

Porosity in electrodeposits can be due to numerous causes, the most common of which are non-conductive inclusions in, and roughness of, the basis metal surface.

2.14 Metal surface preparation for electroplating

In order to be effective, electrodeposits must adhere firmly to the substrate, so that any oxide films and grease must be removed from the basis metal surface prior to plating. Cleaning of the metal surface can be effected by mechanical, chemical and electrochemical methods.

Mechanical methods include grinding, polishing, scratch brushing, grit blasting, etc. Chemical and electrochemical methods include degreasing, etching, pickling and chemical and electrochemical polishing.

Mechanical methods are used to remove scales, burrs, scratches and other defects. Grinding and polishing are cutting processes, the cutting tools being the large number of hard abrasive grains embedded in the grinding and polishing wheels.

In a perfect cutting process the metal should be sheared off by the cutting tool. However, if the rate of cutting is too fast for a given peripheral wheel speed, the surface will partly shear and partly tear. Shearing produces distortion of the metal surface to a shallow depth, but tearing results in a very much greater degree of distortion which penetrates to a far greater depth than shearing. Adhesion of an electrodeposit to a *torn* surface is never satisfactory, however carefully other operations are carried out. The actual appearance of the surface of the metal is not necessarily an indication of its true state, since a surface that has been mopped after bobbing or other cutting operations can hide a deep zone of extreme distortion. The only way to rectify a torn surface and render it suitable for electroplating is to etch off the defective layer and carry out the subsequent operations afresh.

2.14.1 Chemical and electrochemical polishing

The chemical and electrochemical polishing methods have much in common. In both cases a thin oxide film forms on the metal during the polishing operation, which inhibits dissolution of the metal in the bath. The film is thinner on microexcrescences than on microrecesses. Besides, a viscous layer of reaction products between metal and solution is retained in the microrecesses. This favours preferential dissolution of the microexcrescences, thus promoting evening-out of the metal surface microirregularities.

The advantages of chemical and electrochemical over mechanical polishing include greater ease of operation and high output. However, they have a much more limited application.

The brightness attainable by chemical polishing is less than that obtained by electrochemical polishing. Chemical polishing is used in those cases where mechanical polishing is difficult and where the articles do not require a mirror-bright surface, as, for instance, small non-ferrous metal articles of complicated shape.

In electrochemical polishing the articles are made anodic. The effectiveness of this process depends on the composition of the electrolyte, parameters of electrolysis and degree of initial roughness of the metal surface. The electrolyte must be able to promote formation of a passive film on the metal surface, and the latter must not be attacked by the electrolyte in the absence of current flow. The degree of anode polarisation is of fundamental importance.

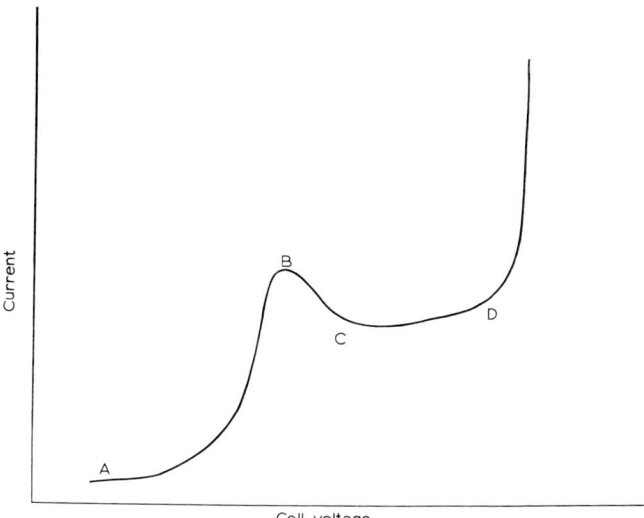

Fig. 6.
Current/cell voltage curve for electropolishing.

In Fig. 6 the portion AB of the curve corresponds to normal anodic dissolution. The rate of diffusion away from the anode of the products of anodic dissolution in this region is faster than their rate of dissolution, the difference decreasing as B is approached. Along BC the rate of dissolution is greater than the rate of diffusion, so that the products of anodic dissolution accumulate on the anode surface. As a result, the electrical resistance increases and the current decreases. The cell voltage continues to rise along CD, whilst the current remains virtually constant. A condition will be established whereby the anode layer is virtually completely saturated. Any further increase in cell voltage will lead to an increase in current, this being accompanied by evolution of oxygen. Polishing occurs along CD and beyond D.

The ability of an electrolyte to even out surface irregularities is confined to very shallow depths; hence, for satisfactory electropolishing the original surface must be relatively smooth. A certain amount of preliminary machining is often required to ensure this.

A large number of electrolytes have been developed for electropolishing, most of them being based on H_3PO_4, H_2SO_4 or CrO_3. Those based on perchloric acid are far less widely used because of the danger of explosion.

2.14.2 Degreasing

The metal surface is often covered with grease or lubricating oil from machining operations or due to handling. These greases or oil films can be either saponifiable or non-saponifiable. Animal and vegetable fats consisting of complex esters of glycerine and long-chain organic acids (*e.g.* stearic, oleic, etc.) are saponifiable. Mineral oils are non-saponifiable. They consist of mixtures of different hydrocarbons (*e.g.* vaseline, paraffin, etc.) and do not react chemically with alkalis.

Both types of grease are water-insoluble, but they can be removed either with organic solvents (*e.g.* trichlorethylene, carbon tetrachloride, etc.) or by means of chemical or electrochemical treatment in alkaline solutions. Saponifiable greases are resolved into water-soluble fatty acid salts and glycerine as a result of chemical reaction with an alkali, *e.g.*

$$(C_{17}H_{35}COO)_3 C_3H_5 + 3NaOH = 3C_{17}H_{35}COONa + C_3H_5(OH)_3.$$

Non-saponifiable compounds may, under certain conditions, form emulsions with alkalis, that is, colloidal solutions of minute droplets of oil which become dispersed in the solution and are easily removed from the metal surface. Emulsifiers are usually added to the alkali solutions in order to weaken the adhesion of the oil to the metal surface. They attach themselves to the boundary between the oil and the aqueous phase, decreasing the surface tension and thereby facilitating destruction of the oil film and promoting formation of an emulsion. Emulsification of oil and its removal is accelerated by increasing the bath temperature. A higher temperature also increases the rate of saponification of greases and prevents hydrolysis of alkali salts. Agitation also accelerates the process.

The concentration of alkali must not be too high, as the solubility of soap then decreases and the emulsion becomes less stable. Common solutions contain 5–10% NaOH or 2–5% Na_3PO_4, together with Na_2CO_3, and are operated hot.

Organic solvents dissolve both saponifiable and non-saponifiable oils and greases, all types of resin, wax, asphalt etc. For this reason metals are subjected to solvent action as the first step in degreasing. The solvent action of organic solvents is higher in the condensing vapour phase than in the liquid phase, and therefore such solvents as trichlorethylene, which have a low boiling point, form a heavy vapour and are non-inflammable, are used as vapours; they are toxic, and hence safety precautions are necessary.

After removal of the bulk of grease or oil from the metal surface by solvents, final removal of any remaining traces of surface impurities is usually effected in alkaline electrolytic cleaning solutions.

2.14.3 *Electrochemical cleaning*

Alkali baths having the compositions indicated above are used. Hydrogen is evolved in cathodic, and oxygen in anodic, cleaning at the metal surface. In cathodic cleaning, due to loss of hydrogen, the catholyte is enriched in OH^- ions, which accelerate saponification of animal and vegetable fats. Anodic cleaning is less efficient, as the anolyte becomes less alkaline, and hence the rate of saponification decreases. Also, oxygen is a less efficient remover of oil drops than is hydrogen, since hydrogen gas bubbles which are formed at the cathode attach themselves to the oil drops and carry them to the surface of the bath, whence they float off into a drainage system.

The baths are normally operated at a temperature of 60–80°C, and the current density used is 3–10 A/dm². The higher the current density, the stronger will be the gas evolution and the faster will cleaning be effected.

2.14.4 Pickling

Pickling usually follows degreasing and cleaning in those cases where the metal is covered with an oxide film.

Ferrous metals are often covered with scales or rust. Scales consist of 85–95% FeO, 5–15% Fe_3O_4 and 0.5–2% Fe_2O_3, whilst rust contains 84% Fe_2O_3, 2% FeO and 14% H_2O.

Sulphuric and hydrochloric acid solutions are used for pickling ferrous metals. Where the metal surface is covered with a scale or rust, dissolution of the metal and oxides occurs simultaneously. The following reactions occur in H_2SO_4.

$$FeO + H_2SO_4 \rightarrow FeSO_4 + H_2O$$
$$Fe_2O_3 + 3H_2SO_4 \rightarrow Fe_2(SO_4)_3 + 3H_2O$$
$$Fe_3O_4 + 4H_2SO_4 \rightarrow FeSO_4 + Fe_2(SO_4)_3 + 4H_2O$$
$$Fe + H_2SO_4 \rightarrow FeSO_4 + H_2$$
$$Fe_2(SO_4)_3 + Fe \rightarrow 3FeSO_4.$$

The time required for removal of the oxides depends on their thickness and on the acid concentration. The optimum concentration range for H_2SO_4 is 20–30% by weight, the rate of dissolution of the oxides dropping rapidly outside this range. An increase in temperature of the acid increases the rate of pickling, and therefore H_2SO_4 baths are usually operated hot.

Scale removal in H_2SO_4 solutions is due largely to the splitting-off of scale lamellae from the metal surface by hydrogen gas bubbles formed when the metal is attacked by the acid, since the rate of dissolution of the iron oxides in H_2SO_4 is slower than the rate of dissolution of the metal itself.

In hydrochloric acid, the mechanical factor is less important, since the iron oxides dissolve much faster in this acid than in H_2SO_4. The disadvantages of HCl as compared with H_2SO_4 are its higher cost and the greater acid consumption. Besides, HCl solutions should not be used at above 40°C, since hydrogen chloride separates and volatilises at higher temperatures.

The use of inhibitors in pickling solutions results in an economy of the acid and of the metal, as well as in a decrease of hydrogen embrittlement of the metal owing to the increase of the overpotential of hydrogen evolution and the ensuing decreased tendency for hydrogen reduction.

For the pickling of copper and its alloys, a mixture of HNO_3, H_2SO_4 and HCl is used.

An alternative method used for etching carbon and alloy steels is electrochemical pickling. This consists of anodic or cathodic treatment in electrolytes of specific composition. Anodic etching proceeds by electrochemical dissolution of the metal, by chemical dissolution and by mechanical detachment of the oxide films from the metal surface by the oxygen evolved.

The electrolytes used consist of acid solutions or solutions of salts of the appropriate metals. In cathodic etching the chemical action of the acids decreases due to polarisation. However, reduction and detachment of the oxide films occurs due to hydrogen evolution.

The advantages of electrolytic pickling are that the process is quicker, the acid consumption is reduced, and steels of various compositions, including those that are not readily etched chemically, can be treated.

Anodic etching is much more widely used than cathodic, since there is danger of hydrogen embrittlement of steels in cathodic etching. However, owing to the poor throwing power of the electrolytes used for anodic etching, this method is unsuitable for the treatment of articles of intricate shape. Baths for anodic etching of steels consist of either H_2SO_4 (200–250 g/l) or acidified solutions of ferric chloride or sulphate. These are maintained at temperatures of 20–60°C and operated at current densities of 5–10 A/dm^2; lead or steel is used as the cathode material.

The final pre-treatment operation prior to electroplating is dipping. This is used to remove the thin oxide films which form on pre-treated metal surfaces during short-time storage or transport, and consists in dipping the articles in dilute acid (or, for aluminium, sometimes NaOH) solutions for short periods of time immediately prior to immersion in the electroplating bath.

Electrodeposition processes

Under this heading the electrodeposition of the more common metals and alloys will be discussed. It is worth noting that for some electroplating solutions, a soluble or sacrificial anode is used, while in other solutions, the source of the metal to be plated comes from the dissolved species alone. Clearly these two cases will require different handling on the shop floor, for in the former case, the composition of the solution should (ideally) remain unchanged, whereas in the latter, the concentration of heavy metal ions decreases smoothly in accordance with Faraday's Laws.

2.15 ELECTRODEPOSITION OF COPPER

Copper as an individual coating is often used for stopping-off specific areas on steel components subjected to carburization, in order to prevent adsorption and diffusion of carbon in these areas. It is not used for protecting steel articles against corrosion because it is cathodic to iron and would accelerate corrosion at any pores or other imperfections in the coating in which iron is exposed. In addition, copper forms a green patina by reacting with the surrounding medium (mainly moisture and carbon dioxide). This is very attractive on domes which are covered with copper sheet, but would be undesirable on engineering components and general articles. Copper is, however, used as an undercoat for nickel and chromium deposits on steel and zinc-base diecastings, as it is soft and hence easily polished, and adheres well to most metals.

The electrolytes generally used for the deposition of copper are the acid sulphate and the cyanide baths, but the pyrophosphate and ethylenediamine solutions are also used for specific applications.

2.15.1 The acid sulphate bath

This bath consists essentially of $CuSO_4$ and H_2SO_4. Deposition proceeds by direct reduction of Cu^{2+} ions, but intermediate reduction to Cu^+ ions can also occur to a small extent. An equilibrium is set up in the electrolyte between Cu^{2+} ions, Cu^+ ions and the copper electrodes, and this is strongly displaced towards Cu^{2+} ions.

The sulphates of Cu(II) and Cu(I) readily hydrolyse with the formation of copper oxide which, if codeposited, would embrittle the copper deposit. For this reason, copper sulphate baths are strongly acidified with H_2SO_4. The latter has the additional advantage of increasing the electrical conductivity of the electrolyte and reducing the activity of copper ions, thus preventing the formation of a coarsely crystalline deposit at the cathode. However, it must be remembered that the solubility of copper sulphate crystals decreases with increase in H_2SO_4 in the solution, and this fact imposes a practical upper limit to the H_2SO_4 content.

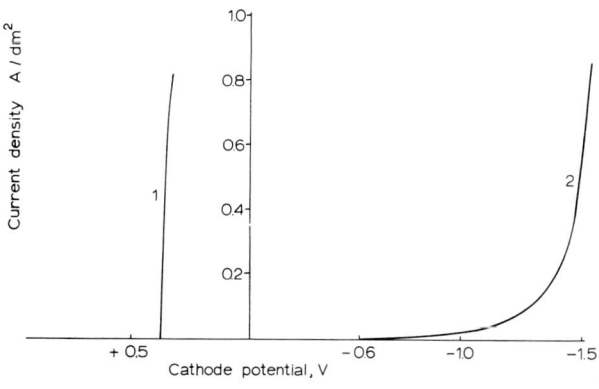

Fig. 7.
Cathode polarisation in acid sulphate and cyanide copper electrolytes. (1) 1.5 N soln. of $CuSO_4$ + 1.5 N soln. of H_2SO_4; (2) 0.5 N soln. of CuCN + 0.6 N soln. of NaCN + 0.25 N soln. of Na_2CO_3.

The cathode polarisation in acid electrolytes is very low (see Fig. 7), but can be increased by addition of small quantities of substances like dextrin, gelatin, disulphonaphthalene, etc. These additives refine the structure of copper deposits and increase the throwing power of the electrolyte.

In acid sulphate baths copper deposition proceeds at a high cathode efficiency. The electrolyte is stable and enables high current densities to be used. Its main disadvantages are the low throwing power and its inability to deposit copper directly on iron and its alloys. On immersing steel articles into the solution, iron will displace copper:

$$CuSO_4 + Fe \rightarrow FeSO_4 + Cu$$

The thus precipitated copper does not adhere to the substrate. For this reason, a thin layer of copper must be first deposited from a cyanide copper bath. The article can then be transferred to the acid bath, where a coating of the required thickness can be plated satisfactorily.

The normal composition range of industrial copper plating solutions is (in g/l):

$CuSO_4 \cdot 5H_2O$	200–250
H_2SO_4	25–50

A bath temperature of between 20 and 50°C, and a cathode current density of from 2 to 5 A/dm² can be used. Under these conditions the cathode efficiency will be 95–99%. By agitating the bath with compressed air or mechanically, a current density of up to 10–12 A/dm² can be used.

The copper anodes need not be of the highest purity, since most of the impurities contained therein are anodic to copper. Rolled anodes are preferred to cast ones, since the latter are liable to contain inclusions of cuprous oxide which form large quantities of anode slime. Electrolytic copper is not very satisfactory as anode material, since adhesion between the individual crystals is weakened during dissolution and the crystals drop out.

2.15.2 Cyanide copper electrolytes

Deposits from cyanide baths are finely crystalline (10^{-5} to 10^{-7} cm grain size), adhere well to the substrate and coat the metal uniformly.

The chief component of these baths is the complex $Na_2[Cu(CN)_3]$. This dissociates into ions as follows:

$$Na_2[Cu(CN)_3] \rightarrow 2Na^+ + [Cu(CN)_3]^{2-}$$
$$[Cu(CN)_3]^{2-} \rightarrow Cu^+ + 3CN^-$$

However, copper ions form a whole range of complexes with the CN^- anion, *viz.* $[Cu(CN)_2]^-$, $[Cu(CN)_3]^{2-}$, $[Cu(CN)_4]^{3-}$, $[Cu_2(CN)_3]^-$ $[Cu_2(CN)_5]^{2-}$, each of which can dissociate.

The degree of dissociation of all these complexes is low, and hence the activity of Cu^+ ions will be low. Accordingly, the equilibrium potential of copper in cyanide electrolytes will be by as much as approximately one volt more negative than in acid sulphate ones. For this reason, copper will not be displaced by iron in such baths.

The potential of copper in cyanide baths is rapidly depressed still further with increase in current density. The higher the concentration of free NaCN, the greater the extent to which the cathode will polarise. An increase in bath temperature will have the opposite effect.

The large cathode polarisation and accompanying decrease in cathode efficiency ensure a good throwing power for cyanide baths.

The solution is normally prepared by dissolving CuCN in a solution of NaCN. However, instead of CuCN either $CuSO_4$ or $CuCO_3 \cdot Cu(OH)_2$ (basic copper carbonate) can also be used.

The following reactions will occur on dissolving the latter two salts in a solution of NaCN:

(a) $CuSO_4 + 2NaCN \rightarrow Cu(CN)_2 + Na_2SO_4$
 $2Cu(CN)_2 \rightarrow 2CuCN + (CN)_2$

(b) $CuCO_3 \cdot Cu(OH)_2 + 8NaCN \rightarrow 2Na_2[Cu(CN)_3] + Na_2CO_3 + 2NaOH + (CN)_2$

In the above reactions, cyanide is lost. In order to avoid this loss, a reducing agent (usually Na_2SO_3) is added to the NaCN solution prior to introducing the copper salt. This reduces Cu^{2+} ions to Cu^+ ions, and the summary reaction can then be expressed by the following equation:

$$2CuSO_4 + 2NaCN + Na_2CO_3 + Na_2SO_3 \rightarrow 2CuCN + 3Na_2SO_4 + CO_2.$$

In order to ensure satisfactory working of the bath, some free (*i.e.* excess) NaCN must be present, as otherwise an insoluble film of CuCN will form at the anode, causing its complete passivation. On the other hand, a large excess of NaCN will result in a considerable drop in cathode efficiency. For this reason, a specific concentration of free NaCN must be maintained for a given electrolyte composition and operating parameters.

Na_2CO_3 will always form and accumulate in the bath due to NaCN reacting with CO_2 gas, which is normally present in very small concentrations in the air. Up to a certain limiting value, the presence of Na_2CO_3 is beneficial, since it increases the electrical conductivity of the solution.

Addition of 0.5–1.0 g/l $Na_2S_2O_3$ to the electrolyte will result in the formation of bright deposits.

The presence of potassium sodium tartrate (Rochelle salt) in the bath facilitates anode dissolution by formation of tartrate complexes, thus enabling electrodeposition of copper to be conducted at higher current densities.

The composition of a basic electrolyte will be as follows (g/l):

CuCN	40–50
Free NaCN	10–15
Na_2CO_3	30

Such an electrolyte is maintained at a temperature of 20–25°C, and deposition is effected at a cathode current density of 0.5 to 0.75 A/dm^2. The cathode efficiency under such conditions will be 60–70%.

On adding 50–60 g/l Rochelle salt and raising the bath temperature to 50°C, the cathode current density can be increased to 5A/dm^2.

Cyanide electrolytes are unstable. At the anode, apart from dissolution of copper and evolution of oxygen, cyanogen is formed:

$$2CN^- - 2e \rightarrow (CN)_2$$

The latter reacts with water, forming HCN and HCNO:

$$(CN)_2 + H_2O \rightarrow HCN + HCNO$$

HCNO hydrolyses as follows:

$$HCNO + 2H_2O \rightleftharpoons NH_3 + H_2CO_3$$

which results in the carbonation of the electrolyte and a loss in cyanide content. HCN may hydrolyse with the formation of formic acid:

$$HCN + 2H_2O \rightleftharpoons HCOOH + NH_3$$

Periodic reversal of current, with the articles to be plated being cathodically polarised for, say, 10 seconds, followed by an anodic period of, say, 1 second, ensures formation of smooth, bright deposits. Higher current densities can be used under such conditions, and no special additives are required.

2.16 Electrodeposition of nickel

Nickel deposits are used for protective–decorative purposes on engineering components, instruments, household articles, etc. They are also used for the protection of chemical apparatus against alkaline solutions, and in the printing trade for increasing the hardness and wear resistance of type-metal stereotypes and blocks. Their widest application, however, is as undercoats for chromium deposits.

Nickel can be deposited from a variety of electrolytes, but the ones in common use are the sulphate and sulphamate baths.

2.16.1 The sulphate bath

The chief constituents of the sulphate bath are $NiSO_4 \cdot 7H_2O$, H_3BO_3 and $NiCl_2 \cdot 6H_2O$.

Nickel sulphate is the main source of nickel ions. Boric acid acts as a buffer, but in addition it forms complexes with $Ni(OH)_2$ of the type $Ni(OH)_2 \cdot 2H_3BO_3$, which lower the rate of $Ni(OH)_2$ formation in the catholyte. Nickel chloride (or other chloride compounds) are added to the bath in order to ensure efficient dissolution of the anode.

The pH of the bath is very important. Since the overpotential of hydrogen evolution on nickel is small, hydrogen tends to be evolved, particularly in low pH solutions. At pH = 0, only hydrogen will be evolved at the cathode at low current densities.

On raising the bath pH, the overpotential of hydrogen evolution increases and at pH = 3 (or higher) the equilibrium potential will be displaced sufficiently to enable nickel deposition to occur. The proportion of nickel deposited will increase with increase in cathode current density, since the slope of dE/di of the potential/current density relation, which is steep at low current densities, flattens out at higher current densities.

Since the polarisation for the reduction of nickel ions decreases with rise in bath temperature to a greater extent than that for the reduction of hydrogen ions, the temperature factor exerts a considerable influence on the relative rates of nickel and hydrogen reduction. The cathode efficiency of nickel increases with rise in bath temperature. The greater the Cl^- ion concentration, the greater will be the increase in cathode efficiency with rise in temperature.

An increase in nickel ion concentration also increases the cathode efficiency by displacing the equilibrium potential of nickel in the positive direction and by decreasing polarisation.

Although the likelihood of hydrogen evolution is greatly reduced by operating the plating bath at a pH of 5, it is impossible to suppress it entirely, even under optimum conditions. Evolution of hydrogen results in a decrease of the H^+ ion concentration in the catholyte, with a corresponding rise in OH^- ion concentration and pH. The limiting pH is that at which $Ni(OH)_2$ begins to be precipitated.

Surface diffusion of nickel adions on the cathode is fast as compared with the actual process by which these ions become discharged. From an addition-free nickel solution, a coarse-grained matt deposit will be obtained, and the growing crystals will tend to develop a preferred orientation of dislocations, resulting in the formation of tensile stresses in the deposit. Furthermore, an uninhibited crystal, due to a more favourable orientation or to the local availability of more nickel ions, will grow a little faster than its surrounding neighbours, and this will result in a non-level coating[10].

In order to produce a bright deposit, the growing cathode surface must be "in-

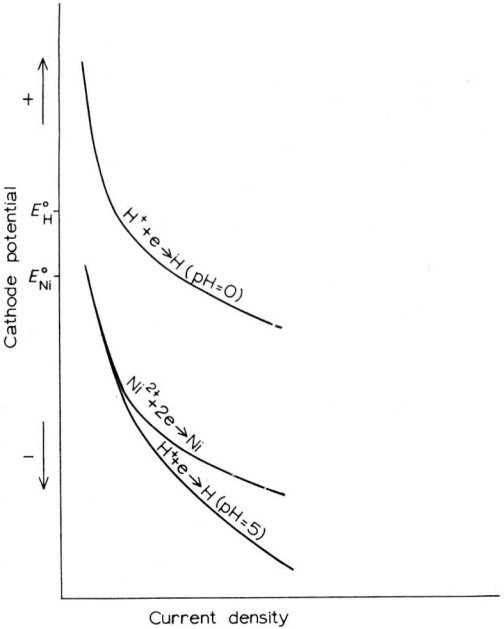

Fig. 8.
Cathodic polarisation of nickel at pH 5.

hibited" by the addition to the electrolyte of specific agents, generally organic surfactants. These invariably interfere with the metal deposition process, as evidenced by the increase in overpotential at any given current density. Some species are adsorbed preferentially at kink sites, inhibiting growth there and forcing fresh nucleation to occur at adjacent sites. This process will result in the formation of a fine-grained, randomly orientated deposit. Other types of additives may be adsorbed along the entire surface. Reduction of metal ions will then occur at any instant at a relatively few sites which constantly change owing to the very liable nature of the adsorption.

Examples of combinations of additives used in commercial electroplating solutions are: coumarin sulphate, benzene sulphonamide and saccharin (B.P. 758, 162 of 1953) or coumarin and aminopolyarylmethane (U.S.P. 2,782,152 of 1954). Each of the above constituents fulfils a definite function or combination of functions, for example brightening, stress-relieving, levelling or making the solution more tolerant towards high concentrations of certain other additives which would, on their own, restrict the operating range allowable for bright plating.

Internal tensile stresses in the deposit increase with rise in Cl^- ion content of the electrolyte.

Tensile stresses can be either reduced or eliminated by molecules that wedge themselves into some of the growing dislocations and invert their orientation. It is either these substances themselves or their decomposition products that act as stress reducers.

Yet other agents that are readily incorporated in the growing lattice, are required to produce good level surfaces.

In very many cases, additives induce some degree of brittleness in the deposit, since the incorporation of these or their decomposition products in the growing lattice greatly interferes with the mobility of dislocations, thus reducing the plasticity of the deposit[9]. The throwing power of the nickel sulphate bath is higher than that of the acid copper, but lower than that of the copper cyanide bath.

The behaviour of the nickel anode during the deposition process has an important bearing on the latter. Anodically polarised nickel easily becomes passive, and at high overpotentials, oxygen evolution will occur, even in solutions of low pH (pH 1 to 2). A loose layer of nickel sulphate, or basic nickel sulphate, is formed at the anode, often followed by the formation of a compact oxide film.

In order to ensure efficient anode dissolution, and hence a continued flow of current through the cell, chloride ions are added to the electrolyte, as already mentioned.

The current density at which anode passivity sets in depends on the ratio between the SO_4^{2+} and Cl^- ion concentrations in the bath. The higher the relative Cl^- ion concentration, the higher the current density that can be used.

Cast anodes dissolve more readily than other forms of anode, but they produce a large amount of slime. Rolled anodes corrode less readily but more evenly. Electrolytic nickel anodes also form slime. In order to prevent the slime from entering the bath, the anodes are put into bags (usually nylon or terylene). The so-called "depolarised" nickel anodes contain small proportions of NiO and carbon in order to promote their dissolution further.

A typical composition of an acid sulphate electrolyte is as follows (in g/l):

$NiSO_4.7H_2O$	250
$NiCl_2.6HO$	45
H_3BO_3	30

Such an electrolyte would be maintained at a pH of 5 and be operated at a temperature of up to 60°C and at a cathode current density of up to $4 A/dm^2$.

An electrolyte containing brighteners would have a similar composition, but a somewhat higher current density can be used. Baths producing semi-bright deposits contain non-sulphur-bearing additives, whereas fully bright deposits result from the addition of specific sulphur-bearing organic compounds to the plating solution.

2.16.2 The sulphamate bath

Most sulphamate electrolytes are similar in composition to sulphate baths, except that the nickel sulphate is replaced by nickel sulphamate. The higher solubility of the sulphamate salt enables baths of higher metal content to be used, with corresponding advantage in rate of deposition. The boric acid and nickel chloride fulfil the same functions as in the sulphate bath.

As in the sulphate bath, the internal tensile stress in the deposit increases with rise in chloride ion concentration of the electrolyte. Therefore, if this stress is to be kept to a minimum, the chloride ion content of the bath should be as low as possible.

Sulphamate nickel baths can be operated successfully without addition agents, in which case matt, relatively soft and ductile deposits are obtained. Additives are, however, sometimes used to eliminate hydrogen pitting, or to induce compressive stresses in the deposit. The agents used for the latter purpose also generally increase the hardness of the deposit.

Naphthalene 1,3,6-trisulphonic acid (NTS) is usually used as the stress-reducing agent. If added in sufficient quantity it will transform internal stress from tensile to compressive to a degree dependent upon the bath temperature and the current density of deposition[36].

2.16.3 Composite coatings

Inert particles are sometimes co-deposited with nickel. So-called "satin" nickel is produced by incorporating inert solid particles in the plating bath, which are kept in suspension by agitation of the bath, and which become mechanically entrained in the nickel deposit.

Whereas in satin nickel the function of the inert particles is primarily to produce a decorative effect, deposits of increased hardness and wear resistance can be obtained by codepositing very much smaller particles of such compounds as Al_2O_3, WC, refractory borides, nitrides, etc.

2.17 ELECTRODEPOSITION OF IRON AND COBALT

Electrodeposits of iron are used in the printing industry for prolonging the life of copper blocks or printing plates, and to a certain extent for repair work of engineering components. In order to increase the surface hardness and wear resistance of iron-plated items, the latter are often given a carburising treatment, followed by quenching and tempering.

The application of cobalt deposits is very limited.

As in the case of nickel, the electrodeposition of iron proceeds with a considerable polarisation. The latter decreases, whilst the cathode efficiency increases, with rise in bath temperature. Owing to the low overpotential of hydrogen evolution on iron, the pH of the electrolyte must be strictly controlled.

The electrolyte is based on ferrous salts, and Fe^{3+} ions should be absent, as they interfere with the quality of the deposit. The bath must therefore not be agitated with compressed air.

Sulphate, chloride and mixed electrolytes are used for the deposition of iron. These can be operated either cold or hot. The mixed sulphate–chloride baths produce hard, brittle deposits, whereas in hot chloride solutions soft, ductile deposits are obtained. Iron plated from the former electrolyte can contain up to 0.1% hydrogen, whereas deposits obtained from hot chloride baths have a hydrogen content of less than 0.002%.

A sulphate electrolyte may have the following composition (g/l):

$FeSO_4.7H_2O$ 420
$Al_2(SO_4)_3.18H_2O$ 100

Such an electrolyte can be operated at any temperature in the range 18–70°C at a cathode current density of 2 A/dm^2 in the lower temperature range and at up to 12 A/dm^2 in the higher range. The pH of the bath must be maintained at between 2.2 and 3.0. The cathode efficiency will be 70–90%.

The higher the bath temperature and the greater the concentration of ferrous sulphate, the faster can the plating process be carried out and the softer and less stressed will be the deposit. The softest deposits are obtained from chloride electrolytes, which may have the following composition (g/l):

$FeCl_2.4H_2O$ 350–500
NaCl   80–100
HCl   0.5–2

Such an electrolyte is operated at a temperature of 90–95°C at a cathode current density of 10–20 A/dm^2 and with a cathode efficiency of 90–98%.

On adding certain specific organic compounds to the bath, iron deposits of increased carbon content can be obtained. Such deposits can be quenched and tempered.

The anodes used are plates of mild steel. Their dissolution is accompanied by formation of slime consisting of carbon, sulphur and other impurities. This must not be allowed to enter the electrolyte, as any suspensions carried over to the cathode would cause formation of rough deposits. The anodes must therefore be bagged.

The physical and chemical properties of cobalt are intermediate between those of nickel and iron.

Electrodeposits of cobalt have no particular advantage over those of nickel, and since the cost of cobalt is much higher than that of nickel, it has so far not been used industrially as an individual coat. Cobalt can be deposited from sulphate electrolytes analogous to those used for nickel deposition.

2.18 ELECTRODEPOSITION OF CHROMIUM

Chromium is an electronegative metal ($E°_{Cr/Cr^{3+}} = -0.74V$) but has a strong tendency to passivity. Its chemical stability therefore approaches that of the noble metals. Not only does it resist the action of organic and oxidizing inorganic acids, alkali solutions, sulphur and H_2S, but it retains its brightness during long-term exposure to moist atmospheres, owing to the resistance of the transparent passive oxide film on the chromium surface to corrosive attack. This film also protects the underlying metal against oxidation at high temperatures.

Decorative electroplated chromium coats are very thin (0.00025 to 0.0005 mm). These are invariably porous, and in spite of their high chemical stability do not, on their own, offer protection to ferrous substrates; on the contrary, they accelerate corrosion of iron at the bottom of pores or cracks in the deposit. For this reason, an impervious undercoat of nickel (or sometimes copper and nickel) is deposited prior to applying chromium.

Owing to the great hardness and low coefficient of friction of electrodeposited chromium, thicker coatings are used to increase the wear resistance of cutting tools (milling cutters, drills, pressing tools, sliding parts of engines, etc.). Such coatings are referred to as "hard" chromium, and their thickness is in excess of 0.0025 mm.

Although influenced by bath composition and operating parameters, coatings thicker then 0.01 mm are generally impervious, since the pores and cracks formed at the earlier stages are bridged over during the later stages of deposition. Hard chromium coats are normally applied directly on to the steel base, *i.e.* without an intermediate nickel layer.

Specific "porous" chromium deposits are produced by etching the steel components after chromium plating. Such deposits are used on cylinder liners for automobile engines, owing to their excellent ability to retain lubricants.

Chromium is deposited from an electrolyte containing CrO_3 anhydride. An aqueous solution of CrO_3 consists of a mixture of polychromic acids, in particular H_2CrO_4 and $H_2Cr_2O_7$, which are in dynamic equilibrium with each other:

$$2H_2CrO_4 \rightleftharpoons H_2Cr_2O_7 + H_2O$$

Thus, the electrolyte contains three species of anion, $HCrO_4^-$, CrO_4^{2-} and $Cr_2O_7^{2-}$.

The $[HCrO_4]^-$ ion predominates in dilute CrO_3 solutions, whereas the $[HCr_2O_7]^-$ ion is formed preferentially in concentrated solutions, such as those that are used for plating baths. Some Cr^{3+} ions are also present. When such an aqueous solution is electrolysed, a strong film forms at the cathode, preventing the formation of a metallic deposit, because it is permeable only to the small hydrogen ions, but not to the larger $[HCr_2O_7]^-$ ions. This film consists of the basic chromium(III) chromate, $Cr(OH)CrO_4$.

In order to effect deposition of chromium metal, some acid must be added which is capable of loosening this film by complex formation. H_2SO_4 forms easily-soluble complexes with Cr^{3+}, dissociating into $[Cr_2(H_2O)(SO_4)_4]^{2-}$, $[Cr_2(H_2O)(SO_4)_5]^{3-}$ and $[Cr_2(SO_4)_6]^{6-}$ ions, and hence tends to prevent formation of this film at the cathode. Thus, during electrolysis a dynamic equilibrium will be established between two processes, namely formation and dissolution of the film.

The presence of Cr^{3+} ions in the electrolyte has an important bearing on the chromium deposition process. They must be present for the production of a satisfactory deposit, but their concentration should not be too great, since otherwise the current density range within which bright deposits are obtained would be greatly restricted and the resistance of the electrolyte would increase. Therefore excess Cr^{3+} ions should be re-oxidised to Cr^{6+} ions, and this process occurs at the anode if lead is used as the anode material, since the oxygen overvoltage on lead is high.

Although Cr^{3+} ions are formed during electrolysis as a side reaction, reduction of chromium to the metallic state proceeds directly from the hexavalent chromium ions.

Chromium deposition is characterised by a very low cathode efficiency, most of the current being used up on side reactions, *viz.* electrolysis of water and reduction of Cr^{6+} to Cr^{3+} ions at the cathode with subsequent oxidation of Cr^{3+} to Cr^{6+} at the anode.

The effect of concentration of CrO_3 and H_2SO_4 on cathode efficiency is shown in Fig. 9, from which it can be seen that the optimum ratio between CrO_3 and H_2SO_4 is approximately 100 : 1.

A high CrO_3 concentration gives a low cathode efficiency and results in a high loss of electrolyte (mainly due to drag-out). The usual concentrations of CrO_3 used

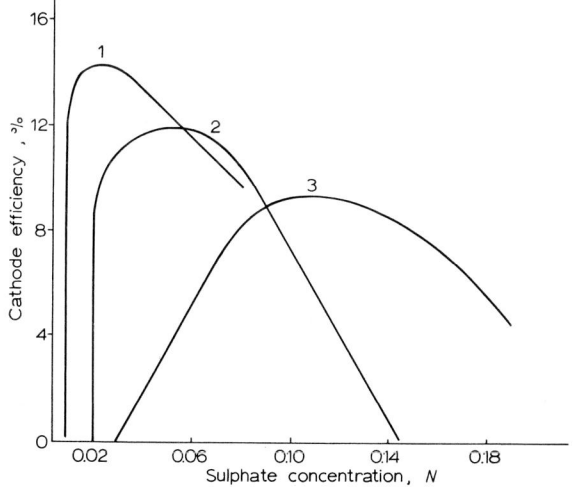

Fig. 9.
Influence of sulphate concn. on cathode efficiency at various CrO_3 contents. (1) 75; (2) 250; (3) 500 g/l CrO_3.

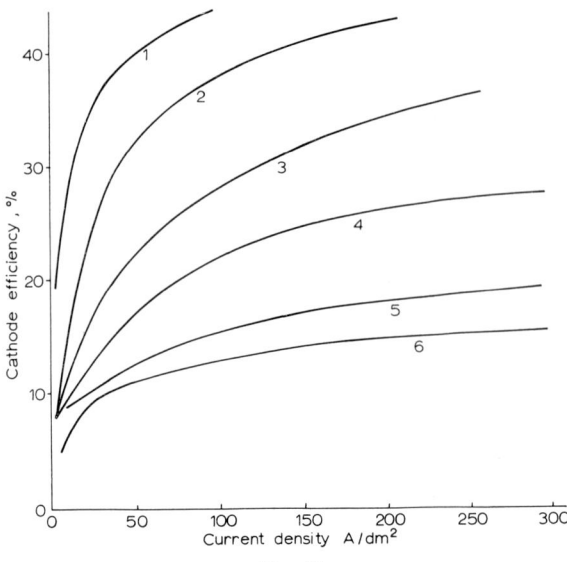

Fig. 10.
Effect of cathode current density and bath temp. on cathode efficiency. (1) 25°C; (2) 35°C; (3) 45°C; (4) 55°C; (5) 65°C; (6) 75°C.

for chromium plating are therefore 150 to 250 g/l, although up to 400 g/l is sometimes used.

Bath temperature and current density exert a considerable influence on the cathode efficiency (Fig. 10). The latter decreases with rise in temperature. These two factors also determine the external appearance of the deposit. For decorative coatings it is important that the chromium coating should be bright since it is difficult to polish electroplated chromium owing to its high hardness.

At relatively low temperatures (35–40°C), matt or grey, highly brittle deposits are obtained over a large current density range. Bright deposits are obtained in the range 45–60°C at current densities of 10 to 55 A/dm². At higher temperatures the deposits obtained tend to be milky. If the current density employed is too low for a given temperature, virtually no deposition at all will occur, the current being used up on side reactions. On the other hand, if the current density exceeds the permissible limit, dark-grey chromium will be deposited.

The throwing power of chromic acid solutions is exceedingly low.

Bath temperature and current density exert a strong influence on the hardness of the chromium deposit. Figure 11 shows the relationship between microhardness of the chromium coating and bath temperature, and it can be seen that all the curves depicted have a hardness maximum which is displaced towards higher temperatures with increase in cathode current density. Thus, by varying the operating parameters, deposits of various hardnesses can be obtained.

The wear resistance of chromium increases rapidly with increase in bath tem-

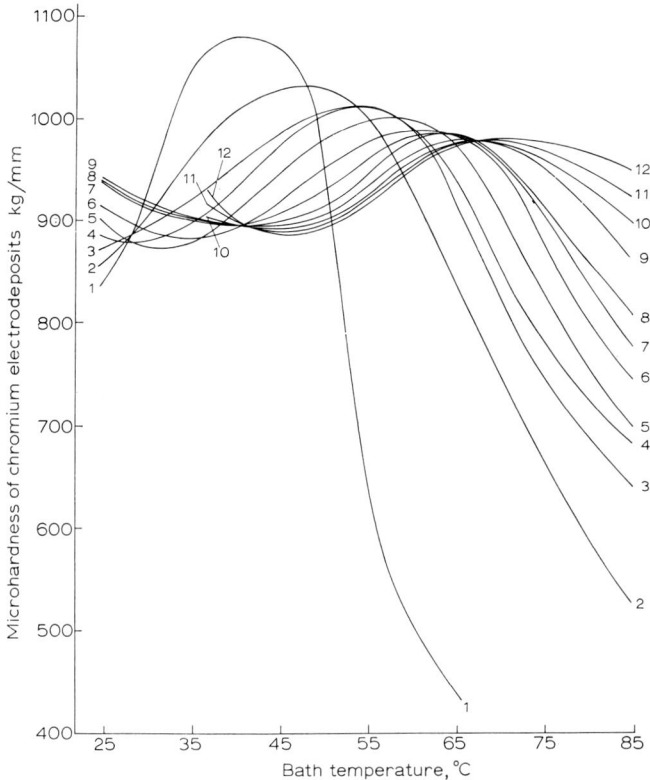

Fig. 11.
Relationship between microhardness of chromium electrodeposit and bath temp. Bath composition (g/l): CrO_3, 250; H_2SO_4, 2.5. Current density (A/dm^2): (1) 10; (2) 20; (3) 30; (4) 50; (5) 100; (6) 150; (7) 200; (8) 250; (9) 300; (10) 350; (11) 400; (12) 500.

perature, reaching a maximum at 55–65°C and thereafter dropping. There is no direct relationship between hardness and wear resistance.

The crystal structure of chromium deposits is so fine that it can be determined only by means of X-ray analysis. According to X-ray data, the grain size of bright chromium is of the order of 10^{-6} cm.

The atomic arrangement of electrodeposited chromium can be body-centred cubic (BCC) or hexagonal close-packed (HCP). Chromium deposits usually contain a considerable amount of hydrogen. In HCP chromium, the hydrogen is present in the form of an interstitial solid solution. In BCC chromium, the hydrogen is accommodated essentially along the grain boundaries. At the instant of deposition, the chromium deposit is HCP, but this lattice, being unstable, is transformed to the BCC form. This transformation is accompanied by a decrease in volume of the deposit. This mechanism may account for the formation of large internal stresses and subsequent relief by crack formation.

According to another theory[11], cracks are caused by formation of unstable chromium hydrides during plating which are spontaneously decomposed to BCC chromium and hydrogen. This is responsible for a contraction in volume and the formation of internal stress, which gives rise to crack formation after a certain thickness, the value of which depends on deposition conditions, is reached.

Another explanation for the origin of tensile stress in chromium deposits has been given by Gabe and West[27]. According to these authors, the high polarisation during chromium deposition gives rise to numerous vacant sites in the lattice which result in microscopic stresses if the vacant sites have a common orientation.

Bright deposits are characterised by a network of cracks and pores, whereas milky deposits are essentially crack-free. It is assumed that in the latter chromium is deposited directly in the BCC form.

The actual thickness of a bright chromium coating at which cracking will occur depends on the following three factors: (1) bath temperature, (2) CrO_3/H_2SO_4 ratio and (3) concentration of CrO_3 in the bath. By raising any one of these, but preferably all of them, the maximum thickness to which crack-free bright chromium coatings can be deposited will be increased. By selecting suitable conditions, films of up to 0.0025 mm in thickness can be built up in the crack-free condition.

The material used for anodes in chromium plating is usually antimonial lead. This is insoluble in chromic acid, and is employed for the following purposes:

(a) to re-oxidise Cr^{3+} ions forming during electrolysis at the cathode to Cr^{6+} (lead peroxide, which is formed, acts as an oxidiser);

(b) to reduce the likelihood of mechanical anode breakage (chromium anodes are brittle and liable to break during use);

(c) to prevent build-up of chromium ions in the bath. The efficiency of chromium anodes is several times higher than the cathode efficiency in the electrolyte, and this would result in a rapid build-up of chromium ions in solution.

During electrolysis, a fine spray of chromic acid solution is thrown up due to intense hydrogen evolution at the cathode and oxygen evolution at the anode. This is harmful to the health of the operator, so that a powerful extraction system must be provided along the upper edges of the plating vat which sucks in the spray and disposes of it. As an extra safeguard, specific additives are sometimes introduced into the electrolyte to minimise this spray.

2.19 PROTECTION OFFERED BY DECORATIVE NICKEL–CHROMIUM COATINGS AGAINST CORROSION OF THE BASIS METAL

2.19.1 Role of the nickel coat

Matt and semi-bright nickel coatings are relatively ductile, whereas bright coatings are far less ductile. Bright nickel-plated articles which invariably contain co-deposited sulphur from brighteners, should not be exposed to elevated temperatures, since at approximately 400°C the sulphur migrates to the grain boundaries, there forming nickel sulphide and thus completely embrittling the deposit and rendering it useless.

The advantages of bright over matt nickel deposits are obvious: no mechanical polishing is required prior to chromium plating, so that not only are time and manpower saved, but also loss of metal due to polishing is prevented. On the other hand, owing to the incorporated sulphur-containing brighteners, bright nickel is electrochemically more reactive, *i.e.* less corrosion-resistant, than matt or semi-bright nickel. For this reason, the direct application of bright nickel to a steel surface, followed by a thin coating of decorative chromium, would give very little protection in a corrosive environment, since the chromium surface would act as the cathode in the cell chromium/moisture/nickel, promoting rapid corrosive penetration of the bright nickel deposit and attack of the underlying basis metal.

This penetration may be considerably retarded by using a double nickel undercoat composed of a thin bright nickel superimposed on a thicker semi-bright (more corrosion-resistant) nickel. Here the bright nickel film, far from escaping attack by a corrosive atmosphere, will act as the anode to the semi-bright nickel substrate when corrosion has penetrated it. Further corrosion of the bright nickel will therefore proceed laterally and penetration to the basis metal will be markedly delayed.

Thus, the main function of any nickel coating is to retard penetration of the corrosive environment to the basis metal. In general, the thicker the coating, the more effective it will be in achieving this. A minimum total thickness of nickel of 0.03 mm on steel for very severe service conditions is specified in the British Standard 1224 : 1970.

2.19.2 Influence of the chromium top coat on the corrosion behaviour of the nickel substrate

If chromium coats of more than approximately 0.0005 mm thickness are to be used, the crack-free variety of deposit should be applied. Deposits of up to 0.0025 mm thick of this kind can be produced. These are far less highly stressed than conventional chromium coats and under normal service conditions give better protection to the substrate than the latter. The reason is that in thicker conventional chromium deposits cracks tend to form spontaneously as a result of the high internal stress, which is thus locally relieved in the top of the chromium coat, but will now consequently be concentrated at the base of the cracks, thus giving a notch effect to the nickel. If the nickel is already highly stressed, it could fail and crack itself, either immediately or after a lapse of time, aided by the joint action of stress and corrosion (nickel is anodic to chromium).

Even if the nickel does not crack, the crack pattern in the chromium may well be reproduced in the nickel by the corrosive attack occurring through these cracks, and once the nickel coat has been penetrated the basis metal itself will be attacked.

However, Safranek and Faust[12] have demonstrated that for the protection of zinc-base diecasting under *severe* service conditions, thick (approx. 0.0025 mm) *microcracked* chromium gives a markedly better corrosion resistance than crack-free chromium of the same thickness. This type of chromium deposit is obtained from the more dilute conventional electrolytes for chromium plating to which a specific concentration of addition agent has been added (*e.g.* silicon fluoride, selenic acid, etc.).

The superior performance of microcracked chromium under such conditions is due mainly to the much larger area of nickel which is exposed to a given area of chromium. This leads to more uniform corrosion and hence to a much slower rate of penetration of the nickel. The distance between cracks should be about twice the thickness of the coating, and for severe exposure, 500 to 800 cracks per cm have been found to give best results.

Microporous chromium coats can offer a degree of resistance similar to that given by nickel and microcracked chromium. Such coats are produced by adding solid particles smaller than those used for satin nickel to the nickel plating bath, and these cause pore formation in the subsequently deposited chromium coat.

Further information can be found in the British Standards Yearbook, 1969 (HMSO), or in the British Standard Specifications (B.S.S.) themselves.

2.20 ELECTRODEPOSITION OF ZINC

Zinc electrodeposits are used for the protection of ferrous metals against corrosion. They are not generally used for decorative purposes, because zinc darkens with time due to formation of basic carbonates on its surface, while under conditions of high humidity, white corrosion products of basic zinc salts form. Dipping into an acid solution of sodium dichromate markedly improves the deposit in this respect. Being anodic to iron ($E°_{Zn/Zn^{2+}} = -0.76V$) zinc protects the latter electrochemically. The rate of corrosion of zinc deposits depends upon service condition. It is relatively slow in rural areas, but can increase five to six-fold in industrial areas where the atmosphere is contaminated by SO_2, SO_3, CO_2, etc. The rate of corrosion increases drastically under conditions of high moisture content of the atmosphere and large temperature variations in which condensation and evaporation follow each other constantly, and for this reason zinc coatings do not offer adequate protection in the tropics. Their resistance to sea water is also low.

The thickness of zinc deposits required for adequate protection depends on the service conditions to which they are exposed, and varies between 0.01 mm and 0.05 mm.

Zinc can be plated from either acid or alkaline electrolytes. The most generally used acid bath is based on zinc sulphate, whereas the alkaline baths are based on zinc cyanide or sodium zincate.

2.20.1 The acid zinc sulphate bath

Owing to the high hydrogen overvoltage on zinc, the reduction of zinc ions

$$Zn^{2+} + 2e \rightarrow Zn$$

proceeds with a high cathode efficiency.

The principal constituent of the bath is $ZnSO_4.7H_2O$. The electrical conductivity of zinc sulphate solutions is low, and hence other salts must be added (*e.g.* sodium sulphate) to increase it. These also increase the cathode polarisation, and with it, the throwing power of the electrolyte.

Satisfactory zinc deposits are obtained in the pH range 3.5–4.5. At lower pH

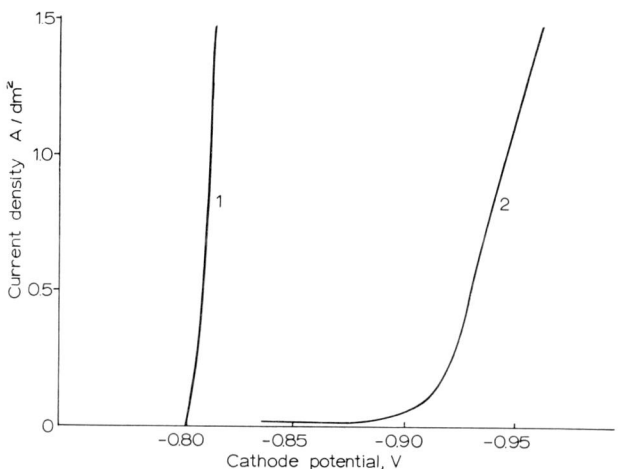

Fig. 12.
Cathode polarisation in zinc sulphate electrolytes: (1) 1.5 N $ZnSO_4$; (2) 1.5 N $ZnSO_4$ + 30 g/l $Al_2(SO_4)_3 \cdot 18H_2O$ + 10 g/l dextrin.

values the anode consumption increases and the zinc deposit may be attacked by the electrolyte, whereas at higher values $Zn(OH)_2$ may precipitate, leading to a rough deposit. In order to stabilise the pH, salts like $Al_2(SO_4)_3.18H_2O$ or $KAl(SO_4)_2.12H_2O$ are often added to the bath.

Addition of surfactants further increases the cathode polarisation and throwing power of the electrolyte, refines the structure and increases the brightness of the deposit. However, the maximum throwing power attainable in acid baths is inferior to that of alkaline electrolytes.

Acid sulphate baths may have the following composition (g/l):

$ZnSO_4.7H_2O$	200–300
$Al_2(SO_4)_3.18H_2O$	30–40
$Na_2SO_4.10H_2O$	50–100
Dextrin	10
pH	3.5–4.5

At 25°C such a bath can be operated at up to 5 A/dm² at a cathode efficiency of 95–98%.

Owing to their low throwing power, acid baths are used for plating products of simple shape, *e.g.* sheet, strip, wire, etc.

2.20.2 *The cyanide zinc bath*

Here, zinc is present in the form of the complex $Na_2[Zn(CN)_4]$ which forms on adding NaCN to any zinc salt. A white precipitate of $Zn(CN)_2$ forms initially, which dissolves in excess NaCN as follows:

$$Zn(OH)_2 + 2NaCN \rightarrow Zn(CN)_2 + 2NaOH$$
$$Zn(CN)_2 + 2NaCN \rightarrow Na_2Zn(CN)_4$$

The latter complex dissociates as follows:

$$Na_2[Zn(CN)_4] \rightarrow 2Na^+ + [Zn(CN)_4]^{2-}$$
$$[Zn(CN)_4]^{2-} \rightarrow Zn^{2+} + 4CN^-$$

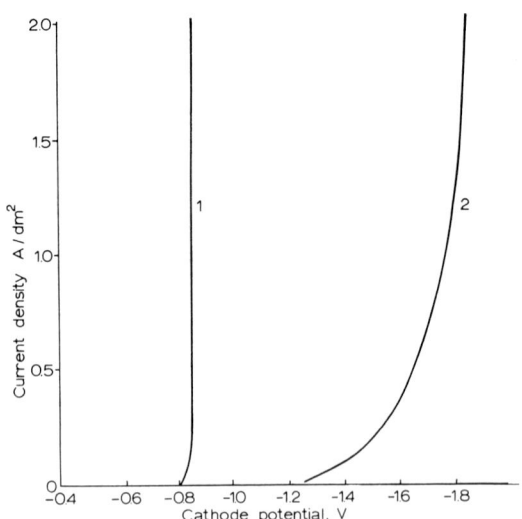

Fig. 13.
Cathode polarisation in acid sulphate and alkaline cyanide zinc electrolytes: (1) 3 N soln. of $ZnSO_4$; (2) 1 N soln. of $Zn(CN)_2$ + 4.3 N soln. of NaCN.

The $[Zn(CN)_4]^{2-}$ anion is very stable, and hence the activity of zinc ions will be extremely low. For this reason, the reduction potential of zinc in such an electrolyte will be far more negative than in acid solutions.

The cathode polarisation increases and the cathode efficiency decreases rapidly with increase in current density, and therefore the maximum cathode current density that can be employed is lower in cyanide than in acid electrolytes.

The concept of free ion discharge from complexes of the above type has met with objections, since the concentration of free ions in such electrolytes must be extremely low. It appears therefore that direct discharge must occur from adsorbed complex anions. However, in the absence of more positive information the above model of double decomposition of the complex with liberation of free metal ions can be conveniently used.

Addition of NaOH to the zinc cyanide bath enables the quantity of NaCN to be reduced. A typical plating bath will then consist of (g/l):

ZnO	40
NaCN	85–120
NaOH	40–60

Such a bath is operated between 20 and 40°C at a cathode current density of 1–4 A/dm². The cathode efficiency will be 70–80%.

Apart from the above constituents, some Na_2CO_3 is invariably present in the bath due to reaction between CO_2 in the air and NaOH and NaCN in the electrolyte.

The NaOH content of the bath must be maintained at the required level, as otherwise hydrocyanic acid will form:

$$2NaCN + CO_2 + H_2O \rightarrow Na_2CO_3 + 2HCN$$

HCN hydrolyses with the formation of formic acid:

$$HCN + 2H_2O \rightarrow HCOOH + NH_3$$

No HCN will form when the NaOH content of the bath is high:

$$2NaCN + 2H_2O + 2NaOH + O_2 \rightarrow 2Na_2CO_3 + 2NH_3$$

In either case, *i.e.* whether the NaOH content is low or high, the NaCN content of the bath diminishes during electrolysis and must be periodically corrected.

Additions of substances like glycerin, sodium hyposulphite, etc., are made to the electrolyte in order to improve the appearance of the deposit.

Since the throwing power of zinc cyanide baths is good, such baths are used for plating articles of complicated shape. However, the solution is highly toxic and precautions must be taken in handling it.

Anode dissolution proceeds freely except at high anode current densities when the NaCN content is low. Under such conditions passivation may occur due to formation of a film of undissolved salts. If the NaOH content is very high, the anodes dissolve unevenly.

2.20.3 The sodium zincate bath

Zincate baths contain basically two constituents, *viz.* sodium zincate, Na_2ZnO_2, and NaOH. Cathodic reduction of Zn^{2+} ions from Na_2ZnO_2 occurs as follows:

$$Na_2ZnO_2 \rightarrow 2Na^+ + ZnO_2^{2-}$$
$$ZnO_2^{2-} + H_2O \rightarrow Zn^{2+} + 4OH^-$$
$$Zn^{2+} + 2e \rightarrow Zn$$

The cathode polarisation of zincate baths is lower than that of zinc cyanide baths, since the ZnO_2^{2-} ion is less stable than the $[Zn(CN)_4]^{2-}$ ion. The maximum current density that will give satisfactory deposits is lower than that used in cyanide baths. Small additions of tin or lead to the bath reduce the tendency to form spongy, dendritic deposits. A typical zincate bath may have the following composition (g/l):

ZnO	4–6
NaOH	65–80
Na_2SnO_3	0.2–0.5

It is operated at a temperature of 45–55°C and at a cathode current density of 0.5–1.2 A/dm². The latter can be increased to 2.5 A/dm² by agitating the bath. The cathode efficiency is 96–98%.

Zincate electrolytes are non-toxic. Their throwing power is lower than that of cyanide solutions, but considerably higher than that of acid baths. They are therefore sometimes used as substitutes for cyanide baths.

2.21 ELECTRODEPOSITION OF CADMIUM

Cadmium deposits, like zinc deposits, are used mainly for protecting ferrous metals against corrosion. However, whereas zinc is invariably anodic to iron, cadmium can be either anodic or cathodic, depending on the environment. It is anodic in solutions containing chlorides, and therefore cadmium deposits are suitable for service under marine conditions. In addition, cadmium is chemically more stable than zinc and resists tropical conditions much better. Dipping into an acid solution of sodium dichromate further improves the resistance of cadmium deposits. The high plasticity of cadmium deposits makes them suitable for coating threaded parts.

However, cadmium is far more expensive than zinc, and this fact limits its application.

2.21.1 Cadmium electrolytes

Like zinc, cadmium is deposited from acid and cyanide electrolytes. In the former, owing to very low cathode polarisation, there is a tendency for formation of spongy deposits, and to obviate this, special organic additives are introduced into the bath.

Fine-grained, well-adhering deposits are obtained from cyanide electrolytes. Here, cadmium is contained in the complex $Na_2[Cd(CN)_4]$, the dissociation of which occurs as follows:

$$Na_2[Cd(CN)_4] \rightarrow 2Na^+ + [Cd(CN)_4]^{2-}$$
$$[Cd(CN)_4]^{2-} \rightarrow Cd^{2+} + 4CN^-$$

Owing to the stability of $[Cd(CN)_4]^{2-}$, the deposition potential of cadmium in such an electrolyte is much more negative than in an acid solution. The throwing power of such a bath is very good owing to high cathode polarisation, good conductivity of the electrolyte and a decrease in cathode efficiency with increase in current density. The deposits obtained are fine-grained, and special additions are made to the bath in order further to improve the quality of the deposits, including their brightness.

Some Na_2CO_3 is invariably present in the bath. As in the case of zinc cyanide baths, this forms during electrolysis. However, its presence is beneficial, since in the absence of this salt, dark, streaky deposits will be obtained at all current densities. For this reason, Na_2CO_3 is added to freshly prepared baths. The stability of the electrolyte can be enhanced by the addition of sodium sulphate.

A typical bath composition is as follows (g/l):

$Cd(CN)_2$	60
NaCN	120
$Na_2SO_4.10H_2O$	50
Additive	5–10

Operating conditions: bath temperature 20–40°C, cathode current density 1–4 A/dm².

A cathode efficiency of 95 to 96% will be obtained. Pure cadmium anodes are used.

2.22 Electrodeposition of tin

Tin is non-toxic, and hence it is used in the food industry (*e.g.* tinplate for canning); it is also used for the protection of household hardware goods, electrical contacts for soldering, for the protection of copper articles against sulphur in vulcanised rubber sheaths, etc.

As compared with hot tinning, electrodeposition gives coatings of more uniform thickness and there is no wastage of tin. Electrodeposition is carried out in acid and alkaline electrolytes. In alkaline solutions tin is quadrivalent, whereas in acid solutions it is divalent. For the same current, therefore, twice the weight of tin will be deposited in acid, as compared with alkaline, solutions. However, the throwing power of acid electrolytes is low and the deposits obtained are coarse-grained so that articles of complicated shape are plated in alkaline, whereas those of simple shape are plated in acid solutions.

2.22.1 Acid tin electrolytes

Sulphate and chloride solutions are used. The cathode reaction is:

$$Sn^{2+} + 2e \rightarrow Sn$$

The standard electrode potential of tin, $E°_{Sn/Sn^{2+}} = -0.136V$, but hydrogen ions are not readily reduced at this potential owing to the high overpotential of hydrogen evolution on tin. In the absence of special additives, the reduction of tin does not proceed with a strong polarisation and the deposit will consist of needle-like crystals. However, addition of such substances as phenol, cresol and its derivatives, glue etc., increases cathode polarisation considerably and refines and brightens the deposit.

Oxidation of Sn^{2+} to Sn^{4+} ions by oxygen from the air is liable to occur in a solution of low acidity:

$$SnSO_4 + H_2SO_4 + \tfrac{1}{2}O_2 \rightleftharpoons Sn(SO_4)_2 + H_2O$$

$Sn(SO_4)_2$ readily hydrolyses as follows:

$$Sn(SO_4)_2 + 4H_2O \rightleftharpoons Sn(OH)_2 + 2H_2SO_4$$

For this reason, a certain minimum concentration of H_2SO_4 must be maintained.

The properties of the deposit depend largely on operating parameters, mainly current density and bath temperature. Coarsely crystalline deposits will be formed at low current densities, whereas an increase in bath temperature will facilitate hydrolysis.

A typical bath composition is as follows (g/l):

$SnSO_4$	30–50
H_2SO_4	50–100
Cresol	6–10
Carpenter's glue	1–4

Such a bath will be operated at a temperature of 18–25°C and a cathode current density of 1–3 A/dm². The cathode efficiency will be 90–95%. Tin anodes under such conditions normally dissolve with an efficiency of approx. 100%.

Chloride solutions are more widely used, since deposition can be carried out at much higher current densities (4–5 A/dm²). They are also more stable and cheaper.

One of the recommended baths consists of (g/l):

$SnCl_2$	75
NaF	37.5
$NH_4F.HF$	37.5
NaCl	22.5
HCl	12.5
NH_4SCN	0.25
Disulphonaphthalinic acid	1.0

It is assumed that the fluoride ions form the complex ion $[SnF_2Cl_2]^{2-}$ with tin, which causes deposition of tin to proceed at a considerable polarisation. Hence, fine-grained deposits are formed, and the throwing power of the electrolyte is satisfactory.

2.22.2 Alkaline tin electrolytes

The chief constituents of these electrolytes are Na_2SnO_3 (sodium stannate) and NaOH. In addition, Na_2SnO_2 (sodium stannite) may form under certain conditions.

Both compounds dissociate as follows:

$$Na_2SnO_3 \rightarrow 2Na^+ + SnO_3^{2-}$$
$$Na_2SnO_2 \rightarrow 2Na^+ + SnO_2^{2-}$$
$$SnO_3^{2-} + 3H_2O \rightleftharpoons Sn^{4+} + 6OH^-$$
$$SnO_2^{2-} + 2H_2O \rightleftharpoons Sn^{2+} + 4OH^-$$

The ratio between the stannite and stannate concentration is determined by the equation

$$2 Sn^{2+} \rightleftharpoons Sn + Sn^{4+}$$

Sodium stannate is more stable, hence the reaction tends to proceed from left to right, but owing to the close proximity of the potentials of Sn/Sn^{2+}, Sn/Sn^{4+} and Sn^{2+}/Sn^{4+}, equilibrium is established slowly. Both Sn^{4+} and Sn^{2+} ions can be discharged at the cathode. Sn^{2+} ions are reduced without much polarisation as a result of which a coarse-grained deposit will be obtained and the throwing power of the electrolyte will be poor. The reduction of Sn^{4+} ions proceeds at considerable overpotentials, and hence in the absence of Sn^{2+} ions the throwing power of the electrolyte will be good and the deposit fine-grained. Hence, during the preparation of the electrolyte, particular care must be taken to ensure that all Sn^{2+} ions are oxidised to Sn^{4+} ions.

NaOH in stannate electrolytes acts as a complex former. An increase in NaOH concentration will reduce the activity of Sn^{4+} and hence depress the electrode potential of tin. Therefore a definite concentration of NaOH must be maintained in the electrolyte. Electrodeposition is carried out at elevated temperatures (65–70°C), since unsatisfactory deposits are obtained at lower ones.

Dissolution of the anode must be controlled so as to proceed entirely with Sn^{4+} ion formation. At low current densities tin tends to dissolve as Sn^{2+} ions. At elevated temperatures, a hydroxide film of golden colour forms on the anode surface, which causes the dissolution potential to be displaced in the noble direction to a value at which Sn^{4+} ions can form.

If the current density is too high, tin dissolution will cease and hydroxyl ions will be discharged at the anode. In the anodic polarisation curve (Fig. 14) the onset of complete passivation will occur at the sharp jump in potential.

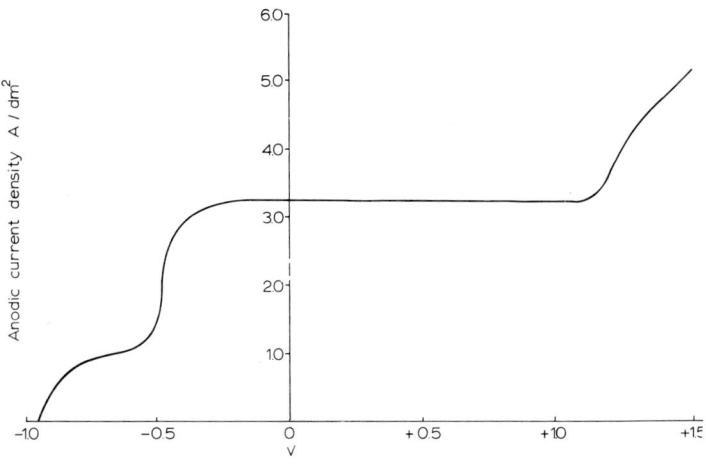

Fig. 14.
Relationship between the anode potential of tin and current density in an electrolyte of the following composition (g/l): Sn, 28; NaOH, 13. Temp. 70°C.

The current density range within which preferential Sn^{4+} ion formation will occur depends on operating parameters (NaOH concentration, bath temperature, anode current density, etc.).

At the commencement of the electrodeposition process the anode must be rendered passive as quickly as possible in order to prevent formation of Sn^{2+} ions. This is effected by passing a high current through the cell until the golden film has formed on the anode. Thereafter the current is lowered to the normal value.

Typical bath compositions are (g/l):

$Na_2SnO_3.3H_2O$ 50–100
NaOH   8–18

The bath is operated hot (70–80°C) at a cathode current density of 1.5–2.0 A/dm^2. The cathode efficiency will be 65–70%.

The deposition process can be speeded up by replacing the sodium stannate with potassium stannate, since the solubility of the latter is nearly twice that of the former.

2.22.3 Flow melting of tin coats

The thickness of electrodeposited tin coats in tin plate used for cans is approximately 1μm. This must be flow-melted in order to close up the pores which are inevitably present in such thin coatings. Deposits obtained from alkaline electrolytes are more amenable to flow melting than those from acid ones, which tend to form individual droplets when so treated.

Flow melting is effected at 260–270°C for a few seconds. An induction heating method is normally used for this purpose. Small articles can be heated in glycerine or in high flash-point oils.

2.23 ELECTRODEPOSITION OF NOBLE METALS

Electrodeposits of gold and silver are used for decorative–protective, as well as for industrial purposes. In industry they are used for lowering the contact resistance of electrical contacts, for increasing the electrical conductivity of current-carrying components of high-frequency electrical equipment, etc. Silver is used for plating reflectors and mirrors.

Electrodeposits of gold and silver are virtually free from internal tensile stress. Since they are cathodic to most other metals their protective qualities increase with increase in thickness. They are applied directly to copper, brass or nickel silver, but in the case of steel and other metals that are considerably more reactive than silver, an intermediate layer of copper or nickel should be applied.

2.23.1 Silver plating

The electrolyte most extensively used for depositing silver is based on silver cyanide. The basic constituents of such an electrolyte are $K[Ag(CN)_2]$ and free cyanide, KCN. The complex decomposes, liberating silver ions, as follows:

$$K[Ag(CN)_2] \rightarrow K^+ + Ag(CN)_2^-$$
$$Ag(CN)_2^- \rightleftharpoons Ag^+ + 2CN^-$$

The activity of silver ions in such an electrolyte is comparatively high, so that unless the free cyanide content is maintained at a high level, silver will be displaced from solution even by metals like copper and its alloys.

An increase in free cyanide content of the bath displaces the potential of silver in the negative direction, increases cathode polarisation with increase in current density and raises the conductivity of the electrolyte. This results in a good throwing power and a fine-grained deposit. A minimum free cyanide content is also required to ensure anode dissolution. Potassium carbonate also increases the conductivity of the bath and hence further improves the throwing power.

A typical electrolyte may have the following composition (g/l):

$K[Ag(CN)_2]$ (calculated as the metal)	30–40
KCN (free) ,, ,, ,, ,,	15–45
K_2CO_3 ,, ,, ,, ,,	25–60

Such an electrolyte is operated at room temperature at a cathode current density of 0.3–0.6 A/dm². By increasing the bath temperature to 30–40°C and periodically reversing the current (PR plating), the current density can be increased to 1.5–2 A/dm².

In order to increase the hardness and wear resistance of silver deposits, salts of various metals (e.g. Cd, Ni, Co, Cu, Pd etc.) can be added to the electrolyte. Vyacheslavov[28] reports good results from a silver bath to which 10–25 g/l K (SbO) $C_4H_4O_6.\tfrac{1}{2}H_2O$ has been added, in addition to 60–90 g/l Rochelle salt.

Bright deposits are obtained by making additions of such compounds as carbon disulphide, hyposulphite, etc. to the bath.

2.23.2 Gold plating

Like silver, gold is normally plated from cyanide electrolytes. The main constituents of a gold bath are AuCN and KCN. Due to the high price of gold, the metal content of such a bath is kept low.

The electrode potential of gold in cyanide solutions is sufficiently negative to prevent contact deposition on the basis metal.

A typical electrolyte may have the following composition (g/l):

$K[Au(CN)_2]$ (calculated as the metal)	4
KCN (free) ,, ,, ,, ,,	15
K_2CO_3 ,, ,, ,, ,,	10–15

Such a bath is operated at a temperature of 60–70°C with a cathode current density of 0.2–0.3 A/dm². The cathode efficiency is 60–80%. Gold anodes are normally used, but insoluble platinum or graphite anodes may be used instead.

By increasing the gold content of the electrolyte to 15–26 g/l, the cathode current density can be raised to 2–3 A/dm².

Leeds and Clarke[13] give the following formula (g/l):

$K[Au(CN)_2]$	11.7,	23.52	or 35.28
corresponding to metal contents of	8,	16	or 24
KCN	90	90	90

Gold deposits are very soft (50 DPN) and in order to increase their hardness and wear resistance, salts of such metals as silver, copper, nickel, cobalt, antimony, etc. can be added to the electrolyte.

The so-called "acid" gold bath is widely used, particularly in the electronics industry. This bath is based on $K[Au(CN)_2]$ and citric acid, and its pH is adjusted to approx. 4 with ammonium hydroxide. This is possible because the complex ion $[Au(CN)_2]$ is stable to below pH 3. Unlike the alkaline solutions, such baths do not attack plastics when plating printed circuit boards. Phosphate baths are also used with pH stabilised by means of a phosphate buffer mixture.

2.23.3 Rhodium plating

Rhodium is plated from an acid sulphate or phosphate/sulphate bath. Electrodeposited rhodium is extremely hard (about 900 DPN) and highly stressed. This high internal tensile stress may cause deposits above 0.0025 mm thick to crack.

Decorative coatings are 0.00025 to 0.0005 mm thick. Owing to their porosity they will not protect base metals, and therefore an intermediate layer of silver or nickel is applied. Thin rhodium coatings are applied, among other things, for the protection of silver articles against tarnishing. Thicker coatings are used as non-tarnishing finishes for lowering the contact resistance of electrical contacts.

2.23.4 Other noble metals

Other noble metals which in their electrodeposited form are of industrial importance, include palladium and platinum.

The hardness of palladium coatings is intermediate between rhodium and gold (200–300 DPN), and since they are much cheaper than the latter metals, they are increasingly used as alternatives to them for electrical contacts.

Platinum electrodeposits are used mainly on titanium for the preparation of inert anodes for electrolytic processes.

2.24 ELECTROPLATING ON ALUMINIUM

Thin deposits of matt chromium can be applied directly to aluminium at high current densities from the normal type of chromium electrolyte. However, for coatings of good durability an undercoat of nickel is required. This cannot be plated directly on to aluminium, and a special preparation of the aluminium surface is required to make it amenable to receive an electrodeposit. A number of processes are available, the most important being the zincate pre-treatment method. More recently, two other processes have been introduced, namely an immersion alloy pre-treatment[14] and a bronze strike method[15].

Prior to the application of any of these pre-treatments, the metal surface must be degreased, cleaned and etched.

In the zincate process the articles are immersed for 1–2 minutes in a warm alkaline sodium zincate solution (75–100 g/l ZnO, 400–500 g/l NaOH). After rinsing, they are ready for plating with copper, brass or silver. Nickel and chromium can subsequently be applied.

In this method, the slight oxide film left after acid etching is removed, and when the bare metal is exposed, some aluminium is dissolved and replaced by zinc.

The replacement zinc deposit is thick and not always sound, owing probably to its mode of growth. In order to control its structure, the process has been modified so as to complex the zinc in an alkaline solution in which nickel and copper ions are also present[14]. A film of controlled thickness is produced, the chemical composition of which changes progressively during its growth. The complexing agents are NaCN and sodium tartrate, and the nickel and copper ions are responsible for arresting the film growth in its final stages, thus preventing filamental growth which would make the film mechanically weak.

The film itself acts as a reducing medium, reducing the metal ions present in the solution to the metallic state. The latter form a barrier between the aluminium and the electrolyte.

In the bronze strike method[15], the articles, after being degreased, cleaned and acid-dipped, are transferred to a stannate-containing activating solution for 30 sec, at 30°C. They are then given a 3 min "strike" in a bronze electrolyte, water-rinsed and plated as desired.

The bronze strike solution contains Cu^+ ions, Sn^{4+} ions, free cyanide, free KOH and certain additives. The parts are plated in this electrolyte at a cathode current density of 1.5–12 A/dm². The corrosion resistance imparted to the basis metal by the bronze strike is claimed to be high.

2.25 Plating of zinc-base alloy diecastings

These are usually nickel and chromium plated. In the mechanical polishing process care must be taken not to cut through the thin sound skin of the diecasting.

These alloys, after appropriate degreasing and cleaning, are given an undercoat of copper either from an ordinary cyanide copper electrolyte or from a Rochelle copper solution. Nickel can then be deposited from any bath, including bright nickel electrolytes. Chromium is subsequently plated in the usual way.

2.26 Electrodeposition of alloys

Many combinations of metals can be codeposited. Alloys have a number of advantages over single metal coatings. Alloying is the most important method of modifying the properties of metals, and therefore alloy electrodeposits are applied when specific properties are required, *e.g.* an increased hardness or wear resistance for electrical contacts (Ag–Sb, Sn–Ni), an increased resistance to high temperatures (Fe–Ni–Cr), specific magnetic properties (Ni–Co, Ni–Fe), etc. In addition, some protective-decorative alloy deposits have a lower porosity than single metal coatings of the same thickness (*e.g.* Cu–Sn, Sn–Ni).

The electrodeposition in alloy form of those metals that cannot as yet be plated singly from aqueous solutions, *e.g.* W–Co, Mo–Ni, Ti–Ni, is of considerable interest.

The laws governing reduction of the individual ions are not obeyed when two or more species of ion are reduced simultaneously. Therefore the equations describing the ion reduction conditions cannot be used in a quantitative study of the kinetics of alloy deposition.

2.26.1 Electrodeposition of brass

Co-deposition of copper and zinc, the standard electrode potentials of which are approx. 1.1 V apart, is possible only from cyanide complex solutions in which the reduction potentials of the two species of metal ions approach each other to such an extent that their simultaneous reduction becomes possible.

The free cyanide content of the electrolyte and the cathode current density used have a strong influence on the reduction potentials of Cu^+ and Zn^{2+} ions as well as on the cathode efficiency with respect to either. The copper ion activity decreases to a greater extent than the zinc ion activity with increase in NaCN concentration. In addition, the cathode polarisation of copper is more pronounced than that of zinc. These facts explain why codeposition of two electrochemically so different metals is possible.

From the polarisation curves for copper, zinc and brass (Fig. 15) it can be seen that brass is deposited at more electropositive potentials than copper at the current densities used in practice, *i.e.* copper exerts a depolarising effect on zinc. This is characteristic of the deposition of a solid solution type alloy.

An electrolyte used for the electrodeposition of brass containing 70% Cu and 30% Zn may have the following composition (g/l):

$K_2[Cu(CN)_3]$	(calculated as the metal)	12
$K_2[Zn(CN)_4]$	,, ,, ,, ,,	16
KCN (free)	,, ,, ,, ,,	12
$KNaC_4H_4O_6$	,, ,, ,, ,,	60

Such an electrolyte is operated at a temperature of 18–60°C at a cathode current density of 1–2 A/dm². The cathode efficiency is 55–60%. The anode material used is rolled brass plates of the same composition as the deposit.

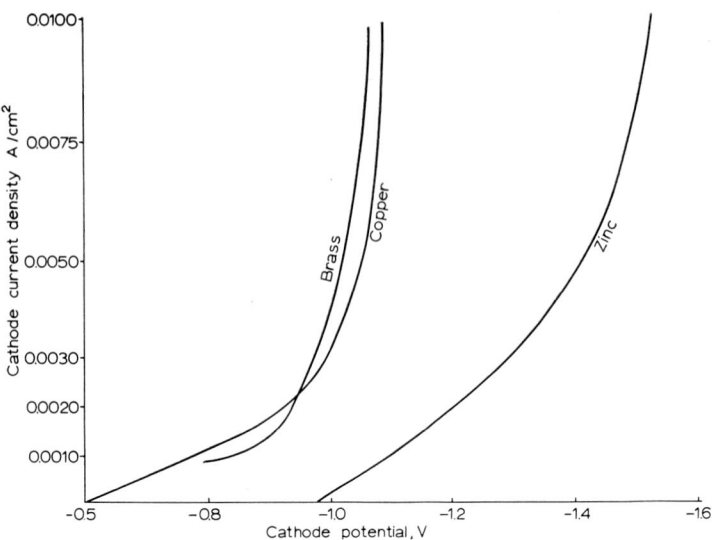

Fig. 15.
Cathode polarisation curves for copper, zinc and brass in cyanide electrolytes.

2.26.2 Electrodeposition of bronze

Electrodeposits of two compositions are used, one containing 10–20% tin, and the other 40–45% tin. The electrolytes used are based on cyanide.

A typical electrolyte may have the following composition (g/l):

		Bath I	Bath II
$K_2[Cu(CN)_3]$	(calculated as the metal)	15–18	8–12
Na_2SnO_3		23–18	40–45
NaOH		9–10	8–20
KCN		26–28	8–15

Bronze is deposited at 60–65°C at a cathode current density of 2–3 A/dm². The cathode efficiency is 70–75%. The deposit from bath I will contain 15–20% Sn, that from bath II, 40–45% Sn. Either bronze or mixed copper and tin anodes are used for both baths.

2.26.3 Electrodeposition of lead-tin alloys

Lead–tin coats have good anti-friction properties; hence they are sometimes used as bearing surfaces. They are widely used for facilitating soldering. Since the standard electrode potentials of lead and tin are close together ($E°_{Pb/Pb^{2+}} = -0.13$V, $E°_{Sn/Sn^{2+}} = -0.14$V), the two species of ion can be reduced simultaneously from simple salt solutions. Alloys containing between 5 and 60% Sn are used industrially.

For the electrodeposition of a coat containing 40% Sn, an electrolyte of the following composition is suitable (g/l):

Lead	(calculated as the metal)	45–50
Tin	,, ,, ,, ,,	40–50
HBF_4	,, ,, ,, ,,	40–75
H_3BO_3	,, ,, ,, ,,	25–35
Glue	,, ,, ,, ,,	3–5

Such an electrolyte is operated at a temperature of 18–25°C at a cathode current density of 0.8–1 A/dm^2. Either lead–tin alloy or mixed tin and lead anodes are used.

2.26.4 Electrodeposition of tin–nickel alloys

Tin–nickel alloy deposits with an undercoat of copper (*i.e.* without an intermediate layer of nickel) have an excellent corrosion resistance, and bright deposits can be obtained.

The electrolyte used for tin–nickel deposition consists of the metal chlorides, to which sodium fluoride and ammonium fluoride are added in order to ensure structurally satisfactory deposits. The composition of the deposits obtained is virtually independent of the metal ion concentration of the electrolyte and remains unchanged within a wide range of current densities (0.5–4.0 A/dm^2) and temperatures (45–70°C). The external appearance of the deposit is greatly influenced by the pH of the electrolyte.

A bath of the following composition (g/l) will give a deposit consisting of 65% Sn and 35% Ni:

$NiCl_2.6H_2O$	250–300
$SnCl_2.2H_2O$	40–50
NaF	28–30
NH_4F	35–38

Deposition is carried out at a cathode current density of 1 A/dm^2, and the cathode efficiency will be 96–98%. The throwing power of such an electrolyte is superior to that of a bright nickel bath.

The Sn–Ni electrodeposit consists of the intermetallic compound NiSn, which can be produced only electrolytically. At 300°C this compound decomposes into Ni_3Sn_2 and Ni_3Sn_4. Cast anodes contain both compounds, and during electrolysis Ni_3Sn_2 dissolves. For this reason, Sn–Ni alloys are not used as the anode material; instead, tin and nickel plates are employed, and the surface area of the nickel plates must be five times greater than that of the tin plates. The average anode current density is 0.5–1.0 A/dm^2.

2.26.5 Electrodeposition of nickel–cobalt alloys

The standard electrode potentials of nickel and cobalt are very close to each other and hence both species of ion can be reduced from simple salt solutions.

Ni–Co electrodeposits are very widely used in the magnetophonic industry. The chief requirements for such deposits are a specific coercive force and residual induction, and this is the case when the cobalt content is 62–85%.

An electrolyte of the following composition will give the required deposits (g/l):

$NiCl_2.6H_2O$	240–250
$CoSO_4.7H_2O$	180–190
H_3BO_3	20–30
KCl	10–15

The electrolyte is maintained at a pH of 4–5 and is operated at a temperature of 50–60°C and a cathode current density of 1–2 A/dm².

On adding hypophosphite to the above bath, a ternary Ni–Co–P alloy will be deposited, the coercive force and residual induction of which are very high.

Fluoboride electrolytes are also used for the deposition of nickel–cobalt alloys.

2.27 ELECTRODEPOSITION OF METALS FROM NON-AQUEOUS ELECTROLYTES

Electrodeposition of such electronegative metals as Mg, Be and Al cannot be effected from aqueous electrolytes, since in such solutions hydrogen is reduced at a very much nobler potential than the metal ions, and hence the entire current will be devoted to hydrogen evolution.

Satisfactory deposits of aluminium have been obtained from $AlCl_3$–$LiAlH_4$–ether[22] and from $AlCl_3$–*n*-butylamine–ethyl ether solutions[23].

Research on the electrodeposition of more noble metals from non-aqueous electrolytes is conducted with the object of ascertaining whether it is possible to attain higher cathode efficiencies and greater coat thicknesses as compared with the deposition from aqueous systems. Also, further applications may arise in the development of methods for obtaining coherent, adherent coatings on active metals such as uranium or titanium.

Copper has been deposited from various solvents, including pyridine, furfural, formamide, formamide–formic acid mixtures, esters, aldehydes, ketones and various ternary systems. Menzies *et al.* have obtained promising results from CuI–10% pyridine, and from copper acetate–acetic acid–20% pyridine solutions[24].

Satisfactory deposits of cadmium have been obtained from CdI_2–acetone solutions[25] and from dimethylformamide solutions of CdI_2–tetrapropyl ammonium iodide[26].

Work has also been done on the deposition of silver, zinc and other metals.

2.28 ELECTRODEPOSITION OF REFRACTORY METALS

Refractory metals are here defined as those of higher melting point than Cr and include Ti, Hf, Zr, V, Nb, Ta, Mo, Re and W. Various processes exist for the electrowinning of these metals, but the product is usually in particulate form, and unsuitable for use as a protective coating. As previously explained on p. 376, these metals cannot be deposited from aqueous solutions, and this section and the following one describe attempts to circumvent this. It is however possible to deposit certain alloys from aqueous solutions containing significant amounts of these metals, such as alloys of Mo and W with the ferrous metals. The content of refractory metal is usually less than 30%. These alloys are not themselves refractory since they are of the eutectiferous type, the eutectic melting point being below that of the iron group metal in the alloy in all cases. They are normally hard, wear resistant and corrosion resistant. Tungsten alloys are readily deposited from alkaline solutions containing hydroxy acids or their salts as complexing agents with ammonia and ammonium salts as additional stabilisers. A typical bath recommended by Brenner et al.[17] for the deposition of Co–W alloys has 25 g/l Co (as $CoSO_4.7H_2O$), 25g/l W as $Na_2WO_4.2H_2O$, 400 g/l Rochelle salt and 50 g/l NH_4Cl. The pH is made up to 9 with ammonia and deposition occurs at 20 mA/cm². Deposits of 20% W are obtained with 93% cathode efficiency. Similar baths have been recommended by Hoar and Bucklow[18] and Rama-Char[19], while similar results for Re alloys have also been reported[20].

2.28.1 Use of molten salt baths

Prior to 1954, the use of molten salt baths produced diffusion alloys with the substrate. These processes were non-Faradaic and suffered limitations of thickness and substrates capable of being specified. Moreover, the rate of deposition decreased with time. Two early Faradaic processes deposited Mo from chloride melts and W and Mo from oxy-ion melts. The results could not be extended to other metals however.

Since then, the work of Mellors and Senderoff[16, 21] using mixed fluoride melts has proved far more successful, although controlled atmospheres and bath conditions are essential. This is the basis of the Union Carbide process. Satisfactory coherent metal coatings are achieved with all Group IVA, VA and VIA metals except for Ti. This is a true Faradaic electrodeposition process. Beryllium can also be plated from melts similar to those of Senderoff and Mellors, although the process is largely a diffusional one (93% cathode efficiency at 1 kA/m²) and details can be found in the work of Fischer[29].

2.29 ELECTRODEPOSITION OF METALS ON PLASTICS

In order to deposit metals electrolytically on a non-conductive surface, the latter must be treated so as to accept a conducting film of metal by a process of chemical reduction, known as "electroless" plating. The metals so reduced are normally either copper or nickel, and once a continuous film has formed, any desired metal can be electrodeposited on it in the usual way.

The surfaces of polymeric materials in the as-received state are hydrophobic, *i.e.* water-repellent. The first step in the treatment of a polymer surface which is to be electroplated consists in submitting it to an etch, usually a chromic acid/sulphuric acid-based solution, to render it hydrophilic. This is followed by sensitising in a solution of stannous chloride in which stannous ions attach themselves to the polymer surface at discrete tiny sites.

When the sensitised surface is subsequently immersed in a noble metal solution such as palladium chloride, the latter is reduced by the stannous ions to form islands of palladium metal of up to 30 Å in diameter. The polymer is now ready to receive a metal coating from an "electroless" plating solution, in which reduction of metal ions on to the palladium islands occurs, palladium acting as a catalyst for their chemical reduction.

The deposit grows radially outwards from each island, so that when the growths have eventually made contact with each other, the metal film will be continuous and conducting. The coatings subsequently electrodeposited are usually nickel and chromium.

The polymer most generally used as a basis for electrodeposition is ABS (acrylonitrile–butadiene–styrene), although many other plastics can also be plated. ABS material consists of a phase of tiny rubber particles (butadiene) dispersed through a matrix of styrene/acrylonitrile. Particles near the surface are etched out by the chromic acid/sulphuric acid solution, and during subsequent plating metal is thrown into the cavities, partly filling them. Thus, a mechanical anchorage is provided within the substrate for the metal deposit, which reinforces metal-to-polymer adhesion, the bonding forces between which are of a chemical nature.

Adhesion of an electrodeposit to a polymer is measured by peeling off a strip of the metal film at right angles to the substrate. Peel values well in excess of 2 kg/cm are obtained for ABS.

3. Anodic oxidation

Anodic oxidation is used to form a protective coating on the metal surface. Though the process can occur by simple chemical oxidation, this gives films of only 1–3 μm thickness (films formed in air are approx. 100 Å thick). Anodic oxidation gives films of thickness 10–25 μm and even, under special conditions, up to

300 μm thick. The latter are known as "hard anodising" conditions producing extremely hard friction-resistant films. Most commonly applied to Al, the technique is also valuable for Ti, Ta, Cu and Mg alloys. In the latter case, the purpose of the treatment is to increase adhesion of paint coatings.

3.1 Aluminium anodising

Though electrolytes including ammonium hydroxides, borates, phosphates, tartrates, oxalates with their corresponding acids, as well as chromic acid, chromates, and sulphuric acid have all been proposed, only the last mentioned acids are used in practice. The films anodically formed are usually colourless with pure aluminium. Films formed from alloys of aluminium, especially those containing copper, tend to be tinted.

A typical commercial bath is 10–50 wt% sulphuric acid operated at 15–20°C at 1.2 A/dm². The cell voltage will be 10–20 V. The lower the temperature, the harder the coating, and films of 15–20 μm are usually obtained after 30 min immersion. This is the bath used for much electrocolouring work. In contrast, hard anodising produces thick, dark-grey films which are not suitable for dyeing. The bath used is often 20% sulphuric acid, refrigerated at 1–3°C.

Sulphuric acid baths are unsuitable for the anodising of fabricated assemblies since it is difficult to thoroughly wash the work after treatment, and trapped acid causes corrosion. For this reason, a chromic acid bath is used for such type of work, this acid not causing corrosion of aluminium. These baths contain 3% chromic acid and the cell voltage is raised to 50 V over the course of an hour. The sole purpose of this type of bath is to enhance corrosion resistance. The work often emerges with a mottled appearance and is sometimes further treated by grease impregnation.

The foregoing techniques apply equally well to pure aluminium, or its silicon, zinc or copper alloys. In the latter case, though, alloys of copper content greater than 5% are not treatable, since the $CuAl_2$ intermetallic compound prevents formation of a continuous oxide film.

3.2 Copper anodising

Copper anodising finds its main use in electrical equipment where the oxide confers a degree of insulation in coils etc. between the adjacent turns of wire. Baths are approx. 20% caustic soda run at up to 100°C. The process differs from aluminium anodising in that an abrupt shut-off of film growth occurs after a short time. Copper dissolves anodically and is converted to the oxide in solution. Cuprates are also involved in the reaction. Mantell[30] illustrates equipment for the continuous treatment of wires etc.

3.3 Theoretical background of anodic film formation

The reader is referred to the work of Vetter[37], and the monograph by Young[31].

4. Electrocolouring

When anodic oxidation of aluminium is followed by impregnation of the oxide layer with a dye, a coloured film results which can be stabilised by immersion in hot water or silicate solution, or in a hot solution of a heavy metal acetate (usually nickel acetate) or by exposure to high pressure steam at 150°C. The dye, trapped in the pores of the oxide film, requires no further lacquering or protective treatment in contrast to most other metal colouring processes.

A black finish can be produced on cadmium by cathodic treatment in the following solution:

Nickel sulphate	75 g/l
Nickel ammonium sulphate	45 g/l
Zinc sulphate	37.5 g/l
Sodium thiosulphate	15 g/l

The bath is operated at 50–55°C at 0.5–2.5 A/dm² and at pH values not in excess of 5.6–5.9 to avoid precipitation of the zinc. By modification of the conditions other colours from black to light green can be obtained. These coloured finishes on cadmium are especially noted for their high corrosion resistance.

Zinc can be coloured black by application of a "black nickel" deposit. The bath consists of nickel sulphate, zinc sulphate and sodium thiocyanate and deposition is effected at low current densities (0.2 A/dm² approx.) the solution being operated cold with nickel or stainless steel anodes.

A black finish on stainless steel is best achieved with the following bath:

Sulphuric acid	250 ml
Sodium dichromate	60 g
Water, to make	1 l

Operating conditions are 70–95°C at 6–8 A/dm², with air agitation. Again, by variation of conditions, blacks, purples and browns can also be obtained.

Gold-coloured films can be obtained by plating from gold solutions to which small amounts of substances, such as silver salts, have been added. The latter will produce a green film. Addition of alkalis such as sodium hydroxide or carbonate give a "rose-gold" colour, while copper salts give a darker hue. The use of alkaline carbonate electrolytes to colour copper substrates has been described[33], while alkaline lactates have been used in the same fashion[34]. Similar cuprous oxide based films have been reported[35], while the colouring of zinc with phosphate or ferro/ferricyanides and similar baths is also described in the literature[32].

In addition to the simple colouring operations here described, the use of masks or photo-sensitive salts in place of dyes permits patterns to be made, in the latter case by what is essentially a photographic process.

5. Plant for electroplating

This falls under the headings:
(1) Automatic Plant
(2) Semi-automatic and mechanical plant
(3) Manually operated plant
(4) Barrel plating plant

In the first three categories, the work is treated individually or at least individually mounted on suspender jigs. In the last case, bulk quantities of small items are handled without being mounted in any way.

Automatic plants process articles mechanically through the various pre-treatment, plating and post-treatment stages, with complete elimination of manual handling. *Fixed programme plants* have a single lift and traverse system so that all the work carriers operate together. Since the operating sequence is fixed, any variation in timing of an individual step within the sequence results in a proportional change in the total times of the other steps. This type of plant is thus best suited to processing of medium to large runs of a given item. A *variable programme plant* has a number of lift and traverse units operating under the instructions of a programmed controller. The result is a more flexible unit, suited to the larger jobbing shop or other plant where a variety of work exists.

In semi-automatic or mechanised plants the lift and traverse units which transfer the work from bath to bath, are manually controlled, though the lift and traverse actions are motorised. In *manual* plants the work is lifted in and out of the bath by hand or by a crane. In countries with low labour costs, the majority of even large production runs is treated in this type of plant.

Barrel plating comes into its own for plating large numbers of small items such as screws etc. These are placed in a rotating barrel which is *either* perforated and immersed in the bath, with external anodes *or* (in the self-contained type) has an

Fig. 16.
Automatic nickel and chromium plating plant for electrical components (Courtesy A.E.B., Belgium).

interior anode, and contains not only the work, but also its own solution. The advantages in the latter case are the small volume of solution required, and also the ability to process really small parts which would drop out of normally perforated barrels. However, the self-contained barrel is seen less and less in normal works, being less efficient than its immersed counterpart.

Bibliography and references

There are, by a wide margin, more titles relating to electrodeposition of metals, than to any other aspect of electrochemistry. The following textbooks and monographs are of general or special interest.

E. RAUB and K. MÜLLER, *Fundamentals of Metal Deposition*, Elsevier, Amsterdam, 1967.
J. M. WEST, *Electrodeposition and Corrosion Processes*, Van Nostrand, New York, 1966.
A. T. VAGRAMYAN and Z. A. SOLOV'EVA, *Technology of Electrodeposition*, Robert Draper, London, 1961.
W. BLUM and G. B. HOGABOOM, *Principles of Electroplating and Electroforming*, McGraw-Hill, New York, 1949.
G. ISSERLIS (Ed.), *Quality Control in Metal Finishing*, Columbine Press, Manchester, 1967.
A. G. GRAY, *Modern Electroplating*, Wiley, New York, 1953.
J. O'M. BOCKRIS and B. E. CONWAY (Eds.), *Modern Aspects of Electrochemistry*, No. 3, Butterworths, London, 1964, chap. 4.
L. BRENNER, *Electrodeposition of Alloys*, Academic Press, New York, 1963.
L. K. GRAHAM (Ed.), *Electroplating Engineering Handbook*, Reinhold, New York, 2nd ed., 1962.
H. SILMAN, *Chemical and Electroplated Finishes*, Chapman and Hall, London, 1952.

W. CANNING AND CO. Ltd., *Handbook on Electroplating*, Birmingham, 21st ed., 1969.
R. A. F. HAMMOND, *Nickel Plating from Sulphamate Solutions*, International Nickel Ltd., London, 1964.
S. WERNICK and R. PINNER, *Surface Treatment and Finishing of Aluminium*, Robert Draper, London, 1964.
W. J. McG. TEGART, *Electrolytic and Chemical Polishing of Metals*, Pergamon, Oxford, 2nd ed., 1959.
N. P. FEDOT'EV and S. YA. GRILIKKES, *Electropolishing and Anodising and Electrolytic Pickling*, R. Draper, London, 1959.
J. O'M. BOCKRIS and G. RAZUMNEY, *Electrocrystallisation*, Plenum Press, New York, 1967.
K. J. VETTER, *Electrochemical Kinetics, 4B, Metal/Ion Electrodes*, Academic Press, New York, 1967.

1 B. E. CONWAY and J. O'M. BOCKRIS, *Proc. Roy. Soc. London*, 248 A (1958) 394.
2 B. E. CONWAY and J. O'M. BOCKRIS, *Electrochim. Acta*, 3 (1960) 340.
3 A. DAMJANOVIC, *Plating*, Oct. (1965) 1018.
4 G. I. FINCH, H. WILLIAM and L. YANG, *Discussions Faraday Soc.*, 1 (1947) 144.
5 K. M. GORBUNOVA and P. D. DANKOV, *Zh. Fiz. Khim*, 27 (1953) 1725.
6 L. I. ANTROPOV, *Zh. Fiz. Khim*, 37 (1963) 965.
7 T. P. HOAR and D. J. ARROWSMITH, *Trans. Inst. Metal Finishing*, 36 (1958).
8 D. R. GABE and J. M. WEST, *Trans. Inst. Metal Finishing*, 40 (1963) 6.
9 T. P. HOAR and D. J. ARROWSMITH, *Trans. Inst. Metal Finishing*, 34 (1957) 354.
10 T. P. HOAR, *Trans. Inst. Metal Finishing*, 39 (1962) 166.
11 C. A. SNAVELY, *Trans. Electrochem. Soc.*, 92 (1947) 537.
12 W. H. SAFRANEK and C. L. FAUST, *Trans. Inst. Metal Finishing*, 40 (1963) 217.
13 J. M. LEEDS and M. CLARKE, *Trans. Inst. Metal Finishing*, 45 (1967) 147.
14 A. E. WYSZINSKI, *Trans. Inst. Metal Finishing*, 45 (1967) 147.
15 J. C. JONGKIND, *Trans. Inst. Metal Finishing*, 45 (1967) 155.
16 S. SENDEROFF, *Electrodeposition of Refractory Metals, Metallurgical Rev.*, 11 (1966) 97.
17 A. BRENNER, P. S. BURKHEAD and E. SEEGMILLER, *J. Res. Natl. Bur. Std.*, 39 (1947) 351.
18 T. P. HOAR and I. A. BUCKLOW, *Trans. Inst. Metal Finishing*, 32 (1955) 186.
19 T. L. RAMA-CHAR, *Electroplating*, 15 (1962) 40.
20 A. R. MEYER, *Trans. Inst. Metal Finishing*, 46 (1962) 209.
21 G. W. MELLORS and S. SENDEROFF, Can. Pat. 688, 546 (1964).
22 D. E. COUCH and A. BRENNER, *J. Electrochem. Soc.*, 99 (1952) 234.
23 I. A. MENZIES and D. B. SALT, *Trans. Inst. Metal Finishing*, 43 (1965) 186.
24 I. A. MENZIES and T. BROUGHTON, *Trans. Inst. Metal Finishing*, 43 (1965) 24.
25 I. A. MENZIES, P. J. MORELAND and S. R. OULSMAN, *Trans. Inst. Metal Finishing*, 45 (1965) 185.
26 I. A. MENZIES and B. SALT, *Trans. Inst. Metal Finishing*, 46 (1968) 65.
27 D. R. GABE and J. M. WEST, *Trans. Inst. Metal. Finishing*, 40 (1963) 197.
28 P. M. VYACHESLAVOV, *Alloy Coatings*, Mashgiz, 6 (1961).
29 H. FISCHER and W. SCHWANN, *Metallwirtschaft*, 12 (1933) 187.
30 C. L. MANTELL, *Electrochemical Engineering*, McGraw-Hill, New York, 4th ed., 1960.
31 L. YOUNG, *Anodic Oxide Films*, Academic Press, New York, 1961.
32 D. R. pat., 626, 502 (1936); 623, 563 (1935).
33 G. L. CRAIG and C. E. IRION, U.S. pat., 1, 974, 140.
34 J. E. STARECK and R. TAFT, Brit. pat., 452, 464.
35 J. E. STARECK, U.S. pat., 2, 081, 121.
36 R. A. F. HAMMOND, *Nickel Plating from Sulphamate Solutions*, International Nickel Ltd., London.
37 K. J. VETTER, *Electrochemical Kinetics, 4B, Metal/Ion Electrodes*, Academic Press, New York, 1967.

JOURNALS. In addition to some of the specialist journals cited above, many others exist. The attention of readers is specially drawn to the bimonthly journal *Metal Finishing Abstracts*.

Chapter 10

Electroforming

R. J. KENDRICK

Director, Electroformers Ltd., Staplehurst, Kent

1. Introduction and historical

Many metal and alloy products may be fabricated by electroplating techniques. The process, known as electroforming, involves controlled electrodeposition onto a mandrel or former of predetermined shape, size, accuracy and surface finish. Subsequently the mandrel is intentionally removed to leave an electroform as a free standing plain or perforated product in its own right.

The word "electroforming" was first coined by Blum and Hogaboom in their book[1] which was one of the classics in the field of electroplating and forming for some thirty years. As they defined it, it was the "production or reproduction of articles by electrolysis", a definition which still holds today nearly half a century later. The word replaced the earlier "galvanoplasty" which had been in use since the inception of the method in the previous century.

The discovery of electrochemical techniques as reproductive tools dates back to 1836, when Jacobi, in Russia, made his first experiments. In 1837, Spencer made similar experiments in the United Kingdom, and the name of De la Rue is also connected with this earliest work. At this time, the future for electroforming was considered to be in reproduction of works of art, and this is where the emphasis lies in the earlier books. But by 1865, R. Hoe of New York were selling machines for electrotyping, a new method for electrochemical reproduction of entire beds of typeface, either with or without illustrative material. This is dealt with later in this chapter. From this time on, electrotyping became the most important sector of the industry, until about the period 1920–1935. At this time, electroforming of iron became increasingly important, and large iron pipes were made (in Grenoble, France, for example) and iron sheet was made in S. Wales while the German industry was equally resourceful, making parts for military and non-military equipment. Most of the parts so made had to be sintered in one way or another, and the process appears to have been entirely superseded. Further details of these early years are beyond the scope of this chapter, but the following works will provide the requisite details: for the earliest years books by Gore[2,3], Wahl[4] as well as De la Rue's paper[5], while for the second phase, the three editions of Blum and Hogaboom[1], the work of McMillan and Cooper[6], and on the electroforming of iron, the monograph of Hughes[7], a pamphlet on the use of nickel to "build-up" worn engineering parts by Faris[8] and for a particularly detailed picture of the German electroforming industry and practice, there is the monumental work of Engelhardt[9]. Silman[26] refers to iron electroforming and other references are given in Mantell[10].

The use of electroformed copper in the reproduction of art wares seems to have reached a peak at the turn of the century, but it is still in use at the British Museum in London for this purpose. At the beginning of the century too, the introduction of the gramophone record served to create another great electroforming industry, the manufacture of record stampers, which is described later. In the USA, there was a steady production of iron moulds and dies by the method, while at the Castner–Kellner Co. works at Runcorn, U.K., during the period 1927 to 1930, quantities of nickel tubing were electroformed. The tubes, $\frac{1}{2}$ inch to $\frac{3}{4}$ inch diameter, were produced in 20 foot lengths using Monel* mandrels arranged vertically in groups of six suspended from rotating planet wheels mounted in a turntable. Electroforming was carried out in a nickel sulphate base electrolyte and the nickel tubes were parted from their mandrel by passage through a reeling machine. Porosity of the nickel deposits was apparently a problem and the plant was closed down.

From 1930 onwards developments in electroforming techniques accelerated rapidly mainly due to the arrival of the motor car. Large automobile factories were created requiring plating plants mass-processing decorative car trims. Techniques were investigated and developed for producing brighter and more corrosion resistant electrodeposited coatings in quicker and cheaper ways. The physical, mechanical and chemical properties of electrodeposits were studied in detail together with the effects on these properties arising from changes in plating conditions and the influence of organic and inorganic addition agents in solution.

It became apparent that under certain plating conditions internal stresses in electrodeposits could be minimised, the use of addition agents could produce deposits with zero stress or reverse the direction of the normal stress, and deposits could be obtained with hardnesses of tool steels. These developments especially gave birth to nickel electroforming which is generally regarded as modern electroforming for the following reasons:

(1) Of all metals which can be electrodeposited with ease the properties of nickel are closest to those of steel, the most universal raw material.

(2) Internal stress, hardness and ductility of electrodeposited nickel can easily be controlled between very wide limits and nickel electroforms are strong, tough, and resistant to corrosion, erosion, and abrasion.

There is now almost no limit to the shapes that may be fabricated by electroforming. Plating conditions can be controlled such that internal stresses in electroforms are frequently lower than those present in metals fabricated by other techniques. Many metals may be used for electroforming including nickel, copper, iron, silver and gold and selection depends on engineering requirements such as me-

---

* Trademark.

chanical properties, electrical or thermal conductivity, reflectivity or corrosion resistance.

2. Advantages of electroforming as a production process

The advantages of electroforming include:

(1) Production in a single stage to very close tolerances, difficult and complex shapes which otherwise by other metal fabrication techniques would require several operations. In some instances components of a particular shape or form can only be made by electroforming, such components include those with internally complex (concave) curvatures, which would be out of reach of mechanical forming tools. Blum and Hogaboom[1] quote work from the U.S. Bureau of Standards where lines smaller than 0.0005 mm could be reproduced with perfect satisfaction.

(2) The process is suitable for both small production quantities and mass production of identical, replicate parts.

(3) Basic plant and equipment is inexpensive for producing precision parts: tooling is relatively cheap and operating and labour costs can be fairly low especially when the process can function for periods unattended.

(4) For certain types of mass production work, electroformed replicate mandrels may readily be made from a single master mandrel.

(5) Dense integral products can be produced free from porosity.

3. Problems encountered during electroforming

The advantages of the electroforming process are fully utilised whenever possible and simultaneously the electroformer exercises his skill and ingenuity to overcome inherent limitations of the process. The main problems encountered are (i) internal stress in the deposit producing distortion, (ii) non-uniform metal distribution arising from variations in current density which may result in variations in mechanical properties within a single electroform, (iii) corner weakness defects which may become prevalent as apparent planes of weakness in deposits formed in creviced regions with sharp root radii, (iv) low deposition speeds leading to inordinate deposition periods when preparing thick electroforms.

It is worthwhile discussing these problems in greater detail with particular reference to nickel because of its commercial use and importance.

3.1 Internal stress

Practically all metals which are deposited from electrolytes without addition agents are formed in a state of tensile stress which for nickel plated at 10 A/dm^2 can vary from 1.4 kg/mm² (2000 lb/in²) in conventional sulphamate solution to 42 kg/mm² (60,000 lb/in²) in all-chloride solution. When electroforming, the need to avoid stress is of major importance as unless it is controlled the electroform will distort and tend to detach prematurely from the mandrel. A controlled level of compressive stress is deliberately induced in some electroforms to facilitate extraction of the mandrel or to promote an increase in fatigue life.

What techniques can one use to control stress? The use of organic addition agents is often the accepted method and their effects are rapidly noticed. To induce zero or compressive stress in nickel deposits aromatic sulphonic acids may be added to the electrolyte. These compounds, besides strongly influencing stress, also harden nickel deposits, sometimes to the level of tool steels, arising from the incorporation of sulphur in the deposit. These types of deposits generally lack ductility and also cannot be exposed to temperatures above 250° because severe embrittlement occurs and such electroforms therefore cannot have appendages welded or brazed on subsequently. When faced with a choice of organic compounds to control stress in nickel deposits it is preferable to use benzene-based sulphonic acids rather than naphthalene derivatives as the latter produce more brittle deposits.

Organic stress controllers, although very effective, do have a number of disadvantages: They can cause temperature embrittlement, they produce rather brittle deposits under general conditions and their use must be under close control and good housekeeping to avoid the deleterious effects of breakdown products arising from their use.

Internal stress in electroforms can be controlled or modified to a lesser extent by making use of other plating variables. In pure copper and nickel deposits internal stress tends to increase in the tensile direction as the solution temperature is decreased and as the current density is raised. In nickel plating solutions internal stress is a minimum in deposits formed at pH values in the region of 4.0. The choice of anion, in solutions free from addition agents, has a profound effect on the internal stress of deposits: tensile stress is considerably lower in copper deposits from sulphate solutions than from cyanide electrolytes and in nickel plating solutions sulphate ions induce moderate tensile stress, sulphamate ions produce very low stress levels and chloride ions act as powerful stress raisers.

Another parameter, which certainly influences stress but has had limited investigation from electroforming aspects, is the use of current wave forms other than direct current. The use of square wave periodic alternating current[11] to deposit nickel from an all-chloride solution has shown that under certain conditions, for example frequencies > 100 Hz coupled with increasing amounts of deplating during each cycle, tensile stress can be reduced by 28 kg/mm² (40,000 lb/in²) which at the same time produces deposits with exceptional ductility, around 40% elongation compared with 25% elongation for a Watts type deposit. The use of certain types of reverse current form for cyanide copper solutions can minimise internal stress.

3.2 Non-uniform metal distribution

The varying ability of different plating processes to produce electroforms of uniform wall thickness upon irregular surfaces is measured by the "throwing power" of the plating solution and depends during deposition upon:

(a) variations in cathode polarisation with current density which alter "primary" to the "secondary" current distribution.

(b) variations in cathode efficiency with current density which modify the "secondary" current distribution.

In general, electroforming from single-salt solutions leads to poor metal distribution as cathode polarisation is low, and relatively independent of current density while cathode efficiency remains high and does not alter with deposition rates. Electroforming from complex solutions does usually produce deposits of relatively uniform wall thickness.

During deposition of thick deposits, as in electroforming, the unevenness of deposit thickness becomes a cumulative problem with time as the "high spots" on the cathode preferentially attract more and more metal, and metal distribution progressively deteriorates at an accelerating rate. If non-uniform metal distribution is likely to present itself as a severe problem the electroformer tries to improve metal distribution by using the following devices.

(i) "Thieves and robbers" in the form of metal strips or wire strategically connected to the cathode around potentially high current density areas. Although useful when electroforming certain types of products the author generally dislikes their employment as although they promote uniform metal distribution "thieves" and robbers" have an insatiable, indeed ever increasing, appetite and they are costly to feed on electricity and metal.

(ii) Shields and screens (non-conducting) of suitable material which withstands operating conditions and is chemically compatible with the electrolyte are placed around high current density areas and designed to restrict current flow to the required level. Very uniform metal distribution can be obtained but success depends largely on the skill and intuition of the electroformer as there are no strict rules which can be followed. A possible disadvantage associated with shields and screens may be interference with solution circulation and agitation leading to variations in properties within a single electroform.

(iii) Auxiliary anode and bipolar conductors placed near the cathode surface where wall thickness requires increasing locally.

Auxiliary anodes can suffer from the following drawbacks (a) products from their disintegration can spill onto the electroform and cause roughness. (b) there can be physical complications in rigidly holding in controlled positions a network of auxiliary anodes which may foreshorten during use. (c) special shapes are frequently required and this adds to cost and delivery dates.

A technique which has been used on a number of occasions, but not extensively, is to progressively stop off areas on the electroform as deposition proceeds. However it is usual for the deposit to overgrow and entrap the stopping off layer. Alternatively the deposit can have its "high spots" machined off periodically. Problems of lamination within the electroform can occur unless, after each stop off layer is applied or after each machining operation, the deposit is satisfactorily treated and activated prior to continuation of the next deposition cycle. Stopping off with material which is not wetted by the plating solutions means that the need for edge machining is subsequently eliminated or drastically reduced. Suitable stop offs during nickel deposition are silicone rubbers, Teflon, Nylon, polythene and pure gum rubber.

High throwing power single salt electroforming solutions have been developed to yield uniformly thick deposits. They rely on a very low metal ion concentration in solution such that at current densities in the region of 1.0 A/dm^2 there is a scarcity of cations at the high current density regions and cathode polarisation promotes relatively good metal distribution but at the expense of very extended deposition periods.

3.3 Corner weakness defects

Corner weakness is one of the major defects restricting the successful electroforming of components containing recessed areas. If one considers an electroform growing on two recessed faces of a substrate meeting at 90° there comes a time, as local deposit thickness exceeds the root radius of the recess, when the deposits growing from each face meet orthogonally and apparently a plane of weakness is formed. This is usually termed corner weakness and the practical effect is fracture of the electroform under negligible load along the "failure to weld" line.

The causes of corner weakness are not fully understood. Recessed areas are low current density regions and therefore local wall thickness in the electroform is usually very thin and unable to withstand applied stress levels that other regions of the electroform can. Because local plating rates are so low it is unknown whether this affects the local integrity of the deposit due to incorporation of metallic impurities. Sharp root radii in recessed areas would act as stress raisers and it would not be too surprising if a moderate load applied to an electroform did not cause the local applied stress level to exceed the local ultimate tensile strength of the deposit. Detailed investigation of corner weakness is required so that the precise reasons for its presence become known and the electroformer can circumvent it and further increase the range of his products.

The practical electroformer has to eliminate or minimise corner weakness either by ensuring that all internal angles on the mandrel have very generous root radii, preferably greater than the deposit thickness called for, or by using certain types of bath additions on the assumption that they help to knit the deposit together along the angle bisecting the crystal growth boundary. It is accepted that the presence of a strongly levelled deposit which produces a series of parallel fillets or radii instead of sharp fissures in the recessed area may considerably reduce the incidence of corner weakness.

3.4 Low deposition speeds

Let us consider the factors that are involved, and which directly influence plating rates. As the current density is raised at a cathode an increasing number of ions are discharged from solution leading to localised impoverishment of metal cations in the cathode layer. When the total rate of ion discharge at the cathode exceeds the rate of supply of metal ions from the bulk solution the limiting current density for metal deposition is exceeded and different cathode reactions with a higher limiting current density proceed. In nickel plating solutions when this stage is reached pronounced hydrogen evolution occurs. The pH of the solution in the cathode region rises rapidly, basic material is incorporated in the deposit, which becomes brittle and highly stressed. Basic green nickel salts are often precipitated at regions where the limiting current density for metal deposition has been exceeded.

There is a simplified relationship between maximum current density and many of the operating variables.

$$\text{Limiting current density} = \frac{\text{(Rate of diffusion of metal ions) (Concentration of metallic salt)}}{\text{(Thickness of cathode film) (Share of current carried by cations other than depositing metal)}}$$

This relationship provides us with a tremendous amount of practical information on how to achieve rapid plating rates.

To obtain higher deposition speeds

(1) Select a salt in which the diffusion rate of metal ions is high.

(2) Increase the concentration of electrolyte bearing in mind drag-out losses.

(3) Minimise the thickness of the cathode film by means of agitation. In a still solution the cathode film is approximately 0.03 cm thick. In vigorously agitated solutions this thickness can be reduced to 0.001 cm.

(4) Minimise the presence of foreign cations so that the depositing metal ions carry all the current.

(5) Raise the temperature of the electrolyte. This increases the rate of diffusion and consequently reduces the thickness of the cathode film. For example, an increase in temperature from 20°C to 70°C reduces the thickness of the cathode film by a third.

Using the above relationship and following the suggestions made, where practicable, significant increases in electroforming rates can be obtained.

4. Electroforming solutions and deposit properties

4.1 Copper

Copper is mainly used in applications where high electrical conductivity is required, such as wave guides, spark machining electrodes, printed circuit foil or for backing up other electroforms, mainly nickel. A number of copper plating solutions have been developed for electroforming purposes. (Typical compositions and preferred operating conditions are given in Table 1).

TABLE 1

TYPICAL COMPOSITIONS AND PREFERRED OPERATING CONDITIONS OF COPPER ELECTROFORMING SOLUTIONS

Bath	Composition	Conditions
Acid	Copper sulphate 150–250 g/l Sulphuric acid 45–110 g/l Composition not critical—higher concentrations used for faster deposition speeds.	Normally at 25°C but can be used up to 50°C
Cyanide	Copper cyanide 30–60 g/l Total potassium or sodium cyanide 40–90 g/l Sodium or potassium carbonate 15–70 g/l Free potassium or sodium cyanide should be maintained at 25–30 g/l and checked daily. Carbonate level must not exceed 75 g/l (precipitate out with calcium hydroxide).	Temperature 60°C pH 11.5–12.5 Vigorous air agitation
Pyrophosphate	Copper 19–22 g/l Pyrophosphate as P_2O_7 150–175 g/l Ammonia 3–8 g/l The ratio of pyrophosphate to copper must be maintained between 6.5 and 7.5 by additions of either copper or potassium pyrophosphate.	Temperature 55–60°C pH 8.5–9.2 Vigorous air agitation

4.1.1 Acid copper

Acid copper is a traditional workhorse. It is a very easy bath to use, operations are unlikely to go wrong, cathode efficiency is high and deposits are seldom pitted. Deposit properties can be selected within the following ranges: hardness up to 100 HV, tensile strength 12.0–42.0 kg/mm² (17,000–60,000 lb/in²), elongation up to 40%. Internal stress in the absence of addition agents is normally tensile and less than 1.4 kg/mm² (2,000 lb/in²). A large number of addition agents can be used to modify and control properties but the choice is not as wide as with nickel.

The main disadvantage of acid copper baths is their low throwing power and the baths do not respond to periodic reverse plating in terms of improving metal distribution. Certain types of periodic reversal of current are used to reduce the severity of nodular growth in thick deposits. All deposits formed without addition agents suffer from corner weakness.

4.1.2 Cyanide copper

Cyanide copper baths have high throwing power and although having a lower efficiency than the acid copper bath, the copper being in divalent form does provide a faster deposition rate, especially in recessed areas which is a very important factor during electroforming. Disadvantages of cyanide baths include its poisonous nature and associated effluent problems and the inability to directly plate onto chemically silvered non-metallic mandrels (see later). The main advantages of cyanide copper baths become apparent when periodic current reversal is used and throwing powers of 100% can be obtained.

Depending on the conditions and the addition agents used, deposit hardness from cyanide baths ranges from 100–250 HV, tensile strength up to 31.5 kg/mm² (45,000 lb/in²), elongations up to 50%. Copper deposited from a cyanide bath has a relatively high internal tensile stress, about 7.0 kg/mm² (10,000 lb/in²). However, by appropriate choice of periodic reversal of current cycle and the use of potassium salts instead of sodium salts, internal stress can be substantially reduced and the author has recorded values of 0.7 kg/mm² (1,000 lb/in²) tensile. Corner weakness still remains a problem.

4.1.3 Pyrophosphate copper

A relative newcomer for electroforming purposes compared to the other solutions but it is becoming quickly established commercially. Throwing power is good and addition agents can be used to obtain brightness and levelling action. Unfortunately there is little data available on published properties except that organic addition agents generally induce compressive stresses. Although pyrophosphate baths are slightly more expensive than cyanide baths they do avoid the problems associated with disposal of cyanide wastes.

4.2 NICKEL

Nickel is extensively used for electroforming purposes because of the wide range of useful engineering properties which can be easily obtained and simply maintained. The solutions used are usually Watts type, conventional sulphamate (containing around 300 g/l nickel sulphamate) and the concentrated sulphamate elec-

TABLE 2

TYPICAL COMPOSITIONS AND PREFERRED OPERATING CONDITIONS OF NICKEL ELECTROFORMING SOLUTIONS

Bath	Composition	Conditions
Watts type	Nickel sulphate 300–450 g/l Nickel chloride 10 g/l Boric acid 40 g/l	Temperature 55°C pH 3.5–4.2
Conventional sulphamate	Nickel sulphamate 300–450 g/l Nickel chloride 10 g/l Boric acid 40 g/l	Temperature 60°C pH 3.5–4.2
Concentrated nickel sulphamate ("NI-SPEED")	Nickel sulphamate 600 g/l Nickel chloride 10 g/l Boric acid 40 g/l	Temperature 60–70°C pH 3.5–4.2 Vigorous air agitation Extremely fast plating rates are possible. If control of deposit stress is important the electrolyte must be subjected to continuous electrolysis using non-activated nickel as anode material at 0.5 A/dm^2 (5 A/ft^2).

trolyte used in the NI-SPEED process (600 g/l nickel sulphamate). Composition ranges and preferred operating conditions are listed in Table 2.

4.2.1 Watts type solutions

These produce matt deposits with a hardness of 150 HV, tensile strength of 42.0 kg/mm² (60,000 lb/in²), elongation of 25% and internal tensile stress levels of 14.0 kg/mm² (20,000 lb/in²) at 5 A/dm^2.

4.2.2 Conventional sulphamate solutions

These produce deposits with a hardness of 180 HV, tensile strength of 42.0 kg/mm² (60,000 lb/in²), elongation of 18–20% and internal tensile stress of 1.4 kg/mm² (2,000 lb/in²) at 5 A/dm^2.

Addition agents are usually made to both these solutions, especially in the case of Watts type solutions, but also in the sulphamate electrolyte at higher current densities. The aims are to lower tensile stress which rises with increased plating rates, to produce zero or compressive stress, or to increase hardness, tensile strength, brightness and prevent pitting. The additions frequently used are aromatic sulphonic acids in quantities to produce zero or compressive stress in deposits with hardnesses ranging up to 600 HV.

Besides giving rise to incorporation of sulphur and causing the electroforms to become heat sensitive, the addition agents may form breakdown products in use and alter the properties of the deposit in an undesirable way. At intervals the solutions may have to be re-purified.

4.2.3 Concentrated nickel sulphamate solutions

Research over several years by International Nickel Ltd. resulted in the evolution of concentrated nickel sulphamate solutions for use in the NI-SPEED process[11, 12]. They are based, normally, on 600 g/l nickel sulphamate and are operated at 60°–70°C and pH 4.0. The NI SPEED process is particularly valuable in the electroforming of many types of article and in heavy nickel deposition, because

(1) Very high current densities can be used—at least twice the maximum value permissible with conventional sulphamate solutions. Substantial reductions in overheads are possible and output capacity per operating vat can be significantly increased.

(2) The process can be operated, without addition agents, to produce ductile deposits which have, as required, compressive, zero or tensile stress. At 60° and 70°C zero-stressed thick deposits are produced at 22 A/dm² (200 A/ft²) and 33 A/dm² (300 A/ft²) respectively. Control of stress is simple provided that the electrolyte is conditioned by continuous low-current density electrolysis at 0.54 A/dm² (5 A/ft²), using non-activated nickel anode materials.

(3) The throwing power of the NI-SPEED process is better than that of conventional sulphamate solutions.

(4) At low current densities, 3.3 A/dm² (30 A/ft²), moderately hard, compressively stressed, lustrous, ductile deposits can be produced without the use of addition agents.

Further development work[13] using sulphur-free addition agents has shown that for engineering applications use of cobalt additions offers several advantages.

(1) The deposits are sulphur-free and can be heated without becoming embrittled.

(2) Control of the properties of the deposit, particularly hardness, is simple: there is no accumulation of breakdown products such as normally occurs in electrolytes containing organic additives.

(3) Hard deposits of the type described are generally more ductile than those hardened by sulphur-containing additives.

4.3 Precious metals

In the silver and jewellery trade there is an increasing shortage of skilled craftsmen such as silversmiths, chasers and engravers, with the result that techniques have been developed for electroforming in silver[14] and gold. Skilled craftsmen are still required for the preparation of master patterns but subsequently perfect replicate reproduction at significantly lower labour cost is obtained by electroforming which can also be used to produce certain decorative effects prohibitively expensive to execute by other techniques. Further developments in the electroforming of precious metals are expected to lead to applications in the electronic and electrical industries.

4.3.1 Silver electroforming

A standard cyanide electrolyte is used containing a relatively high metal and cyanide content and a fairly low carbonate concentration. Although addition agents are used to produce hard fine-grained deposits and minimise nodular growth the deposits are sufficiently pure to receive a Britania hallmark. Current densities of approximately 1 A/dm^2 are used and under certain conditions periodic current reversal is employed.

4.3.2 Gold electroforming

Acid solutions, although of limited application, may be used but electrolytes containing base metal additions, to produce brightness and hardness, are characterised by low efficiency and deposits are highly stressed (21.0 kg/mm^2; 30,000 lb/in^2). Low stressed pure gold deposits are obtained from solutions free from base metal additions.

Alkaline, cyanide-free electrolytes operating around pH 9.0 are finding increased application. Cathode efficiencies of 95% are obtained and stress levels in deposits are less than 3.5 kg/mm^2 (5,000 lb/in^2).

5. Anode materials

For successful electroforming it is imperative that the correct anode material for a particular process is selected and operated correctly under recommended conditions. Failure to do this can result in the electrolyte being thrown out of chemical balance.

Copper anodes. Rolled, cast or extruded bar anodes of electrolytic oxygen-bearing copper (tough pitch) can be used with all types of copper electroforming solutions although formation of anode sludge, carried to the cathode surface despite the use of anode bags, may cause roughness and nodule formation, particularly in acid copper electrolytes. In recent years two relatively new types of copper anode have been introduced, (1) phosphorus deoxidised copper containing at least 0.005% phosphorus and (2) oxygen-free high conductivity copper. Both types appear to prevent the formation of coarse anode particles and promote smoother finer grained deposits with less tendency to nodular formation. Although both these types of copper anode can be used without bags, these are in fact commonly employed and "plastics" type of material such as polyvinyl chloride, "Nylon" and "Terylene" are particularly satisfactory.

Titanium mesh baskets can be used to contain the anode material. Although an initial additional capital expense they are strongly recommended as maintenance costs are reduced, and particularly because anode foreshortening is avoided and a constant anode area is obtained which is of particular importance during copper electroforming as anode current densities are critical. In general, the limiting anode current density in acid copper, cyanide and pyrophosphate copper solutions is 3–4 A/dm^2.

Nickel anodes. These are available in several primary forms for use in titanium mesh baskets as follows:

"Mond" carbonyl nickel pellets, standard electrolyte nickel and sulphur-activated nickel ("S" nickel pieces).

All available non-activated forms of nickel will dissolve satisfactorily up to 15 A/dm^2 in nickel electroforming solutions provided that approximately 5 g/l, preferably 10 g/l, of nickel chloride is present. Increased concentrations of chloride are unnecessary and raise deposit internal stress in the tensile direction. The use of sulphur-activated nickel as anode material does not require the presence of chloride ions and anode current densities in excess of 50 A/dm^2 can be used without anode passivation. Care should be exercised when operating nickel sulphamate solutions if anode passivity occurs as it appears to "anodically oxidise" the electrolyte and the deposits obtained contain sulphur and are characterised by high compressive stress, hardness and brightness; the effects can increase as the electrolyte is aged under these conditions. A controlled degree of anode passivity with nickel sulphamate solutions has been advocated as a means of regulating particular levels of deposit stress[15].

When using titanium baskets it is essential (i) that each container is kept topped up, above the solution level, with nickel anode pieces: failure to ensure this can result under certain conditions in breakdown of the titanium metal; (ii) that activated and non-activated forms of nickel be kept separate.

A variety of materials for use as anode bags is suitable and readily available. The author prefers woven polypropylene and provided that suitable weave is used and air agitation pipes are not located beneath anode busbars there will be no problems of cathode roughness arising from nickel anode residues.

6. Equipment for electroforming

The equipment used is relatively conventional, being similar in many instances to that employed in decorative plating. Wherever feasible, conforming anodes should be used to aid metal distribution.

In electroforming, solution purity is a prime essential for control over properties of deposits. For example, in a correctly balanced nickel plating solution free from addition agents the presence of copper in amounts exceeding 10 ppm has a detectable influence on deposit stress. It is therefore strongly recommended that continuous low current density electrolysis be carried out using a corrugated cathode at 0.5 A/dm^2 (5 A/ft^2) in a separate tank and by this technique metallic impurities are easily reduced to very low and harmless concentrations.

Continuous filtration and vigorous air agitation in most instances helps to produce a better product.

General views of an electroforming plant are shown in ref. 16 and the type of equipment used can be clearly seen.

7. Measuring properties of electroforms

7.1 HARDNESS AND TENSILE STRENGTH

The hardness may be measured by normal indentation techniques either by macro- or micro-hardness tests depending on the thickness of the electroform. Tensile strength may be determined by preparing thin electroformed sheets of metal and testing in equipment designed for handling such material. Alternatively, approximate values of tensile strength can be estimated from the hardness level of the electroform.

7.2 INTERNAL STRESS

A number of techniques for measuring stress can be used, such as the spiral contractometer[17] or the stresometer[18]. The spiral contractometer consists of a long strip wound in the form of a helix, one end of which is clamped, while the free end is attached to a coaxial torque tube which communicates through a system of gears, which magnify any displacement, to a pointer traversing a calibrated dial. The stresometer consists of a flat disc of stainless steel or copper which covers a shallow chamber containing metering fluid connected to a vertical capillary tube. During deposition the flexure of the disc is measured hydraulically due to the displacement of liquid contained in the chamber.

Every electroforming shop should have available an instrument for measuring internal stress. The author uses a very simple instrument consisting of a thick stainless steel plate stopped off on one face with a narrow central channel extending $\frac{3}{4}$ of the length of the plate. Inside the channel a thin stainless steel shim stopped off on one face is fixed rigidly at the closed end and during deposition on the unstopped faces of shim and plate a restraining gate is temporarily fixed across the open end of the channel to avoid movement of the shim. The plate ensures uniform current density along the length of the shim (necessary as stress varies with current density) and after deposition of at least 25 microns thickness of metal the gate is removed and the strip deflects. An attachment working on a lever principle is fixed in position and simply allows the operator to determine the weight (force) required to restore the plated shim to its original position before deposition. The stress level in the deposit is easily calculated from the restoring force and the thickness and area of the deposit on the shim.

7.3 Ductility

In most instances an experienced electroformer will prepare a foil which on bending will give an approximate assessment of ductility. For more accurate work the author forms a deposit on annealed copper rods and notes the extension, during a tensile stretch, at the onset of cracking in the deposit.

8. Preparation of mandrels

The mandrel is the starting point of the electroforming process. It is the matrix, made of conducting materials, or materials coated with a conductive layer, on which the metal deposits from solution, and which, at the end of the run, is parted from the deposit, which will thus have acquired the surface contours of the mandrel.

Choice of materials. The mandrel stage is usually the most important, frequently the most difficult and often the most tedious part of electroforming. The type of material used can influence the life, cost and usefulness of the electroform. There is choice over a wide range of materials that can be used and final selection will depend on:
(i) Material cost.
(ii) Ease of fabrication into the desired shape and whether, if required, accurate replication for mass production is possible.
(iii) Ease of surface preparation prior to plating.
(iv) Inertness towards solution and whether resistant to thermal and deformation stresses.
(v) Ease of removal from electroform and whether re-usable.

8.1 Mandrel types for plain metal products

Mandrel materials can be divided into two types, metallic and non-metallic. Generally, metals offer greater stability and rigidity in the plating bath, they can be machined to more accurate limits, they usually have a more perfect surface finish and surface preparation prior to plating is simple, but they often impose styling restrictions as re-entrant angles are not possible unless the mandrel is expendable each time after use. Non-metallic materials are usually cheaper, more flexible in use and there is greater likelihood of extraction without damaging the electroform. Non-metallic mandrels have their limitations; their surfaces have to be made electrically conducting, usually by coating with a reduced silver film which besides adding another processing stage necessitates initial electroforming at a very low current density until the deposit is thick enough to carry the required final current.

8.1.1 Metals with a natural passive film

These are frequently used where the shape is simple and the mandrel can be simply extracted because there is no adhesion between electroform and mandrel. The electroform should be prepared under conditions of zero or preferably compressive stress to facilitate extraction and the mandrel where practicable should have a definite taper of several tenths of a millimetre per metre. Mandrels of this type are usually made from stainless steel or from chromium plated steel and brass. Repeated use of the mandrel may necessitate periodic reinforcement of the natural oxide film which can be effected by a 30 second immersion in sodium or potassium dichromate solution (1 g/l).

8.1.2 Metals without a natural oxide film

Metals and alloys such as copper and brass which are readily machined are passivated by treatment with a sodium or potassium dichromate solution (10 g/l) or alternatively with a solution of sodium sulphide (8 g/l).

Treatment in dichromate solution gives complete passivation and complete non-adhesion of the electroform. In certain instances where the electroform is being produced on a flat and polished surface, such as the ungrooved region of a record stamper, complete non-adhesion leads in some cases to premature detachment of the electroform. Under this condition, some adhesion is required and the preferred separation technique is treatment of the surface in an albumen solution.

8.1.3 Expendable metals and alloys

Mandrels of this type are used when normal separation is difficult and they are particularly suited for the production of electroforms with re-entrant angles. The use of aluminium, zinc and fusible alloys is well established. With aluminium or zinc the mandrel is removed after the electroforming operation by dissolution in caustic or acid as required.

Fusible alloy mandrels as the name suggests can be melted out from an electroformed object. Alloys with melting points ranging from 50°C to 250°C are made up from casting various combinations and proportions of tin, lead, bismuth, cadmium or antimony. Bismuth is frequently used in the alloys as it expands on solidification and its presence in the alloys counteracts solidification contraction. Unfortunately fusible alloys have two disadvantages: it is difficult, using plain casting techniques, to obtain good surface finish on the mandrel without subcutaneous porosity, and subsequent melting out of the mandrel always leaves traces of fusible alloy fused to the electroform which in the presence of bismuth causes grain boundary embrittlement in nickel and copper electroforms. Various methods of removing final traces of fusible alloy have been suggested but none of them are entirely reliable. The usual approach is to apply a thin parting film of graphite to the mandrel and prevent direct contact between the fusible alloy and the electroform.

8.1.4 Non-metallic materials

There is a wide choice of machinable and castable materials but all types of non-metallic mandrels which are commercially used at present require, prior to electroforming, a pretreatment sequence to produce a satisfactory conducting surface film. This is generally achieved by a reduction silver process whereby after "sensitising" the component it is sprayed with silver nitrate solution and a reducing agent to produce a very thin surface film of silver. The mandrel is then rinsed, preheated in demineralised water at the temperature of the electroforming process, transferred to the electroforming solution and plated, without agitation, at 50–100 A/M² for approximately 10 minutes. Subsequently the rate of agitation and current density are progressively increased to maximum value over a period of approximately 10–20 minutes. The technique is costly, time consuming and reject rates may be high. There are exceptions: (i) automatic "silvering" can reduce reject rates and costs and[19] the price of the silvering stage probably becomes insignificant when producing moulds and dies requiring deposition periods of weeks.

The materials generally used as non-metallic mandrels are:

Perspex. This is an acrylic resin available in sheet or block form. It is relatively brittle, tends to soften at 60–70°C but engraves and machines well to produce a good surface finish.

PVC. This is a relatively soft plastic but is quite widely used. In the form of "slush castings" (hollow) PVC offers an economical technique for the mass production of expendable and easily extractable mandrels with relatively fine detail although overall dimensional accuracy is probably limited. In order to obtain rigidity and avoid buoyancy during deposition some electroformers fill the slush castings with sand or wax but plating solution itself is preferred. Solid PVC mandrels can be used by pouring PVC paste into suitable moulds and curing the plastic.

PVC contains various plasticisers many of which chemically disturb the plating solution, and therefore preliminary tests are advisable before full scale use.

Epoxy resins. There are extremely good casting resins available which are ideal for reproducing identical mandrels in quantity. The material generally machines well, engraves well and accepts a silver film readily unless certain fillers, to reduce casting shrinkage, are used.

Phenolic resins. The material machines well, softens at higher temperatures which is useful for mandrel removal although destructive. The resin is attacked by alkaline electrolytes.

Waxes. Beeswax and other waxes have been used for a long time and although they silver well they tend to have high shrinkage rates, indifferent surface finish and soften in solutions slightly warmer than room temperature. A number of companies now manufacture synthetic polymer waxes which have high softening temperatures, moderate shrinkage rates, excellent surface finish but in the author's limited experience their surfaces cannot as yet be easily conditioned to receive an acceptable conducting surface film.

Plasters, wood. These materials can be impregnated, sealed and shellacked for plating and occasionally are used as temporary mandrels, but generally other mandrel materials are preferred.

8.2 Mandrels for perforated metal products

Special techniques of mandrel production have been developed over many years for the electroforming of perforated metal products, which probably form the most significant growth area of electroforming.

The mandrels are prepared using photosensitive resists which are liquids that are sensitive to various wavelengths of light when dry. The photo-resists are applied to the metal surface and through a high contrast mask of the desired pattern exposed to the appropriate light (*e.g.* ultraviolet) and then developed by means of differential solvents that dissolve away either exposed resist or unexposed resist depending whether it is a positive or negative in action. The result is that chosen areas of the metal surface remain protected by a layer of resist.

In its present form the part can now be used as a temporary mandrel and deposition will take place on the unprotected portions of the mandrel. The resist pattern is usually destroyed or damaged after the preparation of a few electroforms.

Permanent mandrels are made by taking the part after developing the resist, etching the uncoated portions of the metal surface and filling the recesses with dielectric material. The resist is then removed and the metal, the future threads and bars of the mesh product, are exposed. The dielectric is then ground flush with the metal surface to give a permanent mandrel (the shape of the electroform will be a reverse of that from a temporary mandrel).

The preparation of successful mandrels to produce precise perforated products is a combination of considerable skill, expertise, ingenuity, patience and experience. The basis metal may be chosen according to its etching characteristics, the art work has to be modified to take into account undercutting of the resist as etching proceeds, the degree of undercut is influenced by the etching procedure used and is least with spray etching, the uniformity of the photo-resist depends on the method of application and whirl coating is often preferred but this is not always possible with certain shapes, *e.g.* cylinders. Many types of photo-resist are available with different physical and chemical properties, the resolution depends upon the resist material and is normally better than 25 microns, and can be as fine as 1 micron. The adhesion of the dielectric to the etch pits largely determines the life of the mandrel and many techniques have been developed to improve adhesion.

9. Applications of electroforming

The applications of electroforming are rapidly becoming extremely diverse and it is impossible to indicate all the varied products manufactured by the technique. A series of products are described which either illustrate where electroforming is competitive with other methods of fabrication, or it is the only practical method of manufacture, or a particular facet of electroforming can be clearly shown.

9.1 Record stampers

This is one of the largest uses of electroforming. A lacquered aluminium disc is grooved with a stylus to record music or speech and subsequently electroformed with nickel to produce a master which is then passivated and used to electroform a family of "mothers". Each of these is used to produce numerous identical electroformed stampers. Each stamper may produce between 1,000 and 2,000 pressings before damage and therefore hit records require very large numbers of stampers.

Stampers are electroformed from sulphamate or modified Watts type solution under strictly controlled stress-free conditions usually at relatively high current densities (15 A/dm^2; 150 A/ft^2) and the nickel must have sufficient ductility to withstand a coining operation prior to use for pressing operations. The track on

404   ELECTROFORMING

an LP record is approximately half a mile long and any imperfection exceeding 2 microns is unacceptable.

9.2 SEAMLESS PERFORATED TUBES

The use of seamless perforated tubes, especially in the field of rotary textile printing, is expanding rapidly. Electroformed nickel screen printing cylinders are made in a range of mesh sizes, from 40 to 100, in different printing widths and repeat sizes (diameters). Each screen containing several million minute holes of exact size and contour is coated with an emulsion which stops off the majority of holes leaving a pattern of clear holes through which dye is forced onto fabrics, plastics or paper. Multicolour printing onto materials is carried out by using a series of screens in a printing machine, each screen transferring an individual colour to complete an overall pattern (see Fig. 1).

Fig. 1.
Nickel textile printing screen. These screens are electroformed to contain several million holes of exact size and contour. Photo courtesy of Electroformers Ltd., copyright International Nickel Co.

Intricately patterned electroformed nickel screens are also manufactured containing plain and perforated areas in juxtaposition.

Seamless perforated tubes are electroformed over a wide range of current densities from Watts type, conventional nickel sulphamate or concentrated nickel sulphamate solutions containing addition agents to control properties.

9.3 Wave guides

Wave guides must be regarded as one of the most significant industrial uses of electroforming, the components being accurately reproduced and combining strength with internal surface smoothness and lightness, an important factor in missile and aircraft applications. Although simple wave guides are fabricated by other techniques, straight sections can be electroformed on permanent metal mandrels. Complex shapes containing "twists", "bends" and re-entrant angles are electroformed on expendable mandrels such as aluminium or fusible alloy which is subsequently dissolved or melted out. Present trends in wave guides manufacture include the use of a thin copper shell, to provide conductivity, backed by a nickel layer to provide a strong yet thinner and lighter component.

9.4 Plain and perforated metal foils

These products can be made in tremendous lengths by continuous deposition onto a cylindrical permanent metal mandrel with a passivated surface. Perforated mesh is finding increasing application in the electrochemical field as electrode or electrode support material, and a very large potential market exists in the alkaline battery field. The electrodes in nickel–cadmium batteries consist of a support mesh, preferably electroformed nickel, to which a slurry of nickel powder is applied and sintered to produce porous plates of approximately 75–80% porosity. The plates are subsequently impregnated with nickel salts and with cadmium salts which are converted to the hydroxides (active mass) to form respectively the positive and negative plates.

Electroformed foil, mainly copper, is extensively used in the printed circuit industry. Thickness of foil can be accurately controlled with little variation along width and length. At thicknesses less than 0.05 mm (0.002 in) electroforming becomes an increasingly competitive fabrication technique compared with rolling, especially for wide strips. Recently there has been a growing application for electroformed nickel foil for use in "flexible printed circuits".

The electrolytes used for metal foil production are chosen according to the following requirements:

(i) High deposition rates with low power costs to minimise production costs.

(ii) Production of low stressed, ductile deposits.

(iii) Production of sulphur-free copper or nickel in case subsequent annealing, sintering or welding operations are necessary which might embrittle the foil.

9.5 Endless plain or perforated nickel bands

These are manufactured by highly specialised techniques to produce seamless bands without weak spots, characterised by zero stretch, high tensile and fatigue strengths and resistance to abrasion and corrosion. Perforated tapes of any shape can be produced with a pitch accuracy of 0.001 mm and a burr free finish. Deposition of plain bands can be controlled to give a high degree of surface smoothness and insignificant variations in thickness across width and along circumference.

The largest application to date of endless bands is their use as perforated suction tapes in the Molins cigarette making machines. Perforated bands (see Fig. 2) pass round pulleys and a partial vacuum is applied on one side so that tobacco fibres

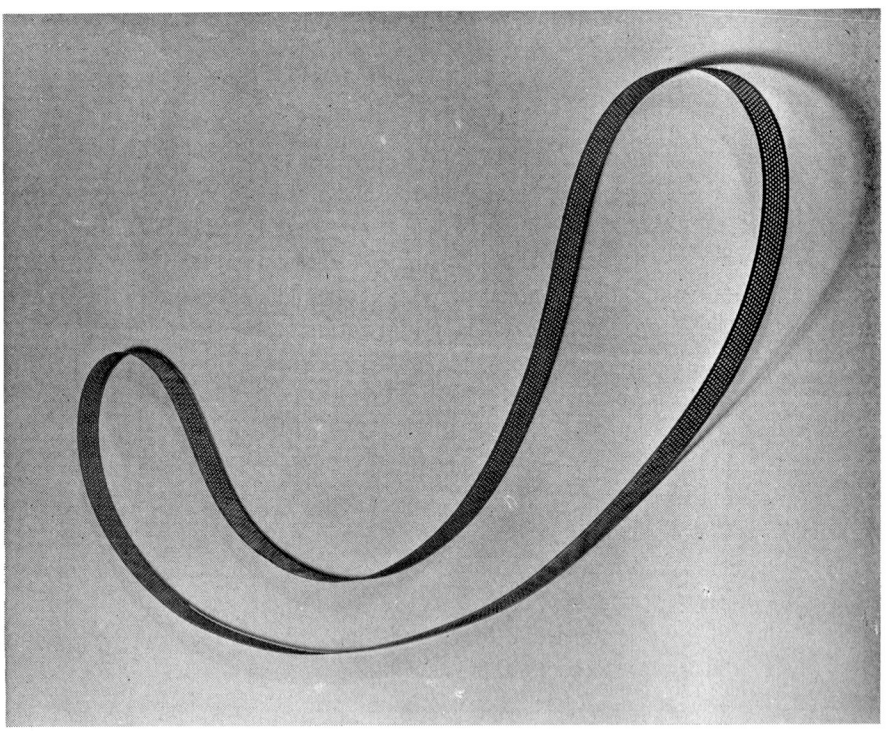

Fig. 2.
Endless perforated nickel band. The product is continuous along its length without seams or joints that could give rise to weak spots.

are picked up on the underside of the band, dropped into a moving band of paper which is rolled to form a continuous cigarette which is subsequently chopped into required lengths.

9.6 Copper spark machining electrodes

The widespread use of electroformed electrodes has to date probably been held back for technical reasons. The required electrodes are usually fairly complex in shape and the acid copper bath is severely restricted by its poor throwing power, while the soft deposits result in fast electrode wear in use. PR cyanide copper can be made to produce hard uniformly thick deposits but recent experiences by the author suggest that sponginess, giving rise to premature electrode breakdown in use, can be a severe problem in deeply recessed areas[20].

Rolls-Royce/EHE Ltd. report that the electroforming of copper satisfies the need for a high degree of repeatable accuracy and relatively low cost in making electrodischarge machine electrodes for aero engine component manufacture. Other materials and methods of electrode manufacture were investigated but it was found that only electroformed copper possessed the necessary properties of low wear rate and dimensional accuracy at an acceptable cost.

The most satisfactory process has been found to be a modified pyrophosphate bath and a patented fusible alloy, both developed[21] by EHE Ltd, Runcorn, Cheshire.

9.7 Razor foils

These are used in the shaver heads of electric razors and merit attention because of the very large quantities produced annually, exceeding 10 million, in competition with etched stainless steel razor foils. Electroformed precision foils are produced with hardnesses of the order of 550–600 DPN.

9.8 Moulds and dies

A wide variety of products are made: moulds with precise parting lines to minimise flashing during accurate moulding of plastics, rubber, type metal, glass, chocolate and other materials readily melted, dies for the pressing and shaping of sheet materials such as steel, brass and copper, and glass reinforced plastics.

Moulds and dies are generally thick walled, complex in shape and their successful manufacture generally demands considerable know-how and skill on the part of the electroformer who has to overcome potential problems of corner weakness and uneven metal distribution. Deposition rates are usually very low and electroformed moulds and dies can be relatively expensive because of extended production times. However, the technique is frequently competitive against other methods of tool making such as machining, engraving or hobbing.

9.9 Metal bellows

Electroformed bellows in nickel metal, nickel–cobalt alloy or nickel–copper, nickel composites are produced on aluminium mandrels which are subsequently dissolved out. The main problem associated with the manufacture of bellows is obtaining a uniform wall thickness and special "high throw" solutions have been developed, generally operating at low current densities (1 A/dm^2; 10 A/ft^2).

9.10 Diamond cutting tools

Electroformed components can be made containing a controlled dispersion of diamond particles. Diamond impregnated cutting wheels, bandsaw blades and drills can be easily used to cut diverse materials ranging from paper and concrete to fibre glass.

Techniques similar to those used to incorporate diamonds can be employed to "co-deposit" a wide range of particles such as carbides, borides and nitrides.

9.11 Aerospace components

Aerospace applications include engine intakes, nose cones, fuel tank bulkheads in rockets, wind tunnel nozzles, venturi tubes, erosion shields and satellite mirrors. The particular type of nickel or nickel alloy used depends on chemical or mechanical requirements. Tensile strengths of 140 kg/mm^2 (200,000 lb/in^2) can be obtained.

9.12 Consumer items

This is a field which probably has an enormous potential market provided that mass production electroforming can be accomplished for this application at significantly lower cost than is possible at present. This demands further development work on new mandrel materials to replace costly re-usable metal masters which severely restrict styling. Flexible or temporary mandrels of conducting "rubbers" or conducting "waxes" are being developed.

A number of consumer items such as coffee percolator bodies, light switch plates, fruit bowls and jugs have been electroformed in America but do not appear to have been economically viable, probably through lack of appreciation of the potentialities of electroforming new shapes and concentration on copying designs easily produced more cheaply by conventional fabrication techniques.

9.13 Apertured components

A very wide range of components can be made for the optical and electronics industries where alternative methods of fabrication would be extremely difficult or too costly. Examples of electroformed material are grids for electron tubes, perforated diaphragms for hearing aids, grid carriers for electron microscopes, shadow masks for colour television and indicator plates for digital computers.

9.14 Metal fibres

There is considerable interest at present in the use of non-metallic fibres *e.g.* carbon fibres, which enable technologists to create new light, strong composite materials. Fine metal fibres made by drawing down wire are finding application in the form of felted materials for use as filtering media, mechanical shock absorbers and metal finishing tools and fibres are also being used in carpet backing to provide anti-static properties.

Recent developments (see Fig. 3) indicate that metal and alloy fibres can be

Fig. 3.
Electroformed nickel fibres. Each fibre is approx. 15 cm long and 25 microns diameter.

manufactured by electroforming. Further development work should result in production of continuous lengths. The properties of electrodeposited fibres can be selected from a wide range; nickel fibres with tensile strengths of 140 kg/mm^2 (200,000 lb/in^2), or soft, extremely flexible, fibres of fully annealed copper or nickel.

9.15 SILVERWARE

The wheel would appear to have turned a full circle in the role of electroforming of silverware items, for whereas firms such as Elkingtons and others produced quantities of such things as dishes etc. up to about 1930, there was a hiatus until

Fig. 4.
A set of cigar jars in sterling silver. Courtesy of Davis and Elson Ltd., copyright Worshipful Co. of Goldsmiths.

recently, when the method is again finding much favour. Much credit for this must go to the Worshipful Company of Goldsmiths and their Technical Development Staff under P. E. Gainsbury, who liasing with such firms as B.J.S. Electroplating Co., have succeeded in achieving this renaissance (see Fig. 4).

9.16 GOLD

Electroforming was the technique employed in making the crown used at the coronation of H.R.H. The Prince of Wales, in 1969. The mandrel was prepared at B.J.S. Electroplating Co. Ltd., and the actual electroforming done by Messrs. Engelhard Ltd.

9.17 ELECTROTYPING

In this large sector of the electroforming industry, reproductions are made of beds of type, with and without illustrative material. The usual practice is to form a thin sheet of copper, which is then given mechanical strength by backing with lead or other materials. Books by Winter[22] and Newell[23] can be recommended for further details, while the chart on pages 412-415 gives a survey of the field. The printing industry is at present evaluating a variety of new methods, and the future of the electrotyping sector of it is very much an open question.

10. The future of electroforming

Improvements in the electroforming process will continue and further increase the number and range of applications. It is difficult, however, to forecast where these future developments will occur. It is possible that main lines of development will be those sketched below.

10.1 METAL AND ALLOYS

Present work with mixed aqueous/non-aqueous electrolytes optimistically suggests that in a few years time it may become possible to electroform chromium with nickel and other metals and introduce a new range of applications particularly at elevated temperatures. Extension of the work may result in commercial usage of electroformed titanium and aluminium products. Aluminium can at present be electrodeposited from an "organic" electrolyte but the process is extremely hazardous in operation.

ELECTROFORMING

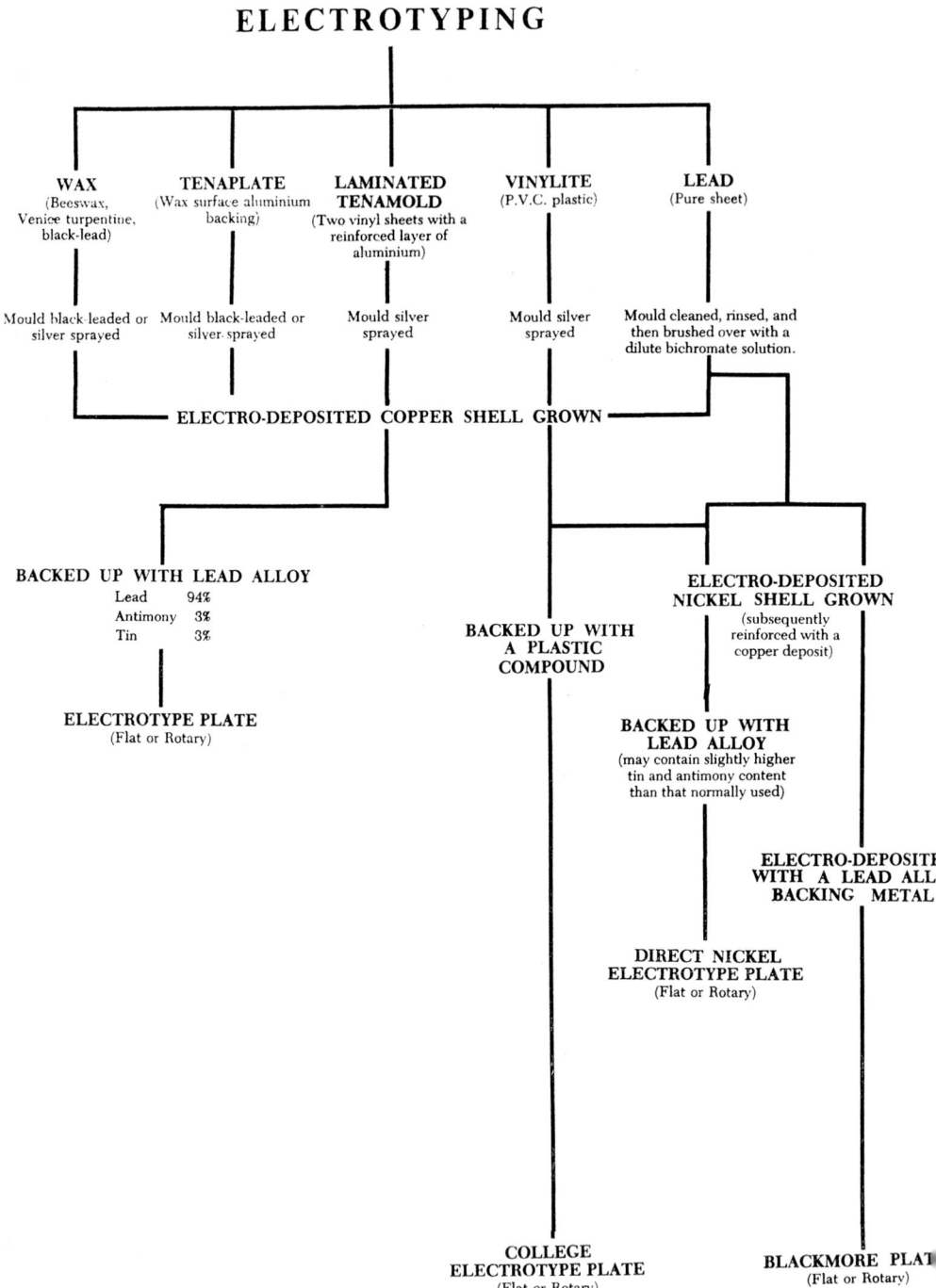

(Designed by H. V. Witherall and produced at Liverpool Regional College of Art, reproduced by permission)

ELECTROTYPE PLATE

(Flat or Rotary)

1. In 1840 the discovery was made that wax, coated with plumbago, became electrically conductive and receptive to a coating of copper. Wax as a moulding medium was messy and unreliable, and has now been replaced by the Vinylite process.

2. Dr. Albert, Munich, introduced lead-moulding in 1905. This medium is recognised as the best available for producing fine-screen and close register colour work, also carton work where the control of shrinkage is essential, as it possesses high fidelity and excellent dimensional stability. It is unsuitable for moulding type forms owing to the high moulding pressures required.

3. Tenaks Products, America, developed Tenaplate in 1935 and the medium is still used in the Trade. It can reproduce any combination of type and blocks and has good dimensional stability, but like wax it has been replaced by Vinylite.

4. Laminated Tenamold, is another product of Tenaks, who claim that it has advantages over the single vinyl sheet for moulding register work. It consists of two vinyl sheets with a reinforced layer of aluminium between. More expensive than Vinylite and has no re-claim value.

5. Vinylite, the thermosoftening plastic moulding medium from the vinyl group, was introduced about 1943, followed in 1946 by Patramold, a plastic material developed by the Printing and Allied Trades Research Association, Silver-spray replaced plumbago as the conducting medium.

6. Lead and Vinylite are the two mediums in general use today. The latter may be Tenamold (single sheet vinyl, purple in colour), Vinylite (a product of B.X. Plastics, marketed by S. D. Syndicate, green in colour), Bakelite 250/2 (black) or Bakelite 250/3 (green), or Mitchell's Polymatt (green and black).

7. Vinylite materials have many advantages over the other mediums; they can be re-deposited to grow further electrotypes; are re-claimable for further use, will reproduce any combination of type and blocks, present less chance of damaging type during moulding, have a uniform shrinkage, and can be filed quite easily or sent through the post with little fear of deterioration.

8. Any number of colour sets can be produced in the lead moulding process each possessing the exact colour values of the original blocks, and it is possible to replace a damaged plate without any difficulty.

9. Electrotypes are nickelled, and often chrome-plated to give runs in excess of a million.

10. Quality magazine work and advertising blocks for newspapers and periodicals are produced in the form of electrotypes.

11. Plates for rotary work can be made flat and then bent to the shape of the printing cylinder by using a bending device, or backed-up on a centrifugal casting machine. The only machine of this type in Britain is at Hazel, Watson and Viney, Aylesbury, and is used to produce Reader's Digest.

12. Electrotypes can have a form of pre-make ready applied to them which reduces "press standing time" and permits the use of a common impression cylinder for colour work.

13. Developments in recent years have seen improvements in both backing-up and the reduction of slabbing with the introduction of tin baths, traying units, pressure casters and consolidators. Pre-press registering devices have facilitated the easier positioning of plates, and the more

general acceptance of rapid plating solution such as copper fluoborate and nickel sulphamate have speeded up shell growing. This operation has also been the subject of experiments in spraying-on the initial coating of copper so that plating could be carried out at maximum current.

14. The decline of the metal-backed plate, with the possible exception of the all-nickel shell electrotypes, in favour of plastic-backed electrotypes has been most marked in America during the last few years, and the examples seen in this country of such plates as Bista, Color-Line, etc., flat and rotary, have been of exceptional quality.

15. Apart from the College and Blackmore electrotypes (see separate sections), very little progress in new duplicate plates has been made here, although it is known that one firm has successfully developed a thermosetting plastic-backed electrotype and has applied for patents.

DIRECT NICKEL ELECTROTYPE PLATE

(Flat or Rotary)

1. Direct-nickel electrotypes, produced by the lead-moulding medium, have a nickel shell of approximately 75 μm grown direct on to the mould, which is then reinforced with a copper.

2. This type of duplicate plate is far superior to the nickel-faced electrotype in that the printing surface, formed by direct contact with the negative, retains the finest details and faithfully reproduces the full tone values of the original.

3. Under normal printing conditions runs in excess of 500,000 are possible.

4. An all-nickel shell electrotype is now being produced in America, claims being made for a longer life, reduced plating costs, and press runs of 2,500,000.

COLLEGE ELECTROTYPE PLATE

(Flat or Rotary)

1. Developed by Mr. W. A. Fry, London, in 1962–3, and is being produced in several firms.

2. This type of plate consists of an electrolytically produced copper shell which is plastic-backed as opposed to metal.

3. The plastic, applied to the shell in paste form whilst it is still in contact with the mould, is used in a cold state and cures at room temperatures.

4. Copper shells as thin as 100 μm may be used, saving valuable depositing time, and the toughness of the backing, combined with resiliency and flexibility, provide a plate of 750 μm upwards (flat or rotary).

5. Any combination of type and blocks can be duplicated by this method.

6. For extra long runs, a nickel or chrome face may be applied.

7. The plastic backing, being minutely porous, contracts on curving, thereby eliminating any stretch of the copper surface.

8. It is not possible to reproduce from this plate using hot-moulding processes, which makes them unsuitable as trade plates for newspapers.

9. Their main asset is high fidelity in fine-screen work a noticeable absence of hard edges. (See Blackmore plate section.)

BLACKMORE PLATE
(Flat or Rotary)

1. Developed in 1963 by the Nickeloid Company, London, and up to the time of writing, produced *only by* them.

2. *First* process ever used in printing by which an *all-metal* duplicate plate is produced *without* any form of heat.

3. Consists of growing a conventional "Direct Nickel" shell (see section on lead-moulding), and then reinforcing it with an electrolytic deposition of a lead alloy backing.

4. Plates supplied, flat or rotary, in 16 gauge (1600 μm) thickness, or laminated to aluminium for a precision lightweight plate of any specified gauge.

5. Being all-metal, these plates are thoroughly satisfactory when subjected to either hot or cold moulding, and they can be repaired.

6. Like the College electrotype, the Blackmore plate is considerably lighter than the conventional electro, and the elimination of heat in the "backing up" operation, any distortion of the face, which is caused when two metals with different coefficients of expansion contract to a different degree, is completely eliminated. This results in a perfect flat surface requiring no "slabbing" with a plate free from high or low spots, hard edges, or dot bruising, and possessing true fidelity with the original.

10.2 REFRACTORY METALS

Thanks mainly to the work of Mellors and Senderoff of Union Carbide, the electroplating and forming of metals such as Nb, Mo, Ta from the so-called group of "refractory metals" is now possible using molten salts and close to being a commercial proposition, and those interested will follow the publications of these authors. Ref. 25 indicates the broad principles. See also Chapter 9, p. 377.

10.3 ENGINEERING DEVELOPMENT

Electrodeposition with associated problems of poor metal distribution, is a slow process in modern times. It is extremely unlikely that electrolyte modifications can overcome this and significant effort should be directed towards contour electroforming for the mass production markets. Contour plating[24] has been used to nickel plate car bumpers at 150 A/dm^2 (1500 A/ft^2). Developments on the engineering side in the design and construction of equipment with high velocity circulation of solution would enable nickel to be deposited at current densities of 200–300 A/dm^2 (2000–3000 A/ft^2) representing forming rates of 50–75 microns thickness per minute.

10.4 Mandrel materials

It is unlikely that many new developments in metallic materials will take place, but in the field of non-metallic mandrels there is tremendous scope for the immediate development of a whole range of new materials such as conducting waxes, resins, plastics and rubbers. Materials will be developed which will give high fidelity reproduction of texture, ease of extraction, and ease of fabrication for immediate deposition.

11. References

1. W. BLUM and G. B. HOGABOOM, *Electroplating and Electroforming*, McGraw-Hill, New York, 1st ed. 1924, 2nd ed. 1930, 3rd ed. 1949.
2. G. GORE, *Theory and Practice of Electrodeposition*, Charles and Co., London, 1st ed. 1856, 2nd ed. 1887.
3. G. GORE, *The Art of Electrometallurgy*, Longmans Green, London, 1877.
4. W. H. WAHL, *Galvanoplastic Manipulations*, H. C. Baird, Philadelphia and London, 1883.
5. W. DE LA RUE, *Phil. Mag.*, 9 (1836) 484.
6. W. G. MCMILLAN and W. R. COOPER, *Treatise of Electrometallurgy*, C. Griffin and Co., London, 1910.
7. W. E. HUGHES, *On the Electrodeposition of Iron*, Bull. No. 6 of the D.S.I.R., 1923.
8. C. H. FARIS, *Uses of Nickel Deposits for Engineering Purposes*, reprinted from *Trans. Inst. Engrs. Shipbuilders in Scotland*, November 22nd, 1927.
9. V. ENGELHARDT (Ed.), *Handbuch der Technischen Elektrochemie*, Band I, Teil 3, Akad. Verlag, Leipzig, 1933.
10. C. L. MANTELL, *Electrochemical Engineering*, McGraw-Hill, New York, 3rd ed., 1961.
11. R. J. KENDRICK, *Proc. 6th Intern. Metal Finishing Conf.*, Institute of Metal Finishing, 42 (1964) 235.
12. R. J. KENDRICK and S. A. WATSON, *Proc. Symp. on Sulphamic Acid and its Electrometallurgical Applications*, Associazone Italiana di Metallurgia, 1966, p. 197.
13. K. BELT, J. A. CROSSLEY and R. J. KENDRICK, *Proc. 7th Intern. Metal Finishing Conf.*
14. *Circuit News*, House J. of British Electricity Authority, December, 1968.
15. A. R. MCCUTCHEON, private communication.
16. *Electroforming as a Continuous Process*, Anon., *Metal Finishing J.*, London, November, 1965.
17. A. BRENNER and S. SENDEROFF, *Proc. Am. Electroplaters' Soc.*, 35 (1948) 53.
17. A. BRENNER and S. SENDEROFF, *Electroplating*, 12 (1959) 207.
18. H. FRY and F. G. MORRIS, *Electroplating*, 12 (1959) 207.
19. J. B. KUSHNER, *Proc. Am. Electroplaters' Soc.*, 41 (1954) 188.
20. P. SPIRO, *Electroforming*, R. Draper, Teddington, 1968.
21. R. FRAMPTON, Rolls-Royce Ltd., private communication.
22. A. WINTER, *Stereotyping and Electrotyping*, Pitmans, London, 1949.
23. L. F. NEWELL, *Electrotyping and Stereotyping*, Pitmans, London, 1962.
24. *Pontiac's New Contour Plating Machine, Product Finishing*, August, 1964.
25. G. W. MELLORS and S. SENDEROFF, *Electroforming of Refractory Metals, Plating*, 51 (1964) 972.
26. H. SILMAN, *Design Engineering*, Nov., 1970, p. 88; *Metal Finishing*, Dec., 1969, p. 36; *Chem. Engineering*, April, 1953, p. 189.

Miscellaneous reading
ASTM Symposium on Electroforming, *Applications, Uses and Properties of Electroformed Metals*, STP No. 318, Am. Soc. Metals, 1962.

Chapter 11

The Electrodeposition of Paint

B. A. COOKE, N. M. NESS AND A. L. L. PALLUEL

Imperial Chemical Industries Ltd., Paints Division, Slough, Bucks

1. Introduction

The application of paint to metal by a process of electrodeposition has gained widespread acceptance during the past ten years on account of the special benefits it offers to a wide range of industries, principally those involved in the finishing of steel fabricated articles. Whilst its main features are now well known in the Paint Industry, it is perhaps not widely appreciated amongst electrochemists in general that such elementary principles as Faraday's laws have, in combination with the modern development of waterborne stoving paint resins, given rise to a uniquely satisfactory method for the priming or one-coat finishing of metal fabrications.

In offering a survey of this rapidly developing technology[*], we feel justified, therefore, in beginning by restating the general features and advantages of paint electrodeposition. We follow this with a survey of waterborne resin systems that are suitable, or have been proposed, for the process; some of this material inevitably consists of a patent review and we have avoided detailed descriptions of specific compositions. It has not been possible to treat the physical chemistry of this subject without reference to original experimental material gathered in our laboratories, simply because the information published to date is small in quantity and fragmentary in nature. In introducing the technology of paint electrodeposition, we have attempted a balanced presentation of the divergent developments which have occurred, especially insofar as they bear on the control of the process in diptanks over an extended time period.

2. General nature of the process

In the application of paint by electrodeposition, the article to be coated is submerged in a bath of a suitable water-borne paint and connected to one pole of a direct current source. An electrode of the opposite polarity is also present in the

---

[*] Between 1963 and 1969 over 400 electropainting production lines were installed throughout the world, mostly in Europe though other important centres of development have been the U.S.A. and Japan. Major installations exist also in Australia, South Africa and South America.

paint tank or the tank may itself act as the counter-electrode. A current passes and deposition of a paint film is induced on the article by the electrode reaction occurring at its interface with the aqueous paint. When deposition is complete, the article may be removed from the bath and rinsed with water to free it of adherent aqueous paint and leave behind the coating induced by the electrode process. The electrodeposition of paint differs characteristically from that of many polymer lattices in that a compact, coalesced film of high electrical resistance is formed. The major feature distinguishing it from other methods of paint application is that the paint film formed on an article differs in composition from the contents of the paint tank from which it originated.

The process of forming an organic film on an electrode in an electrolytic cell can be illustrated by reference to the main class of materials used in current commercially operated processes. These are film-forming resins containing carboxyl groups in sufficient numbers so that when neutralised by a base, stable aqueous dispersions are obtained at relatively low concentrations. The neutralisation may be incomplete, say 50 to 80%, an organic "coupling" solvent* may be present to assist dispersion stability, and the base, which is generally mono-acidic, may be selected from a wide range including amines, ammonia and the alkali metal hydroxides. Anionic resins of this type are insoluble in water in the pure acid condition or if only a small proportion of the carboxyl groups is neutralised by base. If a dispersion of the sort described is electrolysed, a deposit forms at the anode. This consists of the almost base-free resin** if the anode is noble or passive, or of the corresponding soap of the resin in the case of a reactive anode metal. Anodic deposition from this type of self-stabilised dispersion forms the basis of the current commercial process, though there are other instances of electrodeposition of organic film-formers, for example, cathodic deposition from self-stabilised cationic resin dispersions or electrolysis of externally (surfactant) stabilised systems which may be anionic or cationic. Electrodeposition from surfactant stabilised dispersions is somewhat akin to non-aqueous electrophoresis, and indeed paint electrodeposition is frequently referred to as "electrophoresis"†, a term which we prefer to reserve for the motion of the micellar counterion-stabilised particles towards the region of the anode, where they are destabilised by electrolytic action.

In favourable cases (and the exceptions are of no practical interest), precipitation is followed, though not always immediately, by coalescence to yield a continuous film from which phase-segregated aqueous matter is expelled by electro-osmosis. The final film therefore has a high content of involatile matter (*ca.* 90%) and may

---

* For a definition of this term, see 3.2.3.
** We have refrained from using the term "acid resin" although this notion is to be found in simplified accounts of the process, because the anodic deposit is easily shown by analysis to have a small but definite residual base content.
† Other terms in common use are "electropainting" and "electrocoating", the materials used being called "electropaint" or "electrocoat"; there are also several proprietary names.

be of very high viscosity. The deposited film thus differs sharply in both chemical and physical respects from the parent dispersion.

The essentials of the electrodeposition of resins from dispersion were understood by Sumner[1] and Fink and Feinlieb[2] well before the current phase of commercial exploitation.

The presence of pigment particles stabilised in the aqueous resin system does not affect the course of the deposition process: they migrate electrophoretically with the resin micelles (and indeed generally form part of the micelles if the electrodepositable resin has been used as the pigment dispersant) and remain at the anode in the de-stabilised resin film.

2.1 Advantages and limitations of paint electrodeposition

From the considerations just stated, it is possible to deduce the important technological advantages of the process compared with conventional mass-production painting methods.

1. In accordance with Faraday's laws, the thickness of deposit obtained is related to the quantity of electricity passed and not to the concentration and rheology of the dispersion nor to the shape and disposition of the article as is the case in conventional dip painting.

2. Because the deposited film differs sharply from the adherent quantity of dispersion (the "dip-coat" or "drag-out") removed with it when a coated article is withdrawn from an electrodeposition tank, it is possible to rinse the coated article in water, for example in a spray-rinse, leaving the electrodeposit unaffected. This provides a film whose thickness is largely independent of the disposition of the various metal surfaces of which the article is formed.

3. The loss of paint material which occurs on rinsing, which generally is not considered worth recovering*, can be kept small by choosing a low concentration for the dispersion, for example about 10% at which dispersions of many waterborne resins have a viscosity only slightly exceeding that of water, e.g. 1 to 3 cp at room temperature. As a result the tank contents are easily handled, pumped, heat-exchanged, and have a relatively high electrolytic conductivity, typically in the range above 0.5 milliSiemens.

4. The deposited film, being composed chiefly of a highly viscous, poorly dissociated poly-electrolyte, has a high resistance. This, combined with the electrolytic conductivity of the aqueous dispersion, ensures that current shunts from the more accessible regions of the article being coated towards less accessible regions, and so on until if sufficient time is allowed, the article receives an even thickness of

---

* Cf. Addendum, p. 462.

electrodeposit. The isolation of this electrodeposit from the adherent dip-coat by rinsing in water (noted under 2. above) ensures, then, a highly uniform coating of controlled thickness.

5. Because the stabilising base is substantially eliminated from the electrodeposited film during its formation, and therefore does not appear in the composition being stoved or in the final film, a wider choice of stabilising bases is possible than with waterborne dipping resin systems. In particular, there is no objection to the use of relatively involatile amines, which might otherwise interfere with the curing reaction, nor to alkali metals as counterions, which would survive the curing process and be present in the film.

6. The low solids content required for the bath not only reduces the working capital requirement but also the financial risk associated with contamination, overheating or any other accidental occurrence to the bath.

7. In common with other waterborne painting processes, fire and toxic hazards are negligible.

8. Because the electrodeposit has a high solids and low solvent content, there is no need for a protracted period of solvent evaporation ("flash-off") before articles are passed to curing ovens. For the same reason, coatings within confined spaces, such as the box-sections of a motor body, are not subject to the defect known as "solvent-washing" which is experienced with many conventional dipping processes*.

9. Two factors contribute to better coverage of the edges of articles by the electrodeposition method. The high local current densities operating in such areas, which are typical of electrolytic processes, ensure adequate primary coverage of these parts, while the high solids content of the unstoved film makes it possible to have considerably less thermal flow away from edges during stoving. In contrast, surface tension effects in conventional dipping make primary edge coverage difficult, while thermal flow is also a problem with that method.

10. Electrodeposition gives a high material utilisation efficiency and is readily adapted into continuous, automatic processes for the finishing of a wide variety of articles. It wil be pointed out in the section concerned with Technology that it opens the possibility of dispensing with or reducing other operations in complete finishing processes.

The main limitations of paint electrodeposition, or of the majority of systems and processes currently being exploited, are as follows.

1. Being a dipping process, it is limited to paint systems which cure at a relatively high temperature, typically above 150°C, for the simple reason that the autoxidative stability at ambient temperature of lower curing systems would be inadequate for

---

* This arises especially in complex sections in which variations in metal thickness (and therefore heat capacity) exist: solvent refluxes from the rapidly to the slowly heating parts of the interior surface and can entirely remove the paint from the latter, leaving it unprotected.

the prolonged retention of the main inventory of paint in a dip tank. Even so, autoxidation in the tank is one of the sources of instability to be considered in formulating materials for application by this method.

2. Another possible cause of instability is the long-term dispersion behaviour of the pigment/resin system at low solids in water. Clearly, this must not flocculate, yet it is desirable that it should retain its sensitivity to the destabilising effect of electrolysis, *i.e.* quantitatively speaking that the yield of deposit per coulomb of electricity should not alter seriously.

3. Waterborne anionic resin systems are sensitive to heavy metal cations which cause flocculation through soap formation.

4. The electrodeposition process, if not the stability of the dispersion, is interfered with by contamination with many simple mobile anions (Cl^-, SO_4^{2-}, $H_2PO_4^-$, HPO_4^{2-}, etc.), the typical effect being excessively textured or even ruptured films; paints vary in their susceptibility to this contamination, but concentrations above 0.01 equiv.l^{-1} are not generally easily tolerated.

5. The nature of the substrate is highly critical, which is hardly surprising for an anodic process. Steel and phosphated steel behave substantially, though not completely, as passive substrates. Many metals, like zinc and copper, dissolve anodically to form the corresponding soaps of the paint resin, but this does not make the process unworkable: it simply modifies the film growth process and the nature of the film obtained. Aluminium can be coated quite satisfactorily, acting as a passive substrate, though under some conditions much of the resistance increase as deposition proceeds is due to anodisation.

6. The redox action of metal entering the film by substrate dissolution and possibly anodic oxidation of resin constituents can have an influence on the cross-linking reaction by which the system cures. This is not normally a limitation, since the actions mentioned enhance cure, but it is one of the features which can lead to a different result by electrodeposition compared with conventional application of the same material.

7. Metal soaps can contribute colour to the film and there has therefore been a tendency to select dark pigmentations in primer formulations, many of which embody an autoxidising resin system which also contributes colour. This limitation has been overcome in some one-coat electrodepositing finishes, notably those based on acrylic resins incorporating non-autoxidising cross-linking components, which furnish white and pale shades with adequate opacity.

8. A high level of gloss is not easily and consistently obtainable.

Patents on the electrodeposition of dispersed organic film formers appeared as early as 1919 for Japan[3] and others on synthetic resins[4], rubbers[5] and oleo-resinous lacquers[6] followed in the course of the next two decades. Little commercial exploitation of the process took place until the early 1960's when patents[7] appeared covering the anodic deposition of paint, *i.e.* resin plus pigment dispersions. The early adoption of the technique in industrial painting processes was almost certainly handicapped by the limited choice of waterborne resins for paint formulation. The development of waterborne resins of a wide variety of compositions and physical properties is itself a relatively recent advance spurred on by the increasing need to reduce solvent pollution of the atmosphere and to avoid the fire hazard associated with large industrial dip painting tanks.

The advantages of electrodeposition listed above are of considerable technical value, especially the uniformity of coverage and the ready adaptability to mass-production. The special significance of the uniformity of coverage lies in the fact that in conventional painting methods many painted articles have failed by corroding in regions where the coverage is poor; even if the intrinsic corrosion resistance of an electrodeposited primer is not better than that of the conventionally applied material, the performance of the electroprimed article is superior because of the protection afforded in vulnerable regions like seams, edges, hollows and so on. The speed with which the method has been adopted by certain sectors of industry is remarkable. In Europe, for example, a large percentage of the automobiles produced are now being primed by electrodeposition, and several major manufacturers have their entire production on this process. The situation is the same in Japan, where growth has been even faster, 80% of the automotive industry adopting electropriming in 3–4 years. Few electropainting plants for automobiles are operational in the U.S.A., but all the major producers are now installing the process as existing plant is amortised and new production facilities introduced. It is a reasonable prediction that within 5 years, over 90% of the world production of motor vehicles will be electropainted. When it is considered that a complete continuous painting line for the electropriming of car bodies may cost about £300,000 the size of the investment in the process by the motor manufacturers can be seen to be very considerable.

There are many manufactured articles besides motor vehicles for which painting by electrodeposition offers substantial benefits. Indeed many articles which cannot be dip-painted satisfactorily, if at all, with conventional paints are easily dealt with by the process. Complex shaped articles may be impossible to paint on all surfaces by spraying, yet trap too much paint for dipping. Using electrodeposition followed by a suitable rinse stage, such articles can be painted with a uniform and blemish-free film. Even apparently simple articles such as wheels present problems to the dip-painter, as it is virtually impossible to jig a wheel on a production line so that no pool of paint is left after dipping. With electropainting, perfectly uniform films can be obtained without difficulty and many millions of car, truck and tractor wheels are electropainted annually at significantly lower cost than previously. Either a primer or a one-coat finish may be applied.

Petrol tanks, as can be readily understood, are particularly difficult to paint internally and protection of the interior was often achieved by plating the two halves of a tank before welding together. Electropainting provides a ready answer to the problem since the low viscosity electropaint drains freely from the interior on removal from the tank and any residual liquid is so low in solids content as to cause no defects in the interior.

Metal tool handles, switch boxes, meter cases, light fittings, metal furniture, shelving, filters, heaters, horns, fans, are examples of items receiving one coat finishes by electrodeposition. As more experience in colour and gloss control is gained and higher performance one-coat electropaints are developed, one-coat finishing of domestic appliances can be anticipated. Already in Europe several refrigerator, washing machine and dishwasher manufacturers are priming these articles with electropaint and spraying a conventional finish. Electropaints pigmented with aluminium flake are in use in production lines on wheels, tool handles, metal chairs and instrument cases.

3. The composition of electrocoat paints

An electrodeposited paint film, like that of a conventionally applied paint, consists of a relatively small volume fraction of a finely dispersed pigment embedded in a continuous matrix of an organic polymer. In order to be useful as a paint vehicle, the polymer must form a continuous, homogeneous film on casting from solution and be capable of cross-linking (curing) by an autoxidation mechanism or by chemical interaction between reactive groups on the polymer molecule. In the liquid paint, before application, the vehicle is dissolved in a volatile solvent whilst the pigment particles are held in suspension by the action of an adsorbed layer of the vehicle or of an auxiliary dispersing agent.

Most paint vehicles used for the painting of metal substrates are essentially hydrophobic and largely non-polar in character; protection of the metal is ensured not only by the physical barrier interposed between the substrate and the corrosive environment but also by the electrical resistance of the paint film in the electrochemical corrosion circuit. In these essential principles electrocoat paints differ in no way from conventional paints; it will be sufficient therefore to discuss only the features which are special to the electrodeposition process.

1. In electrocoat paints water is invariably used as the preponderant solvent and the polymeric vehicle is held in dispersion by the action of a relatively minor quantity of hydrophilic ionic groups. The essential condition for electrodeposition is, then, that the ionic stabilisation should be inactivated by electrolysis at one or other of the electrodes.

All commercially used electrocoat paints are anionic which implies that the vehicle is acidic in nature and is stabilised in the aqueous medium by salt formation with a base. On passage of an electric current through the bath, deposition occurs at the anode. Cationic vehicles have also been studied; these are essentially basic polymers, stabilised by salt formation with an inorganic or organic acid; deposition occurs at the cathode.

2. Electrodeposition of the vehicle and its associated pigment occurs as a result of a change in the ionic environment in the vicinity of the electrode to be coated when electrolysis commences. In essence, the vehicle for electrodeposition must be stably dispersible as the salt but insoluble in the aqueous environment as the free acid, in the case of anodic deposition, or the free base in the case of cathodic deposition.

3. Because the principal practical function of a cured electrocoat film is the protection of a metal substrate, most vehicles for electrodeposition have been developed from conventional paint vehicles of known protective properties. Sufficient ionic character for efficient electrodeposition can be imparted to an essentially hydrophobic vehicle either by the use of conventional ionic emulsifying agents or by the chemical incorporation of ionic groups into the polymer molecule. A combination of these two approaches has frequently been used, sometimes fortuitously.

The first method tends to give opaque aqueous vehicles of fairly coarse particle size, clearly identifiable as emulsions; the second approach can give optically clear aqueous "solutions" but more frequently gives hazy suspensions of very small particle size, often beyond the scope of optical microscopy. No study of the phase structure of an electrocoat vehicle has been published but it seems certain that even in the "clear solutions" the polymer molecules are associated as micelles: indeed they frequently exhibit the Tyndall effect and show the typical rheological properties of micellar solutions.

4. The satisfactory use of electrocoat paints in large dip tanks requires the highest possible order of dispersion stability. It is well established that the stability of emulsions depends *inter alia* on the particle size, on the concentration of ionic stabiliser, and on the probability of desorption of the stabiliser from the surface of the dispersed particle. Although very good stability can be achieved with emulsions of paint vehicles, using conventional emulsifying agents, experience has taught that the problems of long term stability are on the whole easier to solve with self-stabilised dispersions carrying the stabilising ionic groups on the polymer molecule itself.

5. Uniformity of coating thickness even on the most complex shapes is perhaps the most important technological benefit of the electrocoat process. Uniform coulomb density over the area to be coated is ensured by the deposition of a continuous film of high electrical resistance. This requires the use of polymers of relatively low glass transition temperature (T_g) or the use of solvents or other coalescing aids which have a like effect.

6. Redissolution of the deposited film on interrupting the current is a defect of many electropainting systems which can be inconvenient on a conveyorised painting line where occasional stoppages are inevitable. Such occurrences are often associated with vehicles of excessively hydrophilic character and deposited films of low viscosity.

7. The curing of electrocoat films in the stoving oven is achieved by the conventional autoxidative or chemical cross-linking reactions. Additional "cure" is undoubtedly effected by ionic cross-linking through multivalent cations anodically dissolved from the substrate. The Kolbe reaction and other anodic oxidative reactions have also been reported to be involved in cross-linking of electrodeposited films.

3.1 VEHICLES FOR ELECTRODEPOSITION

Many types of organic polymers have been suggested for use in electrodeposition. The following is a digest of the technical literature and of patents published between 1962 and May 1969.

3.1.1 Maleinised drying oils and modified maleinised oils

Probably the most fruitful method of formulating anionically stabilised aqueous paint vehicles was initiated in 1934 by Clocker[8], who studied the chemical reaction between unsaturated vegetable oils and maleic anhydride. This reaction, which occurs at a practical rate only with multiple unsaturated systems, can take one of two courses according to whether the multiple double bonds in the fatty acid residues are conjugated, as in tung oil, or interrupted by an isolated methylene group, as in linseed or soya bean oil. In the first case, a Diels–Alder mechanism obtains and addition of maleic anhydride results in the formation of a *cyclo*hexene dicarboxylic anhydride; in the second case simple addition occurs at a methylene group adjacent to the double bond and the product is a succinic acid derivative. Fumaric acid can also participate in similar reactions. In either case a dicarboxylic acid or anhydride is formed, linked to the fatty oil by a non-hydrolysable carbon to carbon bond. Provided the molar ratio of maleic anhydride to oil is sufficiently high (*e.g.* greater than 1.5/1) the products are soluble or dispersible in aqueous base and can be used in electrodeposition.

The film properties of the simple maleinised oils are somewhat mediocre, and they are particularly deficient in hardness. More successful paint vehicles have resulted from attempts to increase the molecular weight and the inherent hardness of these products. Four general methods have been used:

(a) thermal bodying of the oil before or after maleinisation[9];

(b) blending or condensation with reactive cross-linking resins, notably phenol/formaldehyde[10] or melamine/formaldehyde[11] condensates;

(c) *in-situ* polymerisation of vinyl monomers, notably styrene[12], vinyl toluene[13] and cyclopentadiene[14] as is commonly practised in the manufacture of solvent borne styrenated alkyds; in such cases, at least part of the resulting vinyl polymers are attached to the fatty residues by grafting or copolymerisation[15];

(d) poly-esterification with di- or tri-hydric alcohols[16] to form condensation polymers akin to the conventional alkyd vehicles.

Aqueous paints based on maleinised oils and incorporating one or more of the additional features listed above were probably the first successful electrocoat paints and are still in quantity production for the coating of car bodies and numerous other fabricated articles.

3.1.2 Alkyds

Alkyd resins—branched polyesters incorporating natural fatty acids, glycerol or other aliphatic polyols, and phthalic anhydride—have been the mainstay of the paint industry since the late 1930's. Anionic solubility or dispersibility in aqueous base can readily be achieved by increasing the relative proportion of phthalic anhydride and leaving some of the carboxyl groups unreacted. In practice, however, a satisfactory compromise between adequate stability in aqueous medium and satisfactory film properties is difficult to achieve. Waterborne phthalic alkyds tend to be excessively low in molecular weight and/or cross-linking functionality and to give films lacking in cohesive strength.

The marketing of newer alkyd raw materials such as trimellitic anhydride[17] and dimethylol propionic acid[18] has facilitated the formulation of waterborne alkyds of adequate molecular weight and of much improved film properties. Such vehicles may well form the basis of successful electrocoat paints. Because the polymeric structure of alkyd resins relies entirely on ester linkages, slow hydrolysis under the slightly alkaline conditions of an electrodeposition bath may result in long term degradation, sufficiently severe to affect film properties.

3.1.3 Epoxy esters and analogous vehicles

Epoxy resins, formed by the poly-condensation of dihydric phenols (notably diphenylol propane) and epichlorohydrin, and marketed under a variety of brand names, have for some years formed the basis of several types of paints noted for their resistance to aggressive environments. In particular the products of their esterification with natural fatty acids have been highly successful as vehicles for primers in the automobile and domestic appliance industries.

That the goal of combining the benefits of electrodeposition with the resistance properties of epoxy esters is a highly desirable one is testified by the fact that such vehicles are featured in the largest group of patents published in the period under review.

Two general methods have been advocated for making conventional fatty epoxy esters anionically soluble or dispersible in water:

(a) The available hydroxyl and epoxide groups in the epoxy resin are completely or nearly completely esterified with fatty acids, including at least a proportion of unsaturated fatty acids, and the latter are subsequently reacted with maleic anhydride as in a maleinised oil[19].

(b) The epoxy and hydroxyl groups are only partially esterified with fatty acids (as in a solvent borne epoxy ester) and some of the residual hydroxyls partially esterified with a di- or tri-basic carboxylic anhydride or acid. The polybasic anhydride and acids used include phthalic[20], maleic[21], succinic[22] and trimellitic[23] anhydride, dimerised fatty acids[24], and maleinised oils or fatty acids[25].

Resinous polyols[26] other than epoxy resins can also be used in a similar way.

The successful use of epoxy ester vehicles in electrocoat tanks has now been established for several years and at the time of writing the growth rate of such products is probably higher than that of any other type.

3.1.4 Acrylic vehicles for one-coat finishes

In the painting of car bodies, the electrocoat primer forms the first coat of a multiple coat paint system and the colour of the paint vehicle is a relatively unimportant factor. In other markets, notably in the domestic appliance and metal furniture industries, where one-coat finishing with conventional paints is an accepted practice, the electrocoat paint may be required to perform both protective and decorative functions. The colour of the deposited film then becomes an important consideration.

Undesirable colour in electrodeposited paint films may originate from three sources, *viz.* inherent colour in the vehicle, chemical degradation during or after curing or from chromophoric metal cations introduced in the film as a result of anodic action during the electrodeposition process.

Although white paints of satisfactory colour characteristics can be formulated for electrodeposition using alkyd and epoxy ester vehicles modified with saturated fatty acids, the patent literature clearly indicates that copolymers of acrylic monomers are the preferred vehicle for such applications. Not only do acrylic vehicles have excellent initial colour and after-yellowing properties, but they can also be rendered water-dispersible at low effective ionic concentrations, thus minimising the extent of substrate dissolution during electrodeposition (because the quantity of electricity required to cause deposition is small).

The range of acrylic type monomers available to the paint industry is very large and a water-borne copolymer for use in electrodeposition commonly contains at least four co-monomers of different characteristics:

(a) an acidic monomer (acrylic acid, methacrylic acid, itaconic acid, etc.) to provide anionic stabilisation in aqueous base,

(b) a reactive monomer (which may carry hydroxyl groups, amide groups or substituted amide groups) to provide sites for cross-linking,

(c) "neutral" monomers selected from the available "hard" monomers (methyl methacrylate, styrene), and

(d) the plasticising monomers (ethyl acrylate, butyl acrylate, 2-ethylhexyl acrylate etc.), the proportions being adequate to give the total copolymer the required mechanical properties.

The acrylic vehicles are almost invariably cured by a chemical cross-linking reaction not involving autoxidation.

Two general approaches to the formulation of anionic acrylic vehicles for electrodeposition, can be distinguished in the patent literature. In the first of these, the acrylic resin carrying hydroxyl or amide groups is co-dispersed in aqueous base with a multifunctional melamine– or urea–formaldehyde condensate[27]; on electrodeposition, both constituents deposit simultaneously and cross-linking of the acrylic copolymers occurs on stoving, through the formaldehyde condensate. In the second approach, the acrylic resin is self cross-linkable by the use of copolymerised hydroxy or alkoxymethyl acrylamide[28].

At the time of writing, acrylic electrocoat paints for one-coat finishes in white and pale colours are being offered by several paint manufacturers, but these products have not as yet achieved wide acceptance in the utilising industries.

The use of the electrocoat process in one coat finishing imposes additional constraints of consistency of appearance, of gloss, and of colour and it is yet to be widely acknowledged that such a process can be as reliable in operation as the electrodeposition of primers.

3.1.5 Emulsion vehicles

The emulsification of conventional paint vehicles by the use of minor amounts of ionic surfactants provides a superficially attractive route to the formulation of waterborne vehicles for electrodeposition. Such vehicles form the subject of a relatively small number of patents[29] and they do not appear to be widely used in commercial operations. In many cases the stability of such emulsions falls short of the stringent requirements for continuous dip-tank operation unless such high levels of surfactant are employed as to cause deterioration in film properties.

On the other hand the "extending" of an anionic vehicle with substantial quantities of a water-insoluble, non-ionic, film forming resin (*e.g.* a hydrocarbon polymer) is widely and successfully practised[30]. In such cases the "extender" resin can indeed be considered to be emulsified in the anionic vehicle and is deposited, like the paint pigment, by virtue of the associated ionic polymer.

3.1.6 Cationic vehicles

Few patents have been published concerning cationic vehicles for electrodeposition; all but one describe acrylic vehicles containing copolymerised basic monomers such as dimethylaminoethyl methacrylate or vinyl pyridine[31]. The polymers are dispersed or dissolved in water by salt formation with an organic or inorganic acid. Cross-linking is most frequently effected by copolymerised alkoxymethyl acrylamide and similar reactive monomers.

The benefits to be expected from cathodic deposition stem from the absence of substrate dissolution at the electrode where film formation occurs. Problems of film discolouration are consequently alleviated. On the other hand, the preferred curing agents, such as copolymerised alkoxymethyl acrylamide or co-dispersed melamine resins, are not usually considered stable under even slightly acidic conditions. Furthermore, the design of electrodeposition plant to withstand pH values below 5 presents some daunting engineering problems.

3.1.7 Cross-linking resins

Where the vehicle is not capable of self cross-linking, whether by autoxidation or by chemical interaction, a separate cross-linking resin must be used in electrodeposition, as in conventional paints.

Most frequently these are condensates of formaldehyde with phenols, urea or melamine and cross-linking occurs by reaction of mutually reactive sites on the multifunctional film formers and the cross-linking resin.

For electrodeposition, four methods of incorporating the cross-linking resin have been advocated:

(1) The use of a water soluble melamine or phenolic resin, simply blended with the aqueous vehicle. Co-deposition of the two constituents then depends on a favourable partition of the cross-linking resin between the aqueous and disperse phases. In many instances such favourable circumstances do not obtain and the deposited film contains less cross-linking resin than is desirable for effective cure.

(2) The use of water-insoluble melamine or phenolic resins, co-emulsified with the principal film former. In this case co-deposition of the two constituents occurs in the same proportions as is present in the aqueous paint.

(3) The use of anionic cross-linking resins, separately dispersed by salt formation in the aqueous paint and electrodeposited independently but concurrently with the principal film forming vehicle. Examples of anionic melamine[32] and phenolic resin[33] have been described in the patent literature.

(4) Prior partial condensation of the film forming vehicle with a cross-linking resin[34]. In this case the combined resins behave as a single constituent.

3.1.8 Other resins

A number of other vehicles for electrodeposition have been disclosed in the patent literature which do not fall neatly into the above classification.

These include: blends of waterborne acrylic copolymers with waterborne alkyds and epoxy esters[35], the products of the maleinisation of polymeric condensates of fatty acids[36], the products of the partial esterification of acidic copolymers with hydroxalkyl fatty esters[37], maleinised hydrocarbon polymers including polybutadienes[38], and acidic copolymers incorporating acidic groups derived from phosphorus and sulphur[39]. The last of these represents an interesting departure from conventional paint technology.

3.2 OTHER FEATURES OF ELECTROCOAT PAINT COMPOSITION

Besides the vehicle, an electrocoat paint contains neutralising base, pigment and, usually, an organic solvent, all of which are important in determining the performance of the dispersion and of the film finally obtained. Although in the selection of these materials much reliance has been placed upon accumulated experience in paint technology, recognition of the special circumstances of the electrodeposition process is leading in directions not contemplated with conventional materials.

3.2.1 *The neutralising base*

Most patents in the field of electrodeposition specify the use of ammonia or of volatile, water-soluble, nitrogenous bases for the neutralisation of the acid groups in the resin. One patent[40] claims benefits from the use of potassium hydroxide and other alkali metal hydroxides whilst a growing number describe the use of either organic or inorganic bases.

While with conventionally applied anionic, waterborne paints the efficient evaporation of the neutralising base from the film on stoving (or its participation in a cross-linking reaction) is an important condition for obtaining satisfactory film properties, in electrodeposition most of the base is removed from the film by electrolytic action. The nature of the neutralising base can then play little or no part in determining the properties of the cured electrocoat film.

The role of the neutralising base is of course fundamental to the physical stability of the aqueous paint. The common practice of partial neutralisation of the available acid groups, for example to a pH range of 7.5–8.5, is consistent with the view that hydrogen bonding between carboxylate ion and undissociated carboxyl groups contributes significantly to the stability of the interface between a dispersed particle and a continuous aqueous phase[41].

The factors affecting the choice of the neutralising base are similar for aqueous paint vehicles to those applying in the surfactant field. The hydrophilic nitrogenous bases carrying hydroxyl or ether groups are often favoured where the hydrophilic properties of the film forming polymer have been reduced to a minimum. The differences between other common organic bases, *e.g.* the alkylamines, and the alkali metal hydroxides are small and can readily be compensated by other formulation variables.

Many electrocoating polymers carry a multiplicity of ester groups which are subject to alkaline hydrolysis; the risk of hydrolytic degradation in the dip tank is, however, usually too small to be significant in commercial operation. Like the hydrolysis of monomolecular esters it is controlled by the concentration of hydroxyl ions (*i.e.* by the pH in the tank), the identity of the neutralising base cation being of no significance. On the other hand, the electrolytic mobility of the base cation plays an important part in determining the conductivity of the paint bath. The patent literature reveals that this factor is still somewhat contentious: whereas one patent[42] discloses means of obtaining very low bath conductivity*, another[40] contends that throwing power is enhanced when using a high conductivity paint neutralised with potassium hydroxide.

The rheology of aqueous anionic paint vehicles is quite unlike that of conventional paints and unlike that of the same resin when thinned in organic solvent without neutralisation. While the viscosity of solvent borne paints is controlled largely by the molecular size of the dissolved polymer and by the possibility of inter-molecular dipole interaction, the rheological properties of anionic aqueous vehicles resembles that of conventional soaps in depending on the size and structure of micellar aggregates. Small changes in pH or solids content can induce rearrangement of the micellar structure resulting in very large changes in viscosity. Many such anionic paint vehicles show a critical range of solids content (chiefly in the region of 30–50%) where very high viscosities and marked thixotropy are observed; below the critical region the viscosity is low and Newtonian whilst above it, where the organic solvent is preponderant, the rheological properties are similar to those of the un-neutralised vehicle.

This anomalous behaviour is of practical importance in determining the mode of addition of the make-up paint to a working tank.

3.2.2 Pigmentation

In electrocoat paints, the pigments perform the same function as in conventional paints, providing opacity, colour, "structure", and a measure of additional hardness, but the use of aqueous anionic vehicles and the mode of application impose additional constraints on both the choice and level of pigmentation.

Few anionic vehicles tolerate significant quantities of soluble multivalent cation without loss of stability. Pigments containing appreciable quantities of water-soluble heavy metal salts should therefore be avoided, so should reactive pigments such as zinc oxide or red lead. This leaves the choice of pigments wide, including all grades of titanium dioxide, iron oxide, carbon black and most organic pigments stable to alkaline conditions. Among extenders, calcium carbonate, china clay, silica, barytes and talc have all been used successfully. Of the anti-corrosive chro-

---

* The conductivity of electrocoating baths has often been used as a means of assessing the concentration of contaminating salts.

mate pigments, lead and barium chromates give stable paints but little anti-corrosive action whilst the use of the more effective zinc and strontium chromates frequently results in instability on storage. Lead silico-chromate (*cf.* Oncor M50—National Lead Co.) is often named as the preferred anti-corrosive pigment in electrocoat paints[43].

Because of the low viscosity of the dilute paint in the tank, pigment settlement can be severe; for this reason "micronised" pigments are often specified and the paints frequently contain mild flocculants to prevent hard settlement.

The anionic vehicles themselves are usually good pigment dispersants and can be pigmented in conventional equipment. Anionic or non-ionic auxiliary pigment dispersants have been found useful in some cases[44]. The dispersion of the pigment in un-neutralised resin before dissolution or emulsification in aqueous base has been advocated[45].

Electrodeposited paint films are usually highly viscous and show relatively little flow during or after deposition. High pigment loadings aggravate this condition and smooth electrocoat films are best obtained at low pigment contents. In primer compositions, pigment contents of between 5 and 15% by volume have been quoted for electrocoat paints[46] whereas in solvent borne primers, pigment volume concentrations in excess of 30% are normal.

In order to achieve opacity at such low pigment loadings, the ratio of prime pigments to extenders in electrocoat paints is often considerably higher than in conventional formulations.

The pigment in the deposited paint contributes significantly to the electrical resistance of the film; the magnitude of this effect increases with pigment volume concentrations to a maximum beyond which film integration suffers. The pigment loading at maximum film resistance is usually well beyond the optimum for film smoothness.

In spite of all these constraints the area for manoeuvre is quite wide enough to allow the formulation equally of smooth, matt, primers or of glossy, white one-coat finishes of high opacity.

3.2.3 *Non-aqueous solvents*

Almost all the electrodepositable vehicles described in the literature contain a proportion of organic solvent. That most frequently named is ethylene glycol monobutyl ether (butyl "Cellosolve" or butyl "Oxitol") but other water-soluble or partially soluble alcohols appear to be equally suitable as are many ketones, hydroxy ketones or ketonic ethers. The effect of adding hydrocarbon solvents to electrodeposition baths has been described[47].

In a vehicle for electrodeposition, the organic solvent may perform one or more of four functions:

(a) It may be necessary or desirable in the manufacture of the resin, for example, to reduce the handling viscosity before dispersion or dissolution in water.

(b) It may improve the stability of the aqueous dispersion or otherwise affect the properties of the liquid paint in the tank.

(c) It may improve the deposition characteristics.

(d) It may improve the properties of the cured film, for example, by acting as a transient plasticiser during stoving.

The first and last of the above functions performed by the solvent are paralleled in conventional paints and need not be discussed further.

Whether the organic solvent is deposited principally with the paint film or retained in the aqueous bath depends on its partition between the continuous aqueous phase and the polymeric film former. Thus, for any given vehicle, the more hydrophilic the solvent is, the less is co-deposited with the paint and the smaller is the effect on the electrodeposition characteristics.

The term "coupling" solvent, frequently used to denote a solvent which materially assists in the dispersion (or "solution") of an essentially hydrophobic paint resin in water, implies a solvent with significant amphipathic properties, compatible with both the polymeric resin and with the aqueous phase, and with a marked tendency to orient itself at the interface between the two. The usefulness of such solvents is entirely consistent with the views of colloid science on the stability of aqueous anionic emulsions[48]. The prime example of a volatile "coupling" solvent is butyl "Cellosolve" but other hydroxylic solvents from the butanols to, say, nonanol can perform the same function.

The presence of organic solvents can have two apparently opposite effects on the electrical properties of the deposited film during the process of electrodeposition.

In cases where integration of the deposited film is difficult at the temperature of the bath, the solvent can act as a coalescing aid, improve the homogeneity of the film, and hence increase its apparent electrical resistance by eliminating conduction through interstices. More commonly, however, the presence of co-deposited solvent reduces the film resistance by reducing the viscosity of the film, thereby enhancing the mobility of residual cations.

In a working electrocoat tank, solvent is added to the bath either with the paint or as an independent addition; it is removed from the tank both by co-deposition with the paint and by natural evaporation. The maintenance of a correct solvent balance in an electrocoat tank is an important factor in ensuring consistency of operation.

Most published papers and patents acknowledge, at least by implication, the importance of the rôle played by the organic solvents in the electrodeposition process[49]. The subject is, however, not beyond controversy since one patent[50] advocates the addition of significant quantities of solvent to an electrodeposition bath whilst another[51], in describing a process for the removal of solvent from the paint before use, implies that its presence can be deleterious.

4. The physical chemistry of paint electrodeposition

A discussion of the physical chemistry of paint electrodeposition might properly commence with a consideration of the electrolytic properties of the aqueous paint dispersion, and progress through the stages in the deposition sequence: the anodic process which leads to precipitation and film formation, the cathodic process which requires compensation if uniform performance is to be maintained during prolonged operation, and the film conduction processes which enable films of sufficient thickness to have protective value to be obtained.

4.1 ELECTROLYTIC PROPERTIES OF THE AQUEOUS DISPERSION

At the concentrations at which they are used in practice, around 10%, aqueous dispersions of electrocoat paints are electrolytes with a normality of the order 0.1 equiv.l^{-1} and rather good electrolytic conductors. An example is provided by the commercially available resin Resydrol P411* which as its ammonium salt at pH 7.8, at 10% solids in water and at 25°C, shows a conductivity of about 4 mSiemens, indicating an equivalent conductance, Λ, of 40 units (Ω^{-1}equiv.$^{-1}$cm^2). Bearing in mind that the viscosity of the dispersion is about twice that of water at the same temperature, which itself suggests that the micelles are not extensively hydrated, it appears that the system has the typically high conductivity of micellar electrolytes despite the high equivalent weight (of the order 1,000 per ionised carboxyl group) of the resin material.

Further information may be obtained by comparing the conductivities of dispersions of a given resin neutralized with a variety of stabilising counterions; we have found that 10% aqueous dispersions of a waterborne vehicle of the epoxy ester type have equivalent conductivities of 50.5, 38.9, 35.8 and 23.5 in the potassium, sodium, lithium and diethanolammonium counterion forms, the viscosities being similar in all these forms, *viz. ca.* 1.5 cp. The similarity in viscosities encourages us to suppose equal contributions to the conductance by the anion in each case, which leads to values of 19 units for the anionic equivalent conductance and 31, 21, 16 and 4.5 for the cations in the order listed above. Evidently, therefore, the cation transference number is high except when amine counterions are used and we have observed a similar level of conductivity to that shown by the diethanolammonium form with a wide range of amines (except ammonia) including di- and tri-methylamine, di- and tri-ethylamine, mono- and tri-ethanolamine, and dimethylaminoethanol.

A comparison of the derived cation conductivities in the dispersions with the values in ordinary solutions suggests a depression of about 2½ times, against a viscosity increase over that of water of 1½ times. If Walden's rule can be supposed to

---

* Vianova Kunstharz A.G., Vienna, and Cray Valley Products Ltd., St. Mary Cray, Kent.

hold over this relatively small viscosity range, there is evidence, therefore, for a certain degree of abnormal interionic attraction attributable to the micellar charge, but the abnormality is not as great as is found in many colloidal electrolyte systems. Summarising, the low viscosity and general conductivity properties of the dilute dispersions of waterborne resins used for electrodeposition suggest that the micellar anions are not extensively hydrated and that abnormal interionic attraction, while present, does not prevent independent migration of polymeric and counter ions.

To assess the fate of dispersed pigment particles in electrocoat paints, electrophoresis studies have been carried out in our laboratories involving direct observation of migration in paints composed of red iron oxide pigment dispersed in an electrocoat resin without any additional dispersant. Both pigment and resin particles were found to migrate at sensibly the same rate of $ca.$ 2.0×10^{-4} cm^{2}s^{-1}V^{-1}; in fact, plant operational experience indicates that the pigment and resin migration rates are close but not exactly equal, but the experimental method would hardly have revealed this. The rate agrees very reasonably with the anionic conductance derived from the conductivity studies mentioned above, for on multiplication by the Faraday constant a conductance of about 19 Ω^{-1}equiv.$^{-1}$cm^2 is obtained.

The discrepancy referred to above between the anodic migration rates of resin and pigment is, in the writers' experience, unimportant from the points of view of either plant control or coating performance.

4.2 THE ANODIC PROCESS

The voltages employed in paint electrodeposition are typically in the range 100–500 V, and most of this acts across the deposited film during the latter stages of deposition when the resistive voltage drop in the dispersion has diminished. The main possibilities for the anodic process are as follows:

4.2.1 Anodic dissolution of metal

This gives rise to soaps of the anode metal with the anionic groups of the resin and no oxygen is evolved. Because the soaps are not self-dispersible, anodic metal dissolution causes destabilisation and, indeed, very satisfactory deposits of this type can be obtained over such anodically reactive metals as zinc and copper. Generally speaking, metal dissolution is a minor, though significant, effect over steel and phosphated steel.

4.2.2 Anion discharge

The direct discharge of carboxyl groups on the resin may well occur at an early stage of the anodic process, but in our view it cannot sustain continuous growth of the film because of the immobility of the carboxyl groups in a coalesced film. No such limitation applies to hydroxyl ions, which are relatively freely available from the dissociation of water; hydroxyl discharge causes the liberation of oxygen and the release of hydrogen ions for conduction through the film, so that the anodic process effects film growth at a distance from the electrode surface.

4.2.3 Side reactions

Little is known concerning the possible side reactions accompanying the primary processes just mentioned. The possibilities include Kolbe type decarboxylation and dimerisation if there is much direct carboxyl discharge, hydroperoxide initiation of vinyl type polymerisation affecting the unsaturated linkage present in any drying oil derivatives in the resin, or simple oxidation of resin components including organic solvents. The latter two possible reactions might well be catalysed by metal dissolved in the film.

It will be appreciated that any combination of the above processes effects destabilisation so that, regardless of the nature of the electrode reaction, the current efficiency of paint electrodeposition is high except when a low current density is used on a paint which redisperses readily in the absence of current.

In depositions on steel and phosphated steel, hydroxyl discharge is the main action (this is sometimes expressed as a pH displacement[52] but iron dissolution may account for up to 10% of the current passed in forming a film[53]: the steel may therefore be said to behave as an imperfectly passive substrate. We have inferred evidence for the oxidative side reactions mentioned from the observation that the quantity of oxygen evolved is smaller than that required by Faraday's Law after allowing for metal dissolution; our experience also provides some evidence for a catalytic action on the part of iron dissolved in the film.

Growth of a continuous film occurs if the destabilised resin coalesces, so that, if the deposition is conducted at constant applied voltage, the current falls steadily as the thickness increases. Perhaps one of the most remarkable aspects of the behaviour of electrodeposited paint films is their capacity for repair of cavities left by gas bubbles passing through the film: the current–time relationship at constant voltage is generally a smooth curve with considerable resistance (of the order $10^5 \Omega cm^2$) being attained in the latter stages of deposition. The field strength acting across the film may be of the order of 100 $kVcm^{-1}$, so it is not surprising that aqueous matter which has phase-segregated from the resin is effectively expelled, although a small but definite fraction of the neutralising base is retained (about 5 to 15% of that present in the dispersion expressed on a basis of solid matter).

In the deposition of paints formulated from electro-depositable resins, the pigment may be said to play a largely passive rôle. It is sometimes found that the proportion of pigment in the deposited film is slightly higher than that present in the dispersion.

4.3 The cathodic process

Deposition consists of the liberation of a substantially acid film and its subsequent removal from the bath. At the cathode, which may consist of the uncoated inner walls of the tank or of special cathode units mounted in a lined or unlined tank, the cation discharge takes a normal course with the release of hydrogen and of free base or alkali. It follows that, if the cathodic products are allowed to mix into the bath, the latter suffers an increase in pH. This disturbs the relation between counterion and paint resin concentration, affecting the stoichiometry and possibly even the course of the anodic process. Clearly, such a change cannot be allowed to continue unchecked and pH control has become a major aspect of the control of the paint deposition process. It is considered in a later section (5.1.3).

4.4 The stoichiometry of deposition

The main anodic process

$$\tfrac{1}{2} H_2O - e \rightarrow H^+ + \tfrac{1}{4} O_2 \tag{1}$$

$$RCOO^- + H^+ \rightarrow RCOOH \tag{2}$$

consists of the release of 1 g equiv. per faraday of carboxylic acid from its counterion stabilisation. If the film is free of retained base and the current efficiency unity, the yield of deposit per coulomb passed, y, is

$$y = \frac{56,000}{xAF} \tag{3}$$

where A = "acid value" of resin or paint, expressed as mg KOH/g involatile matter,
x = the fraction of resin acidity neutralised in making the dispersion,
F = the Faraday constant.

It is common practice in a number of commercial electropainting processes for the quantity $(xA/56)$ to be measured as a part of process control; often referred to as the "M. Eq." value, from the fact that the result is expressed as m equiv./100 g involatile material, it is a measure of the counterion (base) content/g of solid material.

We find the simple expression (3) to be helpful as a guide, but experimental results tend to be higher, generally by 5 to 15%, which is to be expected because the proportion of base retained in the deposited film is of this magnitude. Negative deviations from (3) are generally attributable to either of two causes: delayed initiation or excessive anionic contamination. The first case arises at low initial applied voltage (or current density) and is discussed in the following section, whilst contamination by mobile anions can clearly cause a loss of current efficiency if the corresponding anodic electrolysis products (*i.e.* the acids) do not themselves cause the destabilisation of an equivalent quantity of resin. Most simple contaminant anions do not have a great effect unless present in large amounts (say above 0.01 N concentration in the bath), but an anion derived from a very weak acid, *e.g.* CO_2, may have a marked effect on current efficiency.

4.5 INITIATION OF FILM GROWTH

It has been supposed in the above that destabilisation occurs immediately on the passage of current, that is, that film growth is initiated without delay. In fact, this is not the case: a value of current density can be found below which film formation (as may be recognised by an increase in the resistance of the anode system) cannot be induced. This limiting current density is a function of the geometry of the cell used and of the conditions of mass transfer within it, as exemplified by the work of Beck[54] using a rotating disc anode. In our laboratories, this question has been investigated in an alternative manner, mainly by Strivens[55], using chronopotentiometric and chronoamperometric methods* to indicate an initiation time for film formation.

Beck's approach and ours to this matter focus attention respectively on the steady and the unsteady states; the latter has the merit of a more direct bearing on the practical situation. We are in agreement as to the nature of the processes involved, namely that film formation requires the attainment near the anode surface of a low, critical and probably nearly zero, concentration of stabilising base. As in all electrochemical systems involving mass transfer, the establishment of such a critical interfacial concentration involves the competing effects of electromigrational and convective or diffusive fluxes. At very low current density, convection/diffusion preponderates over migration so initiation can never occur, *i.e.* the initiation time is infinite. Above a limiting current density, initiation occurs after a delay during which the critical interfacial condition becomes established, suggesting a situation to which the Sand equation[56] applies. For the rejection of mono-acidic base cations at a plane electrode, the condition of zero interfacial concentration is predicted to occur after a time t^* given by

$$t^* = \pi F^2 D\, c^2/4\, i^2 \qquad (4)$$

where D and c are the diffusion coefficient and concentration of the electrolyte and i the current density. Inverse proportionality between t^* and i^2 was found to

---

* The two methods prove to be equivalent because in the interval prior to film formation the system has essentially constant resistance once the ordinary electrode overpotentials (which are generally well below the voltages employed) have become established.

hold over a wide current density range from 0.4×10^{-3} A cm^{-2} upwards with t^* values ranging from 700 s to less than 1 s. The proportionality constant corresponding to $\pi F^2 D c^2/4$ in (4) was found to be 1.1×10^{-4} A^{2}s cm^{-4} for deposition on steel from an aqueous waterborne resin dispersion at 10% solids in which the base concentration was 0.10 N. If this can be assumed as the concentration relevant in the Sand equation, the value of D works out to 1.5×10^{-6} cm^{2}s^{-1}.

Although this value for D is low for an electrolyte having ionic conductivities of the order found, we do not consider it unjustifiable. In the mass transfer region, or "Nernst" film, close to the metal interface, the passage of current causes resin to concentrate by the action of concentration polarisation. This causes an increase in viscosity which may become considerable before precipitation commences; as a result the mean diffusion coefficient applicable over the entire concentration range involved in the precipitation process is smaller than that which applies in the mobile aqueous dispersion (to which the conductivities relate). Another possible factor concerns the interfacial base concentration required to induce destabilisation. In applying (4), the value of c used was simply the base concentration present in the dispersion, implying that destabilisation requires a complete absence of base at the interface. In fact, these resins are rather more readily destabilised, so the smaller value implied for c means that our present estimated value of D is too low, but it is unlikely that this is the main source of discrepancy.

With decreasing current density below about 0.4×10^{-3} A cm^{-2}, the system considered above showed increasing values of the product t^*i^2 until initiation failed to occur in unstirred dilute paint below about 0.25×10^{-3} A cm^{-2}. At this limiting current, the steady state may be described in terms of the mean diffusion coefficient, D, "Nernst" film thickness, δ, and equivalent paint concentration, c:

$$i_{\lim} = F D c/\delta \tag{5}$$

Taking δ as 0.05 cm, a commonly found value for natural convection systems, indicates $D = 1.3 \times 10^{-6}$ cm^{2}s^{-1}, in fair agreement with the value calculated from the unsteady state experiments (the same reservations apply concerning the relevant value of c as in that case). The agreement is perhaps somewhat fortuitous in view of the uncertainty in the value of δ, but clearly the two estimates agree as to order of magnitude. The fact that t^* is independent of the conditions of mass transfer (as represented by δ) is a valuable advantage of the unsteady state method of study, for then information obtained in the laboratory is of direct relevance to the practical situation in which mass transfer conditions may be quite different.

We conclude that initiation of film growth is, in general, a diffusion-governed process involving anodic elimination of stabilising base. Two important exceptions to this generalisation have been found, one in which the substrate is reactive and another case in which the resin fails to coalesce immediately on precipitation.

When a reactive metal, *i.e.* one (such as zinc or copper) which dissolves anodically, is used, initiation occurs much more readily than with noble or passive substrates. This is readily understood because the heavy metal soaps of the waterborne resins are strongly ion-paired, so that substrate dissolution causes destabilisation unrelated to the rate of base migration from, or diffusion towards, the interface.

The second exception is illustrated by the effect of anodic electrolysis on dilute dispersions of the alkali soaps of stearic and oleic acids at room temperature. If the applied voltage is kept constant, a virtually time-independent current is observed in the case of the stearate electrolysis while with oleate the current falls steadily with time. In the first case, the deposit consists of crystalline stearic acid which has precipitated but failed to coalesce, whereas coalescence has readily occurred from the oleate bath, oleic acid being liquid at room temperature. In polymeric resin systems, the decisive factor appears to be the relation of the glass transition temperature of the resin material to the temperature at which deposition is carried out; we have observed systems failing to coalesce at a low temperature but doing so at a higher temperature or after prolonged electrolysis during which, it is reasonable to suppose, Joule heating induces coalescence. Alternatively, plasticisers (including resin-compatible organic solvents) may be added to assist coalescence.

Similar observations to the above have been reported by Bruce[57].

4.6 FILM CONDUCTION

After coalescence has occurred, film growth entails conduction through the film which then consists of a highly viscous, mainly non-aqueous material, poorly ionised except by virtue of residual base counterions. We can dispose of the oversimplified notion that the electrodeposited paint film conducts in a simple manner as though comprised of a film of pure acid resin, for on the basis of the specific conductivity of oleic acid (*ca.* 10^{-9} Siemens) we would expect these feebly dissociated and highly viscous materials (viscosities typically above 100 p compared with 0.25 p for oleic acid) to have exceedingly small conductivities, whereas the values calculated for electrodeposited films from observed current/voltage relationships are normally in the range 10^{-9} to 10^{-7} Siemens.

The simple proposition that electrodeposited paint films conduct current uniformly over their entire surface, and therefore that it is meaningful to ascribe to the film material a specific conductivity, is not generally accepted. Thus, Olsen *et al.*[58], in a study of throwing power in an electrocoat system, supposed that current flows between anode and cathode in an amount governed only by the resistance of the intervening dilute paint and that this high current then abruptly ceases, *i.e.* there is no film resistance during the induction period after which, presumably on coalescence, the film conductivity is zero. While some paints approximate to this in behaviour, the current/time relation observed at constant applied voltage invariably gives evidence of continuous film growth; the sharp decrease in current with time is attributable to increasing film thickness which acts not only directly but also by decreasing the mean field strength and thereby the specific conductivity (cf. 4.6.1).

Another view of the film growth mechanism[72] is again based on the precept that the solid paint material in the film has essentially zero conductivity. The observed conduction is supposed to take place through imperfections created by gas evolution. If this were correct, a marked dependence of film conduction on ambient pressure could be expected, but we have found only slight variation in experimental depositions carried out at pressures of from 0.1 to 6 atm.; moreover, film conductivity was found to increase with increasing pressure, which is presumably contrary to the effect expected if the hypothesis holds. Further evidence has been obtained by suppressing gas evolution entirely, either by adding a sufficient quantity of a reducing agent like hydroquinone or by causing the paint electrodeposit to form on the cathode side of a cation-exchange membrane arranged in an electrolytic system to function on passage of current as a source of hydrogen ions. Experiments along these lines revealed completely normal film conduction ("normal" in the sense of section 4.6.1), so we reject the suggestion that gas evolution plays a determining part in the film conduction process. We have already commented (4.2.3) on the remarkable capacity the films have for repair of the cavities left by evolving gas bubbles.

Important aspects of film conduction are:

1. it determines the film thickness developed in a given time in a constant voltage process, or the voltage/time relationship at constant current,

2. it governs the manner in which current is distributed over the surface of an article of complex shape,

3. under extreme conditions, "runaway" deposition, often described as film rupture and characterised by a high apparent film conductivity, can occur.

4.6.1 Deviations from Ohm's law

The complexity of the processes involved in film conduction can be appreciated from experiments in which abrupt changes or interruptions in applied voltage are made. For instance, Fig. 1 shows the effect of abruptly halving the applied voltage in a deposition in which considerable film resistance had already been allowed to build up, the thickness of the corresponding stoved film being some 20 μ. The initial position with 200 V applied (A) is one of stable film conductivity, 0.6×10^{-8} Siemens. An abrupt reduction in applied voltage to 100 V produces a very large fall in conductivity (B) followed by a steady increase until a new more or less stable value is reached (C). Sudden restoration of the original applied voltage causes a rapid rise in conductivity (D) to a value well above that of the original stable condition (A). There follows a decline to about the original level (E).

Evidently, more than one process governing film conduction is affected by voltage changes: a more or less instantaneous effect characterised by a marked positive effect of voltage on conductivity followed by a slow process of adjustment in the

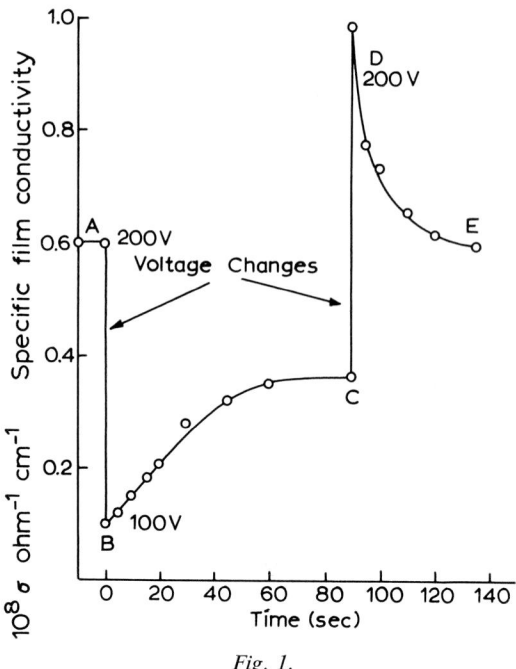

Fig. 1.

opposite direction. We may conveniently consider first the "instantaneous" departure from Ohm's Law, concerning which Beck[59] and Cooke and Strivens[60] agree in finding the magnitude of the effect substantially greater than can be accounted for on the basis of Onsager's theory[61] of field-enhanced dissociation in weak electrolytes, if the calculated value of field strength is assumed to act uniformly across the film. Beck has suggested a mechanism based on space-charge created by proton displacements in a film supposed to consist of pure (base-free) acid resin, but we consider that no account of conduction in electrodeposited paint films can be framed without allowance for the effect of residual base, the analytical evidence for which is unquestionable on the various resin systems studied by us.

The quantities of retained base found are typically between 0.02 and 0.1 equiv./equiv. of carboxyl acidity. As we can expect a very low dissociation constant for a dehydrated polymeric carboxylic acid material (a point made also by Beck), it follows that the retained base supplies, through salt formation, most of the current carriers acting in the film. Only near the anode interface, where protons are being produced and are combining with carboxylate anions, will this not be the case. Then, because of the poor dissociation and therefore low conductivity of the acid material near the metal compared with the relative abundance of mobile cations in the rest of the film, we infer that the electric field is not uniformly distributed throughout the thickness of the film but is concentrated almost entirely over a thin region close to the substrate.

It is suggested, therefore, that residual base plays a dominant rôle in film conduction: it is subject to concentration polarisation causing a gradient of conductivity and non-uniform field strength distribution as a result. On the premise that the maximum field strength acting in the film may considerably exceed the mean value, it is not necessary to look beyond accepted causes of non-ohmic conduction in ionic media, in particular the field-induced dissociation effect. Space permits us only to summarise our confirmatory evidence.

(a) Effect of polarity reversal

Brief polarity reversal, carried out in such a way as not to cause drastic redissolution of the film in the reversed arrangement, caused an almost immediate change to ohmic conduction at a greatly increased apparent film conductivity*.

(b) Impedance studies

Measurements of impedance at a range of frequencies on a DC polarised film gave powerful evidence of a polarised "structure" of the type described.

(c) Time lags

The response to voltage fluctuations is not really instantaneous, time constants of the order 1 ms being observed; these are as expected from Onsager's treatment.

(d) Effect of the counterion

It has been found that films deposited from alkali metal stabilised dispersions show more acutely non-ohmic behaviour than do those from amine stabilised systems.

(e) Modified counterion effect

In depositions from alkali metal stabilised resins, we find that non-ohmic behaviour can be diminished by the addition of amines including those which are too weak to function as stabilisers but which could form salts at the low pH in the interfacial region.

Commenting on (*a*), brief polarity reversal would readily destroy the concentration polarised condition postulated by migration of base from the interior of the film into the interfacial region without substantial ingress taking place from the dispersion. Regarding (*b*), small AC modulations were used and extremely complex behaviour found, both types of reactance occurring depending on frequency; the

---

* In case the effect described was caused by cathodic hydrogen disrupting the film, the experiment was also carried out using as substrate a cation-exchange membrane in the hydrogen form arranged in a suitable electrodialysis train to function on passage of current as a source of hydrogen ions. The same result was observed as with a metal substrate. In both cases the current flowing during the reverse period was, to a first approximation, limited only by the interelectrode resistance of the dispersion (the counter-electrode area was large). Thus, film resistance virtually vanished.

inductive effect became more acute as the DC polarising voltage increased and could be identified with the time lags already mentioned. Concerning the counterion effects, analysis showed that considerably lower base retention occurs in the alkali metal case than with amines and we infer that the high electromigrational mobility of alkali metal cations gives rise to more effective rejection from the interfacial region, intensifying the non-uniformity of field strength distribution and hence the non-ohmic behaviour.

The suggested mechanism of non-ohmic conduction is able to account also for the slow adjustment processes illustrated in Fig. 1. During the recovery period (BC), the film conductivity increased more than three-fold; this could result from the adaptation of the counterion concentration gradient to the diminished electric field present compared with (A) (the residual base content revealed by analysis was only slightly, if at all significantly, greater in condition (C) than in (A). It is noticeable that the ratio of film conductivities between (D) and (C) was smaller than that between (A) and (B), *i.e.* less non-ohmic response to the voltage increase (CD) than to the equivalent decrease (AB). This accords with the view that field strength distribution is more uniform at (C) than at (A) on account of the relaxed concentration gradient.

4.6.2 *Film rupture and the effect of Joule heating*

It is widely considered that every paint or resin is characterised by a voltage above which ruptured or otherwise unsatisfactory deposits are obtained, and it has been suggested that this reflects the intrinsic dielectric strength of the film material. This would imply dielectric strengths of the order of hundreds of $kVcm^{-1}$, but the concept of the concentration polarised film suggests that the field acting is much greater at its maximum value. We are of the view rather that Joule heat production is the main factor governing the rupture phenomenon; insofar as the dielectric strength of a material is frequently governed by Joule heat production within the specimen exposed, the views may be said to converge.

The rate of heat production in an electrodepositing paint film is typically of the order of hundreds of $watts/cm^3$, and thermal runaway is therefore to be expected unless rapid heat transfer occurs away from the film. Heat is readily conducted through the thin film itself so that, if both sides of a thin metal sheet are being coated, the external liquid is the controlling factor*. Thus, agitating the dilute paint makes rupture more difficult to induce**, but perhaps the most effective way of inducing electrodeposited paint films to withstand high voltages is to raise the

---

* For this reason and because of natural convection, it often occurs that vertically mounted surfaces in unstirred dispersions receive a higher thickness of film at the top than at the bottom, *cf.* conventional dip painting.
** So also does electrical insulation and cooling of one side of a metal sheet.

voltage gradually as the thickness increases, keeping the heat production at a steady rate. This can be accomplished either by a sequence of increasing voltage steps or by a resistor in series. In commercial installations, the interelectrode resistance of the dispersion is often sufficient to achieve the required control.

The part played by Joule heating in the integration of some electrodeposited films has already been touched upon in connection with the initiation of deposition. In fact, thermal flow during deposition can be a factor in obtaining smooth blemish-free films, *i.e.* films which have recovered to the necessary degree from anodic gas evolution. In some critical cases, improving heat dissipation by agitating the bath or by depositing on one side only of a thin metal panel is sufficient to inhibit thermal flow and produce flawed films from a bath that would otherwise give satisfactory results. Similar observations have been reported by Göring *et al.*[62] on the coating of those exterior regions of objects which adjoin interior hollow spaces.

4.6.3 The use of impulse current sources

It has been asserted that the use in place of the usual more or less smooth direct current source of a sequence of voltage pulses and interruptions leads to a number of improvements in film properties[63]. To consider this question, a basis of comparison must be decided upon (equal mean voltage or equal mean rate of film growth) because the use of pulsed or other non-smooth current sources affects the conductivity of an electrodepositing paint film. This fact, which is readily confirmed experimentally, follows directly from the positive sense of the effect of voltage on film conductivity. If σ_o is the specific conductivity at zero voltage and σ that at a voltage of V across the film, the non-ohmic effect can, for the sake of illustration, be approximated by

$$\sigma = \sigma_o (1 + a V) \tag{6}$$

where a is a constant. Now, suppose square pulses of height V' volts are used, the mean voltage being $\bar{V}$. The mean current flowing in the case of the pulsed source, I_p, is then given by

$$I_p = \sigma_o \bar{V} (1 + a V') \tag{7}$$

while for the smooth source at the same mean voltage $\bar{V}$

$$I_s = \sigma_o \bar{V} (1 + a \bar{V}) \tag{8}$$

then, because $V' > \bar{V}$, $I_p > I_s$; this can be shown to be general for voltage wave-forms other than the square waves considered here or for other forms of expression for σ than (6). It means that, if pulsed and smooth wave-forms are compared at the same mean voltage, the pulsed source gives a higher rate of film growth. The deviations from Ohm's law are typically so marked (*e.g.* $a > 0.02$ V^{-1}) that, if the wave-form consists of pulses whose duration is, say, less than one-half the cycle time, the rates of film growth in the two cases differ considerably, and thermal runaway occurs readily with the pulsed source. Perhaps the more meaningful comparison from the point of view of the paint user is that obtained when the rates of film growth (mean currents) are equal. Then the mean voltage with the pulsed source is smaller than in the other case and so is the mean rate of Joule heat production. On the other hand, the rate of Joule heat production during the part of the cycle when current flows in the case of the pulsed source is higher than the steady rate with the smooth source; it can occur as a result that a paint which requires the assistance of thermal flow during deposition to produce a flawless film benefits from the use of a pulsed current source. The pulsed source operated at a low mean voltage can therefore give a better integrated film without inducing rupture through the accumulation of Joule heat.

Generally speaking, most electrocoat materials derive little benefit from the use of pulsed or other non-smooth current sources, and the method has the major disadvantage of giving rise to a loss of effective throwing power.

4.7 THROWING POWER

This important property of electrocoat paints, the ability to furnish coating within poorly accessible regions of complex articles, *e.g.* hollow tubes, recesses, seams and box sections, results directly from the (relatively) high conductivity of the dispersions employed allied to the fact that the low conductivity of the deposited films permits the application of unusually high voltages for electrolytic systems. Indeed, one author[64] has suggested characterising the throwing power capability in terms of the ratio of the specific conductivities of dispersion and film. This is not entirely satisfactory, firstly, because the film conductivity of a system is not a constant quantity (*cf.* above), but mainly because it is implicit in the use of this quotient that advantage be taken of a low film conductivity by raising the applied voltage to produce a specified film thickness on freely exposed areas of articles being coated, while it is often the case that advantage cannot be so taken because of film rupture or if there is any arbitrary limit on applied voltage.

Many tests have been put forward for comparing the throwing power of electrocoat paints, most of which emphasise the interior coating of an object, such as a pipe, open-ended narrow rectangular box, crevice or wedge. One which has gained wide acceptance in the motor industry is that described by Brewer *et al.*[65]. A uniform circular pipe, equipped with a steel strip down its centre to serve as an indicator of the interior coating, is coated under standard conditions. The indicator strip is removed and stoved and the performance of the internal coating can be tested. The distance along which coating has occurred is somewhat difficult to assess visually and objectivity can be improved by determining the weight of film on the strip.

As far as throwing power is concerned, the chief requirement of the motor industry is for coating at sufficient thickness to have protective value within the poorly accessible regions of a motor body, in particular the box-sections supplying rigidity to the lower part of the body. This can be accomplished in the electrodeposition process either by the use of a paint of high throwing power combined with a process cycle which enables advantage to be taken of that property or, if the paint has little throwing power or the conditions are unfavourable, by furnishing the difficult regions with additional counter-electrodes. The latter expedient, which follows directly from practice in electroplating, has been a feature of the earliest electropriming installations in the motor industry; it is wasteful of labour besides failing to take the fullest advantage of the capabilities of the process. In both cases, the ideal situation requires correct placing of access and drain holes for the process to be successful, but it is probably true that the counter-electrode method is more successful on a motor body not designed for priming by electrodeposition, hence its widespread use in early commercial developments. The more elegant solution is undoubtedly a high throwing power system combined with purposefully designed box-sections, to achieve which a reliable method of estimating internal coverage through access holes is necessary.

Tests of throwing power such as the one described furnish only a limited amount of information relative to the coating of box-sections. This is because of the geometric difference between a box-section and an open-ended pipe or similar form having a uniform cross-section. In the box-section, the dominant resistance in the system is that of the access hole and this governs the total current entering the box until a significant film thickness has accumulated inside, while with a pipe a continuous and uniform increment in resistance (and therefore decrease in current flowing) occurs as the interior surface becomes coated. It follows that the film thickness yielded in a box-section (which is the quantity of practical importance) is not proportional to the distance of throw observed in a pipe test. In fact, other than experimental depositions on box-sections closely similar or identical to those used in practice, the only satisfactory method of estimating the performance of electrocoat material is by reference to the dispersion and film properties which govern the interior coverage of box-sections. These have been listed by Maisch[66] as including: bath conductivity, film conductivity, deposition yield (per coulomb), limiting current density for initiation and rupture voltage.

Commenting on this list of relevant properties, it has already been pointed out that film conductivity cannot necessarily be varied at will because a mean value of this parameter is implied in the specification of the film thickness required at stipulated voltage and time. Clearly, if the limiting current density for initiation of a paint is not exceeded in the interior of a box-section, no internal coating occurs, but in practice such conditions would be inadmissible (though they could occur readily enough); we consider the more important parameter reflecting initiation to be rather the initiation time applicable to the interior.

The remaining factors are bath conductivity and deposition yield, both of which should optimally be as high as possible. Suppose an "intrinsic throwing power", T, to be defined in terms of the box-section coverage obtained at fixed voltage and time; this would furnish a comparison of paint performances under standard conditions. To a first approximation, T is proportional to the product yK, where y is the yield and K the bath conductivity. Now, K can be expressed in terms of the equivalent conductivity, Λ, of the conducting ions in the dispersion as

$$K = \frac{x A S \Lambda}{56{,}000} \tag{9}$$

where x and A have the meaning previously assigned and S is the fractional concentration of the dilute paint (solids content). Then, if the "intrinsic throwing power"

$$T \approx \text{const.} \; y K \tag{10}$$

it follows from the approximate relation (3) that

$$T \approx \text{const.} \; S \Lambda \tag{11}$$

The solids content S cannot be varied at will because of the economic effect of the losses associated with "drag-out" (*cf.* section on Technology), so that for a stipulated value of this parameter, the intrinsic throwing power is proportional to the equivalent conductivity of the dispersed material. This conclusion, approximate though it may be, provides a useful guide to the maximum attainable intrinsic throwing power, for, as discussed previously, the conductivity of electrocoat dispersions may be of an order approaching that of simple aqueous solutions—a limit which clearly cannot be exceeded.

At a given value of intrinsic throwing power, the box-section coverage or effective throwing power can be varied by varying the voltage and time of deposition. In a general sense, coverage increases, though not proportionally, with increases in voltage and time and it is in these respects that the process conditions as well as the properties of the paint require optimisation if successful coating of poorly accessible regions is to be attained without the use of supplementary counter-electrodes.

Recent activity in this field has included the quantitative treatment of box-section coverage, taking into account all the relevant factors identified, so that the correct number and size of access holes can be specified. In conjunction with paints of high throwing power, this is of value to the motor industry which has accepted that the use of auxiliary counter-electrodes to coat box-sections is an undesirable expedient, to be discarded if possible.

5. The technology of paint electrodeposition

5.1 APPROACHES TO THE TECHNOLOGY

With the rapid and highly competitive development of this method of paint application that has occurred during the past ten years, it is not surprising that distinct and divergent trends have appeared in the technology. Whilst in future developments an incorporation may occur of the best features of each trend, the different solutions found for the underlying problems involve such an interaction of considerations that the distinct trends are likely to persist for some time. The chief considerations to be faced are (1) polarity of the paint vehicle, (2) choice of counterion, (3) method of control of pH, (4) nature of the electrical supply, as to its wave-form and the mode of connection of the plant.

5.1.1 Polarity of the paint vehicle

Hypothetically, the choice as to whether to deposit anodically or cathodically is open to the user of an electrocoat system. In fact, as has already been remarked, cationic vehicles have not at this stage reached commercial exploitation despite the apparent advantage of absence of substrate dissolution during electrodeposition and consequent versatility over a range of metals. In fact, the commercial requirement is dominated by the finishing of steel which is fortunate for the anionic systems because anodic passivation ensures that the main electrode reaction is liberation of the acidified vehicle rather than the production of a metal soap.

5.1.2 Choice of the counterion

As the main class of electrodepositable resins are carboxylic types closely related, if not in some cases identical, to resins developed for waterborne dip application, the choice of neutralising base has been strongly influenced by established practice in that field. In particular, because volatile bases are employed with waterborne dipping resins to ensure elimination of base from the film during the curing process, all the early electrocoat resins were neutralised with bases like ammonia, di- and tri-methylamine, di- and triethylamine, diethanolamine and dimethylethanolamine. The long established stabilisation of carboxylic vehicles by these nitrogenous bases has probably delayed (and perhaps prejudiced against) the recognition that alkali metal hydroxides are also effective stabilisers of these types of resins.

The factor in electrodeposition which distinguishes it from the dipping process is the elimination of base from the uncured coatings. It is therefore of no concern whether an electrodeposited film is derived from an alkali or amine stabilised dispersion. Base elimination is, in fact, not quite complete but the small residual base content does not have any discernible effect on the properties of cured films.* The outstanding advantage conferred by alkali metal counterions is in the conductivity and hence the throwing power of the dispersion; at the same applied voltage the throwing power of dispersions neutralised with potassium hydroxide is an order of magnitude greater than those from organic amines. There has thus arisen one of the main points of technological divergence because with low conductivity dispersions it is necessary to aid the coating of a large complex article like a motor body with appropriately located auxiliary cathodes whereas with alkali metal counterions this may not be necessary. It will be shown in the following section that this difference in approach to the problem of throwing power carries further implications.

5.1.3 Method of pH control

The necessity to prevent or compensate for the accumulation of neutralising base (or acid in the case of a cathodically depositing resin) in the paint tank has already been mentioned (4.3). Limiting discussion to the case of an anodically depositing system, though analogous methods are readily envisaged for cationic materials, the main methods which have been advocated or tried are: (a) base evaporation, (b) replenishment of the dip tank with acid or base-deficient material, (c) cation-exchange, (d) dialysis, (e) electrodialysis using a "semi-permeable" membrane, (f) electrodialysis using an ion-selective membrane. Methods (b), (c) and (d) normally involve items of plant external to the dip tank as do some proposals in the patent literature along the lines of (e), but normally (e) and (f) can easily be incorporated into the tank.

Except in the case of method (b) it is normal for concentrated neutralised paint to be used to replenish the tank as its contents become spent.

(a) Base evaporation

This method is limited to resin systems having highly volatile neutralising bases, such as ammonia and, possibly, trimethylamine. It is necessary to balance the rate of painting (which determines the rate of base production) with the rate of evaporation. The method does not cope easily with fluctuations in painting rate or extended periods of inactivity nor is it possible with the larger, deeper tanks to obtain an adequate rate of evaporation, even of ammonia, to cater for economic throughput rates.

This method is suitable only for small shallow tanks operating at low, reasonably steady, rates of painting.

---

* This holds at all events for the resins within the authors' experience and may not be completely general; similarly, there may well exist resins which are not satisfactorily water-dispersible when neutralised with alkali metal hydroxides. Another interesting fact is that a smaller proportion of base is retained in films deposited from alkali metal stabilised dispersions compared with that found when amines are used.

(b) Acid or base-deficient feed

This method has the attraction of re-use of the stabilising base and of the possible use of the inner walls of the tank as the cathode. Re-use of the base is of economic importance only when expensive amines are employed and the scale of operation is such as to render this item significant. The convenience of using the tank as the cathode, which automatically entails earthing the negative side of the power supply, is of less value than may appear at first sight because of the greater degree of simplification possible if the positive pole is earthed (for it is not then necessary to insulate the conveyor from the jigs, cradles or other metal equipment in electrical contact with, and carrying, the articles being electrocoated, *cf*. 5.1.4).

Because the material removed from the paint tank in the electrodeposition process (deposited film plus adherent dip-coat) is not completely base-free, its compensation requires only a base-deficient rather than an acid material: this fact has doubtless facilitated the development of this method because acid paint carried in organic solvent is more difficult to incorporate than partially neutralised material, or alternatively, it permits the employment of a two-pack paint make-up system in which a neutralised pigmented concentrate is used in conjunction with acid resin. In the latter case, the pigment is not subjected to the difficult transition from stabilisation in solvent at high viscosity to the dilute waterborne condition.

Whether a one- or two-pack feed arrangement is used, efficient mixing is required at the dilution stage, *e.g.* by high speed mixers, colloid mills or ultrasonic homogenisers, and stagewise dilution may be necessary to prevent flocculation.

This method of pH control has found considerable use, notably in Germany and the U.S.A. It has the principal disadvantage of requiring an easily incorporated resin, which probably means a favourable relationship between viscosity and dilution, and not all resins are adaptable without modifications which may prejudice their other properties, that is to say the acceptance of base-deficient feed as a method of pH control removes a degree of freedom from the resin formulator. The method may not compare favourably with an efficient membrane method (*cf*. (*f*)) in respect of small installations because of the regular control necessary to ensure stability; on larger plant the additional supervision or greater complexity of automation are more acceptable.

(c) Cation-exchange

Excess base accumulating in an electrocoat bath is readily removed by cation-exchange using the hydrogen form of either a strong or weak acid exchanger. The latter is rather to be preferred because of the limited effect of a weak acid exchanger when present in excess, such as might occur in a column process in a region of poor mass transfer. The powerful precipitant action of the strong acid exchangers and the virtual irreversibility of flocculation render such materials impracticable, and even with the carboxylic exchangers high flow rates are necessary through packed beds to ensure uniformity of effect.

Perhaps the greatest difficulty with this method arises in connection with the regeneration of the ion-exchange column. Before the introduction of acid regenerant, all traces of paint resin must be eliminated because of its sensitivity to flocculation. After regeneration, it is necessary to rinse all traces of regenerant away, not only on account of the precipitant action but also because anions derived from the regenerant would degrade the paint.

This method is known to have been tried but discarded for commercial use.

(d) Dialysis

The patent literature contains many suggestions for the use of dialysis as a method of treating electrocoat paints after a period of use. The claims as to what is accomplished are not entirely clear: generally amine neutralised systems are used and the free base is stated to dialyse through the semi-permeable membrane. While such a process would effect compensation for base liberated during electrodeposition, it is frequently also claimed that contaminants are separated by this method, *i.e.* anions which may permeate in the form of their salts with the neutralising base.

Dialysis has not become important as a method of pH control, probably because of the large membrane area required to make it workable.

(e) Electrodialysis using semi-permeable membranes

The separation of anode and cathode by a semi-permeable membrane or porous pot is a commonly employed method of isolating the electrode products in an electrolytic process.

To be efficient in controlling the pH of an electrocoat bath, cations must pass into the catholyte at the rate of one or nearly one g equiv./faraday of electricity passed and they should remain there or be eliminated by flushing without returning by diffusion to the anolyte (dilute paint). Now, for a semi-permeable membrane to have sufficient conductivity to function in this application, a relatively open porous structure is required, the material most generally used being canvas or other filter cloth. Such materials are poor barriers to the diffusion of simple monomeric substances* and in electrolysis they permit an anion flux to occur in the opposite sense to that of cations. For the latter reason, the catholyte supplied to installations having semi-permeable membranes is generally demineralised water (an example is one process developed in the U.K.[67]) as otherwise anions from salts in impure water would contaminate the paint. The high resistivity of demineralised water poses an apparent problem in that excessive resistance between the metal cathode and the membrane causes unacceptable voltage drop in the catholyte; this can be overcome by placing the cathode and the membrane in contact or close together (but this reduces the efficiency of flushing). With many semi-permeable membrane

---

* In our experience they do not completely retain paint resin and in some cases readily transmit even pigment particles.

materials the problem of excessive catholyte resistance does not in reality arise because sufficient conducting matter, which may be paint or low molecular weight resin constituents, diffuses into the catholyte from the bath to give it adequate conductivity.

There does not appear to be published information on the current efficiency of the semi-permeable membrane method which is employed in a number of commercial installations in the U.K. and elsewhere on a pilot scale at least. Experimental work in our laboratory indicates that, as expected, the current efficiency is highly sensitive to the rate of catholyte flushing, but that high efficiency at high flushing rates is offset by losses from the paint compartment. The difficulties associated with this method have given rise to criticism of membrane methods in general[68], but it must be assumed that the ion-exchange membrane method was not taken into consideration in reaching this judgment.

(f) Electrodialysis using ion-selective membranes

The deficiencies of the electrodialysis method can be overcome if a cation-exchange membrane is used in conjunction with an anionic paint system (or an anion-exchange membrane with cationic paint). Ion-exchange membranes do not depend on permeation of external electrolyte for their conductivity so that membranes of the required conductivity need not have a highly porous structure. The characteristic property of ion-selectivity (selective permeation of ions of a single sign of charge) is contributed by the highly ionised groups forming part of the polymeric structure of which they are composed, sulphonate groups causing permeation (and electrolytic conduction) to be almost exclusively by cations, and quaternary nitrogen base groups by anions. Cation-exchange membranes effectively resist the natural tendency of anions (which may be hydroxyl or salt anions) to migrate with the electric field or to diffuse from catholyte to paint, so that the base liberated at the cathode can be allowed to accumulate to a high enough concentration to permit very low flushing rates and to keep the voltage loss across the catholyte small. The membranes themselves have resistances low enough to ensure that the voltage loss across the membrane is negligible: less than 1 V at the highest membrane current densities occurring in practice (say 0.05 A cm^{-2}).

The ion-exchange membrane method of pH control[69] has been in operation since June, 1964 without any difficulty of a fundamental nature, a total of about 60 plants with a wide range of production conditions being involved. The current efficiency has proved to be essentially 100% and control is completely automatic with reliability of a very high order. Very infrequent checks of pH are all that are required and it may be necessary occasionally to add small quantities of base to compensate for a slight excess of efficiency attributable to slightly acid material dripping into the tank from articles after painting. The method is conveniently integrated with automatic paint concentration control actuated by a current integrator or a counter sensing the area of metal coated. Because the replenishment paint is neutralised, there is no stability problem associated with incorporation.

The most satisfactory method of building ion-exchange membranes into an electrodeposition tank is in the form of a number of removable units each containing a sheet metal cathode. This arrangement has the advantage over an integral construction that maintenance is easy, it being usually possible to remove a single unit for maintenance without affecting production. It also caters conveniently for the fact that ion-exchange membranes are available in relatively limited sizes. There are commercially available ion-exchange membranes of the requisite electrochemical properties and mechanical robustness and in sufficient size to permit the fabrication of cathode units for large motor body electropainting plants. The overall cathode unit dimensions may be, for example, 0.75×2.5 m. Rigid thermo-

Fig. 2.
A cathode unit constructed according to the recommendations of I.C.I. (Paints Division) by Acalor (1948) Ltd., Crawley, Sussex, England.

plastic materials are favoured for the construction of the units and provision is made for effective support of the flexible membrane. A photograph of a cathode unit constructed according to the recommendations of I.C.I. Ltd (Paints Division) by Acalor (1948) Ltd., Crawley, Sussex, England, is shown in Fig. 2.

With a membrane controlled deposition process, the use of auxiliary cathodes to supplement the coating of difficult regions should be limited to the extent that the auxiliary cathodes (which are presumed not to be shielded by membrane as are the main cathodes) do not carry so high a proportion of the current that pH stability is prejudiced; in practice this may mean up to *ca.* 30%. On economic grounds, it is preferable to dispense with the use of the additional cathodes altogether, but if complex articles are being painted this requires paint of high throwing power or an excessively long deposition time.

These are the facts underlying the distinct approaches to the technology of paint electrodeposition referred to above. On the one hand, we have the process in which an amine neutralised paint is employed and coverage is assisted by means of additional cathodes (which can amount to more than 20 in number on a single motor body). Because base liberated at these cathodes cannot readily be prevented from mixing with the paint, this approach has stimulated the development of the base-deficient feed method of pH control. The other process, which is of more recent development, involves a paint of high throwing power, no auxiliary cathodes, and the automatic pH control offered by the ion-exchange membrane system.

5.1.4 The nature of the electric current supply

Mention has been made of the effect of current supplies which deviate in waveform from smooth direct current. Most generally on the technical scale, the direct current supplies used, obtained by full-wave rectification of 3-phase sources, have a small ripple content and their performance in electrodeposition is not significantly different from that of completely smooth DC. A small number of commercial installations are believed to have been operated with thyristor-controlled supplies having discrete pulses separated by interruptions, but the use of such supplies is not widespread; on plants for the priming of motor bodies it would predictably necessitate the use of auxiliary cathodes because of the loss in throwing power associated with the use of a pulsed source (*cf.* 4.6.3).

The manner in which the installation is connected relative to earth is an important variable in the design of an electrocoat plant, closely related to the selection of the method of pH control. With the base-deficient feed method, it is common to employ the uncoated inner walls of the tank as the cathode and to have the positive pole connected through a bus-bar contacting the jig or other suspending frame carrying the articles being processed: this necessitates thorough insulation of the jigs from the overhead conveyor because both paint tank and conveyor are connected to earth. In the case of membrane-controlled processes, it is clearly not possible to operate with both cathodes and an unlined tank connected to earth, nor is it entirely satisfactory to coat a metal tank with any of the common tank lining materials and to earth both tank metal and cathodes, for then cathodic areas develop at small leaks in the lining and also occur at other points which cannot be protected by lining, such as pump casings, heat exchangers and the like.

One method of operating a membrane-controlled system is to connect neither side of the power supply to earth (the "floating" power supply), in which case an imperfectly lined or even completely unlined tank can be used. There is no significant hazard associated with this method of connection additional to that already present with the voltage employed. If an unlined tank is in use, current paths can occur through the tank if such paths should happen to be shorter (after allowing for the overpotentials encountered at the metal/paint interfaces) than the conduction path directly through the paint, but such "secondary" electrolytic circuits are largely self-limiting in nature because of the resistance of the deposit formed at the region which acts anodically. The "floating" power supply can be employed irrespective of whether the articles being coated are intended to be connected with the current supply at the moment of entry into the tank ("live" or "dead" entry). In general, "live" entry is to be preferred because smaller demands are made on the rectifier installation in regard to initial surge current, but it normally requires that the articles are dry on reaching the electrocoat plant, *i.e.* that water remaining from the last rinse in the pretreatment stages has been dried off. It sometimes occurs even with dry metal entering the tank that "live" entry gives rise to texture defects in the final film (known as "hash-marking"); these occur with some paint systems which react unfavourably to the current density obtaining under these conditions and are accentuated when the conveyor movement is not smooth.

An alternative method of connection, suitable for systems capable of "live" entry, is for the positive pole to be earthed, the tank lined and the negative pole connected to separate cathodes. These may be plain or within membrane compartments. This has the important advantage that the expensive and troublesome insulation of the positive bus-bar from the conveyor is not required, though with the high currents involved in a motor painting plant an earthed bus-bar is necessary to forestall sparking at the conveyor chain links. In this arrangement, the tank lining serves to prevent deposition on the inner walls, but where deposition does occur *e.g.* at small breaks in the lining, it is self-limiting and, unlike cathodic electrolysis, not detrimental to the rest of the lining.

Another aspect of the question of electrical supply is in regard to the variation of applied voltage with immersion time. While, in principle, the alternatives of constant current and constant voltage are both feasible, the change in resistance accompanying deposition makes constant current impracticable or even unattainable on the large scale, especially with articles of complex shape in conveyorised plant. Much the commonest arrangement is one in which a constant applied voltage is used, but it is worth pointing out that the interelectrode resistance (especially with paints of low conductivity) limits the effective voltage during the early stages, thereby minimising the risk of rupture. A widely used variation on the constant voltage arrangement is one in which a sequence of voltage steps is employed, principally with the object of improving effective throwing power by the application of high voltage during the later stages of deposition. A further variation is to employ a resistor in series with the first part of the anodic bus-bar: this has essentially the same effect without requiring multiple rectifiers.

Stepped voltage or series resistor arrangements can be employed with the earthed bus-bar system by making the appropriate connections to the various cathode units rather than to sections of the bus-bar when negative earthing is used. In the case of stepped voltage systems with multiple rectifiers, this requires that the rectifiers delivering current at the lower voltage should be specified as able to withstand the same DC voltage as the high voltage rectifier, because the lower voltage rectifier is polarised at the higher voltage during periods of no throughput in the plant.

All that these variations accomplish is a partial compromise between the extremes of constant current and constant voltage deposition, with the main object being to secure the highest possible mean voltage without inducing rupture of the depositing paint film.

5.2 Ancillary operations

Many of the operations forming part of the complete industrial metal finishing process are standard, but there is a growing recognition that the best results are to be derived from an optimised combination of operations with electrodeposition at its heart. This is especially the case in regard to metal pretreatment, and the usual electrocoat process also entails rinsing operations which have no counterpart in conventional painting.

5.2.1 Cleaning and pretreatment

While electrodeposited paints are more tolerant of greasy substrates than conventional paints, appearance and performance are unsatisfactory unless clean metal is used. Conventional cleaning methods are employed including trichlorethylene vapour degreasing or other solvent cleaning, mild acidic or alkaline aqueous cleaners which may be dipped or sprayed and possibly accelerated by electrolysis or ultrasonics. Care is necessary to ensure that the cleaning agent is thoroughly removed before electrodeposition unless cleaning is to be followed by a metal pretreatment stage.

As with conventional methods of painting, the corrosion resistance and paint adhesion obtained are improved by the use of metal pretreatment. The treatment processes most commonly encountered for steel are those in which a coating of zinc phosphate is applied, usually to reach a coating weight of ca. 2 g m^{-2} of finely crystalline phosphate. Spray application of the phosphate has the advantage of better removal of surface oils and other contamination and generally provides more uniform coatings of finer crystal structure. With motor bodies, combined spray and underbody dip processes, although more expensive, may provide the best solution to the problem of cleaning and phosphating the interiors of the important structural box sections.

One of the currently topical questions in the whole subject of paint electrodeposition is concerned with the effect of electrodeposition on zinc phosphate pretreatment films. An example is provided by the work of Menzer[70], who considered the loss of phosphate suffered by zinc phosphate treated steel undergoing electrodeposition not to have a significant effect on the final performance, nor did he attribute phosphate loss to the acid conditions at the anode but rather to mechanical disruption of the crystalline coating.

Very thorough rinsing is essential after phosphating to eliminate traces of electrolyte that might contaminate the electrocoat bath. Thus, three-stage rinsing is typical, the last two consisting of demineralised water. It is not uncommon for a dilute chromic acid rinse to be interposed in the rinse stages.

Draining and drying before painting are optimal steps with electrocoating, but essential with conventional solvent-borne dipping procedures. If these stages are to be entirely omitted, the wet articles cannot be allowed to enter the deposition tank with the power connected but a short period must be allowed with the power off so that adherent water is dissipated[71]. If entry to the paint tank is to be "live", the standard of dryness required is not as great as is required with solvent-borne dip painting and devices such as a rise and fall in the conveyor (to spill out excess water from motor bodies) or an air-blow may suffice. Wet "live" entry can be facilitated by rapid immersion rates, but these create additional problems in the stable jigging of larger articles; a measure of compromise is possible by immersing at a moderate rate with a reduced voltage applied during the early stages of deposition.

5.2.2 Tank control

Besides chemical control of the tank composition, the main factors in tank maintenance are the agitation required to maintain material in suspension and the heat exchange needed to control temperature. Agitation is especially important because of the low viscosity and consequent speed of pigment settling. It generally takes two forms: circulation by pumps at the rate of several (say 4 to 10) tank turnovers per hour and mixing within the tank using propellor tube stirrers or similar equipment at velocities equivalent to say 10 to 30 tank turnovers per hour. Heat exchange and filtration are usually included in the circulation circuit. The heat exchangers, which are usually of the tube and shell type, are provided to remove the Joule heat created by the deposition process. Filters of many types are used, from coarse strainers to filter batteries removing particles down to 25 μ in size.

In the case of a plant having pH control by means of ion-exchange membranes, the catholyte liquor may to advantage be recirculated. According to their positions in the tank the current passing through individual cathodes varies and hence the ideal flushing rate (for constant outlet conductivity) varies. By recirculating the liquor these differences are rendered unimportant and the conductivity can be kept constant by an automatic conductivity-controlled feed of demineralised water.

5.2.3 Rinsing after painting

The surface of painted articles emerging from the deposition tank is coated with an adherent dip-layer of dilute aqueous material overlying the main electro-coagulated paint film. The dip-layer is essentially in the salt form and variable as to thickness; in most instances it is removed by rinsing though there are several installations, including a substantial one of four motor body painting lines, where rinsing is not practised after electrocoating.

Because an economic method of paint recovery from the rinse waste has not yet been evolved*, the quantity of paint removed in the dip-layer (the "drag-out") is an important factor in the economics of paint electrodeposition. In particular, because drag-out loss increases with increasing paint tank concentration, it governs the solids concentration at which operation is economic. There is some evidence that paints differ in the drag-out obtained at a fixed concentration.

Normally mains water is used in the initial stage after painting, but a final rinse with demineralised water is desirable to obtain the optimum performance from the final stoved film, especially if it is to form the primer in a multiple paint system.

5.2.4 Curing

Only brief draining is required after rinsing, rather than a prolonged period of flashing, because of the small water and solvent content of electrodeposited films. The electrocoated articles can then pass directly to the stoving oven, which can be of conventional types.

5.3 ELECTROCOATING AS PART OF A COMPLETE METAL FINISHING SYSTEM

The main application of electrodeposited paint at the present time is as a primer coat on phosphated steel. In many cases, typified by the motor industry, the electrocoat is the only treatment given to the unseen and inaccessible regions of complex

---

* *Cf.* Addendum, p. 462.

460 ELECTRODEPOSITION OF PAINT

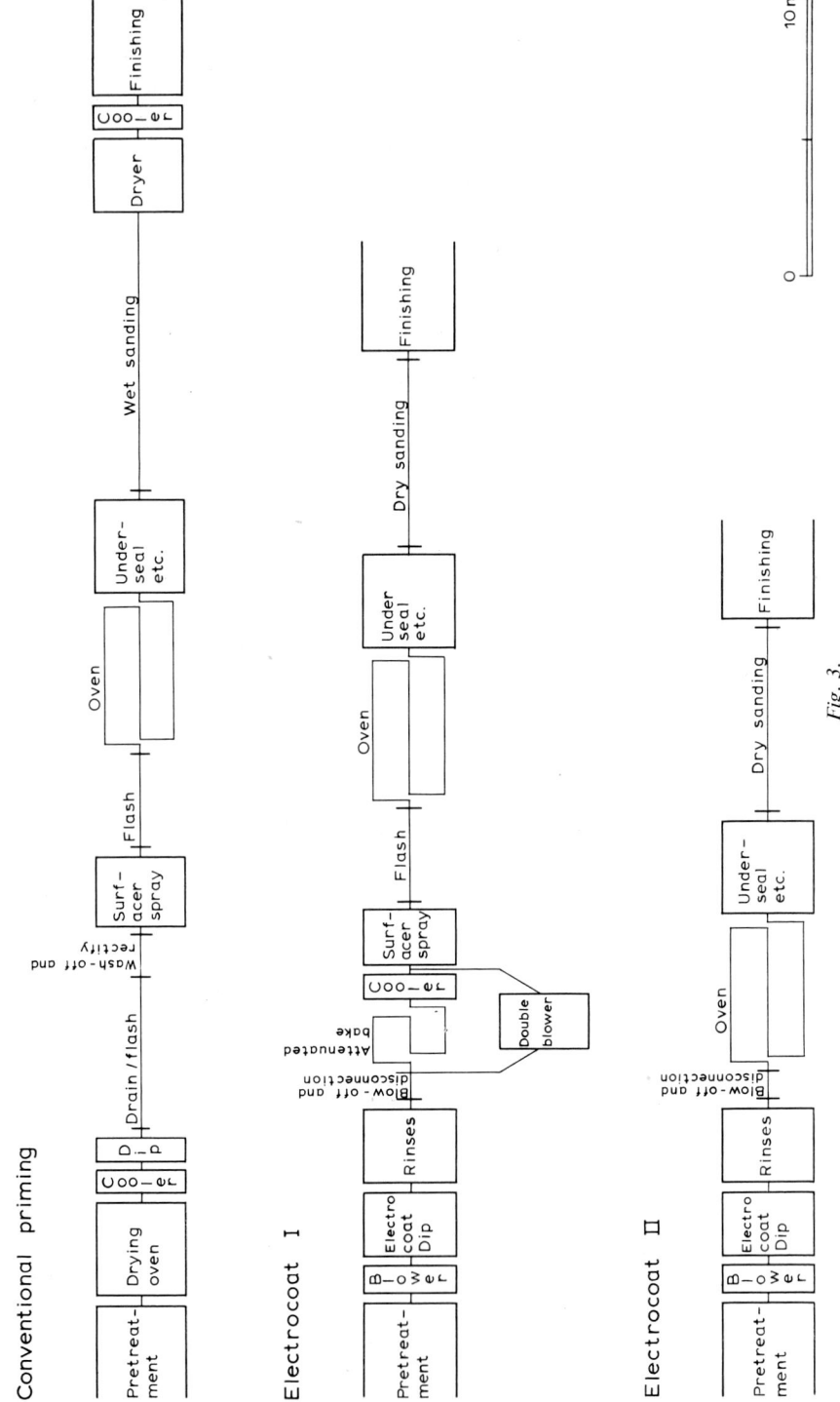

Fig. 3. flow sheets for motor body finishing sequences.

articles but it forms the foundation of a multiple coat finish of high functional and decorative performance on the visible parts. To perform acceptably, therefore, the electrocoat must satisfy requirements not only as to its performance in inhibiting corrosion but also in respect of acceptance of subsequent coats: in particular, intercoat adhesion must be satisfactory and, if a sealing coat is not to be applied, the electrocoat must resist sinkage on overcoating. These considerations are additional factors limiting the range of compositions acceptable for electrocoat paints, yet considerable success has been achieved in formulating satisfactory products for such a demanding outlet as the motor industry.

In discussing the advantages of electrodeposition as a method of paint application, mention was made of the possibility of dispensing with or diminishing other operations forming part of the complete finishing process. This derives from the uniformity of coverage compared with conventional dipping systems (themselves favoured for economy of paint usage and low labour demand) and is illustrated by the flow sheets shown in Fig. 3. These depict three alternative motor body finishing sequences, including a typical process with conventional dip-priming and two which involve electrodeposition. The pretreatment and final colour application steps are common to all processes and are therefore not broken down in the flow sheets which have been drawn with a horizontal scale representing the line length occupied by the various steps.

In the conventional process, which is a typical one though considerable variations are encountered in practice, the dip (in which only the lower part of the body is immersed) is followed by a prolonged period of drainage and flash-off after which all or most of the unstoved primer on the exterior of the bodies is removed by solvent washing. This wasteful step arises because of defects in the dip coat, especially sags and runs, which would make the appearance after finishing unacceptable; it is cheaper to remove the external primer than to correct the defects after stoving. There follows a sprayed primer surfacer coat applied wet on wet (in the sequence shown, though a stoving step is sometimes interposed), followed by flashing and stoving. Wet sanding is normal for the flatting of the primer surfacer coat as dry sanding does not generally give a sufficiently uniform surface to accept the final coats of finish.

The first electrocoat sequence is an example of current practice. Because the bodies to be primed need not be absolutely dry, a blower replaces the drying oven and cooler in the conventional sequence. Primer flashing is replaced by the rinse stages after which a brief stoving or an efficient air-blow suffice to set up the film to accept the sprayed surfacer coat; the latter is itself of reduced thickness compared with the sequence involving conventional priming because of the greater primer thickness present at all points on the body. Because of the relative absence of defects in this case, dry sanding can replace the more costly wet method. A substantial saving in line length is evident from the flow sheet.

The second electrocoat sequence represents an ultimate simplification of the finishing process in which full advantage is taken of the uniformity of film, absence of defects and control of film thickness attainable by electrodeposition. In this case, the film thickness applied in the deposition stage would be some 35–40 μ, i.e. a typical primer surfacer thickness compared with the 20–30 μ required when the electrocoat functions solely as primer. The advantages in process simplification and floor area required are obvious.

The capital cost of the conventional and first electrocoat lines shown are of a similar order, but the simplified electrocoat process offers a substantial saving in this respect. The running costs differ chiefly in respect of the difference in labour requirements. If the common steps of phosphating, undersealing and the final finishing operations are excluded, the conventional process shown has, in typical British practice, a labour requirement of ca. 1.0 man per motor body/h, whilst in a conventional process (e.g. rotor-dipping) entailing priming of the entire body surface the labour required may approach 2.0 man per body/hr. In contrast, the first electrocoat process requires ca. 0.6 man per body/hr, and this could fall to as low as 0.4 in the simplified electrocoat sequence. In addition, although the total thickness of paint applied is the same in all cases, the fact that priming by electrodeposition reduces the wasteful film repair operations gives rise to savings in paint consumption.

In conclusion, the introduction of electrodeposition as a method of priming motor bodies offers substantial manufacturing as well as performance advantages. There can be few developments which have caught the imagination of a major industry to the degree that paint electrodeposition has that of the world's motor manufacturers during the past five years.

Addendum

Since this chapter was drafted, an important development in the technology of the paint electrodeposition process has been the application of ultrafiltration to dilute paint[73]. It has been found that with a suitably selected porous membrane and an operating pressure of the order of 2 atm., an ultrafiltrate is obtained which consists essentially of the continuous phase while the dispersed paint material concentrates on the high-pressure side of the membrane. As with other processes of this kind, rapid circulation of the liquid on the high-pressure side is necessary to limit concentration polarisation.

Several benefits are possible in paint electrodeposition. Firstly, it is claimed that pH control can be accomplished, but this is probably limited to those cases in which weak bases are used and the pH rises to such a high value that a substantial amount of free base is present in the continuous phase. This effect is perhaps valuable to supplement imperfect methods of control rather than as a complete method of control. A second application is as a method of decontamination; soluble salts which might accumulate in the paint are eliminated with the filtrate, so, if a paint is subject or sensitive to such accumulations, its performance can be maintained at a more stable level than without ultrafiltration. A third application, which is of special value to large users of Electrocoat paint, is to dewater dilute paint and return it at a higher concentration to the tank. This makes it possible to recover some or most of the paint lost in rinsing stages whilst maintaining the water balance of the paint tank.

The third application, dewatering so as to effect paint recovery, requires rather larger membrane areas in the ultrafiltration equipment than when the process is used for the other purposes mentioned. It promises, nevertheless, to be attractive economically. Clearly, the rinse recovery process could entail recycle of the ultrafiltrate to the rinse water supply so as to reduce effluent from the plant to a minimum. For decontamination to occur, some or all of the ultrafiltrate would have to run to effluent.

Acknowledgements

We are grateful to Dr. D. B. Bruce of Pinchin Johnson and Associates Ltd., and to Messrs. Berger, Jensen and Nicholson Ltd., for information passed through Dr. A. T. Kuhn and to Mr. B. Jeffrey of Carrier Engineering Ltd. for information relating to plant requirements for paint electrodeposition.

6. References

1 Crosse and Blackwell Ltd., Brit. pat., 455, 810; 496, 945.
2 FINK and FEINLEIB, *Trans. Electrochem. Soc.*, 94 (1948) 309.
3 General Electric Co., Brit. pat., 1, 294, 627.
4 Dunlop Rubber Co., Brit. pat. 291, 477.
5 Sheppard, U.S. pat. 1, 476, 374.
6 Crosse and Blackwell Ltd., Brit. pat., 455, 810; 496, 945.
7 Ford Motor Co., Brit. pat., 933, 175.
8 Clocker, U.S. pat., 2, 188, 882; 2, 188, 885.
9 Pittsburgh Plate Glass Co., Brit. pat., 1, 096, 934.
10 Glidden Corp., Brit. pat., 972, 169; 1, 030, 204. Vianova Kunstharz A.G. Belg. pat., 687, 197.
11 Reichold Chemie A.G., Fr. pat., 1, 541, 164. Vianova Kunstharz A.G., S. African pat. 67/5409.
12 Glidden Corp., Brit. pat., 972, 169.
13 Glidden Corp., Brit. pat., 972, 169; 1, 030, 204.
14 Archer Daniels Midland Co., Brit. pat., 1, 113, 891.
15 BOUNDY AND BOYER, *Styrene, its Polymers, Copolymers and Derivatives*, Reinhold, New York, 1952.
16 Pittsburgh Plate Glass Co., Brit. pat., 1,106,050, Belg. pat., 679,899. Guldenpfennig, S. African pat., 65/2893. Reichold Chemie A.G. Belg. pat., 665,115. Vianova Kunstharz A.G., Brit. pat., 1,104,918. Chem. Ind. Synres N.V., Belg. pat., 711,805.
17 Du Pont, Brit. pat., 1,146,461; see also technical literature of Amoco Chemical Co.
18 Trojan Powder Co., technical literature: dimethylol propionic acid.

19 Celanese Coatings Corp., Belg. pat., 690,753. Reichold Chemie, S. African pat., 66/1399. Shell, Belg. pat., 694, 307.
20 Pinchin Johnson, S. African pat., 66/6315.
21 Pinchin Johnson, S. African pat., 66/6315. Bergers, Australian pat., 6914/66. Celanese Coatings Corp., Brit. pat., 1,099,419. Wülfing, Belg. pat., 656, 503. Berger, Brit. pat., 1, 148,914.
22 Pinchin Johnson, S. African pat., 66/6315.
23 Pinchin Johnson, S. African pat., 66/6315.
24 Pinchin Johnson, Brit. pat., 1,080,172.
25 Pittsburgh Plate Glass Co., Belg. pat., 701,854, Brit. pat., 1,149,153. Celanese Coatings Corp., U.S. pat., 3,305,501. Devoe and Reynolds, Brit. pat., 1,039,793, Belg. pat., 659,863. Reichold Chemie, Fr. pat., 1,472,346, Belg. pat., 693,298, Brit. pat., 1,146,693.
26 Pittsburgh Plate Glass Co., Belg. pat., 701,854. B.P. Chemicals, Brit. pat., 1,127,985.
27 Pittsburgh Plate Glass Co., Brit. pat., 1,096,066. De Soto, Belg. pat., 704,590. B.A.S.F., Brit. pat., 1,146,742; 1,146,858; 1,148,764.
28 Courtalds, S. African pat., 66/149. B.A.S.F., Belg. pat., 681,365, Fr. pat., 1,487,893; 1,485,140; 1,486,213. De Soto, Neth. part., 67/02953. Vianova Kunstharz, Fr. pat., 1,499,083. Coates, Brit. pat., 1,115,130. Courtalds, Brit. pat., 1,127,232.
29 B.A.S.F., Fr. pat., 1,489,548. S.C.E.C.O.A., Fr. pat., 1,531,694. Nippon Paints, Jap. pat., 3096-69.
30 Mobil, Brit. pat., 1,049,025, U.S. pat., 3,340,172.
31 B.A.S.F. Belg. pat., 677,046; 679,845; 683,896; 690,308.
32 Coates, Brit. pat., 1,102,384, S. African pat., 67/4306. Esso, Brit. pat., 1,115,218. British Industrial Plastics, Brit. pat., 1,145,831. Pittsburgh Plate Glass Co., Brit. pat., 1,147,593.
33 Vianova Kunstharz A.G., Belg. pat., 687,197.
34 Vianova Kunstharz A.G., S. African pat., 67/5409; 67/7814. Reichold Chemie, Belg. pat., 710,524, Fr. pat., 1,541,164.
35 Glasurit, Belg. pat., 691,965. De Soto, Belg. pat., 710,605.
36 Bayer, Neth. pat., 67/07358. Berger ,Australian pat., 9402/66.
37 Rheinpreussen, Ger. pat., 1,273,100.
38 Esso, Brit. pat., 1,102,652; 1,107,147.
39 Ford Motor Co., U.S. pat., 3,382,165, Brit. pat., 1,135,404; 1,135,405; 1,135,406.
40 I.C.I. Ltd., Neth. pat., 67/16379.
41 Canadian Industries Ltd., Brit. pat., 973,568.
42 Siemens, Brit. pat., 1,002,958.
43 National Lead Co., technical literature: "Oncor M50", "Oncor S25".
44 Pittsburgh Plate Class Co., Brit. pat., 1,117,875; Nederlands Castor Oliefabriek, technical literature: Necowel L2075.
45 Goodlass Wall, Brit. pat., 1,027,813.
46 I.C.I. Ltd., Brit. pat., 1,030,425.
47 Pinchin Johnson, Brit. pat., 1,046,016.
48 *Cf. inter alia:* SCHULMAN and COCKBAIN, *Trans. Faraday Soc.*, 36 (1940) 651.
49 Distillers Company Ltd., London, "Bisol" Information Sheet No. 605, March, 1966.
50 Pinchin Johnson; Brit. pat., 1,040,016.
51 Ford Motor Co., Brit. pat., 1,100,472.
52 H. SCHENE, 7. Internationale Tagung für Oberflächentechnik der Metalle, Hanover, 1968, p. 87.
53 J. P. WILTSHIRE, *Surface Coatings*, 2 (1966) 88.
54 F. BECK, *Farbe Lack*, 72 (1966) 218.
55 T. A. STRIVENS, to be published.
56 H. SAND, *Phil. Mag.*, 1 (1901) 45.
57 D. B. BRUCE, *Some Mechanisms in Electropainting*, Institute of Metal Finishing Conference, Brighton, May, 1967.
58 D. A. OLSEN, P. J. BOARDMAN and S. PRAGER, *J. Electrochem. Soc.*, 114 (1967) 445.
59 F. BECK, *Ber. Bunsenges. Physik. Chem.*, 72 (1968) 445.
60 B. A. COOKE and T. A. STRIVENS, *J. Oil Colour Chemists' Assoc.*, 51 (1968) 344.
61 L. ONSAGER, *J. Chem. Phys.*, 2 (1934) 599.
62 W. GÖRING, B. ANCZYKOWSKI and H. NOACK, *Farbe Lack*, 75 (1969) 327.

63 P. Stoll Co., S. African pat., 67/1955; Shell Oil Co., U.S. pat., 3,455,805.
64 W. MAISCH, *Ind.-Lack. Betrieb*, 33 (1965) 299.
65 G. E. F. BREWER, M. E. HORSCH and M. F. MADARASZ, *J. Paint Technology*, 38 (1966) 452.
66 W. MAISCH, *Ind.-Lack. Betrieb*, 35 (1967) 3.
67 D. B. BRUCE, personal communication to A. T. Kuhn, June, 1969.
68 G. E. F. BREWER, G. L. BURNSIDE and G. G. STROSBERG, *Solubilizer Balance in the Electrodeposition of Paint*, Amer. Chem. Soc. Meeting, Minneapolis, Spring, 1969.
69 I.C.I. Ltd., Brit. pat. 1,106,979, U.S. pat., 3,419,488.
70 W. MENZER, 7. Internationale Tagung für Oberflächentechnik der Metalle, Hanover, 1968, p. 79.
71 I.C.I. Ltd., Brit. pat., 1,051,801, U.S. pat., 3,420,762.
72 D. R. HAYS and C. S. WHITE, *J. Paint Technology*, 41 (1969) 461.
73 F. C. FORBES, *Prod. Finishing (London)*, 23 (Nov., 1970) 24.

Bibliography

For an earlier review of the subject see R. L. YEATES, *Electropainting*, Robert Draper Ltd., Teddington, 1966.

Chapter 12

Electrodialysis*

G. S. SOLT

Director of Research, Wm. Boby Ltd., Rickmansworth, Herts.

1. Introduction

Electrodialysis as a process differs fundamentally from the great majority of electrochemical processes in that it does not utilise the electrode reactions: it is an ion transport process using selective membranes. Electrodes do, of course, have to be incorporated in electrodialysis plant, but they serve the purely ancillary purpose of applying the EMF necessary for the main object to be achieved. They are in fact a necessary evil, since they involve a great deal of extra complication in the design and operation of plant, and since they tend to be focal points from which operational troubles stem. At least one exasperated research engineer has suggested rotating the electrodialysis stack in a magnetic field in order to create a flow of current without having to use electrodes: unfortunately this is not a useful idea.

The permselective membrane is essentially an ion exchange resin in sheet form; and there are therefore anion and cation selective membranes respectively. Their chemical and physico-chemical properties are described in Chapter 15. For use in electrodialysis, the important features of these membranes are:

(a) They are ion-selective—*i.e.* cation selective membranes will pass practically no anions, and *vice versa*.

(b) They have a low electrical resistance—*i.e.* the electrical conductivity of the membrane material is high and the membrane is thin. This low resistance must be maintained under all conditions of use, especially when passing direct current in a low-salinity environment.

(c) They have the necessary mechanical and chemical stability for practical use.

A useful way of thinking about permselective membranes is as a selective ion conductor. For example, a cation selective membrane is essentially a porous sheet which possesses fixed negative charges due to the electronegative $-SO_3^=$ groups in the resin matrix. Such a sheet promotes the passage of positively charged ions but rejects negative anions.

Suppose a number of these membranes is assembled as shown in Fig. 1. Cation and anion selective membranes are placed alternately and salt solution flows through

---

* This chapter should be read in conjunction with that on Diaphragms for Electrochemical Processes (Chapter 15).

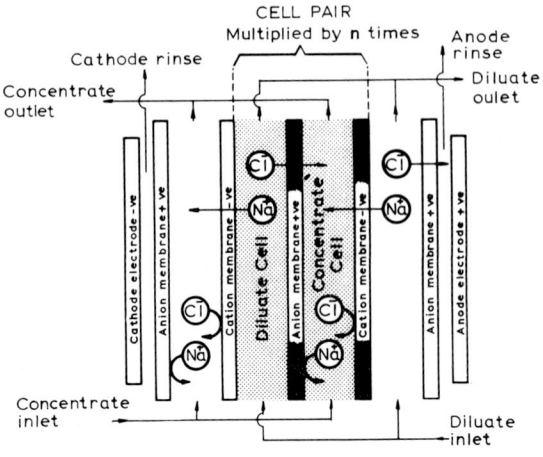

Fig. 1.
Schematic diagram of electrodialysis stack.

the cells formed between the membranes. If an EMF is applied in a direction across the membrane faces, the sodium ions will move towards the negative electrode on the left of the diagram, and the chloride ions towards the positive electrode on the right.

In the cells marked "Diluate Cell", each species of ion finds itself moving towards a membrane which allows it to pass into the neighbouring cell, and so the concentration of electrolyte in these cells falls. In the cells marked "Concentrate Cell", however, each species of ion moves towards a membrane which rejects it, and in these cells the electrolyte concentration therefore increases. As a result, an electric current flows across the cells, which is carried by both ion species in the solution, but only by one species at a time across each membrane.

In such an apparatus, the basic unit is the "Cell Pair", which consists of one cation and one anion membrane and the two cells in which the electrolyte becomes more concentrated and less concentrated respectively. A faraday flowing across such a cell pair theoretically transports a gram equivalent of electrolyte from the "diluate" into the "concentrate" cell. In an apparatus of N cell pairs assembled between a pair of electrodes into a single "stack", the current flows in series through all the cells and one faraday will effect a total transport of N gram equivalents.

This basic process is capable of serving three purposes:

(1) It can be used to reduce the electrolyte concentration in a solution, as in the desalination of water. This is the most important use at present.

(2) It can be used to increase the electrolyte concentration in a solution, as in the production of strong brine from sea water.

(3) It can be used to separate electrolyte from non-electrolyte in a ternary system, as in the de-ashing of glucose solution.

Somewhat different types of membranes and of apparatus construction may be used for each of these types of application.

2. General details of construction

In concept the process is elegant and simple, but its requirements are such that, when their effect on the details of mechanical construction are considered, serious problems immediately become apparent.

For a specific task to be performed in a given apparatus of N cell pairs, the current to be passed is fixed by N and by the mass of salt to be transported. The power consumption is therefore a function of the resistance per cell pair. In each cell pair the resistance of the two membranes is small compared with that of the solution passing through the two cells. The resistance of the concentrate cell can be decreased by recirculating concentrate to raise its conductivity as its electrolyte concentration increases. This, however, increases the danger of scale formation due to supersaturation.

The main resistance in each cell pair is thus due to the diluate cell. In order to keep this to a minimum, the cell must be made as thin as possible. In practice, various plant manufacturers use cell thicknesses between $\frac{3}{4}$ mm and $1\frac{1}{2}$ mm, concentrate and diluate cells usually being the same.

(The argument above is not strictly true for salt production, when the conductivity of both concentrate and diluate are so high that the membrane resistance does become an important factor. Nevertheless, even for this application, the cells must be kept as thin as possible in order to keep the resistance down).

Another way of reducing the cell pair resistance is to increase the membrane area per cell pair. In fact, all other factors being constant, the power required to carry out a specific task varies inversely as the total membrane area utilised. This raises an economic problem in plant design: whether to install a large membrane area with a high capital cost resulting in a plant with a low power consumption, or *vice versa*. It is therefore most important for the membrane area to be cheap. This means that, not only must the membrane itself be cheap, but so must the components which support the membranes; it is also important that gaskets and supports should not obscure a large proportion of the total installed membrane area and so reduce the effective area available for passing current.

Electrodes are troublesome, expensive, and the cathode must be dosed with acid to avoid scaling with inorganic salts. In order to reduce the number of electrodes to a minimum, the number of cell pairs in a single stack should be as large as possible. In order to reduce costs, the area of each cell pair should be as large as practicable. In fact, stacks of up to 300 cell pairs are used, the largest cell in commercial use being 150 $\times$ 50 cms. Even bigger stacks are in the course of development.

With so many cells in a stack, means have to be found for manifolding the incoming and outgoing liquids into the alternate compartments, as shown in Fig. 1. In order to keep pressure losses down, a large number of cell pairs must be supplied in parallel, and a metering orifice has to control the flow entering each cell to ensure even liquid distribution between cells.

Membranes must not be allowed to touch one another, otherwise there is a danger of short circuiting and scorching. A separator therefore has to fill every cell, and its hydrodynamic properties are critical.

The stack in effect comprises two independent circuits, one for concentrate flow and the other for diluate. In practice, it is usual to make these two circuits identical. It is vital that leakage between these two circuits be kept to a minimum, as it leads either to a loss of product by diluate leaking into concentrate or, more seriously, the work done by ion transport is undone by concentrate leaking into diluate. External leakage must also of course be avoided, though a slight weeping is normally experienced and tolerated.

Polarisation criteria, which will be discussed below, dictate that a minimum total liquid path length is necessary for any specific task to be performed. The stack must therefore be built in such a way that this path length can be obtained, either by creating labyrinth flow within each cell, or by internal staging of cells into a multipass system. In larger installations incorporating a number of stacks it is of course possible to connect stacks hydraulically in series to obtain the necessary path length. In all these cases, however, it is necessary to design the cells in such a way that the pressure drop remains as low as possible. A high pressure loss would cause excessive power lost in pumping, and in those cases where a very long path length is necessary, it might lead to inlet pressures so high that the apparatus could not be built to withstand them. The hydraulic design must also reduce the unwanted leakage of DC along liquid conduits, especially in the highly conductive concentrate circuit. This requirement is usually in direct opposition to the need to keep pressure loss low.

Although considerable improvements have taken place in recent years, the ion exchange membrane is not a good mechanical component. Ion exchange resins are, by virtue of their chemical properties, dimensionally unstable and this weakness is reflected in all commercial membranes. Also the need to keep membranes thin makes them intrinsically weak.

The chemical conditions which are created within a stack are as follows: it contains an acid, saline solution with DC flowing. There is danger of scale formation. One electrode produces chlorine. Undesirable DC leakage currents must be avoided by using materials which are good electrical insulators and retain this property indefinitely when immersed in acidified brine. Any fault in operation is likely to cause local overheating. Only noble metals and carefully selected rubbers and polymers will stand up to such conditions.

To summarise, therefore, a stack must be designed to contain hundreds of very large and thin cells, each of which contains a separator and is equipped with a flow metering orifice, but neither of these must create excessive pressure loss. The structure must be cheap, leakproof, it must expose as much membrane area as possible, and it must be easily dismantled for cleaning in case of serious scale formation. The membranes which are available are mechanically rather indifferent, and the choice of materials of construction is limited to those which are expensive, or bad engineering materials, or both. It is hardly remarkable, therefore, that the development of the process has been beset by severe troubles. Indeed its recent successes have been largely due to progressive advances in the chemistry and fabrication of the materials of construction coming to the aid of the Electrodialysis designer.

3. Polarisation

3.1 INTRODUCTION

In electrodialysis, as in other electrochemical processes, polarisation describes the set of phenomena which arises when a current passes across a surface at a current density which is greater than that which the surface will readily transmit. In electrodialysis, the limiting condition is set up when one of the diffusion barriers involved becomes depleted of the transported ion species. Further transport of ions involves either the transport of ions due to dissociation of the medium, or in certain cases additional transport of the counter-ion occurs, but only at the cost of a considerable increase of power consumption.

Figure 2 shows (in a simplified form) how the phenomenon occurs. Figure 2(a) represents a cross-section of an anion-selective membrane and the two cells which border it, with NaCl solution flowing through the cells but with no current flowing. The lines ABC and XYZ represent the concentration of chloride ion in the two cells.

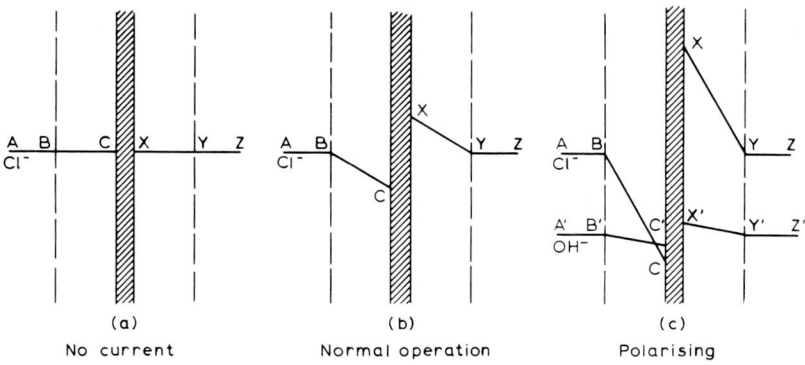

Fig. 2.
Kinetic phenomena on anion membrane. The lines A′, B′ etc. represent concentration of OH⁻ ions.

The regions BC and XY represent the Nernst films on the surface of the membrane, outside which water flows in "turbulent" flow.

If an EMF is applied across this cell, chloride ions begin to move from left to right on the diagram, across the water and membrane. Due to the selectivity of the membrane, the flux of chloride ions across the membrane is momentarily greater than that in the water. As a result, the liquid at the membrane face (at point C) becomes depleted of chloride ions. A similar imbalance in the sodium ion transport from right to left arises from the flux in the solution exceeding that supplied by the anion-selective membrane (which with an ideal membrane is zero). A concentration gradient is thus set up between B and C; while, to the left of point B, any tendency to set up a concentration gradient is nullified by the turbulence supposed to exist there.

By the same token, the region XY becomes enriched with chloride ions, while the bulk liquid from Y to Z remains unchanged.

In this way a kinetic equilibrium is set up whereby the same flux of the transported ion takes place across each plane in the diagram, from A to Z, but three methods of transport are involved. In the turbulent regions AB and YZ, ions are carried to and from the Nernst film by convection; within the membrane, ions migrate solely under the influence of the EMF; while, in the films BC and XY, ion transport occurs by the combined action of diffusion and migration.

The slope of the concentration graph across BC therefore characterises the rate at which chloride ions cross the stagnant film. If the EMF is increased, ion transport increases, and the concentration at point C must fall in order to provide the necessary driving force. Figure 2(b) shows this condition.

The "limiting" condition is reached when the EMF is increased to such a value that the concentration of chloride ions in the water at point C falls to zero: clearly this is a maximum condition beyond which some other phenomenon must set in. Indeed, as soon as this limiting point is reached, a new and disagreeable set of side reactions is observed, which are shown in Fig. 2(c).

We have so far ignored the presence of OH^- ions whose concentration in neutral solution is so low that they do not contribute significantly to the total ion transport. However, as soon as the chloride concentration at point C approaches concentrations in the region of 10^{-7}, the presence of hydroxyl becomes significant. Unlike the chloride ion, the concentration of hydroxyl at point C depends on the dissociation of water and not on transport phenomena, so that its concentration does not fall to zero with increasing EMF.

At and beyond the "limiting" EMF, therefore, significant and increasing quantities of OH^- ion pass across the anion membrane, and the following symptoms are observed:

(a) The apparent resistance of the cell rises sharply. (This is in fact due partly to the back EMF caused by the high concentration potential as well as a rise in ohmic resistance.)

(b) The pH of the diluate falls, and that of the concentrate rises.

(c) The current efficiency of the apparatus falls.

Each of these three symptoms has been used as a criterion to define the point at which polarisation is held to set in. In the early 1960s, there was in fact a spate of research papers on this subject. However, it soon became apparent that work done on cells in which there was turbulent flow of the liquid could not be used to calculate any generally applicable equations. The difficulty of defining the liquid film conditions made it impossible to generalise about cells of different geometry.

The only reproducible work was that which was carried out in stagnant cells, when the Prandtl equations could be used to characterise the liquid film conditions. While this work considerably advanced knowledge of the physical chemistry of polarisation, it did little to help the design of plant, which remains a matter of experience and rule-of-thumb.

In any practical apparatus, indeed, the picture is far more complex even than this. Where there are cells fed in parallel from a manifold, it is impossible to have exactly the same liquid velocity in each cell. Flow velocities within an individual cell cannot be uniform over the whole area through which current is passing. The separator which lies in the cells promotes turbulence so that films are reduced in thickness in some regions, whereas the places immediately adjacent to the points at which the separator touches the membrane inevitably are stagnant. In other words, the hydraulic conditions within the apparatus vary over a wide range. Furthermore the concentration of electrolyte varies from entry to exit. A stack therefore represents a highly heterogeneous set of conditions, in which some proportion of the membrane area is at polarising conditions at all but extremely small EMFs. The proportion of polarised area rises with an increase in EMF. For practical purposes, the designer must be prepared to allow a certain small percentage of the membrane area to be polarised.

While it is impossible to quantify the proportion of area polarising under given conditions (the more so since there is no agreement for the exact criteria at which polarisation is said to have set in), Fig. 3 shows a conceptual picture of the course of events. It plots the percentage of membrane area polarising in a stack at different applied EMFs, all other conditions remaining constant. As this graph shows, there are three main conditions:

(1) At EMF below X, the area under polarising conditions is negligible. As the concentration potential across a single membrane/section interface which accompanies full depletion is of the order of 1 V, a greater EMF must be applied before hydroxyl transport can set in.

(2) Between X and Y a small but significant area is polarising, but the average condition of the whole stack is such that reasonable performance is being obtained.

(3) Above Y polarisation is excessive.

Operation in condition (1) would in many cases be uneconomical, for it implies very large membrane areas and therefore very high capital investment. Condition

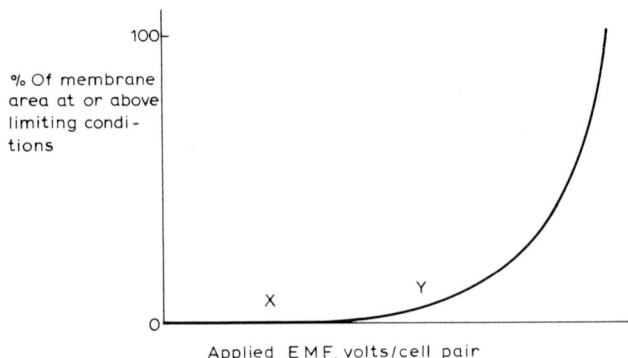

Fig. 3.
Distribution of polarisation (conceptual).

(3) is, in practice, unworkable. The chief interest, therefore, centres on the average condition (2).

The most significant parameter in this context is the applied volts per cell pair, which is of more immediate practical use than the "limiting current density" usually referred to in academic literature. As a criterion of critical conditions, it is roughly independent of temperature and viscosity, and although the critical voltage is theoretically dependent on the liquid velocity, the designer's scope for variation of velocity is usually limited. As a very crude rule-of-thumb, values below 2 V/cell pair give safe operation well within condition (2) (above). If liquid and current distribution were ideal, then this limit would bring operation very near condition (1). In fact, conditions are often very far from ideal.

Two factors must control the shape of the curve in Fig. 3: the mean concentration of electrolyte, and the local Reynolds numbers. The Reynolds numbers, while their overall mean cannot be quantified, depend on the velocity, viscosity and a linear dimension, presumably connected closely to the cell thickness.

3.2 Effect of temperature change

So far, temperature has been considered as a constant. However, changes of temperature have a considerable effect on the operation of the process. In the first place, the ionic mobility rises with rising temperature, as shown by a rise of about 2.3% in the conductivity of saline solutions for a rise of 1°C. Furthermore, the liquid viscosity decreases with rising temperature, with a resulting increase in the Reynolds number. In effect the resistance of a practical apparatus falls by 2% or more for a rise of 1°C. The improved kinetics of the system mean that higher current densities can be employed, and the same limitation of 2 V/cell pair (which is probably conservative) can be applied. The net result is that, at a higher temperature, a smaller cell pair area can perform the same duty at the same power consumption.

In practice, the commercial application of this advantage is limited. The limitations imposed by membranes and materials of construction make it doubtful whether a temperature of 50°C can safely be exceeded without drastically shortening component life. Preheating generally requires a heat exchanger, and in the treatment of brackish water the heat exchanger will either be subject to severe corrosion, or to scale deposition, or both. In treating milk whey, glucose and such solutions, however, pre-heating is normal. This is because the viscosity of these liquids is very high, and their conductivity low in relation to their content of dissolved salts. In the production of salt from sea water, Japanese practice is to allow the temperature to rise to the region of 40°C, but this is the convenient result of conversion of power into heat in the course of the process, and no external source of heat is needed.

3.3 Anomalous polarisation

In some cases, it has been known for electrodialysis stacks to show every symptom of polarisation in conditions which should in theory be well within normal safe operation. This anomalous polarisation indicates that, in addition to the three zones shown in Fig. 2, across which ions are transported, a fourth zone has arisen and is the major hindrance to ion flow. Two different causes have been noted:

If the liquid being treated contains macro-ions too large to pass through the pores in the membrane, these ions will accumulate on or just within the surface of the membrane. Usually such molecules are bound very tightly to the membrane, by a mechanism similar to the familiar fouling of ion exchange resins in conventional water treatment. The removal of the film which is formed in this way is therefore very difficult and may only be possible by scraping away the surface layer of the membrane itself. The macro-ions neutralise the membrane's active groups, and the fouling eventually results in an inert film which covers the surface of the membrane and through which ions diffuse with great difficulty. Within this film depletion takes place very readily, as shown in Fig. 4.

A similar effect has been known to occur with new membranes treating pure solutions which do not cause fouling. On this occasion, it was discovered that, due to a fault in manufacture, the membranes had become covered with an inert polymer film, with similar results. Such membranes are satisfactory in every respect in which they can easily be tested—e.g. ion exchange capacity, selectivity, conductivity (measured with AC in moderate or concentrated solution)—and only fail in respect of their ion transport kinetics.

Worst of all is the result of treating a fouling solution with kinetically faulty membranes.

Because of the difficulty of characterising the liquid film in practical experiments,

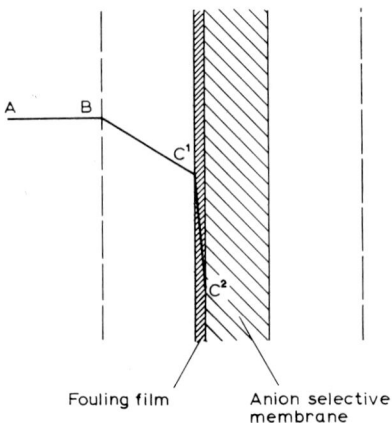

Fig. 4.
Kinetic effect of membrane fouling.

there is no simple way of detecting the presence of a kinetic barrier on a membrane. Normal conductivity measurement of membranes is carried out with AC, in which case little or no depletion of films occurs and a kinetic fault can go undetected. The interesting thing about membranes affected in this way is that at very low DC voltages they also behave normally, and could therefore be used satisfactorily, *e.g.* for salt production.

3.4 POLARITY REVERSAL

Figure 5 shows the same electrodialysis stack operating with the current flowing from right to left, and from left to right respectively. It may be seen that a reversal of polarity, without any other change being made in the stack itself, reverses the

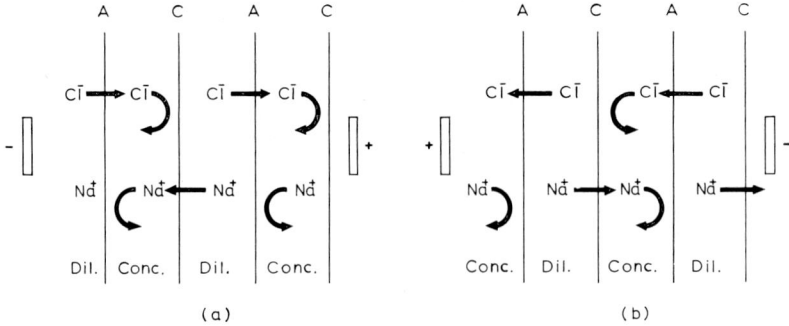

Fig. 5.
Polarity reversal.

function of each cell; the diluate cells in Fig. 5(a) have become concentrate cells in Fig. 5(b).

If the external piping which supplies liquid to the stack is equipped with the necessary crossover valves, then the apparatus can be operated in either direction to yield the identical performance. However, any accumulation of scale or precipitated salts which may arise in the concentrate cells while working on polarity (a) will tend to be dissolved away when the stack is reversed to polarity (b). Polarity reversal therefore acts as a safety device and is often incorporated into actual installations. A plant so equipped is normally operated with reversal being carried out, say, daily, as a regular routine.

Polarity reversal is not, however, the complete answer to scale formation. If the deposit is really serious, then it will tend to prevent any liquid flowing past the affected cell area, and so the hoped-for dissolution of precipitate will not occur. Polarity reversal can therefore only be counted on to remove light and incipient deposits. Another danger is that, if a bulky precipitate has formed, the polarity reversal will tend to dislodge sizable particles which may be carried to the outlet to the cell and lodge in it, so blocking off flow through the cell. (The scale which forms in electrodialysis plants is a complex co-precipitate of calcium and magnesium sulphate, carbonate, hydroxide and, sometimes, silica. It is remarkably insoluble, even in hot hydrochloric acid.) There is, in fact, no satisfactory solution to the problem of removing hard scale once it has formed in a stack, and the proper approach is to prevent its formation in the first place.

4. Construction and design

4.1 INTRODUCTION

The basic requirements of the process leave a good deal of scope in the detail of plant design. There are two main schools of thought in cell design:

(a) The "Tortuous Path" cell is made up with a specially stamped unit which incorporates the outside gasket and a labyrinth flow pattern (Fig. 6). Within the divisions of the labyrinth, the same unit has separator spacers which keep the membranes from touching one another and help in creating turbulence and thus raising the Reynolds number. This design creates a long path within a single cell, so that high desalination ratios can be achieved without multiple staging. The amount of liquid treated by each cell is small, so that the manifold ducts which carry flow to the cells need not be large. On the other hand, this design tends to give rise to large pressure losses, and is not too flexible when small desalination ranges are required.

(b) The "Sheet-Flow" cell consists of a peripheral gasket, within which lies a mesh, weave or grid which serves as separator. The liquid flows straight across the

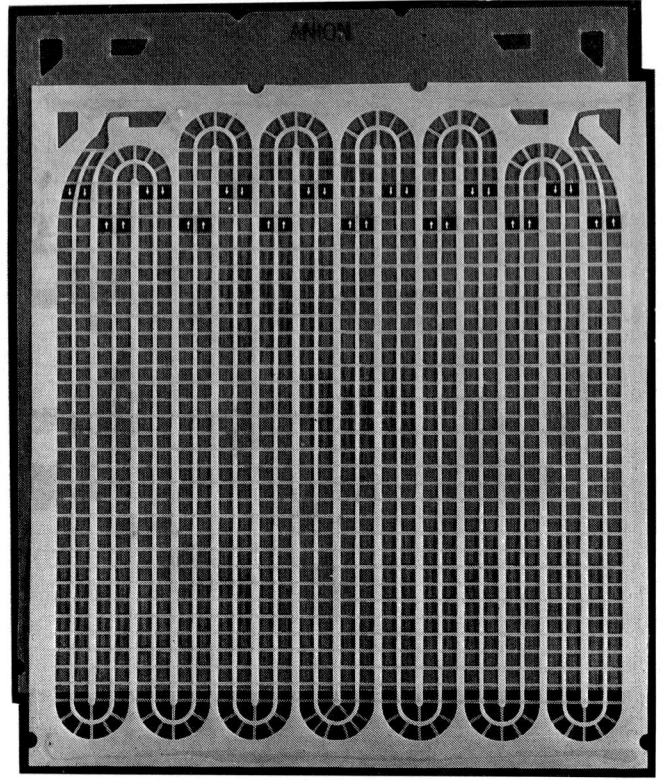

Fig. 6.
Tortuous path cell (by courtesy of Ionics Inc., Watertown, Mass., U.S.A.)

cell, either from corner to corner or from the whole of one side to the opposite side. It is often convenient for this type of cell to be rectangular, with the flow in the direction of the longer axis. This type of cell has just the opposite properties from those listed for the "tortuous path" cell, above. One typical design is shown in Fig. 7.

Several hundred cells are normally incorporated into a single stack and compressed between corrosion-protected steel strongbacks. The sealing pressure must ensure that none of the hundreds of joints made in this way leaks excessively. The precision of each component is therefore of the greatest importance. One design interposes semi-rigid division plates at 100-cell intervals. These plates not only create a fresh flat surface and so iron out any slight cumulative error in the cells beneath, but they also make it possible for the 100-cell "pack" to be handled in one sub-unit. However, in order to give the necessary rigidity, these plates are made of 7 mm thick polypropylene. Either the thick cell thus formed is fed by a separate concentrate flow, or a high local resistance may be expected if these cells are diluate

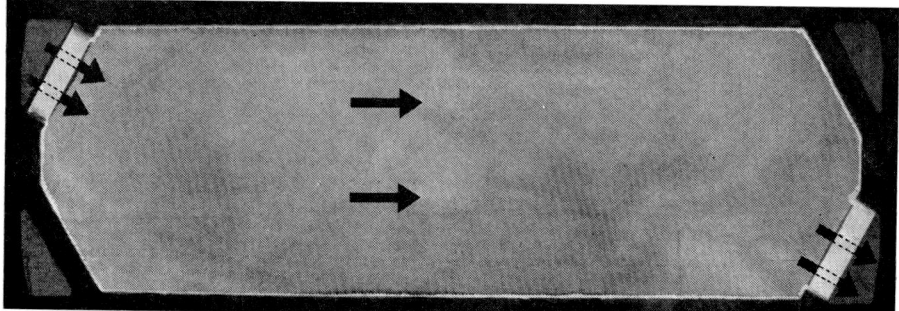

Fig. 7.
Sheet flow cell (photograph by courtesy of William Boby & Co. Ltd., Rickmansworth, Hertfordshire, England).

compartments. Where a plant is built for reversal of polarity, it is impossible to keep the frame cells as concentrate cells, except by supplying them by individual feed pipes—a costly and cumbersome business.

The normal stack arrangement has the cells in a horizontal plane, as it is impossible to lay membranes up vertically. If division plates are incorporated to create a "pack" sub-unit, it is possible (though not universal) to pre-assemble these sub-units and then install the resulting packs vertically in a press frame resembling a filter-press. One of the advantages claimed for this arrangement is that with upward liquid flow, air is less likely to be trapped in vertical cells.

4.2 Electrodes

The stack must be bracketed by a pair of electrodes, and intermediate electrodes are sometimes incorporated within the stack. The electrodes are particularly troublesome components to engineer. At the cathode, there is a release of OH⁻ ions and of hydrogen. In theory, it is possible to terminate the stack with an anion-selective membrane and supply the cathode rinse with a solution containing no hardness and therefore not liable to form precipitates. In practice this has proved quite impracticable. The cathode must therefore be kept acid, the cathode rinse containing at least the electrochemical equivalent of free acid to the current passed. Imperfect distribution of liquid in fact makes it advisable to use a substantial excess of acid.

The anode releases chlorine and oxygen gas. The membranes are susceptible to oxidative degradation, and the spent anode rinse must therefore be flushed to drain at a rate sufficient to keep the concentrations of chlorine low. Even so, the anode membrane has a limited life. Some designers use a porous sheet of inert material in an attempt to keep the chlorine from coming in contact with the membrane, with but moderate success. Probably the simplest protection is to use a double membrane at the anode and build the stack in such a way that it is easily replaced on a routine basis. Anion-selective membranes are less prone to oxidative attack than cation-selective membranes.

The electrode conditions could in theory be met by nickel and stainless steel electrodes respectively. However, if the plant is built for polarity reversal, then each electrode has to be capable of resisting both sets of conditions. Furthermore, the need to conserve water makes it normal for the anode rinse to be spent concentrate liquor, which is acidic. To make a universal electrode, the normal material is therefore a platinised "valve" metal, either titanium or tantalum. Carbon electrodes have been used in the past, but these degrade in oxidising conditions, and their engineering properties are so poor that the high apparent cost of platinised valve metals is preferable.

4.3 THE FLOW SHEET

All cells between a single pair of electrodes must carry the same current. Suppose for example that the greatest reduction possible in a single pass of a given cell design is to halve the incoming salinity, and suppose a water is to be desalinated from 4,000 ppm to 500 ppm: there are four ways of performing this task, as shown in Fig. 8.

(a) The water is passed through three stacks piped in series, in each of which the salinity is halved.

(b) A single stack is internally staged in three passes in each of which the salinity is halved. However, the mass of salt removed in each pass is half that of the previous pass, and therefore the current required is also one-half. It is therefore necessary to insert two intermediate electrodes so that each pass carries the correct current.

(c) A single stack is internally staged in a number of passes, but without resort to intermediate electrodes. The limiting criteria of the last pass imply that the current through it may not exceed that corresponding to a salinity reduction from 1000 ppm to 500 ppm. Because all other passes must carry the same current, they are all restricted to a reduction of 500 ppm, and 7 passes are necessary.

(d) The water recirculates from a batch tank and the current falls as desalination of the batch proceeds. When the contents of the tank reach the desired quality, a fresh batch is treated.

Of these four possibilities, (a) is of course the simplest, but is uneconomic when the flow is too small to justify three separate stacks; (b) is also simple enough, but electrodes are expensive. Method (c), on the other hand, implies that the membrane area in the first few passes is being grossly under-utilised; moreover, a large number of passes may be impracticable because of the high pressure loss involved. The batch process (d) has the advantage of great flexibility in operation, but involves

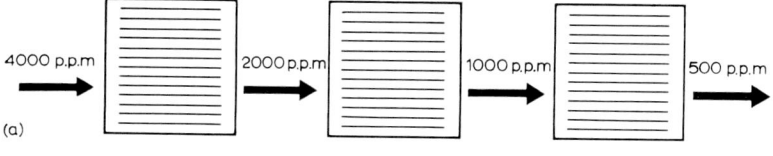

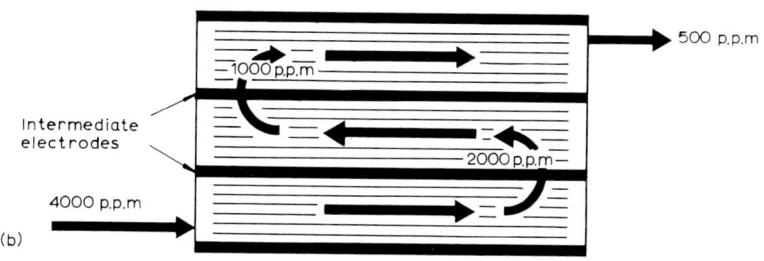

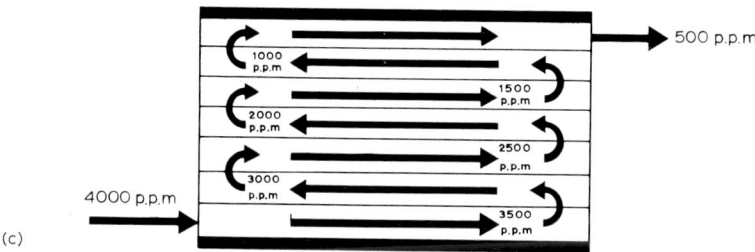

Fig. 8.
Four process flow sheets: (a) three stacks in series; (b) stack staged internally for three passes, with intermediate electrodes; (c) stack staged internally for seven passes, without intermediate electrodes; (d) batch operation. Valves operate automatically to discharge batch when 500 ppm is reached, and to run in fresh batch.

oversize pumps and pipes, and a peaking power consumption at the beginning of every batch which involves oversize DC and power supply. A choice of flow sheet must be made in each individual case, and sometimes it is difficult to make.

These problems of choice are most acute in designing water conversion plants, when the cost of product is very critical. In process applications and salt production, the batch process is usually the most convenient.

Because the membranes are mechanically weak, it is important that they should not be subjected to large hydraulic pressure differentials. Normal practice is therefore to pass the concentrate and diluate streams through identical flow paths, so that if they flow at the same rate and with a similar inlet pressure, the pressures will remain in balance throughout the apparatus. This is particularly important in multi-pass arrangements. Because the main electrical resistance is due to the thickness of the diluate cells, it was at one time suggested that the concentrate should be at a slightly higher pressure and so press the diluate cells to their thinnest. This is successful only if the apparatus is absolutely free from cross-leaks between the two streams—which in practice can rarely be guaranteed. For water conversion it is more usual to keep a slight over-pressure on the diluate, so that any cross-leakage results in a small loss of the product rather than its gross contamination.

In water conversion, the cathode rinse is usually raw water, since concentrate liquor contains so high a proportion of hardness salts that its use would be dangerous. One simple device is to dose enough acid into a stream of raw water to neutralise the cathode and have enough acidity left in it to supply the acid demand of the concentrate circuit. In the anode, on the other hand, there is no danger of scale, and the blowdown from the concentrate circuit may conveniently provide the anode rinse. This arrangement is shown in Fig. 9.

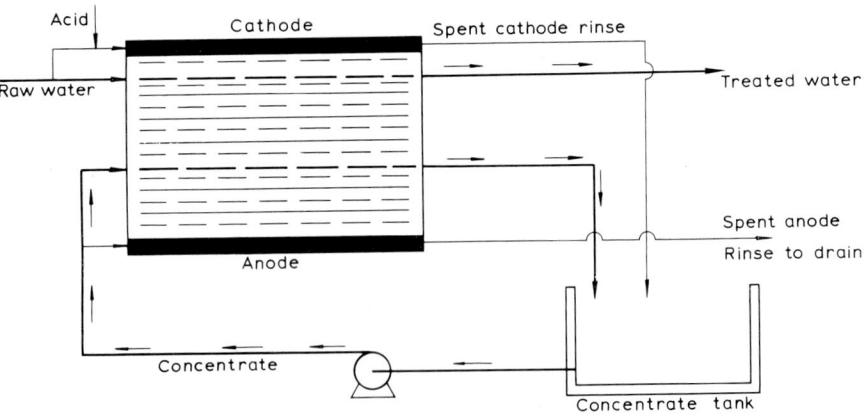

Fig. 9.
Schematic flow sheet (by courtesy of William Boby & Co. Ltd., Rickmansworth, Hertfordshire, England).

For process purposes, brine is usually made up to supply electrode rinses.

Care must naturally be taken to dispose of the gases evolved at the two electrodes separately from one another and so avoid an explosion hazard.

5. Economics and plant design

5.1 Introduction

The process applications are too diverse to allow any general treatment of their economics. On the other hand, water conversion is a field in which rigorous economic studies must be made in each case, so that a great deal has been published. However, the interacting factors are so complex, and some of them so widely misunderstood, that most of the published work is worthless. The costs of the process depend enormously on circumstances. Every project must be examined in detail in order to give even an approximate cost of treatment.

Firstly, there are the usual variables which must be considered for any economic study:

Cost of Power and Acid
Cost of Capital, Amortisation Rates
Cost of Labour
Desired Output
Pattern of Demand for Water

A peculiarity of the electrodialysis process, in addition, is that the costs vary both with the initial salinity of the raw water, and with the desired salt content of the treated water.

In designing the plant, the designer can either aim at a small plant containing little membrane area but in which the membranes are highly stressed, or a large plant operating under less critical conditions. The former design leads to low capital costs, although the power consumption would be high. It would, for instance, be suitable for a plant which is used for only a short season each year. On the other hand, a plant running all year round, especially where power costs are high, will be more economical if its first cost has been increased by using more membrane area to give a lower power consumption.

The thermodynamic theoretical energy required to effect water conversion is so small that it can in practice be ignored. The power consumed in electrodialysis is essentially that which is used in overcoming the resistance of the apparatus. There is also a relatively small loss due to leakage currents, but these are less than 10% of the total current if the apparatus is in good order.

At current efficiencies of the order of 90% about 30 A will transport 1 g equiv./h of salts across a cell pair. Since all the cell pairs in a stack between a pair of electrodes carry the same current in series, the current I needed to transport one g equiv./h in a stack of N cell pairs between two electrodes is $30/N$ A.

The resistance of one cm² of cell pair is roughly dependent only on the concentration of the solution flowing through the diluate cell. Given the inlet and treated water salinities, therefore, the average cell pair resistance per cm² is fixed independent of all other variables except temperature and cell thickness. Suppose this average resistance is r Ω cm² and each cell pair has an available area of A cm², then the resistance of one cell pair is r/A Ω, and the resistance R of a stack of N cell pairs is Nr/A Ω.

Now the power consumption for transporting one g equiv./h will be I^2R. Substituting the values above

$$I^2R = \frac{900}{N^2} \times \frac{Nr}{A} = \frac{900r}{NA}$$

Actually, NA represents the total area of membrane installed, *i.e.* the power consumption varies inversely with the total area of membrane installed, regardless of whether this is done by changing the number of cell pairs or the size of the cell, or both.

The voltage required to cause the current I to flow is RI, which is $30/N \times Nr/A = 30r/A$ V.

This is the voltage over the whole stack. The voltage gradient, which is a measure of the stress under which the membranes are operating, is measured in volts per cell pair, and the voltage gradient is therefore $30r/NA$ V per cell pair.

This factor, like the power consumption, varies inversely with the total membrane area.

Summing up, therefore:

$$\text{Power consumption} \propto \text{volts/cell pair} \propto \frac{1}{\text{total membrane area}}$$

5.2 Optimisation calculations

If we take a typical example of a cell design, 150×50 cm in size, with a cell thickness of 0.13 cm, then its electrical resistance is roughly equal to

$$\frac{0.65}{S} + 0.01 \text{ }\Omega\text{/cell pair} \tag{1}$$

Given an input salinity of S_1 and a treated water quality of S_2 equiv./l, the mean resistance per standard cell pair averaged over the entire plant can be obtained by integration over the range S_1 to S_2 as:

$$\frac{0.65 \ln S_1/S_2}{S_1 - S_2} + 0.01 \text{ }\Omega\text{/cell pair} \tag{2}$$

(This expression is correct only if the applied voltage per cell pair is the same throughout the plant, a condition which gives the lowest power requirement for a given size of plant. There are instances when it is impracticable to design for constant voltage, but even then the error in expression (2) is small.)

The theoretical current is electrochemically equivalent to the number of cell pairs multiplied by the mass of salt to be removed. The equation below assumes a current efficiency of 80%, though somewhat higher values are normally obtained:

$$I = 33.6 \frac{G}{N}(S_1 - S_2)A \tag{3}$$

Suppose the plant contains N standard cell pairs. Then its total resistance will be:

$$R = N\left(\frac{0.65 \ln S_1/S_2}{S_1 - S_2}\right) + 0.01 \; \Omega \tag{4}$$

The power consumption is $I^2R/1000$ kW, and amalgamation of eqns. (3) and (4) gives:

$$\text{Power} = 0.0112 \frac{G^2}{N}(S_1 - S_2)[65 \ln S_1/S_2 + (S_1 - S_2)] \tag{5}$$

(The above expression includes an allowance for losses due to current rectification.)
If the cost of electricity is inserted, the power cost per cubic metre of product is:

$$\text{Cost} = k_5 \frac{G}{N}(S_1 - S_2)[65 \ln S_1/S_2 + (S_1 - S_2)] \; £/m^3 \tag{6}$$

5.3 Capital costs

5.3.1 Plant cost

The capital cost of the plant roughly follows the equation:

$$\text{Capital cost} = £k_1N + k_2 \tag{7}$$

The true relationship is of course not linear, but the above equation involves a relatively small error. It is taken for convenience, as on differentiation it yields an economic yard-stick independent of the scale of operations.

The sum of capital, depreciation and maintenance charges per cubic metre of treated water then becomes:

$$\frac{k_4}{G}(k_1N + k_2) \; £/m^3 \tag{8}$$

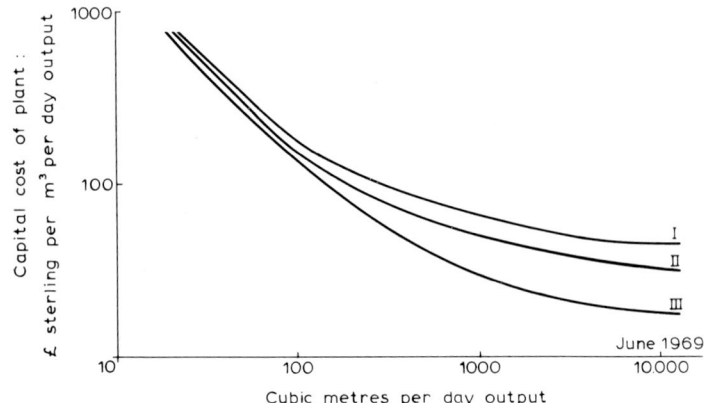

Fig. 10.
Capital costs of electrodialysis (by courtesy of William Boby & Co. Ltd., Rickmansworth, Hertfordshire, England).

Typical conditions:

			Inlet		Outlet	Power consumed
Curve I	overall reduction	6:1	3000 ppm	to	500 ppm	3½ kWh/m³
Curve II	overall reduction	3·3:1	3000 ,,	to	900 ,,	3 ,,
			1700 ,,	to	500 ,,	1¾ ,,
Curve III	overall reduction	1·8:1	3000 ,,	to	1700 ,,	1¾ ,,
			1700 ,,	to	900 ,,	1¾ ,,
			900 ,,	to	500 ,,	¼ ,,

Note: Costs depend on full specification and on site conditions.

5.3.2 *Transformer rectifier*

Direct current is normally obtained by a transformer rectifier, whose cost is dependent on the power consumption. In the author's experience, the capital cost in the size range which is normally encountered roughly follows a linear relationship:

$$\text{Capital cost} = £k'_1 \text{ kW} + k'_2 \tag{9}$$

where kW represents the maximum rating of the transformer rectifier. This is normally about 25% more than the actual DC output in order to give the necessary operational margin. The cost of current rectification per cubic metre of treated water then takes the form:

Specific capital cost:

$$= k_4 \left(k''_1 \frac{G}{N} (S_1 - S_2) \left[65 \ln \frac{S_1}{S_2} + (S_1 - S_2) \right] + \frac{k''_2}{G} \right) £/m^3 \tag{10}$$

5.4 OTHER COSTS

5.4.1 Membrane replacement cost

In normal conditions this is so low that with a minimum error it can be included in the factors for eqn. (8).

5.4.2 Labour costs

These are hardly likely to vary enough between alternative designs to affect the optimisation of plant design.

5.4.3 Total costs

Summing the three cost items set out above, we can differentiate the total with respect to N, and equate the differential to zero:

$$N_{optimum} = K.G \sqrt{(S_1 - S_2)\left[65 \ln \frac{S_1}{S_2} + (S_1 - S_2)\right]} \text{ cell pairs} \quad (11)$$

The treatment above is somewhat theoretical unless it is possible to substitute for the constants k_1 to k_5. In practice, these constants are subject to such enormous variations between individual cases that it would be quite impossible to generalise without misleading. Indeed, most of the published cost information must be viewed with the utmost suspicion, since it is at best meaningless and at worst positively misleading. Much of it is commercially oriented to prove that desalination in general, and electrodialysis in particular, are economically attractive. In order to prove this, writers have frequently used amortisation and interest rates which are unrealistically low, have neglected civil and ancillary costs, and have taken low electricity costs. The vital factor of percentage utilisation over the year has been almost universally neglected. In actual practice, desalination plant often has a utilisation factor below 75% and often far lower. This is because

(a) The plant specification, at the time of purchase, must cater for the peak load at some future date.

(b) Water consumption tends to vary seasonally with a peak load in the summer.

(c) Where desalination plant is used in conjunction with a natural water source, the operating cost of the desalination plant is higher than that of the natural water source and its use will be restricted to meeting any demand for which the natural water source is inadequate.

Clearly, if utilisation is only, say, 50%, then the factor k_4 is doubled.

However, the graph shown in Fig. 10 has been published by one commercial supplier of equipment to show the pattern of capital costs in different applications. The capital cost quoted is for the plant installed and commissioned on site, but excludes civil work and supply and distribution of water.

Taking the limitation of 2 V per cell pair as being a standard, the power consumption in electrodialysis is equivalent to $\Delta S/15$ kWh/m³, where ΔS is the salinity reduction in mequiv./l. At high and low voltages the power cost varies *pro rata*.

5.5 Fields of economic application

The desalination of sea-water, together with the purification of "brackish" water, has tended to dominate the field of electrodialysis. (Brackish water has dissolved salts other than NaCl). A variety of views of these processes are given in refs. 1–6. Much of this material is however not scientific, in the sense that it seeks to promote the activities of a company or group of companies. The existence of a "desalination lobby" has been suggested, especially in the U.S.A. In addition to the above two applications of "primary" water treatment, it is widely assumed that in the more populous parts of the world, water will have to be re-used many times over. For the removal of dissolved salts from this "secondary" water, electrodialysis would seem to be a natural contender.

There are a number of applications for the process in the food industries[7], and the plants for desalination of cheese whey and de-acidification of orange juice are both operating at large pilot scale. The former plant is shown in Fig. 16, though few details are known. Other applications in the food industry relate to the treatment (mainly de-ashing) of sugars and syrups and related compounds. Other food products in which the use of the method has been studied include wines, yeasts and the up-grading of cattle forage crops, though this does not in any way imply that these are all destined for scale-up. A review of the more advanced projects is found in ref. 8 and the reader may consult ref. 9 for details of the remainder.

Other applications outside the food industry concern the treatment of antibiotics, well-drilling fluids, paper pulp, leach solutions in metallurgy, fission products of nuclear reactors, clays, alkaloids and blood of patients with certain conditions. The automatic regulation of pH, and the manufacture of HCl from salt solutions are also reported.

Two applications which are well advanced in the field of effluent control relate to the scrubbing of flue-gases to remove SO_2 and the regeneration of chromic acid liquors in the plating industry[10, 11].

Generally speaking, electrodialysis is the most economical process for the conversion of brackish waters, up to 5,000 ppm, to potable water of 500 ppm. Above this salinity, evaporative methods tend to become competitive, but only on the large scale. Below 500 ppm., ion exchange demineralisation is less costly but on the large scale the case can be made for electrodialysis, down to 200 or 300 ppm.

Reverse osmosis is, at the time of writing, about to become a commercial process, but the costings published so far do not inspire much confidence in their reliability. As far as one can judge, reverse osmosis appears to be attractive on the small scale because of its intrinsic simplicity, and in those cases where brackish water is to be converted to boiler feed water. Where, for instance, a 3,000 ppm water is to be converted, electrodialysis would go to, say, 300 ppm, leaving the rest to be dealt with by ion exchange. A reverse osmosis system, with at least 95% rejection, would go down to 150 ppm or less, thereby halving the load on the ion exchange plant.

Specific conditions can create cases where electrodialysis is quite unexpectedly economical. For example, there seems to be a large potential demand for sea water conversion on the 1–2 m³/h scale for tourist resorts. In these cases, the amortisation and interest charges tend to be very high and the utilisation factor is often as low as 33% per annum. Capital cost is therefore overwhelmingly the most important factor, and it appears, at least on paper, that electrodialysis provides the cheapest capital plant for meeting these requirements, although the power cost is significantly higher than the nearest competitive processes which, in this instance, are vapour re-compression and vacuum freezing. On the other hand, an oil-rich Middle Eastern location at which steam is almost free of charge yields costs according to which sea water conversion is cheaper than electrodialysis of brackish water.

The Benghazi illustration (mentioned below), which is under construction at the time of writing, was ordered after being tendered in competition with a continuous counter-current ion exchange system. The ion exchange plant was in fact slightly cheaper in first cost, but its running cost was higher by an order of magnitude and it was clearly uneconomical by comparison.

TABLE 1

RELATIVE EFFECT OF VARIOUS FACTORS ON PROCESS ECONOMICS

	Factors favourable to:		
	Electrodialysis	Reverse Osmosis	Evaporators
(i) Raw water salinity	Low	Low	High
(ii) Scale of operations	Medium/large	Small	Large
(iii) Power/heat cost	Cheap power	Cheap power	Cheap heat
(iv) Desired treated water quality	High t.d.s.	Low t.d.s.	Very low t.d.s.
(v) Amortisation, interest, utilisation	Varies	Varies	Varies
(vi) Site conditions	Needs good supervision	Needs moderate supervision	Needs good supervision

Table 1 shows the general inference of the six cost factors on the competiveness of electrodialysis, as compared with the alternative processes.

6. Variations on the process

Two variations have been explored to avoid some of the short-comings of the process.

6.1 THE FILLED CELL

The main electrical resistance in the electrodialysis stack arises in the diluate cell. In the brackish water range, the conductivity of the diluate is less than the conductivity of an ion exchange resin. A number of workers have therefore attempted to devise plants in which the diluate cell is packed with ion exchange resin, either cationic or anionic, or mixed. None of these trials has yielded a commercially attractive process. The primary problem is the mechanical one of introducing the resin into the cells and keeping it evenly distributed within the cell. This has been the main stumbling block. Furthermore, the reduction of cell resistance has never been as high as had been expected in theory, due no doubt to kinetic problems at the relatively small areas of contact between resin particles. The prospect of a successful filled cell design in the future still exists, however, and no doubt development in this direction will be resumed some day.

6.2 TRANSPORT DEPLETION

This process is essentially electrodialysis, but using selective membranes of one sign only, the other membrane being non-selective. In practice, this has meant an apparatus of cation membranes using a porous non-selective membrane instead of the anion membrane. This device completely overcomes the disagreeable consequence of polarisation on the anion membrane. If the current density of such a stack is increased beyond the limiting point, an increased electrical resistivity will become apparent, but there will be no pH change and therefore no danger of scale formation. Scale can only form if the concentrate is allowed to become supersaturated, and would then form as a soft precipitate rather than a hard scale growing from the membrane.

The penalties for this agreeable state of affairs are that the current efficiency falls to about 50% and, all other things being equal, the power cost of the process is therefore about three times that of true electrodialysis. Development continues in the belief that the simplicity of the plant and its freedom from trouble in the field will out-balance the increased power cost. On the face of it, this assumption is perfectly sound. Unfortunately, transport depletion also has to use electrodes for providing the EMF and, since these are inevitably a source of complication and potential trouble, it is doubtful whether the process will win through in competition with reverse osmosis, which also holds out prospects for relatively trouble-free operation in the field.

7. Examples in the field

7.1 Desalination for potable water

7.1.1 Buckeye, Arizona, USA (Fig. 11)

Built	1962
Rated output	100 m³/h potable water
Waste water	15% of output
Feed water	2,076 ppm total dissolved solids
Power consumption	3.2 kWh/m³ product
Cell type	Tortuous path
Flow sheet	Continuous flow in three parallel trains each of two stacks in series

Fig. 11.
A desalination plant at Buckeye, Arizona (by courtesy of Ionics Inc., Watertown, Mass., U.S.A.).

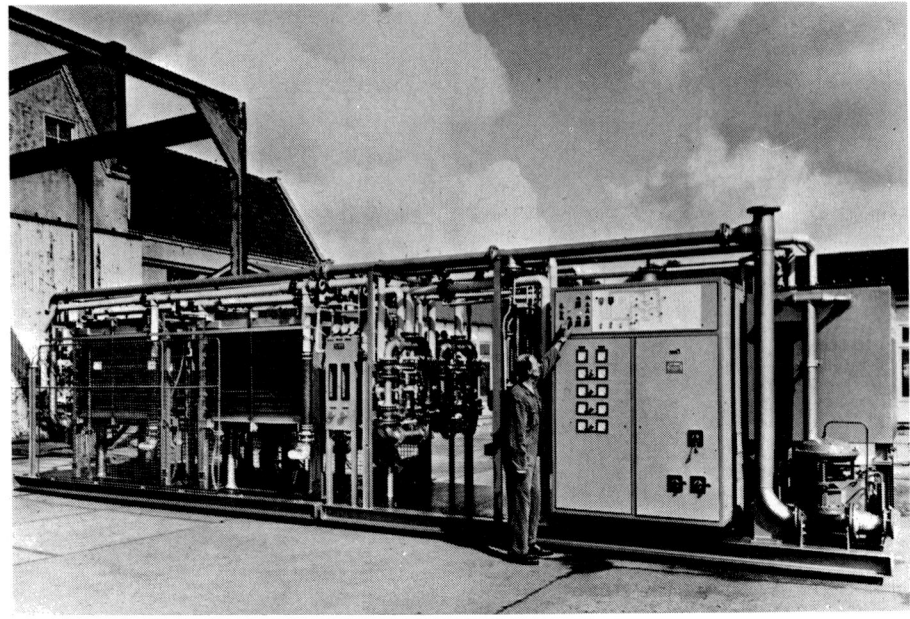

Fig. 12.
A desalination plant for Vieste, Italy. The photograph shows the plant being shipped to site (by courtesy of William Boby & Co. Ltd., Rickmansworth, Hertfordshire, England).

7.1.2 Vieste, Italy (Fig. 12)

Built	1968
Rated output	45 m^3/h potable water
Waste water	10% of output
Feed water	2,200 ppm total dissolved solids
Power consumption	3.2 kWh/m^3 product
Cell type	Sheet flow
Flow sheet	Continuous flow in two parallel trains each of two stacks in series. Each stack is internally staged in two passes with an intermediate electrode, making four passes in series. This is necessary as specification calls for 300 ppm solids in treated water, as compared with the normal requirement of 500 ppm.

7.1.3 Plants under construction

At the time of writing (1969) two plants are under construction which will be by far the largest ever installed. Figure 13 shows the proposed plant at Benghazi and

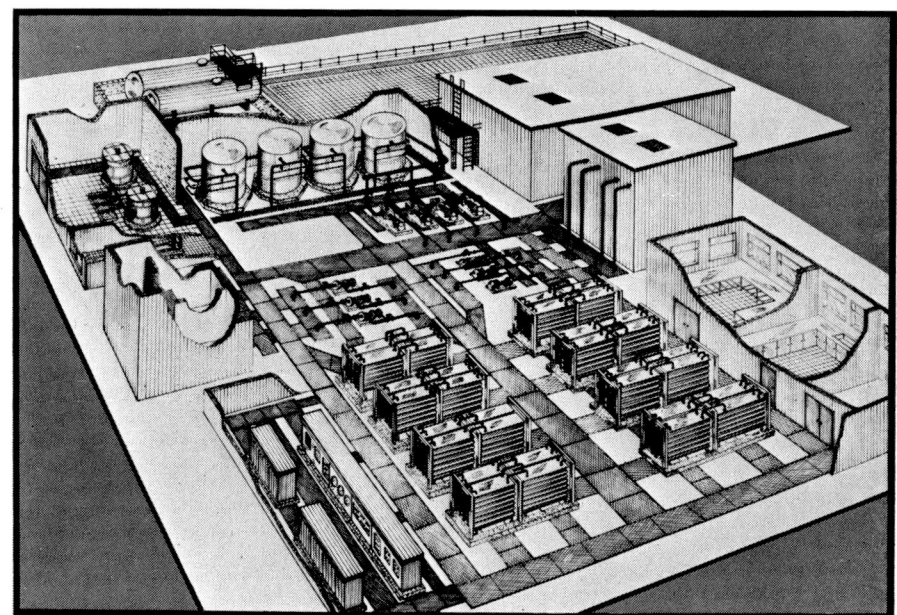

Fig. 13.
Diagram of a desalination plant under construction at Benghazi, Libya (by courtesy of William Boby & Co., Ltd. Rickmansworth, Hertfordshire, England).

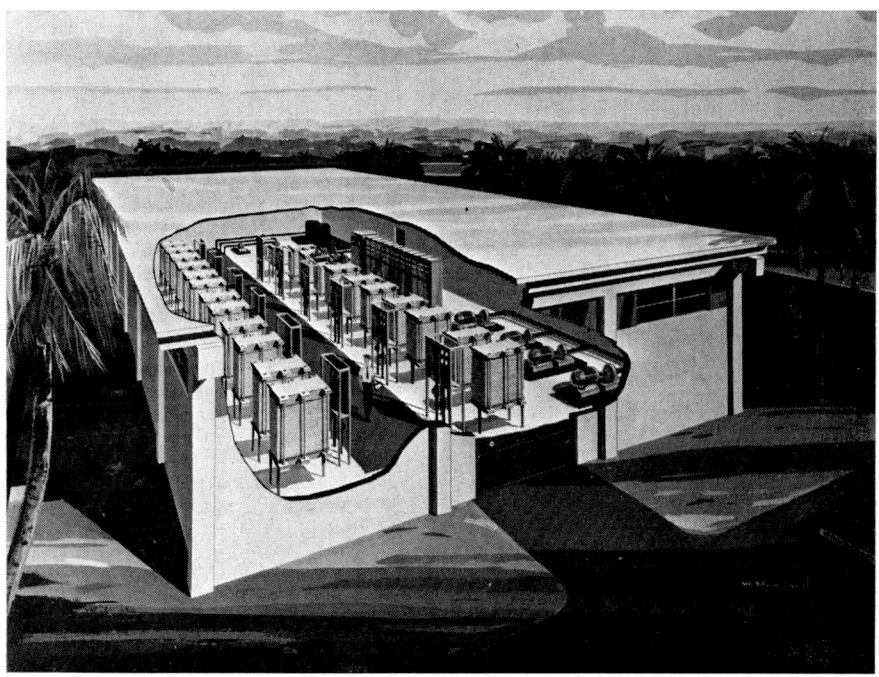

Fig. 14.
Diagram of a desalination plant under construction at Siesta Key, Florida (by courtesy of Ionics Inc., Watertown, Mass., U.S.A.).

Fig. 14 that at Siesta Key. The latter is being built in two stages. Technical details are set out below:

	Benghazi	Siesta Key Stage 1	Stage 2
Rated output, m³/h net	800	188	310
Waste water, % of output	8	17	
Raw water, ppm total dissolved solids	2,000	1,300	
Number of parallel trains	16	7	12
Stacks/train	2		2
Total number of cell pairs	9,600	4,200	7,200
Total membrane area, m²	15,000	3,900	6,700

7.2 SALT PRODUCTION FROM SEA WATER

Figure 15 shows a plant at Sakaide, Japan. It consists of 8 stacks each containing 817 cell pairs and produces 10,000 tons of NaCl per annum in the form of 17° Bé brine, which is subsequently processed to table salt. The plant was built in 1960.

Fig. 15.
Salt production plant at Sakaide, Japan (by courtesy of Asahi Glass Co. Ltd.).

Fig. 16.
Desalination of cheese whey (by courtesy of Ionics Inc., Watertown, Mass., U.S.A.).

7.3 DESALINATION OF CHEESE WHEY

Figure 16 shows a plant which reduces the inorganic salt content of cheese whey. No details are available regarding this plant.

8. References

1 J. R. WILSON (Ed.), *Demineralisation by Electrodialysis*, Butterworth, New York, 1960.
2 L. H. SHAFFER and M. S. MINTZ, *Electrodialysis*, in *Principles of Desalination*, K. S. SPIEGLER (Ed.), Academic Press, New York, 1966, Chap. VI.
3 *Süsswasser aus dem Meer, (Fresh Water from the Sea)*, Dechema Monographien Vol. 47 (in 2 parts), Papers 781–834, Verlag Chemie, Berlin, 1962.
4 *Proc. 1st Intern. Symp. on Water Desalination*, U.S. Office of Saline Water, Washington D.C., 1965.
5 *Süsswasser aus dem Meer* (2nd Symposium, 1967), to be published.
6 S. N. LEVINE (Ed.), *Selected Papers in Desalination and Ocean Technology*, Dover Publications, New York, 1968.
7 J. A. ZANG, R. J. MOSLEY and R. N. SMITH, *Electrodialysis in Food Processing*, Chem. Eng. Progr. Symp. Ser., 62 (69) (1966) 105.

8 H. J. COHAN and R. W. KENNEDY, *A Survey of the Electrodialysis Industry in the USA*, 2, 389 in *Proc. 1st Intern. Symp. Water, Desalination* (see ref. 4), 1965.
9 *Chemical Abstracts*. Readers with special interests see under "Dialysis" (electro) or "Dialysers" (electro) for the process or compound in question.
10 *Chemical Week*, August 10th, 1968, pp. 51–2.
11 U.S. Pat., 1,102,899 (1965).

Further reading

W. MCRAY and E. G. PARSI (Eds.), *Electrodialysis*, The Electrochemical Society, New York, 1969 (collected papers given at the meeting of the Society, May, 1968).

Chapter 13

Miscellaneous Industrial Processes

C. JACKSON

Imperial Chemical Industries Ltd., Mond Division, The Heath, Runcorn, Cheshire

A. T. KUHN

Department of Chemistry and Applied Chemistry, University of Salford, Lancashire

The aim of this chapter is to present a picture of the electrolytic processes which have been, or still are, commercially successful and which are not dealt with elsewhere in this book. Some attempts are made to analyse the reasons for the failure of those processes no longer operated.

It is not proposed to describe at length the various cell designs and process details except where not described in the literature. Rather it is the aim to provide up-to-date information on the various processes, together with full literature sources should the reader require further information.

General reading on the electrochemical industries can be found in refs. 1–11.

1. Inorganic oxidation and reduction

1.1 HYDROGEN PEROXIDE AND PERSALTS

1.1.1 Hydrogen peroxide

Hydrogen peroxide can be made by the hydrolysis of peroxycompounds formed by the electrolysis of sulphate at a platinum anode.

$$2\ HSO_4^- \rightarrow H_2S_2O_8 + 2e$$
$$H_2S_2O_8 + 2\ H_2O \rightarrow 2\ H_2SO_4 + H_2O_2$$

(a) Process

Because of military demand, the electrolytic hydrogen peroxide processes, first developed by Degussa[12] in Germany and Austria, were expanded in these countries during World War II to capacities much greater than any other. Three processes were used.

(i) The persulphuric acid (Weissenstein) process[13, 14] operated by Degussa, consisted of the electrolysis of sulphuric acid between platinum anodes and lead cathodes separated by a porous ceramic diaphragm. The resulting persulphuric acid was then hydrolysed by heating.

(ii) The potassium persulphate (Pietsch and Adolph) process[15–21] operated by

E.W.M., Otto Schickert AG and Henkel, when a solution of ammonium sulphate in sulphuric acid was electrolysed. Potassium bisulphate was added to the electrolyte to precipitate potassium peroxydisulphate which was separated and hydrolysed. Platinum anodes and graphite cathodes were used, the latter being wrapped with asbestos cord which acted as a diaphragm.

(iii) The ammonium persulphate (Lowenstein/Laporte) process[22, 23] operated by E.W.M. and Becco in the U.S.A. The electrolysis of ammonium sulphate in sulphuric acid with platinum anodes and graphite or lead cathodes yielded ammonium persulphate which was subsequently hydrolysed. Being an "all liquid" process with a higher yield than the Weissenstein or Pietsch and Adolph route, this method has found the most favour.

Full cell and process details, together with comparisons of the above process can be found in refs. 24 and 25 and the physical properties in ref. 26. Later development of the process has resulted in high amperage cells, *e.g.* 8 kA cell by Becco, with reductions in the energy requirement to 11–12 KWh/kg H_2O_2 (100%) and large plants can be operated with fewer personnel[27]. More modern cell designs are discussed in ref. 28 and a rarely seen photo of a large Pt anode is shown in ref. 112.

(b) Remarks

Although the use of peroxide as a bleach, oxidising agent and source of persalts is expected to increase[29], virtually all new plant utilises the chemical autoxidation of anthraquinone[30]. The electrolytic process has the disadvantage of high plant capital, high cost of electricity and necessary high purity of the reactants. The autoxidation process is always cheaper to operate and capital costs are lower until a crossover point is reached at low tonnage plants. This crossover point has been quoted at 3–4 tonnes 100% H_2O_2/day but even 30–50 tonnes 100% H_2O_2/day is economic if the price of electricity is 1–2 pfg/kWh[27]. Indeed, DeNora have recently supplied a 6 tonne 35% H_2O_2/day plant to Egypt[31]. A world survey of hydrogen peroxide production as it stood in July 1968 appears in Table 1 and Laportes' total capacity and European capacity of peroxide and perborate in ref. 32.

TABLE 1

WORLD SURVEY OF HYDROGEN PERIXODE MANUFACTURE, JULY 1968
by courtesy, Laporte Industries Ltd.

Country–Company	Location	Process
Argentina		
Duperial	San Lorenzo	Elect.
Atanor	Gral	Elect.
Australia		
LCL	Botany	Elect.
LCL	Botany	AO

TABLE 1 *(continued)*

Country–Company	Location	Process
Austria		
Alpine	Schaftenau	Elect.
Oesterreichische Chem.	Weissenstein	AO (Degussa)
Werke GmbH	Carinthia	
Belgium		
Solvay	Jemeppe	AO
Degussa	Antwerp	AO
Brazil		
Montiqueira	Torena, San Paulo	Elect. (FMC)
Rio Cotia	San Paulo 50%	Elect.
Britain		
LIL	Warrington	AO
Canada		
CIL	Hamilton	Elect.
Du PONT	Maitland	AO
Chile		
Farmo Quim. del Pacif.	San Joachim	Schmidt
Colombia		
Electroquimica	—	Schmidt.
Colombiana	Medellin	Elect.
Egypt		
Misr Chemicals	—	Elect. 50%
France		
Oxysynthese	Jarrie	AO
Ugine	Pierre-Benite	Elect.
Solvay	Tavaux	AO
Germany (E)		
Eilen Celluloid	Eilenburg	35%
Germany (W)		
Degussa	Rheinfelden	Elect.
Degussa	Rheinfelden	AO
EWM	Munich	AO
Kali Chemie	Honningen	AO (Solvay)
Greece		
Chromat Pireos	—	—
Hungary		
Ungaria	Budapest	—
Holland		
Solvay	Holland	AO
India		
Arat Elect	—	—
Nat. Perox. Ltd.	Bombay	Laporte
Nat. Perox. Ltd.	Bombay	AO
Israel		
Israel Oxydon	Holon	Electrolytic persulfate (5000 A cells)

TABLE 1 *(continued)*

Country–Company	Location	Process
Italy		
Montecatini-Edison	Bussi	AO
Solvay	Rossignano	AO
Japan		
Mitsubishi-Edogawa	Yokkaichi	AO Expanding to final target
Tokai Electrochemical	Yoshiwara	Elect. (FMC)
Nippon Peroxide	Koriyama	AO
Mexico		
Cia Electroquimica	—	Elect. (FMC)
Mexican SA	Mexico City	AO
Pakistan		
Dawood Cotton Mills	Karachi	35%
Poland		
Zaklady Elekt	—	—
Portugal		
Quijabel S.A.R.L. Aq. by Solvay	Lisbon	Abello-E-35%
Rumania		
Indust. Import	—	AO LIL
South Africa		
Dundee Chem. Co. AE and LCL	—	—
Spain		
Foret SA	Barcelona	LCL
Abello	Madrid	—
Peroxidos	La Zaida	AO
Solvay	Torrelavega	—
Sweden		
Elektrokemiska AB	Bohus	USSR AO
Elektrokemiska AB	Bohus	Elect. (to be phased out?)
Switzerland		
Neher	—	Elect.
Oxysynthese	Aarau	FMC
U.S.A.		
Allied	Syracuse N.Y.	100%
Du Pont	Memphis	AO 100%
FMC	Buffalo	Elect. 100%
FMC	S. Charleston	AO 100%
FMC	Vancouver	Elect.
Pennsalt	Wyandotte	—
P.P.G.	Barberton	—
Shell	Norco	—
Yugoslavia		
Belinka	Ljubljana	35%

The foregoing discussion describes the manufacture of hydrogen peroxide by anodic oxidation, as it were, of water. There is, of course, a second electrochemical route to H_2O_2, by reduction of oxygen. The idea of basing an industrial process on this was first conceived by Traube in 1882, and the idea has been taken up at various times since. The work of Berl in the 1940's and others is described in Billiter's book[8] but it is worth quoting that 2–5% aqueous solutions were obtained at 90% current efficiency, with up to 25% solutions being formed at lower efficiencies. The idea has recently been revived as shown by two patents assigned to Kimberley Clarke of the U.S.A. (U.S. Pat., 3,454,477 and 3,462,351) where air is reduced at a porous electrode, containing a hydroquinonoid-type catalyst.

1.1.2 Peroxydisulphates

The electrolytic route is the only serious route to ammonium and potassium persulphates. As mentioned above, platinum anodes, together with a base metal e.g. Ta, current carrier, are used but the cost of these is insignificant in the cost of manufacture of the product[33]. Laporte Industries and Noury and Van Der Lande (Roermond) appear to be the only European producers of these compounds. At Roermond[34], platinum anodes and carbon cathodes with asbestos diaphragms are used at an anodic current density of about 1 A/cm². Use is mainly as bleaching and oxidising agents and as a copper etching agent in the electronics industry[35].

1.1.3 Perborate

Sodium perborate can be manufactured by electrolysing a sodium carbonate solution containing borax, between platinum anodes and mild steel cathodes.

(a) Process

Full details of the semi-continuous process can be found in the descriptions of the 14,000 tonne/annum Degussa plant[36] and the 3,650 tonne/annum Henkel plant[37, 38] which also give details of the platinum loss rates of the anodes. It is interesting to note that the Henkel electrolytic plant operated at 60% of the cost of the chemical process, but had to be closed down during the war because of a shortage of borax. The perborate electrolysis carried out by Noury Van Der Lande is operated on a suspension of 100 g perborate crystals in a solution containing 150 g of soda and 20 g borax/l. The current density at the Pt anode is 0.5 A/cm² and at the iron cathode is 0.25 A/cm². The current yield is 55–60%[34].

(b) Remarks

With the increased use of perborate as a soap powder additive, production should increase. However, the electrolytic process is not generally regarded as best except for small tonnage plants[33], though DeNora have recently supplied a 12 t/day electrolytic plant to Montecatini at Terni[31]. Degussa for instance have discontinued electrolytic production since enough H_2O_2 is available from the A.O. process to enable perborate to be made chemically[12]. A list of producers appears in Table 2.

TABLE 2

WORLD SURVEY OF SODIUM PERBORATE MANUFACTURE
by courtesy, Laporte Industries Ltd.

Country		Company	Process	Location
Australia	1967	LCL	Chemical	Botany
Austria	1966	Treibacher	Electrolytic	Treibach
Belgium	1966	Solvay	?	Jemeppe
Brazil	1966	Elect Rio Rotia	?	
U.K.	1966	ICI	Chemical	Runcorn
U.K.	1967	LCL	Chemical	Luton
U.K.	1967	LCL	Chemical	Warrington
France	1966	Air Liquide	Chemical	Chalon Sâone
France	1966	Solvay	?	Tavaux
France	1966	Ugine	Electrolytic	Pierre Benite
Germany	1964	Degussa	Electrolytic	Rheinfelden
Germany	1966	Henkel	Chemical	Dusseldorf
Germany	1965	Kali Chemie	Chemical	Honningden
India	1966	Nat Peroxide	Chemical	Bombay
Israel	1959	Oxidon	?	Holon, Telaviv
Italy	1964	Montecatini	Electrolytic	Terni
Italy	1964	Montecatini	Chemical	
Italy	1964	Solvay	Chemical	Rossignano
Netherlands	1966	Noury Lande	Electrolytic	Deventer
Spain	1966	Abello	?	Leon
Spain	1966	Foret	Chemical	Barcelona
Spain	1966	Solvay	?	Santander
Sweden	1966	Elektrokemiska	Electrolytic	Bohus
Switzerland	1966	Henkel	Electrolytic	Pratteln
Switzerland	1966	Neher	Electrolytic	Mels
USA	1966	J. T. Baker Chemical	?	Phillipsburgh
USA	1966	Du Pont	Chemical	Niagara Falls
USA	1966	Du Pont	Chemical	Memphis
USA	1966	FMC	Chemical	Buffalo
USA	1966	S.B. Penick	?	Newark
Yugoslavia	1964	Belinka	?	Ljubljana

1.2 MANGANESE DIOXIDE AND PERMANGANATE

1.2.1 Manganese dioxide

Manganese dioxide can be made by the anodic oxidation of manganous sulphate solution with inert electrodes. The anode reaction is:

$$2\ Mn^{2+} \rightarrow 2\ Mn^{3+} + 2\ e$$
$$2\ Mn^{3+} \rightleftharpoons Mn^{4+} + Mn^{2+}$$
$$Mn^{4+} + 2\ H_2O \rightarrow MnO_2 + 4\ H^+$$

and the manganese dioxide formed adheres to the anode to form a permeable diaphragm from where it is periodically removed.

(a) Process

A detailed description of the process operated by the Western Electrochemical Corp. is given in ref. 39 together with details of the cell design and performance. The Mitsui process is detailed in ref. 40 and this method is believed to be similar to that operated by other U.S. manufacturers[41]. Factors influencing current efficiency, cell voltage, energy consumption and the composition of the deposit are given in ref. 42. Western Electrochemical use a non-diaphragm type cell with graphite anodes but diaphragm-type cells have been used both commercially and in laboratory installations[43-46] to minimise cathodic reduction of the product. The anode can also be lead, lead alloy or titanium[47-52]. The use of lead anodes can result in a contaminated product[53] but this can be reduced to acceptable levels by careful operation. Russian practice appears to favour the production of the manganese dioxide as a slurry in the cell rather than as a deposit on the anode[54-57]. This is achieved by electrolysing at a low rather than a high temperature when the Mn^{4+} ions have time to diffuse out from the anode into the bulk of the solution before they are hydrolysed to MnO_2.

(b) Remarks

Use is almost entirely as a depolarizer in Leclanché and alkaline MnO_2 cells, where high purity and available oxygen content make it superior to the impure natural ores. Electrolytic MnO_2 is thus employed in the high performance varieties of MnO_2 cells. It is however, four to five times more expensive than natural ores selected for their battery activity[58] and for many applications the latter material is quite satisfactory. Thus although the proportionate consumption by the battery industry of electrolytic MnO_2 is probably increasing, the major material used is still selected natural ore. It is estimated that world consumption of selected ore for batteries is 60–70 thousand tons[59].

The major electrolytic producers in decreasing order of output are detailed in Table 3.

TABLE 3

MAIN PRODUCERS OF MnO_2 IN DECREASING ORDER OF MAGNITUDE (ref. 59)

Tekkosha	Japan
Mitsui	Japan
American Potash	U.S.A.
Hanshin	Japan
Knapsack	W. Germany
Lavino	U.S.A.

WORLD PRODUCTION FIGURES (ELECTROLYTIC)

Japan*	34,400 t (3990 in 1960)
U.S.A.	12,600 t
W. Germany	6,000 t
Belgium**	4,000 t
France	600 t
Italy	300 t

* See also ref. 182.
** Doubtful[59], process believed to be chemical[181].

1.2.2 Permanganate

Potassium permanganate can be produced by the electrolytic oxidation of potassium manganate solution (formed by the roasting of pyrolusite ore with potassium hydroxide solution) at a nickel anode, the anodic reaction being:

$$2\ MnO_4^{2-} \rightarrow 2\ MnO_4^- + 2\ e$$

(a) Process

The wartime plant of I.G. Farbenindustrie, Bitterfeld contributed much to the knowledge of the process. This 200 tonne/month plant is described in ref. 60 and a 400 tonne/month extension in ref. 61. The cells were of the non-diaphragm type, stirred and fitted with iron cathodes. Further developments are given in ref. 62.

Details of the Boots–Ketjen plant operated by Boots Pure Drug Co., Nottingham[63] are given in ref. 64 whilst the process operated by the Carus Chem. Co., Illinois is fully detailed in refs. 65–66. A review of early developments of the cells can be found in ref. 67.

Methods for producing permanganate from ferromanganese[68], silicon–manganese or electrolytic manganese are summarised in ref. 69 and although some progressed through pilot plant stages, none is now used commercially[70].

(b) Remarks

In addition to the major plants mentioned above the Societé des Usines Chimiques Rhône-Poulenc also utilise electrolysis[71]. It is known that $KMnO_4$ is also produced in the U.S.S.R. and Czechoslovakia, probably *via* electrolysis. China, Spain and South Africa may well have plants but no information has been obtained[41].

In view of the fact that there are relatively few manufacturers in the world, competition is fierce and no details of process economics or production levels have been published[72].

1.3 Metal oxides

1.3.1 Cuprous oxide

Cuprous oxide can be made by the anodic oxidation of a copper anode in a diaphragm cell fitted with a copper cathode, the electrolyte being a hot solution of NaCl + NaOH (or carbonate).

The four European producers by electrolysis are Metal Powders Ltd., Chemical Supply Co. (on a very small scale), Nordax A/S Oslo, and Outokumpu Oy (Helsinki)[73]. The Nordax A/S production amounts to about 200 tonnes/annum[74]. The cell operating voltage is about 2.5 V per cell and product cost £7.5/tonne in electricity. Outokumpu production is 300 tonnes/annum but can easily be increased[75]. Röhm and Haas in the U.S.A. are also believed to operate an electrolytic plant, brief details of which appear in ref. 1.

1.3.2 Mercuric oxide

Mercuric oxide, a catalyst for the preparation of acetaldehyde was made electrolytically from mercury by Dr. Alexander Wacker G.m.b.H. at a rate of 4.6 tonne/month/cell. The plant, containing twelve cells, is fully described in ref. 76. Each cell contained a continuously scraped mercury anode and a cast iron cathode, and although the process was continuous, power consumption was high, the cell voltage ranging from 12 to 17.

There do not appear to be any commercial electrolytic processes for mercuric oxide now in operation. One of the difficulties with the method is that the product always contains free mercury metal.

While many different reaction conditions have been recommended[77-79], some early work done by the Wood Ridge Chemical Corp.[80] used a mercury anode and a nickel cathode in a 15% solution of caustic soda maintained at 60°C. The current density used was 24 A/dm^2 at 5 V. Yield was 600 g HgO/kWh but the product contained 1–15% free mercury.

Complete oxidation can be obtained by the addition of NaCl to the cell, the oxidant then being NaOCl. Unless concentration, temperature and pH are carefully controlled, basic chlorides of mercury are formed, and the alkalinity required ensures that only yellow mercuric oxide will be produced.

1.4 Miscellaneous processes

1.4.1 Potassium ferricyanide

Potassium ferrocyanide may be electrolytically oxidised to potassium ferricyanide with a high current efficiency in a diaphragm cell

$$Fe(CN)_6^{4-} \rightarrow Fe(CN)_6^{3-} + e$$

Eastman–Kodak operate a plant in which stainless steel anodes and nickel cathodes are used, the available details of which are given in ref. 1.

1.4.2 Basic white lead carbonate

White lead carbonate may be made by the anodic oxidation of a lead anode in a suitable electrolyte.

(a) Process

The continuous Sperry Process[81] employed a cell fitted with lead anodes, a linen diaphragm and an iron cathode. The electrolyte consisted of a sodium acetate and carbonate mixture, and anolyte and catholyte were individually recirculated through the cell, white lead being removed from the anolyte as a slurry.

$$Pb \rightarrow Pb^{2+} + 2e$$
$$Pb^{2+} + 2 CH_3COO^- \rightarrow (CH_3COO)_2Pb$$
$$3 Pb(CH_3COO)_2 + 4 Na_2CO_3 + 2 H_2O \rightarrow 2 PbCO_3.Pb(OH)_2 + 2 NaHCO_3 + 6 CH_3COONa$$

The process was operated by Anaconda White Lead Products, until abandoned because of loss of markets, to give 26.4 tonnes/day. A full description of plant and process appears in ref. 82.

All the white lead manufactured was used as a pigment. The outstanding characteristics of that produced electrolytically were its exceptional purity (only lead being dissolved in the cell), uniformity and brilliant whiteness.

In about 1925, experimental work on electrolytic white lead production was done at the Brimsdown research laboratories of the Associated Lead Manufacturers Ltd.[83]. A pilot plant was operated at the company's Bootle Works. The product was considered too fine for normal pigmentary use and the economics of the process were unattractive.

In 1927 the Lonabarc Co. Ltd., marketed an electrolytic white lead. In spite of ts high price it was favoured for a time by some printing ink manufacturers[83].

(b) Remarks

Electrolytic white lead was never economically competitive with ordinary process white lead but the criticism of its small, uniform particle size must be seen in the context of a time when stack white lead, with its wide range of particle sizes, was regarded as the norm. The electrolytic process may have held a special attraction for Anaconda in its ability to utilise primary lead having a high bismuth content, the bismuth being retained in the anode slimes.

1.4.3 Aluminium hydroxychloride

Aluminium chloride solution was electrolysed in a diaphragm cell by I.G. Farben, Ludwigshafen to give chlorine, hydrogen and any of the three compounds:

$Al(OH)_2Cl$ and $Al_2(OH)_3Cl_3$—tanning agents
$Al_2(OH)_5Cl$ —water repellent for textiles.

Descriptions of the cells, which comprised graphite electrodes separated by crocidolite (blue) asbestos cloth diaphragms, appear in refs. 84 and 85.

2. Organic oxidation and reduction processes

2.1 ELECTROHYDRODIMERISATION OF ACRYLONITRILE TO ADIPONITRILE

Nylon 66 is obtained from a polymer formed by reacting adipic acid with hexamethylenediamine. The latter is obtained by hydrogenating adiponitrile, $(NC(CH_2)_4CN)$ and the adiponitrile was usually formed from adipic acid.

In 1965 the Baizer process[86,87] for electrohydrodimerisation of acrylonitrile to adiponitrile came on stream at Monsanto's Decatur Alabama Plant, thus freeing considerable amounts of adipic acid for reaction with the diamine. The plant has since received widespread publicity[1,88,89] as one of the few examples of a modern, commercially successful electro-organic process.

(a) Process

The anolyte is a dilute aqueous solution of mineral acid, the catholyte an aqueous solution of acrylonitrile, quaternary ammonium salt and reaction products. The quaternary ammonium salt is added in order to increase the electrical conductivity of the solution and also the solubility of acrylonitrile and adiponitrile in water. Recycling of the electrolyte makes the process continuous, adiponitrile being removed outside the cell.

Lead cathodes and anodes separated by cationic exchange membranes are used in a bipolar filterpress cell, the materials of construction of which are restricted to non-conducting polymers. The published details of the process can be found in ref. 90 together with details of operating problems and process performance[91].

Other companies have been active in the field including Asahi (using emulsion electrolyte)[92], U.C.B.[93, 94], B.A.S.F.[95] and also non-diaphragm cells have been employed[96, 97]. An excellent summary of processes and process economics appears in ref. 98 and also in ref. 137 which compares this and the amalgam routes.

(b) Remarks

That the process is a commercial success is due in no small measure to the fact that continuous operation is achieved and also the cell design has benefited from modern technology and materials of construction. Continuous operation of the process enables costly labour requirements to be reduced to a minimum whilst modern technology enables maximum efficiencies to be attained. The process has also benefited from continuous design development, an absolute "must" if initial successes are to be maintained, (*c.f.* sorbitol/mannitol process) and the plant is now reported to be operating at better than design yields and has been operated at 150% of design capacity. An anode corrosion problem appears to have been successfully solved[99].

2.2 ELECTROLYTIC LEAD ALKYLS

The continuous electrolytic process for the production of tetramethyl and tetraethyl leads for use as petrol anti-knock additives which is operated by Nalco[71] has gained as much publicity as the Monsanto Process (above)[1, 88, 89].

On stream in Freeport, Texas since early 1964 (after an explosion and fire held up commercial operation) the plant can produce 15,400 tonne/annum TML or 18,100 tonne/annum TEL and is described in detail in refs. 100–102.

Organomagnesium halides (Grignard Reagents) ionise in organic solvents (usually ethers) and the resulting solutions can be electrolysed.

$$R.Cl + Mg \xrightarrow{ether} R.MgCl$$
Alkyl halide   Grignard reagent
$$R.MgCl \underset{\rightleftharpoons}{\overset{ether}{}} R^- + MgCl^+$$
At anode: $4 R^- + Pb \rightarrow R_4Pb + 4 e$
At cathode: $4 MgCl^+ + 4 e \rightarrow 2 Mg + 2 MgCl_2$

Magnesium deposition on the cathode is prevented by using excess alkyl halide. Operating procedures can be varied to provide plant flexibility for the preparation of related compounds or mixtures of related compounds. As the rate of reaction falls with time and the internal cell resistance increases, the applied cell voltage is increased to counteract this effect and maintain a constant reaction rate.

The ten 36.4 m³ cylindrical cells are of novel design utilising the steel cell wall as cathode separated from the anode by means of a non-conducting, low permeability diaphragm. The anode consists of lead pellets constrained within the cylindrical diaphragms and hence as the lead is used during the reaction, the anode can be continuously "made up" and the inter-electrode gap maintained, simply by feeding fresh lead pellets into the anode assembly.

Remarks

The process was long used for tetramethyllead production only, but NALCO are now understood to be making tetraethyllead by this route also.

2.3 METHYL ETHYL KETONE

Methyl ethyl ketone used as a solvent for vinyl and acrylic coatings and nitrocellulose lacquers, may be prepared by the electrolytic oxidation of a mixture of 1- and 2-n-butenes in dilute sulphuric acid by a platinum anode.

The continuous process patented by Esso Chemical Ltd.[103] utilises a stream of 1- and 2-n-butenes in preferably 6–11.5 moles H_2SO_4/litre at a temperature of 48°–85°C and an anodic oxidation potential of 0.5–1.65 V (with respect to standard hydrogen reference). A full list of the effects of process variables appears in the patent.

The cell utilises a platinum coated anode and either a power-generating oxygen fuel cell or a power-consuming hydrogen evolving cathode, anode and cathode being separated by an ion exchange membrane. Maximum selectivity of the reaction is achieved by rapid removal of product, product yield being 75–80%.

It has been stated that Esso are to construct a 45,000 tonne/annum plant to be on stream in 1971[104], but confirmation of this is lacking.

2.4 PROPYLENE OXIDE

This compound may be made by anodic oxidation of propylene. The Kellogg Company appear to have been most active. There are two distinct routes. The first is an electrolysis in the presence of an aqueous brine. The propylene chlorhydrin is formed as intermediate and this compound is then recycled to the cathode where "P.O." and HCl are formed. A pilot plant is believed to have been constructed by Kellogg and BASF in Germany, probably fitted with precious metal anodes. Scepticism as to the process viability has been expressed in *European Chemical News*[105] and by Ibl[172] and elsewhere. Elegant though the process may seem, Ibl's analysis shows how current densities must be kept down to preserve efficiency. In such an analysis, the economics of such a plant might be seen as a simple diaphragm cell, operating some five times less hard (on a capital invested basis) than normal chlorine cells, the main saving being one of chlorine transport, and some overvoltage saving.

Kellogg later published details of a "direct" process in which propylene was oxidized at a porous silver electrode to propylene oxide in the absence of Cl⁻ ions. Alkaline and weak electrolytes such as borates were specified. Little is known of this, but unpublished work by one of the authors indicates that the exchange current on a laminar electrode appears to be so small that it is on the threshold of detectability by GLC methods. This process is described in U.S. Pat., 3,427,235 while the chlorhydrin process was disclosed in Brit. Pat., 1,049,756, 1,090,006; Belg. Pat., 637,691 and 677,299 and Fr. Pat., 1,540,800, while three Bayer Patents are Brit. Pats. 1,176,649, 1,176,650 and 1,176,669.

2.5 SORBITOL AND MANNITOL

Glucose and mannose can be reduced to the corresponding alcohol using an agitated mercury or a solid lead or zinc amalgam cathode separated from a lead anode by a porous ceramic diaphragm[106].

$$CHO.(CHOH)_4CH_2OH + 2 H^+ \rightarrow CH_2OH.(CHOH)_4CH_2OH$$

(a) Process

The 113,400 kg/month plant built by the Atlas Powder Co. in 1937 and utilising lead amalgam cathodes is fully described in refs. 107 and 108. The batchwise process employed an anolyte of dilute sulphuric acid and a catholyte of glucose solution with added sodium sulphate for conductivity. The early work on the process has been published[109] as has the cell performance[110].

(b) Trends

The electrolytic process was supplanted in 1948 by one of catalytic hydrogenation because the latter enabled manufacture of the reduction products at a lower cost and was more versatile than the electrolytic process[111]. Diaphragms were frail and expensive and the process was slow and limited by the conductivity of the electrolyte and the cell size. No details were published of the lead content of the product.

2.6 REDUCTION OF NITROBENZENE

2.6.1 *Benzidine and analogues*

Benzidine, an intermediate in the production of azo dyes, can be made by the reduction of nitrobenzene in caustic soda solution (containing *o*-dichlorobenzene or dichloronaphthalene[173] as solvent) with subsequent rearrangement in acid solution of the resultant hydrazobenzene.

$$2\,C_6H_5NO_2 \longrightarrow C_6H_5\text{-NH-NH-}C_6H_5 \longrightarrow H_2N\text{-}C_6H_4\text{-}C_6H_4\text{-}NH_2$$

(a) Process

The electrolytic process (Rohner Process[174]) was operated at the Rohner factory at Pratteln, Switzerland and also by the Gesellschaft fur Chemische Industrie, Basle*. Diaphragm cells were used, utilising mild steel anodes and cathodes and concrete diaphragms. A non-diaphragm cell was also employed in Basle. All the cells were fitted with agitators and means for heating and cooling. The reduction time depended on temperature and current density but could take several days at 80–90°C with a current density of 300 A/m². Cell voltages were 3.5–4.0 V during the fast stage of the reduction (nitrobenzene to azobenzene) and then increased to 4.0–4.5 V in the azobenzene to hydrazobenzene stage. The formation of 0.5–1% aniline was unavoidable in the first phase but not in the second. Current efficiencies were approximately 100% and 50% respectively for the two phases.

The same plants were also used for the manufacture of o-tolidine and o-dianisidine and full plant and process details are to be found in ref. 175.

(b) Remarks

The Rohner Process was not really suitable for the electrolytic processing of low cost products, although it was shown to be economic[176]. Poor cell geometry and unfavourable diaphragm material caused an excessive cell voltage for a reaction potentially capable of proceeding at about 1 V. Other unfavourable factors were:
   (a) Low current efficiency in the later stages of reaction.
   (b) Small, batchwise operation necessitating excessive handling of materials.
   (c) High toxicity of products and the frequency of removing inspection samples.
   (d) Excessive plant maintenance.

2.6.2 p-Aminophenol

p-Aminophenol sulphate can be made by the cathodic reduction of nitrobenzene[177] in strong sulphuric acid using platinum electrodes.

$$C_6H_5NO_2 \rightarrow C_6H_5NH.OH \rightarrow C_6H_4\begin{subarray}{l}NH_2\\ \\OH\end{subarray}$$

phenylhydroxyl-   p-aminophenol
amine

(a) Process

A small batch process (60 kg/week) was operated by Eastman–Kodak in 1914

---

* *Editorial note:* the process is also known to have been operated in the U.K. in Huddersfield prior to 1955.

because of the wartime shortage of photographic developers. The cell was a diaphragm cell fitted with an unglazed porcelain diaphragm and is described, together with the process in ref. 113.

(b) Comments

The process was not very economic because of the high cost of manpower and maintenance of electrodes which had to be kept clean and free of poisons (Fe or Cu) otherwise the process was irreproducible. The cost of the sulphuric acid used was also high.

The control of the electrode potential at a specific value by a potentiostat has been reported[114] to give a marked improvement in current efficiency which could enhance the attractiveness of the process.

2.7 Reduction of Salicylic Acid

A solution of sodium salicylate may be readily reduced at a mercury cathode to salicylaldehyde which is extensively used in coumarin production.

Indian workers have successfully operated a pilot cell to produce 0.25 kg/hr of salicylaldehyde and are considering a 50 kg/day semi-continuous plant[178]. The catholyte consists of a mixture of salicyclic acid and aqueous sodium sulphate or chloride in the presence of boric acid. A microporous rubber diaphragm separates it from the 20% sulphuric acid anolyte. The catholyte pH is maintained at about 5.5 and the temperature $< 18°C$. The anodes are perforated cylinders of lead/lead–antimony or lead–silver alloys, the cathode amalgamated copper or brass. The cathode is rotated to enable high current densities to be used for the reduction whilst keeping the pH of the cathode diffusion layer very near to the pH of the bulk electrolyte. The current density range is 12–15 A/dm².

2.8 Piperidine

Piperidine sulphate is made by the electrochemical reduction of pyridine in sulphuric acid[115].

The process operated by Robinson Bros. uses a lead-lined cell and lead cathodes separated from the anode by a porous pot diaphragm. The anolyte is sulphuric acid and the catholyte contains about 5 equivalents of sulphuric acid and 10% by volume pyridine. The preferred working temperature is 30–40°C, when the total yield of piperidine is about 90%. The current density is in the range 13–15 A/dm². The process is still operated by Robinson Bros.[116] with an output of the order of 20 tonne/week[117], but most producers now manufacture piperidine by catalytic hydrogenation[118].

2.9 IODOFORM AND CHLOROFORM

The electrolysis of an iodide solution results in free iodine formation at the anode. Reaction with hydroxyl ions forms IO^-

$$2\ I^- \rightarrow 2\ I + 2e$$
$$2\ I + 2\ OH^- \rightarrow IO^- + I^- + H_2O$$

If alcohol or acetone is present, the IO^- reacts to form the iodine-substituted product, for example:

$$CH_3COCH_3 + 3\ IO^- \rightarrow CHI_3 + CH_3COO^- + 2\ OH^-$$

or

$$CH_3CH_2OH + 5\ IO^- \rightarrow CHI_3 + H_2O + I^- + 3\ OH^- + CO_2$$

A commercial installation was operated by Chemische Fabrik Schering[8] to produce 180 kg/day of very pure iodoform. The electrolyte was 40–50 kg KI (or NaI), 40 kg soda and 8 l of 96% ethanol in 400–500 l of solution. Electrolysis was carried out at 50–70°C and 4 V with a current efficiency of 93–95%.

Porcelain or iron vessels were used with platinum gauze anodes and nickel cathodes separated by a parchment diaphragm. The cell is fully described in ref. 9.

Chloroform could be similarly prepared but neither process has been used by Schering for the last 30 years[119].

2.10 DIMETHYL SULPHOXIDE

The formation of this compound, by anodic oxidation of dimethyl sulphide is understood to have been brought to pilot plant scale both by Glanzstoff Chemie in W. Germany and by Petroles d'Aquitaine in France. The work of Worbs *et al.* at Glanzstoff appears to be going well. The electrolyte used, a sulphuric acid / DMSO mixture, is described in Brit. Pat. 1,177,308 and Belg. Pat. 737,577 with other process details.

3. Processes involving electrolytic regeneration

3.1 DIALDEHYDE STARCH

Dialdehyde starch can readily be produced by the chemical oxidation of starch by periodic acid, the resulting iodic acid being reoxidised to periodic acid electrochemically, thus achieving good utilisation of expensive periodic acid.

(a) Process

The process originally developed to pilot plant stage by the U.S. Government involved the simultaneous oxidation of starch and rejuvenation of the periodic acid in a cell containing a lead dioxide coated lead anode, steel cathode and alundum diaphragm. The cell is described in ref. 120 and the process details together with results of experimental work and the effect of variables are given in refs. 121–123. Because of the irreproducibility of this process, due to the passage of iodide across the diaphragm to the anode and subsequent reaction with the starch, a two-stage process was developed where the oxidation of the starch occurred in a different vessel[88]. This process was experimentally investigated and costed[124] by the U.S. Government and a 113,400 kg/annum plant was built by Miles Laboratories Inc.[125].

(b) Trends

Dialdehyde starch is finding increased usage in leather tanning[126], as a wood adhesive[127] and for increasing the wet strength of paper[128].

Expected trends are to be found in refs. 129 and 130. Development work on the process is detailed in refs. 131–133. The plant built by Miles is unknown but believed to be similar to the filter press type designed by Mantell at Newark College[125]. The Mantell cell which was shipped to the U.S. Department of Agriculture, Peoria, has now been dismantled[134].

3.2 Chromic acid regeneration

Chromic acid (in sulphuric acid) is employed as an oxidising agent being itself reduced to chromic sulphate; regeneration of the expensive chromic acid is carried out electrochemically.

$$2\ Cr^{3+} + 3\ O + 3\ H_2O \rightarrow 2\ Cr^{6+} + 6\ OH^-$$
$$2\ Cr^{6+} + 16\ OH^- \rightarrow 2\ CrO_4^{2-} + 8\ H_2O$$

This is exemplified in four commercial processes[143].

3.2.1 Oxidation of fatty acids

Oleic acid was oxidised to azelaic and pelargonic acid by means of chromic acid, with regeneration of spent chromic acid, in a process employed by Emery Industries Inc., detailed in refs. 3 and 135. The highly corrosive nature of the process meant costly maintenance, and lack of development of the process after the initial success led to its abandonment in favour of ozone oxidation. Energy-wise, economies favour the electrolytic process, however the other economic factors favouring the ozone process are adaptability to continuous processing, lack of serious corrosion problems, better reaction control, better products and no need for the monitoring of cell operation[136].

3.2.2 Oxidation of anthracene

A suspension of anthracene in sulphuric acid can be oxidised to anthraquinone at a platinum anode[183]

$$C_6H_4(CH)_2C_6H_4 + 3O \rightarrow C_6H_4(CO)_2C_6H_4 + H_2O$$

but this has never formed the basis of an industrial process[11].

The use of chromic acid as the oxidising agent has been (and still is) widely used, the spent oxidant being electrochemically regenerated. A detailed description of the type of cell used is given in ref. 3.

L. B. Holliday are large producers known to use this method, and although cell details have never been released, they are in all probability similar to those described in ref. 7. Electrolytically regenerated ceric sulphate or manganese dioxide may also be used[3] but the development of processes for the production of anthraquinone by the vapour phase oxidation of anthracene and by the phthalic anhydride method offer viable alternative routes. The use pattern is given in ref. 139.

3.2.3 Oxidation of toluene-o-sulphonamide to saccharin

Chromic acid oxidation of toluene-*o*-sulphonamide to saccharin was tried on the pilot scale by the Boots Pure Drug Company, Nottingham, with subsequent electrolytic regeneration of the chromic acid. The following problems have been encountered: (a) the accumulation of ammonium salts causes crystallisation of ammonium chrome alum; (b) diaphragms for the electrolytic cells are difficult to make and are usually too porous and not sufficiently robust; (c) anodes require constant scraping to keep clean, they become buckled if overheated and sulphated if left unused; (d) cell temperature is difficult to control; (e) purification of saccharin produced is more difficult; (f) running costs are not attractive when cost of cell repairs is taken into account[63]. More recent literature references to this method include refs. 140–142.

3.2.4 Oxidation of raw montan wax

Chromic acid is used as an oxidising agent in the purification of montan wax by Farbwerke Hoechst A.G. The electrolytic cascade process used to regenerate the chromic acid is fully described in ref. 143.

3.3 BROMIDE REGENERATION

3.3.1 Oxidation of dextrose to calcium gluconate

Dextrose can be electrolytically oxidised to the calcium salt of gluconic acid in the presence of calcium carbonate and a bromide catalyst[144], using graphite electrodes.

$$2 C_6H_{12}O_6 + CaCO_3 + H_2O \rightarrow Ca(C_6H_{11}O_7)_2 + CO_2 + 2 H_2$$

The bromide ion gives free bromine at the anode which reacts with the sugar to give gluconic and hydrobromic acids which are neutralised by the carbonate. The bromide is constantly regenerated, hence a small quantity acts catalytically to facilitate the oxidation of large quantities of sugar without the inconvenience of handling elemental bromine.

The batch process details together with a historical review, costs of production and uses are given in refs. 145–149. Medically important calcium lactobionate was also produced[150, 151].

(a) Remarks

The process is still in use in the U.S.A. and India despite competition from biological and catalytic methods[89]. In Europe, calcium gluconate and lactobionate are manufactured electrochemically by Chefaro Rotterdam and by Sandoz, Switzerland[151]. A pilot plant for calcium lactobionate was set up by the National Dairies Res. Dept. on Long Island but the lack of a strong enough market did not warrant continued operation[152].

3.4 CAUSTIC SODA REGENERATION

3.4.1 Sodium sulphate electrolysis

During the manufacture of rayon, cellulose is precipitated from alkaline solution by the use of sulphuric acid, resulting in the formation of sodium sulphate as a by-product. The sodium hydroxide and sulphuric acid may be regenerated by the electrolysis of the sodium sulphate.

(a) Process

The process, as operated by I.G. Farbenwerke, Bitterfeld, is fully discussed in refs. 153, 154 and 4, which also describes other work. A graphite anode and a flowing mercury cathode were used, the sodium ions dissolving in the mercury to form sodium amalgam which was subsequently decomposed to caustic soda. As in chlor/alkali cells the graphite anodes corroded and were subsequently replaced by a 10% silver/lead alloy in smaller units. For large units, its durability was not good enough. In order to prevent mixing of any hydrogen produced at the cathode and oxygen at the anode, and also to prevent excessive dilution of the sulphuric acid formed at the anode with sodium sulphate, a rubber latex diaphragm was used. To reduce constructional difficulties (because of bulging of the diaphragm), floor-space and energy consumption, a vertical cell was designed and the idea was developed by DeNora after the war[1], however this work has now ceased[31].

The electrolysis can also be carried out in a double diaphragm cell fitted with lead/silver alloy anodes and steel cathodes. The effect of variables on the performance of such a cell are fully described in ref. 155 together with a proposed commercial scheme.

(b) Remarks

This process is not believed to be currently operated by any rayon producer in the world because it is not economical at the present time[156] considering the prevailing prices of caustic soda and the state of the market for caustic, as well as the buoyant market for Glauber's salt. The power consumption for this process would be large and it would be more economically feasible to use the calcination process for the sodium sulphate. In the U.S.A. there is a good market for anhydrous sodium sulphate in the detergent and paper industries, among others.

During the war years, and later, a considerable amount of work was done on the process in Holland but the work never progressed beyond pilot plant stage. The conclusion reached was that if the sodium sulphate were absolutely unsaleable, the electrolysis process might be economic if the power were sufficiently cheap. An interesting observation from the Dutch work was that sodium sulphate, as recovered from the rayon industry, was impure, containing a significant quantity of zinc. It was found that the salt must be purified extremely well to make it suitable for electrolysis, otherwise the metal impurities were discharged on the cathode forming amalgams. These formed solid isles floating on the mercury which clogged pipes and pumps[157].

3.4.2 Caustic regeneration in the petroleum industry

One stage of petroleum refining is the removal of mercaptans by caustic scrubbing. The mercaptans, dissolved in the caustic, are then oxidised to insoluble disulphides and separated from the solvent, the caustic being recycled. In one version of the process, the caustic is regenerated in hydrogen/oxygen cells and several small commercial installations exist in the U.S.A. and the Near East[88]. In order to minimise floorspace, fibre diaphragms of controlled porosity are used, which enable mercaptan oxidation cells 760 cm high by 30.5 cm diameter to be designed.

4. Use of electrolytic amalgam as reducing agent

The sodium amalgam formed in the mercury cell for the electrolytic production of chlorine from brine can be used as a reducing agent for a variety of compounds. Although amalgam does not appear to have been widely used commercially, the situation where the market for chlorine is far larger than that for the co-produced hydrogen, may provide the incentive to industry to use the amalgam as a reducing agent in the existing plants.

An excellent review of processes and a list of various reductions together with the equivalent weight of chlorine appears in ref. 158 whilst commercially applied processes are detailed below.

4.1 BENZIDINE AND ANALOGUES

Benzidine can be produced by the reduction in an aqueous alcoholic medium of nitrobenzene with sodium amalgam in the Bonelli process[179].

(a) Process

I.G. Farben at Leverkusen[159, 160] use sodium amalgam to reduce about 3000 kg/day nitrobenzene but stop the reduction at the halfway stage producing azoxybenzene in an 8 hour batch process. The azobenzene is further reduced to hydrazobenzene by metallic zinc. The amalgam reduction is stated to be cheaper than the zinc reduction since chlorine and 50% caustic soda are simultaneously produced and the reason for using zinc at all is due to the explosion hazard between the hydrazobenzene left in the reactor and a fresh charge of nitrobenzene.

Solvent-free hydrazobenzene, substituted nitroanthraquinone, nitrophenols and 1,1,-diamido-3,3,-disulphodiphenylethylene were also produced during the war, but there was no demand for the latter products.

The Bonelli process[179] was at one time operated by I.C.I. General Chemicals Division. Batches of 128 l of nitrobenzene were reduced in a suitable agitated reactor to give hydrazobenzene which was converted to benzidine hydrochloride by dilute hydrochloric acid. The plant capacity was 10.7 tonne/week and in addition the plant was used to prepare dianisidine dihydrochloride, *o*-toluidine dihydrochloride and dichlorobenzidine sulphate before being closed down in 1965.

4.2 SODIUM SULPHIDE

Sodium sulphide was produced in a continuous process by I.G. Farben at Höchst and Ludwigshafen by the reduction of sodium polysulphide solution with sodium amalgam[161, 162]. Either 30% or 60% sodium sulphide could be made at a rate of 2250 kg/day.

$$Na_2S_x + (2x - 2)Na/Hg \rightarrow xNa_2S + Hg$$

4.3 SODIUM HYDROSULPHITE

Sodium bisulphite may be reduced to sodium hydrosulphite by sodium amalgam according to the equation

$$2\ NaHSO_3 + 2Na/Hg \rightarrow Na_2S_2O_4 + 2\ NaOH + Hg$$

The process, initiated by the high price of zinc, was operated by I.G. Farben at Leverkusen[163] where the plant capacity was 30–50 t/month; however, the use of amalgam was discontinued when the price of zinc dropped.

The process is currently operated by B.A.S.F. at Ludwigshafen[159, 163], the N.V. Koninklijke Nederlandsche Zoutindustrie[164] and is to be operated by Olin Mathieson[165].

A process involving the reduction of sulphur dioxide dissolved in sodium sulphite solution was also employed at Leverkusen[166] and was re-examined[167] with a view to large scale manufacture in the Netherlands in the event of a zinc shortage.

4.4 SODIUM ALCOHOLATE

The corresponding alcoholate may be prepared by reduction of methanol or ethanol by sodium amalgam.

$$2\ Na(Hg) + 2\ ROH \rightarrow 2\ NaOR + H_2 + (Hg)$$

Sodium methylate, important in the synthesis of sulpha drugs, was manufactured thus by Olin Mathieson[158] during World War II.

In the Lülsdorff works of Dynamit Nobel, some 10 of the 65 ka mercury cells are fitted with tower denuders for the above process[168].

4.5 HYDRODIMERISATION OF ACRYLONITRILE TO ADIPONITRILE

The use of alkali metal amalgam by U.C.B. to produce adiponitrile from acrylonitrile appears to be going beyond pilot plant stage[180]. In one version of the process acrylonitrile is dimerised to adipic acid with alkali metal amalgam in tert.-butanol, tetrahydrofuran or dioxane containing 0.5–2% trimethylbenzylammonium-p-toluene sulphonate, perchlorate or phosphate at a temperature below 0°C.

A similar process has been patented by I.C.I. and full details of the various processes are discussed in a recent article in *European Chemical News*[137].

5. Electrochemistry and glass manufacture

Pilkington Brothers Limited have recently patented an electrochemical process for the surface modification of glass manufactured by their float process[169]. The essence of the float process for continuous glass manufacture is that a ribbon of molten glass is flowed across the surface of a molten metal (usually tin). The new process[170] is an electrochemical system which drives metallic ions into the glass to a controlled depth and intensity whilst the ribbon is advancing. These are quickly reduced by the atmosphere of hydrogen over the bath, to give metal particles (about 400 Å in diameter) concentrated immediately below the glass surface. The resulting glass is coloured (the colour depending on the metal used) and has outstanding heat and light reflecting properties.

The molten tin of the float glass is made the cathode and a pool of molten metal (again usually tin) introduced onto the surface of the glass acts as the anode. Metal is driven into the glass from the anode under the influence of an applied voltage. The temperature of the bath (600°–750°C) ensures that the glass is sufficiently conductive. A high operating temperature also means that a low voltage input—necessary to avoid surface damage to the glass—can be used. A current of 50 A at an applied voltage of 50 V, for example, would drive about 0.16 mg/in^2 of tin into the glass.

Permutations of metals and temperatures can be used to produce glasses with varying heat and light transmissions and it is even possible to create changes in these properties simply by altering the input current.

The surface modification process, costing only £15,000 can be installed in two hours. Changes from clear glass to modified glass and back again can then be effected in minutes. This flexibility makes short yet economic runs of different types of glass possible on even the largest continuous production line. It is of particular importance in manufacturing special glasses for small markets. It has so far been used to produce three coloured glasses in the range, grey–bronze to copper–bronze, whose main applications will be to limit solar heat gain and glare in motor cars and buildings. Further development work will undoubtedly lead to production of glasses with other properties.

Acknowledgements

The authors would like to thank the following for their help in the preparation of this work.

J. G. Marshall (Laporte Ind.), Drs. Werner and Lange (Degussa), H. J. Van der Lee (Noury and Van der Lande), J. O. Hay (Harshaw Chem. Co.), F. L. Tye (Ever-Ready Battery Co.), C. J. Brodrick (Boots Pure Drug Co.), A. H. Reidies (Carus Chem. Co.), A. P. Knowles (Metal Powders Ltd.), J. Brun (Univ. Norway, Trondheim), J. Larinkari (Federation of Finnish Chem. Ind.), E. H. Amstein (Assoc. Lead Manufacturers Ltd.), M. Baizer (Monsanto), M. S. Buckingham (Editor E.C.N.), M. E. D. Windridge (Robinson Bros. Ltd.), R. R. S. Smutney (Abbott Labs.), D. C. Miller (U.S. Dept. Ag. Peoria), V. J. Muckerheide (Emery Ind.), A. J. Cofrancesco (G.A.F. Corp.), F. H. Lentz (Enka Corp.), F. Beck (B.A.S.F.), M. A. Van Damme-van Weele (N.V.K.N.Z.), A. S. Robinson (Pilkington Bros. Ltd.), E. Cadmus (Wood Ridge Chem. Corp.), M. T. Sanders (Atlas Chem. Corp.), H. S. Isbell (The American Univ., Washington D.C.), J. Lango (Nalco), and P. Gallone (Oronzio de Nora).

Also G. E. Edwards (I.C.I. Mond Division) and U.C.W. Kopsch for their assistance, and the management of I.C.I. Mond Division for permission to prepare and publish the work.

6. References

1 C. L. MANTELL, *Electrochemical Engineering*, McGraw–Hill 4th ed., 1960.
2 A. REGNER, *Electrochemical Processes in Chemical Industries*, Artia, Prague, 1957.
3 C. L. MANTELL, *Electro Organic Chemical Processing*, Chem. Proc. Rev. 14, Noyes Development Corp., 1968.
4 C. HAMPEL, *Encyclopaedia of Electrochemistry*, Reinhold, New York, 1964.
5 R. E. KIRK and D. F. OTHMER, *Encyclopedia of Chemical Technology*, Interscience, New York, 2nd ed., 1964.
6 *Ullmann's Encyklopädie der technischen Chemie*, Urban und Schwarzenberg, Berlin, 1953.
7 V. ENGELHARDT, *Handbuch der technischen Elektrochemie*, Akademie Verlag, Leipzig, 1932.
8 J. BILLITER, *Die technische Elektrolyse der Nichtmetalle*, Springer-Verlag, Wien, 1954.
9 V. GAERTNER, *Praktische Elektrochemie*, Jugend und Volk, G.m.b.H., 1952.
10 W. L. FAITH, D. B. KEYES and R. L. CLARK, *Industrial Chemicals*, Wiley, New York, 3rd ed., 1965.
11 W. A. KOEHLER, *Principles and Applications of Electrochemistry*, Vol. II, *Applications*, Wiley, New York, 2nd ed., 1944.
12 L. WAGNER and W. LANGE, Degussa, private communication, 16 June 1969.
13 B. E. A. VIGERS and W. S. WOOD, B.I.O.S. Final Rept. 683, Item No. 22, 1945.
14 V. W. SLATER and W. S. WOOD, B.I.O.S. Final Rept. 921, Item No. 22, 1946.
15 M. WOLDENBERG and L. M. WHITE, C.I.O.S. Final Rept. XXIII-18, Item No. 21, 1945.
16 J. MCAULAY, D. B. CLAPP, V. W. SLATER and K. A. COOPER, C.I.O.S. Final Rept. XXXIII-42.
17 A. DAVIDSON, T. N. BUKLEY and B. VIGERS, B.I.O.S. Final Rept. 294, Item No. 22.
18 B. E. A. VIGERS and W. S. WOOD, B.I.O.S. Final Rept. 599, 1945.
19 B. E. A. VIGERS and W. S. WOOD, B.I.O.S. Final Rept. 886, 1945.
20 H. L. HULLAND, W. S. WOOD, J. MACGREGGOR and F. W. FAIRLIE, B.I.O.S. Final Rept. 1381, 1946.
21 ANON., *Chem. Eng.*, 55 (1948) 102 (August).
22 J. MCAULAY, D. B. CLAPP, V. W. SLATER and K. A. COOPER, C.I.O.S. Final Rept. XXXIII-43.
23 W. S. WOOD, *Chem. Ind. London*, (1953) 2.
24 W. C. SCHUMB, C. N. SATTERFIELD and R. L. WENTWORTH, *Hydrogen Peroxide*, Am. Chem. Soc. Monograph Series, Rheinhold Corp., 1955.
25 W. S. WOOD, *R.I.C. Monograph No. 2, Hydrogen Peroxide*, 1954.
26 V. W. SLATER and W. S. WOOD, *J. Brit. Interplanet. Soc.*, 7 (1948) 137.
27 A. H. HEINRICH SCHMIDT, *Chem. Ing.-Tech.*, 37 (1965) 832.
28 B. E. A. VIGERS, *Proc. Inst. Elec. Engrs. London*, 107 (1960) 463, 475.
29 See ref. 5, 11 (1966) 391, and ref. 10, p. 458.

30 R. Powell, *Hydrogen Peroxide Manufacture*, Chem. Proc. Rev. 20, Noyes Development Corp., 1968.
31 P. Gallone, de Nora, private communication.
32 *European Chemical News*, 15 (1969) 14 (7 March).
33 J. G. Marshall, Laporte Ind., private communication, 14 May, 1969.
34 H. J. van der Lee, Noury and Van der Lande, private communication, 1 Oct., 1969.
35 T. O. Schlabach and B. A. Diggory, *Electrochem. Tech.*, 2 (1964) 118.
36 H. Shaw and O. Whitston, B.I.O.S. Final Rept. 854, Item No. 22.
37 J. Young and R. G. Tongue, B.I.O.S. Final Rept. 887, Item No. 22, 1946.
38 J. Lawrence, R. B. Peacock and C. W. Lefeuvre, B.I.O.S. Final Rept. 1367, Item 22, 1946.
39 E. Schrier and R. W. Hoffmann, *Chem. Eng.*, 61 (1954) 152, 372.
40 See ref. 5, 13 (1967) 30.
41 J. O. Hay, Harshaw Chem. Co., private communication, 17 June, 1969.
42 G. W. Nicholls, *Trans. Electrochem. Soc.*, 57 (1932) 393.
43 R. I. Agladze, Russ. Pat., 59, 273 (28 Feb., 1941).
44 T. Banerjee, *J. Proc. Inst. Chemists India*, 26 (1954) 127.
45 R. Dufour, M. Verron and A. B. Lemagoarou, Fr. Pat., 1, 112,567 (15 Mar, 1956).
46 Cartoucherie Francaise, Brit. Pat., 797,993 (9 July, 1958).
47 C. Drotschmann, *Batterien*, 20 (1967) 990.
48 V. Aravamuthan and T. R. Venkatasubramanian, Ind. Pat., 60,333 (1959).
49 O. W. Storey, E. Steinhoff and E. R. Hoff, *Trans. Electrochem. Soc.*, 86 (1944) 337.
50 P. Marx, *U.S. Bur. Mines. Inform. Circ.*, 7464 (1948).
51 T. Tsuruoka, K. Shiroki and R. Asaoka, *Denki Kagaku*, 27 (1959) 229.
52 T. Ishino, H. Tamura and M. Yanokawa, *Technol. Rept. Osaka Univ.*, 6 (1956) 359.
53 E. H. Ungiadze, *Tr. Inst. Prik. Khim. i Elektrokhim., Akad. Nauk Gruz. SSR*, 2 (1961) 161.
54 S. A. Zaratskii, *Gosudarst. Inst. Prik. Khim., Sbornik Statei*, 1919–39 (1939) 174.
55 See ref. 2, page 441.
56 S. A. Zaratskii, *Elektrokhim. Margantsa Akad. Nauk Gruz. SSR*, 3 (1967) 229.
57 S. A. Zaratskii and E. I. Antonovskaya, *Elektrokhim. Margantsa Akad. Nauk Gruz. SSR*, 3 (1967) 232.
58 *Industrial Minerals*, July (1968) 10.
59 F. L. Tye, Ever-Ready Battery Co. Ltd., private communication, 26 March, 1969.
60 A. H. Loveless, B.I.O.S. Final Rept 964, Item No. 22.
61 T. Hagyard and W. Meachen, B.I.O.S. Final Rept. 1577, Item No. 22, 1947.
62 See ref. 6, 12 (1960) 230.
63 C. J. Brodrick, Boots Pure Drug Co. Ltd., private communication, 3 June, 1969.
64 *Brit. Chem. Eng.*, 9 (1964) 383.
65 Carus. Chem. Co., U.S. Pat., 2,843,537; 2, 908,620.
66 See ref. 10, page 642.
67 *Chemische Technologie*, Winnacker–Weingartner, Vol. 2 (1950), C. Hanser, München, p. 506.
68 R. E. Wilson and W. G. Horsch, *Trans. Am. Electrochem. Soc.*, 35 (1919) 371.
69 See ref. 6, 12 (1960) 225.
70 See ref. 5, 13 (1967) 35.
71 Rhône Poulenc, U.S. Pat., 1,826,594 (1931).
72 A. H. Reidies, Carus Chem. Co. Inc., private communication, 20 May, 1969.
73 A. P. Knowles, Metal Powders Ltd., private communication, 9 June, 1969.
74 J. Brun, Tech. Univ. Norway, Trondheim, private communication, 9 June, 1969.
75 J. Larinkari, Federation of Finnish Chem. Ind., Helsinki, private communication, 19 June, 1969.
76 W. G. Gardiner, FIAT Final Rept. No. 822, 1946.
77 A. Lowy and A. R. Ebberts, *Trans. Am. Electrochem. Soc.*, 45 (1924) 49.
78 H. Hibbert and R. R. Read, *J. Am. Chem. Soc.*, 46 (1924) 983.
79 Brit. Pat., 115,140 (1918); 335,638 (1929).
80 E. L. Cadmus, Wood Ridge Chem. Corp., private communication, 10 June, 1969.
81 E. W. Sperry, Brit. Pat., 144819 (1919); U.S. Pat., 1452620 (1922).
82 R. G. Bowman, *Trans. AIME*, 73 (1926) 146.

83 E. H. Amstein, Assoc. Lead Mfs. Ltd., private communication, 20 May, 1969.
84 J. Neubauer and Z. G. Deutsch, FIAT Final Rept. 431, 1945.
85 W. C. Gardiner, H. E. Hauser, C. E. Lyon and K. E. Rule, FIAT Final Rept. 788, 1946.
86 M. Baizer, *J. Electrochem. Soc.*, 111 (1964) 215.
87 M. Baizer and Monsanto, U.S. Pat., 3,193,476; 3,193,477; 3,193,481 all of 6 July, 1965.
88 C. L. Mantell, *Chem. Eng.*, 74 (1967) 128 (5 June).
89 C. J. Wilson, *Int. Sci. Tech.*, (1966) 82 (March).
90 J. H. Prescott, *Chem. Eng.*, 72 (1965) 238 (8 Nov.).
91 D. E. Danly and J. F. Giblin, Paper presented at meeting of Am. Chem. Soc., April 3, 1968, San Francisco, California.
92 Asahi Kasei Kogyo, Neth. Pat., 67,08254 (1956); Fr. Pat., 1503182; 1503244 (Oct. 1967); Jap. Pat., 4774/69 (1969); Brit. Pat., 1,169,525.
93 Y. Arad, M. Levy, I. R. Miller and D. Vofsi, *J. Electrochem. Soc.*, 114 (1967) 899.
94 u.c.b., Neth. Pat., 67,04123.
95 B.A.S.F., Neth. Pat., 67,04132; Belg. Pat., 679514.
96 A. P. Tomilov et al., Neth. Pat., 66,10378 (1967).
97 Rhône-Poulenc, Fr. Pat., 1,491,516 (1967).
98 P. L. Morse, *Rept. 31*, Process Economics Program., Stamford Res. Inst. California, 1967.
99 Monsanto, Brit. Pat., 1,150,303 (1969).
100 L. L. Bott, *Hydrocarbon Process. Petrol. Refiner* 44 (1965) 115.
101 E. Guccione, *Chem. Eng.*, 72 (1965) 102 (21 June).
102 Anon., *Chem. Eng.*, 72 (1965) 249 (8 Nov.).
103 Esso Res. and Eng. Co., Brit. Pat., 935236 (1963); U.S.Pat., 3,247,084.
104 *European Chemical News*, 15 (1969) 10 (9 May).
105 *European Chemical News*, 14 Jan., 1968; 17 May, 1968; 10 Oct., 1969. Ibid 9 (221) 30; 9 (223) 36.
106 H. J. Creighton, Atlas Powder Co., U.S. Pat., 1,612,361 (1926); 1,653,004 (1927); 1,712,952 (1929); 1,990,582 (1935).
107 R. L. Taylor, *Chem. Met. Eng.*, 44 (1937) 588.
108 D. H. Killeffer, *Ind. Eng. Chem. News Ed.*, 15 (1937) 489.
109 H. J. Creighton, *Trans. Am. Electrochem. Soc.*, 75 (1939) 389.
110 M. T. Sanders and R. A. Hales, *J. Electrochem. Soc.*, 96 (1949) 241.
111 M. T. Sanders, Atlas Chem. Ind. Inc., private communication, 23 May, 1969.
112 *International Nickel* magazine, 1970/1, p. 17.
113 A. S. McDaniel, L. Schneider and A. Ballard, *Trans. Am. Electrochem. Soc.*, 9 (1921) 441.
114 H. C. Rance and J. M. Coulson, *Electrochim. Acta*, 14 (1969) 283.
115 Robinson Bros. Ltd., Brit. Pat., 395,741 (1933).
116 M. E. D. Windridge, Robinson Bros. Ltd., private communication, 23 April, 1969.
117 M. S. Buckingham (Ed.), *European Chemical News*, private communication, 22 July, 1969.
118 R. R. Smutney, Abbott Labs. Inc., private communication, 20 June, 1969.
119 Schering, A. G., private communication, 14 July, 1969.
120 H. F. Conway and V. E. Sohns, *Ind. Eng. Chem.*, 51 (1959) 637.
121 W. Dvonch and C. L. Mehltretter, *J. Am. Chem. Soc.*, 74 (1952) 5522.
122 C. L. Mehltretter, J. C. Rankin and P. R. Watson, *Ind. Eng. Chem.*, 49 (1957) 350.
123 H. F. Conway, E. B. Lancaster and V. E. Sohns, *Elect. Tech.*, 2 (1964).
124 V. F. Pfeifer, V. E. Sohns, M. F. Conway, E. B. Lancaster, S. Dabic and E. L. Griffin, *Ind. Eng. Chem.*, 52 (1960) 201.
125 Anon., *Chem. Eng.*, 67 (1960) 64 (22 Feb.).
126 J. F. Wagoner, J. C. Stemoroski, W. Windus and W. C. Witham, *J. Am. Leather Chemists Assoc.*, 57 (1962) 302.
127 F. B. Weakley and C. L. Mehltretter, *Forest Prod. J.*, Jan. (1965) 8.
128 T. E. Yeates and C. L. Mehltretter, *J. Tech. Assoc. Pulp and Paper Ind.*, 48 (1965) 655.
129 Anon., *Chem. Eng.*, 64 (1957) 71 (14 Dec.).
130 Anon., *Chem. Eng.*, (1960) 44 (3 Oct.).
131 E. B. Lancaster and H. F. Conway, *Electrochem. Tech.*, 1 (1963) 253.
132 H. F. Conway and E. B. Lancaster, *Electrochem. Tech.*, 2 (1964) 46.

133 C. L. MEHLTRETTER, *Die Stärke*, 18 (1966) 208.
134 D. L. MILLER, U.S. Dept. of Agriculture, Peoria, private communication, 12 June, 1969.
135 J. D. FITZPATRICK and L. D. MYERS, U.S. Pat., 2,450,858 (1943).
136 V. J. MUCKERHEIDE, Emery Ind. Inc., private communications, 5 June, 1969, and 16 July, 1969.
137 *European Chemical News*, 17 April, 1970, 47; 24 April, 1970, 42; 1 May, 1970, 32.
138 A. J. CONFRANCESCO, G. A. F. Corp., private communication, 11 June, 1969.
139 See ref. 10, p. 108.
140 J. MIZUGUCHI, *J. Pharm. Soc. Japan*, 67 (1947) 90.
141 K. SUGINO and J. MIZUGUCHI, Jap. Pat., 176,554.
142 J. MUZUGUCHI, *J. Electrochem. Soc. Japan*, 19 (1951) 16.
143 M. KAPPEL, *Chem. Ing.-Tech.*, 35 (1963) 386.
144 H. S. ISBELL, U.S. Pat., 1,976,731 (1934).
145 H. S. ISBELL and H. L. FRUSH, *Natl. Bur. Std. U.S. J. Res. Paper*, 328, 6 (1931) 1145.
146 H. S. ISBELL, H. L. FRUSH and F. J. BATES, *Natl. Bur. Std. U.S. J. Res. Paper* 436, 8 (1932) 571.
147 C. G. PINK and D. B. SUMMERS, *Trans. Am. Electrochem. Soc.*, 74 (1938) 625.
148 H. S. ISBELL, H. L. FRUSH and F. J. BATES, *Ind. Eng. Chem.*, 24 (1932) 375.
149 H. S. ISBELL, *Natl. Bur. Std. U.S. J. Res. Paper*, 618, 11 (1933) 713.
150 H. S. ISBELL, *Natl. Bur. Std. U.S. J. Res. Paper* 914, 17 (1936) 331.
151 H. J. VAN DER LEE, Noury and Van Der Lande, private communication, 19 Sept., 1969.
152 H. S. ISBELL, The American University, Washington D.C., private communication 28 Aug., 1969.
153 Z. G. DEUTSCH and J. NEUBAUER, FIAT Final Rept. 429, 1945.
154 W. C. GARDINER, FIAT Final Rept. 831, 1946.
155 W. W. STENDER and I. J. SEERAK, *Trans. Electrochem. Soc.*, 68 (1935) 493.
156 H. COWLINE, F.M.C. Corp., private communication, 10 June, 1969.
157 F. H. LENTZ, American ENKA Corp., private communication, 18 June, 1969.
158 R. B. MACMULLIN, *Chem. Eng. Progr.*, 46 (1950) 440.
159 F. BECK, B.A.S.F. AG, private communication, 7 Feb., 1969.
160 W. C. GARDINER, FIAT Final Rept. 818, 1946.
161 W. C. GARDINER, H. E. HAUSER, E. H. KARR, C. E. LYON and K. E. RULE, FIAT Final Rept. 790, 1946.
162 J. S. SMATKO, FIAT Final Rept. 797, 1946.
163 J. A. A. KETELAAR, *Chem. Ing.-Tech.*, 35 (1963) 372.
164 M. A. VAN DAMME-VAN WEELE, N.V.K.N.Z., private communication, 23 Sept., 1969.
165 *European Chemical News*, 16 (1969) 31 (3 October).
166 B.I.O.S. Final Rept. 1373, 1946.
167 T. DIJS, J. G. HOOGLAND and H. I. WATERMAN, *Chem. Ind. London*, Nov. (1952) 1073.
168 A. T. KUHN, private communication.
169 PILKINGTON BROS. LTD., Fr. Pat., 1,486,271 (1967).
170 A. S. ROBINSON, Pilkington Bros. Ltd., private communications, 9 Oct., 1969, and 17 Oct., 1969.
171 J. LANGO, Nalco Corp., private communication, 12 Nov., 1968.
172 N. IBL, *Chem. Ing.-Techn.*, 4 (1970) 180.
173 D.R. Pat.,, 297,019 (1915).
174 D.R. Pat., 181,116 (1903).
175 See ref. 7 page 289.
176 F. FICHTER, Congr. Intern. Electrochim. Paris, Sect. 7, rapport 19, 1932.
177 B. B. DEY, T. R. GOVNDACHARI and S. C. RANGOPLAN, *J. Sci. Ind. Res. India*, 4 (1945) 559; *J. Electrochem. Soc.*, 99 (1952) 289.
178 K. S. UDUPA, G. S. SUBRAMANIAN and H. V. K. UDUPA, *Ind. Chem.*, May (1963) 238.
179 D. R. Pat., 94736 (1896).
180 *European Chemical News*, 51 (1969) 19 (26 Dec.).
181 Brit. Pat., 855,810.
182 K. TAKAHASHI, *J. Electrochem. Soc. Japan*, 37 (1969).
183 N. WEINBERG and H. R. WEINBERG, *Chem. Rev.*, 68 (1968) 461.

Chapter 14

Electrodes for Industrial Processes*

A. T. KUHN and P. M. WRIGHT

Department of Chemistry and Applied Chemistry, University of Salford, Lancashire

1. Introduction

The functions required of an electrode are fairly easy to define. The electrode should be:
(i) A good conductor.
(ii) An efficient electrocatalyst for the reaction in question.
(iii) Corrosion resistant and otherwise free from chemical attack.
(iv) Mechanically rigid, (in most cases).

In practice, one tends to separate possible anode materials from possible cathodes for the simple reason that anodic dissolution rules out all but a very few materials for the former. Of these few, some such as magnetite are only poor electrocatalysts. While the scope of this chapter is restricted to those materials which have proved their worth in actual processes, there is no doubt that the advent of new anode materials would prove of much value to the technologist. In this context, the promise of boride, nitride and carbide type materials as well as spinel structured semiconductors, and pseudo-metals, such as the tungsten bronzes, should be constantly appraised, as indeed they are. Furthermore, the novel electrodes based largely on ruthenium oxide, described here, have themselves a high content of non-noble valve metal oxides. There is doubtless a considerable effort aimed at increasing this and simultaneously reducing the precious metal content. We feel that such work is bound to be partially successful, but for the present the main anode materials are those described in this chapter.

2. The lead dioxide anode

The suitability of lead dioxide as an anode material had been known for many years[1-3], but because of practical difficulties its introduction into commercial applications was delayed until recently[4].

Pacific Engineering and Production Co. of Nevada[4], and Sanwa Chemical Co.

---

* Section 6 was contributed by P. C. S. Hayfield and M. A. Warne, Imperial Metal Industries Ltd., Witton, Birmingham.

Ltd. of Tokyo, Japan[5] are, to our knowledge, the only producers of lead dioxide anodes in the world.

The high cost of the precious metals and their alloys as anode materials had for many years been a stimulant for research into new materials[4, 6-7]. The potential of lead dioxide for such a purpose was realised shortly after the beginning of the century[3]. With a resistivity of 40 to 50 $\times$ 10^{-6} Ω cm[8], it is a better electrical conductor than most metals, including mercury, and a much better conductor than carbon or graphite. It is relatively hard (about 5 on the moh scale), and therefore resists abrasion[9]. It is chemically inert to most oxidising agents and strong acids. It possesses an oxygen overvoltage of the same order of magnitude as that of platinum[10-12].

In spite of all these favourable properties, difficulties were found in the manufacture of the anodes. Ferchland[3] had deposited lead dioxide from lead nitrate solutions and found the deposits to be very brittle and easily friable. Angel and Mellquist[6] investigated the reasons for the fragility of the deposits. They attributed the brittleness to inhomogeneity of the deposit caused by fluctuations in deposition current and electrolyte composition (especially pH and lead ion concentration), the latter fact being confirmed by Japanese workers[10, 13].

Several types of bath have been used for the deposition of lead dioxide[6, 14-16], and of these the lead nitrate bath is found to give the best deposits[8]. With the use of modern electroplating techniques and of suitable additives, deposits of high strength, high density and surface smoothness are obtained[8]. Shibasaki[17] has shown that bright smooth deposits of PbO_2 are the strongest, and consequently the most desirable.

Angel and Mellquist[6] deposited PbO_2 on iron, copper and nickel but found that such electrodes passivated. A massive lead dioxide anode was prepared from a neutral lead nitrate bath by Japanese workers[10], using the inside of iron or nickel cylinders as the substrate. The deposit was easily detached from the substrate by cutting the metal cylinder. It was not possible to deposit PbO_2 on the external surface of the cylinders as the deposit easily cracked. This was attributed to the differences in the thermal coefficients of expansion of the deposit and substrate. A technique for depositing the oxide on the outer surface of a non-metallic cylinder[18] was developed, which, after removal of the substrate, resulted in a hollow cylinder of PbO_2. Grigger and Miller deposited the oxide from an acid nitrate bath onto Ta and Ni[19, 20]. Difficulty was experienced in depositing the oxide onto thin attackable base materials because of the serious anodic dissolution that took place[8]. Schumacher et al.[7] deposited PbO_2 onto nickel and platinum plated titanium wires from an acid lead nitrate bath. The problem with these, and the previous PbO_2 electrodes, lay in the difficulty of applying an electrical connection[7]. With a conventional Cu contact, severe heating is observed in the contact area. Only a silver contact was found to be suitable[7].

The development of lead dioxide coated graphite anodes[21-23] solves the problem of the electrical connection, which can be conveniently effected through the graphite. These anodes developed by Pacific, primarily for perchlorate manufacture[4], have an additional contact advantage in that silver perchlorate is explosive and would constitute a hazard in the other current contact procedure.

The replacement of platinum by lead dioxide anodes in perchlorate manufacture had been investigated prior to 1950[6, 10, 24-26], but with the use of ammonium perchlorate as oxidant for solid propellants in rocket motors, research received a new stimulus[7, 9]. Angel and Mellquist[6] obtained a current efficiency (C.E.) of 72–79% with no additive; however in the presence of chromate, normally used with platinum anodes to prevent the reduction of chlorate at the cathode, a drop in C.E. was observed. This is probably due to interaction of the chromate with the electrode to form lead dichromate. Japanese workers[24-26] found that 2 g/l of sodium fluoride added to the electrolyte improved C.E.'s by over 40%, resulting in performances similar to the platinum anode. Schumacher et al.[7] investigated further the effects of the NaF additive and modified the addition to 0.5 g/l. Grigger and coworkers[8, 27] have used $Na_2S_2O_8$ as the additive. Udupa et al.[28] have found that addition of fluoride with sulphate, as sulphuric acid, has the maximum effect on C.E. Table 1 compares the results of perchlorate manufacture, using lead dioxide anodes, by various research workers. In addition to the effects of additives on C.E., pH, current density (C.D.), temperature[28], anode life[29], and cathode materials[7], have also been investigated. More recently [5, 30, 31], comparisons between platinum and lead dioxide anodes have been made, for most parameters, in perchlorate production from sodium chlorate.

Lead dioxide anodes are in use for perchlorate manufacture at Henderson, Nevada[4], and Tokyo, Japan[5]. The Pacific plant utilises, in one area, 1440 flat 3.18 × 51.2 cm lead dioxide anodes (Figs. 1 and 2) and has a production capacity of 22 tonnes of sodium perchlorate/day. The plant has been in operation for over 10 years. Another section uses 2400 10.9 × 110 cm anodes and has a capacity of about 62 tonnes/day of $NaClO_4$. Pacific claim a big cost saving over a similar plant with platinum anodes, complete cost for a 33 tonnes/day plant being about $13,000/ daily tonne compared with $75,000. Not all the saving is made from the use of the less expensive PbO_2 anode, but also in purification and recovery equipment. PbO_2 cells are operated at low chlorate concentrations and therefore it is not necessary to remove residual chlorate and chromate, (not needed in the PbO_2 cell), before using the $NaClO_4$ as ammonium perchlorate feedstock. Table 2 compares the two-processes.

In addition to the production of perchlorate, Pacific use their anodes to generate the chlorate feedstock from chloride solution. Plants using 5000 lead dioxide anodes have been built with a capacity of about 14 tonnes of sodium chlorate/day, and Weyerhaeuser Co. have a captive plant at Longview, Wash.

Sugino and Yamashita[32] had shown that a massive PbO_2 anode[33] was in fact superior to graphite or magnetite for chlorate production, and a C.E. of 81% was

TABLE 1

COMPARISON OF RESULTS FOR PERCHLORATE MANUFACTURE USING LEAD DIOXIDE ANODES OBTAINED BY VARIOUS WORKERS

	Japanese workers[24-26]	Angel and Mellquist[6]	Grigger et al.[8]	Schumacher et al.[7]	Udupa et al.[28]	Udupa et al.[28]
Anode	massive PbO$_2$	PbO$_2$ on iron	PbO$_2$ on Ta	PbO$_2$ on platinum clad Ta	PbO$_2$ on graphite	PbO$_2$ on graphite
Cathode	stainless steel	mild steel	stainless steel	stainless steel	stainless steel	stainless steel
Electrolyte	satd. soln. of NaClO$_3$	satd. soln. of NaClO$_3$	600 g/l NaClO$_3$	505 g/l NaClO$_3$	600 g l NaClO$_3$	600 g/l NaClO$_3$
Additives	2 g/l NaF	—	2.08 g/l K$_2$S$_2$O$_8$	0.5 g/l NaF	2 g/l NaF & H$_2$SO$_4$	2 g/l NaF
Temperature	29–30°	25–35°	25–35°	30–45°	30°	40°
pH	—	—	—	6.0–6.9	6.4–6.8	8.0–10.8
Bath voltage	4.5	5.05	5.0–6.5	4.75	4.3	4.7–5.1
Anodic C.D.	20 A/dm^2	20 A/dm^2	30 A/dm^2	15.5 A/dm^2	23.2 A/dm^2	40 A/dm^2
Cathodic C.D.	20 A/dm^2	—	—	7.25 A/dm^2	—	—
C.E.	82.9%	68.7%	68–74%	91.5%	82.8%	88.6%
Energy consumption	—	—	—	2.29 kWh/kg of NaClO$_4$	2.22 kWh/kg NaClO$_4$	2.37 kWh/kg NaClO$_4$

Figs. 1 and 2.
Lead dioxide anodes (courtesy of Pacific Engineering Co.).

obtained. The production of chlorate using lead dioxide coated graphite anodes has also been investigated by Udupa *et al.*[34-36]. A direct comparison of graphite and PbO$_2$ anodes in chlorate manufacture[34] indicated that greater C.E.'s were obtainable with PbO$_2$, 80%, as opposed to 70% with graphite. It was also found that the PbO$_2$ anode had a much longer life, being intact after the test, whereas the graphite disintegrated in 152 days. Subsequent work[36] showed the PbO$_2$ anodes to

TABLE 2

COMPARISON OF THE ELECTROLYTIC PRODUCTION OF SODIUM PERCHLORATE USING PT AND LEAD DIOXIDE ANODES

	Typical data for Pt	Typical data for Pt	Data with Pacific PbO_2	Data with Pacific PbO_2
Cell:				
Material of construction	steel	steel	stainless steel	stainless steel
Length × width × height	2.59 × 0.127 × 0.61 m	2.59 × 0.127 × 0.61 m	1.22 × 0.228 × 0.61 m	12 each 0.127 m diam. × 1.067 m
Electrolyte (g/l):				
$Na_2Cr_2O_7$ (g/l)	600 $NaClO_3$	90–110 $NaClO_3$	600 $NaClO_3$	600 $NaClO_3$
NaF (g/l)	5	5	nil	nil
Circulation (g/min)	—	—	0.5	0.5
Temperature (°C)	2	2	10	10
Current per cell (A)	40–45	40–45	55–60	55–60
Current density (A/cm² anode)	5000	5000	5000	5200
Current density (A/cm² cathode)	12.9	12.9	6.45	6.45
Current efficiency (%)	3.225	3.225	4.63	4.63
Voltage/cell	90	70	85	85
Energy (kWh/kg)	6.8	6.8	4.75	4.64
	0.65	0.865	0.5	0.49
Anode:				
Material	perforated Pt sheet	perforated Pt sheet	PbO_2 coated graphite	PbO_2 coated graphite
Consumption	6.32 g/t	6.32 g/t	—	—
Life (months)	—	—	6	12
Cathode:				
Material	steel	steel	stainless steel	copper, stainless steel
Raw materials:	$NaClO_3$	$NaClO_3$	$NaClO_3$	$NaClO_3$
Concentrations:				
Perchlorate, final (g/l)	500	580–600	950–1050	950–1050
Chlorate, initial (g/l)	600	90–110	600	600
Chlorate, final (g/l)	90–110	5	3–5	3–5

be still intact after 24 months of continuous operation. An investigation of chlorate manufacturing parameters[35] showed that it was possible to use PbO_2 anodes under conditions too drastic for the conventional graphite anode. Pacific have also found the PbO_2 anode to be remarkably versatile in chlorate manufacture[4]. The PbO_2 anode will operate effectively in chloride concentrations from near zero to saturation, unlike processes using graphite or magnetite where final chloride concentration is maintained above 100 g/l.

Pacific estimates that a complete plant having a capacity greater than 5 tonnes/day, costs about $40,000/daily tonne of capacity. Purity of the product, in terms of low final chloride concentration and the absence of anode sludge, results in a cost saving of about 25% on purification equipment.

Other production processes in which Pacific have found application for their anodes are hypochlorite, iodates and bromates.

The primary application of the hypochlorite cell is for the production of low concentration (up to 30 g/l) solutions of hypochlorite. These solutions are then used, in plant, for oxidation of some in-process constituent or to destroy the undesired further effect of some treatment chemical. Southern Peru Copper Corp. use the cell to kill the action of certain flotation agents. The cell also has applications for water treatment where it can serve as a direct replacement for the usual chlorination or hypochlorination treatments.

Some work has been carried out on the production of bromates using PbO_2 anodes[13, 37, 38]. A current efficiency of 90–92% was obtained in commercial production by Osuga and Sugino[13]. The lead dioxide electrode underwent some loss, 53–56 mg/1000 A h being observed. Lead however was not detectable in the product. As with chlorate production the main advantages of the process are the anode life and the product purity, no filtration being necessary. Sundararajan et al.[37], obtained 99% pure potassium bromate using lead dioxide anodes in addition to a current efficiency of 91–96% which was obtained over a C.D. range of 10–40 A/dm². The use of lead dioxide anodes for bromate production has also been investigated by Dzhafarov and Efendieva[38].

The use of PbO_2 anodes for the production of periodic acid has been known since the beginning of the century[39]. Willard and Ralston[40], used PbO_2 on Pt to convert iodic acid to periodic acid in HCl. More recently, Mehltretter[41] oxidised iodine in NaOH directly to periodate at a lead dioxide on lead anode, in the presence of sulphate or borax, whilst Torigai and Ishii[42] have used a massive PbO_2 anode for the direct production of periodic acid from iodine in HCl. Osuga and Sugino using the same type PbO_2 anode converted iodine to periodic acid in bromine water[43]. Using this electrolyte, anode losses were approximately halved to 0.07–0.15 g/100 A h, whilst maintaining a C.E. of 70–75%. In addition an investigation of other possible anode materials was carried out. Platinum and graphite were found to be ineffective for the production of periodic acid. Massive PbO_2 and PbO_2 coated lead were found to be the best materials. The PbO_2 coated lead anode could only be used in the presence of sulphate however, and in this case anode loss was ten times that observed with the massive PbO_2 anode. The use of H_2SO_4 also makes the separation of the HIO_4 difficult. Recently Russian workers[44] have patented a process for periodic acid manufacture from iodine using lead dioxide anodes.

Deep Water Chemical Co. Ltd., Compton, Calif. is using Pacific lead dioxide coated graphite anodes for the production of iodates. Osuga and Sugino[43] observed that this type of PbO_2 anode was the least efficient of massive PbO_2, PbO_2/Pb, PbO_2/Pt, and $PbO_2/graphite$ anodes, and that graphite itself gave product discolouration which was difficult to remove. Discolourations have been observed with Ni and stainless steel as well as graphite[45]. The usefulness of the PbO_2 anode in iodate manufacture has been confirmed by Ching Fa Teng and Yuan-P'u Lee[46], using a $PbO_2/graphite$ anode.

In evaluating possible uses of the lead dioxide anode some general principles may be applied.

(1) It should never be used cathodically. As a cathode the PbO_2 is reduced to metallic lead or else goes into solution.

(2) As a chemical it is quite active. While it is insoluble in water it slowly dissolves in HOAc and NH_4OAc. It dissolves more readily in HCl and HNO_3/H_2O_2 mixture. In general it is a moderately strong oxidising agent and will consequently react with oxidisable substances in acid solution.

(3) Anode life is markedly dependent on the usage to which it is put. In general, anode life increases as the voltage increases up to the discharge potential of the oxidisable substance. Increases in voltage above this level gradually decrease anode life. However the greater current density at higher potentials frequently offsets increased anode replacement costs.

(4) The lead dioxide anode is essentially an oxygen anode. That is, the configuration of bound atoms at the electrode surface presents positively charged oxygen atoms to the electrolyte. The function of the lead atoms is to provide the foundation for the charged oxygen atoms at the surface. As long as the transfer of electrons from the substance being oxidised *via* the lead dioxide coating to the power source is sufficiently rapid, the anode is inert. The lead is held at its highest valency state and does not dissolve.

(5) The lead dioxide anode exhibits some unique characteristics. This may be due to the spatial arrangement of the positively charged oxygen atoms at the electrode surface and may explain its extraordinary applicability to iodate production and the differences observable between platinum and PbO_2 anodes[47-50].

Some research has been carried out into processes other than oxyhalogen compound formation[51-59]. The use of lead dioxide/graphite anodes to replace costly platinum anodes has been reported for the production of bromoform[51], and iodoform[52, 53]. The intermediate in the synthesis is however the hypohalite ion, an oxyhalogen compound, which is then used as a halogenating agent. In bromoform synthesis a C.E. of 92.5%[51] was obtained, (Pt (87%) and graphite (86%)). PbO_2 anodes have been shown to be the best choice for iodoform production[52, 53]. Anode loss is almost negligible and a C.E. of 90%[52] is reported. Other syntheses in which an investigation into the possible use of lead dioxide anodes has been carried out are production of manganese dioxide[54], manganic sulphate from manganous sulphate[55], chromic acid from chromium sulphate[10], magnesium bleaching powder[56], oxidation of quinoline[57], oxidation of isobutyl alcohol to isobutyric acid[58], electrowinning of chromium[59], persulphates from sulphates[60], oxidation of phenol[61], and reduction of nitric acid to hydroxylamine[62].

Since the breakthrough by Pacific with the lead dioxide coated graphite anode for commercial use, greater interest has been shown in processes for the electrodeposition of lead dioxide and possible ways of anode manufacture[63-75].

Pacific have developed a technique for electrodepositing PbO_2 on other substrates[72]. Ti, Zr, Co, Hf, Ta and others have been plated and the PbO_2/Ti anodes are available for commercial use.

The PbO_2/Ti anodes have two distinct advantages over the PbO_2/graphite anode:
(1) Increased anode surface to volume ratio permitting construction of a more compact cell.
(2) Reusability of the Ti substrate at the end of anode life.

In addition Pacific manufacture powdered lead dioxide[72]. This material may find use in a recent patent using powdered PbO_2 for compacting lead dioxide anodes[71].

Two crystalline modifications of electrolytic lead dioxide have been identified by X-ray diffraction. The tetragonal β PbO_2 is the common modification and is the form found in the commercial anodes produced from acid lead nitrate baths. The α form is less common and has an orthorhombic structure. Some work has been carried out on this form because it is the most stable form[67, 68]. It is deposited from alkaline and neutral solutions.

The lead dioxide anode has in many ways shown itself to be a most suitable replacement for the costly platinum, and in some processes (notably chlorate production) strongly challenges the graphite anode.

3. The magnetite anode

Commercial use of magnetite has been limited to the production of alkali chlorates and chlorine. The first use of magnetite anodes is attributed to Oscar Carlsson[76], about 70 years ago in chlorate production. According to Billiter[76], and Jeitner[77], magnetite anodes have been used at the Griesheim Elektron works, following work by Specketer[78, 79] but the practice is now discontinued. At that time Griesheim Elektron kept a monopoly but this was terminated with the entry of Radocha (Poland) and Société d'Electrochimie (Switzerland). I.G. Farbenindustrie[80] employed magnetite anodes at a plant built in 1915 at Bitterfeld, Germany.

Japan Carlit still continue to use the anodes at their plant at Shibukawa, Japan[81], and it is reported that the anodes are in use for chlorate production in India[34].

The advantages of the magnetite anode for chlorate manufacture stem from its resistance to corrosion and wear. Owing to its dimensional stability the cell anode cathode distance is kept practically constant during electrolysis[81]. A loss of 5 mm on the external diameter of 60 mm anodes after 4 years continuous operation is reported[81]. This loss represents the very small cell voltage increase of 0.01 V under normal operating conditions[81]. In addition temperature has little effect on the wear rate, permitting high temperature operation (70°) and consequently improved C.E.[80, 81]; graphite anodes can only stand 40° and undergo excessive consumption[80]. Chemical, as opposed to electrochemical, formation of chlorate and minimisation of cooling water requirements are additional advantages of high temperature operation in chlorate cells[81].

The limitations of the magnetite anode arise from difficulties in fabrication, non-machinability, fragility and low conductivity, 20 times poorer than graphite[82, 34, 77].

Japan Carlit feel that the disadvantages of the magnetite anodes are more than offset by the dimensional stability of the anodes and the high operating temperature of the cells[82]. They claim the following operating conditions for their magnetite cells: cell voltage 3.2–3.5 V, final chlorate concentration 400–450 g/l, a C.E. of 83–85% and an electric power consumption of 6500–7000 kWh/tonne of $NaClO_3$[81]. Figure 3 shows chlorine overvoltage on pure magnetite in 5.3 M NaCl at 25°.

The Japan Carlit anodes are prepared from hematite[81] which is smelted in an electric furnace and then cast into hollow cylinders with one end closed. The inner

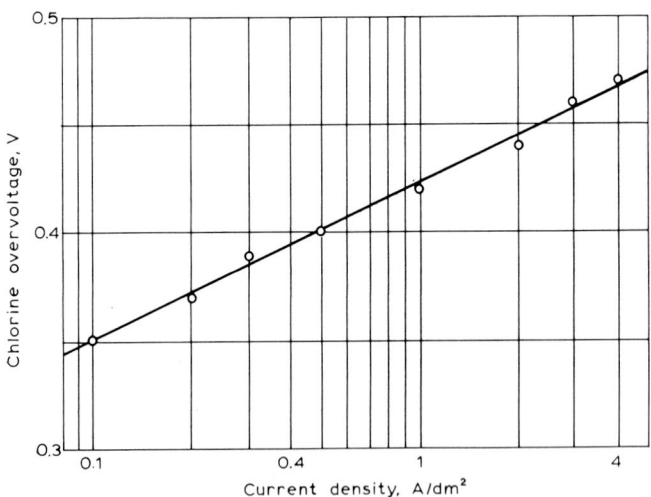

Fig. 3.
Chlorine overvoltage on pure magnetite in 5.3 M NaCl at 25°C, 1 atm. Cl_2 pressure[81] (by kind permission of the Electrochemical Society).

surface is then coated with a layer of electrodeposited copper. This procedure seems to have been adopted by previous manufacturers[76-79].

Other methods have however been claimed for anode manufacture[83-93]. Among the more interesting of these is a sintering technique[77, 91, 92] which enables the production of various anode shapes[77].

Attempts have been made to improve the physical properties of magnetite anodes by addition of other oxides[86, 89, 91, 93, 96]. It has been found that anode strength is greatly increased by the addition of CuO[97], whereas it has been suggested that the presence of Al_2O_3 and SiO_2 weakens the anodes[89]. Addition of up to 10% of metal oxides or SiO_2 and up to 1% V_2O_5 have been found to improve the anodes with respect to oxidation resistance when sintering[91]. Nagai et al.[94-96], have carried out a systematic examination of the effects of various oxide additives on magnetite anodes. The Fe_3O_4–SiO_2–Na_2O system was found to have equal conductivity and erosion resistance but much higher strength[94]. Some anode variations involving magnetite have been claimed in the literature[98-103]. Among the most recent of these is a lead or lead alloy[98] into which a piece of solid magnetite is embedded. Earlier, a magnetite anode into which a piece of noble metal was embedded was reported[99]. The use of magnetite on Pt for protection against anodic attack in chlorine production is a further modification[100, 101] and Japanese workers have recently been investigating the deposition of platinum on magnetite for use as the anode in perchlorate production[102, 103].

The non-applicability of the magnetite anode for perchlorate production is attributed to the much lower oxygen overvoltage[104] on magnetite in chlorate solution. Use of the magnetite anode, besides the commercial chlorate application, has been proposed elsewhere. Electrodeposition of metals from ammonium metal complexes, electrodeposition of chromium[105], electrolysis of sewage[106], and the decomposition of fatty acids[107, 108] have all been investigated using the magnetite anode.

Chlorate production remains however as the single commercial application of the anodes. Japan Carlit seem optimistic about the continued use of the anodes and, with present work showing that resistance to corrosion and wear can be coupled with increased strength, there may still yet be a place for the magnetite anode in the electrochemical manufacturing industry.

4. Lead and lead alloy anodes and the "Chilex" anode

Lead and its alloys find application as "insoluble" anodes in solutions relatively free from halide ions. At an insoluble anode in halide-free solution the anodic process is invariably oxygen evolution. Under these circumstances lead forms an adherent protective layer of lead dioxide which enables it to resist corrosion.

The major application of the alloys is in the electrowinning industry, especially zinc and copper electrowinning. Other applications are electrolytic deposition of manganese dioxide and chromium plating, and an application did exist at one time in sodium sulphate electrolysis.

In zinc electrowinning the recommended alloy, often referred to as the "Tainton" anode, consists of lead alloyed with approximately 1% silver. Originally electrolytic lead was the recommended material[109], however pure lead suffered from certain disadvantages[109, 110]. The anode could not tolerate the presence of chloride ions in the electrolyte[109], there was a tendency to disintegrate resulting in the zinc being contaminated, and also to bend and buckle, attributed to intercrystalline oxidation[110]. Fink and Pan[111, 112] showed, during some work to develop an anode for the chlor–alkali cell, that lead alloyed with small amounts of silver resulted in greatly reduced corrosion in chloride solutions. On investigation this transformation in the corrosion resistance was found to be due to the protective layer of PbO_2. Tainton et al.[110] in attempts to improve the corrosion resistance of lead anodes investigated many binary alloys of silver but found the lead silver series to be superior, provided that the system contained no free silver, i.e. at or below the eutectic of 2.6% Ag. An alloy containing 1% Ag, the "Tainton" anode, was investigated in zinc electrowinning and after a month's operation the terminal voltage was 3.37 V and C.E. 91.5%. The amount of lead in the Zn was determined at 0.004%. These results were confirmed by later workers[113–120]. In addition it was observed that lead–calcium and lead–tellurium alloys[113, 114] showed similar characteristics and that the ternary alloy lead–silver–calcium intensified the individual effects of the binary alloys[114]. However the lead–silver–calcium ternary alloy was subsequently shown to be unsuitable because of low stability[115]. In some instances quaternary alloys have been investigated[121–123], but the Tainton anode has yet to be superseded.

All copper electrowinning plants on line today use lead alloy anodes containing 6–15% antimony and 0–1% silver. Antimony is added to give the so-called "hard" lead which has superior mechanical properties over pure lead. The silver is added, as in the case of the Tainton anode, to increase the corrosion resistance.

An interesting feature in the history of copper electrowinning was the specific use of "Chilex" (Chile Exploration Co.) anodes at Chuquicamata, Chile. This anode was developed[124] because of the severe corrosive actions of the leached Chilean ores and consisted of many alloying metals but was essentially copper silicide. The anode however raised power consumption by virtue of its greater electrical resistance and oxygen overvoltage, and was not used later than 1933.

The Tainton anode is also used in chromium and manganese electrowinning. Fink and Kolodney[125] found that lead anodes were unsuitable in manganese electrowinning because of loss of Mn by oxidation to MnO_2. They developed an alloy anode of lead antimony and cobalt. Schlain et al.[126] found that lead alloyed with 1% silver and 1% arsenic corroded only 12% as rapidly as the Tainton anodes and that this alloy, as an anode, was as least as good with respect to cell voltage, cathode C.E., physical properties, etc.

The lead 1% silver anode is also used for the production of electrolytic manganese dioxide[127-131]. The application is however confined to countries outside the U.S.A., where graphite is favoured as there is no risk of lead contamination of the product[132]. Russian workers have investigated the use of lead alloy anodes for cadmium electrowinning[133, 134]. In a pilot plant they observed that the Tainton was more corrosion resistant than the "hard" lead anode. It was also found that whereas "hard" lead resulted in lead contamination of the product, equivalent to that of pure lead, the Tainton anodes resulted in diminished contamination.

The lead 8% antimony "hard" lead anode has found further application, other than copper electrowinning, in cobalt electrowinning and chromium plating[135, 136]. Some research has been carried out into the possible use of other alloys, based on lead antimony and tin, for chromium plating[137], but without success. An additional advantage of the hard lead anode in chromium plating is in its ability to maintain bath stability by reoxidising reduced chromic acid.

Attempts have been made to use lead alloys as anodes for sodium sulphate electrolysis but, to our knowledge, no full scale plant is in operation at the present time. I.G. Farbenindustrie[138] operated a plant during the second world war using lead–silver (7.5–10%) anodes. Various alloys were investigated during preliminary research but were not suitable. Lead and "hard" lead were attacked too much, the Tainton did not under test come up to requirements, and alloys of lead with Te, As, Se, and Cu, as proposed in the literature, were not satisfactory either. Research into the favourable lead–silver system resulted in the adoption of the 7.5–10% silver alloy, however this also had certain disadvantages,

(1) Composition of the alloy is above the eutectic and special precautions had to be observed in production and treatment.

(2) The anodes were extremely sensitive to chloride ions which rapidly brought about corrosion.

(3) Durability was confined to small units.

Japanese workers[139-145] have developed anodic materials for sodium sulphate electrolysis based on lead silver and tellurium alloys. On a pilot plant scale a Pb 2% Ag 1% Te alloy operated with 95–98% C.E. and a cell voltage less than 5 V. Under pilot plant operation anode consumption was 0.2–0.5 mg/Ahr. Tsuiki and Ueno[146] have investigated the effect of the Te additive, as well as Sn, to Pb and Pb Ag alloy. The addition of Te to Pb was found to greatly reduce the grain size of the lead alloy and decrease anodic corrosion. The addition of Te to the Pb Ag alloy made the alloy crystals very fine, dispersed segregated Ag alloy crystals more uniformly, and reduced anode consumption by as much as 90%.

Sato et al.[147-151] have investigated various lead alloys for corrosion resistance in sea-water and some other alloy modifications have been suggested in the literature[152-159].

The use of lead alloy anodes is unquestionably established in the electrowinning industry and for chromium plating. Application in the production of electrolytic

MnO_2 is confined to plants outside the U.S.A. Application of lead alloy anodes for Na_2SO_4 electrolysis, whilst the next logical step from brine electrolysis, is unlikely to find commercial applications, although large quantities are produced cheaply as a by-product from the viscose rayon industry.

5. Carbon and graphite anodes

Carbon and graphite have been the main "insoluble" anodic materials in industrial electrochemical processes for a very long time[159, 160]. One of the major applications of graphite anodes, in which almost exclusive usage is found, is in the chlor–alkali industry. In addition graphite anodes are used for the production of chlorates from chlorides, bromates from bromides, iodates from iodides and in manganese dioxide production. Metal electrowinning processes are further applications for graphite and in particular sodium, magnesium and lithium. Carbon is not as generally used as graphite as an anode material owing to its inferior properties, but it still finds application in the generation of fluorine and in the aluminium electrowinning industry, where, incidentally, more carbon is consumed as electrodes than in any other electrochemical process[159].

The world wide acceptance of graphite as an anodic material is due entirely to its excellent properties[159-163]. It has high chemical resistance, good conductivity, considerable mechanical strength, high surface to volume ratio and is cheap. Although graphite possesses good corrosion resistance when it is used anodically, some consumption does in fact take place, primarily due to the chemical attack by "active" oxygen generated during electrolysis, but also to the highly oxidising nature of the products[162, 164, 165].

Graphite was first used as anodes in the production of chlorine by the Brown Co. in Berlin, New Hampshire[165]. Prior to this, carbon or magnetite anodes were used[160]. Carbon anodes are inferior to graphite with respect to mechanical and chemical resistance[166]. In the chlor–alkali industry graphite anodes have the additional advantage of low chlorine overpotential. As can be seen from Fig. 4, graphite is superior to both platinum and magnetite as an electrocatalyst for chlorine evolution.

Vetter[167] has pointed out that graphite is essentially a passive electrode. Possible corrosion reactions, whose reversible potentials are much lower than the chlorine reversible potential, must be strongly inhibited for graphite to be able to function as an anode in chlor–alkali cells[162]. However, as mentioned previously, some anode consumption does occur, approximately 3 kg/tonne of chlorine in diaphragm cells and 2 kg/tonne in mercury cells[162], and much research has been carried out to determine the corrosion mechanism[161, 163-165, 168-183], the effect of electrolysis variables[161, 165, 172, 174, 184-199], and possible ways of reducing the consumption[170, 173, 178, 185, 200-203].

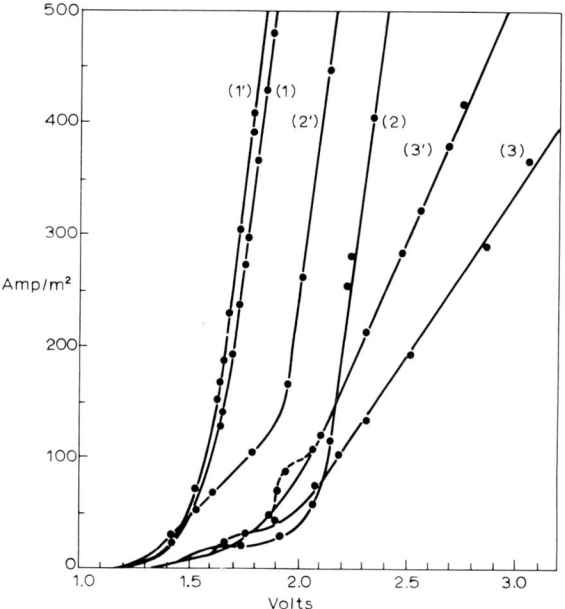

Fig. 4.
Comparison of chlorine overvoltages at 25°C and 70°C. (1,1') Graphite; (2.2') smooth Pt; (3,3') magnetite.

The presence of carbon dioxide and, to a lesser extent, carbon monoxide in the anode gases in chlorine production, indicated that some form of chemical attack of the graphite anodes was occurring. In addition a graphite sludge formed showing that anode attack was not entirely chemical but also mechanical. The presence of oxygen in the anode gases suggested that "active" oxygen was the corroding species, a generally accepted view[161, 163, 169, 171, 176, 183]. However there may be some contribution from highly oxidising products, e.g., hypochlorous acid[161, 164, 174, 183, 186, 204]. The nature of the graphite anodes is the important factor in mechanical wear. Graphite anodes are porous disperse bodies which consist of interconnecting carbon crystals, 70–80% of whose volume consists of solid material, the remaining volume consisting of pores[161]. It is this pore structure which plays a significant part in the mechanical consumption of the anodes[161, 169, 176, 205]. Chemical attack within the pores weakens the intercrystalline bonds causing graphite particles to spall off.

The mechanism of anode consumption is not simple due to the porous nature of the graphite, and it is easier to consider external and internal attack separately, internal attack being that occurring in the graphite pores and resulting chiefly in mechanical consumption.

Vaaler[174] suggests that external attack of graphite anodes occurs by rapid adsorption of oxygen followed by a conversion to chemisorbed oxygen, which is then rapidly desorbed, breaking a carbon bond, to give carbon monoxide. Carbon monoxide is then mostly oxidised to carbon dioxide by dissolved oxygen or chlorine. Bulygin[176] decided, as previous workers had[179, 184, 189] that to investigate the corrosion of graphite it was essential to work in an electrolyte in which the only anodic product would be oxygen. From this approach he deduced a similar corrosion mechanism to Vaaler. He also showed that external attack could result in crumbling of graphite *via* a "swelling" mechanism, first observed by Sirak[184], and Thiele[179]. However this phenomenon has only been observed in strong oxidising acids and not under conditions typical for chlorine production.

Internal attack of the graphite was attributed, by Sproesser[205] to hydroxyl ion discharge resulting from the depletion of the chloride ion by internal electrolysis. Stender[163] showed however, that electrolyte circulates continuously through the pores of a working graphite anode and therefore chloride ion impoverishment does not occur. The internal attack was explained by virtue of the lower potential within the pores favouring oxygen evolution. In view of the lower potential within the pores Ioffe[181] has pointed out that internal oxidation should be less intense (lower C.D.). According to Ksenzhek[206], only 2% of the C.D. is involved in internal electrolysis. Bulygin[176] investigated further the C.D. within the pores of a graphite anode and found that electrolysis within the pores could not cause any significant destruction of the anode. In agreement with an earlier postulate[175, 186] he showed that the attack was chiefly due to the chemical action of hypochlorous acid within the pores. Kokoulina and Krishtalik[169] have recently confirmed Bulygin's work.

In the production of chlorine, anode consumption is to some extent dependent on cell operating variables. Among the variables that have been investigated are feed brine concentration, cell operating temperature, anode current density, brine feed rate[165], pH[171, 174, 207], and sulphate concentration[163, 187, 190-196, 207, 208]. In studying the effects of operating variables difficulties are encountered in reproducing commercial conditions in the laboratory. To some extent these difficulties have been overcome. Johnson[165] used a laboratory scale type of diaphragm cell to investigate the effects of feed brine concentration, temperature, anode C.D. and brine feed rate. It was found that temperature increase resulted in an increase in anodic attack of the order of 1.15 g/1000 A h/10°C rise above 50°C. The variation of anode consumption with increasing C.D. was found to be a slight decrease in terms of quantity of current carried, but almost a linear increase with time. Even a small lowering of the sodium chloride content of the feed brine, from saturation, resulted in a significant increase in the rate of anode attack, which then rose rapidly as the brine concentration further decreased. At constant temperature and C.D., brine feed rate was found to be of major importance for controlling anode corrosion, zero corrosion being indicated for infinite rate of flow. These determinations took no account of pH variations within the anolyte[171], however, a factor that has subsequently been shown to be of great significance[171, 174, 207]. Filippov et al.[188, 189] studied anode wear under constant pH conditions by maintaining a fast flowing electrolyte. However mechanical wear was not taken into account and, in view of the short nature of the experiments, it is doubtful whether steady state conditions were achieved. Krishtalik et al.[171] using a double diaphragm cell, with neutralisation of any migrating alkali within the diaphragms, successfully controlled anolyte pH and investigated its effect on anode wear in a low pH range. They showed that anode wear, in a solution containing 300 g/l of sodium chloride at 80°C with a C.D. of 800 A/m², was independent of acidity above approximately 0.01 N. At lower acidities the wear increased with fall in acid concentration, slowly at first, but then more rapidly. In the pH range 3.0–4.2, Vaaler[174] has shown, by analysis of the anode gases, that anode consumption is doubled at the higher pH. More recently Wallen and Wranglen[207] have developed a technique for studying graphite anode corrosion over a wide pH range and have also investigated the effect of chloride concentration. The results obtained are not typical of those used in commercial chlorine production but show that maximum corrosion is obtained at intermediate pH's. These workers also used their technique to study the effect of sulphate ion on anode corrosion[207]. It was found that at low chloride concentrations the effect of sulphate was very small, but in solutions of high chloride concentration the effect was very pronounced, anode wear being more than doubled with an addition of 40 g/l of sodium sulphate.

Owing to work by Murray and Kircher[190], Gardiner[191], and Jacopetti[187], the detrimental effects of sulphate ion in brine electrolysis have been known for some time. Work by Krishtalik et al.[192] has shown that graphite anode corrosion shows a roughly linear increase with increasing sulphate content. Flisskii et al.[193–195] have attributed the corrosion increase to the specific adsorption of the sulphate ion, resulting in weakening of the carbon bonds in the lattice and culminating in the discharge of CO and CO_2. Japanese workers[208] have found that in saturated brine sulphate has no effect in concentrations less than 3 g/l.

The effect of chlorate ion on the corrosion of graphite anodes seems to be much smaller than that of sulphate[192]. Fukunaga et al.[209] claim a reduction in anode losses in diaphragm cells of 20% by incorporating 100 mg/l of cobalt in the brine.

In addition to the effects of operating variables on anode consumption, work has also been carried out to determine the effects of anode characteristics, such as pore structure, degree of graphitisation, etc., on anode life. Foerster[185] observed that artificial graphite electrodes were broken down much less than charcoal electrodes under the same conditions, a result confirmed by Arndt and Fehse[200]. The increase of stability with graphitisation is attributed to crystal growth, which also results in increased conductivity[161].

Flisskii et al.[201, 203] have recently investigated the relationships between anode stability and electrode structure. Recent work on the surface area of graphite anodes indicates that the specific surface of graphite is of the order 1 m²/g[210]. Flisskii[203] obtained similar data for fresh anodes but, after several months use in chlorine electrolysis, the surface area increased approximately tenfold. However no connection between surface area and stability of the anodes was found. This was explained by the fact that the specific surface of graphite is determined primarily by the number of micropores, and only a small proportion of the total surface corresponds to the macropores, and evidently micropores do not play any significant part in brine electrolysis[203]. Marek and Heinz[211] have, however, found increased corrosion with increasing surface area. Studies of the relationship between anode stability and pore volume[201, 203] indicate that the pores having radii within the range 7,500–75,000 Å are the ones susceptible to attack, whilst the greatest number of the pores are concentrated in the radius range up to 7,500 Å. Kuchinskii et al.[201] showed a 30% corrosion increase for a pore volume increase of 0.032–0.089 cm^3/g for pores in the diameter range 2,500–70,000 Å.

Bulygin[178] has studied the density distribution in the surface layer of a graphite anode at high and low C.D.'s. At high C.D. the electrolysis process was found to occur in only a small number of large pores resulting, after a time (4 hours in Bulygin's case) in abrupt crumbling of the surface layer. At low C.D.'s however, the anodic destruction was of a surface character. Bulygin suggested that in this case electrolysis took place in the finest of the pores. Nystrom[183] has shown that some pore enlargement occurs in electrolysis resulting in, with time, progressive enlargening of the pores enabling internal electrolysis to occur at the anode.

Methods of improving the corrosion resistance of graphite anodes are based on impregnation. Many methods employing numerous impregnants have been reported in the literature[170, 181, 212–254].

The main concepts on the mechanism of impregnant action were formulated by Ioffe[202]. More recently Krishtalik et al.[173, 255, 256] have carried out a detailed investigation into the mechanism of impregnants in reducing anode consumption. The impregnant was found to protect the regions of grain contact thereby minimising mechanical losses. Sjodin and Wranglen[170] have carried out an analysis of impregnation procedure and impregnants. Their work indicates that use of impregnants may result in increased anode corrosion if the impregnant is not removed at the same rate as the external surface of the anode. It is believed that sodium chromate assists in this respect by oxidising away the impregnant. Linseed oil, properly dried, is accepted as the best impregnant of those in use today.

Impregnation of graphite anodes for diaphragm cells is essential if reasonable life is to be obtained. Mercury cell anodes are rarely impregnated however. This is partly due to the fact that under mercury cell conditions, i.e. lower pH (approx. 2.0), and lower temperature (10–20° lower than in diaphragm cells), the rate of anode attack is much lower[162], but may also be attributed to the tendency of impregnated anodes to "shatter" in mercury cells.

Various workers[164, 168, 257–267] have investigated the overpotential of anodic evolution of chlorine on graphite. Several papers refer to overpotentials for low C.D.'s only[257, 259], and those at high C.D.'s usually refer only to a single temperature with considerable variation in results between authors[168, 260]. Some authors have tried to express the relationship between overpotential and C.D. in the form of a Tafel equation[261-264]. There is considerable variation in the Tafel constants however[261, 262]. Krishtalik and Rotenburg[263, 264] identified two pH regions in which chlorine overpotential differed considerably[263], and subsequently showed that the polarisation curves in acid solutions were comprised of two sections with Tafel slopes of about 60 and 110 mV[264]. Japanese workers have considered the effect of chlorine bubbles on the electrode potential[262, 265], whilst Kubasov and Volkov[266] have expressed Tafel type equations incorporating temperature dependence. Recently Drossbach et al.[267] have determined kinetic data for chlorine evolution and oxygen evolution, the latter in aqueous and non-aqueous media.

The lack of agreement in the kinetics of chlorine evolution on graphite has been attributed to changes in the electrode characteristics as a result of "ageing" of the electrode[292, 293]. Flisskii[203] has shown that the surface area of working graphite anodes increases with time and recent work has indicated that changes in the type of oxide film on graphite anodes also occur[292].

Krishtalik and Rotenburg[293] assumed that new graphite electrodes were covered with a stable oxide which was replaced, when the electrode was polarised anodically, by a layer of anodic oxides with properties different from that of the stable oxide. Binder et al.[294] have distinguished between two types of oxide on graphite anodes.

Janssen and Hoogland[292] have carried out a detailed investigation into the ageing of graphite anodes. Their work agrees with the surface area changes observed by Flisskii, showing a maximum increase of 17 times the original surface area, and provides evidence for the changes in the nature of the oxide layer assumed by Krishtalik and Rotenburg. They found that new graphite anodes were covered with a stable oxide which protected the graphite against attack. This oxide disappeared under continuing anodic polarisation at 1.72 V resulting in an increase in the roughness and surface area of the electrode.

The investigation of anode consumption in the manufacture of chlorates has not been as thorough as that for chlorine production. The same principles apply however. Janes[186] investigated the effects of chloride concentration, temperature and anode C.D. on the anode consumption. In view of the operating conditions favouring oxygen evolution the corrosion of graphite anodes is consequently more intense. Filippov et al.[182, 199] investigated the effects of chromate, pH and C.D. on the behaviour of graphite anodes under the conditions of chlorate manufacture. Recently, Jaksic and Csonka[254] have postulated that chlorate anode consumption may be kept to the level of that experienced in the chlor-alkali industry by controlling, to a minimum, the available chlorine in the cell.

The use of graphite anodes in bromate, iodate and manganese dioxide production suffers from the same disadvantage as in chlor-alkali and chlorate manufacture, namely anode consumption. In bromate manufacture, anode sludge results in discolouration of the product[268] and necessitates the incorporation of filtration equipment in the process. Polish workers have carried out a systematic investigation of the anodic consumption of graphite in various electrolytes[269-273] including potassium bromate and manganese sulphate[269].

Although graphite and carbon are used extensively as anode materials in fused salt electrowinning, very little has been published about their performances in such processes. In fact with the exclusion of aluminium electrowinning and fluorine manufacture the literature is almost non-existent.

In aluminium electrowinning two types of carbon anodes are in general use; prebaked and Soderberg[274]. Soderberg anodes were developed by Electrokemisk A/S of Norway after the invention of Soderberg[289-291] and were adapted for use in aluminium electrowinning in the early twenties[275]. Oehler[276] has compared the two types of anode and shown that prebaked are more suited to large plants because of their superior properties, *e.g.*, C.E., anode consumption, power consumption, etc., whilst the Soderberg anodes are more suited to smaller plants. The prebaked anodes have a low percentage of binder material (coal tar pitch) and the Soderberg anodes have a higher percentage[274].

The use of carbon as an anode in aluminium electrowinning has several advantages[277]:

(a) It is virtually insoluble in molten Al.
(b) It is able to withstand the high operating temperature (980°C).
(c) The anode products, CO_2 and CO, are expelled from the system and do not contaminate the electrolyte.
(d) It is cheap.

The main disadvantage is the anode consumption. This is due to attack of the carbon by oxygen liberated at the anode which reacts to form the primary product, CO_2[278, 279], CO being produced by carbon reduction of the carbon dioxide. The thermal effect of oxidation is useful to some extent in maintaining the bath in a molten state, thereby saving power consumption[274]. As in aqueous electrolysis some mechanical consumption of the anodes also occurs[280, 281].

Hollingshead and Braunworth[281] have investigated the effects of operating variables on anode consumption. Increasing C.D. and baking temperature, and decreasing bath temperature decreased anode consumption. Addition of NaCl and increasing the NaF/AlF_3 ratio increased anode consumption.

The behaviour of anode carbon in cryolite Al_2O_3 melts was investigated by Antipin and Dudyrev[282]. They suggested anode consumption proceeds in a 3-stage process;

(i) At C.D.'s 0-0.1 A/cm^2, C-O compounds are formed on the electrode surface.
(ii) 0.1-0.3 A/cm^2, O ions penetrate the carbon lattice.
(iii) Above 0.3 A/cm^2, CO_2 is formed on the surface.

More recently the kinetics of the anode process in Al electrowinning have been investigated by Thonstad and Hove[283]. Welch and Richards[284] had found that the relationship between overvoltage and C.D. followed the classical Tafel equation and Thonstad and Hove tentatively suggested the following mechanism;

O^{2-} (in melt) → O^{2-} (on surface) (1)

$O^{2-} + C → C_xO + 2e$ (2)

$2C_xO → CO_2$ (adsorbed) $+ C$ (3)

CO_2 (adsorbed) → CO_2 (g) (4)

and showed that the overpotential may be due to slow transport of the oxygen ions through the double layer (1), or to slow reaction between chemisorbed oxygen and carbon (3). Stern and Holmes[285] from analysis of industrial potential decay curves concluded that the slow step was a chemical reaction of approximately second order (3), whilst Mashovets and Revazyan[286] suggests that the rate is determined by (1) and (3).

The investigation of catalysing (Fe_2O_3), and inhibiting (H_3BO_3), additives in the anodes gave weight to reaction (3) being rate controlling, as small but significant changes in the overpotentials were observed.

Robozerov and Vetyukov[287, 288] have investigated the effect of H_3BO_3 additions on the anode characteristics. Additions were found to markedly decrease oxidation and crumbling of the anodes whilst increasing mechanical strength and electrical resistance without affecting the density or the porosity of the anodes.

Recently Kronenberg[277] has investigated the possibility of using gas depolarised graphite anodes for Al electrowinning. His work has shown that gas depolarisation of the anode is possible with hydrogen or natural gas but serious fluoride losses were observed.

The use of carbon anodes for fluorine generation is adequately dealt with in Chapter 2.

6. Noble metal coated anodes

6.1 INTRODUCTION

The inertness of noble metal coated anodes is undoubtedly their major advantage. However, even the noble metals when made anodic in aqueous media undergo some corrosion. This corrosion is normally only a surface phenomenon resulting in the formation of noble metal oxides, but it is sufficient to ensure that the oxides, and not the massive metals, are the electrocatalysts.

A survey of the use of solid platinum and platinum-clad metals is provided by Wranglen[295] but the use of solid platinum for all but a very small number of specialised industrial cells can be neglected because of the excessive capital costs. This situation has prevailed for more than half a century, but as early as 1911 Stevens patented[296, 297] noble metal coated tungsten and tantalum electrodes for use in electroplating, electrometallurgy and for electric batteries and cells. The non-exploitation of this form of anode is possibly associated with the cost and lack of availability of tungsten and tantalum at the time, and an inadequate technology to make the most economic forms of anode shapes. Subsequently platinised tungsten was evaluated for perchlorate electrolysis using platinum electrodeposited to 5 microns in thickness, but at moderately elevated temperatures the tungsten dissolved through pores in the coating.

Cladding of platinum foil on substrates such as copper and silver has been practised, especially for per-oxidation reactions such as persulphate, perchlorate and perborate, and these are in fairly wide use. The platinum foil thickness is in the region of 70 microns, but even so it is difficult in practice to ensure completely sealing off of the substrate and in any event the capital cost of the platinum in large plant is still very considerable.

The advent of titanium in commercially available quantity brought with it the very real possibility of employing an economically attractive inert substrate that would not corrode even when coated with a porous conducting film of noble metal. The first published information upon the use of titanium in this sort of application appeared in short notes by Cotton in 1958[298, 299]. The present article deals with the developments in noble-surface coated titanium stemming from that point. Firstly however, it should be realised that, prior to being coated, titanium is covered with a passive oxide, or in some cases a hydride film[300]—films which give titanium its remarkable corrosion resistant properties.

The first patents in the field, resulting from independent research in Holland[301] and in the U.K.[302], were made at much the same time, and essentially covered the construction of titanium-based electrodes coated with electrodeposited platinum, rhodium and other noble metals, and in particular directed attention[302] to the use of such composite anodes in cathodic protection. The first commercial production of such anodes began in U.K. in 1959 and for several years the commercial use was limited to such applications as cathodic protection, electroplating and electrodialysis, although prototype cells for generating sodium hypochlorite from seawater were in operation using platinum coated titanium anodes. From the beginning, however, there was an obvious incentive to replace graphite in the electrochemical industry.

Smooth platinum, and the not greatly different surface produced by electrodepositing platinum, is characterised by a chlorine overpotential of several hundred millivolts at current densities in the 2–10 kA/m² current density range. These values are higher than those obtained with graphite and, coupled with other factors such as initial cost, high wear rate, etc., this was sufficient to preclude platinum electrodeposit from these applications.

Advantage was, however, soon taken of the existence of the technique of coating ceramic and glass with noble metal, by thermal decomposition of noble metal compounds in an organic base[302, 303]. When applied to titanium at decomposition temperatures lower than those used for ceramics, this type of surface was found to have many of the attributes of platinum black, but with the advantage of much better durability[302, 304-307]. In particular, the chlorine overpotential at current loadings of 2–10 kA/m² was found to be of the order of 25–100 mV—significantly less than the best attainable figures for plain graphite.

There seems little doubt that the platinum dissolution rate is related to the total current passed, and, for example, in near neutral saline environment, Wranglen[295] reports a loss rate of 1 μg/A h. In contrast to this, practical experience of plated platinum used in bipolar cells producing sodium hypochlorite from seawater have given a rate of less than 0.05 μg/A h[308], while in laboratory diaphragm cells operated at high current densities losses of about 0.2 μg/A h were recorded.

There is some evidence that platinum dissolution can be explained in terms of a potential-dependent change in surface structure[309], and there is little doubt that the imposition of AC on to the DC source can give rise to increased dissolution[310, 311]. The behaviour of platinum under the influence of impressed current can thus be quite complex, but in many environments the rate of dissolution is sufficiently low as to enable thin platinum coatings to be used economically in such applications as cathodic protection, electrodialysis, electroplating, etc. For example, practical experience in the cathodic protection field indicates that lives of 10 years are achieved with a coating thickness of 5 microns.

The noble metals of Group VIII are quite variable in their behaviour as electrodes under impressed current conditions and, for example, under such conditions iridium is much more durable than platinum. It is not therefore very surprising that very early in this development, alloys of noble metals were tested. As a result, a 70/30 wt.% platinum/iridium alloy was selected as representing the best compromise between technical superiority, availability, cost and ease of application. For the first serious attempts to employ noble metal surfaces in caustic chlorine cells, therefore, 70/30 platinum/iridium was applied to titanium by a thermal route. This type of surface had an adequately low overvoltage for chlorine release and under chlorine diaphragm cell conditions at least, the rate of wear appears to be economic. Indeed, in spite of further developments described later, this type of surface can be regarded as eminently suitable for use in chlorine diaphragm cells, sodium chlorate cells and possibly most chlor-alkali cells not involving liquid mercury. It has certainly been incorporated in several commercial chlorate cells[312, 313].

Chlorine cells employing mercury carry the almost inevitable hazard that very occasionally there will be accidental shorting between the sodium amalgam cathode and the anode. Depending on the type of anode employed and its wettability in sodium amalgam, shorting can result in, in addition to the physical effects of arcing, an amalgamation and reduction of the anode. It soon became evident that the physical effects of shorting with a metal anode could be more serious than with graphite and, in particular, this could adversely affect the wear rate of the noble metal coating—even that of 70/30 platinum/iridium. This hazard was eventually countered by the disclosure[314] that certain noble metal oxides could function electrochemically as efficiently as the metals themselves, and that they behaved much better, upon shorting with sodium amalgam.

This has been attributed, in part, to their poor wettability in the sodium amalgam. Further developments of this type of surface led to the production of mixtures of titanium oxide and noble metal oxides with a further variant in the form of a "mixed crystal" oxide containing essentially titanium oxide, TiO_2, with ruthenium oxide, RuO_2[314-316]. There seems little doubt that these developments foreshadow a new era in the design and operation of caustic chlorine cells.

Thus noble metal coated titanium anodes now cover a very wide range of coatings and forms of anode. Basically these are divided into the high and low overpotential groups, with further subdivisions depending upon the particular electrolyte conditions to be encountered, whether subjected to AC or DC conditions, to strong acid environments, and so on. In the present article some general characteristics of the noble metal coated anode will be described, and others more particular to special types of coating, and finally some assessment given of the present state of application of this class of electrodes in a range of industrial electrochemical processes.

6.2 PROPERTIES OF NOBLE METAL COATED TITANIUM ANODES

6.2.1 Titanium substrate

Although titanium forms only one part of the noble metal coated titanium anode, it is important to emphasise that the rapid development of this form of electrode is a result of a combination of factors, of which a principal one has been the availability of titanium in a range of forms such as sheet, rod, tube, wire, mesh, etc. that could be readily fabricated and welded into a variety of shapes to conform to any particular design of cell. For the majority of electrodes commercial purity titanium is employed. Alloyed material, apart from being more expensive than commercial purity titanium, usually has a lower anodic breakdown potential in chloride solutions and is thus unsuitable, but a notable exception to this is the titanium–0.2% palladium alloy. Residual impurity content of the titanium affects the residual anodic current[317], but there is no sound evidence that this increased activity affects the ultimate durability of coatings and there is certainly no economic incentive to employ ultra pure titanium.

6.2.2 Mechanical properties and corrosion resistance

The mechanical and corrosion resistant properties of titanium have been described on a number of occasions[318, 319]. Suffice it to state here that commercial purity titanium has a tensile strength similar to that of mild steel and that it is usually quite suitable for the design of anode frames of adequate rigidity. For the applications in which noble metal coated anodes are used the corrosion resistance of the titanium is of a high order, but this generalisation needs qualification because of the very wide range of environments in which the material might be exposed. Even the very small corrosion rates of exposed titanium[409] quite insignificant in terms of normal usage of uncoated titanium, can contribute to loss of adhesion of coating by undermining. In general terms, the higher the acidity of hydrochloric or sulphuric acid, and also the higher the temperature, the greater the residual dissolution and the more likely any coating is to become undermined. Of a wide range of applications in which anodes have so far been employed, including electrolysis of brine to form chlorine, chlorate and hypochlorite, the only instances established where this has become serious are in the metal finishing field[300], where coating undermining has occurred in strong hot acid solutions. Restriction of circulation of electrolyte, such as occurs when anodes are buried in deposit, may also produce sufficiently acid conditions to promote attack of the substrate. As a general rule, all forms of noble metal coated anodes exhibit best durability when operated in well irrigated cells.

Except for the most simple types of electrode, such as a rod dipping into an open tank of solution, crevices inevitably exist in some part of the electrode structure, such as between the anode stem connection to the busbar and the sealant/insulation material which surrounds it. Crevice corrosion of the titanium in the presence of chloride has been reported very occasionally, and although titanium is normally resistant to attack in wet chlorine, it seems that in crevices, conditions can exist in which titanium is no longer passive. In laboratory experiments using brine, crevice attack has not been detected under heat transfer conditions below 140°C and the few examples in electrolytic cells in which this form of attack is believed to have occurred are probably associated with local overheating due to high resistance connections.

Another important parameter concerning titanium is the value of the anodic breakdown voltage in the electrolyte. In a wide range of non-halide electrolytes titanium can be made anodic up to high voltages, *e.g.* in excess of 100 V in sulphuric acid at 20°C, the overall reaction being the formation of an anodic film of the anatase modification of titanium dioxide which is entirely resistant to the electrical stress imposed upon it. In halide bearing solutions titanium is appreciably less resistant[320]. For most purposes, however, the electrical stress imposed on non-conducting areas of noble metal coated titanium anodes (see Fig. 5) is below the breakdown voltage of titanium in brine solution, and for this reason titanium is a suitable substrate, whereas many other materials are not. The breakdown voltage of titanium is not greatly affected by salt concentration and pH over fairly wide ranges, but falls appreciably with temperature, see Fig. 6. In diaphragm-type chlorine cells, operating at up to 90–95°C, there is still a margin between the anode

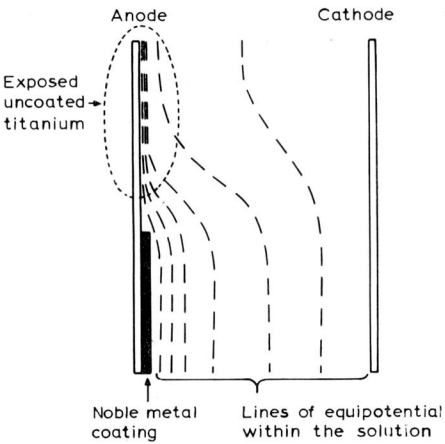

Fig. 5.
Diagram to illustrate the formation of electrochemical stress on uncoated areas of noble metal coated type anodes. Hence the importance of choosing a substrate resistant to anodic breakdown, particularly in the presence of halide ions.

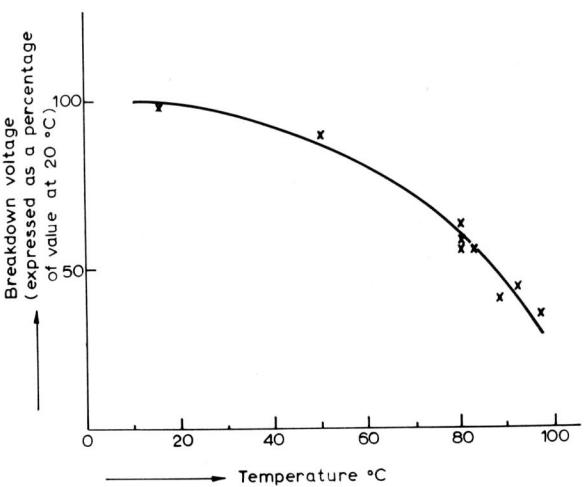

Fig. 6.
Influence of temperature on the anodic breakdown potential of commercial purity titanium in brine.

electrode potentials and anodic breakdown voltage and no instances of failure have been reported. Breakdown is usually of greater importance in the cathodic protection of large structures, such as pipelines. Because of the area to be protected, large areas of uncoated titanium have to be used to make the protection economic. Consequently a large voltage has to be applied to impose the desired throwing power. Even here however the problem can be overcome by attention to design.

The breakdown voltages of niobium and tantalum offer a much greater freedom of use as substrates, but the cost of the anode is then appreciably higher.

6.3 ANODE DESIGN

Anode design is important for a variety of reasons. A first consideration, in order to promote the highest electrochemical efficiency and uniform wear of the coating, is to ensure a uniform working electrode potential across the operating part of the surface. Often this is difficult to achieve because of the shape of the electrode and its position with respect to the counter electrode, *e.g.* a rod-shaped anode in a tank will usually exhibit a higher current density at the anode tip. In more symmetrical situations, such as in chlorine cells, a guiding first rule is to ensure the resistive path between the anode stem or feeders to all parts of the coated anode surface is low enough for variations of only a few tens of millivolts across the working surface, with maximum passage of current. A common form of chlorine anode operating at 10 kA/m² or more consists of a massive central stem which is welded to thick bar or plate sections acting as current distributor. Attached to the distributor are a number of less massive pieces, such as rods, blades or mesh, which support the noble metal coating. Because the resistivity of titanium is relatively high for a metal, 60×10^{-6} Ω cm at 85°C, use is sometimes made of copper-cored or aluminium-cored bars for principal current-carrying sections.

A quite separate issue in design of electrodes for chlorine cells is that of chlorine bubble release, and attention to this factor has resulted in significant decreases in cell voltage compared with the use of graphite anodes, and is additional to the decrease brought about by a lower chlorine overpotential, which is discussed in a later section. A range of industrial anode designs are shown in Figs. 7 and 8.

A further feature of anode design involves the actual manufacture. For plant requirements of several thousand anodes, the process of construction requires as much simplification as possible, consistent with meeting the various electrochemical factors and mechanical considerations such as rigidity and flatness. Flatness is an essential factor for electrodes where anode/cathode separation may be as small as 2 to 3 mm only.

6.3.1 Anode coatings

An essential requirement of noble metal coated titanium is that of adhesion of the coating, which must be maintained throughout the useful working life of the electrode. Coatings are mostly applied to a roughened titanium surface, roughening being effected either mechanically or chemically[321-323].

For the deposition of platinum by electroplating, a wide variety of bath com-

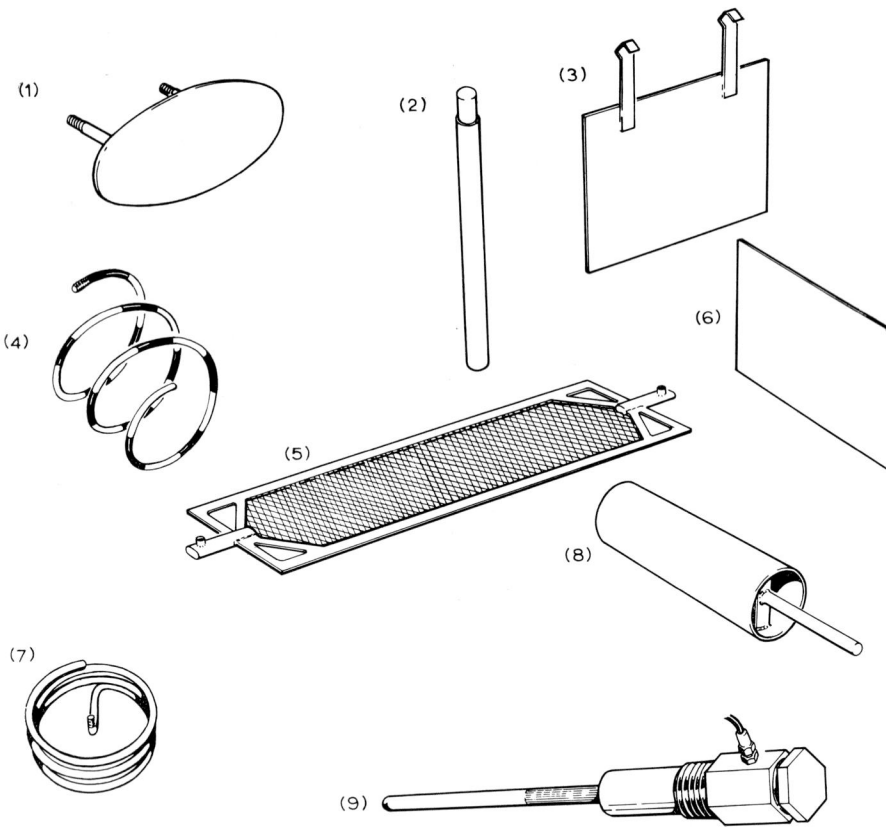

Fig. 7.
Illustration of a range of anode shapes used in industrial electrochemical processes. (1) Ship anode; (2) copper-cored rod for peroxidation reactions; (3) anode used in gold plating; (4) copper-cored skip plated anode for cathodic protection of water boxes; (5) Expamet coated anode from electrodialysis cell (by courtesy of Wm. Boby and Co. Ltd.); (6) coated plate anode from bipolar hypochlorite cell; (7) auxiliary anode used in nickel plating of the inside of kettles; (8) anode used in recovery of silver from spent photographic solution (by courtesy of Photomec Ltd.); (9) standard cantilever rod anode used in cathodic protection of *e.g.* steel ducting.

positions have been formulated[324-327], and some of the principal ones known to have been used in commercial production of electrodes include:

(i) Diammino-dinitrito-platinum (platinum "P" salt)
(ii) Chloroplatinic acid
(iii) Alkali hydroxyplatinates
(iv) Bromides[326, 327]

Other proprietary baths based upon these or alternative chemical compositions would doubtless be satisfactory. As with plating of other metals, the type of bath and conditions of plating impart different characteristics. Some deposits are much

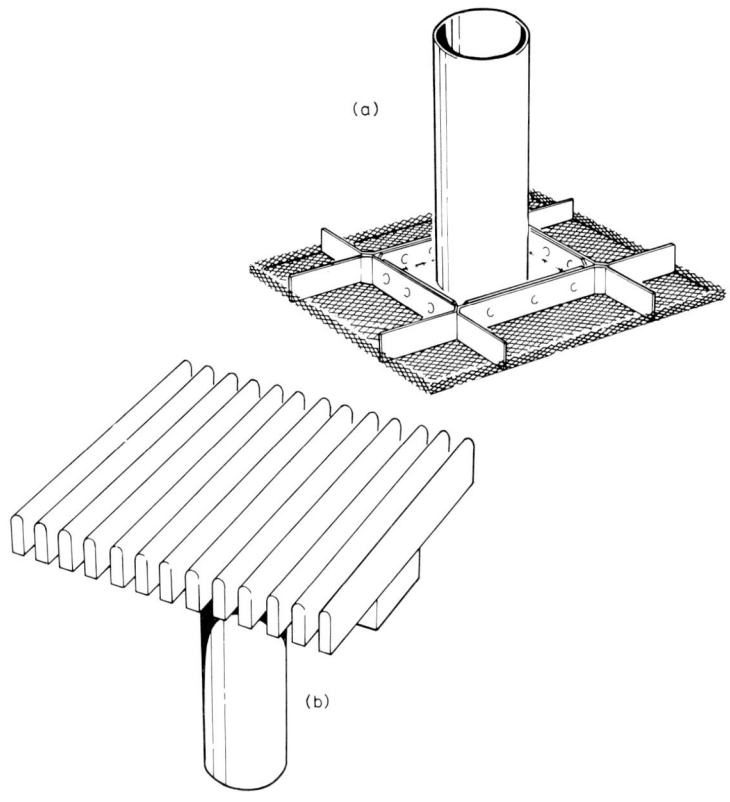

Fig. 8.
Anode designs for use in mercury-type chlorine cells, incorporating noble metal coated titanium.
(a) Expanded mesh; (b) strips.

harder than others, some are more stressed than others. Some plating baths are much more manageable to operate over long periods than others, although all need close control. The "P" salt solution bath leads to a relatively soft unstressed deposit while that from the alkali hydroxyplatinates contains more dissolved hydrogen* and is consequently harder—up to 400 V.P.N.[300]—an attribute which is favoured in applications where the anode needs a measure of resistance to erosion as well as electrochemical forms of wear. With most of these baths and a suitable choice of conditions, thicknesses of up to 25 microns can often be deposited, but an economic life is usually obtained at a much lower value of 2 to 5 microns.

Other noble metals deposited by electrodeposition include rhodium, palladium, ruthenium and iridium. The efficiency of deposition is in some instances as low as

---

* The authors find this surprising in view of work by Schuldiner and Warner[328] who showed that dissolved hydrogen, termed "dermasorbed" hydrogen, readily diffused to the surface of a platinum electrode under anodic conditions and was subsequently oxidised away.

10% even in well regulated baths. A notable step was made recently by Tyrrell[326, 327] with the introduction of noble metal plating from bromide-based baths, which allowed improvements in the deposition of iridium in particular. By suitable attention to bath composition and temperature, alloy plating of different compositions, such as platinum–iridium and platinum–ruthenium, can be achieved.

Method of deposition of coatings other than electrodeposition include metal spraying, vacuum evaporation, electrophoresis and thermal decomposition of solutions or paints containing finely dispersed or soluble noble metal compounds. The more recently perfected application methods vary from one manufacturer to another, but some insight into the types of procedure involved can be gleaned from patent literature. Undoubtedly the most significant method of deposition is based upon thermal decomposition of paints to provide a coating of high surface area. The patents by Angel and Deriaz describe the application of noble metals in the form of a metal-organic formulation which is applied to the surface of the titanium by brush, or by spraying, roller coating, etc. The solvents in the paint are dried off by a preliminary heating in air, and subsequently the organo complex is carbonised by heating further to higher temperature. A carbon residue of the organo complex may remain but it is less than 1% of the final deposit. Exact times and temperatures of the air stoving treatments depend upon the composition of the paint. Actual composition of such coatings is difficult to define, reflection electron diffraction results not conforming to the solid solution of elemental platinum but rather to that of oxide. By the thermal decomposition route it is readily possible to make a wide range of alloy coatings by mixing the required proportions of organo-metal salts prior to application of the paint[307].

Alternative methods of producing high surface area coatings have been suggested. In one such route the platinum layer is alloyed with mercury[329] and the layer subsequently heated in air to volatilise the mercury. The non-uniform penetration of the mercury thus leaves a finely divided surface structure. In other methods, less noble metals such as copper are diffused into the platinum layer, or alternatively coplated with the platinum[330] and subsequently leached out chemically or electrochemically.

6.3.2 Overpotential characteristics

The literature on the anodic polarisation characteristics of platinum in a range of electrolytes is considerable and outside the scope of the present chapter. Some of the features relating to the range of noble metal coated titanium anodes are embodied in Fig. 9a, in which the relationship between electrode potential and logarithm of current over the range of current of commercial interest is indicated. The data in the first of these Figures shows the essential difference between the so-called "high" and "low" overpotential type surfaces, which amounts to as much as 500 mV. Surface area determination on a range of materials[331], using a method

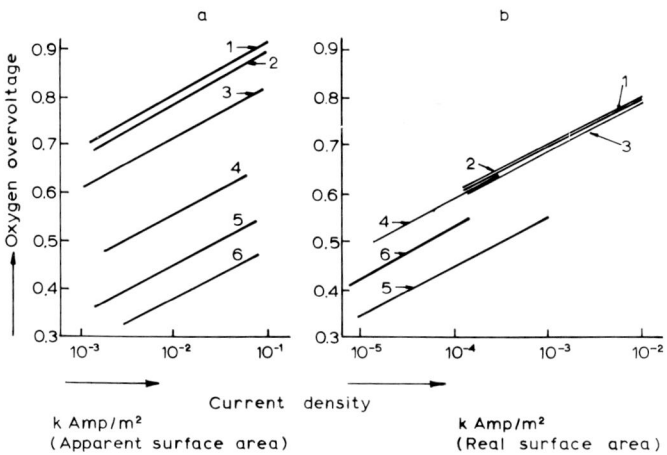

Fig. 9.
Oxygen overvoltage on various platinum and platinum coated electrodes in 1 M sulphuric acid at 25°C.

	Surface roughness[331]
(1) Pt wire	2.0
(2) Pt foil	2.7
(3) Electrodeposited Pt on titanium	4.3
(4) Pt deposited by thermal decompn. of paint soln.	190
(5) Pt + 0.05% Rh deposited as above	97
(6) High surface area Pt on Ti, prepared according to ref. 330	2740

involving the measurement of the capacitance of the electrical double layer between the metal and the electrolyte, shows that most of these overpotential differences can be accounted for by area of the working surface, see Fig. 9b. An exception to this rule is the addition of inherent low overpotential metal, such as ruthenium, to the coatings.

It must be made clear that the present data on overpotential represents a considerable simplification of the behaviour of platinum in aqueous electrolytes. In raising the electrode potential of clean platinum, the metal gradually covers with an adsorbed layer of oxygen, and at potentials as low as + 0.9 to + 1.0 V(NHE) an oxide or hydroxide layer is beginning to form. Such passive film is detectable by ellipsometric methods[332-336] and, depending on electrolyte composition and temperature, may reach several monatomic layers in thickness. As potential of the electrode rises further, the composition of the passive layer changes to a higher oxide level[337] with an accompanying change in Tafel relationship[338-340] to yield higher working electrode potentials. It is an objective in choosing low overpotential electrodes to select a composition and physical form of electrode in which rises of overpotential of this nature do not take place, and in this respect 70/30 platinum/iridium allows a greater range of working electrode potential before this takes place, compared with pure platinum[340].

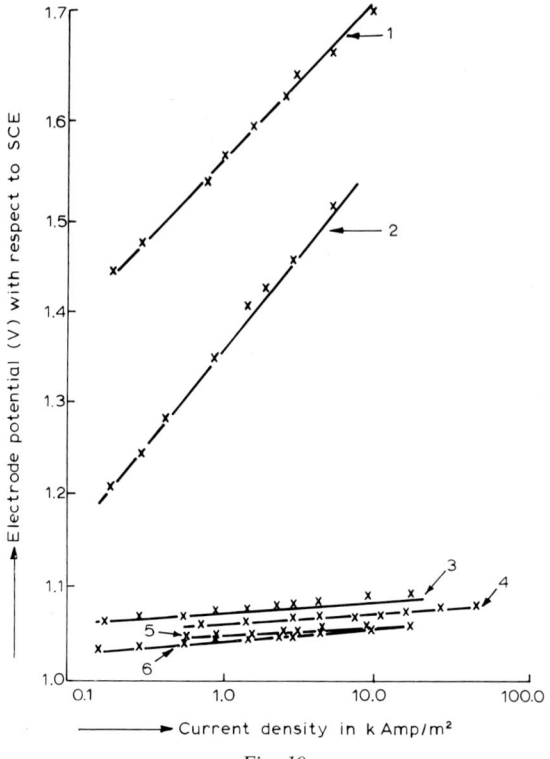

Fig. 10.
Tafel plots for various anode surfaces in acidified, flowing, 22% brine at 70°C (uncorr. for surface area). (1) Electrodeposited Pt on Ti; (2) graphite; (3) Pt coating on Ti, thermal decompn. of a paint; (4) 70/30 Pt/Ir paint deposit on Ti; (5) oxidised Ru coated on Ti.; (6) bulk Ru.

In Fig. 10 are illustrated a range of Tafel characteristics for graphite, platinum electroplated onto titanium, and also some of the more recently developed coatings suitable for operation at very high current densities in mercury-type chlorine cells, and it will be noted that overpotential levels of the latter remain well below the 100 mV level, even at current densities of 10 to 30 kA/m^2.

6.4 Application of noble metal coated titanium anodes in particular industrial electrochemical processes

6.4.1 Chlorine

Since the formulation of the patents in 1958 on platinised titanium-type anodes, much thought has been given to such anodes in both mercury and diaphragm-type cells, and frequent mention has already been made to particular anode requirements for the different types of cell. Low overpotential types of pure platinum have been found wanting for all types of cell, both in respect of rising overvoltage, high platinum wear rate and, in the instance of mercury-type cells, a lack of durability during adventitious electrical shorting to sodium amalgam.

6.4.2 Diaphragm cell

For diaphragm-type cells much improved and economically acceptable characteristics are derived from using paint-deposited 70/30 platinum/iridium coatings, and laboratory and plant trials at current densities of 2 kA/m² provide evidence of wear rates in the region of 55 mg of coating per tonne of chlorine produced. This represents a wear rate in terms of micrograms per ampere hour of about 0.067 and means that, in theory, an anode of initial loading of 10 g/m² can be expected to operate at 2 kA/m² in excess of six years.

The introduction of noble metal coated titanium anodes is limited not merely by the need to build up experience on consistency of anode behaviour and optimisation of operating conditions, but by other factors, notably for diaphragm cells the choice of a stable diaphragm material to cope with the use of a dimensionally stable anode. In comparison with mercury-type cells, the full transition to titanium based anodes in diaphragm cells is more complicated. The full advantage of the metal anode for diaphragm cells will only be realised with a complete redesign of such cells. This redesign stage has already been achieved for chlorate cells. Despite this the economic advantages are such that replacement of graphite even in existing diaphragm cells is worth while and has already commenced.

It is interesting to note that "ruthenium oxide" type anodes are stated as not being used in diaphragm cells as conditions permit the use of 70/30 platinum/iridium anodes. From this it can be inferred that ruthenium oxide type anodes are more expensive. However, it is not, as yet, possible to quote a price for these anodes because of the unsettled nature of the market.

6.4.3 Mercury cells

Noble metal coated titanium anodes for mercury cell operation has awaited the development of surfaces showing improved wear characteristics and a variety of other attributes already discussed, such as resistance to amalgam shorting and current reversals. Evidence accumulating from laboratory and plant trials suggests that oxidised ruthenium coatings and various compositions of mixed titanium and ruthenium oxides meet these requirements, and the way now seems clear to a very considerable installation of this type of titanium-based anode. The benefits to be derived from this change-over are mainly concerned with a decrease in operating cell voltage with K factor, expressed in the formula $V = A + KI$, where $V =$ cell voltage, I is the current in kA/m² and A is a constant, in the region of 0.08–0.09.

Current densities in this type of chlorine cell may be as low as 4 kA/m², but new cells are being considered with current densities of 10 kA/m² and higher. As the percentage of anode working surface per geometrical surface area may only be a half or less, depending upon the design which allows optimum bubble release, the actual current density on the coated metal surface may rise in excess of 20 kA/m². For coatings of oxide in the region of 10 to 20 g/m², this means a wear rate at such high current densities of 20 to 40 mg/ton of chlorine (0.029 µg oxide/A h) and the evidence accumulating is that these wear rates are maintained over long periods and may even be of a lower level.

6.4.4 Chlorate cells

Platinum electroplate and 70/30 platinum/iridium paint deposited coatings have been considered for chlorate electrolysis[312, 313, 339, 340]. Krebs et Cie S.A.[312] are reported to observe loss rates of 500 mg/tonne of sodium chlorate produced for 70/30 platinum/iridium coatings in preliminary trials, but 2 years protracted experience with this same type of coating has led to estimates of 220 mg/tonne (approaching 0.1 µg/A h). At least two large chlorate plants are now equipped with noble metal coated titanium anodes.

6.4.5 Hypochlorite cells

Some of the first commercial electrodeposited platinum coated titanium anodes were fitted into small hypochlorite dosing cells used for sterilisation purposes with capacities of a few kg/h of available chlorine. These were simple monopolar cells, and overvoltage was of secondary importance compared with the ability to operate at high current density, with long life, robustness and compactness. No data are available on platinum wear rate, but it is perhaps not without significance that this type of cell continues to be used. A later version of cell using platinised titanium was of bipolar construction[308, 341, 342]. These cells were regarded as sufficiently successful for the building of cells having outputs of up to 9.07 kg/h of available chlorine. Wear rates for one such unit operating at 5.0 kA/m² over 2 years, incorporating electrode plates of size approx. 10 × 7.5 cm has been reported as not greater than 0.05 µg/A h. A number of problems, mainly in design, have slowed down the rate of development in this field, and it is suspected that the rate of loss of platinum rises as the temperature falls. This loss at low temperatures is a matter for conjecture, because the effect has not been observed, for example, in cathodic protection situations. A recent article by Marshall and Millington[342] puts forward the view that loss in hypochlorite cells at low temperature is a consequence of increased anode potential resulting from operating cells at a given current density. Platinum wear rates reported by Marshall and Millington were in the region of

0.02–0.05 μg/A h over the current density range 1.7–4.9 kA/m² in 3% sodium chloride flowing at 7 cm/sec at 25°C.

Despite this slow development, there is now every reason to expect the use of hypochlorite cells to expand, particularly for plant using seawater at relatively inaccessible areas, *e.g.* where desalination is most required.

6.4.6 Peroxidation reactions

Smooth platinum clad copper or silver rods have been employed for some considerable time in the electrolytic oxidation of perchlorate, persulphate and perborate[295]. As mentioned in the introduction, even the use of foil down to 70 microns thick represents a considerable capital investment in noble metal, and the changeover to less costly and consistently durable anodes is a considerable attraction. Platinum has much to commend itself as a first choice of coating because platinum has functioned adequately before and, with a smooth finish, has the requisite high overpotential required for this type of electrochemical reaction. Metals with inherently low overpotentials such as ruthenium and iridium would have no merit either as coatings in their own right or as an alloying constituent. Wear rate of platinum in these reactions is relatively high at 0.4–0.5 μg/A h, and it follows that the thickness of platinum coating requires to be greater than for most other applications and typically $12\frac{1}{2}$ to 25 microns.

6.4.7 Metal finishing and electrowinning

A wide range of electrolytic conditions is encountered in this category of electrochemical cells, ranging from the very acid to strongly alkaline. Most electroplating and electroforming baths, however, are based upon acid solutions in the 50 to 70°C temperature range, and this sets some limit upon the use of electrodeposited platinum on titanium, not only because of the possibility of undermining of coating through slow corrosive attack on titanium through pores, but because of direct dissolution of the platinum.

The resistance of noble metal coated titanium and tantalum[343] to attack in 5% hydrochloric acid at 75°C, and other acid situations, has been described by Haley[344, 345]. Tantalum, however, remains a better base metal than titanium for highly corrosive situations, but is a good deal more costly.

In strong acid situations the platinum dissolution rate increases[346–348] and each particular metal finishing situation needs to be judged upon its own merits. Unexpectedly the addition of one particular organic agent to a nickel plating solution increased the dissolution rate from 5 to 150 μg/A h.

Despite the complexity of the electrolytes, noble metal coated titanium anodes have been quite extensively used in the metal finishing industry[300] for the plating of precious metals (gold in particular) and other metals, such as copper, nickel and chromium. The introduction of anodes into continuous nickel/chromium plating systems is just beginning.

6.4.8 Electrode boilers

The use of electrical AC heating of solutions has frequently been considered and massive cast iron electrodes are used in some industrial boilers. Platinised titanium subjected to a superimposed 50 Hz AC deteriorates with the formation of layers of non-adherent PtO_2[311]. Alternative coatings for titanium well suited to this application include iridium, 70/30 platinum/iridium, rhodium and oxidised ruthenium[349]. An explanation for this choice, based upon the affinity of precious metals to hydrogen, has since been published by Hoare[350]. No data are available for coating wear rates under power conditions of water heating, but it can be reported that prototype anodes of low loading, 5 g/m², have withstood three month long test periods. Anodes coated with iridium and rhodium to loadings of 55 g/m² operated continuously at 1 kA/m² for 2 years with no visible signs of wear. Subsequent testing of a 70/30 platinum/iridium coated anode with a loading of 4 g/m² has now proceeded satisfactorily at 2 kA/m² for over 1 year, which indicates a wear rate better than 0.5 μg/A h.

6.4.9 Cathodic protection

Systematic evaluation has been conducted of electrodeposited platinum coatings operating at a range of current densities in flowing seawater, with monitoring of anode thickness by *beta* backscatter gauge every six months. Wear rate has been found to decrease linearly with time, and to increase with current density over the range 0.3 to 1.5 kA/m². Wear rates for a range of preparation techniques have been found to lie between 0.3 and 1.0 μg/A h. Various types of power source have also been employed in order to examine the influence of AC current ripple upon activation of platinum, including unsmoothed three phase full wave rectification, unsmoothed single wave full wave rectification and various types of automatic cathodic protection controllers, such as the silicon controlled thyristor system[351], without any evidence of excessive loss rates from the platinum surface. The situation regarding activation of platinum by superimposed AC ripple remains unchanged from the review by Juchniewicz and Hayfield[310] in which, from the practical viewpoint, anodes subjected to ripple of higher frequency than 100 Hz were unlikely to be significantly activated. Certainly no incidence of premature failure of platinised titanium anodes has been reported which may be attributable to such high frequency ripple. Superimposed lower frequency ripple, repeated on/off switching and pure AC of 60 Hz and lower frequency are all unacceptable if persistently applied. In such unusual circumstances the alloy plating of 70/30 platinum/iridium would be the obvious choice, and long term monitoring of the wear rate of such an alloy put down as an electrodeposit has not shown wear rates significantly different from that of pure platinum.

6.4.10 Organic electrosynthesis

Types of anode system likely to be used are under review[352, 353] but the materials of construction are uncertain. Noble metal coated titanium is likely to find use but choice of the most suitable coating probably requires considerable investigation. Experience in other fields of usage of anodes involving organic additives suggests that the wear rate of pure platinum may be unacceptably high because of complexing reactions. Mention has already been made of the activating influence of one organic additive to a nickel plating solution[300], giving rise to wear rates of many μg/A h. Experience in the cathodic protection of a mild steel vessel containing a fermenting solution of organic sugars in $\frac{1}{4}$% sodium chloride produced unexpectedly high wear rates of hundreds of μg/A h. Later, more systematic experimentation revealed that electrolysis of saccharose gave rise to the strong activation of platinum anodes, with alloys of 70/30 platinum/iridium, iridium and rhodium, both coated on titanium and in wrought form, being attacked equally fast. As in the metal finishing field, careful consideration needs to be given to selection of anode material to meet the needs of particular circumstances.

6.5 Conclusions

From the contents of this review chapter, it will be appreciated that the subject of noble metal titanium anodes, which had its origin little more than 10 years ago, has already widened into an industry involving large capital sums. So eager is the expanding electrochemical industry to absorb the advantages accruing from permanent anodes, that empirical development has far outstripped the scientific basis for many of the coatings and particular applications to which they are best suited. The list of possible coating types is far from exhausted, and the field remains one open to vigorous development in both the commercial and academic sense. Two recent publications[407, 408] give some idea of the interest generated by noble metal coated anodes.

7. Cathode materials

While having to satisfy all the criteria laid down at the beginning of the chapter, the conditions where anodic dissolution are concerned are far less stringent, and there is a wide choice of possible cathode materials. Having regard to anodic dissolution, the process designer frequently opts for a cathode material which corrodes (though only slightly) at open circuit, but which is "safe" under the cathodic overpotential of the operating process. Mild steel cathodes form the best example of this as used in the chlorine industry. Wherever the cathode reaction is hydrogen evolution, the designer is wary of hydrogen embrittlement. This not only reduces the mechanical strength of the electrode, but also its catalytic activity. Even so, it is an acceptable penalty.

Where the cathodic reaction is hydrogen evolution in an aqueous electrolyte, the work of Bockris and Conway[354] enables us to see which materials are the best electrocatalysts, and by optimising this factor on a cost basis, the most viable material can be selected. Platinum is the best electrocatalyst, though on price considerations it is excluded. Its near neighbour in the Periodic Table is iron and this is often the preferred choice. Where a specific cathodic reaction other than hydrogen evolution is desired, we are fortunate in having metals of the Pb–Hg–Tl types, with extremely high overvoltages for hydrogen evolution. As in polarography, this gives a potential range in which other reactions can be allowed to occur. The use of mercury cathodes is dealt with in the Chlor–Alkali Industry chapter. Piperidine manufacture is an example of the use of lead cathodes although the reader should also refer to the chapter on miscellaneous processes. Molten lead has been used as a cathode, as have its alloys (see Philblack or Szechtmann processes) but these are now of historical value only. In electrowinning of metals, the product metal acts as a cathode itself, though "starters" have to be used.

Considering the principles of cathode selection just outlined, it is hardly surprising to find that iron, mild steel and the stainless steels find extensive application as cathode materials in electrochemical manufacturing processes. Iron is used in sodium electrowinning (Down's cell), water electrolysis and periodate manufacture. Chlor–alkali, chlorate, perchlorate manufacture and magnesium electrowinning utilise mild steel cathodes, although in the latter a modification to improve wettability has been proposed[355]. Some attempts have been made to improve the electrocatalytic properties of cathode materials for aqueous processes, i.e. to lower the hydrogen overvoltage by surface roughening[356-361]. Stainless steel is used for cathodes in perchlorate manufacture, using lead dioxide anodes[362], and manganese, chromium and titanium electrowinning.

Pure aluminium cathodes are used in cadmium and zinc electrowinning. Recently however aluminium alloy cathodes have been reported with superior properties[363-365]. Watanabe and Karashima[363] have patented an aluminium alloy containing 0.3–3.0% magnesium and 2.0% silicon, which is annealed at 150–350°. They claim good mechanical properties. Znamenski and Stender[364] claim better resistance to corrosion in zinc electrowinning for their alloy containing 0.08–0.15% each of Ti and Zr. Euchev[365] reports that an alloy containing 0.1% Ti, 0.05% Cr and 0.05% Si in zinc electrowinning enables easy separation of the zinc deposit from the cathode.

Ware[366] has carried out an investigation into cathode materials for zinc electrowinning. He found that zinc, which eliminates the need for stripping the deposit, could be used but it underwent extensive corrosion at solution level and it would be necessary to use a thick lead-in. Titanium was found unsuitable but zirconium cathodes decreased the corrosion of the deposit and resulted in poorer deposit adherence, factors which lend themselves to easier mechanical stripping of the deposit.

Tember and Gladyshev[367] have investigated the possibility of using a pure graphite rod in the production of high purity metals by electrolysis of the sulphates, chlorides and perchlorates of a series of non-ferrous and rare metals. Graphite is occasionally used as a cathode in chlorate manufacture. The use in aluminium electrowinning is however the major application of carbon cathodes.

The carbon cathode forms the cell lining and is installed either as prebaked blocks, of a type described by Bacon[368], or as a rammed monolithic lining, depending on the size of the cell[369]. The trend is to the former however[370], as they offer higher density, higher operating strength, lower porosity and lower resistance over the latter.

In operation the carbon cathodes are not completely inert and are slowly consumed: by sodium attack; by mechanical disruption, caused by electrolyte penetration into the cathode pores followed by subsequent crystallisation; by aluminium carbide formation; and by slow oxidation, due to traces of air diffusing through the insulation.

Sodium attack of the cathode material stems from the simultaneous liberation of sodium metal with the aluminium metal. The liberated sodium permeates the cathodes quite rapidly[371, 372], although the rate and degree of penetration is greatly dependent on the cathode pretreatment and the type of carbon employed, graphite and electrically calcined anthracite being more resistant to the initial impregnation. Dewing[373] has proposed that the sodium metal permeates the cathode material by transportation through the layer planes of the carbon crystallite and not by diffusion through the pores. Whatever the mechanism of the penetration there seems little doubt that the migration of the sodium precedes the absorption of the cryolite into the cathode and it is believed that the combined effects of the sodium penetration and the subsequent cryolite absorption cause the formation of sodium fluoride and aluminium carbide within the cathode according to the reaction:

$$12 \, Na + 4 \, Na_3AlF_6 + 3 \, C = 24 \, NaF + Al_4C_3$$

Carbon cathodes absorb up to nearly a third their own weight of cryolite in their lifetime[374], most of which is believed to occur in the first few months of cell life. The fluorine content of this absorbed electrolyte is mostly recovered by leaching the old cathodes but the recovery costs and the added cryolite inventory add to the aluminium production costs. The crystallisation of the absorbed cryolite in the cracks and voids of the cathodes has been discussed by Waddington[371]. It is suggested that crystallisation results in structural disruption in a manner analogous to the "frost heave" phenomenon that occurs during the formation of ice crystals in certain soils. Panebianco and Bacchiega[375] have shown the consequences of absorption in the cathode lining pores to be expansion and disintegration. Rapoport and Samoilenko[376] showed the occurrence of a similar phenomenon with carbon blocks and in addition that the deformation could be reduced by the addition of 10% of calcium fluoride. Watanabe et al.[377] have studied the various types of cathode lining failure resulting from deformation. They conclude that failure is due to a thermal expansion gradient and swelling occurring from the top of the lining to the bottom.

Various workers[376, 379, 381] have rated cathode materials for aluminium electrowinning. Rapoport and Samoilenko[376] rated cathode materials by a laboratory test of their ability to resist deformation when subjected to alkali metal attack during electrolysis in molten cryolite. They accelerated the deformation with potassium fluoride and found that graphite, calcined anthracite and their mixtures were the best materials. Modifications of the above test have been reported as being used by both carbon manufacturers and aluminium producers[368, 378]. Muller and Shea[379] confirmed Rapoport and Samoilenko's findings when determining the resistance of cathode materials to particle degradation in molten sodium at 900–1000°C. In addition they observed that graphitisation of all calcined materials resulted in improved resistance to the sodium attack.

Balazs and Orkenyi[380] whilst reviewing methods of determining carbon cathode breakdown have also observed the beneficial effects of graphitisation. Best results were obtained with 100% dry graphite. Assumptions developed from observations of plant cells and the results of laboratory tests also indicate the superiority of graphite for aluminium cell cathodes. Bullough et al.[370] have recently discussed the advantages and disadvantages of graphite cathodes for aluminium cells. Advantages claimed are:

(a) A reduction in the rate of cryolite absorption, attributed to the low wettability of graphite in cryolite alumina melts, and a reduction in the total cryolite absorption, due to the increased density of graphite. The absorption of cryolite has been related by Beljajev[382] to the porosity and pore size distribution of cathodes. Graphite samples taken from pilot plant cells have shown a decrease of up to 30% in cryolite absorption compared with monolithic carbon.

(b) A reduction in the likelihood of the "frost heave" phenomenon resulting from the high thermal conductivity of graphite[383] which causes a more uniform temperature distribution throughout the cathode.

(c) A reduction in the initial cathode voltage drop of about 80 mV due to the low electrical resistivity of graphite. Graphite cathodes have been in use in large reduction cells since 1961[370], however there are some disadvantages.

(a) Graphite is between two or three times more expensive than carbon black.

(b) The brittle nature of graphite can cause problems in the connection of the current collector bars.

(c) Graphite is relatively soft and cell life may be limited by mechanical abrasion.

Balazs and Orkenyi[380] also observed that graphite and graphite skeleton electrodes were resistant to the effects of lithium salt and potassium chloride additions where carbon electrodes were destroyed. Pre-electrolysis of samples in the presence of lithium salt resulted in immunity to attack and it was inferred that lithium forms interstitial compounds with carbon. Rapoport et al.[384] showed that lithium, as all alkali metals added as salts to the aluminium electrolytic cell, penetrated into the cathode lining. The deforming effect of the penetration however decreased with increasing temperature and ultimately ceased. Additions of lithium fluoride were found to reduce deformation considerably. The basic difference between lithium and all other alkali metals was attributed to its capacity for spontaneous formation of lithium carbide, Li_2C_2, conditions of 500–800°, relatively high lithium content, and a relatively long reaction time being essential. Zhemchuzina and Nyurenberg[385] have investigated the effect of additions of 10 mole% of BeF_2, MgF_2, BaF_2, NaCl, $MgCl_2$, and KCl on cathode stability. The presence of the fluorides of Be, Mg and Ba had a protective action, probably due to carbide formation on the cathode surface, whilst KCl had a harmful effect.

The rates of diffusion into the cathode linings of various elements were studied by Dell[372]. Sodium was found to diffuse rapidly whereas the diffusion of fluorine, calcium and aluminium occurred to a lesser extent. Sodium diffusion increased with current density but if silicon nitride bonded silicon carbide cathodes were used, diffusion rates were found to be independent of the current. However, only alkali metals and calcium diffusion were found to produce swelling.

The use of silicon carbide type refractories instead of carbon has come into limited use in recent years, the disadvantage of greater initial cost being offset by improved heat dissipation and elimination of sodium diffusion. The borides and carbides of transition elements, and in particular titanium and zirconium borides and carbides, have recently been suggested as alternatives to the carbon cathodes in aluminium electrowinning.

Ransley and co-workers[386-390] have pioneered the boride-carbide cathodes and their work has generated interest by other workers[391-396]. Ransley[389] has pointed out the advantages of the boride-carbide cathodes over conventional carbon cathodes. With the boride-carbide cathodes, the voltage drop in aluminium cells can be considerably reduced, new types of cell become possible, the cell lining can be made neutral and magnetic fields may be reduced or eliminated. Voronin et al.[397] have investigated the corrosion resistances of pressed specimens of powdered ZrB_2, TiB_2, TiC and B_4C as cathodes in aluminium cells. The corrosion resistance was dependent on the powder purity and the density of the pressed product. The compositions of the most resistant powder mixtures were reported. A possible use of the boride-carbide cathodes has been reported for uranium electrowinning,[398] however, the cathodes would have to be buried in the cell linings to avoid contamination.

Although the aforementioned are the main industrial cathodes, with the inclusion of mercury cathodes in chlor–alkali manufacture, the use of a few other materials[399–403], mostly liquid metals (Cd, Zn, Pb and Hg), have been recorded. Titanium cathodes[404] have been suggested for cobalt electrowinning from cobalt sulphate solutions, a patent has recently been taken out for tungsten and titanium carbide cathodes[405], formed by plasma arc spraying, for HCl electrolysis, and bronze cathodes[406] are reported for perchlorate manufacture.

In general, the use of titanium as a cathode material is only now developing. The two problems of anodic dissolution at fairly cathodic potentials following the removal of the protective film, and hydrogen embrittlement, though real, can usually be avoided. Electrocatalytically the metal and even its hydrided surface appear to be about as active as iron.

8. References

1 P. FERCHLAND, Ger. Pat., 206,329.
2 Konsortium für elektrochem. Ind., Ger. Pat., 195,117.
3 P. FERCHLAND, Z. Elektrochem., 9 (1903) 670.
4 ANON., Chem. Eng., 72 (1965) 82.
5 T. OSUGA, S. FUJII, K. SUGINO and T. SEKINE, J. Electrochem. Soc., 116 (1969) 203.
6 G. ANGEL and H. MELLQUIST, Z. Elektrochem., 40 (1934) 702.
7 J. C. SCHUMACHER, D. R. STERN and P. R. GRAHAM, J. Electrochem. Soc., 105 (1958) 151.
8 J. C. GRIGGER, H. C. MILLER and F. D. LOOMIS, J. Electrochem. Soc., 105 (1958) 100.
9 Private communication with Pacific Engineering and Production Co. of Nevada.
10 K. SUGINO, Bull. Chem. Soc. Japan, 23 (1950) 115.
11 E. MULLER and M. SOLLER, Z. Elektrochem., 11 (1905) 863.
12 E. MULLER, Z. Elektrochem., 10 (1904) 61.
13 T. OSUGA and K. SUGINO, J. Electrochem. Soc., 104 (1957) 448.
14 Y. KATO, Brit. Pat., 456,082.
15 Y. KATO and K. KOIZUMI, J. Electrochem. Assoc. Japan, 2 (1934) 309.
16 F. MATHERS, Trans. Am. Electrochem. Soc., 17 (1910) 261.
17 Y. SHIBASAKI, J. Electrochem. Soc., 105 (1958) 624.
18 K. SUGINO and Y. SHIBASAKI, J. Electrochem. Soc. Japan, 16 (1948) 10.
19 H. C. MILLER and J. C. GRIGGER, U.S. Pat., 2,872,405.
20 J. C. GRIGGER, U.S. Pat., 2,945,790, July 19, 1960.
21 H. V. K. UDUPA and K. C. NARASIMHAM, Ind. Pat., 66,195, Dec. 22, 1958.
22 F. D. GIBSON, JR., U.S. Pat., 2,945,791, July 19, 1960.
23 K. C. NARASIMHAM and H. V. K. UDUPA, Proc. Symposium on Electrolytic Cells, Central Electrochemical Research Inst., Karaikudi-3, (1961) 22.
24 K. SUGINO and M. YAMASHITA, J. Electrochem. Soc. Japan, 15 (1947) 61.
25 S. KITAHARA and T. OSUGA, J. Electrochem. Soc. Japan, 10 (1942) 409.
26 Y. KATO, K. SUGINO, K. KOIZUMI and S. KITAHARA, Electrotechnical Journal, Japan, 5 (1941) 45.
27 H. C. MILLER and J. C. GRIGGER, U.S. Pat., 2,813,825, Nov. 19, 1957.
28 K. C. NARASIMHAM, S. SUNDARARAJAN and H. V. K. UDUPA, J. Electrochem. Soc., 108 (1961) 798.
29 K. C. NARASIMHAM, S. SUNDARAJAN and H. V. K. UDUPA, Bull. Nat. Inst. Sci., India, 29 (1965) 279.
30 S. OBARA, I. OHARA and T. SEKINE, J. Electrochem. Soc. Japan, 36 (1968) 291.
31 O. DE NORA, P. GALLONE, C. TRAINI and G. MENEGHINI, J. Electrochem. Soc., 116, (1969) 147.

32 K. Sugino and M. Yamashita, *J. Electrochem. Soc. Japan*, 16 (1948) 123.
33 K. Sugino, K. Shirai, Y. Aiya and S. Fujii, *J. Electrochem. Soc.*, 109 (1962) 419.
34 H. V. K. Udupa *et al.*, *Chem. Age India*, 16 (1965) 491.
35 N. Ramachandran, V. Dhuruvan, S. Sampath and H. V. K. Udupa, *Ind. Chem. Eng.*, 8 (1966) 6.
36 H. V. K. Udupa *et al.*, *Ind. J. Technol.*, 4 (1966) 305.
37 S. Sundararajan, K. C. Narasimham and H. V. K. Udupa, *Chem. Process. Eng.*, 43 (1962) 438.
38 E. A. Dzhafarov and M. Sh. Efendieva, *Azerb. Khim. Zh.*, 5 (1967) 166.
39 E. Muller, *Ber.*, 35 (1902) 2655.
40 H. H. Willard and R. R. Ralston, *Trans. Electrochem. Soc.*, 62 (1932) 239.
41 C. L. Mehltretter, U.S. Pat., 2,830,941, 1958; C. L. Mehltretter and C. S. Wise, *Ind. Eng. Chem.*, 51 (1959) 511.
42 E. Torigai and E. Ishii, *Bull. Osaka Ind. Res. Inst.*, 7 (1956) 195.
43 T. Osuga and K. Sugino, *J. Electrochem. Soc.*, 104 (1957) 448.
44 Sh. Sh. Khidirov, D. P. Semchenko and V. I. Lyubushkin, U.S.S.R. Pat., 217,384.
45 E. A. Dzhafarov, M. Sh. Efendieva, F. G. Bairamov and A. M. Musaev, *Azerb. Khim. Zh.*, 2 (1966) 125.
46 Ching Fa Teng and Yuan-P'u Lee, *Hua Hsueh Tung Pao*, 1 (1962) 51.
47 N. Sato, T. Sekine and K. Sugino, *J. Electrochem. Soc.*, 115 (1968) 242.
48 A. Kunugi, H. Urata and S. Nagaura, *Denki Kagaku*, 36 (1968) 237.
49 N. Sato, T. Sekine and K. Sugino, *Denki Kagaku*, 34 (1966) 119.
50 C. A. Brockman, *Electro-Organic Chemistry*, John Wiley and Sons, Inc., London, 1926, p. 14.
51 S. Chidambaram, M. S. Pathy and H. V. K. Udupa, *Ind. J. Technol.*, 5 (1967) 346.
52 R. Ramaswamy, M. S. Venkatachalapathy and H. V. K. Udupa, *J. Electrochem. Soc.*, 110 (1963) 294.
53 E. A. Dzhafarov, Sh. M. Efendieva, F. G. Bairamov and A. M. Musaev, *Azerb. Khim. Zh.*, 4 (1966) 105.
54 M., G. Potdar and H. V. K. Udupa, *Bull. Acad. Polon. Sci., Ser. Sci. Chim.*, 16 (1968) 39.
55 M. S. Venkatachalapathy, R. Ramaswami and H. V. K. Udupa, paper presented at the symposium on *Electrolytic Cells*, Karaikudi, Dec. 1958.
56 A. Fukusawa, *Tokyo Kogyo Shikenso Hokoku*, 57 (1962) 111.
57 V. G. Khomyakov *et al.*, *Tr. Mosk. Khim. Tekhnol. Inst.*, 32 (1961) 349.
58 N. G. Bakhchisaraits'yan *et al.*, *Zh. Prikl. Khim.*, 35 (1962) 1643.
59 J. A. Whittaker, *J. Electrochem. Soc.*, 109 (1962) 986.
60 K. Sugino, J. Mizuguchi and M. Yamashita, *J. Chem. Soc., Japan*, 67 (1946) 108.
61 A. I. Gladysheva and V. I. Laurenchuk, *Uch. Zap. Tseut Issled. Inst. Olovyan Prom.*, 1 (1966) 68.
62 T. Sekine, *J. Electrochem. Soc. Japan*, 20 (1952) 390.
63 L. Wasilewski, A. Korczynski, B. Kot and E. Mieczkavska, *Zeszyty Nauk. Politech. Slask. Chem.*, 24 (1964) 69.
64 E. A. Dzhafarov, N. G. Bakhchisaraits'yan and M. Ya. Fioshin, U.S.S.R. Pat., 154,247.
65 E. A. Dzhafarov and N. G. Bakhchisaraits'yan, U.S.S.R. Pat., 154,462.
66 H. M. Zimmermann, U.S. Pat., 3,087,870.
67 W. G. Darland Jr., U.S. Pat., 3,033,908.
68 N. G. Bakhchisaraits'yan and E. A. Dzhafarov, *Dokl. Akad. Nauk Azerb. SSR*, 17 (1961) 785.
69 N. G. Bakhchisaraits'yan, E. A. Dzhafarov and G. A. Kokarev, *Tr. Mosk. Khim. Tekhnol. Inst.*, 32 (1961) 243.
70 Brit. Pat., 850,380.
71 S. Kiyohara and Y. Shibasaki, U.S. Pat., 3,318,794.
72 Fr. Pat., 1,483,489.
73 Y. Shibasaki, K. Kujiwara and T. Shimojoh, *Denki Kagaku*, 36 (1968) 59.
74 Y. Shibasaki and K. Fujiwara, *Denki Kagaku*, 36 (1968) 725.
75 S. Obara and T. Sekine, *Denki Kagaku Oyobi Kogyo Butsuri Kagaku*, 37 (1969) 252.
76 J. Billiter, *Die Technische Elektrolyse der Nichtmetalle*, Springer-Verlag, Wien, (1954), p. 358.

77 F. Jeitner, *Chem. Ing.-Tech.*, 34 (1962) 353.
78 D.R. Pat., 157,122.
79 D.R. Pat., 193,367.
80 R. Bauer, *Chem. Ing.-Tech.*, 34 (1962) 376.
81 T. Matsamura, R. Itai, M. Shibuya and G. Ishi, *Electrochem. Technol.*, 6 (1968) 402.
82 Private communication with T. Matsamura, Japan Carlit.
83 Brit. Pat., 16,902.
84 U.S. Pat., 1,079,079.
85 U.S. Pat., 1,226,121.
86 N. V. Kuronova, *Sb. Tr., Vses. Nauchn. Issled. Inst. "Goznaka"*, 76, No. 5, (1967).
87 M. Yasuda, *J. Tokyo Chem. Soc.*, 40 (1919) 413.
88 F. Vogel, *Metallboerse*, 19 (1929) 761.
89 G. A. Volin and D. D. Kaganov, *J. Chem. Ind. (U.S.S.R.)*, 16 (1939) 37.
90 Dutch Pat., 85,519.
91 Ger. Pat., 1,091,545.
92 Ger. Pat., 1,068,675.
93 U.S. Pat., 3,232,858.
94 T. Nagai, T. Kishi and T. Takei, *Denki Kagaku*, 28 (1960) 448.
95 T. Nagai, T. Kishi and T. Takei, *Denki Kagaku*, 28 (1960) 213.
96 T. Nagai, T. Ito and T. Takei, *Denki Kagaku*, 28 (1960) 149.
97 M. DeKay Thompson and T. C. Atcheson, *Trans. Am. Electrochem. Soc.*, 31 (1917).
98 U.S. Pat., 3,294,667.
99 U.S. Pat., 3,305,539.
100 U.S. Pat., 3,103,484.
101 Brit. Pat., 902,023.
102 H. Yamamoto, Y. Kokubu, K. Nezu, T. Nagai and T. Takei, *Denki Kagaku*, 33 (1965) 561.
103 H. Yamamoto, Y. Kokubu, K. Nezu, T. Nagai and T. Takei, *Denki Kagaku*, 33 (1965) 501.
104 H. C. Howard, *Trans. Am. Electrochem. Soc.*, 43 (1923).
105 T. Yoshida, *J. Chem. Soc. Japan, Ind. Chem. Sect.*, 53 (1950) 3.
106 N. A. Maslenikov, *Nauchn. Tr. Akad. Kommun. Khoz.*, 20 (1963) 97.
107 T. Eguchi, *Denki Kagaku*, 29 (1961) 19.
108 T. Eguchi, *Denki Kagaku*, 28 (1960) 664.
109 H. Hock and F. Klawitter, *Metall u. Erz*, 22 (1925) 377.
110 U. C. Tainton, A. G. Taylor and H. P. Erlinger, *Am. Inst. Mining Met. Engrs., Tech. Pub.* No. 221, (1929) 12.
111 C. G. Fink and L. C. Pan, *Trans. Am. Electrochem. Soc.*, 49 (1926).
112 C. G. Fink and L. C. Pan, *Trans. Am. Electrochem. Soc.*, 46 (1924).
113 H. R. Hanley, C. Y. Clayton and D. F. Walsh, *Am. Inst. Mining Met. Engrs., Tech. Pub.* No. 321, (1930) 3.
114 V. G. Ageenkov and S. L. Sosunov, *Tsvetn. Metal.*, 7 (1934) 61.
115 S. A. Pletenev and A. L. Soboleva, *Tsvetn. Metal.*, 8 (1935) 48.
116 P. S. Titov and I. N. Nikonov, *Tsvetn. Metal.*, 7 (1934) 53.
117 T. Ishida, *J. Soc. Chem. Ind. Japan*, 39 (1936) 484.
118 L. Cambi and R. Piontelli, *Chim. Ind. ,Milan*, 20 (1938) 649.
119 T. Shibano, *Nippon Kogyo Kaishi*, 78 (1962) 199.
120 A. E. Koenig, J. U. MacEwan and E. C. Larsen, *Trans. Electrochem. Soc.*, 79 (1941).
121 J. L. Bray and F. R. Morral, *Trans. Electrochem. Soc.*, 80 (1941).
122 U.S. Pat., 2,602,775.
123 G. Z. Kiryakov and V. V. Stender, *Izv. Akad. Nauk Kaz. SSR, Ser. Khim.*, 5 (1953) 91.
124 Brit. Pat., 212,875.
125 C. G. Fink and M. Kolodney, *Trans. Electrochem. Soc.*, 76 (1939).
126 D. Schlain, J. D. Prater and B. Lukens, *U.S. Bur. Mines Rept. Invest.*, No. 3863 (1946).
127 M. I. Kurashvili, *Elektrokhim. Margantsa Akad. Nauk Gruz. SSR*, 2 (1963) 375.
128 E. M. Ungiadze, *Tr. Inst. Prikl. Khim. i Electrokhim. Akad. Nauk Gruz. SSR*, 2 (1961) 161.
129 S. Nishihara, *Rept. Chem. Research. Inst. Kyoto Univ.*, 16 (1947) 10.
130 P. Marx, *U.S. Bur. Mines Inform. Circ.*, 7464 (1948).

131 E. UNGIADZE and D. BOGVERADZE, *Elektrokhim. Margantsa Akad. Nauk Gruz. SSR*, 2 (1963) 367.
132 W. C. AITKENHEAD, Wash. State Inst. Technol., *Bull.* No. 219, June, 1953.
133 O. A. KHAN and G. N. SOSNOSVKII, *Tsvetn. Metal.*, 36 (1963) 32.
134 O. A. KHAN and G. N. SOSNOVSKII, *Tsvetn. Metal.*, 34 (1961) 35.
135 E. M. BAKER and P. J. MERKINS, *Trans. Electrochem. Soc.*, 61 (1932).
136 H. ITO and T. SHIBANO, *Nippon Kogyo Kaishi*, 75 (1959) 183.
137 T. YOSHIDA, K. HARA and T. ARAI, *J. Chem. Soc. Japan Ind. Chem. Sect.*, 56 (1953) 826.
138 Fiat Final Report No. 831.
139 Jap. Pat., 204,853.
140 Jap. Pat., 210,923.
141 S. OKADA and S. YOSHIZAWA, *J. Electrochem. Soc. Japan*, 20 (1952) 471.
142 S. OKADA, M. TSUIKI and Y. UENO, *J. Electrochem. Soc. Japan*, 21 (1953) 568.
143 Jap. Pat., 6602 ('53).
144 S. OKADA, S. YOSHIZAWA, F. HINE, H. HARA, S. ANDO and I. KIMURA, *Men. Fai Eng. Kyoto Univ.*, 24 (1962) 112.
145 Ger. Pat., 1,161,037.
146 M. TSUIKI and Y. UENO, *Gifu Daigaku Kogakubu Kenkyi Hokoku*, 11 (1961) 54.
147 E. SATO, S. SUZUKI and S. YOSHIDA, *Boshoku Gijutsu*, 11 (1962) 436.
148 E. SATO, *Boshoku Gijutsu*, 10 (1961) 2.
149 E. SATO, *Boshoku Gijutsu*, 9 (1960) 436.
150 E. SATO, *Boshoku Gijutsu*, 8 (1959) 367.
151 E. SATO, *Boshoku Gijutsu*, 9 (1960) 152.
152 V. G. BUNDZHE, YU. D. DUNAEV, G. Z. KIRYAKOV and V. F. KOZIN, U.S.S.R. Pat., 217,648.
153 Brit. Pat., 811,404.
154 Jap. Pat., 2959 ('53).
155 Can. Pat., 411,573.
156 Can. Pat., 430,269.
157 U.S. Pat., 2,340,400.
158 U.S. Pat., 1,437,507.
159 C. L. MANTELL, *Carbon and Graphite Handbook*, Interscience, New York, 1968, Chap. 15, p. 249.
160 J. S. SCONCE, *Chlorine, ACS Monograph* No. 154, Reinhold, New York, 1962, Chap. 5, p. 85.
161 V. V. STENDER and O. S. KSENZHEK, *Zh. Prikl. Khim.*, 32 (1959) 110.
162 L. E. VAALER, *Electrochem. Technol.*, 5, (1967) 170.
163 V. V. STENDER, *Electrochemical Production of Chlorine and Alkalies*, Leningrad, ONTI, Chem. Theoret. Press, 1931.
164 O. S. KSENZHEK and Z. V. SOLOVEI, *Zh. Prikl. Khim.*, 33 (1960) 279.
165 N. J. JOHNSON, *Trans. Electrochem. Soc.*, 86 (1944) 127.
166 A. REGNER, *Electrochemical Processes in Chemical Industries*, ARTIA, Prague, 1959, p. 181.
167 K. J. VETTER, *Chem. Ing.-Tech.*, 34 (1962) 362.
168 V. V. STENDER, *Applied Electrochemistry*, Izv. Khar'kovsk Univ., Khar'kov, 1961.
169 D. V. KOKOULINA and L. I. KRISHTALIK, *Electrokhimiya*, 3 (1967) 848.
170 B. SJODIN and G. WRANGLEN, *Electrochim. Acta*, 10 (1965) 203.
171 L. I. KRISHTALIK, G. L. MELIKOVA and E. G. KALININA, *Zh. Prikl. Khim.*, 34 (1961) 1537.
172 R. H. GEISE, L. E. VAALER and A. J. KALLFELZ, *J. Electrochem. Soc.*, 111, 73C, Abs. 170, (1964).
173 F. I. MULINA, L. I. KRISHTALIK and A. T. KOLOTUKHIN, *Zh. Prikl. Khim.*, 39 (1966) 1338.
174 L. E. VAALER, *J. Electrochem. Soc.*, 107 (1960) 691.
175 T. S. FILIPPOV, *Trans. Ukrainian State Inst. Appl. Chem.*, 1 (1936).
176 B. M. BULYGIN, *Zh. Prikl. Khim.*, 31 (1958) 1832.
177 B. M. BULYGIN, *Zh. Prikl. Khim.*, 32 (1959) 121.
178 B. M. BULYGIN, *Zh. Prikl. Khim.*, 32 (1959) 521.
179 H. THIELE, *Z. Elektrochem.*, 55 (1951) 193.
180 T. S. FILIPPOV and N. I. NECHIPORENKO, *Trans. Ukrainian State Inst. Appl. Chem.*, 3 (1937) 104.

181 W. S. IOFFE, *Z. Elektrochem.*, 42 (1936) 71.
182 V. I. EBERIL and T. S. FILIPPOV, *Zh. Prikl. Khim.*, 40 (1967) 2482.
183 W. A. NYSTROM, *J. Electrochem. Soc.*, 116 (1969) 17.
184 I. YA. SIRAK, *Zh. Prikl. Khim.*, 6 (1933) 808.
185 F. FOERSTER, *Elektroch. Wassriger Losungen*, Barth, 4th ed., 1923.
186 M. JANES, *Trans. Electrochem. Soc.*, 92 (1947) 23.
187 M. JACOPETTI, *Rend. Accad. Sci. Fis. Mat. Soc. Napoli*, 10 (1939–1940).
188 T. S. FILIPPOV, A. V. MOTSAREVA and V. A. GRINEVICH, *Abstracts of papers at the 4th Conf. on Electrochem.*, 1956, p. 98.
189 N. I. NECHIPORENKO, Candidates Dissertation, (Khar'kov).
190 R. MURRAY and M. KIRCHER, *Trans. Electrochem. Soc.*, 86 (1944) 83.
191 W. GARDINER, *Chem. Eng.*, 54 (1947) 11, 108.
192 L. I. KRISHTALIK, G. L. MELIKOVA and E. G. KALININA, *Zh. Prikl. Khim.*, 34 (1961) 1543.
193 M. M. FLISSKII, I. E. VESELOVSKAYA and R. V. DZHAGATSPANYAN, *Zh. Prikl. Khim.*, 33 (1960) 1901.
194 M. M. FLISSKII, I. E. VESELOVSKAYA, R. V. DZHAGATSPANYAN and O. V. CHERNYAVSKAYA, *Zh. Prikl. Khim.*, 34 (1961) 2483.
195 I. E. VESELOVSKAYA, M. M. FLISSKII, R. V. DZHAGATSPANJAN and L. V. MOROTJKO, *Zh. Prikl. Khim.*, 36 (1961) 2179.
196 S. TOSIYUKU, *R. Zh. Khim.*, Abstr. No. 57786 (1960).
197 L. I. KRISHTALIK, *Abstracts of Comm. and Papers, Electrochem. Sect., 8th Mendeleev Congress*, 1959, p. 70.
198 L. I. KRISHTALIK, *Zh. Prikl. Khim.*, 34 (1961) 1807.
199 V. I. EBERIL and T. S. FILIPPOV, *Zh. Prikl. Khim.*, 40 (1967) 2488.
200 K. ARNDT and W. FEHSE, *Z. Elektrochem.*, 28 (1922) 376.
201 E. M. KUCHINSKII, N. P. LIPIKHIN and M. M. FLISSKII, *Zh. Prikl. Khim.*, 37 (1964) 460.
202 V. IOFFE, *Zh. Prikl. Khim.*, 13 (1936) 668, 784.
203 M. M. FLISSKII, *Zh. Prikl. Khim.*, 38 (1965) 2815.
204 O. S. KSENZHEK and V. V. STENDER, *Tr. 4-go Chetvertogo Sovesh. po Electrokhim. Moscow*, (1956) 823.
205 L. SPROESSER, *Z. Elektrochem.*, 7 (1901) 971, 1012, 1027, 1071, 1083.
206 O. S. KSENZHEK, Authors Summary of Candidates Dissertation, (Dnepropetrovsk, 1956).
207 B. WALLEN and G. WRANGLEN, *Electrochim. Acta*, 10 (1965) 43.
208 S. OKADA, S. YOSHIZAWA and T. ISHIKAWA, *Zairyo Shiken*, 7 (1958) 111, 250.
209 T. FUKUNAGA, O. SUZUKI and T. MATSUNO, Abs. No. 248, *J. Electrochem. Soc.*, 112, 85C (1965).
210 P. L. WALKER, R. J. FORESTI and C. C. WRIGHT, *Ind. Eng. Chem.*, 45 (1953) 1703.
211 R. W. MAREK and E. A. HEINZ, Abs. No. 141, *Conf. on Carbon*, Case Inst., Cleveland, Ohio, June, 1965.
212 D. G. FITZ-GERALD and B. C. MOLLOY, Brit. Pat., 1376.
213 J. C. BURNS JR., U.S. Pat., 2,820,728.
214 E. KIEFER and W. KRELLNER, U.S. Pat., 2,368,306.
215 C. C. HARDMAN, U.S. Pat., 2,902,386.
216 R. HUNTER, L. STEWART, H. HOUSER and L. DEPREE, U.S. Pat. 1,861,415.
217 G. J. ATKINS, U.S. Pat., 75,114.
218 W. W. GLEAVE, U.S. Pat., 2,433,212.
219 C. FRANK, K. DIETZ, F. PRIVINSKY and E. THIEL, Ger. Pat., 610,652.
220 W. WRIGG and R. ROWLEY, U.S. Pat., 2,685,533.
221 A. R. DE VAIN, Fr. Pat., 567,925.
222 H. W. NICOLAI, Ger. Pat., 845,038.
223 J. PARKER, Brit. Pat., 813,515.
224 S. SUZUKI, Jap. Pat., 2,663 (50).
225 A. PLANCHEN et al., Fr. Pat., 492,735.
226 E. SZARVASY, Ger. Pat., 319,087.
227 A. V. ANTROPOFF, Ger. Pat., 342,794.
228 C. LINDEMANN, Ger. Pat., 187,029.

229 C. Higgins and D. Pritchard, Ger. Pat., 167,041.
230 E. S. Mkrlchyan and G. A. Sarksyan, U.S.S.R. Pat., 201,341.
231 H. Shibata, Y. Yamazaki and K. Sugihara, Brit. Pat., 1,074,042.
232 B. L. Bailey and B. Best, Brit. Pat., 953,504.
233 J. Sucharda and A. Cernihorsky, Czech. Pat., 100,959.
234 A. Heymann, W. Zierner and J. Schuecker, Ger. Pat., 1,112,048.
235 R. Mader and B. Vesily, Czech. Pat., 95,463.
236 W. W. Carlin, U.S. Pat., 2,920,004.
237 C. C. Hardman, U.S. Pat., 2,881,100.
238 D. D. Kaganov and G. A. Volin, U.S.S.R. Pat., 43,876.
239 R. M. Hunter and L. E. Ward, U.S. Pat., 1,927,661.
240 O. A. Laubi, U.S. Pat., 1,779,242.
241 C. S. Lowe, U.S. Pat., 3,046,216.
242 A. Ritter, Swiss Pat., 73,797.
243 H. Sjodzi and J. Muramacu, Jap. Pat., 23,616.
244 B. L. Bailey, U.S. Pat., 3,120,454.
245 B. Morris and J. Schempf, *Anal. Chem.*, 31 (1959) 286.
246 A. Korshunov, *J. Chem. Ind.*, 12, (1935) 384.
247 Nederlandse Zoutindustrie, Brit. Pat., 677,491.
248 Great Lakes Carbon Corp., Fr. Pat., 1,335,152.
249 Kureha Chemical Industry Co. Ltd., Belg. Pat., 634,242.
250 Le Carbone-Lorraine, Fr. Pat., 999,968.
251 G. Eger, *Handbuch der Techn. Elektrochem.*, Vol. I, Part 1, (1961) p. 103.
252 J. Billiter and F. Fuchs, *Handbuch der Techn. Elektrochem.*, Vol. II, Part 1, 1933, pp. 233, 234, 245, 279.
253 W. W. Gleave, Swed. Pat., 128,823.
254 M. M. Jaksic and I. M. Csonka, *Electrochem. Technol.*, 5 (1967) 473.
255 F. I. Mulina, L. I. Krishtalik and A. T. Kolotukhin, *Zh. Prikl. Khim.*, 38 (1965) 2808.
256 F. I. Mulina, L. I. Krishtalik and A. T. Kolotukhin, *Zh. Prikl. Khim.*, 38 (1965) 2812.
257 V. V. Stender, P. B. Zhivotinskii and M. M. Stroganov, *Trans. Electrochem. Soc.*, 65 (1934) 189.
258 T. Inove and K. Sugino, *J. Electrochem. Soc. Japan*, 27 (1958) E55.
259 O. Suzuki, A. Ikeda and S. Abe, *J. Electrochem. Soc. Japan*, 27 (1959) E38, E40.
260 L. S. Genin, *Electrolysis of Sodium Chloride Solutions*, Goskhimizdat, Moscow, 1960, p. 73.
261 H. Smidt and F. Holzinger, *Chem. Ing.-Tech.*, 35 (1963) 37.
262 S. Okada, S. Yoshizawa, F. Hine and Z. Takehara, *J. Electrochem. Soc. Japan*, 26 (1958) E55.
263 L. I. Krishtalik and Z. A. Rotenburg, *Zh. Fiz. Khim.*, 39 (1965) 328.
264 L. I. Krishtalik and Z. A. Rotenburg, *Zh. Fiz. Khim.*, 39 (1965) 907.
265 F. Hine, S. Yoshizawa and S. Okada, *J. Electrochem. Soc. Japan*, 24 (1956) 375.
266 V. L. Kubasov and G. I. Volkov, *Electrokhimiya*, 1 (1965) 1395.
267 P. Drossbach, H. Hoff, P. Schmittinger and J. Schulz, *Chem. Ing.-Tech.*, 37 (1965) 639.
268 D. T. Ewing and H. W. Schmidt, *Trans. Am. Electrochem. Soc.*, 47 (1925) 117.
269 A. Korczynski and R. Dylewwki, *Zeszyty Nauk. Politech. Slask. Chem.*, 35 (1967) 63.
270 A. Korczynski and R. Dylewski, *Chem. Stosowana* Ser. A, 11 (1967) 141.
271 *Ibid.*, 10 (1966) 369.
272 A. Korczynski and R. Dylewski, *Zeszyty Nauk. Politech. Slask. Chem.*, 30 (1966) 65.
273 *Ibid.*, 29 (1966) 67.
274 C. L. Mantell, *Carbon and Graphite Handbook*, Interscience, New York, 1968, p. 310.
275 M. Sen, J. Sejersted and O. Bockman, *J. Electrochem. Soc.*, 94 (1948) 220.
276 R. E. Oehler, *Extractive Metallurgy of Aluminium*, Vol. 2, *Aluminium*, Interscience, New York, 1963, p. 231.
277 M. L. Kronenburg, *J. Electrochem. Soc., Electrochem. Technol.*, 116 (1969) 1160.
278 T. G. Pearson and J. Waddington, *Discussions Faraday Soc.*, 1 (1947) 307.
279 A. J. Beljajev, M. B. Rapoport and L. A. Firsanova, *Metallurgie des Aluminium*, VEB Verlag Technik, Berlin, 1956.

280 R. Schadinger, *Alluminio*, 21 (1952) 252.
281 E. A. Hollingshead and V. A. Braunworth, *Extractive Metallurgy of Aluminium*, 2 (1963) 31.
282 L. N. Antipin and V. K. Dudyrev, *Zh. Fiz. Khim.*, 31 (1957) 2031.
283 J. Thonstad and E. Hove, *Can. J. Chem.*, 42 (1964) 1542.
284 B. J. Welch and N. E. Richards, *AIME Intern. Symp. Extractive Met. Alum.*, New York, 1962.
285 H. Stern and G. T. Holmes, *J. Electrochem. Soc.*, 105 (1958) 478.
286 V. P. Mashovets and A. A. Revazyan, *Soviet Electrochem.*, 2 (1961) 185.
287 V. V. Robozerov and M. M. Vetyukov, *Khim. Tverd. Topl.*, 3 (1968) 128.
288 V. V. Robozerov and M. M. Vetyukov, *Tsvetn. Metal.*, 41 (1968) 6.
289 C. W. Soderberg, U.S. Pat., 1,440,724.
290 C. W. Soderberg, U.S. Pat., 1,441,037.
291 C. W. Soderberg, U.S. Pat., 1,670,052.
292 L. J. J. Janssen and J. G. Hoogland, *Electrochim. Acta*, 14 (1969) 1097.
293 L. I. Krishtalik and Z. A. Rotenberg, *Zh. Fiz. Khim.*, 39 (1965) 168.
294 H. Binder, A. Kohling, K. Richter and G. Sandstede, *Electrochim. Acta*, 9 (1964) 250.
295 P. G. Wranglen, *Tekn. Tidskr.*, H20 (1960) 551.
296 R. H. Stevens, U.S. Pat., 1,077,894.
297 R. H. Stevens, U.S. Pat., 1,007,920.
298 J. B. Cotton, *Chem. Ind. London*, April 28th, (1958) 68.
299 J. B. Cotton, *Platinum Metals Rev.*, 2 (1958) 45.
300 M. A. Warne and P. C. S. Hayfield, *Trans. Inst. Metal Finishing*, 45 (1967) 83.
301 N.V. Curacaosche Exploitatie, Brit. Pat., 885,107.
302 J. B. Cotton, E. C. Williams and A. H. Barber, Brit. Pat. 877,901.
303 G. F. Taylor, *J. Opt. Soc. Am.*, 18 (1929) 138.
304 C. H. Angel and M. G. Deriaz, Brit. Pat., 885,819.
305 C. H. Angel and M. G. Deriaz, Brit. Pat., 984,973.
306 Amalgamated Curacao Patents Co. N.V., Brit. Pat., 869,865.
307 H. B. Beer, Brit. Pat., 964,913.
308 A. F. Adamson, D. G. Lever and W. F. Stones, *J. Appl. Chem., London* 13 (1963) 483.
309 C. Marshall and J. P. Millington, *J. Appl. Chem. London*, 19 (1969) 298.
310 R. Juchniewicz and P. C. S. Hayfield, *Proc. 3rd Intern. Congr. on Metallic Corrosion, Moscow, 1966*, Vol. III English Ed., p. 73.
311 J. P. Hoare, *The Electrochemistry of Oxygen*, Interscience, New York, 1968, p. 173.
312 Anon., *Platinum Metals Rev.*, 13 (1969) 103.
313 Solvay et Cie., to be published.
314 H. B. Beer, Brit. Pat., 1,147,442.
315 H. B. Beer, Brit. Pat., Case No. 649–67.
316 J. B. Cotton, W. R. Bennett, J. A. Bell and P. C. S. Hayfield, Fr. Pat., 1,583,370.
317 V. V. Andreeva and V. I. Kazarin, *Dokl. Akad. Nauk SSSR*, 121 (1958) 873.
318 J. B. Cotton, in L. L. Shrier (Ed.), *Corrosion*, Vol. 1, George Newnes Ltd., London, 1963, Sect. 5.32.
319 B. H. Hanson, *Corrosion Resistance of Titanium*, Imperial Metal Industries (Kynoch) Ltd., Brochure.
320 I. Dugdale and J. B. Cotton, *Corrosion Sci.*, 4 (1964) 397.
321 W. A. Marshall, *Trans. Inst. Metal Finishing*, 44 (1966) 111.
322 C. H. Angell, Brit. Pat., 1,105,388.
323 Ajinomoto Co. Inc., Brit. Pat., 998,709.
324 S. D. Crammer, C. B. Kenaham, R. L. Andrews and D. Schlain, *U.S. Bur. Mines Rept. Invest.*, 7016.
325 F. H. Reid, *Met. Rev.*, 8 (1963) 167.
326 C. J. N. Tyrrell, Brit. Pat., 1,108,052.
327 C. J. N. Tyrrell, *Trans. Inst. Metal Finishing*, 45 (1967) 53.
328 S. Schuldiner and T. B. Warner, *J. Electrochem. Soc.*, 112 (1965) 212.
329 Johnson Matthey and Co. Ltd., Brit. Pat., 957,703.

330 J. B. COTTON and P. C. S. HAYFIELD, Brit. Pat., 1,113,421.
331 L. COWLEY, unreported work.
332 A. K. N. REDDY, M. A. GENSHAW and J. O'M. BOCKRIS, *J. Electroanal. Chem.*, 8 (1964) 406.
333 A. K. N. REDDY, M. A. GENSHAW and J. O'M. BOCKRIS, *J. Chem. Phys.*, 48 (1968) 671.
334 S. SHIBATA, *Bull. Chem. Soc. Japan*, 40 (1967) 696.
335 W. VISSCHER, *Optik*, 4 (1967) 402.
336 P. C. S. HAYFIELD, *Symposium on Recent Developments in Ellipsometry, Nebraska, 1968*, North Holland, Amsterdam, 1969, p. 126.
337 J. SCHMETS, J. VAN MUYLDER and M. POURBAIX, *Atlas d'Equilibres Electrochimiques*, Chapitre IV, Section 13.6, *Platine*, pp. 378–383.
338 E. L. LITTAUER and L. L. SHREIR, *Electrochim. Acta*, 11 (1966) 527.
339 P. VAN LAER, CEBELCOR, Reports CEFA/R.64 and CEFA/R.65.
340 G. FAITA, G. FIORA and J. W. AUGUSTYNSKI, *J. Electrochem. Soc.*, 116 (1969) 928.
341 J. B. COTTON, *Trans. Inst. Chem. Engrs. London*, 41 (1963) 355.
342 C. MARSHALL and J. P. MILLINGTON, *J. Appl. Chem. London*, 19 (1969) 248.
343 E. F. ROSENBLATT and J. G. COHN, U.S. Pat., 2,719,797.
344 A. J. HALEY, Engelhard Industries Inc., *Technical Bulletin*, 7 (1967) 157.
345 M. A. WARNE and P. C. S. HAYFIELD, patent pending.
346 J. LLOPIS and A. SANCHO, *J. Electrochem. Soc.*, 108 (1961) 720.
347 A. N. CHEMODANOV, I. K. MOREZAVA, V. V. GOREDETSLEII, M. A. DEMBROVSKII, V. V. LOSEV and YA. M. KOLOTYRKIN, *Protection of Metals*, 1 (1965) 433.
348 A. N. CHEMODANOV, Y. M. KOLOTYRKIN, M. A. DEMBROVSKY and T. V. KUDRYAVIN, *Proc. U.S.S.R. Acad. Sci.*, 171 (1966) 1384.
349 J. B. COTTON, Brit. Pat., 1,068,732.
350 J. P. HOARE, *Electrochim. Acta*, 9 (1964) 599.
351 W. MATTHEWMAN, J. MALSTER and A. D. WALLACE, *Corrosion Technol.*, 10 (1963) 92.
352 Brit. Pat. Application 23070/66.
353 F. GOODRIDGE, *Chem. Process Eng.*, 49 (1968) 93.
354 J. O'M. BOCKRIS and B. E. CONWAY, *J. Chem. Phys.*, 26 (1957) 532.
355 I. A. BARRANIK, V. P. SHAPOVALOV, L. N. ANTIPIN *et al.*, U.S.S.R. Pat., 203,920.
356 B. N. KABANOV and S. A. ROZENTSVEIE, *J. Phys. Chem. U.S.S.R.*, 22 (1948) 513.
357 E. M. KUCHINSKII and G. N. KOKRANOV, *J. Phys. Chem. U.S.S.R.*, 36 (1962) 251.
358 K. M. GORBUNOVA and L. I. LIAMINA, *Electrochim. Acta*, 7 (1962) 251.
359 C. D. STOCKBRIDGE, P. B. SEWELL and M. COHEN, *J. Electrochem. Soc.*, 108 (1961) 928.
360 J. BRENET, T. MARKOVIC and E. ATLIC, *Werkstoffe Korrosion*, 17 (1966) 2.
361 S. A. ROSENTSVEIE and B. N. KABANOV, *Zh. Fiz. Khim.*, 22 (1948) 1214.
362 D. R. STERN and J. C. SCHUMACHER, U.S. Pat., 2,840,519.
363 H. WATANABE and M. KARASHIMA, Jap. Pat., 1412 ('57).
364 G. N. ZNAMENSKI and V. V. STENDER, U.S.S.R. Pat., 128,145.
365 I. EUCHEV, *Rudodobiv Met.*, 23 (1968) 44.
366 G. C. WARE, *U.S. Bur. Mines, Rept. Invest.*, 6301 (1963).
367 G. A. TEMBER and V. P. GLADYSHEV, *Poluch. Anal. Veshchestv. Osoboi Chist., Mater Vses. Konf., Gorky U.S.S.R.*, 27 (1963).
368 L. E. BACON, *Extractive Metallurgy of Aluminium*, Vol. 2, *Aluminium*, Interscience, New York, 1963, p. 461. Published in *Khim. i Khim. Tekhnol.*, Alma-Ata, Sb. 2 (1964) 221.
369 R. A. LEWIS, *Chem. Eng. Progr.*, 56 (1960) 78.
370 V. L. BULLOUGH, L. O. DALEY and C. J. MCMINN, *Electrochem. Technol.*, 5 (1967) 182.
371 J. WADDINGTON, *Extractive Metallurgy of Aluminium*, Vol. 2, *Aluminium*, Interscience, New York, 1963, p. 435.
372 M. B. DELL, *Extractive Metallurgy of Aluminium*, Vol. 2, *Aluminium*, Interscience, New York, 1963, p. 403.
373 E. W. DEWING, *Trans. Met. Soc. AIME*, 227 (1963) 1328.
374 D. D. BEATTIE, *Extractive Metallurgy of Aluminium*, Vol. 2, *Aluminium*, Interscience, New York, 1963, p. 453.
375 B. PANEBIANCO and R. BACCHIEGA, *Met. Ital.*, 52 (1960) 539.
376 M. B. RAPOPORT and V. N. SAMOILENKO, *Tsvetn. Metal.*, 30 (1957) 44.

377 T. Watanabe, H. Hayashi and F. Mochizuki, *J. Electrochem. Soc. Japan, Overseas Ed.*, 36 (1968) 123.
378 J. J. Vadla, *Proc. Conf. Carbon, 4th, Buffalo, 1959*, Pergamon Press, Oxford, 1960, p. 169.
379 N. W. Muller and F. L. Shea Jr., *Extractive Metallurgy of Aluminium*, Vol. 2, *Aluminium*, Interscience, New York, 1963, p. 417.
380 E. Balazs and J. Orkenyi, *Freiberger Forschungsh.*, B113 (1967) 951.
381 M. B. Rapoport, *Tr. Vses. Nauchn. Issled ovatel Alyumin Magn. Inst.*, 39 (1957) 357.
382 A. I. Beljajev, *Met. Al. Band I*, Veb Verlag Technik, Berlin, 1956, p. 110.
383 D. D. Beattie and M. K. B. Day, *Extractive Metallurgy of Aluminium*, Vol. 2, *Aluminium*, Interscience, New York, 1963, p. 387.
384 M. B. Rapoport, V. I. Kudryavtsev and G. A. Shifman, *Izv. Akad. Nauk SSSR Metal.*, 5 (1967) 151.
385 E. A. Zhemchuzina and G. Ya. Nyurenberg, *Izv. Vysshikh Uchebn. Zavedenii, Tsvetn. Met.*, 8 (1965) 90.
386 C. E. Ransley, Brit. Pat., 784,696.
387 C. E. Ransley, Brit. Pat., 826,635.
388 C. E. Ransley, U.S. Pat., 3,028,324.
389 C. E. Ransley, *J. Metals*, 14 (1962) 129.
390 C. E. Ransley, *Extractive Metallurgy of Aluminium*, Vol. 2, *Aluminium*, Interscience, New York, 1963, p. 487.
391 Ger. Pat., 1,130,607.
392 D. W. Morgan, U.S. Pat., 3,081,254.
393 Brit. Pat., 1,065,792.
394 U.S. Pat., 3,274,093.
395 Neth. Pat., 6,510,275.
396 Ger. Pat., 1,251,962.
397 N. I. Voronin et al., *Tsvetn. Metal.*, 40 (1967) 63.
398 P. D. Piper and R. F. Liefield, *Ind. Eng. Chem.*, 1 (1962) 208.
399 A. Sh. Avaliani, *Tr. Inst. Metal. i Gorn. Dela Akad. Nauk Gruz. SSR*, 7 (1956) 183.
400 T. Batvecas Rodriguez et al., *Cellini Proc. U.N. Intern. Conf. Peaceful Uses of At. Energy, Geneva*, 4 (1958) 73.
401 H. V. K. Udupa, *Bull. Acad. Polon. Sci., Ser. Sci. Chim.*, 9 (1961) 51.
402 H. V. K. Udupa and G. S. Subramanian, *Chem. Ing.-Tech.*, 38 (1966) 868.
403 N. E. Khomutov and T. N. Skornyakova, *Zh. Prikl. Khim.*, 36 (1963) 1521.
404 E. S. Lekht and L. S. Ivanova, *Tsvetn. Metal.*, 39 (1966) 37.
405 Ger. Pat., 62,308 (East).
406 A. Legendre, *Chem. Ing.-Tech.*, 34 (1962) 379.
407 *Chemical Week*, Feb. 15th, 1969.
408 *European Chemical News*, Oct. 31st, 1969.
409 N. T. Thomas and K. Nobe, *J. Electrochem. Soc.*, 116 (1969) 1748.

Chapter 15

Diaphragms and Electrolytes

C. JACKSON*, B. A. COOKE** AND B. J. WOODHALL***

1. Diaphragms

1.1 Introduction

Diaphragms serve to keep apart the products and/or reactants in an electrochemical process, or to maintain a pH gradient. The first-named author describes those diaphragms which might be termed "mechanical" in their action, while the second-named author discusses "chemical" diaphragms, *i.e.*, ion-exchange membranes. Apart from these two categories, or possibly in the latter one, there is the mercury diaphragm in which metal cations of certain types can move. This has already been referred to in the discussion on Chlorine Diaphragm cells. Another example of this type of diaphragm may be found in two ICI Patents relating to adiponitrile manufacture. (Brit. Pat. 1,076,783 and 1,067,447), but apart from these anomalies, other diaphragms appear to fall squarely into one category or the other.

1.2 Mechanical diaphragms

All ions in solution participate in conducting the current in proportion to their concentrations and mobilities. For concentrated electrolytes, the diaphragm presents no barrier to the migration of ions but its introduction increases the electrical resistance of the electrolyte. This is because current can only pass through the thin streams of liquid filling the capillaries of the diaphragm, so the greater the path length (or tortuosity) of electrolyte within the diaphragm, the more efficient the diaphragm as a separator but the greater the potential drop across it. Consequently a compromise must be reached between the current efficiency and the voltage drop tolerable.

Of course, the electrical resistance may be reduced by ensuring the diaphragm

---

* ICI Ltd., Mond Division.
** ICI Ltd., Paints Division.
*** ICI Ltd., Petrochemical and Polymer Laboratory.

has as high a void fraction (or percentage of holes) as possible thus maximising the volume of electrolyte contained within the diaphragm structure. However whilst it is desirable for the void fraction (or porosity) to be high, the actual size of the individual interconnected pores (often best regarded for mathematical treatment as parallel capillaries) must be small to ensure that no passage of gas bubbles occurs through the diaphragm. Simple measurements usually give the mean pore size but in practice a pore size distribution occurs unless careful steps are taken to prevent this during diaphragm manufacture. This must be uniform over the entire diaphragm area to ensure good current efficiency and current distribution, *i.e.* the diaphragm must be homogeneous.

As electrolysis proceeds, the concentration of non-gaseous products will increase in the electrolyte. An example is the increase in the concentration of caustic soda in the catholyte of a brine cell. Conditions are thus established for the diffusion of the product across the diaphragm with consequent loss of efficiency. Diffusion will be greater the greater the porosity, the size of the pores, and the concentration gradient across the diaphragm, and also the thinner it is. Diffusion presents another reason why the diaphragm must be homogeneous. It will be seen later how diffusion is to a large extent overcome in the chlorine cell when the special case of the filtering diaphragm is considered.

Diaphragms are often charged with respect to the electrolyte and may be positive, negative or neutral depending on their position in the cell. It is thus important that the diaphragm material is non-conducting to prevent it from acting as an electrode. Russian workers suggest[1] that although the diaphragm is usually charged, the electro-osmotic flow of liquid across it in a commercial electrolyser is insignificant since commercial electrolysers operate with concentrated electrolytes and with a low potential difference across the diaphragm.

The properties of an ideal diaphragm may now be summarised. The diaphragm must be:

1. permeable to ions but not molecules
2. of high void fraction to minimise electrical resistance
3. of small mean pore size to prevent the passage of gas bubbles and minimise diffusion.
4. homogeneous to ensure good current efficiency and even current distribution
5. non-conducting to prevent action as an electrode
6. chemically resistant to the reactants and products
7. resistant to cell operating conditions of temperature, pH, etc.
8. of some mechanical strength and rigidity
9. cheap (unless the lifetime is so long that the cost can be treated as plant capital).

Several of these properties are in conflict in practice and generally a compromise must be made. Chemical stability is a constant problem especially if the anolyte is acid and the catholyte alkaline and most diaphragms have a relatively short life requiring frequent renewal. Many scores of materials have been used in the past[2-12] of which porous concrete, asbestos and asbestos mixtures (all alkaline resistant), ceramics, crocidolite (blue) asbestos, quartz and fire clay (all acid resistant) are but a few. Microporous rubber[60], linen and metal gauzes[61] have also been commercially applied. Many of these materials have not achieved one or all of the above criteria and many potential electrolytic processes have failed for lack of a suitable diaphragm material.

The most widely used materials have been asbestos, either in the form of fibre, cloth or paper, and ceramic, the properties of these materials best approaching most of the criteria laid out above. Neither material is ideal, asbestos being slowly attacked in acid conditions and ceramic having a high electrical resistance whilst at the same time being brittle and thus unstable to temperature change and mechanical shock. The manufacturing techniques of ceramic diaphragms, because of the nature of the material, prevent the production of large diaphragms and thus place a constraint on cell size. Also, the larger the piece, the greater the thickness and consequent electrical resistance.

A special case of diaphragm is the filtering diaphragm as exemplified in the chlor/alkali diaphragm cell. In this case diffusion of hydroxyl ions across the diaphragm from cathode to anode compartments is counteracted by maintaining a flow of electrolyte in the opposite direction *i.e.* from anode to cathode compartments. This is achieved by maintaining a higher anolyte level than catholyte level, thus applying a hydrostatic head of liquor against the diaphragm. Hence another diaphragm characteristic must be considered, namely the permeability of a diaphragm to liquid flow. The quantity of liquid flowing will be proportional to the hydrostatic pressure and surface area of the diaphragm but inversely proportional to its thickness. A useful measurement of the permeability is given by:

$$K_1 = \frac{m^3 \text{ cell liquor/h}}{(m^2 \text{ area diaphragm}) (m \text{ head of brine})} \times 10^3$$

Care must be taken if the diaphragm material is compressible (*e.g.* asbestos since the flow rate will only be proportional to the hydrostatic head over a small range of values. This fact places a constraint on the height of anolyte level in industrial chlor/alkali cells, all of which use asbestos diaphragms. Thus in brine electrolysis, as well as withstanding the effects of chlorine saturated brine of pH 3–4 on the anode side and caustic liquor pH $>$ 14 on the catholyte side, the diaphragm must sustain a flow of electrolyte. It is not surprising therefore, that magnesium and calcium are leached from the anode face and deposited as hydroxides in the more alkaline regions of the diaphragm[13]. The net result is a gradual loss of permeability, which is increased by any impurities in the feed brine. Figure 1 shows the rate of extraction at 25°C of magnesium and calcium from one sheet of Turner's Electrolytic Asbestos Paper 0.5 mm thick, and with acid brine at pH 3 flowing at 15 ml/

hr/cm² from a hydrostatic level of 20 cm. In a working cell this effect is counteracted to a certain extent by the rise of anolyte level which occurs (brine feed rate being constant) but this cannot be allowed to rise indefinitely. Rejuvenation techniques are operated which involve the washing out of impurities with acidified brine[14] or water[15] but there is a limit to the number of such washings a diaphragm will stand and the lifetime of an asbestos diaphragm is usually of the order of 100–120 days.

Since the reason for flowing electrolyte through the diaphragm is to counteract back diffusion of hydroxyl ions, another compromise must be made, this time between current efficiency and outflowing caustic strength, since the greater the diaphragm permeability, the greater the current efficiency but the weaker the caustic solution leaving the cell. Evaporation of the caustic liquor to form a saleable product is expensive and this fact is the reason why the preference, at least in the U.K., has been mercury cells for brine electrolysis since they produce caustic liquor essentially free from unchanged sodium chloride. With the present trends of chlorine demand exceeding the demand for co-produced caustic however, new outlets for caustic must be found, e.g. carbonation to soda ash, and industry is taking a new look at the cheaper diaphragm cell.

Although the presence of calcium and magnesium in the feed brine is detrimental to diaphragm life, they can be useful doping agents for diaphragms whose permeability is too high and as such, occasionally find a use.

Although methods of measurement of porosity[16], pore size, electrical resistance[17–19] and permeability[20] of unused diaphragms may be devised[21, 59] they generally only give a rough indication of subsequent diaphragm performance. The properties of an unused diaphragm and those of one during operation differ greatly, especially when the physical character of the diaphragm changes during its life, and bench testing is of value only for giving "orders of magnitude".

Full theoretical treatments of diaphragms for electrolysis, together with their physical properties can be found in refs. 22–25. Extensive treatments of diaphragm performance with special reference to ion-exchange membranes can be found in refs. 26 and 27.

In order to improve the properties of asbestos diaphragms, many attempts have been made to modify the structure of the diaphragm by impregnating or combining the asbestos with, for example, rubber[28], carbon[29, 30], halogenated drying oils[31, 32], calcium fluoride or titanium oxalate[33], asphalt or sulphur[34] and a polymer of an unsaturated ethylenic compound[35]. The latter is claimed to act as an ion exchange membrane, allowing passage of sodium ions but not chloride ions so that caustic liquor essentially free of unchanged sodium chloride results.

Many new materials have been fabricated into porous sheets with controlled properties for use as diaphragms for electrolytic processes. Examples of these are polymeric acrylic resins[36], silica acid impregnated P.V.C.C. felt[37], supplementary chlorinated P.V.C.[38], porous P.V.C.[39], P.V.C. flexible foam[40], microporous P.V.C.[41], porous polyolefin[42], and glass fibre fabric coated with a copolymer of vinylidene fluoride and hexafluoropropene[43].

A table of copolymers useful as diaphragm materials together with their acid resistance appears in ref. 44. Most of these materials when fabricated into diaphragms, fail in chlorine electrolysis due to the high surface area exposed to the corrosive nature of the anolyte which enhances chemical degradation.

Polytetrafluoroethylene will withstand the cell conditions but porous P.T.F.E. fails as a diaphragm because of its non-wettability[61]. To a certain extent this can be overcome by the addition of wetting agent to the anolyte liquor but because of the reactive conditions relatively large concentrations are required and the method is uneconomic as well as resulting in the contamination of the cell liquor. Fuel cell technologists utilise P.T.F.E. powder[62] to limit the wettability of fuel cell electrodes and the right combination of P.T.F.E. and hydrophilic solid (*e.g.* titanates[45]) may one day yield a resistant, wettable P.T.F.E. diaphragm.

1.3 EXAMPLES OF COMMERCIALLY AVAILABLE ELECTROLYTIC DIAPHRAGMS

1.3.1 Asbestos

Turner's Asbestos Cement Company Limited, manufacture an Electrolytic Grade asbestos paper for use as a diaphragm in the Gibbs chlorine cell[13] and the Knowles cell for water electrolysis[46]. Details of the chemical resistance were given earlier whilst porosity data for a 0.5 mm sheet appears in Table 1. Figure 1 shows the rate of extraction of calcium and magnesium from the paper. The permeability of a Gibbs diaphragm composed of four 0.5 mm sheets is about 25.

For the Knowles cell, a medium weight (1.9 kg/m²) organic-free chrysotile asbestos cloth is used as diaphragm. A service life of upwards of 10 years is expected but in some instances as much as 40 years service has been obtained.

TABLE 1

POROSITY DATA FOR 0.5 MM TURNERS ELECTROLYTIC GRADE ASBESTOS PAPER

Porosity	36.14%
Surface area	8.9 m²/g
Total pore vol.	0.4 cm³/g
Pore vol. in pores of diameter 300 Å	0.0228
Pore vol. in pores of diameter 200 Å*	0.0147
% Total pore vol. in macropores (diameter > 300 Å)	96.3

* The amount of pore volume due to pores < 200 Å is very small and fairly evenly distributed over the pore size range 18–200 Å.

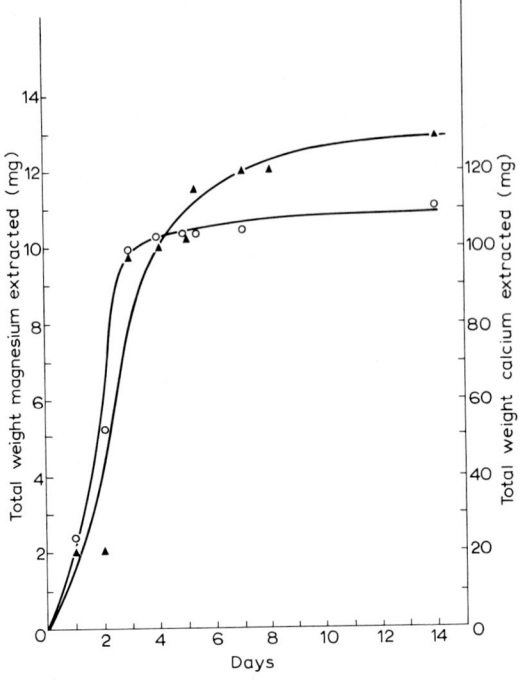

Fig. 1.
Rate of extraction of calcium and magnesium from Turner's Electrolytic Grade asbestos paper. (▲) magnesium; (O) calcium. Weight of paper exposed was 0.74 g.

1.3.2 Ceramic

The Aerox Filtration Division of Doulton Industrial Products Limited manufacture ceramic diaphragms in three standard grades, CG5, CF18 and H125. Of these, Celloton H125 is most commonly used since its characteristics are more ideal for most electrolytic applications[47]. Celloton H125 has an extremely good mechanical strength, is highly resistant to most acids (except fluorides) but is attacked by strong caustic in particular at elevated temperatures. It may be used in contact with weaker alkaline solutions but normally the application has a shortening of the life of the diaphragm which in some cases may be acceptable. Celloton is ideal in processes requiring a non-corrosive diaphragm to act as a diffusion membrane for an osmotic process (for example in the manufacture of hydrogen peroxide). Plating processes use diaphragms to protect the anodes during electrolysis and to revitalise chrome plating solutions. A newer process is the recovery of precious metals after they have been used for other processes such as coating on printed circuits and catalytic screens.

The following indicate the maximum dimensions of Celloton diaphragms and Table 2 the physical characteristics of the three grades.

TABLE 2

PHYSICAL PROPERTIES OF CERAMIC DIAPHRAGMS

	Aerox (Celloton)			Schumacher
	CF5	CF18	H125	
Max. pore size (μ)	3.5	4.3	1.0	0.8–1.5**
Porosity	33.8	41.3	48.0	42–53**
Permeability*	1.1	3.0	0.8	—
Electrical resistance factor+	26.2	6.5	3.0	—
Modulus of rupture (kg/mm^2)	4.4	1.87	2.84	1.3
Test piece thickness (mm)	3	3	3	—
Density (g/cm^3)	—	—	—	1.45

* Permeability = vol. in ml passing by 1 cm^2/h at 20°C and 10 cm head of water.

+ Electrical resistance factor = $\dfrac{\text{Resistance of impregnated diaphragm}}{\text{Resistance of electrolyte of same dimensions}}$

** Depending on type

Flat diaphragm plates: 76.2 cm long × 76.2 cm wide × 0.6 cm thick

Cylindrical diaphragms: Up to 35.56 cm diameter or more and 91.44–101.6 cm high.

Rectangular diaphragms: Many sizes to requirements within the limits of ceramic practice.

Thickness: Can be as low as 0.15 cm but depends on size.

Schumacher also manufacture a ceramic diaphragm for electrolyte use under the trade name Diapor[48]. The wall thickness is about 0.5 cm for smaller sizes but for larger sizes can be up to 1.2 cm for greater mechanical strength. Diapor is manufactured in a variety of shapes: as plates up to a size of 50 × 35 × 1.2 cm, cylindrical closed or open tubes up to a diameter of 20 cm and length 65 cm or rectangular boxes up to a side length of 40 cm. Porosity characteristics are given in Table 2.

1.3.3 Polyvinyl chloride, polypropylene and Terylene

Porvair Limited market "Vyon" which is a sintered high density polythene that is finding increased use in electrolytic applications[41, 49]. It is used as a diaphragm in electroplating and electrowinning to prevent contamination of the cathode deposit by anodic corrosion products, and being robust and easily fabricated, is finding uses in the "porous pot" field conventionally held by ceramics.

A microporous P.V.C. has also been developed, the major application being in primary cells. It is possible to use the P.V.C. for the above applications where a finer porosity is required. Having an 80% void volume it is rather flimsy for these applications but lamination with Vyon affords some rigidity, at the same time

TABLE 3

PHYSICAL PROPERTIES OF PORVAIR DIAPHRAGMS
(Vyon)

Thickness (cm)	Air flow (m³/min at 2.54 cms W.G.)		Water perm. (m³/min/m² at 70g/cm²)	Tensile strength (min.) (kg/cm²)	Elongation (%)	Density (g/cm³)	Vol. porosity (%)	Max. pore size (μ)	Max. Working Temp. (°C)
	Min.	Typical							
0.0762	1.71	2.72	2.24	10	10	0.56	41	90	80
0.102	1.45	2.26	1.71	15	10	0.56	41	85	80
0.158	0.85	1.36	1.47	25	10	0.57	40	80	80
0 204	0.43	1.02	1.22	30	10	0.58	39	75	80
0.254	0.34	0.57	0.98	35	10	0.60	37	70	80
0.318	0.28	0.51	0.73	40	10	0.61	36	65	80
0.476	0.28	0.51	0.49	40	10	0.65	33	60	80

Microporous PVC

Thickness (cm)	Min. tensile strength (kg/cm²)	Elongation (%)	Permeability (cm³/cm²/min at 50 cm w.p.)	Pore size (μ)	Volume porosity (%)
0.019	10	20	25	5–9	80
0.025	12	20	20	5–9	80
0.050	17	20	5	5–7	80
0.076	10	20	8	5–7	80

providing an extremely fine diaphragm. Physical characteristics are given in Table 3.

Schumacher Filters Limited market a polyolefin diaphragm free from fillers or plasticiser under the name of Filtroplast[48]. Diaphragm tiles are made in sizes up to 50 × 50 cm, in thicknesses 0.2, 0.4 or 0.8 cm. Diaphragm cylinders range from 15 cm length with 1.2 cm diameter to 50 cm length with 12 cm diameter. They do not stress, corrode or absorb moisture, and are resistant to alkali, most acids and a variety of organic solvents. Strong oxidising agents, especially at higher temperatures can cause attack. The diaphragms can be supplied in four pore sizes (different granulation) and in electrolysis, the poor wettability can be overcome by immersing in electrolyte or by adding wetting agent. When electrolysis commences, no further difficulties are met with regard to this hydrophobic character. The pore volume is about 45% and the density 0.5 kg/dm².

P. and S. Textiles Limited supply Terylene and propylene cloth diaphragms to nickel refineries throughout the world for the electrowinning of the metal[50]. If lead anodes are used, the diaphragm, as well as preventing contamination of the cathode compartment by anode slimes, must also prevent migration of sulphuric acid into the cathode compartment where it would cause deposition difficulties. The properties of the most successful diaphragms supplied by P. & S. are given in Table 4.

TABLE 4

PHYSICAL PROPERTIES OF P. & S. NICKEL ELECTROWINNING DIAPHRAGMS

	No. 488 Terylene	No. 14 Propex
Composition	100% multifilament Terylene	100% multifilament polypropylene
Weight (g/m^2)	450	390
Air permeability: (m^3/m^2/min at 25 mm w.g.)	0.4	0.3
Weave	Plain	Plain
Finish	Heat stabilised	Heat stabilised

1.3.4 New materials

Although not used as electrolytic diaphragms, there are several new materials which could have interesting applications.

(a) Polytetrafluoroethylene

P.T.F.E. (Teflon or Fluon) is unaffected by all organic solvents, concentrated acids and alkalies. It is stable at temperatures in excess of 260°C, is chemically inert and pure. Millipore (U.K.) Limited market a P.T.F.E. filter under the name of Mitex[51] which is fabricated by a process which gives a continuous mat of P.T.F.E. fibres fused together at each intersection. Two pore sizes are currently available, 5 and 10 μ, and the porosity is in the range 60–68%. However, its use in aqueous electrolysis, especially as a filtering diaphragm would only become practicable if the hydrophobic character, which would cause occlusion of gas bubbles within the pores with a resultant increase in electrical resistance, could be overcome. Success with this problem is advertised by Chemplast Inc. who market a fibrous P.T.F.E. material "Zitex"[52].

Filter aids, stable inorganics and other substances that may increase the hydrophilicity may be incorporated in the pores of the filter to give one solution to the problem, or the surface of all the fibres throughout the filter may be etched (by sodium/liquid ammonia) to give several gradations of wettability.

(b) Graphite

Interest in graphite fibre as a reinforcing material in laminates is high[53, 54]. Carbon and graphite cloths, consisting entirely of flexible filaments, are produced by the pyrolysis of rayon cloth by Union Carbide[55]. The cloth made at present would be too porous for diaphragm use but the technique, if applied to a rayon cloth of suitable permeability, might provide a useful electrolytic diaphragm, being much more flexible than conventional porous graphite filters.

(c) Stainless steel

Porous stainless steel strip with pore sizes variable from < 1 to 500 μ and porosity variable from 0–40% has been developed by the Mallory Metallurgical Company, Indianapolis[56]. An electrochemical process creates separation between the metal crystals without destroying the metallic strength. The material, which may contain up to 23,600 holes/cm is available in thicknesses up to 0.381 cm and currently in continuous lengths 15.24 cm wide. A possible use might be as a diaphragm in water electrolysis. Another technique for producing fine holes in thin metal sheets is spark erosion (15 μm)[57].

General Electric market a plastic filter material under the name of "Nuclepore" which is produced by the bombardment of thin plastic film with fragments from disintegrating radioactive atoms followed by etching out the tracks left by the particles. The result is a filter with holes ranging from 1 to 10 μ according to the length of the etching process[58].

2. Ion-Exchange Membranes

Ion-exchange membranes, which are ion-exchange polymers in sheet form, have as their most characteristic property the ability to discriminate between permeating or migrating ions by the sign of their charge. As normal conditions of service entail ionic fluxes through membranes of steady composition rather than cyclic use involving separate absorption and regeneration stages, it is perhaps more accurately descriptive of their action to call them ion-selective (or in one variant "permselective") membranes. Cation-exchange membranes, *i.e.* those selectively permeable to cations, can be synthesised from strong and weak acids, and anion-exchange types from strong and weak bases, but those derived from weak acids and bases have thus far not become commercially important, all current commercially used materials being of the type which dissociate over almost the entire pH range.

To take the case of a cation-exchange membrane; this consists, like the corresponding ion-exchange resin, of a polymeric salt (or acid when in the hydrogen form) of which the anions, typically sulphonate, are covalently attached to the polymer. Such a membrane when immersed in electrolyte-free water contains counterions (in this case cations) in equivalent quantity to the fixed ions, though when an external electrolyte is present some anions from the latter (called "co-ions") enter the membrane phase together with an equivalent number of additional counterions. The internal co-ion concentration is always smaller than the external electrolyte concentration and, at low external concentration, is very small in accordance with the Donnan exclusion principle. It follows that, unlike non-selective barrier materials, ion-exchange membranes do not rely on the ions furnished by a permeating external electrolyte in order to carry electrolytic current: provided the external electrolyte concentration is not very high compared with the fixed ion concentration in the membrane, current is carried mainly by the mobile counterions. Because the counterion concentration and mobility are properties essentially of the membrane, the conductance of ion-exchange membranes is relatively insensitive to the concentration of external electrolyte*, the only qualifications being that the external concentration is not too high (with typical commercial membranes this means up to at least a few tenths of an equiv.l.$^{-1}$) and that the external electrolyte does not form strong ion-pairs or poorly dissociated complexes with the fixed ions in the membrane.

A property closely connected with the selectivity of ion-exchange membranes is their ability to suppress free electrolyte diffusion, that is to say, the electrolyte diffusion flux observed through such a membrane under given conditions is considerably smaller than would be displayed by a hypothetical membrane of the same structure as the swollen membrane but without the ion-exchange groups present. Ion-exchange membranes therefore form highly effective barriers in electrolytic processes involving electrolytes only. This special action, which derives from the ion exclusion effect, does not operate against non-electrolytes which it may be desired to retain on one side of a barrier in an electrolytic cell, but diffusion rates in these membranes are generally much lower than in non-selective barrier materials because denser, less porous, continuous or reinforced gel-like structures can be employed in their fabrication, a fact permitted by the high counterion conductance. A compromise between control of diffusion (requiring a dense structure) and high conductance (favoured by an open structure) can therefore be established at more favourable values of conductance and backdiffusion flux than is the case with non-ionic materials.

The electrochemical properties of ion-exchange membranes are such that the majority of electrolytic processes requiring barriers can be expected to benefit from their use unless exceedingly low barrier resistance is necessary (say, below 1 ohm cm^2). Their main limitations besides high price (currently from the region of U.S. \$10/m^2 upwards), are in respect of mechanical strength, dimensional stability, resistance to solvents and solvent mixtures, reduced effectiveness at high external concentration, and chemical stability. The difficulty in producing a robust, dimensionally stable sheet material with sufficient ion-exchange groups to perform the required electrochemical function derives from the fact that the fully swollen membranes contain about 50% of water and that the water content (and therefore the dimensions of a sheet) varies with the ionic environment, especially with the nature of

---

* If this is not the case in any particular instance, it suggests the presence of a discontinuity in the membrane structure (*e.g.* a region of low or zero fixed ion concentration located within the path of electric current). This could affect the use of the membrane in electrolytic processes.

the counterion. In some cases, membranes must be kept wet throughout their useful life to prevent the development of cracks due to shrinkage on drying. Approaches towards the improvement of mechanical properties have been made by the incorporation of reinforcing fabrics as well as the inclusion of plasticising constituents, *e.g.* co-monomers or "backbone" polymers, in the polymer composition. The long-term dimensional stability of membranes without reinforcement is often poor, with a tendency to develop a protrusion in the direction of any hydrostatic head applied during use or, in the absence of a head, in the direction of ion transfer. In water demineralization by electrodialysis, the indifferent dimensional stability of many membranes has led to the development of apparatus constructions which subject the membranes to the smallest possible pressure differences or other mechanical stress; it follows that such membranes may not have the robustness necessary for use in many industrial applications unless special constructions are adopted. Membranes incorporating reinforcing fabrics have been known to fail on exposure to organic solvents or aqueous solvent mixtures; this is attributable to differential swelling of the fabric compared with the ion-exchange component.

The chemical stability of ion-exchange membranes is difficult to discuss in terms of a definite life expectancy, although that is a question which is frequently asked. If the useful life of a membrane is not terminated by mechanical failure, it frequently ends with the accumulation of foreign matter, especially precipitates (scale) or ion-pair forming counterions (*e.g.* polyvalent or polymeric ions), to such a degree that the performance has declined unacceptably. It is thus frequently the case that no fundamental change has occurred in the polymeric structure of the membrane or in the ionisable groups. In fact the chemical stability of the ion-exchange component is generally considerable, especially of the cation-exchange types most of which have sulphonated cross-linked, grafted or copolymerised styrene as the active constituent. Such materials may withstand temperatures of 100°C over considerable periods but the plasticising, binding or reinforcing constituents may alter in structure or lose adhesion, causing the membrane to fail in service. The course followed in developing membranes of enhanced chemical stability has therefore been to employ substrate or backing materials of high chemical or thermal stability, such as polypropylene, or for extreme service, fluorocarbon polymers, rather than modifications in the ion-exchange constituents. Anion-exchange membranes, being derived from quaternary ammonium salts, are much less stable and are not normally recommended for service above 50°C; in common with the corresponding anion-exchange resins, the free-base form is even less stable. Many standard grades of ion-exchange membrane available from commercial suppliers are sensitive to oxidising conditions, but some manufacturers offer products which tolerate these conditions to some degree.

An extensive literature exists on the behaviour of ionic membranes, much of which has been contributed by biologists inspired by the ideal of understanding the functioning of membranes present in living tissue. This work attained considerable momentum during the 1930s and 1940s before ion-exchange membranes of commercially interesting properties existed; fundamental work has continued at an increasing pace with the methods of irreversible thermodynamics brought to bear on the multiplicity of fluxes (*i.e.* those of counterions, co-ions and of solvent) which are induced by electrical or diffusional driving forces. A substantial review covering this work as well as a source of reference on the earlier phase has been published by Lakshminarayanaiah[63]. A valuable brief survey of membrane phenomena is that by Läuger[64].

Juda and McRae[65] in 1948–49 are generally credited with the first successful development of ion-exchange membranes having low electrical resistance combined with adequate mechanical strength for use on the technical scale, while during the period since 1950 synthetic work on membranes intended principally for use in desalination has been pursued by several groups, notably in the U.S.A., Netherlands, U.K., Japan, South Africa and Israel. From the outset, two approaches to membrane preparation were distinguished: the incorporation of finely divided ion-exchange resin particles into a binding matrix material, yielding an obviously heterogeneous product, and the formation of a continuous sheet of ion-exchanger, giving a supposedly homogeneous membrane.

The first method was the basis of products* which remained in commerce until about 1968; the preparation of successful membranes of this type is difficult because of the need for contact between the swollen resin particles, so that the binder content is typically only 30–40% by volume. Although the mechanical strength could be improved by the incorporation of woven fibrous supporting material, these membranes were characterised by a high ion-exchange capacity relative to the electrochemical performance obtained. Following the second general line of approach, the methods used have included:

1. The casting of styrene/divinylbenzene/radical catalyst compositions and their subsequent polymerisation and chemical reaction (sulphonation or amination).

2. The incorporation of plasticising co-monomers into styrene-containing polymers which are subsequently reacted.

3. The grafting of reactive branches, usually of polystyrene, onto inert "backbone" materials either in sheet form or which can be so fabricated.

4. The direct reaction of flexible sheet materials, like polyethylene or plasticised polyvinyl chloride, with liquid or even gaseous reactants.

5. The impregnation of sheet materials with monomeric or partially condensed liquid compositions which yield ion-exchange polymers on curing.

6. Casting from a common solvent a mixture of a film-forming polymer and a linear polyelectrolyte, the film-former being able to restrain the tendency of the polyelectrolyte to leach out during service.

---

* Permaplex A and C series membranes (The Permutit Co. Ltd., London).

In a number of cases, an open-structure woven fabric (which may be in glass fibre or synthetic polymers, e.g. nylon, polyester, polypropylene or vinyl chloride, acrylonitrile or vinylidene dichloride polymers or copolymers) is incorporated to contribute dimensional stability and mechanical strength. A high proportion of the current commercial products are of this type. Functionally, the inclusion of the supporting fabric does not invalidate the classification of reinforced membranes as "homogeneous" because some continuity of ion-exchange activity still exists between the faces of the membrane.

The "homogeneous" type of membrane characteristically has a relatively low ion-exchange capacity, generally 1–2 m equiv./g dry membrane, while the balance of electrochemical properties, selectivity, conductance and electrolyte diffusion flux, is generally superior to that of the heterogeneous type. The commonly used description "homogeneous" may be open to question even in the case of unsupported membranes. For example, graft copolymers used to make membranes of type 3 typically contain inclusions of homopolymer which appear as phase-segregated regions in the micro-structure of the final membrane. Evidence has been advanced that even membranes made from random copolymers (styrene/divinylbenzene) exhibit spatial variations in the density of crosslinks, so it might be questioned whether complete homogeneity is attainable. Nevertheless, as far as performance is concerned, the main contrast is between the compositions made from particulate ion-exchange resins and those with an essentially continuous ion-exchanger phase.

An ion-exchange membrane operating in an electrolytic process in which a flux of counterions occurs in one direction forms a system liable to exhibit film control as does any other system with mass-transfer at a solid/liquid interface. This arises because the discrepancy between the fluxes of individual ionic species in the membrane and the liquid phases must be compensated by processes like convection and diffusion taking place in the liquid within a short distance of the interface—within the notional boundary film. On the side of the membrane which donates counterions, the concentration of electrolyte falls, the limit not being reached until the interfacial concentration is essentially zero. On the receiving side, the rise in concentration can cause precipitation of, or supersaturation in, sparingly soluble constituents. The major limiting factor in the electrodialytic treatment of natural saline water is in fact this phenomenon of concentration polarization (cf. the chapter on *Electrodialysis* by G. S. Solt). In this case, the depletion on the donating side of the anion-exchange membranes progresses to a stage at which the solution/membrane interface is starved of electrolyte and a further increment in current is carried by ions generated by the dissociation of water. Hydroxyl ions then pass through the membrane and the consequent rise in alkalinity on the receiving side causes the precipitation of calcium carbonate if calcium and bicarbonate ions are present and of magnesium hydroxide if magnesium is present; precipitation is encouraged by the excess electrolyte concentration present on the receiving sides of these membranes. Curiously, water-splitting is much less evident at the cation-exchange membrane interfaces, so it is practicable to operate cation-exchange membranes under limiting current conditions, *i.e.* with the donating side essentially depleted in electrolyte. In general, whenever a choice is open, a cation-exchange membrane is to be preferred for use as a barrier: it is likely to be more chemically stable and to retain its efficiency over a wider range of current density.

A phenomenon of great fundamental interest and of practical significance when relatively concentrated electrolytes are in use is the electro-osmotic water flux accompanying ion transfer through ion-exchange membranes. Although with modern commercial membranes water transfer seldom exceeds 10–15 moles/Faraday, the effect of this on the mass-balance of an electrolytic operation can be important at high concentrations. The water transfer number is to some degree a function of the membrane structure, especially of the cross-linkage, and membranes showing small electro-osmotic fluxes have been prepared and made available commercially; this usually entails a somewhat increased resistance. The rate of permeation of water under a hydrostatic head is normally very small, for example $< 2 \times 10^{-7}$ cm³/sec/cm² exposed membrane surface at an applied pressure of 1 atmosphere on an unflawed homogeneous membrane. The older type of heterogeneous membrane exhibits significantly higher water fluxes under a pressure head.

A general comment on the question of membrane supply and availability might not be out of place in introducing the list of membranes from well-established suppliers given in Table 5. With the exception of the Japanese effort, ion-exchange membrane development has, in the past, been aimed principally at the desalination market, the prospects for which were highly rated some years ago. With the possible exception of the processes involving casting, the methods of manufacture are not intrinsically costly so that recovery of research and development expenditure is an important factor in determining selling prices. An optimistic view of the prospects of the desalination market may have induced some companies to spread this expenditure over larger potential sales than have indeed materialised, so that there have been disappointments and several instances of actual withdrawal from the field; it is still probably true that most manufacturers are operating on a scale, and indeed with equipment, that is closer to the pilot plant scale than to commercial operation as normally understood within the chemical industry. The general commercial position is therefore somewhat unstable, though at the present time the small user (as most who employ ion-exchange membranes for purposes other than desalination are likely to be) still enjoys relatively low prices which could not be supported by chemical process applications alone.

The entry of several Japanese companies into the field derives from their country's lack of indigenous supplies of salt and work on the concentration of sea-water as a substitute. The brine concentration outlet has been quite limited and this has led Japanese manufacturers to take an interest in the general desalination market. Although these ventures have been successful, there is no reason to believe their

TABLE

ION-EXCHANGE MEMBRANES AVAILABLE (EARLY 1970)

Manufacturer	Code No.	Selectivity	Type	Reinforcing fabric	Thickness (mm)	Sizes available
American Machine and Foundry Co. (1)	A60	A	Homogeneous type 3 on polyethylene (2)	None	0.3	Rolls *ca.* 1 m wide
	C60	C	—	—	0.3	—
	A100	A	Homogeneous type 3 on polyethylene (4)	—	0.2	—
	C100	C	—	—	—	—
	C322	C	Homogeneous type 3 on fluorocarbon polymer	—	0.2	—
Asahi Chemical Industry Co. (5)	A	A	Homogeneous type 1 or 2	None	0.2	1 m square pieces
	C	C	—	—	—	—
	CA1	A	—	—	—	—
	CK1	C	—	—	—	—
Asahi Glass Co. (6)	AMT	A	Homogeneous probably type 2	Polyester	0.2	1 m × 1 m, 1 m × 1.5
	CMG	C	—	Glass fibre	—	—
	CMV	C	—	PVC	—	—
Ionac Chemical Corporation (7)	MA3148	A	Heterogeneous (8) (9)	Yes (10)	0.2	0.9 × 3 m
	MA3236	A	—	—	0.3	—
	MA3475	A	—	—	0.3	—
	MC3142	C	—	—	0.2	—
	MC3255	C	—	—	0.3	—
	MC3470	C	—	—	0.3	—
Ionics, Inc. (12) (13)	111BZL183	A	Homogeneous type 1	Dynel (14)	0.6	0.5 × 1 m
	111BZLO65	A	—	Dynel (15)	1.4	0.5 × ?
	61AZL183	C	—	Dynel (14)	0.6	0.5 × 1 m
	61AZLO65	C	—	Dynel (15)	1.2	0.5 × ?
	61DYGO67	C	—	Glass fibre	0.6	0.5 × ?
Tokuyama Soda Co. (18)	AV4T	A	Homogeneous type 2	PVC (19)	0.2	1 m × 1.5
	CL-2.5T	C	—	PVC (19)	0.2	—

1 689, Hope Street, Springdale, Conn., U.S.A. Products are called "Amfiion".
2 Believed to be low-density polyethylene.
3 Developed by the Central Technical Institute, T.N.O., Holland and frequently referred to as TNO A60 and C60 membranes.
4 Believed to be high-density polyethylene.
5 Hibiya-Mitsui Building, 12, 1-chome, Yurakucho, Chiyoda-ku, Tokyo, Japan. Products are sometimes called "Aciplex".
6 2-14, Marunouchi, Chiyoda-ku, Tokyo, Japan. Products known as "Selemion".
7 Birmingham, New Jersey, U.S.A. Products known as "Ionac".
8 The Manufacturers' literature states these membranes to be of heterogeneous type, though the general electrochemical properties are closer to those of homogeneous membranes. This may be due to discrepancies in understanding of the term "homogeneous".
9 Ionac membranes as received are coated with a protective film of a water-soluble polymer.
10 Nature not disclosed, but believed to be different for the various membrane types.
11 High thermal and chemical stability specifically claimed.

...OM WELL-ESTABLISHED MANUFACTURERS

Burst strength (approx.) atm. over-press.)	Dimensional stability	Ion-exchange capacity (m equiv./g dry membrane)	Resistance (ohm cm²)	Specific resistance (ohm cm²)	Price range	Remarks remarks
3.5	L	1.5	5	170	M	Note (3)
3.5	L	1.5	5	170	—	Note (3)
3.5	LM	1.5	8	400	M	
3.5	LM	1.5	8	400	—	
7.0	LM	0.9	2	100	VH	
—	MH	2.0	2	100	L?	Poor flexibility
—	MH	2.5	1.5	75	L?	—
1.4	MH	2.0	2	100	L?	
2.0	MH	2.5	3	150	L?	
> 7.0	LM	2.0	4	100	L	
ca 7.0	LM	1.5	6	300	L	
ca 7.0	LM	1.5	6	300	L	
13	H	1.0	10	500	M	
13	H	0.8	35	1200	M	
14	H	0.7	20	700	M	Note (11)
13	H	1.1	10	500	M	
13	H	1.3	20	700	M	
13	H	1.1	10	500	M	Note (11)
8.5	H	1.8	14	230	MH	
24	H	1.6	30	210	H?	Note (16)
8.0	H	2.7	12	200	MH	
26	H	2.6	30	400	H?	Note (16)
12	H	2.5	15	250	MH	Note (17)
7.0	LM	1.6	3	150	L	
3.5	LM	1.6	3	150	L	

12 65, Grove Street, Watertown, Mass. U.S.A. Products known as Nepton.
13 Ionics Inc., offer a wider variety of membranes than those listed, which illustrate those having the more distinctive features.
14 4 oz./sq.yd. grade fabric.
15 15 oz./sq.yd. grade fabric. Products with 8 oz./sq.yd. backing are also offered.
16 The manufacturer specifically recommends this product for use in electrolytic cells requiring a robust separator.
17 A membrane of reduced swelling which may be preferable when polar organic solvents are involved.
18 Supplied through Iwai and Co. Ltd., Chemicals Department, C.P.O. Box 226, Tokyo, Japan. Products known as "Neosepta".
19 Tokuyama Soda Co. offers also anion and cation exchange membranes AVE and CLE, stated to be formed on polyethylene gauze, for which reduced sensitivity to a wide range of organic solvents is claimed.

commercial position to be fundamentally more stable than those already discussed.

Table 5 gives a selection of currently available ion-exchange membranes from six manufacturers who may be said to be well established in this field; those offering development products only for evaluation, or whose position in the field could be said at this time to be tentative, have not been included. The list is not complete as regards individual manufacturers' published product ranges, but an attempt has been made to cover a wide range, especially in regard to materials which might serve as barriers in industrial electrochemical processes. The quantitative information given, which is derived mainly from manufacturers' published data, should be viewed with some caution, as individual samples often deviate considerably from the mean, especially in electrical resistance. The property of ion-selectivity has been dealt with qualitatively only (A = anion-selective, C = cation-selective), although manufacturers' bulletins usually contain a good deal of quantitative information on this. The varying methods of measurement make it difficult to compare the reported values and, in any event, they are usually so high that selectivity is not the criterion of choice between competitive products. This may not be the case where high external electrolyte concentrations are used and some products have been developed specifically to function efficiently under such conditions: a suggested guide in this case is either to select membranes of unusually high specific resistance (indicating a highly crosslinked structure) or those of high ion-exchange capacity (provided such membranes do not have abnormal water contents in the fully swollen state).

The indication given in the Table as to membrane type or method of manufacture is largely the author's own surmise; the type numbers refer to the categories listed on p. 587. The thicknesses are those of fully swollen membranes in water. As regards the sizes available, this is frequently an important consideration in choice and it will be seen that some products which have attractive properties are limited in this respect. Dimensions are approximate only. The bursting strengths quoted are measured by the Mullen test as widely used in the paper industry. The qualitative appraisal offered of dimensional stability (L = Low, M = Moderate, H = High) is quite subjective; consideration has been given to the dimensional change from the wet to the dry state, from one counterionic state to another and to the effect of prolonged use at room temperature or slightly higher. The ratings are to be considered relative to the generally low degree of dimensional stability possessed by ion-exchange membranes. The first column of resistance values refers to the resistance presented by 1 cm^2 of swollen membrane in a dilute aqueous electrolyte at 25°C; anion-exchange membranes are in the chloride form and cation-exchange membranes in the sodium form for the purpose of this measurement. The concentration of external solution is usually 0.1 N or less, *i.e.* within the range in which concentration has little influence. The specific resistance values are simply the quotients of area resistance and thickness values.

The price ranges are L = up to 20, M = 20–60, H = 60–150, and VH = over

150, all in U.S. \$/m² membrane area. The information is given as a guide to order of magnitude, but it cannot possibly be accurate for all commercial conditions. Thus, there is a marked tendency for prices to be negotiated in relation to the size of order. This and indeed all other information in the Table is given in good faith to assist potential users to decide whether ion-exchange membranes are suited to any new application which may be under consideration, rather than to give guidance on the selection of a specific product. In a field such as this, as yet technically and commercially immature, extensions and alterations to product ranges occur frequently and many manufacturers are prepared to adapt their products to novel applications. Though one fluorocarbon membrane is listed in Table 5, mention

TABLE 6

OPERATIONAL RANGES FOR DU PONT FLUOROCARBON MEMBRANES

Agent	Concentration (%)	Temp. (°C)
KOH	25–40	150
HNO_3	70	100
H_2O_2	30	80
H_3PO_4	85	150
H_2SO_4	50	150

should also be made of the Du Pont XR perfluorosulfonic acid membranes, which though commercially less advanced than others, have nonetheless recorded some remarkable performances, including continuous operation for nearly two years in a fuel cell at approx. 85°C. These have a K^+ ion resistivity of 3 ohm cm² and the manufacturers indicate the operational ranges given in Table 6 with an estimated absolute temperature maximum for operation of 225°C.

3. Electrolytes for electrochemical processes

3.1 THE CRITERIA FOR A SUITABLE ELECTROLYTE

Among the criteria for a suitable electrolyte, we must list the following.

3.1.1 Good electrical conductivity

While molten salt media satisfy this *par excellence*, and many aqueous electrolytes do not lag far behind, the same cannot be said of non-aqueous media. The conductivity of solvents such as ethers, dimethylformamide, dimethylsulphoxide containing dissolved salts, is poor. Of the many processes based on this type of electrolyte, only the NALCO process (q.v.) appears to be established. Even in the best designed cells, power costs can multiply up to tenfold, and few processes or products can bear this cost.

3.1.2 Solubility of products and reactants

The reactant must always be soluble in the electrolyte, either in its massive or its ionised state. The product may sometimes be soluble—which makes for a smooth running process, but raises problems of extraction or separation from the electrolyte—or it may be insoluble, which makes for ease of separation, but can cause clogging of the cell or obscuring of the electrodes.

3.1.3 Corrosion problems

At the temperature and concentration of operation, the electrolyte should not corrode cell or electrodes.

Having established certain criteria, one may consider the three main types of electrolyte, namely aqueous, non-aqueous and molten salts.

3.2 AQUEOUS ELECTROLYTES

So much is known about these, that little can be usefully said. Complex inorganic ions can be advantageous, and the reader is referred to the chapters on electrorefining in aqueous media, and on electrodeposition of metals. Turning to the use of aqueous electrolytes in organo-electrochemistry, one meets the problem of poor solubility of reactants in so many cases. This has been largely overcome by use of "salting-in" procedures with a solubilising McKee salt[66], such as an alkyl or aryl sulphonate. The uses and advantages of such salts in electrochemical processes, are nowhere better described than in the literature (including patent literature) surrounding the ADN process (q.v.). Thus the solubility of acrylonitrile (ACN) can be increased from approx. 6% in aqueous tetramethylammonium chloride to a value of 20% in aqueous tetramethylammonium p-toluene sulphonate[67]. The quaternary ammonium salts have a separate advantage in that they repress the onset of cathodic hydrogen evolution to values of -2.0 V vs. R.H.E. and thus make possible a wider potential operating range on lead and mercury electrodes. Moreover, the presence of such salts can also affect the nature of the reduction product, as exemplified again by the patent literature surrounding the ADN process, and also the rather remarkable claim[68] recently made, whereby CO_2 is reduced to malic or glycollic acids. Having quoted the virtues of these salts, two caveats should be added. Firstly, they are costly, and little wastage of them can be tolerated. Secondly, they can be anodically oxidised under many conditions, and this calls for a cell with an ion-exchange membrane or some other solution to the problem of oxidation. It has been suggested that use of platinum-type anodes in certain potential regions (anodic to 1 V *vs.* R.H.E.) might impede oxidation.

3.3 Molten salts

The extensive use of these in electrowinning and electrorefining processes requires little further comment. In each case, the salt itself, without support electrolyte, suffices, and in each case, the product is the ionic species present in the salt after charge transfer has occurred. The fluorine and aluminium processes (*q.v.*) are also useful models. It is not generally known that molten salts can operate below 100°C, and this appears to be a field with much promise. Among the many books on the electrochemistry of molten salts, that by Delimarskii[69] is worth mentioning. Organic compounds are poorly soluble in melts, as a rule, and above a given temperature—which may be as low as 250°C, depending on the compound—pyrolysis commences.

The behaviour of various sodium salt melts including NaCN, $NaNO_3$ and sodium borates, is described in ref. 71.

Organic molten salts have been little studied, but in the writer's laboratory their stability is not found to be good, though they are satisfactory solvents for many organic reactants.

3.4 Non-aqueous media

As already stated, these have not found much application in industrial electrochemistry. An excellent review of their electrochemical properties is found in ref. 70 and little can be added to this. One industrial electrochemical process for which non-aqueous media have been repeatedly considered (but apparently with no success) is the extraction of metallic sodium. Sittig's monograph[71] gives full references and bibliography for both organic and liquid ammonia electrolytes.

Acknowledgement

C.J. would like to thank the following for their help in the preparation of this work: B. L. Green, (T.A.C. Ltd.); W. McClure, (T.B.A. Ltd.); N. S. Hutchinson, (Doulton Ind. Prods. Ltd.); C. Kreuser, (Schumacher Filters Ltd.); R. H. Ashbolt, (Porvair Ltd.); J. Smith, (P. and S. Textiles Ltd.); J. G. Mulvany, (Millipore (U.K.) Ltd.); R. B. Forsyth, (Union Carbide); and also ICI (Mond Division) Ltd. for permission to prepare and publish the work.

4. References

1 V. V. STENDER, *Asbestos Diaphragms*, Rept. of Govnt. Inst. of Appl. Chem., Leningrad, 22nd ed., 1944.
2 *Gmelins Handbuch der Anorganischen Chemie*, 8th ed., Verlag Chemie, Weinheim, 1969, p. 69.
3 C. J. THATCHER, U.S. Pat., 1,393,467 (11 Nov., 1921).
4 SIEMENS and HALSKE A. G., Brit. Pat., 309,316 (11 April, 1929).
5 J. Y. JOHNSON to I.G. Farben, Brit. Pat., 321,394 (29 Oct. 1929).
6 M. LEBLANC, *Z. Elektrochem.*, 7 (1900) 290.
7 C. J. THATCHER, *Met. and Chem. Eng.*, 13 (1915) 336.
8 M. BUCHNER, *Z. Angew. Chem.*, 7 (1904) 985.
9 J. R. CROCKER, *Electrochem. and Met. Ind.*, 6 (1908) 153.
10 P. ALEXANDER, Wacker Gesellschaft für Elektrochem., Brit. Pat., 312,713 (6 June, 1929).
11 SIEMENS and HALSKE AG., Brit. Pat., 309,316 (11 April, 1928).
12 G. F. JAUBERT, U.S. Pat., 1,837,050 (1 Dec., 1926).
13 B. L. GREEN, Turners Asbestos Cement Co. Ltd., private communication (2 April, 1969).
14 HOOKER CHEM. CORP., Brit. Pat., 1092,167 (22 Nov., 1967); Dutch Pat., 66,04238 (1 April, 1966).
15 S. YA. FAINSHTEIN, *Production of Chlorine by the Diaphragm Electrolysis Method*, Izdatel'stvo "Khiniya", Moscow, 1964.
16 R. M. WERNER and L. A. CLARENBURG, *Ind. Eng. Chem., Proc. Design and Development*, 4 (1965) 288.
17 R. E. MEREDITH and C. W. TOBIAS, *J. Electrochem. Soc.*, 110 (1963) 1257.
18 W. S. SPIEGLER, *J. Electrochem. Soc.*, 113 (1966) 161.
19 R. S. ALWITT, *Electrochem. Tech.*, 6 (1968) 172.
20 L. I. KRISHTALIK, *Zh. Prikl. Khim.*, 36 (1963) 1776.
21 V. V. STENDER, *Trans. Electrochem. Soc.*, 67 (1936) 51.
22 V. V. STENDER, *The Electrolytic Production of Chlorine and Alkalies*, Onti. Khimteoret, Leningrad, 1935.
23 T. MUTSUNO, *Kogyo Kagaku Zasshi (J. Chem. Soc. Jap. Ind. Chem. Sect.)*, 60 (1957) 951.
24 A. KORCZYNSKI and R. DYLEWSKI, *Chem. Stosowana*, 12 (1968) 183.
25 T. MUKAIBO, *J. Electrochem. Soc. Japan*, 20 (1952) 482.
26 N. LAKSHMINARAYANAIAH, *Chem. Rev.*, 65 (1965) 491.
27 Various authors, *Discussions Faraday Soc.*, 21 (1956).
28 V. V. STENDER, J. G. JORNITSKY and B. G. SABO, *Trans. Electrochem. Soc.*, 71 (1937) 565.
29 C. P. LEBLANC and L. W. HENSLEE to Pittsburgh Plate Glass Ind., U.S. Pat., 3,022,244 (20 Feb., 1962).
30 M. C. HUTCHINS to Johns-Manville Corp., U.S. Pat., 3,269,889 (30 Aug., 1966).
31 W. W. CARLIN to Pittsburgh Plate Glass Ind., U.S. Pat., 3,057,794 (9 Oct., 1962).
32 COLUMBIA SOUTHERN, Brit. Pat., 907484.
33 J. BILLITER to Siemens and Halske A.G., Brit. Pat., 262,470 (31 Mar., 1927).
34 J. MÜLLER, Brit. Pat., 235,557 (5 Aug., 1926).
35 D. R. NIELSON and W. W. CARLIN to Pittsburgh Plate Glass Ind., Can. Pat., 737,916 (5 July, 1966).
36 H. RUDOLPH, Brit. Pat., 498,293 (2 Feb., 1939); 534618 (21 Mar., 1941).
37 K. WEBER and G. BODE, U.S. Pat., 2,342,230 (22 Feb., 1944).
38 M. KALTWASSER, K. H. BERNDT and E. FISCHER to V.E.B. Electrochem. Komb., Fr. Pat., 1,450,872 (18 July, 1966).
39 D. SATAS, *Ind. Eng. Chem.*, 57 (4) (1965) 38.
40 K. M. DEAL, D. C. MORRIS and R. R. WATTERMAN, *Ind. Eng. Chem., Product Res. and Dev.*, 3 (1964)209.
41 PRITCHETT and GOLD and E.P.S. CO. LTD., Brit. Pat., 565,022 (24 Oct., 1944).
42 G.E.C. Co., Brit. Pat., 980,229 (13 Jan., 1965).
43 K. E. THOMPSON to E.I. DuPont, U.S.Pat., 2,983,624 (9 May, 1961).
44 E. C. HEMES, *Fibres Plastics*, (Jan. 1961) 13.

45 R. G. HALDEMAN, *Cyanamid Electrode Matrix Materials*, Stamford Res. Labs. Rep. (16 May, 1967).
46 W. MCCLURE, Turner Brothers Asbestos Co. Ltd., private communication (8 Aug., 1969).
47 N. S. HUTCHINSON, Doulton Industrial Products Ltd., private communication (28 Feb., 1969).
48 C. KREUSER, Schumacher Filters Ltd., private communication (19 Dec., 1968).
49 R. H. ASHBOLT, Porvair Ltd., private communication (25 Feb., 1969).
50 J. SMITH, P. & S. Textiles Ltd., private communication (23 Oct., 1969).
51 J. G. MULVANY, Millipore (U.K.) Ltd., private communication (11 Mar., 1969).
52 ANON., *Chem. Eng.*, 72 (2 Aug. 1965) 66.
53 W. T. GUNSTON, *Science J.*, 5 (1969) 39.
54 *New Scientist*, 41 (No. 636) (1969) 398.
55 R. B. FORSYTH, Union Carbide Corp., Carbon Products Divn., Private Communication (7 May, 1969).
56 *Metals and Materials*, June (1969) 154.
57 B. J. STANIER and C. H. B. MEE, *J. Sci. Instr.*, 41 (1954) 51.
58 *Financial Times*, 28 Jan., 1965, p. 10.
59 T. MATSUNO, *J. Electrochem. Soc. Japan*, 20 (1952).
60 W. C. GARDINER, FIAT Final Rept. 831 (17 June, 1946).
61 E. A. CHAPMAN, *Chem. Process Eng.*, Aug. (1965) 387.
62 H. A. LIEBHAFSKY and E. J. CAIRNS, *Fuel Cells*, Wiley, New York, 1968.
63 N. LAKSHMINARAYANAIAH, *Chem. Rev.*, 65 (1965) 491.
64 P. LÄUGER, *Angew. Chem. (Intern. Ed.)* 8 (1969) 42.
65 W. JUDA and W. A. MCRAE, U.S. Pat., 2,636,851; U.K. Pat., 720,002.
66 R. H. MCKEE, *Ind. Eng. Chem.*, 38 (1946) 382.
67 M. M. BAIZER, *J. Electrochem. Soc.*, 111 (1964) 215.
68 A. T. KUHN, *Brit. Chem. Eng.*, 17 (1971) 39.
69 I. K. DELIMARSKII and B. F. MARKOV, *Electrochemistry of Fused Salts*, Sigma Press, Washington D.C., 1961.
70 A. J. BARD (Ed.), *Electroanalytical Chemistry*, Marcel Dekker, New York, Vol. III, 1969, Chap. 2, C. K. MANN, *Nonaqueous Solvents for Electrochemical Use*.
71 M. SITTIG, *Sodium, ACS Monograph Series*, 133, Reinhold, New York, 1956.

Chapter 16

Cell Design

C. JACKSON

Imperial Chemical Industries Ltd., Mond Division, Ltd., The Heath, Runcorn, Cheshire

1. General discussion of cell design

A cell design is always a compromise between capital cost and power consumption (unless the electrical power is cheap). By far the greatest amount of development work has gone into cells for the electrolysis of brine to chlorine and caustic, and water to hydrogen and oxygen as the extensive patent literature shows. Most of this work has been aimed at up-grading the capacity of the cells by increasing the current density whilst eliminating unnecessary power losses. To this end, much attention has been paid to improving bus-bar connections (*e.g.* ref. 1), electrode designs and to minimising anode/cathode gaps. It is worth noting that although individual improvements are often small, the resultant savings are large over a period of time because of the tonnages involved. This is especially the case in the U.K. where power costs amount to about 50% of the total running costs.

Many electrochemical processes have failed in the past because of a lack of continued cell development after the initial installation. Far too often, the units installed were larger versions of laboratory cells operating as batch units. This entailed high manpower usage and poor control over the reaction. Cells were not optimised for best yields of products or minimum power consumption and it is not surprising that many of the processes were replaced by cheaper chemical methods.

The aim in designing a cell should be to design a cell to fit the process, rather than to design a process to fit a particular cell design. Very often the cell is only one stage of a multistage process and should be considered as such at the design stage. Installation of a slightly more expensive cell may well eliminate the need for a possibly expensive piece of equipment later in the process.

The heart of an electrochemical process lies at the electrode–reactant interface, and it is only when the nature of the electrodes and the reactants is known, that the cell design can be started. The cell is the box holding electrodes and reactants together in a manner which will allow the electrode reactions to proceed without hindrance under optimised conditions. The electrodes are almost invariably solid metal or graphite. The main exceptions to this are found in the chapter describing electrolysis of chlorine, where mercury cathodes are widely used. Liquid lead cathodes were also used in the Szechtmann (Philblack) process which was used to make Na–Pb alloys in anti-knock manufacture in Germany during the 1940's (see Chapter 3). Other molten metal cathodes have been used (on pilot plant scale) in various high-temperature electrowinning processes for metal extraction, and the appropriate chapter should be consulted. In all these cases, the product of the electrode reaction dissolved (amalgamated) with the liquid metal, this amalgam requiring removal from the cell, treatment in some exterior vessel and recycling.

In the mercury cell the cathode is approximately horizontal, but vertical mercury cathodes have been used, notably in connection with fuel cells (see Chapter 3) and sodium sulphate electrolysis[2] (see Chapter 14).

Leaving aside, therefore, these very few instances of non-solid electrodes, we turn to the next problem in the design of a cell. This revolves around the nature (gaseous or liquid or solid) of both reactants and products, at both working and counter electrodes. The cell design must clearly allow access and egress of these. Failure to do so will result in concentration polarisation, gas-locking or fouling of the electrode with solid, insoluble products. During gas-locking, the gases drive the electrolyte out of the cell, wholly or partially taking the electrodes out of service. Fouling the electrodes has a similar effect and insufficiently supervised, a cell in this condition may suffer hot spots and similar damage. Next, the materials of construction of the cell must be chosen to withstand the chemical attack of the electrolyte conditions employed. These days (at least if cost is disregarded) there are materials to match most conditions. When the materials suggested are non-metallic their suitability can be tested by simple life tests, with the proviso that if these materials are to be used in an extended (*i.e.* high area) form, such as diaphragms, the rate of chemical attack will be many times higher than on a massive smooth-surfaced sample. Where the materials of construction are metallic, a simple life test by immersion will not normally serve to evaluate their suitability. This is because in the finished cell design, whether by accident or design, the components by virtue of their position in the cell, will either attain a certain potential value, or be directly tied to part of the electrical cell circuit. Thus any testing without consideration of this factor is open to doubt. Nor should it be assumed that only anodic conditions jeopardise the suitability of metallic components. Titanium is only one of several metals used in cell construction which will fail under certain cathodic conditions.

Now, having a "box" to hold the electrodes, and capable of containing the electrolyte, as well as supplying and removing reactants and products, the electrochemical performance must be optimised. Cell performance is basically evaluated on two criteria: (i) current efficiency (C.E.); (ii) cell voltage.

1.1 CURRENT EFFICIENCY

Let it be assumed that no side-reactions proceed on the electrode in question (C.E. is measured with respect to one electrode only), when run under ideal conditions. The C.E. in a working cell is usually a measure of the extent to which the products formed at the opposite electrode migrate (under potential or concentration gradients) to the electrode under consideration, where they will react a second time, at the expense of the C.E. There are highly efficient methods of avoiding this type of current inefficiency. In the main, these consist of arranging reactant flows through the cell in such a manner as to minimise migration; secondly, the most important method—is the use of diaphragms. These are treated elsewhere in this work, but the most basic truth can be stated here. The more efficient the working of these diaphragms, the higher the ensuing cell voltage.

1.2 Cell voltage

This is the measured voltage which the cell operates under load. It is normal to define the point at which this measurement is made, *i.e.* at bus-bar, at electrode stem, etc., for differences of 0.1 V or more are commonplace between these various points. The cell voltage can, and should be broken down into its component parts[3]. Without knowledge of these[4], it is impossible to know where the design may be improved.

$$C_v = E_{rev} + E_{ohm} + E_{cont} + \eta_1 + \eta_2$$

C_v the total cell voltage, is the sum of the reversible potential (sometimes, but not always, close to the rest potential or open circuit potential), the ohmic drop across the electrolyte (which may itself be the sum of the resistance of the electrolyte and the contribution from gas bubbles at the electrodes), the contact resistance (at points where conductors are joined together) and lastly, the overvoltages at the two electrodes. Where a diaphragm is used, there is a voltage drop across its two faces, which can be included in the E_{ohm} term, and indeed is hard to separate from it by normal techniques of measurement.

2. More detailed consideration of cell design

2.1 Electrode design

Having left aside liquid metals as a special case, a second special case may also be considered, this is the situation where the reactant(s) is/are gaseous. This situation will rapidly lead to concentration polarisation (mass transport) problems, especially where they are of limited solubility. This problem is one that has confronted fuel cell designers for some time, and the modern fuel cell books such as Berger[5] or Liebhafsky and Cairns[6] contain a wealth of theoretical and practical details on this one problem. It should be noted that no industrial process has yet required this technique of gas electrodes.

The physical form of the electrode has been endlessly researched, in an attempt to decrease both activation and concentration polarisation and cells have been designed with fast flowing electrolytes[25, 27], with rotating electrodes, and other mechanical devices. Some more recent approaches have included the "wiped electrode"[7] which has a "windscreen wiper" passing across the face of the electrode. The enhancement in currents, for certain reactions, is believed to be due not to actual "wiping" but to the creation of a turbulent layer close to the electrode. The problem is largely one of maximising the surface area per unit volume, though this approach has its limits, for the point is reached where ohmic drops in the electrode itself outweigh the savings resulting from high areas. From time to time, attempts have been made to construct particulate electrodes. One example of these will be found in the Braithwaite (NALCO) process[8] where the electrode is pelleted lead. A series of papers on powdered carbon electrodes is referred to in the chapter on water electrolysis. Gerischer[9] developed the slurry electrode, as did Sun Oil of Canada[10] and Shell[31], while recently rather ambitious claims have been made for a "fluidised bed" electrode[11]. A truer picture is presented by Smith[12] whose data indicate that fluidisation of particulate electrodes can actually be disadvantageous, and that "restrained bed" electrodes give superior performance. Other indications are that this electrode design is not specially suitable for reactions involving gases, or indeed anodic reactions, where film formation can raise cell voltage. A rather critical theoretical treatment is given in ref. 13 where the authors endeavour to show the system is not intrinsically superior to the porous electrode. Another analysis is found in refs. 28 and 29.

But apart from these development systems, and the unique case of the NALCO process, cells tend to use rigid, massive electrodes. Rigidity is required to maintain a constant inter-electrode gap, while massiveness reduces the ohmic losses in the electrode metal itself. Thus constrained, there remain few changes which can be rung. These are best exemplified in the two industrial cases which have been the most developed, namely the chlorine industry and the water electrolysis industry. In the main, the developments have been the perforation of electrodes (described in the patent literature as "foraminated") and their corrugation or similar deformation. Both techniques aim at easing the escape of gaseous materials, and further details can be found in the appropriate chapters. Another approach to these problems has been sought in the use of mechanical (usually rotary) motion of the electrodes, and pre-war German workers used rotating electrodes (Tainton process and amalgam cell)[14]. No examples of this technique are to be found in use to-day, with the sole exception of the Asahi chlorine cell[15].

Again, because of the problems of gas removal, horizontal electrodes are rarely used, vertical electrodes having the additional advantages of savings in floor area and absence of fouling by anode slimes. Economy in floor area is also the reason why cylindrical electrodes seldom find an application nowadays except where different current densities at each electrode with uniform current distribution is required.

It was noted earlier that cell voltage is an important criterion in cell performance. Whilst the cell designer cannot improve the reversible potential, he can improve the internal resistance of the cell and the contact resistance term. The internal resistance of the cell is composed of several contributions: the resistance of the electrolyte, bubble layer resistance, resistance of the diaphragm and the resistance of the electrode itself. Bubble layer resistance can be minimised by careful design of the electrodes to enable free escape of gas as previously noted. The electrical resistance of the electrode material may considerably influence the cell design, since the current distribution is affected[16, 17]. In the monopolar Hooker chlorine cell where current is fed along the anode from one end, a voltage drop of up to 100 mV at the beginning of the run and 400 mV at the end of the run (when the anode is partially worn away) can occur along the graphite. This cell design would not readily accept simple replacement of the graphite by, say, sheets of platinised titanium because of the low electrical conductivity of titanium metal. Expensive massive current carriers would have to be installed.

The conductivity of the electrolyte may be increased by additives in certain cases, thus decreasing the cell voltage, or it may be more desirable to work at an elevated temperature to achieve the same effect. The resistance of the diaphragm is dealt with later.

Contact resistance, as the name implies, arises where bus-bars are joined together or connected to the electrodes. Imperfect connections in a cell room can contribute up to 20% of the total energy required for the process, and consequently care must be taken in their design. The nature of the materials to be connected, *e.g.* graphite to copper, will determine the method used but consideration must be given to the nature of the process. Frequently removed electrodes will be connected in a different manner to permanent electrodes. Monopolar cells are more demanding in terms of bus-bar requirements and, needing electrical linkages to each pair of electrodes, have more points of electrical contacts. Bipolar cells are more economic in bus-bar requirements, and need only two electrical connections, besides generally being more economic in floor area.

The other aspects of facilitating reactant and product flow is largely a matter of manifolding and the tightness with which the electrodes and diaphragms are packed into the "box". This is very important if bipolar electrodes are to be used (unless arranged in a filter press manner) to minimise electrolyte seepage and hence current leakage around the edges of the electrodes.

2.2 Materials of construction

Cell bodies are now seldom made of concrete (*pace* the Hooker chlorine cell) (though this was popular) nor are earthenware bodies now used. Wood lined with lead may be used and this has certain advantages when the cell body is itself used as an electrode. Cast iron is used for high temperature processes, while mild steel is supreme wherever it can be held at cathodic potentials. When rubber lined, its range of usefulness is further extended. The use of plastics and glass-fibre resin-bonded materials is slowly growing[18]. Seals can be made with plastic gasket materials, replacing tars and pitches. Bus bars are increasingly made of aluminium which has the highest electrical conductivity per unit of purchasing power.

At roughly this point, where one has a workable, perhaps even efficient cell, the concept of value-engineering enters, *e.g.* solutions to some of the problems of marrying dissimilar materials have given materials experts new fields to explore in the search for new cell materials.

All the above remarks apply to aqueous cells. For various reasons cells operating with molten salt electrolytes are still little more than giant "pots".

2.3 CELL HEATING AND COOLING

Due to the internal resistance of the cell, a proportion of the input power will be dissipated as Joules' heat. This may be desirable, or not, depending on the process. If the electrolysis is required to proceed at an elevated temperature, Joules heating might be sufficient for this temperature to be attained. Otherwise it is usual to pass the incoming electrolyte through some form of heat exchanger prior to entry into the cell. Lagging may also be needed to prevent excessive heat loss (consult the chapter on chlorine for a reference to heat balance on a sodium cell).

Cell cooling may also be achieved by passing the incoming electrolyte through a heat exchanger although actual cooling of the electrodes themselves has been tried in several cell designs. This is best achieved by using hollow electrodes through which coolant passes[30], describing a chlorate cell.

Discussions of the essential features of a cell design together with the problems of scale-up are given in the two papers by MacMullin[19, 20] and the paper by Agar and Hoar[21]. MacMullin points out that it is often more desirable to scale down an envisaged industrial cell to laboratory size rather than the reverse. Other useful descriptions of cell designs are given by Adamson *et al.*[22] and Chapman[23] whilst Beck and Guthke[24] gives a description of novel cell designs applicable to electroorganic synthesis including those with vibrating electrodes.

Chlorine cell designs can be located in refs. 25 and 26 or Chapter 3.

3. References

1 S. J. S. PARRY, J. R. ANTON and M. W. VOELKER, *Electrochem. Techn.*, 6 (1968) 73.
2 W. C. GARDINER, FIAT Final Rept. 831 (17 June, 1946).
3 V. V. STENDER, P. B. ZIVOTINSKY and M. M. STROGANOFF, *Trans. Electrochem. Soc.*, 65 (1934) 189.

4 L. H. Wolgast, Brit. Pat., 1,091,920.
5 C. Berger, *Handbook of Fuel Cell Technology*, Prentice-Hall Inc., Englewood Cliffs, New Jersey, 1968.
6 H. A. Liebhafsky and E. J. Cairns, *Fuel Cells and Fuel Batteries, a Guide to their Research and Development*, Wiley, New York, 1968.
7 I.C.I. Ltd., Brit. Pat., 1,071,923.
8 L. L. Bott, *Hydrocarbon Process. Petrol. Refiner*, 44 (1965) 115.
9 H. Gerischer, *Ber. Bunsenges. Physik. Chem.*, 67 (1963) 164–7.
10 Sun Oil, Can. Pat., 700,933, Dec. 29, 1964.
11 F. Goodridge, *Chem. Process. Eng.*, Feb. (1968) 93.
12 D. H. Smith, *Chem. Process.* (London), Sept. 15 (1969) 45.
13 A. R. Brown and N. Hiddleston, *Nature*, July 6 (1968) 94.
14 *Gmelins Handbuch der Anorganischen Chemie*, Bd. 21, Verlag Chemie, Weinheim, 8 Aufl. 1964, p. 69.
15 M. Murozumi, *Electrochem. Tech.*, 5 (1967) 236.
16 C. W. Tobias and R. Wijsman, *J. Electrochem. Soc.*, 100 (1953) 459.
17 I. Rousar, *J. Electrochem. Soc.*, 116 (1969) 676.
18 R. R. Dukes and C. H. Schwarting, *Chem. Eng.*, March 11 (1968) 206; April 8 (1968) 172.
19 R. B. MacMullin, *Electrochem. Tech.*, 1 (1963) 5.
20 R. B. MacMullin, *Electrochem. Tech.*, 2 (1964) 106.
21 J. N. Agar and T. P. Hoar, *Discussions Faraday Soc.*, 1 (1947) 158.
22 A. F. Adamson, B. G. Lever and W. F. Stones, *J. Appl. Chem. London*, 13 (1963) 483.
23 E. A. Chapman, *Chem. Proc. Eng.*, August (1965) 387.
24 F. Beck and H. Guthke, *Chem.-Ing.-Tech.*, 41 (1969) 943.
25 J. S. Sconce, *Chlorine: Its Manufacture, Properties and Uses*, Reinhold, New York, 1962.
26 See ref. 14, p. 41.
27 Carter, to ICI, Brit. Pat., 597, 387.
28 P. Le Goff, F. Vergnes, F. Coeret and J. Bordet, *Ind. Eng. Chem.*, 61 (1969) 8–17.
29 J. N. Hiddleston and A. F. Douglas, *Electrochim. Acta*, 15 (1970) 431.
30 Hooker Corp., Neth. Pat., 6,510,232.
31 Shell, Brit. Pat., 1,098,837.

Chapter 17

An International Survey of Industrial Electrochemical Processes

A. T. KUHN

Department of Chemistry and Applied Chemistry, University of Salford, Lancashire

D. W. LAWSON

European Chemical News, 33 Bowling Green Lane, Londen E.C. 1

The chapters preceding this one describe individual processes or groups of processes, mainly from a scientific or technological aspect. The purpose of this chapter is to give an insight into the economic importance of the various sectors of the industries, and to discuss methods for obtaining economic information. It can be stated for example that chlorine production, in particular, is a good index of the size and rate of expansion of a national economy. Such information can be obtained in two ways.

1. Economic information on a product group

For certain groups of electrochemicals, economic information collected by various organisations including the product associations is available.

1.1 Chlorine

The Chlorine Institute Inc., at 342 Madison Ave., New York 10017, publishes a valuable series of pamphlets. These include No. 16, *Chlor-Alkali Producers outside North America* and No. 10, *North American Chlor-Alkali Industry*. Similar information is contained in the series published and continuously updated by Chemical Data Services at 33–39 Bowling Green Lane, London, E.C. 1. Lastly, the *Chemical Guide to Asia, Africa, Australasia*, published by Noyes Development Corp. in 1966, indicates plants which all not found in the sources mentioned above. It is accepted that none of these sources gives a complete picture, for at least some manufacturers are reluctant to disclose their resources. World chlorine production figures are given in ref. 36. Most chlorine cells are supplied by outside contractors, such as those referred to in Chapter 3. These contractors publish details of plants installed by them. Similarly, manufacturers of heavy electrical equipment also disclose some of the equipment installed by them, with its power rating and place of installation. Much valuable information can be extracted from such data, though it should be borne in mind that at least some new equipment goes to replacement of existing plant rather than to extending it.

1.2 ALUMINIUM, TIN, COPPER, ZINC, MAGNESIUM

These metals all have Associations, to which most producers and refiners of them belong, and which supply certain amounts of information. Magnesium is the exception and no Association appears to exist. However certain figures for electrolytic magnesium production can be found in the publications of the Chlorine Institute referred to above.

1.3 HYDROGEN PEROXIDE

The Table shown in Chapter 13 was compiled by an industry in this field.

1.4 PERCHLORATES

It is believed that a complete listing of these is given in Chapter 3.

1.5 MANGANESE DIOXIDE

It is believed that a complete and accurate list is given in Chapter 13, kindly supplied by F. L. Tye of Ever-Ready Battery Co.

1.6 METALS (MISC.)

It is believed that a reasonably complete listing of all metals electrowon is given in Chapter 6. The compilation of a complete list of electrorefiners including the many small ones, is beyond the scope of this work.

2. Economic information on a national basis

Certain countries publish detailed information giving annual production figures for various electrochemical products. In addition the International Organisations, including the organs of the U.N.O., publish limited economic surveys. There follows below, information on certain countries which has been obtained by direct approach. The list does not claim to be complete, but may be more accurate than most other published information of this kind.

2.1 THE SOVIET UNION AND COMECON GROUP COUNTRIES

The book *Soviet Chemical Industry* by Y. Meltzer, Noyes Development Publications, gives an overall picture. The United States Congress publish an annual volume *Soviet Economic Performance* from which the size of the aluminium, copper and chlorine industries can be estimated. Apart from these "staple" electrochemical industries, the USSR appears to have developed various electroorganic synthetic routes and pilot plants more extensively than the West.

2.1.1 Developments in industrial electrolysers in the Comecon area

In general most of the development work in this field has taken place in Russia, and to a lesser extent in Poland, Hungary and Czechoslovakia. The 50th Anniversary of the Russian revolution occasioned the publication of some useful review articles on the electrochemical industry and electrochemical research[1-4]. A good introduction to this technology is given in several reviews by prominent Russian electrochemists[5-11, 18].

The total amount of power generated in the USSR in 1965 was 507,000 million kWh and in 1980 it is estimated it will reach 2.7–3.0 billion kWh. The price of electric power in the USSR (around 1 kopeck/kWh) is relatively cheap and this fact, together with a strong tradition of electrochemical research in Russia, has led to a large amount of work and consequently a large number of publications on electrochemical research and development, and the industrial operation of several new electrochemical processes.

However, it is not intended that this Appendix should be a comprehensive literature survey, instead developments which are known to have reached the pilot plant or semi-technical stage are reviewed. Most of these developments are in organic chemistry, since in the field of inorganic chemicals most of the work has been concerned in improvement of existing processes such as those for chlorates, chlorine, per-compounds, etc. Nevertheless some promising laboratory research in electrochemical synthesis is reviewed at the end of the paper, where this work is thought likely to lead to industrial development.

Tomilov and co-workers[10, 12] have described a pilot plant process for the electrolytic synthesis of hexamethylenediamine and aminocapronitrile from adiponitrile. The electrolytic cell had a rated load of 2000 A and was designed for electroreduction in an alkaline medium. The cell consisted of a rubberised steel vessel hermetically sealed by its lid. The electrode pack consisted of sponge copper covered steel cathodes and hollow magnetite anodes. No diaphragm was necessary since it was found that neither products nor starting material were oxidised on the magnetite anode. Electrolyte was circulated through the cell and through a heat exchanger by means of a centrifugal pump to maintain a cell temperature of 3–5°C. For effective emulsification of the adiponitrile the velocity of the emulsion in the interelectrode space had to be 20–50 cm/sec. The optimum rate depends on the difference between the specific gravities of the electrolyte and the organic substance; the smaller this is, the lower the velocity for effective emulsification. A gas separator was employed to separate gaseous products and reaction products were separated from the alkali solution in decanters. The organic products were then extracted with toluene and the hexamethylenediamine layer was decanted off and distilled. The toluene extract was rectified and the product fraction was rerectified at reduced pressure to obtain an aminocaprolactam fraction.

At a loading of 2000 A, over a period of 100 hours, a chemical yield for amines of 73% was obtained at a current efficiency of 51.5%. A hexamethylenediamine: aminocapronitrile ratio of 4.4 : 5.6 was obtained at a current density of 60 mA/cm^2 and a ratio of 3 : 7 at 30 mA/cm^2 at a loading of 1000 A for 130 hours. Materials and power usage per ton of product mixture is given in Table 1 for operation at two different current densities.

TABLE 1

Material	Materials factor	
	30 mA/cm^2	60 mA/cm^2
Adiponitrile (ton)	1.15	1.165
NaOH (ton)	0.24	0.52
Copper acetate (ton)	0.003	0.003
Electrical energy (kWh)	8650	14000
Toluene (ton)	0.02	0.02

Thus, economically it appeared to be more expedient to operate at the lower current density, at lower power consumptions.

This reaction has also been combined with a useful anode reaction[8, 11]. Hydrochloric acid (25–30%) was used as the electrolyte and a diaphragm cell was used with a spongy nickel cathode. The anode product is chlorine which is thereby recovered from hydrochloric acid, a by-product from chlorination of hydrocarbons. Practical operation of a cell at 1000 A showed that 6 tons of chlorine were obtained per ton of hexamethylenediamine. Some operating parameters of the process are given in Table 2.

An economic evaluation of this process showed that hexamethylenediamine obtained by the electrochemical method had a 20% lower cost than the catalytic hydrogenation product. Also, electroreduction gave a higher quality product than that obtained by the chemical route, which could be directly converted to nylon salt without further purification.

The Monsanto electrochemical adiponitrile process is now well known. Knunyants[13, 14] has developed an alternative process, which uses alkali as the solu-

TABLE 2

Parameter	Average value
Current density	100 mA/cm²
Voltage	5.5 V
Adiponitrile concentration	96.9 g/l
Material yield	
hexamethylenediamine	60%
aminocapronitrile	12.2%
HCl content in catholyte	
free	15.4 g/l
combined	59.9 g/l
Catholyte temperature	17.4°C
Content of anode gas	
chlorine	99.2%
hydrogen	0.21%
Content of cathode gas	
chlorine	0.21%
hydrogen	97.3%

bilising electrolyte. The main products of the reduction are again propionitrile and adiponitrile, and the ratio between them is highly dependent on the nature of the cathode. The maximum current yield of adiponitrile (60%) was obtained on a graphite cathode. Another factor determining the product ratio is the current density. The maximum adiponitrile yield (60% electrically and 82–84% chemically) was obtained at a current density of 10 mA/cm². The yield of adipontrile tended to fall off with increasing alkali concentration, 1–2 mole/litre being an optimum. The nature of the alkali (LiOH, KOH, NaOH) had no great influence on the reduction process. Table 3 summarises the power costs and acrylonitrile conversions for the three electrical methods for adiponitrile synthesis.

Satisfactory yields of adiponitrile have never been obtained by simply hydrodimerising acrylonitrile in the presence of hydrogen; the yield is no more than 20%. Cathodic reduction and alkali metal amalgam hydrodimerisation can be seen from Table 3 to compare closely in performance. However, the use of amalgams leads to product contamination with mercury, although amalgam reactors are simpler to construct than are electrolysers, and there are no acrylonitrile losses due to oxidation processes. Nevertheless mercury is toxic, and therefore an electrochemical process is preferable. This cathode dimerisation process for adiponitrile represents a technological and economic advance because of the low production costs of acrylonitrile from propylene and ammonia.

An industrial electrolyser has been described[10, 15] by Fioshin and co-workers for the synthesis of carboxylic acid esters, such as the di-ester of sebacic acid or the ester of ω-hydroxypentadecanoic acid, by means of anodic condensation. A special feature of the electrolyser was a phase separator which allowed the condensation

TABLE 3

ENERGY REQUIREMENT AND ACRYLONITRILE CONSUMPTION PER TON OF ADIPONITRILE

Method	Consumption		Process reference
	Acrylonitrile (ton)	Power (kWh)	
Reduction with potassium amalgam	1.35	9700	ICI Fr. Pat., 1,385,906 Brit. Pat., 1,067,447
Cathodic reduction in a quaternary ammonium base, salt solution	1.12	12000	Monsanto Fr. addition Pat. No. 82,951 (1964)
Cathodic reduction in alkaline solution	1.20	5000	Knunyants Fr. Pat. 1,401,175 (1965)

and recovery of the highly volatile solvent which was used as the medium for the anodic reaction. Cooling pipes for the phase separator served as the cathode and cylindrical platinised titanium anodes were situated between the stainless steel cooling pipes. A gaslift was created by the hydrogen and carbon dioxide formed during the electrolysis and this caused the electrolyte, together with the electrolysis products, to be ejected into the phase separator. The gases were then vented to waste from the phase separator. Thus the phase separator can be integral with the electrolysis cell if current supply is at the cell base, or it may be a separate unit in the case of current supply to the top of the cell. This arrangement enabled continuous electrolysis to take place at temperatures close to the boiling point of the electrolyte. The condensation of the monomethyl ester of adipic acid to sebacic acid gave sebacic acid in 84–86% chemical yield with a current efficiency of 60–65%. The product was much purer and the costs less than half compared with the process based on castor oil, which is expensive and in short supply.

In Poland a pilot plant has been operated for the electrochemical oxidation of glucose or lactose with formation of calcium gluconate or lactobionate respectively[10, 16, 17]. Copper cathodes were arranged on both sides of graphite anodes, which makes it possible to use both sides of the anode, thereby increasing the productivity of the cell. The cell was a rectangular steel vessel lined with PVDC, and electrolyte was continuously circulated through the cell. The cell had a rated load of 400–600 A and operated at an anode current density of 14–16 mA/cm² for the oxidation of lactose and 20–22 mA/cm² for the oxidation of glucose.

An application of an electrolytic method was revealed by work described by Sergunina[19] for the decontamination of drinking water. Sodium chloride solution was electrolysed in a cell containing magnetite or graphite granules, with stainless steel current conducting electrodes. The granules operated as discrete bipolar cells.

The effect of the electrolysed solution, containing hypochlorite, chlorate, and perchlorate species, on autoclaved mains water contaminated with *Escherichia coli* and on untreated river water, was measured. It was found that the bactericidal action of the electrolysis products was not inferior to hypochlorites obtained by chemical means from chlorine and in certain cases it was better than the chemical product. The electrolyser was shown to be reliable under experimental production operating conditions.

Other reactions have been described in the literature, but it is not certain whether they have progressed to the pilot plant or production scale. For example, Tsodikov (quoted in ref. 11) has described an electrolytic process for the simultaneous production of nicotinic acid and tetrahydroquinoline, in a cell with a ceramic or plastic diaphragm. Quinoline was oxidised at a lead dioxide anode in an ammonium sulphate/sulphuric acid electrolyte, the quinolinic acid so formed being decarboxylated to nicotinic acid, which was obtained in 40–43% chemical yield. Quinoline was reduced at a lead cathode in sulphuric acid solution to 1,2,3,4-tetrahydroquinoline, in a chemical yield of 70–75%. The overall reaction was:

$$11 \text{ [quinoline]} + 16 \text{ H}_2\text{O} \xrightarrow{36e} 2 \text{ [nicotinic acid]} + 9 \text{ [tetrahydroquinoline]} + 6 \text{ CO}_2$$

8.5 tons of tetrahydroquinoline were obtained/one ton of nicotinic acid for an energy consumption of 15,000–20,000 kWh. This affects the economics of production, since the same power consumption would be required for the synthesis of nicotinic acid alone.

Another reaction described by Fioshin[9] as of major practical interest is the electrolytic reduction of 3-nitrobenzene sulphonic acid to 3-aminobenzene sulphonic acid (metanilic acid). This is a dyestuffs intermediate and also it is used in the synthesis of *p*-aminosalicylic acid, an anti-TB drug. Khomutov and co-workers have published considerable detail on this electrolytic reduction reaction.

Other reactions of commercial interest for which few details of actual or possible scale-up exist are iso-butyric acid[20]; hydroxylamine[11, 21–23] for cyclohexanoneoxime formation for caprolactam manufacture; 4-methyl-1,4-pentanediol[24], olefin oxides[25]; *p*-aminophenol[26]; sodium dithionite[27–29], sodium sulphate electrolysis[30–32] and the use of bipolar powder electrode cells in electrosynthesis[33–35].

2.2 UNITED STATES

Overall production is listed annually in the Bureau of Mines Yearbook. Additionally, the paper *Report on Electrolytic Industries in the USA* appears to have become an annual feature in the *Journal of Electrochemical Technology*, now printed with the *Journal of the Electrochemical Society*.

2.3 CANADA

The Science Council of Canada commissioned a study of chemistry in this country, and a sub-group reported specifically on the electrochemical industries (E. J. Casey, DRB ref. DCBRE 551). The journal *Canadian Chemical Processing* is a valuable source of information, and the Editor, (T. E. Buck) endeavours to provide the most up-to-date information possible on all electrochemical processes.

2.4 AUSTRALIA

Chlorine production is recorded in the sources quoted. A 40,000 tonne/annum aluminium plant is planned for 1970 in New Castle NSW, by Alcan, while Comalco at Bell Bay, Tasmania plant plan 80,000 tonnes annually, and Alcoa at Geelong, Victoria are now operating at this level. Copper refining is done at Mt. Isa Mines, Townsville, Queensland (70,000 tonnes) and Electrolytic Refining and Smelting Co. are at Melbourne, 25,000 tonnes. Electrodialysis—a 230,000 l/day plant has been installed by Ionics at the USN Base at North West Cape, W. Australia. The authors are indebted to Dr. D. F. A. Koch of the Australian CSIRO for this information.

2.5 INDIA

This country possesses a very active Central Electrochemical Research Institute (Director H. V. K. Udupa, D.Sc.) at Karaikudi-3, S Railway, from which much information can be obtained. Among processes which have been brought to pilot scale there, are the electrochemical manufacture of succinic acid, calcium gluconate and benzidine. Indian chlorine capacity is shown in Table 4, and actual consumption in 1967 was 284,000 t, with a forecast of 400,000 t in 1974. India has two electrolytic zinc plants: Cominco Binani Zinc in Kerala and Hindustan Zinc in Rajasthan, with joint capacity of 38,000 t/annum. Four aluminium reduction plants in W. Bengal (Aluminium Corp. of India), Orissa (Indian Aluminium), Renukoot (Hindustan Aluminium), and Tamil Nadu (Madras Aluminium Co.), have a joint capacity of 120,000 t/annum with further capacity of 55,000 t planned. Electrolytic silver refining is done at the India Government Silver Refinery and by Hindustan Photo Films Mfg. Co. with joint capacity of 130 t/annum. Copper refining at the Chosi Metal refinery in Surat runs at 230 t/annum while chlorates are made by Mettur Chemicals (Tamil Nadu) (2 t/day), W. India Match Co. (Maharashtra) and the Travancore Chemical Co., Kerala. Lastly, India possesses one of the largest electrolytic hydrogen plants in the world at Nangal in the Punjab, with a 25,000 m^3/hr hydrogen capacity.

TABLE 4

ELECTROLYTIC CHLORINE MANUFACTURERS IN INDIA AND THEIR ANNUAL CAPACITY

Name of unit	Location	Annual capacity in tonnes (as on Dec. 31, 1968)
Mercury cells		
1. Dhrangadhra Chemicals	Tuticorin, Tamil Nadu	54,000
2. Standard Mills	Bombay, Maharashtra	40,000
3. Travancore Cochin Chemicals	Alwaye, Kerala	33,000
4. National Rayon Corporation	Bombay, Maharashtra	20,000
5. Mettur Chemicals	Mettur Dam, Tamil Nadu	20,000
6. Shri Ram Vinyl	Kota, Rajasthan	16,500
7. Kanoria Chemicals	Renukoot, Uttar Pradesh	16,500
8. Jayshree Chemicals	Chatrapur, Orissa	16,500
9. Hukumchand Jute Mills	Amlai, Madhya Pradesh	13,000
10. Century Chemicals	Bombay, Maharashtra	13,000
11. Calico Mills	Bombay, Maharashtra	12,500
12. Durgapur Chemicals	Durgapur, West Bengal	10,500
13. Andhra Sugars	Tanuku, Andhra Pradesh	10,000
14. Nepa Mills	Nepanagar, Madhya Pradesh	4,000
15. Rohtas Industries	Dalmianagar, Bihar	4,000
16. Orient Paper Mills	Brajrajnagar, Orissa	3,000
17. All Products	Bulsar, Gujarat	3,500
18. Hindustan Heavy Chemicals	Calcutta, West Bengal	2,500
19. J.K. Chemicals	Bombay, Maharashtra	1,000
	Total	293,500
Diaphragm cells		
1. Mettur Chemicals	Mettur Dam, Tamil Nadu	20,000
2. D C M Chemicals	Delhi	13,200
3. Alkali and Chemical Corp.	Rishra, West Bengal	9,200
4. Tata Chemicals	Mithapur, Gujarat	7,000
5. Titaghur Paper Mills	24-Parganas, West Bengal	6,400
6. Sirpur Paper Mills	Sirpur, Andhra Pradesh	6,000
7. Calico Mills	Ahmedabad, Gujarat	2,300
8. Rohtas Industries	Dalmianagar, Bihar	1,700
9. Shree Gopal Paper Mills	Yamunanagar, Punjab	700
10. Mysore paper mills	Bhadravathi, Mysore	600
	Total	67,100
	Grand Total	360,600

2.6 ISRAEL

Electrochemical Industries (Frutarom) Ltd. have a 30 t/day chlorine plant, using cells of their own design. Plans are in hand to start up an HCl electrolysis plant which will add a further 30–40 t/day mercury cell plant, while persulfate (and possibly hydrogen peroxide) is made in 5000 A cells by the Oxidon Co. at Holon. In Haifa, Messrs. Electroclor make hypo by direct injection of chlorine into caustic soda. An electrolytic copper refining plant is under construction at Timna Copper Mines, near Eilat, and Makhteshim have plans for an electrolytic ADN plant. Lastly the Ministry of Defence is believed to operate plants, presumably for chlorate production (information supplied by Dr. A. Cantoni and R. Wachs).

2.7 Japan

Dr. F. Hine has provided the following information.

An estimated 271 billion kWh load is forecast for 1969. Of this, 20% will be consumed by the electrochemical and electrothermal sector of the chemical and metallurgical industry. Chlorine consists of the major part of this load, with a rated capacity of 2.4 million tonnes, making Japan the second largest producer after the 6.5 million tonne US capacity. 90% of this production is in mercury cells, of which some produce at up to 500 t/day, with 300 kA cells operating at 8 kA/m². Extensive use of automation has reduced manpower requirements to 4–5 man/shift including brine purification, liquefaction stage and rectifier plant. Installation of DSA anodes in several plants is reported. Chlorate and perchlorate—Japan Carlit have supplied the following information—annual production in 1968 was 43,000 t, and production has been expanding annually by 2–4000 t. Manufacturers (electrolytic route) are Japan Carlit, Hodogaya Kagaku Co. and Showa Denko KK while a chemical process is operated by Toyo Soda and Osaka Soda. Perchlorates are made only by Japan Carlit, with capacity of 5000 t/annum. Production of chlorates is divided approximately 50–50 between the sodium and potassium salts. Most cells use magnetite anodes. Bromates and iodates are made in reagent quantities only. Periodic acid dialdehyde starch process is operated by Japan Carlit Co.

Magnesium—the author estimates 10,000 t/annum. An estimate for electrorefined copper is 550,000 t, with zinc (electrolytic) at 500,000 t. Japanese copper refining cells run at 1.5 A/dm² with 2000 kWh/tonne power consumption. The electrolytic zinc route is not expected to expand. Aluminium—production is increasing at 20%/annum with 600,000 t made in 1969. Power consumption is 16,000 kWh/t but will be reduced to 15,000 or even 14,000 kWh/t. Cells of 100 kA size are now being installed. These are fitted with pre-baked anodes. Labour requirements are as low as 25 men/shift in a plant of 130 cells, 20,000 t/annum including rectifier room, handling and casting plants.

Water and heavy water electrolysis—this area is insignificant, though plants for both processes are operating (2 heavy water plants). Sodium—estimated 8–9000 t/annum. A succinic acid plant is believed to be operational. Fluorine is understood to be produced, though no information is available.

2.8 SWEDEN

The authors are indebted to G. Karlberg of the Dept. of Applied Electrochemistry, Royal Institute of Technology, Stockholm 70, for the following list of plants operating in Sweden today, with notes.

Plant at	Owner
Chlorine	
Skoghall	Uddeholms AB
Bohus	Elektrokemiska AB
Skutskar	Stora Kopparbergs Bergslags AB
Domsjo	Mo och Domsjo AB
Ostrand	Svenska Cellulosa AB
Korsnas	Korsnas-Marma AB
Stromsbruk	Strom-Ljusne AB
Kopmanholmen	Fors AB
Stenungsund	Fosfatbolaget
Chlorate	
Mansbo	Stabindustrier AB, Alby klorat
Alby	—
Trollhattan	Fosfatbolaget
Stockvik	—
Domsjo	Mo och Domsjo
Oxygen–hydrogen	
Ljungaverk	Fosfatbolaget AB
Domsjo	Mo och Domsjo AB
Copper, silver and gold refining	
Ronnskar	Boliden
Aluminium	
Sundsvall	Svenska Aluminiumkompaniet
Electrolytic iron powder	
	Husqvarna Vapenfabriks AB

*1945:*In the years 1929–1945, aluminium sulphate solutions were purified from iron by electrolysis over a mercury cathode in a plant at Helsingborg, now owned by Boliden AB.

1968: Production of hydrogen peroxide at Elektrokemiska AB, Bohus, earlier based on electrolytically manufactured ammonium persulphate, shifted to the organic method, based on anthraquinone.

2.9 NORWAY

Professor Jomar Brun and Mr R. Tunold have supplied the following details.

Hydrogen/heavy water	Norsk Hydro	25 t/annum D_2O, 1000 t H_2
		400,000 t/annum NH_3
(will probably switch to reformer hydrogen shortly)		
Chlorine	Hydro, Borregaard	62,000 t/annum
Chlorates	Vadheim	3000 t/annum
Electrorefining	—	Nickel 33,000 t/annum
		Copper 15,000
		Cobalt 800
		Selenium 9
		Silver, gold, platinum 4
Zinc	Norzink, Odda	Zinc 60,000 t
		Cadmium 80
Aluminium	5 firms	400,000 t
Aluminium refining	Vigeland Brug	3200 t
Magnesium	Norsk Hydro	33,000 t

2.10 FINLAND

Professor J. Larinkari, of Kemian Keskusliitto has provided the following information.

Chlorine	Finnish Chemical Oy, Aetsa	80,000 t/annum
		(40,000 t/annum expansion planned)
	Oulu Osakeyhtio	37,000 t/annum
	Kymin Osakeyhtio	23,000 t/annum
Chlorates	Finnish Chemical Oy	16,000 t/annum
	Oulu Osakeyhtio	6,000 t/annum
Cuprous oxide	Outukumpu Oy	300 t/annum
Hydrogen	AB Woikoski Oy	—

2.11 THE BENELUX COUNTRIES

Professor M. Pourbaix, of CEBELCOR, Av. Paul Heger, Grille 2, Brussels 5, has assisted the authors in collating the following information. In the Netherlands, a list of electrochemical industries is published by the *Vereniging van de Nederlandse Chemische Industrie*, 2 Javastraat, 's-Gravenhage.

Chlorine	Zoutchemie Botlek (100% KZO)	Rotterdam (Botlek)
	K.N.Z. (100% KZO)	Hengelo and Delfzijl
	Natronchemie (100% Solvay)	Herten/Roermond
Sodium bromate	Chefaro (100% KZO)	Dordrecht
Potassium bromate	,,	,,
Potassium bromate	Noury and Van der Lande	Roermond
Ammonium and potassium persulphates	Chefaro and Noury and Van der Lande	Dordrecht/Roermond
Calcium bromolacto	Chefaro	Dordrecht
Calcium gluconate	Chefaro and Noury and Van der Lande	Dordrecht/Roermond
Sodium perborate	Noury and Van der Lande	Roermond
H_2–O_2	Oxygenium	Schiedam
H_2O_2	Natronchemie (100% Solvay)	Herten
In Belgium,		
Chlor-alkali	Solvay and Cie SA	Jemeppe s/Sambre
	Produits Chimiques de Tessenderloo S.A.	Tessenderloo
Cu(refining)	Métallurgie Hoboken SA	Hoboken

(this company is believed to produce other metals by electrochemical reduction)

2.12 FRANCE

Little information was obtainable from this country. The following organisations would seem to cover the industries in question. Chambre Syndicale de l'Electrométallurgie et de l'Electrochimie and Fédération Nationale des Industries Electrométallurgiques Electrochimiques et Connexes, both at 33 rue de Lisbonne 75, Paris.

Apart from these organisations, the French electrochemical industry is covered in the twin publications *Chimie et Industrie* and *Génie Chimique*, both edited by Professor P. Le Goff. The latter kindly agreed to seek out information, but the response was disappointing. Solvay state that their Tavaux plant for H_2O_2 and perborate is not electrochemical. Their chlorine production is at Tavaux and Dombasle. Potasse et Produits Chimiques have a chlorine plant at Thann (Ht-Rhin) where all production is stated to be in mercury cells.

2.13 Austria

Aluminium	Ranshofen, Lend
Copper refining	Montanwerke, Brixlegge
Zinc	Bleiberger-Bergwerksunion, Gailitz
Mischmetall	Treibacher Chem. Co.
Perborates	Treibacher Chem. Co.
Chlorine	Solvay (Hallein), Donauchemie, Brückl/Kärnten

(The Weissentstein Chem. Co. have abandoned their electrolytic perprocesses)
(Information kindly supplied by Dr. G. Jangg of the Vienna T.H.).

3. Bibliography (U.S.S.R.)

V. G. KHOMYAKOV, V. P. MASHOVETS and L. L. KUZ'MIN, *Electrochemical process technology*, Goskhimizdat, 1949.

M. YA. FIOSHIN, S. S. KRUGLIKOV and A. P. TOMILOV, *Electrochemical methods of making organic and inorganic substances*, Viniti Press, 1958.

ZH. BILLITER, *Industrial electrolysis of aqueous solutions*, Goskhimizdat, 1959.

V. V. STENDER, *Applied electrochemistry*, Izd. Khar'kovskogo Gos. Univ., 1961.

N. P. FEDOT'EV, A. F. ALABYSHEV et al., *Applied electrochemistry*, Goskhimizdat, 1962.

M. YA. FIOSHIN and V. N. PAVLOV, *Electrosynthesis of organic compounds*, Znanie Press, 1965.

E. A. DZHAFAROV, A. P. TOMILOV and M. YA. FIOSHIN, *The electrosynthesis of organic and inorganic substances*, Azerneshr Press, Baku, 1965.

L. M. YAKIMENKO, *Electrolytic cells with solid cathodes*, (UDC 048 : 541.13) Khimiya Press, 1966.

A. P. TOMILOV and M. YA. FIOSHIN, *Progress in the electrochemistry of organic compounds*, Nauka Press, 1966, pp. 256–277. (UDC 541.13 : 661.7) (available as translation in RTS 5500).

References

1 N. P. FEDOT'EV, *50 Years of the electrochemical industry in the USSR*, Zh. Prikl. Khim., 40 (1967) 2179–2191 (2103–2112 English pagination).

2 N. M. EMANUEL, *Electrochemistry*, Usp. Khim. *(Russ. Chem. Rev.)* 36 (1967) 858–866 (English pagination).

3 A. N. FRUMKIN, *Electrochemistry, Razv. Fiz. Khim. SSSR. 1917–1967 Akad. Nauk. SSSR Inst. Istor. Estest. Tekh.*, (1967) 149–176 (available as translation RTS 5163).

4 G. M. KAMAR'YAN, *Scientific and technical basis of the development of electrolytic manufacture of chlorine and caustic soda*, Zh. Prikl. Khim., 40 (1967) 334–341 (319–324 English pagination).

5 V. G. KHOMYAKOV and A. P. TOMILOV, *Examples of possible applications of electrolysis of organic compounds in industry*, Khim. Prom., (1959) 566–73 (Available as translation TT 62-20371).

6 V. G. KHOMYAKOV, M. YA. FIOSHIN and A. P. TOMILOV, *Electrochemical methods of synthesis of some starting materials for high polymers*, Khim. Prom., (1959) 16–20 (available as translation TT 61-10756).

7 V. G. KHOMYAKOV and M. YA. FIOSHIN, *Innovations in the field of electrochemical synthesis of oxidising agents*, Khim. Prom., (1962) 30–37 (available as translation AD 602527).

8 M. YA. FIOSHIN, *Achievements in the field of electrochemical synthesis of monomers*, Itogi Nauki Electrokhimii, (1965) 152–168 (available as translation RTS 4960).

9 M. YA. FIOSHIN and A. P. TOMILOV, *Progress in the electrochemical production of organic compounds*, Khim. Prom., 43 (1967) 243–252 (available as translation RTS 4748).
10 A. P. TOMILOV and M. YA. FIOSHIN, *Industrial electrolysers for the synthesis of organic compounds*, Khim. Prom., 44 (1968) 84–91 (available as translation RTS 4799).
11 M. YA. FIOSHIN, *The feasibility of combining various electrochemical synthetic reactions*, Khim. Prom., 44 (1968) 882–885 (available as translation RTS 5178).
12 A. P. TOMILOV et al., *Electrochemical synthesis of hexamethylenediamine and aminocapronitrile*, Khim. Prom., (1965) 329–333.
13 A. P. TOMILOV et al., *Production of adiponitrile by the hydrodimerisation of acrylonitrile*, Khim. Prom., 42 (1966) (12) 892–896 (available as translation RTS 4747).
14 I. L. KNUNYANTS et al., *The electrochemical reduction of acrylonitrile*, Brit. Pat., 1,014,428 22, Dec. 1965 (C.A., 64 (1966) 7683 c).
15 G. M. KAMARYAN et al., *Electrolyser for the synthesis of carboxylic esters*, Russian Inventor's Certificate No. 185,853; Oct. 19, 1966 (C.A., 66 (1967) 1110729).
16 Polish Pat., 46836 (1961).
17 Polish Pat., 55009 (1969) (C.A., 70 (1969) 102545).
18 A. P. TOMILOV and L. F. FILIMONOVA, *New means for the possible use of electrochemical processes in industrial organic synthesis*, Zh. Vses. Khim. Obshchestva im D. I. Mendeleeva, 14 (1969) 328–338.
19 L. A. SERGUNINA, *An effective electrolytic method for the decontamination of drinking water*, Gigiena i Sanit. 33 (1968) 16–21; 33 (1968) 22–27.
20 M. YA. FIOSHIN et al., *Electrochemical synthesis of iso-butyric acid on an industrial scale*, Khim. Prom., 45 (1969) 496–8 (RTS 5474).
21 A. KORCZYNSKI et al., *Electrochemical synthesis of hydroxylamine sulphate. Preparation of crystalline products in a continuous process*, Przemysl Chem., 48 (1969) 274–6; 48 (1969) 207–10; 48 (1969) 94–98.
22 N. F. STAROSTENKO et al., *Electrochemical preparation of hydroxylamine*, Zh. Vses. Khim. Obshchestva im. D. I. Mendeleeva, 13 (1968) 599–600.
23 A. KORIZYNSKI, *Crystalline hydroxylamine sulphate*, Polish Pat., 54,799 (C.A., 70 (1969) 25240).
24 A. P. TOMILOV et al., *4-methyl-1,4-pentanediol*, Russian Author's Certificate 233,651 (C.A., 70 (1969) 120,598).
25 A. P. TOMILOV and M. YA. FIOSHIN, *Use of electrolysis for synthesising olefine oxides*, Khim. Prom., 45 (1969) 413–416 (RTS 5473).
26 A. KORCZYNSKI et al., *Electrochemical preparation of p-aminophenol from nitrobenzene*, Przemysl. Chem., 48 (1969) 156–9.
27 I. P. SENTYUREVA, *Effect of cathode material on the electrochemical preparation of sodium dithionite*, Zh. Vses. Khim. Obshchestva im. D. I. Mendeleeva, 14 (1969) 228–9.
28 I. P. SENTYUREVA and N. S. FEDEROVA, *Kinetics of the electrochemical preparation of sodium dithionite*, Izv. Vysshikh. Uchebn. Zavedenii Khim. i Khim. Tekhnol., 11 (1968) 573–576 (available as translation RTS 4961).
29 E. E. KRAVTSOV and YU. A. LYSENKO, *The possibility of utilising certain chemical industry waste products. Parts 2 and 3*, Nauchn. Zap. Lugansk Sel. Inst., 8 (1961) 257–260, 261–4.
30 Chemolimpex Magyar, *Improvements in or relating to the electrolysis of sodium sulphate*, Brit. Pat., 938,111 (25 Sep., 1963) (C.A., 60 (1964) 5080 d).
31 A. K. GORBACHEV et al., *Electrochemical preparation of caustic soda and sulphuric acid by the use of oxygen depolarisation on porous cathodes*, Khim. Prom. Ukr., (1968) pp. 3–5 (available as translation RTS 5105); (1969) pp. 11–43 (available as RTS 5529).
32 V. K. BEIDIN et al., *Electrolysis of sodium sulphate aqueous solutions*, Russian Author's Certificate No. 227,308 (C.A., 70 (1969) 92692).
33 W. TOMASSI et al., *Application of a bipolar powder electrode in the coupled electrochemical preparation of zinc and manganese dioxide*, Przemsyl Chem., 47 (1968) 692–3.
34 W. TOMASSI et al., *Application of activated carbon with a catalyst deposit in a bipolar powder electrode*, Przemsyl Chem., 47 (1968) 495–497.
35 W. TOMASSI et al., *Application of a powder cathode to the electrochemical preparation of manganese dioxide. II. Pilot plant*, Przemsyl Chem., 43 (1964) 608–9.
36 ANON., *Chemische Industrie International*, 1 (1969) 1–4.

Index

accuracy of ECM, 273, 300
Acker process, 103, 508, 599
active sites in electrodeposition, 328
addition agents in refining, 235
additives for electroplating, 334, 336, 351, 352
adhesion of electrodeposits, 327
adions in electrodeposition, 328
adiponitrile, 507, 519, 610
agitation of electrolyte, 332
air-agitation in electroforming, 398
air cathodes, 92, 106, 107, 509
alkaline tin refining, 220
alloys, electrodeposition of, 372
—, electroforming of, 411
alternating current, effect on Pt dissolution, 547
— current, effect on wear rates, 560
— current in electroforming, 388
β-alumina, use of, 103
aluminium, anodising of, 379
—, electroplating on, 371
—, information on production of, 608
—, refining of, 246
— alloy cathode, 562
— hydroxychloride, 507
— manufacture, anodes for, 194, 544
— manufacture, gas-depolarised electrodes for, 545
— sulphate, electrochemical purification of, 618
amalgam, energy in, 109
— cell for Be manufacture, 208
— method for Zn recovery, 186
— reactors, 611
— reductions, 518
amalgams, hydrodimerisations on, 611
—, sodium manufacture from, *see* Acker process
aminobenzenesulphonic acid, 613
aminocapronitrile, 609
aminophenol, 511, 613
ammonium persulphate, 498
anode, graphite crucible, 210
—, niobium carbide, 252
—, sacrificial lead pellet, 508
—, slimes and sludges, 183, 190, 230
—, wear, 20, 74, 91, 95, 101, 110, 543, 547
— baskets, 397
— cathode gap, 197

— coatings, 551
— designs, 552
— effect, 2, 14, 19, 194, 196, 197, 198
anodes, auxiliary, in electroforming, 389
—, bagged, 183, 190, 227, 232, 396
—, beryllium, 252
—, carbon or graphite, 2, 5, 90, 93, 100, 121, 123, 194, 206, 207, 208, 209, 538
—, composite, 101
—, connection to, 44, 118
—, continuously cast, 230
—, copper, 122, 505
—, cylindrical Pt/Ti, 612
—, design of, 118, 119, 551
—, Downs cells, 100, 105
—, impregnation of, 122, 542
—, lead, lead alloy, lead dioxide, 93, 97, 121, 186, 210, 503, 512, 525, 613
—, life of, 233
—, magnetite, 93, 95, 111, 533, 609, 616
—, manufacture of, 194
—, mild steel, 93, 97, 115, 133, 501, 511
—, mixed crystal, 548
—, nickel, 3, 71, 74, 152, 165
—, nickel sulphide, 182, 232, 241
—, Ni plated, 139
—, platinised Ti, 93, 95, 97, 98, 121, 545
—, platinum, 2, 93, 97, 98, 191, 497, 498, 501, 509, 513, 515, 545
—, plutonium, 250
—, prebaked carbon, 194
—, resin-coated, 230
—, slotted, 100
—, Söderberg, 194, 544
—, stainless steel, 506
—, structure of, in F_2 manufacture, 21
—, titanium, 503
— for electroforming, 396
— for electrowinning, 182
— for Hg cells, 118, 119
— with wells, windows, etc., 100
anodic condensation of hydroxypentadecanoic acid, 611
— film formation, 380
— oxidation of coatings, 378
anodising, 327, 378
anthracene, oxidation of, 515
anthraquinone, 515

anti-knocks, *see* Acker process
antimony, 176, 210, 254
applications of ECM, 304
arcing in ECM, 271, 272
— in Hg cells, 547
Asahi cell, 115, 119
asbestos cloth, reinforced, 130
— diaphragms, porosity of, 579
asphalt in cell construction, 226
Australia, information on, 614
Austria, information on, 620
automatic plants for electroplating, 381
azelaic acid, 514

bags for anodes, *see* anodes, bagged
Baizer process, *see* adiponitrile
Balbach-Thum cells, 219
Bamag cell, 143, 147
barium, 176, 209
barrel-plating, 381
baskets, anode, *see* anode baskets
bauxite, treatment of, 192
Bayer process, 192
bellows, metal, 408
bells for gas collection, 133
belts, endless, electroforming of, 406
benzidine, 510, 518, 614
beryllium, 208
—, refining of, 251
— powders, 243
beta-ray backscatter, 560
Betts process, 219, 233, 237
bichromate as additive, 96, 98, 121, 122
bismuth, separation from Pb, 250
—, winning of, 176, 210
blanks, stainless steel, 226
Bonelli process, 518
borides as electrodes, 525, 565
boron, 209
brackish water, electrodialysis of, 475, 488
Braithwaite process, 508, 602
brass, electroplating of, 373
Breteque process, 191
bright metal electroplating, 335
brine-purification, 110, 118
bromate manufacture, anodes for, 122, 543
bromates, electrochemistry of, 121
—, manufacture of, 531, 616
bromide regeneration, 516
bromine, 122
bromoform, 532
bronze, electrodeposition of, 374
— cathodes, 98, 566
bubble-layer resistance, 90, 156, 603
bubbles, effect of, 18, 37
burrs, electrochemical removal of, 264

busbar connections, 599
busbars, Al, 604
butene, oxidation of, 509

cadmium, electroplating of, 364
—, manufacture of, 186, 537
—, manufacture of cathodes for, 562
caesium, 176, 209
calcium, manufacture of, 176, 209
— gluconate, 516, 612, 614
— lactobionate, 516, 612
Canada, information on, 614
carbides in cathodes and anodes, 565
carbon cooling elements, 186
— dioxide, 539
— fibres and cloth, 117, 583
carbonation of NaOH, xii, 108
carboxylic acid esters, 611
cascade processes and systems, 160, 163, 227, 515
cast iron electrodes, 560
cathode, air, 117
—, amalgam, 200
—, amalgamated Cu, 512
—, bagged, 190
—, bronze, 98, 566
—, carbon and graphite, 90, 93, 123, 227, 498, 501, 563, 611
—, cooling pipes as, 98, 612
—, Cu, 2, 122, 505, 612
—, dummy, 332
—, gallium pool, 191
—, impure molten Sn, 255
—, iron and steel, 71, 95, 165, 206, 207, 209, 210, 250, 505, 506, 514
—, lead (solid, liquid, amalgamated), 103, 497, 507, 562, 600
—, liquid metals, 566
—, manganese, 251
—, mercury, 191, 208, 512
—, mercury (flowing/vertical/agitated), 115, 510, 516, 600
—, mild steel, 93, 97, 115, 133, 139, 501, 511, 562
—, molten metal, 600
—, molybdenum, 206, 251, 254, 255
—, nickel, 152, 191, 505, 513
—, nickel alloy or spongy, 98, 610
—, niobium tube, 252
—, perforated, Ni or W, 250, 252
—, platinum, 191, 545
—, rotating, 103, 115, 512
—, scraped, 227
—, sectioned, 198
—, silicon carbide, 198
—, silver, 6, 98

cathode, sponge Cu-covered Fe, 609
—, stainless steel, 93, 122, 191, 208, 210, 226, 562
—, starter, 226, 228, 229, 562
—, tantalum, 251
—, tin-plated, 200
—, titanium, 566
—, tungsten, 206
—, zinc amalgam, 510
—, zinc plated, 200
cathodes for industrial processes, 561
cathodic etching, 344
— protection, electrodes for, 546, 547, 560
— protection of machine tools, 266
caustic-chlorine balance, 108
caustic soda, carbonation of, 108
— soda, electrolysis of, 102
— soda, regeneration of, 517
cavitation of electrolyte in ECM, 271, 278
cell construction and design, 34, 35, 90, 93, 156, 599, 609
— heating and cooling, 98, 138, 178, 604, 612
— lining in Al manufacture, 195
— linings of Pb-Sb, 180
— linings, steel, ceramic, etc., 93
— pairs in electrodialysis, 468
— voltage, 601
cells, bipolar graphite, 90, 95
—, ceramic, 206
—, double diaphragm, 144, 517
—, filterpress, 90, 133, 507
—, materials of construction, 266, 603
—, Mg alloy, 5
—, monopolar, 115
—, pressure, 155
—, refining, design of, 226
—, sheet-flow, 477
—, Tank type, 133
—, tortuous path, 477
—, use of PVC in, 93
—, vertical Hg, 516
— for perchlorate manufacture, 97
— for zero-G use, 155
cementation, 179, 181, 187, 191, 234
centrifuging of electrolytes, 267
ceramic diaphragms, 580
— materials in Downs cells, 105
ceric sulphate, 515
Chemech cell, 95
Chilex anodes, 535
chlor-alkali cells, cathodes for, 562
— industry, 89
chlorate cells, anodes for, 547, 558, 604
— cells, anode wear in, 526, 543
— ions, effect on graphite wear, 541
chlorates, 92

chlorine, drying and handling of, 119
—, evolution on graphite, Pt, Ru, etc., 539, 543, 556
—, manufacture of, 106, 556
—, production information on, 607
— bubble release, 551
chloroform, 513, 532
chromates as additives with PbO_2 electrodes, 527
chromic acid, 514, 533
chromium, anodes and cathodes for manufacture of, 188, 533, 535, 562
—, electrodeposition of, 354
— plating, anodes for, 358, 536, 537
circuit breakers, 118
CJB electrolyser, 147
coated electrodes, 545
cobalt, electrodeposition of, 352
—, manufacture of, 181, 536, 566
—, refining of, 224
combined anode and cathode processes, 610, 613
Comecon area, processes in, 609
computerised prediction (of reactions), 96
computers in cell control, 111, 198
concentrate cell, electrodialysis, 468
concentration polarisation, 602
concrete in cell construction, 226, 603
conductors, bipolar, in electroforming, 389
construction materials for cells, 226, 603
contact angle, electrode/electrolyte, 15
— resistance, 603
contractometer, 398
cooling in cells, coils, 93, 98, 138, 178, 604, 612
— in cells, carbon elements, 186
— in cells, cathodes for, 98, 612
copper, anodising of, 379
—, electrodeposition of, 344
—, electroforming of, 392
—, electrolytic removal of, 181
—, information on production of, 608
—, manufacture of, 178, 536
—, refining of, 219, 224
— powder, 242
— silicide as anode, 536
cresyl sulphonic acid as electrolyte, 219
cryolite, 194
cuprous oxide, 505
current density, distribution of, 96
— efficiency, 600
— efficiency in Downs cell, 102
— reversal, 123, 348, 388
cyclohexanone-oxime, 613

DC, unsmooth, 156
deburring, *see* burrs, electrochemical removal of

decomposer, *see* denuders
decontamination of drinking water, electrolytical, 612
deformation of cathode linings, 564
degreasing, 341
Demag electrolyser, 141
dendrites, electrodeposition of, 177
De Nora cell, 109, 111, 144, 163
denuders, 107, 111, 119
deuterium oxide, 159
dextrose, oxidation of, 516
dialdehyde starch, 513, 616
diamond cutting tools, 408
dianisidine dihydrochloride, 511, 518
diaphragm cells, 107, 115, 119, 164
— cells, anodes for, 557
diaphragms, 129, 575
—, additives to, 117
—, alumina, 202
—, alundum, 514
—, asbestos, 115, 133, 139, 150, 152, 501, 507, 579
—, bag type, *see* anodes, bagged
—, canvas, 227
—, capillaries in, 575
—, carbon cloth, 117, 583
—, cells and, 134
—, ceramic, 123, 497, 510, 512, 580, 613
—, ceramic clad, 101, 105
—, "chemical", 575
—, clogging of, 123
—, cloth, 188, 581
—, commercial available, 579
—, concrete, 511
—, doping agents for, 578
—, double, 144
—, Downs cell, 101
—, fibre, 517
—, filtering, 577
—, iron mesh, 101
—, life of, 117
—, linen, 506
—, mechanical, 575
—, mercury, 109, 575
—, molybdenum, 252
—, novel or potential, 583
—, parchment, 513
—, permeability of, 576
—, polymeric, 581
—, polypropylene, 581
—, p.t.f.e. as, 583
—, p.v.c. as, 90, 91, 581
—, reinforced, 91
—, rejuvenation of, 578
—, rubber, 156, 512, 517
—, seals and, 144

—, shaking of, 101
—, stainless steel, 207, 584
—, structural modification of, 578
—, titanium, 202
—, tortuosity of, 575
—, various, 98, 578 et seq.
—, wetting of, 117, 579
dichlorobenzidene sulphate, 518
dichromate as additive, 96, 98, 121, 122
diluate cell, 468
dimethyl sulphide, oxidation of, 513
— sulphoxide, 513
dissolved hydrogen in electroplating, 337
Downs cells, 20, 99 et seq.
— cells, anode effect and, 20
— cells, electrodes for, 562
— cells, literature on, 104
drilling, electrochemical, 264, 306
ductility of electrodeposits, 399

earthenware for cell construction, 603
ECM, *see* electrochemical machining
economic information, 607 et seq.
economics of Cl_2 manufacture, 119
— of ECM, 301
— of H_2 manufacture, 155
efficiency of F_2 production, 45
electrical circuit for ECM, 270
electrocatalysts for cathodes, 91, 92
electrochemical cleaning, 340, 342
— drilling, 306
— forming, 305
— grinding, 298
— machining, 263 et seq.
— polishing, 340
— trepanning, 308
— turning, 306
electrocoat painting, 417, 423
— painting, anodic process in, 435
— painting, pigments for, 431
— painting, vehicles for, 425
electrocolouring, 327, 380
electrocrystallisation, 330
electrode activation, 91, 133, 156
— boilers, 560
— design, 601
— feeders, stainless steel, 612
— gap, 110, 111
electrodeposition, co-deposition of foreign matter, 334
—, corner defects and, 387
—, kinetics and mechanism of, 328
—, non-uniformity of, 389
— of alloys, 372
electrodeposits, stresses in, 386, 387
electrodes, bipolar, 133, 134, 156, 602, 613

—, carbon powder, 156
—, construction of, 155, 156
—, corrugated, 130, 602
—, cylindrical, 602
—, electrodialysis, 479
—, expanded metal, 130, 133, 552
—, fluidised bed, 602
—, foraminated, 602
—, gauze, 156
—, horizontal, 602
—, industrial materials for, 525 et seq.
—, mercury, 113, 118, 517
—, molten tin, 502
—, monopolar, 133
—, nickel, 133
—, particulate, 602
—, pelleted lead, 602
—, perforated, 130, 133, 602
—, perforated Ni, 150
—, porous, 154, 156, 501, 510
—, Raney, 156
—, refining, 228
—, restrained bed, 602
—, rotating, 103, 602
—, roughened, 133
—, scraping or brushing of, 321, 505
—, silver-Pd, 155
—, slurry, 602
—, steel mesh, 147
—, vertical, 113, 118, 130, 517, 602
—, vibrating, 604
—, "wiped", 602
electrodialysis, 467 et seq.
—, applications of, 488, 491
—, costs of, 483 et seq.
—, electrodes for, 479
—, plant design, 477
electrofluorination, mechanism of, 74
electroformed nickel, 386
electroforming, 385 et seq.
—, alternating current and, 388
—, anodes for, 396
—, applications of, 403
—, control of stress in, 387
—, copper products, 242, 392
—, corner weakness in, 390
—, electrodes manufactured by, 156, 310
—, equipment for, 397
—, production by, 387
—, solutions for, 392
— of alloys, 411
— of refractory metals, 415
— of tools, 310
electroforms, properties of, 398
electrohydrodimerisation, *see* adiponitrile
electrolyte, ammonium sulphate/H_2SO_4, 613

—, cavitation of, 271, 278
—, circulation, 227
—, clarification of, 267
—, flow of, 238
—, liquid NH_3, 103
—, NaAc + carbonate, 506
—, purification of, 234
— for refining, 234
— pumps for ECM, 268
— supply for ECM, 266, 295
electrolytes, borates as, 510
—, fast flowing, 115, 602
—, molten salt, 7, 103, 104, 177, 192, 245, 377, 415, 595
— for ECM, 295
— for miscellaneous processes, 103, 593
electrolytic pickling, 344
— regeneration, 92, 513, 514, 516
electrophoresis for Pt deposition, 554
electroplating, 327 et seq.
—, additives for, 350, 351
—, permanent anodes for, 546
electropolishing, 327, 340
electrorefining, aqueous, 219
—, copper, 179
—, history of, 219
—, molten salt, 245
electroslag refining, 246
electrotransport, 246
electrotyping, 411, 412
electrowinning, anodes for, 535, 559
epitaxial growth, 331
etching, cathodic, 344
explosion bonding, 115
extraction of Cu, solvent, 180

fatty acids, 514, 535
ferrocyanides, oxidation of, 506
ferrotitanium as anode, 253
filterpress cells, 134, 138, 144, 156, 163, 164
filtration, continuous, 398
Finland, information on, 618
flocculating agents in cells, 187
flotation concentration of Cu ores, 178
flow-melting of Sn coats, 369
flowsheets of tankhouses, 179
fluocarbons as ion-exchangers, 593
fluorination, electrochemical, 71 et seq.
fluorine, 1 et seq.
fluosilicic acid as electrolyte, 219
foaming, prevention of, 130, 156
"fogs", metallic, 196
—, removal of, 156
foils, perforated, 405, 407
forced circulation, 129
— flow of brine, 115, 117

foreign bodies, 334
forming, electrochemical, 305
fouling of electrodes, 600
France, information on, 619
frost heave in Al cells, 195, 564
frothing in cells, 130, 156
fuel cells, *see* air cathodes
—, amalgam, 113, 118

gallium, 190
galvanoplasty, 385
gaps, anode–cathode, 198
gas-depolarised electrodes for Al manufacture, 545
gas domes, 139
— evolution in ECM, 278
— lift in cells, 130, 612
— locking of cells, 600
gaskets, asbestos, 139
Gaspack H cell, 153
gas-stirring in cells, 122
germanium, 176, 210
glass manufacture, electrochemical tinting in, 520
gluconates, *see* calcium gluconate
gluconic acid, *see* calcium gluconate
glucose, electrodialysis of, 475
—, oxidation of, 612
—, reduction of, 510
gold, electroforming of, 395, 411
—, electroplating of, 370
—, refining of, 219, 225, 255
gramophone records, electroforming, 386, 403
graphite, overvoltage on, 539
— anode corrosion, 541
— diaphragms, 117, 583
— (as) electrocatalyst, 539
— electrodes, 90
— wear, 91, 112, 540
grinding, electrochemical, 298
growth sites in electrodeposition, 328

hafnium, 176
half-crystal position in electrodeposition, 328
Halkyn process, 210
hardness, tests for electroforms, 398
— of electrodeposits, 339
Haring cell, 332
— and Blum formula, 333
Hazelet cast anodes, 230
heat balance in F_2 cell, 34
heaving of cell linings, 195, 564
heavy water, 159
hexamethylene diamine, electrosynthesis of, 609
Hoechst cell for HCl electrolysis, 90
Hooker S-4 cell, 116

hot-spots in cells, 600
Hybinette cells, 228
— process, 224
hydrochloric acid, electrolysis of, 89, 566
hydrodimerisation, *see* adiponitrile
hydrogen, manufacture of, 127
—, overvoltage on Zn, 185
— electrolysis, 155
— embrittlement, 562
— evolution, 110, 119
— fluoride, 71
— peroxide, 497, 499, 501
— reduction of metals, 182, 183
— removal in ECM, 268
hydroxylamine, electrosynthesis of, 553, 613
hydroxypentadecanoic acid, electrosynthesis of, 611
hypochlorite cells, 99, 546, 558

impregnation of graphite anodes, 542
impulse current sources, 445
India, information on, 614
indium, 191
insulation of ECM tools, 282
internal stress in electrodeposits, 338, 398
iodates, 120, 531, 532, 543
iodine, electrochemistry of, 120
iodoform, 513, 532
ion-exchange membranes, 507, 584, 590, 593
— membranes in Cl_2 manufacture, 109, 117, 120
— membranes in electrocoat process, 452
iridium, electroplating, 553
iron, electrodeposition of, 352
—, electroforming of, 385
— electrodes in Downs cells, 562
— pipes, electroforming of, 385
— powder, manufacture of, 242
isobutyl alcohol, oxidation of, 533
isobutyric acid, electrosynthesis of, 533, 613
isotopic separation, 158
Israel, information on, 615

Jamin effect, 117
Japan, information on, 616

Kesting process, 96
kinetics of F_2 evolution, 27
kink-sites in electrodeposition, 330
Knowles cells, 136
Kossuth process, 123
Krebs NC 12 cell, 93, 94
Kroll-Betterton process, 225
Kroll process, 203
Kureha cell, 113, 119

labour costs in refineries, 240
labyrinth flow, 470
lactobionate, *see* calcium lactobionate
lactose, electro-oxidation of, 612
laminates, Ti/mild-steel, 115
leaching of Cu ores, 178 et seq.
— of ores, 176
lead, oxidation of, 506
—, refining of, 219, 248
—, winning of, 210
— alkyls, *see* Acker process
— alloys as anodes, 525
— as cell construction material, 177, 603
— carbonate, basic white, 506
— cathodes, 562
— dioxide, 533
— dioxide anodes, 562
— electrode, 525 et seq.
— tetraalkyls, *see* Acker process
— tin alloys, 374
levelling in electrodeposition, 333
Liberator cells, 234
limiting current density in electroforming, 391
liquid metals, 566
literature on Cl_2 industry, 120
— on D_2O manufacture, 170
lithium, winning of, 207
Lowenstein Laporte process, 498

McKee salts for solubilising, 594
macrothrowing power, 333
magnesium, information on, 608
—, manufacture of, 199
—, refining of, 255
— alloy, cell bodies, 5
— bleaching powder, 533
— manufacture, cathodes for, 562
magnetite electrodes, 525, 533
— for manufacture of anodes, 93, 533
mandrels for electroforming, 385, 399, 402
manganate, oxidation of, 503
manganese, 189
— dioxide, 186, 503, 504, 515, 533, 536, 537, 543
manganic sulphate, 533
manganous sulphate, 533
manifolding, 470
mannose, reduction of, 510
mass-transport, 601
membranes, asbestos, 107
—, ion-exchange, 155, 584
—, semi-permeable of Cr_2O_7, 96, 98, 122
mercuric oxide, 505
mercury anode, sacrificial, 505
— cathodes, 618
— cells, anodes for, 118, 119, 538, 545

— cells, bipolar, 115
— inventory, 108, 112
— loss, 108, 112
— pumps, 113
metal distribution formula in electroplating, 332
metal fibres, electroforming of, 409
— finishing, anodes for, 545, 552
— powders, 242
— spraying for Pt deposition, 554
metanilic acid, 613
methyl ethyl ketone, 509
4-methyl-1,4-pentanediol, 613
microcracking of Cr deposits, 360
microphotographs of anodes, 91
microporosity of Cr deposits, 360
microsteps in electrodeposition, 328
micro-throwing power, 333
milk-whey, electrodialysis of, 488, 495
Miller process, 225
Mischmetall, 205
mixed crystal anodes, 548
models, mathematical, 96
Moebius cell, 227
molten salt electrolytes, 7, 103, 104, 177, 192, 245, 377, 415, 595
molybdenum, 176, 210, 254
monatomic steps, 329
monopolar cells, 115, 164
Monsanto process, *see* adiponitrile
Montan wax, 515
Moritz cell, 144
moulds and dies, 407
moving burden, Hg as, 109
multi-stage cells, 160

Nalco process, 508, 602
nickel, electroforming of, 393, 404
—, electroplating of, 348
—, refining of, 219, 224
—, winning of, 182
nickel–cobalt alloys, electrodeposition of, 376
nickel sulphide anodes, 232
— tubing, 386
nicotinic acid, electrosynthesis of, 613
niobium, 203
— powder, 243
— refining, 252
nitric acid, reduction of, 533
nitrides, 525
nitrobenzene, reduction of, 510, 518
— sulphonic acid, reduction of, 613
noble metal anodes, 95, 547
— metal electrodes, 545
— metals, electroplating of, 369
— metals, overvoltages on, 555, 556

non-aqueous electrolytes, 103, 376, 593
Norway, information on, 618

Oerlikon cells, 142, 164
olefin oxides, electrosynthesis of, 509, 613
oleic acid, oxidation of, 514
Olin-Mathieson cells, 93
optima, determination of, 96
organic electrode reactions, electrodes for, 561
— electrolytes, 175, 177, 376, 595
— oxidation processes, 507
— reduction processes, 507, 518
overvoltages at molten metals, 245
oxygen, 127
— overvoltage, 555
ozone, formation of, 98, 122

paint, electrodeposition of, 417
painting and firing of noble metal coatings, 554
palladium, electroplating of, 371
Pechkrantz cells, 160
pelargonic acid, 514
perborates, 501
perchlorate cells, cathodes for, 97, 98, 562
— manufacture, 97, 98, 528, 530, 535, 546, 562
perchlorates, 97
perchloric acid, 98
periodate process, electrodes for, 562
"per" reactions, 497, 498, 559
Philback process, see Acker process
plutonium, refining of, 250
polarisation, 2, 471, 475
polarity reversal, 443, 476
polypropylene, diaphragms of, 581
pore-size distribution in diaphragms, 576
porosity, measurement of, 577
— of diaphragms, 576
— of electrodeposits, 339
porvair diaphragms, properties of, 582
post-ECM treatment, 298
potash compounds, 88
potassium, winning of, 176, 209
potassium-ferricyanide, 506
potassium persulphate, 501
powders, electrodeposition of, 177
power costs in refineries, 239
power-shut-off systems in ECM, 272
power supplies for ECM, 269
pre-baked anodes for Al manufacture, 194, 544
precious metal plating, anodes for, 559
— metals, electroforming of, 395, 410
— metals for activation of electrodes, 91, 133, 156
pre-enrichment, electrolytic, 162
pressure electrolysers, 132, 133, 147, 155
pre-treatment for electroplating, 328, 339

printed-circuit foils, electroforming of, 392
printing, rotary textile, 404
—, see electrotyping
propylene chlorhydrin, 509
— oxide, 509
pulsating current, 168
pulsed electrolysis, 75, 82
pumps, electrolyte, for ECM, 268
purity of refined Al, 247
PVC diaphragms, 581
Pyne-Green cells, 234
pyridine, reduction of, 512

quaternary ammonium salts, 507, 594
quinoline, electro-oxidation of, 533, 613

rare earths, 205
razor foils, 407
reaction modelling, 96
recirculatory brine systems, 108
record stampers, 403
redox systems, 92, 513 et seq.
reduction of metal ores, 175
refineries, list of, 222
refinery costs, 239
refractory metals, electrodeposition of, 377
— metals, electroforming of, 415
— metals, refining of, 245
reversible voltage of F_2 cells, 26
rhenium, 176
rhodium, electroplating of, 371, 554
Rohner process, 511
rotary textile printing, 404
roughening of electrodes, 133, 562
rubber in cell construction, 226, 604
rubidium, 176, 209
ruthenium, electroplating of, 554
— oxide anodes, 555, 556, 557

saccharin, 515
safety circuits for ECM, 271
— in plant operation, 135
salicylaldehyde, 512
salicylic acid, reduction of, 512
"salting-in" of reactants, 594
scale formation, 469
Schroeder process, 92
scrapers, mechanical, 227, 505
scrap recovery by refining, 220
screen, Ni mesh, 202
screw dislocations, 329, 330
sealing of cells, 132, 604
seawater, electrolysis of, 99, 108
sebacic acid di-ester, electrosynthesis of, 611
semi-permeable membrane from silicates, 122

separation factor, isotopic, 159, 163
sewage, electrolysis of, 535, 613
shields and screens in electroplating, 332, 389
short-circuits in refineries, 238
shorting in Hg cells, 547
silicon carbide cathodes, 565
silver, electroforming of, 396, 410
—, electroplating of, 369
—, refining of, 225, 255
Simons process, 71 et seq.
skirts, 37
slag production from ores, 178
slimes, adherent, 231
—, anode, 227
—, falling, 232
—, floating, 232
—, recovery of, 230
sludge, red nickelliferrous, 4, 74
— disposal from ECM, 268
solvent extraction, 180
spiral contractometer, 339, 398
— growth, 330
stainless steel, see cathodes, stainless steel, and, diaphragms, stainless steel
starter cathodes, 226, 228, 229, 562
steel in cell construction, 226, 604
stopping off, 344, 390
stressometer, 339, 398
strontium, 176, 209
Stuart cell, 138
succinic acid, 614
sulphate ions, 541
sulphates, electrolysis of, 516
surface area measurement of electrodes, 554
— diffusion of adions, 329
— finish of ECM, 273
— preparation for electroplating, 339
surfactants, 84
Sweden, information on, 617
Szechtmann cell, see Acker process

Tainton anodes, 536, 602
Tank electrolysers, 133, 136
tantalum, winning of, 203
— as electrode substrate, 545
— powder, 243
Teflon as diaphragm, 583
temperature in refining, 236
tensile-strength, 398
tensile stress in electrodeposits, 339
terylene, 183
—, diaphragms, 581
tetraethyl lead, see Acker process
tetrahydroquinoline, 613
tetramethyl lead, 508
thallium, 191

thermal decomposition of Pt containing paints, 554
"thieves and robbers" in electrodeposition, 389
thorium, 176, 211
three-layer cells for Al manufacture, 247
throwing power, 333
— power in electrocoat, 446
— power in electroforming, 391
Thum cell, 227
ticklers, auto, 101, 105
tickling of Downs cells, 101
tin, electroplating of, 365
—, production facts of, 608
—, refining of, 219, 221, 254
—, separation from Pb, 250
tin–nickel alloys, electrodeposition of, 375
titanium, anodic breakdown voltage, 550
—, corrosion resistance of, 548
—, Cu or Al cored, 551
—, mechanical properties of, 548
—, properties of, as electrode substrate, 548
—, refining of, 254
—, resistivity of, 551
— as electrode substrate, 548
— cathodes, 566
— cathodes in Zn manufacture, 563
— manufacture, cathodes for, 562
— powders, 243
toluene-o-sulphonamide, oxidation of, 515
toluidine, 511
— dihydrochloride, 518
tool design for ECM, 280 et seq.
— insulation for ECM, 282
Treadwell generator, 152
treatment, post ECM, 298
trepanning, electrochemical, 308
tubes, seamless perforated, 404
tungsten, manufacture of, 176, 211
—, refining of, 254
— as electrode substrate, 545
— bronzes as electrodes, 525
turning, electrochemical, 306
two-stage refining processes, 252

United States, information on, 613
upgrading of D_2O, 166
uranium, 176, 211, 254, 565

vacuum evaporation for Pt deposition etc., 554
vanadium, 176, 253
void fraction, diaphragms, 575, 576
voltage regulation, ECM, 269

water electrolysis, 127 et seq.
— electrolysis at 200°C, 156
— electrolysis, electrodes for, 562

— electrolysis, thermodynamic and experimental data, 156
wave-guides, electroforming of, 392, 405
wear-rates of Pt electrodes, *see* anode wear
wear-resistance of electrodeposits, 339
Weissenstein process, 497
Westvaco system, 92
wettability of anodes by Hg, 547, 548
wetting agents, 84, 117, 156, 579
— of electrodes, 15
wiping of electrodes, 242, 602
Wohlwill process, 219, 225
wood in cell construction, 226, 603
Wunsche process, 123

Zdansky-Lonza cell, 147
zinc, electroplating of, 360
—, hydrogen overvoltage on, 185
—, information on production of, 608
zinc-alloy castings, plating on, 372
zinc cathodes, 562
— electrowinning, anodes for, 536
— manufacture, 183
— powder, 242
zirconium, manufacture of, 176, 210
—, refining of, 254
— cathodes, 563
— powder, 242